COMPUTER APPLICATIONS FOR ENGINEERS

THOMAS K. JEWELL
UNION COLLEGE

WILEY

JOHN WILEY & SONS, INC.

NEW YORK • CHICHESTER • BRISBANE • TORONTO • SINGAPORE

ACQUISITIONS EDITOR   Charity Robey
PRODUCTION MANAGER   Katharine Rubin
COVER/TEXT DESIGNER   Karin Gerdes Kincheloe
PRODUCTION SUPERVISOR   Linda Muriello
MANUFACTURING MANAGER   Lorraine Fumoso
COPY EDITOR   Richard Blander
COVER PHOTO   Marjory Dressler

Recognizing the importance of preserving what
has been written, it is a policy of John Wiley &
Sons, Inc. to have books of enduring value
published in the United States printed on
acid-free paper, and we exert our best efforts to
that end.

Library of Congress Cataloging in Publication Data:

Jewell, Thomas K.
      Computer applications for engineers / Thomas K. Jewell.
         p.   cm.
      Includes index.
      ISBN 0-471-60117-9
      1. Engineering—Data processing. 2. Microcomputers. I. Title.
   TA345.J48   1991
   620′.00285′416—dc20   90-15471
                                                               CIP

Printed in the United States of America

10 9 8 7 6 5 4 3 2 1

TO PAM

YOU HAVE MADE THIS PROJECT POSSIBLE THROUGH YOUR HELP
AND UNDERSTANDING; AND TO JIM, KEITH, AND JOHN FOR ALL
OF YOUR UNDERSTANDING AND SUPPORT

STUDENT'S PREFACE

The computer is an essential tool for engineers, much as a hammer is an essential tool for a carpenter. Part of your engineering education will be learning how to use the computer. This text will give you a foundation in effective computer use. You will continue to practice and will become more proficient with the computer in your later courses as you work toward becoming an engineer. Your achievement of an acceptable computer proficiency level will provide you with a powerful analytical tool.

Using the computer will encourage a logical approach to problem definition, problem solution, and engineering design. The computer will free you from having to do tedious, repetitive calculations when you are examining alternative solutions to problems. Because of these advantages, you will have more time to be creative in your problem solving.

It has been my experience that the computer proficiency level of most incoming engineering students remains rudimentary. This is not because secondary schools are failing to introduce students to computers. Incoming engineering students are more comfortable with computers today than they were a few years ago. However, they are not computer literate in the engineering context, because they have not had sufficient exposure to problem solving. I see weaknesses in problem definition skills, which are an essential element in any computer application. I also see weaknesses in the skills needed to compose and document a well-structured program. Properly structured programs exhibit good programming style and observe a prescribed set of rules. Such programs are easily transferred and understood. This text is designed to help you develop your problem-solving skills while simultaneously becoming proficient in computer usage.

What is involved in computer literacy? If you are to be truly computer literate, you must be able to recognize problems for which computer solutions are appropriate. You must also be able to use the computer to implement your solutions. You should have the ability to develop, test, and document a well-structured program using a high-level language such as FORTRAN or PASCAL. You should also be able to develop engineering applications for generic microcomputer packages such as spreadsheets and equation solvers. You may not use all these skills all the time as an engineer. However, you will need them in reserve for analysis and design problems.

We are in a period of rapid growth in the availability of microcomputers and associated software. At the same time, prices for larger computers continue to go down. As an engineer, you will need to know how to choose from among many options to select the preferable computer resource for a particular application or problem. In the past, graduates were expected to know how to use a slide rule or calculator. Graduates are now expected to know how to use a computer with proficiency.

Thomas K. Jewell

INSTRUCTOR'S PREFACE

Engineering graduates need to be comfortable with, and proficient in, using the computer as a tool. Faculty of undergraduate engineering programs, as well as engineering program accrediting agencies, realize the need for computer literacy. They are taking the necessary steps to ensure that graduates have the required skills. The Accreditation Board for Engineering and Technology (ABET) has included the following requirements in its "Criteria for Accrediting Programs in Engineering in the United States":

> Appropriate computer-based experience must be included in the program of each student. Students must demonstrate knowledge of the application and use of digital computation techniques to specific engineering problems. The program should include, for example, the use of computers for technical calculations, problem-solving, data acquisition and processing, process control, computer-assisted design, and other functions and applications appropriate to the engineering discipline. Access to computational facilities must be sufficient to permit students and faculty to integrate computer work into course work whenever appropriate throughout the academic program.

A 1986 survey of civil engineering departments shows the general state of educational computing at that time. These statistics are from a questionnaire returned by 93 out of 212 civil engineering programs in the United States. Nine percent of the responding schools required students to purchase microcomputers. Thirty-two percent were considering making it a requirement. Thirty-two percent of the programs reported that students were computer literate when entering the institution. However, the questionnaire offered no explanation about what computer literacy entailed. In addition, ninety-seven percent of the programs required at least one course in high-level programming, of which:

83% required FORTRAN

19% required BASIC

16% required PASCAL

5% required C or APL

This list adds up to more than 100 percent because some programs required multiple languages.

The survey reported that 39 percent of the course work at the undergraduate level required use of the computer. Twenty-two percent of course work required students to write high-level programs. Twenty percent of the surveyed programs required a course in computer technology other than language training. Respondents had no strong preference between focusing this course on computer literacy, operating systems, system software, hardware interfacing, and so forth, or on an advanced course in programming. Word processing, statistics, and spreadsheets were the most often cited software being used. Following these, CAD/CAM (Computer-Aided Design/Computer-Aided Manufacturing), database management, project management, simulation, optimization, data acquisition, and presentation graphics were pretty evenly split among respondents. These statistics are believed to fairly represent the other engineering disciplines. The data show that the computer is an important resource in undergraduate engineering programs. They also show that FORTRAN is preferred as a programming language.

This text presents an integrated approach toward achieving computer literacy and learning to use computers as productivity tools. Topics covered include:

Computer architecture and operating systems

Problem formulation and algorithm development

Programming language (FORTRAN 77)

Microcomputer applications

  Equation solvers

  Spreadsheets

  Database management

Numerical methods

Development of engineering applications packages

Most computer texts cover one programming language or microcomputer applications package. They do not integrate that particular language or applications package into the context of using the computer as a productivity tool. This text is designed to fill that gap.

Using this text will help students develop an understanding of what computers are and how they operate. They will understand the capabilities and limitations of computers. Students will gain an appreciation of the value of the computer as an analytical tool, and will be comfortable in using it. Among other things, they

1. Will understand the need to confirm the correctness of answers;
2. Will have the ability to develop, test, and document a structured computer program;
3. Will recognize how computers can be effectively applied in engineering problem solving;
4. Will be able to choose the most appropriate software and hardware resources to use for a particular application;
5. Will be able to recognize when the computer is not the proper tool to use to solve a particular problem;
6. Will have an ability to use, verify, and modify applications software packages;
7. Will be able to communicate with expert programmers or lay people about computer usage;

8. Will learn to appreciate the types and magnitudes of errors that can develop through computer calculations; and
9. Will have the ability to become proficient in additional languages or software packages on their own as the need arises.

A course based on this text will provide students with a basic competency in computer applications. This course could replace an introduction to engineering course. However, there should not be a significant lag between the end of the course and the next time students will be required to use the computer in course work. They will need to increase their proficiency by continuing to use the computer as an effective tool in their other courses.

The material is designed to be covered in a one semester or two-quarter course at the freshman or sophomore level. I have used all of the elements of the text in a 10-week, 70-contact-hour course. Elements of all chapters should be included if students are to become proficient in using the computer as a tool. However, they do not all have to be presented in the same course. The applications chapters, for instance, could be included as an adjunct to later courses.

The text assumes a basic knowledge of physics, calculus, and mechanics. Students should have completed a first course in these areas before using the text. They will need to have some exposure to college-level work before they can appreciate the value of computer applications. To provide realistic examples, I have included some concepts in the text to which freshmen or sophomore engineering students may not have been exposed. Sufficient background information for these concepts is included in the body and the appendices of the text. A lack of theoretical background should not be a problem.

Students must develop many original programs and applications if they are to understand the material in the text thoroughly. Developing original programs will also help them become effective engineering computer users. I have found that frequent exams or quizzes are important to ensure that students are keeping up and are understanding the material presented. Some type of term project is also an important ingredient in mastering the subject matter. First, a project presents a comprehensive application that uses previously developed procedures. Second, it gives students experience with the documentation and communication of results.

The use of a computer laboratory is important for learning and teaching operating systems, high-level languages, and applications programs efficiently and thoroughly. It is very difficult to teach computer programming and applications skills solely through classroom lectures. The laboratory environment will improve group dynamics by showing everyone that they are experiencing the same difficulties and it will provide nearly instantaneous feedback and correction of errors that students commonly make when they are learning computer applications. The laboratory optimizes time and allows the instructor to make sure that the whole class has reached some minimum competency level before continuing to the next topic. With the instructor acting as a coach, the frustration and anxiety levels in students will be significantly reduced. The computer laboratory should include individual work stations and a large monitor screen for the instructor that can be seen by all students. Bringing a computer projector into the classroom will also help in demonstrating concepts, especially when alternative approaches are available for accomplishing the same end.

To accomplish the goals of this text, students must have microcomputer applications programs available to them. The programs can be the ones illustrated in the text, or they can be programs with similar capabilities. Buying student versions of multiple packages can be expensive. In some cases, site-licenses may be purchased that allow unlimited copying of the program at a reasonable price. Another approach is for engineering departments to buy sufficient packages for the computers available to students. The software can either be installed in the machines, or checked out to students through a library.

After the students complete the introductory computation course, the faculty and the curriculum should require the students to continue to use the computer in an appropriate manner during course work.

Why does the text use FORTRAN 77 as a high-level language? Several languages have been presented as a replacement for FORTRAN. FORTRAN, however, is still with us, and is doing admirably. It is the language of choice of engineers and scientists. Even if a better language for these applications comes along, it will take a long time for it to supplant FORTRAN, because of the large body of FORTRAN software already available. FORTRAN 77 incorporates the latest official enhancements to the language, many of which have been adapted from more structured languages such as PASCAL. FORTRAN 77 removed many of the earlier weaknesses of FORTRAN. It has developed into a robust language that should be useful to engineers in the future. An updated FORTRAN standard is now being reviewed for adoption.

I have not tried to present a comprehensive users guide for the microcomputer applications packages used in the text. My objectives are to introduce students to the applications, and to help them develop an appreciation for which product is most appropriate for a particular application. They will develop increased competence and confidence through practice and reference to the documentation for the products.

Since software products are continually being updated, newer versions of some of the packages described in the text may become available while the text is being used. It has been my experience that new versions of applications packages are enhanced older versions. Newer versions will add competitive features or remove some previous limitations. Therefore, capabilities and procedures described in the text should still be correct. I will make every effort to update the package descriptions in a timely fashion. Instructors can add details of updated versions in the meantime.

Mention of commercial hardware and software products does not constitute endorsement of one product over another by either the author or the publisher. Comments on the performance of hardware and software are derived from my experience with a certain product.

Programs and applications in the text have been selected for their educational value. They have been carefully developed and verified, but are not intended to be used for any purposes other than education.

FEATURES OF TEXT

Several features of the text help to make it a more effective learning tool. The most significant feature is its integration of various computer software and hardware tools into the context of engineering problem solving. This allows students to develop an appreciation of how to select the best tool for a particular application.

The text emphasizes the communications aspects of computer applications through structured program design and program and applications documentation. A separate chapter covers applications development and documentation.

Emphasis is placed on making computer programs and applications as general as possible. The dangers of making programs too general are also discussed.

The text presents a generic form of flowcharting and pseudocode writing that is applicable to all types of computer applications. FORTRAN and applications package examples illustrate more specific forms.

Realistic and practical applications are emphasized in examples and exercises, making allowance for students' limited backgrounds. Many examples and exercises develop as the chapters progress. These use FORTRAN in increasingly sophisticated applications as students acquire more knowledge of FORTRAN. They also illustrate alternate solutions by using different microcomputer applications packages. The continuity in examples and exercises will help students determine which computational tools may be more appropriate for general classes of problems.

Examples and chapter-end exercises are drawn from the engineering disciplines of chemical engineering, civil engineering, electrical engineering, industrial engineering, and mechanical and aerospace engineering. Examples and exercises also cover the general concepts of data analysis and statistics, economic analysis, engineering mathematics, numerical methods, and probability and simulation. This text provides a good introduction to the numerical methods that are essential to engineering analysis. Examples illustrate the differences between analytical and numerical solutions for calculus problems. This text will provide students with an understanding of numerical precision and accuracy. To aid the student in locating referenced examples, a List of Examples follows the Table of Contents.

Chapter-end exercises cover a range of difficulty and sophistication to accommodate student needs during the first two years of the engineering curriculum. Exercises are separated according to discipline, and several exercises for each discipline are continued within all chapters. Exercises are identified by topic. Numbers in brackets before an exercise title indicate chapters in which related exercises may be found. [ALL] indicates that related exercises are contained in Chapters 2 through 9. Separate exercises develop syntax knowledge or develop modifications of text examples. These exercises make good exam questions. Appendix A contains selected solutions for the syntax exercises.

Examples in the chapters are structured to show students how to approach problem solving with the computer. Elements include:

Statement of problem

Mathematical description

Algorithm development

  Input/output design

  Numerical methods

  Computer implementation

Program development

Program testing

Several appendices are included. Some of these develop theoretical concepts used in examples and problems throughout the text. They provide students with a ready source of information, minimize duplication of theoretical concepts, and allow chapters to focus on computer applications.

Additional appendices contain solutions to selected problems; standard ASCII codes; FORTRAN intrinsic functions; and information on dimensions, units, constants, and conversions.

A diskette containing examples, exercise solutions, and pertinent data files is available to instructors. These files can be used in any way the instructor sees fit. Some could be used as subprograms for development of advanced applications. Some could be used as class examples to show the effects of program or data modifications. Some might be given to students as program files to be modified for new applications. The purpose of the diskette is to save time for both the student and the instructor, so more time can be spent on learning important concepts.

INTEGRATION OF COMPUTERS INTO THE ENGINEERING CURRICULUM

A course based on this text is not sufficient to ensure the computer proficiency of engineering graduates. Use of computers has to be integrated into the curriculum to develop students' competency. This will also give them the familiarity with computer operations that they need, and will make them more comfortable with the computer. They do not have to program in every course, but they should use the computer in every course where it is appropriate. They have to be able to sense the natural use of the computer as a tool. They must also have sufficient computer resources available.

Placement of the first computer applications course in the curriculum is open to debate. Many schools put the introductory programming course in the freshman year, primarily because it is hard to find engineering courses to include in the first year. But since a majority of the computer applications that students will undertake do not start until the junior year, they may have forgotten what they learned about programming in their freshman year.

I believe that the computer applications course should be delayed until the sophomore year, closer in time to when students will actually use most of the applications. They will also have already taken some of the introductory engineering courses, and thus will be more prepared to develop the problem definitions, mathematical models, and algorithms necessary for computer applications. Development of computer algorithms depends on both proper problem definition and on development of realistic models. Therefore, students will profit more from a computation course when they are better prepared for it.

Students can still be introduced to computers in the freshman year. They can use previously developed programs, laboratory data reduction and analysis, graphics, and word processing. They can also learn about the operating system for their computer. All of these interactions with the computer will help them get over their trepidation of "the machine." These early applications will introduce students to the value of the computer as a problem-solving tool and will also improve students' performance in their first computer applications course. However, proper problem definition requires some degree of engineering sophistication on the part of the student engineer. Development of realistic models is facilitated by work in college level mathematics and physics courses.

COMMUNICATIONS ASPECTS OF COMPUTER APPLICATIONS

Engineers must clearly communicate concepts and results to all types of audiences. Structured computer programs allow them to communicate the logic and syntax of a computer application to the intended audience. Carefully designed output, whether in tabular or graphic form, will communicate results. Documentation for programs will communicate to other engineers how to use and check the programs. As students develop their expertise in computer use, they should also develop their ability to communicate computer programs and results.

T. K. J.

ACKNOWLEDGMENTS

STUDENTS

I express my appreciation to the many students who have helped develop and refine problems over the years. Students who helped develop specific programs are identified in the title block for the program.

REVIEWERS AND CONTRIBUTORS

The following people reviewed the manuscript: Shan Somayaji, California State Polytechnic University, San Luis Obispo; Norman Laws, University of Pittsburgh; Michael G. Zabetakis, University of Pittsburgh; George F. Engelke, California State Polytechnic University, Pomona; Theodore J. Weidner, Rensselaer Polytechnic Institute; Robert T. Alguire, University of Arkansas; Mardith B. Thomas, Iowa State University; Barbara Ann Sherman, University of Buffalo; Joseph E. Saliba, University of Dayton; Jerry R. Bayless, University of Missouri-Rolla; Richard P. Ray, University of South Carolina; James M. McDonough, University of Kentucky; and Jon K. Jensen, Marquette University. They made many fine suggestions that enhanced the final product.

Mr. Todd M. Piefer, Product Manager, Universal Technical Systems, Inc., reviewed the section on equation solvers. His careful analysis contributed a great deal to the final development of the material in that section. He also provided considerable technical support and software for TK SOLVER applications.

Mr. Barry Briggs, Lotus Development Corporation, provided technical support on product enhancements for LOTUS 1-2-3.

JOHN WILEY & SONS, INC.

Ms. Charity Robey Forman, as sponsoring editor, has contributed a great deal of support and encouragement, and many helpful suggestions to this project. Mr. Richard Blander, Senior Copy Editor, significantly improved the manuscript through his careful review and editing.

MANUSCRIPT PREPARATION

The manuscript was prepared by using an IBM PS/2 Model 50 microcomputer and the LOTUS MANUSCRIPT word processing package. Commercial software packages used to develop applications include TK SOLVER, LOTUS 1–2-3, and ENABLE.

T. K. J.

ABOUT
THE
AUTHOR

Thomas K. Jewell is a Professor of Civil Engineering at Union College. He graduated from the United States Military Academy at West Point in 1968. He received his MS degree in environmental engineering and his Ph.D. in civil engineering from the University of Massachusetts at Amherst. His training in computers began at West Point. During his graduate education, he used computers extensively in his research on the modeling of storm water runoff from urban areas. Since joining the faculty at Union, he has taught programming and computer applications courses for engineers and has integrated computer use into all of his undergraduate engineering courses. He has stressed effective use of the computer for engineering analysis and design, and the importance of choosing the best computer resource for a particular application. He has published several articles on curriculum and computer applications development, and is the author of a text on systems analysis and design for civil engineers.

CONTENTS

Appendix H: Probability, Statistics, and Simulation   737

Appendix I: Electric Circuit Analysis   745

LIST OF
EXAMPLES

This list of examples is designed to help you quickly locate referenced examples in different chapters.

CHAPTER ONE

INTRODUCTION

1.1 INTRODUCTION

The computer is one of the most powerful tools available to an engineer. A knowledge of computers, how they work, and how to use them is essential to your career as an engineer. In this text you will be presented with an integrated approach to learning about various computer tools that are available to the engineer. You will gain an understanding of how to choose the best tool for a particular application.

Presentation of these computational tools separately, in different courses, or on a "need to know" basis does not give you a sense for how to use the set of tools effectively. This is similar to learning about different hand tools, but not knowing how to use the tools together to build something useful. It is also similar to knowing how to use different saws but not knowing what jobs each is most useful for. You would not be proficient tool users because of your lack of ability to accomplish practical applications. This text's integrated approach is meant to eliminate some of the problems associated with specializing in just one facet of computer usage.

If this is your first computer course, do not feel intimidated by classmates who appear to be computer experts. You will all be together in your struggles with defining engineering problems and using the computer to assist in solving them. Once you have mastered the techniques presented in the text, you will feel comfortable with using the computer as a tool. You will find that the computer can help you both understand and solve engineering problems. Your first course will not make you an expert in all aspects of computer applications. However, it will give you a good base from which to continue your development.

If you have not done so already, take a few minutes to read through the Preface to the text. The Preface provides some interesting background information on the role of computers in engineering education. It also gives you an overview of what you will be learning in this course. Chapter 1 will introduce you to computers and will help you understand their capabilities and limitations.

1.1.1 What is a Computer?

Basically, a computer is any machine that automates computation. However, we often limit our perception of a computer to electronic machines that are programmable, will accept input data, operate on that data, and produce output. By programmable, we mean that the machine is capable of accepting and acting on a series of commands given to it by the user. Input refers to any data that the user wishes to enter into the computer system, whereas output refers to data that the computer has generated or manipulated through its operations, and that the user wants either to print out or save for future use. Computer systems consist of several modules, or devices, that accomplish specific tasks. A central processing unit (CPU) accomplishes computations and coordinates and controls the operation of the other parts of the computer system. Input and output devices allow you to input data and output results of computer operations. Data-storage devices allow you to save data and programs for later use. Some devices may serve all three purposes. The CPU may also be connected to other computer resources through a network. The equipment part of the computer system is known as the *hardware*. Figure 1.1 shows a schematic of the hardware components of a computer system.

Software is the name applied to the sets of instructions, or programs, that are used with the computer system. Programs allow a computer to accept input data, to operate on that data to produce useful information (more data), and to output both the input and generated data. Computer programs and data are organized into files.

A *file* is a related set of data that you want to be able to save and use as a separate entity. A file has to have a beginning and an end and must be assigned a unique name to identify it. Normally, program and data files will be separate. In order to use files you have to have some way of creating, editing, saving, and manipulating them. The *operating system* is the set of software programs that allows you to use files. The operating system also coordinates the various data processing activities of the computer hardware.

The computer would be useless without programs. Engineering applications software are the programs that are used to accomplish engineering analyses and

FIGURE 1.1   Computer system hardware components.

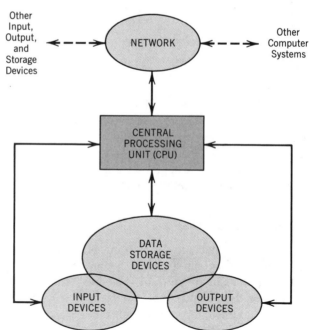

designs. These include, but are not necessarily limited to, high-level languages, spreadsheets, equation solvers, database management systems, and graphics packages. Programs can take on many different forms, as you will see as you progress through this text.

You will quickly realize that the computer is merely a piece of inanimate hardware, like any other tool. You, as the user, have to make the computer do its work. This is what programs are for. The computer cannot think. It merely carries out instructions provided through programs. You will begin to write simple programs in Chapter 3 and will continue to expand on the complexity and sophistication of your programs throughout the rest of the text.

1.1.2 Computer Applications

Computers can be used to automate many engineering applications. When they are used effectively, they produce results that demonstrate an increase in productivity and a reduction in numerical errors. The results of improper computer applications can be disastrous, particularly in engineering, where people's lives may be at stake. This text will help you to judge when computer applications are appropriate and will place particular emphasis on checking the accuracy of your results. This section will describe some of the areas in which computers can help you become more efficient and accurate in your engineering work.

1.1.2.1 Analysis and Design

Computers have facilitated the automation of many previously tedious and/or repetitive computational design procedures. Even when they are assisted by a calculator, humans are prone to making simple mathematical or entry errors that can render a lengthy computational sequence worthless. Once the computer is correctly programmed to complete a sequence of operations, it will continue to repeat this same series of operations flawlessly as many times as is necessary. Some of the analysis and design problems for which computers have been used effectively include the following:

Analysis of stresses in frames and machine parts

Analysis of complex chemical reactions and equilibria

Design of complex microelectronics circuits

Modeling of scheduling and manufacturing optimization

Modeling of unsteady flow of fluids through conveyance systems

The design of a new integrated circuit illustrates the value of the computer. Integrated circuits are the miniaturized electronics circuits used to form the heart of digital computers. Once engineers decide precisely what they want to create, they use specialized computer tools to make schematic drawings of the integrated circuit. The drawings are then subjected to computer-aided circuit simulation routines that check the logic of the design. If there are problems, then the schematic drawings are modified until everything works. The computer is then used to convert the schematic drawings into multilayered physical layouts that can be used in the manufacturing of the silicon chip that contains the integrated circuits.

Computers have also allowed the development of new procedures that, because of their computational complexity, were not possible before digital computers. The theory for many of these procedures was known prior to the implementation of large-scale computers, but the computational processes necessary for solving them were too involved to undertake by hand. There are other procedures that have been developed largely because computers are available. For example, methods for finding approximate solutions for complex differential equation mathematical models require the computational power of computers.

1.1.2.2 Database Management

Databases and database management are used by almost all engineers. A database is any large body of data that is used for some specific purpose or purposes. The data gathered by the United States Census Bureau is an example of a database. A database management system provides the means of accessing a subset of data within the database and using the data to produce useful information. The database has to be organized in such a manner that the database management system can work efficiently with the data within it. The database management system will often be used to create the organization of the database, and then to enter and store, or save, the data in the desired format. The management system should be able to edit and correct data as necessary, and be able to add more data as they become available. After accessing the data, the management system may use them to produce reports, combine and use the data for further analysis using its own programming language, or feed the data to another program for further computations.

The database management system may accomplish one or many tasks related to the database, depending on the uses desired. You may only want to extract data from an established database, with no need for modifying any of them. A database management system could extract selected data from the database and feed them into the appropriate spot in an analysis or design program. For example, you might want to extract data on material and/or section properties from a database for use in a structural analysis and design program. If there is an established database from which to draw information, the use of database management will free you from having to input data yourself. Other types of databases include the following:

Graphics information and data

Water quantity and quality data

Personnel and finance records

Industrial production data

Library card catalogs and information retrieval indexes

Mailing lists

A database management system will allow you to use, modify, and replace data in these databases.

1.1.2.3 Computer-Assisted Data Acquisition and Analysis

Computer-assisted data acquisition and analysis can save a great deal of time and money. It can facilitate the collection of large amounts of data which could not be accomplished efficiently by manual methods. When you use it properly, computer-assisted data acquisition can automate the entire process of data gathering, transmission, processing, and placing information into a database.

Transducers are the primary interface between the acquisition system and the physical system for which data are being collected. Typical transducers measure pressure, strain, temperature, or pH. Pressure and strain transducers work on the same principle. As the gage is deformed, its electrical resistance changes. The change in resistance is measured and correlated with the pressure or stress being applied. Temperature transducers are thermocouples that change resistance with increasing or decreasing temperature. A pH meter measures the hydrogen ion concentration in a solution. All these transducers have to be calibrated using known standards, and all of them output a continuous, or analog, voltage. These continuous voltages have to be changed into equivalent digital pulses for computer processing. This change is accomplished by an analog–digital converter. The digital signal can be used by any digital computer. Data can be transferred to the computer over dedicated telephone lines, over telecommunications links, or over long distances by satellite relay. Software within the computer interprets the digital signals as numerical values. The numerical values can be stored in a database, and any necessary analysis can be performed. Graphical and/or tabular output can be generated.

1.1.2.4 Process Control

Computers can be used to control processes that require monitoring and adjustment of the process to maintain output quality. Processes requiring this type of control are especially prevalent in chemical or manufacturing plants. However, the methodology is equally applicable to control of a hydroelectric generation plant or a water supply system. Process control is also called real time control because it is accomplished simultaneously with the process. The control systems are often referred to as supervisory control and data acquisition systems, or SCADAs. The SCADA involves database, data acquisition, and control applications.

An application of this type of system is used to automatically control the pumping and distribution of water from the Rio Grande aquifer used by the city of Albuquerque, New Mexico (*Insight*, July/August 1988). Microwave radio equipment is used to gather data on pressure, flow, and reservoir levels from 89 wells, 27 pump stations, and 41 reservoirs dispersed throughout a 200-square-mile area. These data are checked against specified operating limits. Alarms are sounded for out-of-tolerance data, so that operators can check them. The data are logged and processed, and new data are calculated for use in control. The SCADA system determines an optimized pumping strategy that saves the city $700,000 annually in energy bills alone. Pumps are started up or stopped to meet the demands of the system. Pumps are stopped if emergency conditions are detected. Valves are opened, closed, or set at intermediate points to control the flow in the system.

The computers used in the application are fairly inexpensive, so a completely separate system is kept for backup in case the primary system should fail. Thus, a computer failure will not interrupt the flow of water. The SCADA system reports the status of the city's wells, pump stations, and reservoirs to the operator and alerts the operator to dangerous conditions, such as chlorine leaks, high reservoirs, and power failures at any of the remote sites. The system can also produce a variety of reports showing historical trends and pumping activity.

1.1.2.5 Expert Systems

Expert systems involve the development of a knowledge base derived from past experiences. This knowledge base is then used to select the operating policy for the system based on the state of several key parameters. There are separate computer languages, including LISP and PROLOG, that are specifically designed to set up the structures to catalog the knowledge that is acquired. They also have structures to manipulate the knowledge to help in making decisions. Expert systems are sometimes referred to as artificial intelligence, but that is really a misnomer. There is nothing artificial about it. The programmer establishes what knowledge will be stored, and then develops the rules that will be applied to the saved knowledge. Based on information supplied by the user, the rules establish how the knowledge base will be manipulated and what information will be given to the user. Typical applications of expert systems are construction claims evaluations, construction risk analysis, and bridge evaluation and selection.

1.1.2.6 Robotics

Robots are being used extensively in manufacturing processes. The construction industry is developing new robotics applications. Robots are also useful for performing tasks in hazardous environments such as nuclear containment vessels or sewers. They are excellent tools for performing repetitive tasks that require close tolerances. The ability of a robot to perform tasks depends on the sophistication of the electronic and mechanical components used in its construction and the quality of the software used to run it. Some sophisticated systems will use data from a computer-aided design to develop the instructions for an industrial robot to manufacture the part or assembly. This is referred to as Computer-Aided Design/ Computer-Aided Manufacturing, or CAD/CAM. As an engineer you may become involved in the design or programming of robots or in the design of processes using robots.

1.1.2.7 Graphics

Graphical representation of objects and systems is an essential element of engineering design and production processes. Without graphical representations, an engineer's conceptual design could not be translated into reality, unless the engineer did all the work from start to finish. This would be impractical for any but the simplest projects. The computer is a natural adjunct to the construction and storage of graphics images and databases. The usefulness of computer graphics as an engineering design tool depends on the ability of engineers to develop the graphics software to represent the system being designed.

All graphics are based on coordinate geometry and the ability to express lines and curves as either mathematical functions or loci of points. The computer can store information on the points, lines, and curves in a database. Computer programs can modify, add to, or delete from the database, or output the graphics image. Often-repeated shapes and details can be stored and recalled as needed. You will probably take some type of computer graphics course in your undergraduate education. This will give you a feel for the methodology of using computer graphics systems and will help you master any system that you may use in practice.

Conventional drafting involves the two-dimensional representation of three-dimensional objects. Computer-aided drafting systems significantly increase the productivity of draftsmen because of the ease of making changes in the drafting database and the automation of the actual drawing process. The computer can also be used to store three-dimensional graphics databases. All points in these databases have x, y, and z coordinates. With a three-dimensional database, the computer can be instructed to generate any view of the object, whereas with a two-dimensional database, you are limited to the views stored in the database. You can also specify the angle at which the object will be observed. Three-dimensional computer graphics can help you visualize objects and spatial relations. Some three-dimensional graphics programs will accomplish shading, cuts through an object at any desired angle, hidden-line removal, and even animation.

Computer-aided graphics does not replace the necessity for you to study conventional graphics. You must still know how to sketch objects and draw plans, and how to apply spatial relations to the laying out of drawings. As an engineer you may not actually enter and manipulate computer graphics data. However, you will have to interact with computer-aided drafting technicians. You will have to prepare preliminary sketches for the draftsman who will input the data to a computer-aided graphics program. You will also have to be able to interpret the plans produced.

In your engineering career you may accomplish design work with the assistance of a graphics workstation that requires extensive interaction with the computer. If you are involved in machine or electronics component designs, you may use computer graphics as a primary design tool. You may use graphical simulation and animation to model dynamic processes, or to analyze interactions and identify conflicts among system components.

You will almost certainly use computer-generated graphics to illustrate your problem definitions and solutions. As an engineer you will be expected to write many reports for, and give many presentations to, a variety of audiences. In your reports and presentations, you will have to include data pertinent to the concepts involved. Graphical representation of these data will help your audience assimilate the information and will assist them in making comparisons among the data presented.

Later in this text we will explore the use of computer-aided graphics in your presentation and analysis of data. These applications will introduce you to the use of special graphing routines provided by several of the popular microcomputer generic software packages. The presentation graphics modules of these packages offer several options for displaying data. Pie charts are effective for viewing the fraction of a total data set with a particular attribute. Vertical bar charts or *X-Y* plots illustrate the variation of data over time or with any independent variable. Several dependent variables can be included in the same plot for comparison. Histograms are specialized bar charts for illustrating frequency distributions. Each bar represents the number of times some value or range of values occurs in a sample. Scatterplots are specialized *X-Y* plots that show the relative location of the data points without joining one point to another. Many times the best fit line through the data group is superimposed on the plot. Your choice of graph type has to depend on the type of data to be presented and on the audience that will be using the information. You should make your graphs as simple as possible to avoid confusion, while at the same time you must include sufficient information to make the intent of the graph clear.

Other typical applications for computer graphics include surveying and mapping, structural design, microprocessor design, and computer-aided design and manufacturing (CAD/CAM). Application of the computer to surveying and mapping allows the engineer to go from field data or database to finished contour map in one step. However, frequent updating of mapping databases may be necessary. In structural design, computer-aided graphics can be integrated with analysis and design. Computer programs can reduce misaligned or erroneously placed members and structural components by checking for interference among the elements of the graphical database. Computer graphics can assist in the layout of microprocessors. Other programs can use the resulting images to produce the circuits. CAD/CAM programs can integrate the graphical analysis and design process with manufacturing. Sophisticated packages allow the user to go straight from drawings to instructions that will control production machines.

Computers can store any type of image that appears on a television screen. The screen is divided into hundreds (or thousands) of tiny rectangles called pixels. Phosphors coat the inside of the screen. A cathode-ray tube (picture tube) aims a beam of electrons at each pixel. The electron beam governs the intensity and/or color (in the case of a color tube) of the phosphor within a particular pixel. The beam moves very quickly over all the pixels on the screen, and the composite of all the pixels forms the image that we see. In television, the image is updated as the action progresses. The speed of updating is such that we see it as a continuous process, although the updating is actually done in discrete increments.

A VCR tape stores the information required for image updating. A computer can also store the data for a screen image and can instruct the cathode-ray tube to reproduce the image. The computer can store a series of related images and output them in rapid succession to form a type of animation. Computers can also update the instructions to the cathode-ray tube using three-dimensional graphics databases. These instructions could produce an animated scene such as a space ship docking with a space station. Motion picture producers have taken full advantage of the graphics capabilities of computers to produce some of the spectacular space and animated movies of the past few years. This animation capability can also test the interaction between two parts in a mechanical system, and has many other engineering design applications.

Some confusion exists over nomenclature in the computer-aided graphics field. You will often hear the acronym CADD (Computer-Aided Drafting and Design), which really does not fit the process for which it is intended, the use of graphics to assist in design. Computer-aided drafting is really a separate entity. As discussed previously, it is the automation of the drafting process to produce blueprints or other drawings that can be used to develop an end product. Computer-aided drafting may be linked interactively with design, but it may also be driven by design input. On the other hand, computer-aided design can be effectively undertaken in many situations without any reference to computer-aided graphics.

1.1.2.8 Word Processing and Publication

Whether we like it or not, writing occupies a significant portion of an engineer's time. We may have to write a report for a project we have been working on for a year, a letter to a client on a problem just uncovered, a memo to our colleagues asking for input on a particular problem, or a laboratory report for experimental work done as part of a project. It has been estimated that 10 to 20 percent of an engineer's time is spent in writing activities. This amounts to 800 to 1,000 pages per year (Eisenberg 1982).

The text editor that you will use to produce your FORTRAN programs later in this text accomplishes many of the tasks of a word processor. You will have to know the commands for your particular text editor before you will be able to enter and run any programs. Your text editor will allow you to input text, save and modify that text, and move blocks of text around. For FORTRAN, the text will be in the form of computer instructions that will be interpreted by the computer. The computer will then accomplish the desired tasks or perform the desired calculations.

Word processing involves using the computer to input, store, modify, and output textual information. Word processing probably does not speed up the initial typing of a manuscript, since typing is typing. However, it does allow you to start with an outline and expand on that outline as you develop your thoughts and logic. Once your text has been entered, it can be edited, text deletions or additions can be made quickly, and a new copy of the manuscript can be printed out. You can easily move portions of the outline or manuscript around as you see better ways of organizing your thoughts. You can move individual words or large blocks of text. Most word processing programs allow you to search for a particular sequence of characters, or character string, and either replace that string with a new group of characters or mark the text for further editing.

Most word processing programs will give you a fair amount of latitude in how your text is output. Different character shapes or fonts, character sizes, and margins can emphasize desired portions of your text. You can easily change text formats. These formats provide information on how the text will appear when it is printed. Facilities are usually available for importing data from text files or applications package output files into your word processing documents. These files may include addresses from a database, spreadsheet data tables, or graphics images. The incorporation of one address from an address list database into each copy of a document is called mail merge and is the source of the "personalized" computer letters that you receive in the mail.

When you choose a word processing system, you need to make sure that the package will accommodate documents of sufficient length for your uses. Also, consider the speed of moving through the document. Some packages accommodate documents of almost any length by storing only a portion of the document in the active computer memory. Shifting from the top to the bottom of the document may proceed slowly as the computer reads in succeeding portions of the document from the data storage device. If the whole document is kept in computer memory, shifting from one point in the document to another will proceed quickly.

Many word processing packages provide you with spelling checkers. Although these are valuable, they will not catch all errors. If you should incorrectly type the word "their" for the word "there," the computer would recognize "their" as a properly spelled word and would not flag it to be changed. Some packages also have a built-in thesaurus.

A good word processing program will support output to a variety of printers. The different types of printers will be discussed in the computer hardware section of this chapter. The recent advances in the quality of output devices have spawned the concept of desktop publishing. Desktop publishing provides a means of going straight from a word processing document to a publication quality manuscript using only your personal computer. It takes a practiced eye to spot the difference between documents produced by high-quality desktop publishing systems and documents produced by commercial typesetting systems.

This text will not attempt to teach you word processing. There are too many different packages with different capabilities to do that. Almost any word processing package supported by a university or college computer center will meet your needs. Most computer centers also conduct classes on the particular package supported by the college. When you take a job with a firm, you may find that a specific package with specialized capabilities is available. Fortunately, the transition from one package to another is fairly easy, as most accomplish the same basic operations.

Once you have mastered a word processing package you may find that you will become a better writer. It will be simple to change or rearrange elements in a document and to get a new copy printed out whenever you need one. You will be free to concentrate on getting your thoughts down and can improve your presentation through several easily accomplished revisions.

It should be noted that the ability to type well is an invaluable skill for word processing and computer programming in general. The old hunt-and-peck method will slow you down tremendously. A computer can process information only as fast as the information is input. If your hand-eye-typing coordination is poor, you might as well write everything out longhand and give it to a typist for input. If you are not proficient in typing skills, it would be worth your time to improve them. Many high schools provide continuing education courses in typing. Or the computer can be used as a tool to teach typing skills. There are several good typing tutorial programs that you can use. Your reward for the time you spend on improving your typing skills will be increased productivity over your years of computer use.

1.1.3 Engineering Computer Literacy

The applications already discussed should indicate to you that there is a need for engineers to be familiar with the computer if they are to be able to harness its potential value for computer-aided analysis, design, and graphics.

What then is computer literacy? Many people use the term "computer literacy" to refer to some minimum level of competence that any engineer should have with respect to computer applications. However, there is no absolute definition of what computer literacy is. It may vary considerably depending on the discipline, and even the type of work that an engineer is doing. There are still some engineers who are getting along nicely without using the computer at all. There are many others who could not possibly do their jobs without using computers. It is my experience that new engineers looking for their first jobs are at a distinct disadvantage if they are not conversant with computer applications. Many firms look to the young engineers to strengthen their computer applications abilities.

The following is a list of what I feel are the essential elements of computer literacy needed by engineers. You may not actually use each of these elements in your work, but you will not really understand computers until you are familiar with each. You could compare computer literacy to driving a car. You do not have to parallel park on a hill each time you drive, but you have to be able to do it when the need arises. The elements of computer literacy are as follows:

1. The ability to recognize problems for which computer solutions are appropriate
2. The ability to use the computer to implement these solutions
3. The ability to program in high-level languages (FORTRAN, BASIC, PASCAL, etc.)
4. The ability to use generic microcomputer packages (spreadsheets, equation solvers, etc.)
5. The ability to develop, test, and document a structured program application

6. The ability to determine which computer hardware and/or software products are best for a certain application

A thorough knowledge of both computer hardware and software is essential to developing engineering computer literacy.

1.1.4 Considerations in Using the Computer as an Engineering Tool

When you decide whether or not to use the computer as an engineering tool, first consider if the benefits will outweigh the costs. Next, evaluate the state-of-the-art in computer technology as it relates to the application at hand.

1.1.4.1 Costs vs. Benefits

Computers are extremely powerful tools, but they can be used improperly. Engineering firms are in business to make a profit. For computers to have a place in everyday operations, the benefits generated by computer usage have to outweigh the additional costs involved. This is an especially important consideration in computer usage because it is easy to overlook some of the costs. Also, there is a tendency on the part of engineers to experiment with the computer and undertake tasks that do not produce monetary returns.

Increased productivity and reduction in errors are two of the major benefits that should come through computer usage. However, you have to identify who in the organization achieves increased productivity, and by how much their productivity is increased. You also have to determine how much the firm is likely to save through reduction of errors. You may find it necessary to have a certain level of computer capability in order to be competitive in bidding for contracts or in applying for grant work. More and more requests for proposals are requiring computer capabilities. Sometimes clients will go so far as to specify what computer resource is to be used.

Costs for computers include much more than just the capital cost of the equipment. You have to include operating, maintenance, and personnel costs, as well as costs for supplies, printer ribbons, paper, and so on, that the computer will use. Employee time spent on actual projects can be billed directly to clients. However, there will also be training and development time that must be charged to overhead. Software costs can be greater than the actual cost of the computer system. If you are going to produce any of your own software, the costs can be very high. As you progress through the text you will find that production of quality software is a time-consuming process. One long-time computer user in engineering education (Fenves 1982) estimates that up to 75 percent of total computer costs can be associated with software purchase, maintenance of programs, installation of program updates, and software development. Up to 80 percent of the software cost can go toward the upkeep and maintenance of existing programs. Branscomb (1982) traced the increase in the proportion of total computer costs that can be attributed to software. In the 1950s, software was 30 percent of the total cost, and hardware accounted for 70 percent. In the 1960s, the portion attributable to software rose to 65 percent. In the 1970s, it rose to approximately 80 percent; and Branscomb predicts that software will rise to about 85 percent of total costs in the 1980s.

Personal computers, or microcomputers, have been hailed as the low-cost solution to many firms' computational needs. However, many firms are looking only at the capital and software costs when they are considering microcomputers. Mr. Glen Orenstein (1986), manager of microcomputers for Stone and Webster Engineering Corp., a large engineering consulting firm, has estimated that hardware and software cost per microcomputer system typically runs between $5,000 and $10,000. Three years is a commonly accepted useful technical life for computer equipment. Thus, the annual costs are between $1,700 and $3,300 per microcomputer. However, when operating costs, support labor, training, maintenance, and software development are added, the total annual cost is estimated to be between $4,000 and $12,000, with $6,500 being typical. Orenstein also conducted an informal poll of computer managers in several large engineering firms. They indicated that the annual costs ranged from $10,000 to $18,000 per microcomputer.

1.1.4.2 Innovations and Advances in Technology

As you begin to use computers, you will be both impressed and awed by the fast pace of computer hardware and software development. You will also realize that most of the advances in technology have been coming from the computer industry. We, as engineers, have been relegated to a reactionary mode, trying to apply this new technology effectively. Many engineers' anxieties about computers are heightened by their perception of the rate of development. They fear that even if they gain some level of competence, it will be impossible for them to keep up without sacrificing too much of their primary function as engineers. However, this fear is largely unfounded. You will soon find that with a minimum level of training you will begin using computers effectively as tools while utilizing the prevailing state-of-the-art technology. As you continue to use the computer, your knowledge will increase. You will realize that it is not necessary to grab every new advance that manufacturers offer. You have time to evaluate new technology (both hardware and software) to see if it will increase the productivity of your firm enough to justify its purchase. As part of your decision-making process you will have to evaluate which applications can continue to be run on older equipment and which applications should be converted to the newer equipment.

1.1.5 Reasons for Using Computers in Education

Although the following discussion is oriented toward you as students, it has a great deal of carryover into your later professional practice.

1.1.5.1 Productivity

As students you are learning how to use computers so you can increase your productivity as engineers after graduation. However, the computer can also increase your productivity as students, allowing you to learn more in a given amount of time. The computer can speed up your calculations and can increase your precision in computations. People tend to make errors in tedious repetitive calculations. If you properly program the computer you can eliminate these errors. You should use the computer to increase your productivity in assignments, even when you are not required to do so by your instructor. Many of the items discussed in the subsequent sections also relate to increased student productivity.

1.1.5.2 Enhance Understanding of Theory and Solution Techniques

The computer can be used to reinforce concepts learned in class. Computer applications can also help you increase your understanding of calculation methods. However, computer applications should not be used in lieu of learning the theory and solution methods for problems. In order to develop a computer application, you must thoroughly understand the theory of the problem, as well as the numerical methods and approximations needed to arrive at a solution. The computer will allow you to investigate alternatives quickly. You can be asked to answer "What if" questions that would be impossible to formulate owing to time constraints if solutions were accomplished by hand. Values of input variables can be changed during successive runs and the effects of the changes on the output observed and interpreted.

1.1.5.3 Make Analyses and Designs Otherwise Impossible

The computer will allow you to undertake realistic problems while the theory behind the application is still fresh in your mind. You will be able to deal with many variables and much larger bodies of data. You will be able to look at many alternatives in a given amount of time. You will be able to search for an optimal, or near optimal, solution in the same amount of time it would take you to find one valid solution by conventional methods.

1.1.5.4 Experience the Acquisition, Processing, and Analysis of Data

It is likely that you will be exposed to computer-assisted laboratory data acquisition during your undergraduate education. The computer has been used as a valuable adjunct to graduate and research laboratory data acquisition and analysis for some time and is finding its way down into undergraduate programs. Use of the computer to process and analyze data will relieve you of the need to make tedious calculations and manipulations of data. It also permits the analysis of realistic data sets and will teach you about the need for attention to detail in any data-gathering program.

Although the computer should not be used to replace all your hands-on laboratory experience, it provides the ability to augment or replace expensive physical models in laboratories. Hands-on use of laboratory equipment helps you gain a feel for the accuracy and precision of laboratory results. Accuracy is the deviation of laboratory results from the true results, and precision is a measure of the reproducibility of results from one experiment to another. Laboratory experience will give you a much better understanding of the accuracy of computer results. You will not automatically accept the computer's output of seven significant figures as representative of the accuracy of the data.

1.1.6 Possible Pitfalls in Using Computers in Education

As is the case with any tool, a computer can be misused. Probably the greatest danger in using computers in both education and engineering practice is the possibility of becoming dependent on others' engineering competence. Engineering applications programs will allow you to find answers for problems that may be outside your area of expertise. However, if you use one of these applications programs, you may not understand how to check the correctness of your results. This would be a clear violation of good engineering ethics principles.

It is important not to treat a computer as a black box that produces output when you provide input. You need to understand what is going on within the computer. Even competent engineers should not use computers if they do not understand how the computer operates and transforms data and if they have not studied the computational processes used in any particular application program. You also have to be careful not to attach unwarranted significance to the number of digits that the computer puts out. Interpretation of computer output must follow the same rules for significant digits applied to hand-calculated results.

The more immediate danger to you as a student is that a computer may allow you to arrive at correct answers without understanding the methods and physical principles involved in solving a problem. Since learning engineering skills is a cumulative process, it is important for you to understand each step of the process as it is taught to you. To do anything less is shortchanging yourself and your profession.

Another important concern is that you must be careful not to depend on computer tools to such an extent that you ignore other computational tools. At times you will have to apply your engineering judgment and make field decisions when you do not have ready access to a computer, and will also have to be able to perform on-the-spot order of magnitude checks to confirm that calculated results are reasonable. Therefore, you will still have to understand the conventional solution techniques. You still need to develop the feel for problems that translates into good engineering judgment. You should use the computer as an adjunct to conventional methods, not as a replacement for them. Maintain your appreciation for the ''rules of thumb'' and common sense that are used by experienced engineers.

1.1.7 Types of Problems Best Solved with Computers

Any problem that requires a large number of repetitive calculations, or that will be accomplished multiple times, is a candidate for computer implementation. A problem that requires a large number of computations in a sequential manner, but will only need to be solved once, is probably not. It would take more time to develop the computer application than it would take to do the problem by hand. When several different alternatives are to be investigated for a problem, the computer can be of assistance. You can solve the problem once, then generate the different alternatives by changing the computer input data. If a problem requires any sort of repetitive calculations to find its solution, you are probably wise to use the computer. The repetitious parts of the problem solution may be common procedures for which computer applications have already been developed, such as solving systems of linear equations. These previously developed applications can be adapted for the problem under consideration. Almost any trial-and-error solution or approximation technique will lend itself to computer solution. The computer can also efficiently deal with large bodies of data such as text, numerical data, and graphical data.

The types of problems that you will attempt to solve with a computer will often be determined by the hardware that you have available to you and by the availability of appropriate software to use with the hardware. If you have a personal computer on your desk and have equation-solving software available, you will be more likely to use the computer for routine daily work than if you have to submit a computer job to a service bureau in a building separate from your office.

1.1.8 Steps in Solving a Problem Efficiently with a Computer

In any computer application, the necessity of understanding the problem and the engineering principles involved in it cannot be overemphasized. All of Chapter 2 will be devoted to these concepts, and they will continue to be emphasized throughout the text. As an engineer, you must search for an understanding of the whole problem and decide how the computer can assist in its analysis and solution. Be inquisitive, and never be afraid to ask for help should you need it.

As you read through the following steps, you will see that only a few of them actually involve hands-on use of the computer. Those that do are identified by an asterisk (*).

1.1.8.1 Problem Identification and Goal Definition

First you have to establish that a problem really exists and determine what you want to evaluate through your problem solution. Determine if there is a need for your solution.

1.1.8.2 Problem Statement

Define the problem by writing out a problem statement. To properly define the problem you will need a thorough understanding of the scientific and engineering principles involved in the problem. You also must carefully define assumptions required to develop a solution.

1.1.8.3 Mathematical Description

Develop a mathematical description (mathematical model) that can be evaluated by the computer. To do this, you need to understand the mathematical descriptions of the engineering processes involved in the problem. You will also need to synthesize these descriptions through substitution and mathematical manipulation to apply them to the problem at hand. You must be able to ascertain that the physical dimensions of the terms of the mathematical model are consistent.

1.1.8.4 Numerical Analysis Techniques

Choose numerical analysis techniques that will allow you to proceed from a mathematical model to numerical answers. Numerical analysis techniques may involve substitution, approximation, and/or repetition.

1.1.8.5 Algorithm Formulation

Develop the computer algorithm. Development of the computer algorithm involves the transformation of the mathematical model and numerical methods into a form that the computer can interpret. It necessitates the development of the logical organization of the computer application. You can accomplish this through decomposing the problem into smaller steps, then successively refining the steps until sufficient detail is present to solve the problem. This is called *top-down design* and will be discussed in more detail in Chapter 2 and in examples in subsequent chapters. Flowcharts can also be used to help clarify the logic of the computer application. Flowcharts, graphical representations of the sequence of steps required to solve the problem, will be discussed in detail in Chapter 2.

1.1.8.6 * Computer Programming

Write the computer program. The computer program is a detailed sequence of instructions prepared in the syntax of the particular language or computer application at hand. The writing of the program will be a straightforward task if the previous steps have been successfully accomplished.

1.1.8.7 * Program Testing

Test the program. Program testing involves the verification of the program output against known answers. If the answers do not match, go back through the program to search out and correct any errors. If you find no errors in the program, you will have to question the correctness of steps leading up to the actual programming. You should be prepared to return to and correct any of the previous steps.

1.1.8.8 * Production

Output your results. Producing useful results is the objective of the computer application. Once the program has been verified, it can be run using the problem input data to generate the desired output.

1.1.8.9 Interpretation of Results

Interpret your results. We have a tendency to accept computer output without careful analysis. Even after the program has been verified against known data, we still need to double check and verify the original problem theory and solution logic. We need to look beyond numerical answers. Even at this stage, we must be willing to recycle to any previous step if it is necessary to do so.

1.1.8.10 Documentation

Document your results. You will have to be able to convince others of the correctness, accuracy, and worth of an application program that you have developed. You also have to give others instructions on how to use your program. This is what we mean by documentation. We will cover documentation in more depth in Chapter 9.

1.1.9 Don't Reinvent the Wheel

The procedures just outlined are for developing a computer application from start to finish. If you have already accomplished similar analyses in the past, you may not have to go through the whole outline. In fact, your last choice should be to develop a completely new program. For cost effectiveness and ease of development, your priorities in choosing applications should be as follows:

1. *Old Solutions.*   Use an existing program or applications model.
2. *Similar Problems.*   Find a problem similar to the one you are working on and modify the program for its solution to conform to the present problem.
3. *Available Resources.*   Write a new program using available procedure libraries or segments of old programs.

4. *Completely New Program.*   Use your resources to develop a completely
   new program.

These are listed in order of increasing cost and increasing difficulty of development.

1.2 HISTORY OF COMPUTING

1.2.1 Technical Developments

Engineers have always tried to model the real world to facilitate creating products
that will work and that also are of reasonable cost. It is generally cheaper and
easier to work with a model than with a final product. Modeling allows engineers
to look at more possible alternative designs before choosing the best one. Mathematical models are one type of model. They represent the product or system being
analyzed by a set of mathematical equations and relationships. In order for the
mathematical model to be useful, an engineer has to be able to obtain a quantitative result. Development of electronic computational devices, or computers, over
the past 50 years has increased the engineer's ability to obtain quantitative results
with mathematical models.

The first electronic computers used vacuum tubes as their primary logic and
memory elements. The Atanasoff-Berry Computer (ABC) was developed at Iowa
State University to solve simultaneous equations in the late 1930s. This was a
special purpose, one-of-a-kind machine. A more general computer was the Electronic Numerical Integrator and Calculator (ENIAC), which was developed during World War II. It was 100 feet long and contained thousands of vacuum tubes.
It was the first successful general-purpose electronic computer, but it had the
disadvantage of having to be rewired every time a new application was undertaken.

In the late 1940s, computers were developed that could read and store programming instructions without having to be rewired. In 1951 Remington-Rand Corporation produced UNIVAC I, a first-generation general-purpose computer. One of
these was purchased and used by the Census Bureau. The vacuum tubes used in
these first-generation computers produced a great deal of heat, were unreliable,
and took up a great deal of space. Input and information storage for these machines was by punched cards. Most applications were limited to business-oriented
data processing, such as payrolls and accounting. The programming languages
used were obscure and complicated. The computer could be run only by a trained
operator. These first-generation computers demonstrated their utility for large-
scale data processing but were not yet ready for the general use of professionals
such as engineers.

The second generation of computers appeared in the early 1960s. These used
transistors in place of tubes for logic elements. Transistors were smaller, faster,
and more reliable, generated less heat, and had lower power requirements. Thus
machines could have more memory, and larger applications could be developed.
Easier-to-use programming languages, such as FORTRAN, started to be developed during this period.

The third generation of computers appeared in the late 1960s with the introduction of integrated circuits. Integrated circuits contain thousands of transistors on a
tiny silicon chip. They led to the development of larger, faster, cheaper, and more

reliable computers. The increased capacity led to the development of time-sharing systems. Time sharing allowed several users to access a large computer simultaneously, with an accounting of how much time and computer resource each application used. This encouraged smaller firms to use the computer. The Beginner's All-purpose Symbolic Instruction Code (BASIC) language was also introduced. BASIC was designed to be a teaching language and was especially suited to the interactive computing environment. Also during this period Digital Equipment Corporation (DEC) introduced minicomputers. These were scaled-down versions of larger computers, which brought computational power to firms that could not previously afford it. Many engineering firms either started using time sharing or bought minicomputers.

The fourth generation of computers was introduced in the early 1970s. These employed large scale and Very Large Scale Integration (VLSI) of circuits. The improved technology produced a single silicon chip about the size of a postage stamp. The chip contained the complete circuitry of the computer's control unit. These were named microprocessors and made possible the development of the microcomputer. The first microcomputer based on a microprocessor was developed in 1975. It was followed two years later by the introduction of the Apple II computer, which was the first mass-produced and mass-marketed microcomputer. It opened up computing power to virtually anyone who needed and wanted it. In 1981 the IBM PC (Personal Computer) was introduced. Developments in both computer hardware and software since that time have been nothing less than phenomenal.

1.2.2 Cost vs. Computational Power

Over the years the capital and operational costs for computers have continued to go down. Mass production of very-large-scale integrated circuits and peripheral hardware have produced lower capital costs. The newer machines also require less space. Operational costs have gone down because of reduced climate control and power requirements, and ease of operation and maintenance. At the same time, the computational and storage capacities of the machines have been increasing. Costs for large computers have dropped at a slower rate than smaller computers. This is because the largest computers offer increased computational and processing speeds in conjunction with very large computer memory. The net result for engineers has been a large reduction in costs per instruction processed. Practical computing power is now affordable by any engineering firm, and by many individuals.

1.2.3 Personal Computers

Since the introduction of their personal computers, IBM and Apple have been vying for control of the personal computer market. IBM seems to have the edge, at least in the engineering field, but Apple is still actively pursuing its market share. There are also several companies that produce microcomputers that are compatible with IBM PCs. Over the past 12 years, the cost of an average microcomputer system has remained approximately the same, but the power of that average system has increased tremendously. The newer machines have much faster processing speeds, can use 20 to 100 times more main memory, and will store 200 to 400 times as much data on a single disk. Users now have access to much more sophisticated software. The interface between person and machine has become more natural. The mystique of the specialized computer operator and programmer has been supplanted by the concept of having a personal computer on an engineer's desk for everyday as well as specialized tasks. In the personal computer market, the price/performance ratio has been improving by approximately 15 to 20 percent per year (Bender 1989).

1.2.4 Academic Computing Environment

Engineering computer usage was in its infancy in the mid-1960s. Computers still used transistors and filled up whole rooms to provide the power comparable to the first Apple II. One job could be processed at a time. Users could actually toggle some of the switches on the main computer to control programming options. The face of the computer contained a bank of flashing lights, which nobody except the computer scientists could interpret. Programming was in obscure languages. Program and data input were accomplished using mark-sensed cards. The user put pencil marks in columns on the cards, and the marks were read by an interpreter. Use of the computer was generally limited to specialized computer courses, and most faculty computer users were specially trained. In the late 1960s, punched cards and FORTRAN were introduced into the curriculum.

By the early 1970s, time sharing had become more prevalent. Instructors increased the use of computers in a broad spectrum of courses. FORTRAN was the language of choice for engineers and scientists. Time sharing provided a more efficient interface between the user and the computer because of its capability for entering, running, and modifying programs from a terminal. Output was sent back to the terminal immediately. This was referred to as *interactive computing*. However, interactive computing also required that many users be able to use the main computer simultaneously. To prevent system overloads, much of the student computer work was still accomplished through batch processing. In batch processing a student's computer instructions, program, and data were submitted as a card deck. A card reader interpreted the card deck and routed the instructions to the processor. The student waited for printed output to come back from the run. Corrections were made by typing new cards for the lines that needed correcting, and the deck was resubmitted. This limited the student to one or two runs per hour, and then only when someone was on duty to run the machine and distribute output. An advantage of batch processing was that it did not require much storage of data on the computer system.

In the early 1980s, the technology for storage of data on magnetic disks improved, and the number of users able to use systems interactively increased. Therefore, most colleges and universities switched from batch to interactive computing for all students. Card decks faded into the storage drawers. The structured high-level language PASCAL increased in popularity. With the increased use of personal computers, BASIC also gained more favor as an engineering language. This was primarily because the early personal computers did not have the capacity to run more sophisticated languages. Developers also continued to improve FORTRAN. With three languages vying for the limelight, educators were left with the question of which language should be taught to students. This question still has not been settled, although FORTRAN is probably the language that has been used in a majority of the applications over the years. All three of these popular languages are now available in microcomputer versions.

Computers have been more thoroughly integrated into engineering curriculums in the 1980s. Today there are appropriate uses for the computer in most engineering courses. They are increasing your learning productivity. Computer use is also driven by pressure from industry and engineering practice. Employers expect graduates to be computer literate.

1.2.5 Future Outlook

Personal computers will continue to become smaller, more powerful, and cheaper. They will soon replace calculators as the everyday computational tool of choice for engineers. There are already hand-held computers on the market that are not much bigger than a calculator. Engineers will be able to use these computers in the field. When they return to the office they will be able to plug the portable computer into computer networks. They will have access to any level of computing resource necessary for further analysis of the field data.

1.3 COMPUTER SYSTEM HARDWARE

Hardware is the actual equipment of computer systems. As you know, the technology behind computers is evolving rapidly. The rapid change has led to the assumption by many computer users that the useful technical life of computer equipment is about three years. However, technical obsolescence has very little relation to mechanical life. It has been my experience that computers, especially microcomputers, are amazingly dependable pieces of equipment. This is not to say that they do not require maintenance, but their service to down-time ratio is very large. Also, machines that have been replaced by more state-of-the-art equipment can still be used effectively to do specialized tasks.

1.3.1 How Computers Store and Process Data and Instructions

Computers and calculators accomplish arithmetic operations in different ways. Calculators use analog circuitry. Arithmetic operations are accomplished by measuring the flow of current through circuits. Digital computers, on the other hand, use counting procedures to accomplish arithmetic operations. The basic building block of the computer memory is a microtransistor that acts as an on–off switch. The on–off can be used to represent zero or one in the binary number system. Therefore, data are stored in the computer in binary format. Each zero–one switch represents a "bit" of data in computer parlance. These bits are joined together to represent numbers and computer instructions. The computer takes data and instructions in decimal and text form, converts them to the binary format that the computer can use, operates on them, converts the results back to decimal and text form, and outputs results. The process is controlled by the programs and data that we provide. The computer accomplishes these tasks in a matter of microseconds for programs of moderate size. Calculators provide fast and accurate computations for straight arithmetic operations, but they lack the digital computer's flexibility and storage capabilities in working with large quantities of numerical and text data.

1.3.2 Classification of Computer Systems by Type

Computers are generally divided by type into mainframe computers, minicomputers, and microcomputers. There used to be rather distinct separation among the three as to the type of equipment, software, and processing each was best for. This distinction has blurred with the development of more powerful minicomputers and microcomputers. Computational power that was available ten years ago only with large, mainframe computers, is now available with microcomputers.

1.3.2.1 Mainframe Computers

Mainframe computers are the computational workhorses of computer centers and research facilities. They have large memory capacity and can process arithmetic operations very quickly. Many large simulation programs will run efficiently only on mainframe computers. Mainframes can accommodate many users simultaneously in an interactive mode and can schedule and run batch jobs at the same time.

1.3.2.2 Minicomputers

Minicomputers are basically scaled-down versions of mainframe computers. Although they are smaller, they can still run the same types of programs that the mainframes do. Their size makes them less expensive, and therefore affordable for many engineering firms that could not justify buying and maintaining mainframes. Firms can do much of their own processing on the minicomputers. They can contract with computer centers to run large jobs that are beyond the capabilities of the minicomputer. Minicomputers can accommodate multiple users simultaneously, but not as many as the mainframe computers. Minicomputers also permit lower-cost expansion of capability as needs increase. Two or more minicomputers can be linked together in what is called a network, for the purpose of sharing data and tasks. Individual components of the minicomputer network can be set up to do certain computational tasks more efficiently, such as statistical analysis. Any statistical analysis would be routed to that machine, with other tasks routed to other machines in the network. A minicomputer network offers the opportunity to upgrade a portion of the system as the state-of-the-art advances, without having to replace the whole system. It is also possible for mainframe computers and microcomputers to be linked to the network, so that work can be routed to the most appropriate level of resource. Computational resources of any type can be connected to the network through communications technology, so that data and resources can be shared with almost any facility in the world.

1.3.2.3 Microcomputers

Microcomputers, because of their proliferation in secondary schools and homes, are the computers with which most students are familiar. Microcomputers contain smaller or modified forms of the same components found in larger computers. Most of the same operations are also carried out by the microcomputer system. Until quite recently, one of the big differences between microcomputers and their larger relatives was that the microcomputers were designed to be used by a single user. Now there are some microcomputers that will allow more than one user to access them simultaneously, but the number of users is less than allowed by minicomputers. There also are some microcomputers that will enable a single user to undertake more than one task at once. This is called *multitasking* and allows execution of one or more tasks by the computer while the user is inputting instructions or data for a separate task. Multitasking is useful for accomplished users who want to get the maximum performance from their machine.

The computational speed of microcomputers varies widely, but most are slower than the larger computers. Some have special microprocessors, called math coprocessors, that can be installed to speed up arithmetic operations. Microcomputers also offer a good method for communicating with other computing resources. Using the proper programs and communications technology, a microcomputer can interact with much larger computers. Users can send data back and forth between the machines. In some cases, each machine can do part of the processing involved in an application.

1.3.3 Components of Computers

Computer systems consist of numerous components that act together to process data. Figure 1.2 shows a schematic diagram of typical computer system components. These will be described in subsequent sections.

FIGURE 1.2   Computer system components.

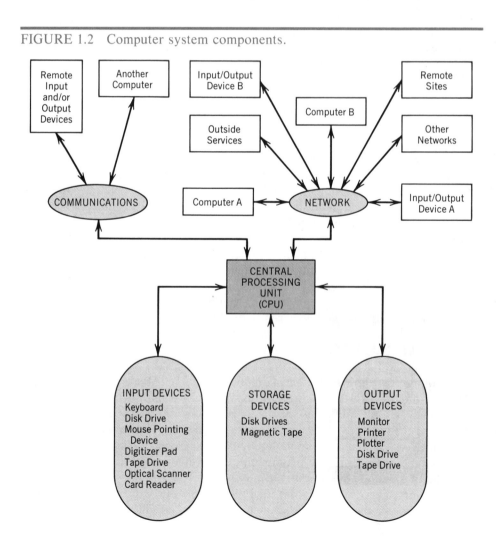

1.3.3.1 Central Processing Unit (CPU)

The central processing unit (CPU) is the principal hardware component of any computer. It consists of a control unit, an arithmetic/logic unit, and the primary computer memory. The control unit coordinates transfer of information among the various components of the computer system, such as memory and peripheral storage devices. The arithmetic/logic unit allows the actual processing of data to produce results. Memory is made up of random access memory (RAM) and read only memory (ROM). From RAM, memory can be accessed for storage and retrieval of information. From ROM, information can only be retrieved. ROM is used mainly to provide program information to the computer.

Although the CPU is the heart of the computer, it could not accomplish any meaningful tasks without relying on the various peripheral devices available. These peripheral devices input data and programs into the computer, output computed data, and store data or programs for later retrieval, modification, or use. Certain peripheral devices also communicate with other computer resources.

1.3.3.2 Input Devices

Input devices accommodate the input of various types of data to the CPU, including numeric and text data, programs, graphics data, and bar code information. The devices can vary from keyboards, to magnetic tape readers, to the bar code readers that you see in grocery stores.

Terminal or Microcomputer Keyboard   Figure 1.3 shows a keyboard for an IBM microcomputer, which is typical of microcomputer keyboards. Keyboards for minicomputer and mainframe computer terminals are similar. The keyboard looks like a typewriter keyboard, with the addition of several function keys. The

---

FIGURE 1.3   Typical microcomputer keyboard.

Keyboard Guide for:
IBM Enhanced Keyboard

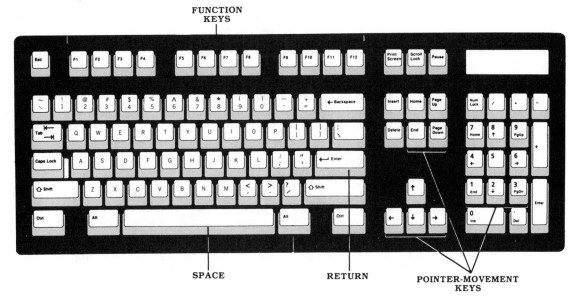

function keys may be in different positions on the keyboard, but they are usually labeled F_1 through F_n. Different software packages program the function keys to accomplish different tasks. Commercial software vendors usually supply a template showing the commands executed by the function keys for their program. Each function key may accomplish two or three commands by use of the **CTRL** or **ALT** keys simultaneously with the function key. Simultaneously means that you hold down the **CTRL** or **ALT** key while you press the function key. Many keyboards also have a numeric keypad that can be used for entry of numeric data. It is generally placed on the right-hand portion of the keyboard. The numeric keypad has an alternate mode that can be programmed to accomplish other functions. The **NUM LOCK** key toggles the keypad between the numeric and function modes.

The main keyboard can also be programmed to execute commands. For example, pressing the **CTRL** and **S** keys simultaneously might be programmed to save the file that is being worked on. The **CAPS LOCK** key produces all capital letters for the alphabetic keys, but does not affect other keys. To type the alternate characters for these keys, you have to press the **SHIFT** key and then the desired key. The **ENTER** key works much the same as the carriage return key on a typewriter. It is named the **ENTER** key because it enters the latest information typed and causes the computer to process the information. The arrow keys are used to move the cursor around the screen. Software documentation for a particular package being used should be studied to find out which functions particular keys perform.

Disk Drives   Disk drives are both input and output devices for computer systems. You must have a means of placing the data on the disk drive before it can be input to a program. Disk drives for the different types of computers do not all look the same. However, they all serve the same purpose, the storage of large amounts of data. The mechanics of how the data are stored will be more thoroughly discussed in a later section. The medium that data are stored on is similar to the material on cassette tapes: metal particles (usually iron oxide) imbedded in plastic. Two types of disks are used: floppy and hard. Floppy disks are made of flexible plastic and are enclosed in a protective cardboard or plastic envelope. Hard disks are made of rigid plastic and can store much more data than floppy disks. Read and write operations are much faster with a hard disk than with a floppy disk. The larger storage capacity and increased speed make hard disks ideal for storage of applications software that must be read into memory each time you want to run the program.

The record/playback heads of a floppy disk drive work the same way that the record and playback heads work on your cassette deck, except that the heads move back and forth to position themselves over different portions of the disk medium. Floppy disk drives are often installed in the main enclosure of a microcomputer, but they can also be found as separate units connected by cable to the control unit of the CPU in the main enclosure.

The data are stored in a much more compact format on hard disks than on floppy disks, so the record/playback heads have to be much closer to the hard disk surface. The heads actually float on a cushion of air whipped up by the rapidly rotating disk medium. The gap is only about 10 millionths of an inch. Therefore, the disks have to be enclosed in a protected environment to keep dust away from the disk surface. A particle of dust would seem like a boulder to the head when the gap is that small.

Most hard disks for microcomputers are installed in the main enclosure of the microcomputer. A typical hard disk pack for a minicomputer will have several circular hard disk media. The disk pack is usually stored in a separate cabinet from the CPU. The capacity of the hard disks in microcomputers is smaller than the capacity of the hard disks in minicomputers. However, even the smallest ones can store about 5,000 pages of text. The advancements in hard disk technology have contributed significantly to making the microcomputer a viable engineering and scientific computational device. Operation of the hard disk in the microcomputer is essentially the same as the operation of the larger hard disk drives in minicomputers. Hard disks are also called fixed-disks because the disk medium is usually fixed in the computer, as opposed to floppy disks that can be taken out and stored or transported to other machines. However, there is at least one manufacturer that is producing portable hard disks. These incorporate the hard disk medium and the read/write heads in a cassette that can be removed from a receptacle that is connected to the computer. Multiple hard disk cassettes can be used with a machine, or the cassettes can be moved to another machine that also has the receptacle.

Mouse Pointing Device   A mouse pointing device is a puck-shaped device that has a roller-ball on the bottom and one or more buttons on the side or top. As the device is rolled back and forth across a flat surface, the roller-ball turns and sends electrical impulses to the computer. These impulses are translated into coordinates that are normally shown on the computer screen display as some type of cursor. As the device is moved, the cursor moves on the screen. Pressing one of the buttons sends a signal to the computer indicating the present coordinates of the cursor. Many graphics packages use the mouse pointing device for fast entry and editing of graphical database information. Another use that is gaining favor is for selecting commands from a display on the screen. Key words for commands are shown on the screen. When you move the cursor to the command and push the button, the computer uses the coordinates to choose which command to execute, then executes that command.

Digitizer Pad   The digitizer pad works much the same way as the mouse pointing device, except that the movement of the puck is limited to the surface of the digitizer pad. There is a grid of fine wires embedded in the surface of the digitizer pad. The position of the puck is interpreted as coordinates by measuring its location in reference to the grid. The buttons on the puck have the same effect as the buttons on a mouse device, and the digitizer pad can be used for the same purposes as the mouse.

Magnetic Tape Drive   Tape drives and magnetic tape can accomplish the same tasks as hard disks. The tapes used in these drives are similar to the tapes used in stereos and recording studios. However, the drives are designed for very rapid fast forwarding or rewinding to place the read/write head at a certain point on the tape. At one time tapes were used as the primary storage devices for larger programs and databases. The advantage of tapes is that they can be taken off the drive and stored, and different tapes mounted for processing. The disadvantages are the handling involved and the slow speed of data recovery as compared with disk drives. As hard disk capacities have increased, they have largely replaced tapes for primary data storage. However, tapes still have an important function for backup of information on disks. Most computer centers back up modified portions of their hard disks to tape every day and back up the entire hard disk about once a week. Thus, if a sudden power failure or a severe software or hardware malfunction occurs that corrupts the data on the hard disks, the tapes can be used to reload all the information that was on the hard disk before the problem developed. Users would lose only modifications they had made since the last backup. Tapes are also used to transmit programs and data from one computer center to another. Many government databases, such as weather data, can be purchased on tape for loading onto your computer. Tape drives are also available for microcomputers, but they are used almost exclusively to back up hard disk information.

Optical Scanner   There are several types of optical scanning devices. The two types that are most important to engineering computer users are bar code readers and text scanners. Bar code readers interpret the parallel lines such as those seen on items in stores and convert this information into data for input to the computer. In stores, the bar code contains identification information about the product. The scanner extracts the information from the bar code and automatically enters it into the register. The register is connected to the store computer system, which provides the price for the item and updates the inventory or other stock control data. For other computer applications the bar codes can represent either numeric or text data. Some computer magazines are providing programs in bar code form for readers to load into their computers. Text scanners convert text or graphics from a page into computer-compatible data.

Card Punch and Reader   Prior to widespread adoption of interactive computing, punched cards were the primary means of inputting both programs and data into computers. A typewriter keyboard was used to drive the key punch. The punch made one or more holes in a column to correspond to a code for the number or character to be placed in that column. The number or character was also printed at the top of the column so you could see what had been entered. Each card contained 80 columns. When the cards were run through a card reader, it interpreted the punched codes and sent the corresponding numbers or characters to the computer. You will still find a few programs, and some data, available on punched cards.

1.3.3.3 Output Devices

Monitor (Screen)   The monitor is the primary output device for routine interaction with the CPU. The monitor screen echoes input as you enter it from the keyboard and displays messages sent by the CPU. It can also be used to display output data, or graphics output if the computer has that capability. The distinctness of the image on the screen, or resolution, is a function of the number of pixels in the screen display. A pixel is the smallest area on the screen that can be given a particular degree of luminescence or color by the picture tube electron beam. The more pixels there are on the screen, the more distinct output images can be.

Printer   A printer can output printed copies of computer programs or data. Many printers can also output graphics images. The printed copies are useful for study, reports, or archives. Several types of printers are available for output of data. This section will discuss the general types. There are a number of subdivisions within each of the types.

Dot matrix printers use a grid of tiny cylinders to produce output. The printer forces the cylinders corresponding to the character to be output forward. These cylinders strike a ribbon and make an imprint on the paper behind the ribbon. The distinctness of the output is a function of the number of individual cylinders contained in the print head grid. The advantages of dot matrix printers are that they are relatively inexpensive and can easily be programmed to produce different character fonts (shapes) and sizes. Bold type can also be accomplished by double striking characters with a slight offset. Graphics can be output as a series of related dots. The disadvantage is that the print quality is not as good as that produced by typewriters.

An impact printer is essentially a typewriter with additional circuitry to allow it to output characters sent to it by the computer. Large, high-speed, impact printers were the main output devices for minicomputer and mainframe computer systems prior to the development of practical laser printers. The output quality of these printers is not as good as typewriter output. There are also slower impact printers that produce typewriter, or letter-quality, output.

Laser printers combine copier and computer technology to produce high-quality text and graphics output. The computer electronically produces an image on the copier drum. The image is then transferred to any medium that a conventional copier could use. Laser printers have become the mainstay for high-quality output for computer centers. Some laser printers are capable of producing output similar in quality to commercial typesetting equipment. Development of smaller, less expensive laser printers and sophisticated programs to run them has brought the laser printer within the reach of many personal computer users. These laser printers have facilitated the development of desktop publishing. Publication quality output can be produced directly from your microcomputer on a device that can be placed next to it on your desk.

Plotter Plotters provide high-quality output of graphical and related data. Engineering offices use them extensively for production of plans and blueprints. Plotters can produce plans much faster than a draftsman. They also produce uniformly high-quality output. Pen plotters are the oldest plotting technology. They make prints by drawing an ink-filled pen across the media surface. Pen plotters can be purchased in either drum or flat-bed configurations. Drum plotters are used to produce larger plots. The drum rotates under the plotter pen to produce movement along one axis, and the plotter pen moves parallel to the axis of the drum to produce movement along the other axis. Coordination of the two movements can produce any shape, including accurate circles. The width of the plot is limited by the width of the drum, but the plot can be of any convenient length. Flatbed plotters will produce plots up to a certain size on a piece of plotting paper laid flat under the plotter pen. The pen moves along both axes to produce the plot. High-quality flatbed plotters have been developed for use with microcomputers. Some of these are a cross between a flatbed and a drum plotter. Rollers control the movement of the paper along one axis, and the pen moves along the other axis.

Pen plotters remain the most popular type of plotters because they can work with a variety of media, such as regular paper, translucent plastic sheets, vellum, overhead transparencies, and drafting paper. They offer a wide variety of paper-feed and size options and can produce color output. The line quality of plots is excellent. Pen plotters are relatively inexpensive. However, they are slower than other types of plotters.

Thermal transfer plotters operate by passing a wax-coated film between a heated printing element and the medium, either paper or film. The heated element melts tiny dots of the ink onto the medium precisely where printing is needed. Numerous tiny dots eventually form the completed plot. Thermal plotters are much faster than pen plotters but do not offer as high quality output. They are presently available in widths up to only 24 inches. Electrostatic plotters work like copiers, except the image is created by the computer rather than being taken from the page to be copied. Electrostatic charges are placed onto a receptive medium, then the charged medium is passed over a writing station with liquid toner. The toner is drawn to the charged portions of the medium. Electrostatic plotters provide excellent line quality and shading of three-dimensional drawings. Inkjet plotters produce a drawing by spraying tiny drops of ink onto a sheet of medium. The excellent color saturation provided by inkjet plotters produces drawings that have almost photographic quality.

Output to Storage Devices   Outputting to storage devices is a way of saving computer programs or data for future use. Several types of storage devices will be discussed in the following section.

1.3.3.4 Information Storage Devices and Media

Disk Drives   Both floppy and hard disk drives are information storage devices. If information is put on a disk, it will remain there when the computer is turned off and will be there the next time you turn the computer on. A set of related data that you want to keep together for future use is called a file. When you send a file to a disk, you assign a name to it. The data are identified by the name when they are placed on the disk. The space the file occupies on the disk is reserved so no other files can replace it without your instruction to do so. You can read the data back into the computer by referencing the file name. You can modify the data and put it back on the disk. Enough space will be made available for the modified file. *Unless you have saved the new file under a different name, only the latest version will be stored on the disk.* Saving the old file under a different name or on a different disk is called backing up your data. If something happens to corrupt the data in the original file, you still have the old version to fall back on. Backing up data files is important insurance against accidental loss of data. It should be noted that when you read data from a disk into the computer, nothing happens to the data on the disk. A copy is sent to the computer, but the data are still on the disk. If you do not want a certain file anymore, you can instruct the computer to delete it. The delete command does not really remove the data from the disk; it just allows data from a new file to be written onto the same space to replace the existing data. It is important to remove files that you do not use regularly to avoid clutter on your working disks. If the file contains information that you might possibly need at some later time, you can make an archive copy.

As we have discussed before, floppy disks do not store as much data as hard disks. Transfer of information to or from a floppy disk is also slower than transfer-ring the same data to or from a hard disk. However, the floppy has the advantage of portability. You can remove the disk from the disk drive and take it to another machine, or even send it to a colleague in another state. You can also put a new floppy disk in the drive when one becomes full. You have to remove some files from a hard disk to make room if it becomes full. Usually you will store programs and data that you use frequently on the hard disk, and store data that you use occasionally on floppy disks. You can always transfer data from a floppy to the hard disk when you want to work with it, and then transfer it back to the floppy when you are done. Floppy disks are also ideal for making backup copies of all important files.

Magnetic Tape   As was stated previously, magnetic tapes are used quite often to back up all the files on a hard disk system. They can also be used to store or transmit large amounts of data. Rather than storing all the data on the hard disk system, you can use the computer to extract the desired information from the tape and store only the needed subset. Magnetic tape storage devices are available for all types of computers.

Other Storage Devices   It is possible to store information from the computer on punched cards. The computer drives a card punch, which produces the same type of cards as the typewriter card punch. The cards produced could be stored or sent to a colleague. Cards, however, are a cumbersome way of storing and transmitting data. You can also store data on a paper tape medium. Again, the computer drives a punch that puts coded punches on the paper tape to represent characters. The tape can be saved and read back in at a later time. Today these methods of data storage are rarely used. Most computer centers still have card readers to use when necessary.

1.3.3.5 Communications and Networking

It is often necessary to transmit computer data from one location to another. The transmitted data may be data from a remote sensing site that will be input to a computer analysis program. It may be instructional data sent to a remote site to control some process. It may also be computer program or data files that are being transmitted from one computer system to another.

The transmittal of files from one computer system to another is gaining popularity as a means of sharing data or programs with colleagues. It is also becoming a popular method of interaction among microcomputers, minicomputers, and main frame computers. Data communication allows you to use the microcomputer to create and edit files and accomplish preliminary analysis. You can then send the files to the larger computers for computation-intensive analyses. Transmittal of files from a microcomputer to a larger computer is called uploading, and the opposite process downloading. Files could be sent to a computer in another department in your building, or they could be transmitted to another part of the country. Two methods for communicating between computers will be discussed below.

Modems   Modems are devices that allow one computer system to send information to another system over standard telephone lines. They allow you to connect a microcomputer or terminal to a minicomputer or mainframe computer located some distance away. The computer you are communicating with may be across campus, or it may be across the country. Modems are often used by business people to communicate with their home office computer, either while talking with clients or from their hotel room in the evening. You could use a modem while you were at home to communicate with the computer system at your college or university. Modems can also be used to connect two microcomputers, allowing them to send data back and forth.

The earliest types of modems were acoustic modems. The connection between the computer and the telephone was made by placing the telephone handset in a cradle that cupped the ear and mouth pieces. The tones that represented the data transmission were transmitted through the handset. This method was subject to outside noise interference. Often the communication link between the two computers would be broken if the handset were bumped.

Improved models allow the modem to be spliced into the telephone cord. This eliminates the outside noise problem and allows the modem to accomplish tasks such as automatically dialing the telephone number of the other computer. Even in-line modems have problems with spurious noise in the telephone lines. For example, if you have multiple phone extensions in your home and someone picks up one of the other phones while you are connected with an outside computer, you will most likely lose your connection. Also, the call waiting service offered by most phone companies will interrupt your connection if a second call comes in while you are using the phone for computer communications.

Some modems are housed in external boxes and are connected to the host computer through a cable. Other modems are on printed circuit boards that can be placed inside the main enclosure of the computer. For microcomputer users, the advantage of the modem on a card is that it is out of the way and does not take up additional desk space. The disadvantage is that it takes up one of the microcomputer's expansion slots.

Networks Networks and local area networks (LANs) allow various computer resources to communicate with each other over direct connections. They are used mainly to connect resources at one location, but can also have connections to communicate with remote locations. Users can often enter the network through modems. Networks can be used to connect several minicomputers, microcomputers, and mainframe computers together. They can also be used to connect shared peripheral equipment that can be accessed by various computer resources in the network. Some networks have been set up to allow similar groups of computer users to communicate with each other from different parts of the country. A network of academic computer users is an example.

1.4 SOFTWARE

1.4.1 Definition and Types

Software is the other half of what makes computers useful machines. The hardware is a system of physical components that are connected together and are designed to interact with each other in a specific manner. However, they cannot interact without some sort of instructions. This is where software comes in. It provides the instructions that tell the hardware components how to interact. Software can be divided into two main types; operating systems and applications software. Operating systems coordinate the actions of the computer hardware. They govern actions such as how data are put onto disks and how files are manipulated in the system. Applications software provides the problem-solving part of the computer instructions. Applications software must interact with the operating system to accomplish its tasks.

1.4.2 Operating Systems

Operating systems provide various housekeeping functions for computer systems. These include signing on, creating files, editing files, saving files, manipulating files, and running programs. Signing on to a microcomputer may be as simple as turning the machine on and waiting for the operating system to automatically load. Signing on to a network may require several steps. The steps may include identifying the computer resource desired, specifying a user code, and providing a security password. Working with files can involve numerous operations that will be discussed in a separate section. Running programs involves commands that will execute the appropriate applications software. For multiuser systems, the operating system must also schedule and allocate the various computer resources to efficiently accomplish the tasks required by each of the users. The operating system has preassigned, or default, input and output devices. The user can change or add to these as the application requires. The operating system includes the main operating system program and several utility programs and files for accomplishing specific tasks. These utility programs and files include programs for formatting and backing up disks, system configuration files, and output device drivers. The utility files can be accessed when needed. They do not take up available computer memory when they are not needed.

Since different types and brands of computers may use different operating systems, this text will be as general as possible when referring to operating systems. However, it will occasionally be appropriate to refer to a particular operating system to illustrate a point. Where this is necessary, the VAX/VMS operating system will be referenced for minicomputers and IBM DOS Version 3.3 for microcomputers. Although your computer may have a different operating system, the tasks involved are basically the same for any system.

1.4.3 Compatibility Between Applications Software and Hardware

Software programs are generally written in a language that can be understood by any computer that has the capability of interpreting that language. However, different computers have to use a compatible data storage technology, or they must be able to send the program back and forth in a form understandable to both. Minicomputer and mainframe computer manufacturers have established standards that facilitate the transfer of data and program files among computers. A FORTRAN program that is written for an IBM computer can be stored on tape, taken to a facility that has a Control Data Corporation (CDC) computer, and loaded and run on the system. Occasionally there are minor inconsistencies between the way languages are implemented on different machines, but these do not present major problems in compatibility. This level of compatibility does not exist among microcomputers.

Incompatibilities among microcomputers exist for several reasons, all of which are related to the absence of standards that are accepted by all manufacturers. The way the operating system and hardware save data on a disk varies from one manufacturer to another. Also, applications programs often take advantage of unique features of a microprocessor that are available only on a particular micro-computer. Thus, files written to a disk on an IBM system may not be able to be read by an Apple disk drive. Even if the files can be read, the program may not run on the Apple because of hardware inconsistencies. The lack of a standard operating system for microcomputers is also a big impediment to the transfer of applications from one microcomputer to another. Until some consensus is reached, transferability of microcomputer applications from one system to another will be limited.

1.4.4 Applications Software

Applications software can be thought of as any programs that interact with the computer hardware and the operating system and that produce useful results when used properly. Applications software can be divided into several types, based on ease of usage and the way the user interacts with the software.

1.4.4.1 Machine and Assembly Languages

Machine language is the binary language understood by computers. It is possible to write programs in machine language, but it is impractical for most users because of the complexity of the instructions necessary. Therefore, some method is needed to allow input of instructions in a form more decipherable by humans, but which can still be efficiently translated into machine language. Assembly language is the first step in this direction. Programs written in assembly language run efficiently, but the commands and syntax are still somewhat cryptic. It is also difficult to interpret programs that another person has written.

1.4.4.2 High-level Languages

High-level languages provide the next step beyond assembly language. High-level languages allow the user to develop the logic of the program using commands that are easily understood by humans, such as READ, WRITE, WHILE i > 10 DO, and IF(A.GT.33) THEN. This is not to say that mastery of any of these languages does not take a good deal of study. It is much like learning a foreign language. You have to memorize the syntax and grammar, then use them in practical applications to develop your expertise. However, as is the case with foreign languages, once you have learned one it is much easier to learn others.

After you write a high-level language program, it must be translated into machine language so the computer can use it. The statements written in the high-level language are called the source code. Two methods are used by software to translate the source code into machine language. The simplest is called interpretation. Interpretation software converts one line of the source code at a time into machine language. This uses the least amount of computer memory to implement and offers the advantage of identifying syntax and language errors as you enter the program. The disadvantage of interpretation is that the program has to be interpreted each time it is executed, which extends the time necessary to run the program. Compilation is the second type of translation software. Compilers operate on the whole source code file after it has been entered. They provide syntax and language error messages if there are errors in the source code. Once the errors are eliminated, a successful compilation results in generation of an object file that contains the machine language for the program. This object file can be run any number of times without recompilation. Several high-level languages popular with engineers will be described as follows.

FORTRAN   FORTRAN stands for "formula translation" and was originally developed in 1957 by IBM. It introduced an efficient user-oriented way of expressing formulas for translation into machine language. It soon became the language of choice for engineers and scientists. FORTRAN is a compiled language. Since its introduction, FORTRAN has undergone several revisions and upgrades. Its present standard form, FORTRAN 77, allows users to develop structured programs that are easy to understand by other users. An updated standard is now being reviewed by FORTRAN users. Many FORTRAN application programs have been developed. FORTRAN will remain an important language as long as these applications continue to be used. As microcomputer capabilities have increased, efficient FORTRAN compilers have been developed for them.

BASIC   BASIC stands for "beginners all-purpose symbolic instruction code." It was developed in the mid-1960s as an instructional language. BASIC was designed to be easier to learn than FORTRAN, yet incorporated many of its key features. BASIC underwent a big increase in usage with the introduction of microcomputers. Many computer graphics programs were written in BASIC during this time. Both compiled and interpreted versions of BASIC are now available.

PASCAL   PASCAL is named after Blaise Pascal, the French mathematician, physicist, and philosopher. It was developed as a structured language that implements strict organizing principles. This feature makes development and maintenance of larger programs much easier. Many of the structured programming aspects of PASCAL have been incorporated into FORTRAN and BASIC. These languages were not originally developed as structured languages. Therefore, the implementation of structured programming with FORTRAN or BASIC is not as natural as it is with PASCAL. Some engineers and scientists feel that PASCAL will become the language of choice.

C   C is a language developed by Bell Laboratories for writing systems software. It has many of the features of a high-level language and facilitates development of well-structured programs. It has gained favor with many microcomputer software developers. For example, the equation solver TK SOLVER is written in C.

ADA   ADA is a structured language supported by the Department of Defense. All defense contracts that require programming have to be done in ADA. It is designed to facilitate the development, maintenance, and change of large programs by teams. ADA has many similarities with PASCAL.

1.4.4.3 Command Languages
Command languages are actually part of the operating system, but they are also an integral part of using applications programs. Sometimes called job control language, these are the commands that allow the user to create, manipulate, and save files and to run programs. The commands of MS-DOS and the commands of VAX/ VMS form command languages. For example, the steps involved in compiling and running a high-level language program include compilation, linking, file assignment, and execution. Compilation has been discussed previously. It is the translation of the source program into machine language. However, if separate modules that represent individual procedures are used in the development of a structured application, these modules will be contained in separate source files to facilitate their use in other applications. Each module that has not previously been com-

piled will have to have a separate command to compile it. The linking stage combines all the compiled object files that will be used in the application into a single executable file. File assignment links any input and output files to the application. It is essentially a bookkeeping process that tells the computer what file to use when a certain input or output statement is encountered in the running of the application. Execution is the actual running of the execution file generated in the linking stage.

On most systems it is possible to enter commands one at a time from the terminal or microcomputer keyboard. However, when you are developing applications or will run an applications program a number of times, you can save the commands in a command file. One instruction will execute all the commands in the command file. Most systems also have numerous options that you can include in command files to increase their generality.

As with high-level languages, different command languages may differ in details, but they all accomplish essentially the same tasks. This makes it easier to move to a new type of system after you have mastered one. If you have not already learned the command language for your computer system, you will have to learn it before being able to apply the concepts you will be learning in this text.

1.1.4.4 Generic Software

Generic software refers to a whole family of applications packages that were not developed for a certain discipline. They have a wide spectrum of uses. Included in this category are spreadsheets, equation solvers, database management systems, and word processing programs. These generic software packages have many of the features of high-level programming languages. They also have features that make them much easier to use for particular types of applications. For example, spreadsheets are particularly useful for processing any problem data that can be arrayed in a tabular format. Equation solvers are specifically designed to solve systems of mathematical equations without having to write a detailed program. Since microcomputers have come into common usage, the generic software packages have gained stature as engineering and science tools. Some engineers and educators have gone so far as to say that these generic software packages will make the teaching of high-level languages obsolete. It has been my experience that the generic software packages are outstanding tools, but each has its limitations. The high-level languages still have the edge for large-scale applications. The key to being an effective computer user is in knowing when each type of package should be used.

1.4.4.5 Applications Packages

Applications packages are tools developed for a specific engineering application. These may be locally produced or may be purchased from a firm that specializes in developing engineering software. The complete package will consist of not only the program or programs but also detailed instructions for their use. The developers of the programs may also provide continued support. Any of the languages and generic software discussed above could be used in an applications package. You may be contractually restricted from making any changes in the software. In that case, you must be doubly sure that a package will do everything you want it to before purchasing it. If you are allowed to make changes and desire to do so, you will have to be familiar with the development language or software used in the package. Development of applications packages will be discussed briefly in the next section and in greater detail in Chapters 2 and 9.

1.4.5 Development of Applications Packages

As engineers you may be involved in the development of computer applications packages. You will almost certainly be involved in their use and will evaluate new packages for possible purchase. In any of these roles, it is important that you understand what goes into making a good applications package.

The first requirement in the development of applications packages is the writing of clear computer programs that can easily be understood and modified by other users. This is true even if the source code for the package is not to be transmitted to clients. Your firm will likely want to make changes and additions to the package in the future. If you have moved on to another job, whoever becomes responsible for the program will have to become familiar with its inner workings.

Structured programming provides the wherewithal to develop easily understood and transferable programs. You will learn how to develop structured programs beginning in Chapter 3. The rules of structured programming prescribe good programming style. Structured programming languages either require or facilitate the implementation of these rules. The concepts of structured programming are useful when you are developing applications for any of the high-level languages or generic software packages.

Structured programming stresses modular development of programs. Each module has a specific purpose and can often be used in more than one application. Within a particular module, indention is used to identify the hierarchy of logic blocks. Comments are used in the program to describe what is being accomplished. Top-down design procedures are used in developing the logic of the application. These procedures involve the decomposition of the problem into general tasks and then the successive refinement of the logic of each task until sufficient detail is achieved.

Once the structured application is developed, it must be tested and any errors in logic corrected. This process is called debugging. In debugging it is important that you test all possible options. The program may run fine for some options, while another option could still have errors in it.

Documentation is an important part of the development of applications packages. Whoever uses your package must be able to understand and use it without having to continually call you for instructions. Documentation is often the weakest link in the applications development chain. Without good documentation, even an excellent program will not be used. Documentation could be in the form of a formal printed report. It could also consist of instructional prompts presented on the screen as the program is used. Different types of documentation may be used in the same program. The elements of good documentation will be discussed in Chapter 3 and in more detail in Chapter 9.

After developing a computer application, it is important to consider licensing and legal requirements. These are designed to protect you from unfair use of your package and from court suits resulting from improper use of the software. You must also consider how you are going to distribute your package. Is it a package that will be used by only a few engineers in your office, or is there a possibility that it might be used nationwide? The legal requirements would be vastly different.

1.4.6 Availability

Software is available from many sources. Some of it is free, and some of it is very expensive. However, the value of software does not always correlate with its cost. That is why you must carefully evaluate new software before purchasing and using it. Computer-related magazines often compare and evaluate programs that are written for similar purposes. You may also be able to gain valuable insight by talking with people who have used a certain package.

Once you have defined what you want to do with a software package, you can begin to shop around to see what is available to meet your needs. You may find free software that meets your needs. You may find two packages that claim to do the same thing, with one costing $500 and the other $200. The $200 package may do as much as or more than the $500 package.

Public sector software is software that has no copyright. You can legally copy it and use it on your system. The documentation of the program may be nonexistent, or it may be thorough and well presented. Universities and governmental agencies are two of the primary sources of public sector software. There are some extremely sophisticated packages that are available through these sources. Some have extensive documentation and are supported for revisions and corrections by the sponsoring agencies.

There are also many software programs, especially for microcomputer users, that have been written by individuals and placed in the public sector for copying by anyone who has access to computer bulletin boards. Computer bulletin boards are systems that are available to anyone who has a modem and calls up the system. Individuals are allowed to put programs into the bulletin board and to copy them from the bulletin board into their own system. There may be someone who controls what goes into the bulletin board, or it may be wide open. This software is generally not documented, except through information stored in the files of the packages. Recently some problems have cropped up from the widespread use of bulletin boards. Mischievous or malicious users have placed files that contain damaging programs called viruses in the bulletin boards. You will not be able to tell that the viruses are there when you copy the files over to your system. Sometime later they will make themselves known, just like a human virus after its incubation period. The incubation period is usually measured in a number of sign-ons or a number of runs. The effect of the virus may be a harmless message on the screen. However, some of the viruses can go so far as to erase a whole disk. If you should happen to have infected software copied from a bulletin board on your hard disk, you could lose everything on your hard disk. There are some programs now available that claim that they can eradicate viruses. Your best defense is to use software only when you are sure of its source. Or if you do copy any software from a bulletin board, keep it on a separate floppy disk until you are sure that it does not contain any viruses.

Commercial software is available from a number of sources. There are many specialized firms that make a business of developing and marketing engineering software. Their ads can be seen in most of the engineering journals. Large publishing companies are also getting into the business of marketing software. There are also computer service bureaus that maintain engineering software packages, and will make them available on a rental basis to you, usually through time-sharing services with their computers.

Any software that is going to remain useful over a long period of time must have regular maintenance. Maintenance will correct errors found by users. It will also institute changes in procedures that will add capabilities or make the software more effective. If the original developer is unavailable to accomplish maintenance, someone else must do it. Who will accomplish maintenance and how it will be accomplished are important considerations when acquiring software. You do not want to do it yourself if you can help it.

1.5   MECHANICS OF INFORMATION STORAGE AND RETRIEVAL

If computers are to accomplish anything useful, they must be able to store information while processing it in the CPU. They must also be able to store information on peripheral devices for later use. This section will briefly discuss the mechanics of how this is done.

1.5.1   Information Storage in the Computer

As you have learned, computers use binary (on–off) logic. Each binary switch is called a bit, and represents either 0 or 1. Fixed-length groups of bits, called bytes, are used to represent each character for data transmission and storage. Usually either seven or eight bits make up a byte. A seven-bit byte gives 128 possible binary combinations, so up to 128 characters can be used. The eight bit byte allows up to 256 characters. Each lowercase letter, each uppercase letter, each number, and each symbol is assigned a particular code. The American Standard Code for Information Interchange (ASCII) is a popular code, available in both seven- and eight-bit versions. It was developed by the American National Standards Institute in an effort to standardize the various codes. Another popular eight-bit code is the Extended-Binary-Coded Decimal Interchange Code (EBCDIC) developed by IBM, which is used on most of their computers. However, the IBM PC uses ASCII. A complete set of ASCII standard codes can be found in Appendix B.

Computer memory is measured in the number of bytes of storage available, since a byte is the smallest element of memory that can store a single character. One kilobyte (K) is equal to 1,024 (2^{10}) bytes. Therefore, 48K represents 48 × 1,024 bytes of storage. Another way of representing units of computer storage is by "words." Early personal computers used words that were one byte, or eight bits, long. Today, many personal computers use 16-bit words, and some even use 32-bit words. Thirty-two bit words are the rule for minicomputers and mainframe computers. Some supercomputers use 64-bit words. The length of the words in a computer system is important, because information is transmitted a word at a time. Therefore, larger words result in faster data transmission.

1.5.2   Information Storage on Disks

Both floppy and hard disks store information in basically the same way. A microcomputer floppy disk uses a flexible medium with a metallic covered surface. The floppy disk medium is enclosed in some type of protective covering. Hard disks use a rigid medium that is housed in a protected shell to keep dust away from the surface.

The two most common types of floppy disks in use today are either 5.25 or 3.5 in. in diameter. The disk drive head is located above a slot in the floppy disk cover. The head either reads or writes data as the medium spins underneath it. This slot has to be protected when the diskette is not in the computer. Even the sweat and oils from a fingerprint can damage the disk medium. The 3.5-in. disks have a cover that automatically slides over to cover the slot when the disk is taken out of the computer. Removable covers protect the slot for 5.25-in. disks.

Microcomputer floppy disks also have a means of preventing accidental erasure or writing over of files. There is a notch in the side of the 5.25-in. diskette. If this notch is covered, the disk drive will not allow data to be written to the disk, but will allow information to be read from it. When the notch is covered, the diskette is said to be write-protected. Removable tabs are used to cover the notch. The 3.5-in. diskettes have a more rigid cover than the 5.25-in. diskettes. Write protection for 3.5-in. disks is accomplished by sliding a movable tab one way or the other.

The hardware mechanics for reading from and writing to disks were discussed in Section 1.3.3.2 under Disk Drives. Figure 1.4*a* shows a schematic of a floppy or hard drive medium and the storage method for data. In order for a disk to accept data, it has to be formatted for the particular system. Formatting divides the disk into a series of circular strips called tracks. Some disk drives can write and read information on both sides of the disk. The two tracks above and below each other form a cylinder. Each track is divided into a series of sections called sectors. Each sector occupies a fixed amount of space on the track. For example, a sector in the IBM PC environment will store 512 bytes. When a file is written to a disk, the data are placed in the first available sector. As a sector is filled, the write head moves on to the next available sector. If it encounters sectors that already have data stored in them, the head skips over those sectors. If the last part of a file occupies only part of a sector, the remainder of that sector is unavailable to data from another file. Each sector occupied by the file is marked with a label so the disk drive can identify the parts of the file when it comes back later to read the file. Another section of the disk contains a File Allocation Table, or FAT, that keeps track of all the occupied sectors on the disk. Still another section of the disk stores the directory of files on the disk. The directory contains the file names and other information on each file. Data are actually stored on the disk in pulses, either a maximum signal or nothing, as opposed to the continuously variable signal that you would have in an audio recording. Figure 1.4*b* illustrates the relationship among the directory, the file allocation table, and the sectors for storing a typical file.

1.5.3 Safeguarding Data

Software and data can often be worth much more than the computer system hardware that you are using. Therefore, the safeguarding of data must be a prime consideration in the day-to-day use of computers. If you do not take the time to safeguard your data, you are giving yourself no insurance against the occasional incidence of data loss due to power failures, system failures, or disk failures. The media can wear out or the disk can be physically damaged. You may not be able to replace the lost data. Even if you can, the cost may be prohibitive.

Saving your work often when using the computer is the first line of defense against data loss. Some programs accomplish this automatically, saving the work

FIGURE 1.4   Methods of data storage on floppy and hard disks.

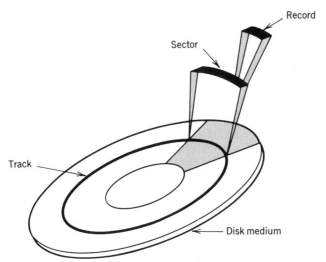

(*a*) Tracks, sectors, and records.

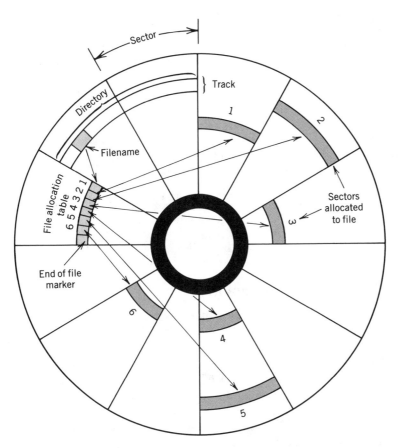

(*b*) Directory, file allocation table, and sectors.

after a selected amount of time. However, many programs still require the user to do the saving. You should periodically back up the file you are working on to guard against power failures or other problems that cause you to lose what is presently in computer memory. At least at the end of each session you should back up your saved file, in case the saved file becomes damaged. This backup should be done on a separate disk from the primary file, to guard against damage such as spilling coffee on the disk. It is best to store the backup disk in a location separate from where you store the primary disk. Store it in a separate building if possible. This guards against catastrophic losses from emergencies such as fire or theft.

1.6 ACCURACY AND NUMERICAL PRECISION

Accuracy and precision are important considerations when determining the validity of computer-generated results. The basic concepts of accuracy, precision, and error will be discussed here and their application demonstrated throughout the text.

1.6.1 Approximations and Errors

In engineering we almost always work with estimates or approximations for values. That is why safety factors are used to compensate for unavoidable errors caused by uncertainties in data or incomplete knowledge of the processes involved. This is not the same as errors due to mistakes or lack of engineering judgment.

Almost any engineering measurement will exhibit some irregularity, even if the process being investigated has predictable results. Data acquisition involves some amount of instrument or human error. Error is the difference between the true value of something being measured, and the measured value.

true value = estimated (approximate) value + error

If true values are not known, there are statistical means of dealing with the uncertainty in test results. More discussion of statistical analysis and error is included in Appendix H.

1.6.2 Accuracy vs. Precision

Accuracy refers to how closely a computed or measured value approximates the true value. Thus, it is a measure of the error of a value. Inaccuracy in results is also called bias. Precision, on the other hand, has to do with the reproducibility of results for several estimates of a value under similar conditions. If you are target shooting, and you put a series of shots in a tightly packed group around the bull's-eye, your results are said to be both accurate and precise. They are accurate because the center of the group is in the bull's-eye, and they are precise because the group is tightly packed together. If you have a tightly packed group of shots that is above and to the left of the bull's-eye, your shooting is precise, but not accurate. The sites on the rifle are probably set wrong. If your shots are centered around the bull's-eye, but they are not closely grouped, your shooting is accurate but not precise. You probably need to check your technique. If your group of shots is centered above and to the right of the bull's-eye, and the grouping of shots is scattered, your shooting is both inaccurate and imprecise.

1.6.3 Significant Figures

The term *significant figures* is used to state your confidence in, or the reliability of, a numerical value. For example, if you are measuring the diameter of a steel rod using a ruler with inches and tenths, you might see that the diameter is between 1.1 and 1.2 in. You are confident, or certain, of the first two digits, while the third digit would have to be an approximation. On the other hand, if you use a micrometer that is accurate to 0.001 in., you could see that the diameter is between 1.145 and 1.146 in. In this case, you would be certain of the first four digits, while the fifth digit would be an approximation.

The number of significant figures is set by the weakest link in the data-gathering process. For example, if you time a process using a hand-held stopwatch that is accurate to 0.001 s, you cannot justify reporting the elapsed time to thousandths of a second. The response time of your eye–finger system is only about a tenth of a second. Therefore, a time on the stopwatch of 3.145 s would have to be reported as 3.1 s.

Significant figures are the digits in a number that can be used with confidence. The number of significant figures is defined as the number of certain digits plus one estimated digit. It is customary to set the estimated digit at one-half the smallest increment on the measurement device. Thus, the diameter of the steel rod as measured with the ruler would be reported as 1.15 in., whereas the diameter as measured by the micrometer would be reported as 1.1455 in.

One way of expressing numbers so there is no confusion as to the intended number of significant figures is through scientific notation. The main part of the number gives the significant figures and is then multiplied by some power of 10 to show where the decimal place is to be put. Therefore, the number 3,465,000 with four significant figures can be represented as 3.465×10^6, while the number 0.00000003465 would be 3.465×10^{-8} in scientific notation.

When you are accomplishing numerical computations, it is standard practice to carry as many digits as possible through the computations to minimize round-off errors. The final answer should be reported with the minimum number of significant figures of any of the values used in the computation. Numbers to the right of the last significant digit should be rounded up if placing a decimal point before them gives a value ≥ 0.5. Otherwise, they are rounded down.

1.6.4 Effects of Computer Hardware and Software

Computers can store only a limited number of significant digits. For example, pi = 3.14159265358979 would be stored in a computer that uses seven significant figures as 3.141592. Errors caused by the limited number of digits carried through computations are called computer round-off errors.

Most computers are capable of outputting more digits than they can actually carry through computations for a number. For example, the default output for a FORTRAN program on a VAX minicomputer uses nine digits, although the computer can normally carry only seven significant figures through computations. Also, the number of digits output has nothing to do with the actual significant figures of the value. You should be consistent in your treatment of computer values and not attribute more significant figures to the results than are warranted. Most computer languages and applications packages allow you to format output to give an appropriate picture of the accuracy of results.

Precision does not have as much application when discussing computer results as it does when discussing the results of measurements of a physical phenomenon such as a chemical reaction. Computers are very precise in their computations. You will get the same results each time you run a computer program. However, the accuracy of the answers will be influenced by the accuracy of the input data, the accuracy of the mathematical model being used, and any computer round-off errors that have been propagated through the computations. Some applications allow you to adjust the number of digits that are carried through computations. For example, in FORTRAN you can use DOUBLE PRECISION storage to more than double the number of digits that the computer can store and carry through computations. Using DOUBLE PRECISION will result in reduced round-off error. This will provide more accurate results, but not more precise results.

Computer round-off errors are of concern if you are using programs that arrive at answers after repeating iterative calculations many times, or that require the calculation of small differences between large numbers. Sometimes these errors will be large enough that it is obvious that there is a problem. At other times the answers may appear to be reasonable unless they are checked carefully. When using programs in which round-off error may be a problem, the user has to make spot checks to ensure that all results are within allowable tolerances and do not violate any required specifications. Sometimes these checks will be built into programs, and error messages are printed out if errors exceed a specified tolerance.

1.7 ETHICAL CONSIDERATIONS IN COMPUTER USE

There are several important ethical considerations when using computers. These include using software packages for applications that are outside your area of expertise, using unauthorized copies, and gaining unauthorized access to files and programs.

Software packages are available for a wide variety of engineering applications. You may be tempted to use some of these packages to develop solutions to problems that are outside your primary area of expertise. The problem with this is that you may be unable to interpret and evaluate the answers that the computer puts out. You have professional, ethical, and legal responsibilities for the results of your engineering designs. You must understand the theory on which your designs are based. You must also understand the limitations of the software being used, the assumptions that were made during its development, the mathematical model incorporated into the software, and the solution method used to solve the mathematical model. The computer is just a tool, and you are responsible for its proper use. You, as the engineering designer, will be held primarily responsible for the correctness of the results.

Much software is easy to copy. Even software packages that use protection schemes to prevent copying have been subverted by knowledgeable computer software developers. However, any software package that has been developed for commercial distribution is copyrighted, and it is illegal to use unauthorized copies. When you use unauthorized copies, you are depriving the developers of the compensation that is due them. As a software purchaser, you do have the right to make an archive copy of the software. Most software developers encourage you to use this backup copy as your normal work copy. If something happens to corrupt the copy, you have the original from which to make an additional copy. Some developers of copy-protected software provide a backup copy with your purchase.

Some software developers will grant what is called a site license, which allows the licensee to make a number of copies of a software package for use in an office or corporation. There may be limits placed on the number of copies that can be made, or on where and how these copies can be used. Within the restrictions imposed, site licensing eliminates the concern about unauthorized copies being used.

Other software developers distribute their software through a procedure called "Shareware." This is a pay-if-you-like-it concept. The developers give users permission to copy and try their programs without charge. But they retain their copyrights and expect satisfied users to pay them. Many shareware developers will provide full documentation or user support only to registered users who have paid the expected fee. The fees charged for shareware are generally quite reasonable.

Some computer users, called hackers, have made it a practice to try to figure out security codes in order to access computer systems without proper authorization. Some of these breaches have been purely mischievous, while others have had serious national defense implications. Hackers have decoded long distance access codes that have let them make long distance telephone calls for free. Some have even broken the security code for certain Department of Defense computers and have made unauthorized entry into restricted databases. Responsible computer users do not partake in these types of activities.

SUMMARY

The computer is an extremely powerful tool when it is used properly. However, to use it properly you have to understand both the computer's strengths and weaknesses. You have to be able to properly define a problem and get it into a form that can be analyzed with a computer. You need to be familiar with the various kinds of computer hardware and software available. This will enable you to make sound decisions as to when the computer is the appropriate tool to use for a particular problem. It will also help you choose which computer resource should be used. As an engineer, you will typically use the computer in analysis and design, in database management, in graphics applications, and in preparation of reports. You will need to stay abreast of the rapidly advancing technology of computers. Your use of the computer should be predicated on the ability of the computer to analyze a given problem for less money than manual or other solution techniques.

Loss of data is one of the most frustrating failures when using computers. It can be catastrophic if the lost data are irretrievable. At a minimum, you will have to go back and retrieve or reenter the lost data. System malfunction, disk malfunction, physical damage to disks, or power loss may all cause data loss.

You should save and back up your files frequently when you are modifying them.

You are an engineer, not a computer scientist. As an engineer, you need to use the computer as an effective engineering tool. You should not expect to know everything there is to know about computer programming and computer systems. You are bound to be confronted with computer questions for which you do not have the answer, or for which the answer is not readily available in the documentation. That is the time to seek the help of experts. It is not efficient use of your time as an engineer to be researching answers to obscure computer questions. When seeking help you have to consider the most efficient way of getting the answers you need.

REFERENCES

Bender, E., "Industry Outlook," *PC World*, PCW Communications, Inc., San Francisco, May 1989, p. 51.

Branscomb, L. M., "Electronics and Computers: An Overview," *Science*, Vol. 215, American Association for the Advancement of Science, Wash., D.C., 12 February 1982, pp. 755–760.

Chapra, S. C., and Canale, R. P., *Introduction to Computing for Engineers*, McGraw-Hill, New York, 1986, Chapters 1–5.

Chapra, S. C., and Canale, R. P., *Numerical Methods for Engineers*, 2nd. ed., McGraw-Hill, New York, 1988, Chapter 3.

Eisenberg, A., *Effective Technical Communication*, McGraw-Hill, New York, 1982, p. 1.

Fenves, S. J., "Computers in Civil Engineering Practice," *Journal of the Technical Councils of ASCE*, Vol. 108, No. TC1, American Society of Civil Engineers, New York, 1982, pp. 44–52.

Godfrey, K. A. Jr., "Computers: What Do Students Need to Know?" *Civil Engineering*, American Society of Civil Engineers, New York, June 1987, pp. 72–75.

Insight, "Microvax Reliability Helps Albuquerque Conserve Precious Resource," Vol. 8, No. 6, Digital Equipment Corporation, Maynard, MA, July/August 1988, p. 5.

Norton, P., "The Norton Disk Companion: A Guide to Understanding your Disks," Peter Norton Computing Inc., Santa Monica, CA, 1987.

Orenstein, G. S., "The True Cost of Microcomputers," *Civil Engineering*, American Society of Civil Engineers, New York, November 1986, pp. 57–59.

Red, W. E., and Mooring, B., *Engineering: Fundamentals of Problem Solving*, Brooks/Cole Engineering Division, Wadsworth Inc., Monterey, CA, 1983.

CHAPTER TWO

PROBLEM FORMULATION AND REPRESENTATION

2.1 INTRODUCTION

Proper problem definition and representation are essential ingredients for effective computer applications. Problem definition involves the complete statement of all the conditions of the problem, plus the development of the necessary mathematical models and solution techniques required to arrive at a solution (or solutions) to the problem. Representation involves any graphical depiction of the problem as well as the conversion of the problem definition into a form that is amenable to a computer-assisted solution.

Problem definition and representation present the greatest challenge to students starting out in computer applications. Once the problem is properly defined and represented, the actual application of the computer-assisted solution follows well-defined rules and procedures. Problem definition and representation for computer solutions are not as straightforward as most textbook problems that you have solved in the past. In order to properly define and represent the problem, you will have to integrate the technical knowledge that you have learned in your previous courses. Learning how to use the computer as an engineering tool will help you to develop a logical approach to all problem definition and decision making.

Problem definition requires the identification of all the factors that influence a problem. First you must ask yourself a series of questions about the problem at hand. Are there constraints on the problem? In other words, do the problem solutions have to remain within an identifiable set of limits, or can they vary in any way the model allows them to? What are the input variables that influence a problem? What are the relationships among these input variables? Can a mathematical model be developed that models the interactions among the input variables to produce usable problem solution output? What is the problem solution output that is desired? Once you have answered these questions, you will be well on your way to having the problem defined.

Good graphical representation is a necessary adjunct to all problem definition. To be able to solve mechanics problems, you have to be able to sketch and properly define the parameters of the free body diagram. Similarly, you need to be able to identify a control volume for fluid flow problems. The old adage that a picture is worth a thousand words is certainly true in regard to problem definition. A sketch helps you understand the problem and the systems and components involved in it. It also helps someone else understand how you went about your solution. All the usual graphical representations of problems, such as plan views and orthographic projections, are important in computer applications. It is also important to be able to visualize the problem solution process in a graphical form. This graphical form is known as the *flowchart*. It will be discussed in some depth in this chapter.

Proper representation of the problem in a manner conducive to computer-aided solution requires both proper definition of the problem and an understanding of the mechanics of the particular computer tool to be used. Since there are multiple computer tools available, you must be able to decide which tool is the best for the job at hand, even though more than one could do the job. The importance of being able to choose the proper computational tool is one of the primary reasons for the integrated approach used in this text. Once you have chosen which tool is appropriate, you can transform the problem definition into a form that can be used with that particular tool. You must also be open to the possibility of discovering that you have not chosen the best tool for the problem at hand. This sometimes becomes apparent only when you have started to apply the tool. In this case, you will have to backtrack and search for a better tool. It may also become apparent that your problem definition is faulty or insufficient. In that case, you must also be willing to go back and refine your definition. If you fail to heed either of these needs for feedback, your final solution may be inefficient or, worse, incorrect.

2.2   TOP-DOWN DESIGN

To assist in problem definition and representation, structured programming uses a procedure called *top-down design*. Top-down design is a systematic approach to problem definition that lends itself to computer applications. It is very similar to the approach applied by engineers in design. Developing a computer program is a form of design. You are synthesizing your knowledge of engineering principles and programming syntax to produce a software system that will respond to your inputs in the desired way. The software system serves the purpose of providing information that will be helpful in design or decision making. The system is made up of various components that all work together to achieve that purpose.

The first step in top-down design is to develop a general problem definition. This is what you might find in a textbook problem statement. In real world problems, you may find that one of the most difficult tasks is to come up with a concise problem statement. There will be no book for you to look it up in. Various people associated with a problem may have quite different ideas of how it should be defined. As part of the problem statement, you should identify what input data you have, what the desired output will be, and how the output should be presented.

Next, you have to define and understand the mathematics of the problem. One of the primary engineering uses of the computer is to help solve mathematical problems quickly, especially problems that require many repetitive calculations. Unfortunately, a computer will usually solve an incorrectly posed problem just as quickly as it will solve one that is posed properly.

A general solution technique has to be developed to solve the mathematical representation, or model, of the problem. This could simply involve the substitution of values into a series of equations to estimate desired values for dependent variables. It might also involve complex numerical techniques that require many repetitions. The latter case leads to the definition of an algorithm as a series of steps, often iterative in nature, that are used to arrive at the solution for a mathematical model.

The development of a solution algorithm is not sufficient in itself. It must be refined into a computer algorithm. To do this, you must interweave the logic of the numerical algorithm with the logic structures available in the computer tool being used. You must also include provisions for data input and output. One approach to developing a solution algorithm is to begin by decomposing the problem into a series of smaller problems. Then each of the smaller problems can be refined as much as is necessary to provide the detail necessary to generate the computer program. This process is termed decomposition and successive refinement. Decomposition should break the problem down into logic blocks, or modules, each of which accomplishes a specific task or procedure. In so doing, you may be able to identify procedures that have been used for previous problems and saved as program modules or that are available as library modules on your computer system. You will not have to refine these portions of the present problem decomposition. All you will have to do is access the appropriate module as the need arises. The other portions of the decomposition can be successively refined until the necessary detail is present for writing the computer code for each module. Your successive refinement should also include liberal use of error diagnostics.

Top-down design adds a considerable amount of structure to a program. This structure makes it easier for other users to understand and interpret the program. It also makes it easier for the original author to maintain and modify the program in the future. Top-down design also facilitates the isolation and saving of procedures developed for one problem, allowing them to be called back and used for similar applications. Each module can be developed, tested, and verified separately, then combined with other modules to form the application program. A library of useful modules can be developed and saved for use in subsequent applications.

Top-down design helps structure the application, regardless of the computer software resource being used for a particular problem. It is an especially important procedure for high-level languages such as FORTRAN and PASCAL. You should avoid the temptation to sit down and start writing code after a cursory problem definition. Time spent on top-down design prior to writing computer code will result in much better applications programs that will be usable by a larger audience of engineers.

Top-down design will be illustrated through the various examples in this chapter and subsequent chapters. Two helpful tools, flowcharting and pseudocode, can document your top-down design and provide a bridge from the problem definition to program writing. This chapter will develop examples of each. Additional discussion of structured programming and top-down design is included in Chapter 3.

2.3 PROBLEM DEFINITION (STATEMENT OF PROBLEM)

Problem definition involves several steps. In order to establish that there really is a problem to be analyzed, you have to be able to identify the problem needs. You have to ascertain what is known and what it is you have to estimate. You have to define the goals for the computer application and identify any constraints or limitations imposed on the problem. Careful identification of assumptions is as important here as it is in any other type of engineering application.

Problem identification requires that you establish needs. If there are no needs, there is no problem. There is no reason to produce an elegant computer program that is not going to produce useful results and satisfy some need for information. Many engineering problems have to be solved by iterative procedures. A computer program that will automate the solution procedure could be used many times and would save the engineer a great deal of time. Thus, his or her productivity would be increased. There is a real need for such a program. Conversely, a great deal of time and money have been spent to produce programs that will solve partial differential equations. Some of these programs have little practical value because it is difficult or impossible to accurately estimate the boundary conditions necessary to produce an accurate solution for the differential equations. Thus, it is questionable if there was a real need to develop these programs.

Finding out what is known will require identification of the types and sources of data for problem input. How hard will it be to get the data into a form that the computer can interpret? If there is a need to use a large amount of data and the data are in a form that is incompatible with the computer system being used, it may not be feasible to try to analyze the problem using the available computer resource.

Your output requirements have to be related to the needs of the problem. What information has to be generated, and how should it be presented? Does the information have to be stored for future use, or is a printout of the results sufficient? Is it to be used for design purposes, or is it only to provide a quick estimate for a planning study? Who is the intended audience?

Identification of the needs and the output requirements for the problem will help you define what you plan to accomplish with the program. You could refer to what you plan to accomplish as the goals of your program. You will also need to define intermediate steps, or objectives, that you will have to accomplish to reach your goal. Identification of goals and objectives will assist you in your development of both the mathematical model and the computer algorithm for the problem. A simple example might be a program to find the total distance along a path joining several sets of X-Y coordinates. You could define the output requirements as the coordinates of the points, the distance between each successive two points, and the total distance. Your intermediate objectives would be to find the distance between each successive set of points, and then sum all the distances.

Any constraints and limitations on the problem solution have to be identified, so that the computer application does not violate key constraints. Limitations on the total amount of money available to implement a solution, or on the yield stress of steel, should be included in the computer application to prevent the mathematical model from exceeding practical limitations. Any assumptions that have to be made in order to develop a mathematical model should be included as part of problem definition, since they place constraints on the use of the program output.

Also, the required accuracy of the computations will form a constraint on the problem. Less accurate results are generally cheaper to produce, so it is advantageous to use a level of accuracy that is appropriate for the problem at hand.

2.4 MATHEMATICAL MODEL(S)

Engineering is the application of mathematics and physics to solve practical problems. The engineer must be able to predict the behavior of the solution, whether the solution applies to a building or to an automobile engine. One of the tools that the engineer uses to predict this behavior is the mathematical model. A mathematical model is an abstraction, or approximation, of the actual system that will be associated with the solution for the problem at hand. How good the mathematical model is will depend on the quality of problem definition, the mathematics and physics knowledge of the engineer developing the model, and the state-of-the-art of technology with respect to the phenomena being modeled. When using mathematical models, it is important to identify, and remember, what assumptions were made. You have to understand what limitations might be inherent in the mathematical depiction of the problem. You have to make sure that units are consistent through all terms of a particular equation.

In developing a mathematical model, you need to determine the mathematical expressions that will relate what is known to what you want to find. You are developing a mathematical system that models the physical system. When you input values to the model, it will act on this input and produce an output. Your goal is to have this output be a reasonable approximation of the corresponding response, or output, of the actual system. You can produce alternative solutions for the model by providing alternate input. You also need to continue to evaluate the reasonableness and accuracy of the mathematical model throughout the engineering process by gathering whatever data you can to verify the output.

Many mathematical models that are difficult or tedious to solve by normal hand calculations can be solved efficiently with the computer. However, the solution will still be only as good as the mathematical model. This is why the computer user must be particularly attentive to the development of the mathematical model, the assumptions made, and the limitations of the model.

2.5 SOLUTION TECHNIQUE(S) (ALGORITHM DEVELOPMENT)

It is not sufficient merely to develop a mathematical model. You have to be able to solve the mathematical model if you are to generate any useful information from it. This may seem trivial for simple problems, but with more complex problems it is critically important to consider what procedure you will have to follow to evaluate the mathematical expression or expressions.

2.5.1 Input/Output Design

You need to determine what variable values will have to be provided to the computer to solve a certain problem. You also have to decide the best way to enter the data. Using a data file may be advantageous for applications that require a large amount of input data, or it may be advantageous to enter and change data from a terminal during an iterative design process.

Output should include all input data. This allows you to check for data entry errors. Other output must be designed so it displays all necessary output data with clear labels, but does not include unnecessary intermediate data that will distract the reader. Tables should be used to show how dependent variables react to changes in independent variables. Graphical displays should also be included when appropriate.

2.5.2 Algorithm Formulation

An algorithm consists of a detailed sequence of simple steps used to solve complex problems. It is often iterative, in that more than one pass is required through the series of steps or portions of them. However, the algorithm must end in a finite number of steps. Either the solution must converge to an answer, or you must determine that the solution will not converge. The algorithm must contain some sort of mechanism to determine when it should be terminated. Changing the input to the algorithm can generate alternative solutions.

Development of an algorithm can be accomplished only if you first understand the mathematical model posed and have a general understanding of the steps involved in arriving at a solution. You may find that there are existing algorithms that can be used for or adapted to your present problem. For example, if you have a system of linear equations, you have the choice of solving the equations by substitution or by employing one of several established techniques for solving linear systems. Implementations of most of these methods are available in library routines maintained for minicomputer and mainframe computer systems. You must understand the rules for using the routine, the assumptions used in its development, and the limitations of its operation.

You may also find that a new problem can use a procedure that you have programmed previously. This is frequently the case if you are doing a series of similar, but not identical, analyses.

2.5.3 Numerical Methods

Numerical methods are an integral part of the development of your algorithm. They are also a key ingredient in the computer implementation of algorithms. You will use numerical methods when successive approximations are required to approach the correct solution to a problem. A trial-and-error solution is a type of numerical method. However, there are more sophisticated methods available that employ computed correction factors that speed up the convergence to the final solution.

Whenever you are using repetitive calculations to find an approximate solution, the accuracy of the answers provided is a real concern. A measure of the accuracy is the error, the difference between the estimated and true values of the answers. The algorithm error can often be estimated through error functions that have been derived for particular numerical techniques. However, you also have to be concerned with the numerical accuracy of the computational device being used. Computers have to convert decimal numbers into binary format for computations. Computers also retain only a certain number of significant figures throughout computations. Therefore, rather large numerical errors can build up after thousands of repetitive calculations. You also have to be alert to the possibility of convergence criteria giving a false positive (indicating that the algorithm has converged to a solution when it actually has not). Sometimes you can evaluate numerical error of algorithms by checking your answers against known correct answers. If you have no known answers, you must use extra care in making your algorithm as accurate as possible. Minimizing the number of calculations required and checking for operations that may cause errors, such as adding or subtracting very large numbers and very small numbers, are two ways for you to increase accuracy. When using FORTRAN to implement your algorithm, you can use a data storage option called DOUBLE PRECISION, which allows the computer to store additional significant figures.

2.6 TRANSLATION TO COMPUTER ALGORITHM

Translation of the solution algorithm into a computer algorithm requires that you apply the structures of a computer language or software package to implement the logic of the solution algorithm. This will require writing some sort of computer program. Any computer program consists of four main elements. These are input, computational sequences, control structures, and output. A program may require several implementations of each element. The interactions among the elements may be fairly complex. Engineers use flowcharts or pseudocode to assist in applying programming structures to the algorithm logic. Flowcharts and pseudocode are also useful tools for modular design of structured programs.

The best time to structure your program is during the development of the logic of the computer algorithm. Application of top-down design using flowcharts and pseudocode helps identify modules that accomplish specific procedures. Any of these modules that have been developed and saved previously can be used for the present application. You will not need to include their detailed logic structure in the current design. You can develop detailed flowcharts or pseudocode for new modules. You will also have to develop logic for the main program module that will coordinate the old and new procedure modules.

2.7 FLOWCHARTS AND PSEUDOCODE

Flowcharts and pseudocode provide means of expressing design in a general way without using exact language syntax. They display the logic of the various modules to be used for a computer application. Modules are identified through the top-down design procedure of decomposition. Successive refinement adds the details of the logic for a particular module. Development of a flowchart or pseudocode helps the programmer analyze and develop the logic of the application thoroughly before writing program code.

Both flowcharts and pseudocode can be developed in a generic format that is language-independent. Thus, they can be used for any application. Examination of the structure of the flowchart or pseudocode may indicate the best computer resource for the problem. Examples of both flowcharts and pseudocode will be used in the text. Whether you use flowcharts or pseudocode is a matter of personal preference. You may choose one or the other, depending on the problem. At times, you may feel that it would clarify a problem if you used both.

2.7.1 Flowchart

Flowcharts provide a graphical representation of problem solution logic, computational sequence, and data input/output sequences. They are written in a somewhat standard form, which helps other users of an application interpret the logic used in the problem solution. Flowcharts will frequently be included with documentation. A generic form of the flowchart will be developed in this chapter. More specific forms may be used for specific applications or languages. However, the basic components of logic representation will be the same.

2.7.1.1 Representation of Control Structures for Programming (Logic)

In developing a structured computer algorithm, you will want to avoid unnecessary and indiscriminate branching, as that obscures the logic of the program. You should use three basic control structures in any representation of problem solution logic. These are sequential, selection, and repetition control. Graphical representation of these structures is somewhat standardized. I have chosen what I believe to be the most representative symbols for use in this text.

Sequence   Sequential control means that instructions or operations are completed in the order that they appear. There is no branching or looping. Sequential control is an essential structure in any computer application and is the only structure that can produce a complete program without any other type of control. However, without selection and repetition control, programs would be limited to simple once-through logic. Figure 2.1 shows the symbols that can be used to depict sequential control logic.

Selection (Branching) Control   Selection control is a representation of the decision-making process in problem solution logic. Given some defined set of alternative conditions, the solution process will branch to different subsequent processing based on which condition exists. The alternative branches usually rejoin later in the solution process. However, there may be instances where they do not. Figure 2.2 shows a representation of elementary selection logic structures. Note the use of sequential control within the branches of the logic structure. As we start to develop applications, you will see that the branches of one logic structure may also contain additional logic structures. This is called nesting. To properly depict program logic, a nested selection control structure must be contained within the branch of the higher order control structure.

Repetition (Looping) Control   Repetition, or looping, control allows a portion of the solution process to be repeated a number of times. If the number of times the solution process subset is to be repeated is set prior to starting the repetition, it is referred to as an absolute repetition structure. Some sort of counting mechanism must be employed to keep track of how many times the solution process has looped. Each time the loop is executed, the counter is checked, and if it has

FIGURE 2.1   Symbols representing sequential control.

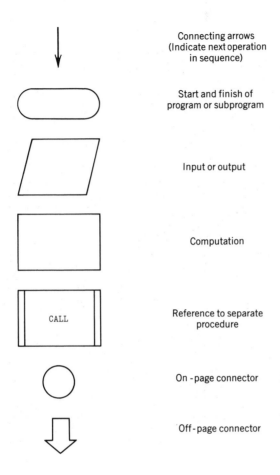

Connecting arrows
(Indicate next operation
in sequence)

Start and finish of
program or subprogram

Input or output

Computation

CALL

Reference to separate
procedure

On - page connector

Off - page connector

reached the desired number of repetitions, the repetition ceases. Another type of repetition structure allows the repetition to continue as long as some condition holds. As soon as the condition no longer holds, the repetition ceases. This is referred to as a conditional repetition structure.

Figure 2.3 shows the schematic depiction of both absolute and conditional repetition structures. Note that as with selection control, additional repetition structures can be nested within a repetition structure. Also, repetition and selection structures can be combined to form complex logical structures.

2.7.2 Pseudocode

Pseudocode is a tabular representation of program logic, written in a form that is understandable to the programmer. There is no standardized format for pseudocode. Pseudocode does not have to follow language syntax rules, but you should be able to develop the language code from the structure of the pseudocode. To help identify the logic structure, it is a good idea to indent subservient levels of

FIGURE 2.2   Symbols representing selection control.

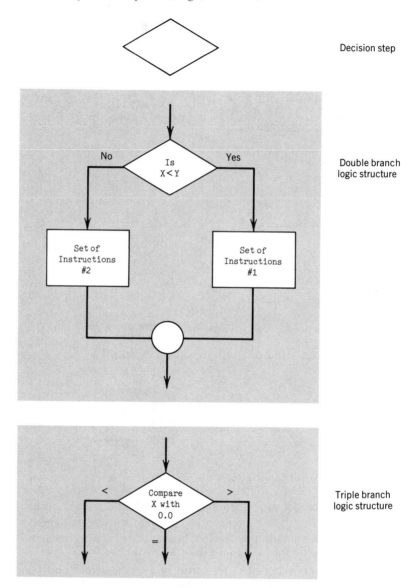

Decision step

Double branch
logic structure

Triple branch
logic structure

logic. Thus, the sequential logic within a repetition loop would be indented, as would a nested selection structure within the repetition structure. The sequential logic within the nested selection structure would be indented twice. For example, pseudocode for the double branch logic structure of Figure 2.2 would be

```
IF X < Y THEN
     ACCOMPLISH SET OF INSTRUCTIONS #1
OTHERWISE
     ACCOMPLISH SET OF INSTRUCTIONS #2
END OF IF STRUCTURE
```

FIGURE 2.3   Symbols representing repetition control.

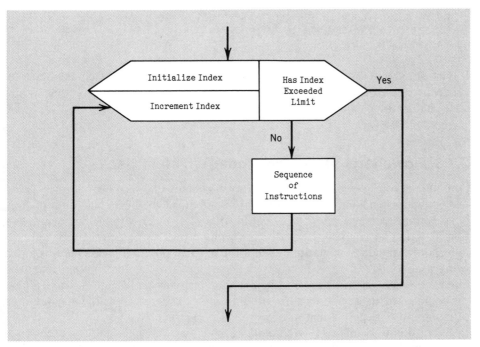

(*a*) Absolute repetition.

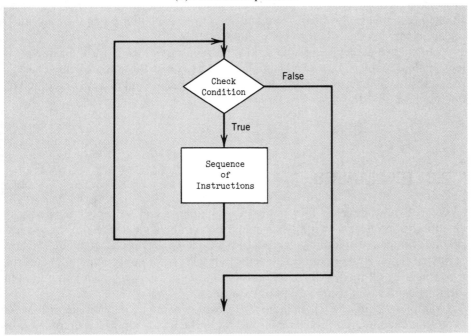

(*b*) Conditional repetition.

The pseudocode for the looping structure of Figure 2.3*a* would be

```
    DO TO END OF LOOP 10 FOR INDEX FROM INITIAL TO LIMIT
        SERIES OF STATEMENTS TO BE REPEATED
10 END OF LOOP
```

It is assumed that the computer will increment the index by one each time the loop is executed. If you want to increment the index in steps other than 1, you should state that in the pseudocode.

2.7.3 Flowcharts vs. Pseudocode: Which to Use?

Sometimes you may prefer to use both a flowchart and pseudocode to develop and show the logic and structure of a program before you begin computer coding, or sometimes you may prefer one over the other. If you are producing documentation that another engineer is to use, you would probably be better off including a flowchart in the documentation, even if the original program had been developed using pseudocode. This is because engineers are used to interpreting graphical information, and even engineers who may not be computer literate can understand the logic used in developing the program. However, in developing your own programs, it is largely a matter of personal preference.

According to Scanlan (1988), computer scientists have shifted to a preference for documenting algorithms for structured programs with pseudocode rather than flowcharts. This is despite research indicating that students prefer having algorithms presented in flowchart form and seem to learn better when they are introduced to the flowchart before the pseudocode.

You will see examples of both a flowchart and pseudocode for the same problem in this chapter. In the remaining chapters, generally either a flowchart or pseudocode will be used. This is done to save space, not out of any preference for one method over the other.

2.8   EXAMPLES

We will now develop four computer applications of varying complexity to illustrate the concepts of problem definition, mathematical modeling, and algorithm development. Example 2.1 will be a simple application to compute the area of any circle given the radius. The algorithm will require only sequential control. Example 2.2 will modify the circle area program to allow for area computations for a series of radii, using repetition control.

Examples 2.3 and 2.4 will demonstrate development of computer algorithms for two often encountered problems in physics and engineering, the quadratic equation and a particle trajectory problem. The quadratic equation example will require use of selection control, while the particle trajectory problem will require use of all three control structures. Decomposition and successive refinement will also be illustrated. Example 2.3 will illustrate use of both pseudocode and flowcharts for the quadratic formula.

EXAMPLE 2.1 □ Area of Circle of any Radius

STATEMENT OF PROBLEM

Develop a computer algorithm that will compute and output the area of a circle, given the radius.

MATHEMATICAL DESCRIPTION

$$\text{Area} = \pi r^2 \tag{2.1}$$

ALGORITHM DEVELOPMENT

Input/Output Design

The radius will be input at the beginning of the program. Output should include both the radius and the area, with appropriate labels.

Numerical Methods

No special numerical methods are required for this problem.

Computer Implementation

Figure 2.4 shows the flowchart for this program.

FIGURE 2.4   Flowchart for area of circle of any radius.

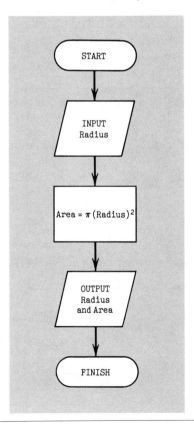

EXAMPLE 2.2   □   **Area of Circle for Radius 1,2, . . . , 12 ft**

STATEMENT OF PROBLEM

Develop a flowchart for a computer application that will compute and output the area of a series of circles with radii from 1 to 12 ft.

MATHEMATICAL DESCRIPTION

$$\text{Area} = \pi r^2 \tag{2.1}$$

FIGURE 2.5   Flowchart for area of circles of selected radii.

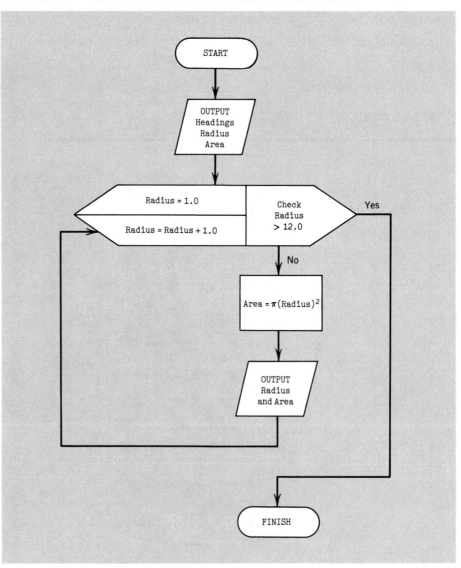

ALGORITHM DEVELOPMENT

Input/Output Design

The radius will be input at the beginning of the program. Output should include both the radius and the area, with appropriate labels. This is the same as Example 2.1.

Numerical Methods

A repetition structure will be used to increase the radius from 1 to 12 ft in specified increments.

Computer Implementation

Figure 2.5 shows the flowchart for this problem.

EXAMPLE 2.3 □ Quadratic Equation

STATEMENT OF PROBLEM

Many phenomena in engineering and physics can be approximately or exactly modeled by a second-order equation of the form

$$y = f(x) = ax^2 + bx + c \qquad (2.2)$$

Equation 2.2 is the formula for a parabola. It is also called a quadratic equation because the highest order term in the equation is a squared term. An example of an exact application of a parabolic equation would be the modeling of a parabolic arch used to form the support structure for a bridge.

Develop a computer application that can be used to find the roots of the quadratic equation. The roots of the equation are the values of x that satisfy the condition $f(x) = 0.0$.

MATHEMATICAL DESCRIPTION

Equation 2.2 could be solved by trial and error until a value of x that satisfies the condition $f(x) = 0.0$ is found. However, this is not the most efficient way of accomplishing the task. Mathematicians have developed a direct solution technique that can be used to find the roots. It is the quadratic formula, which has the form

$$X = \frac{-b \pm \sqrt{b^2 - 4ac}}{2a} \qquad (2.3)$$

Equation 2.3 can be used to solve for either real or complex roots of the quadratic equation. If the portion under the radical evaluates to a positive value, there are two real roots, whereas a negative value under the radical indicates two imaginary roots. When $b^2 = 4ac$, the two roots are equal.

(EXAMPLE 2.3 □ Quadratic Equation □ Continued)

ALGORITHM DEVELOPMENT

There will be no need for repetition in this problem, but there will be a need to make decisions as the computations progress, depending on the outcome of evaluating the variable values under the radical.

If $b^2 - 4ac$ is positive, you can compute the two real roots using Equation 2.3. However, if it is negative, you will have to take the square root of the absolute

FIGURE 2.6   Flowchart for solution of quadratic equation.

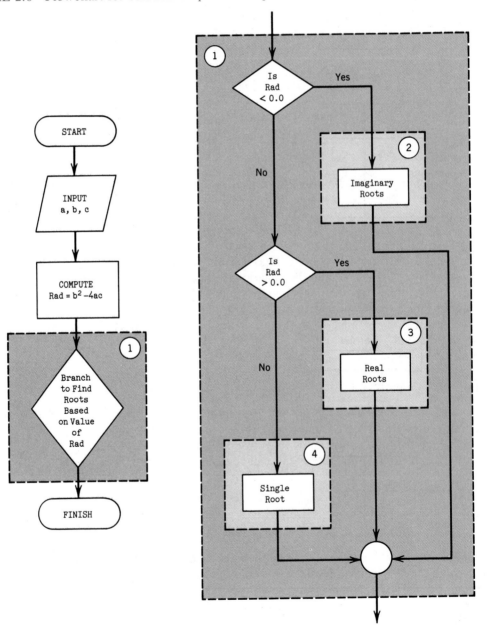

(*a*) Decomposition of first-level refinement.

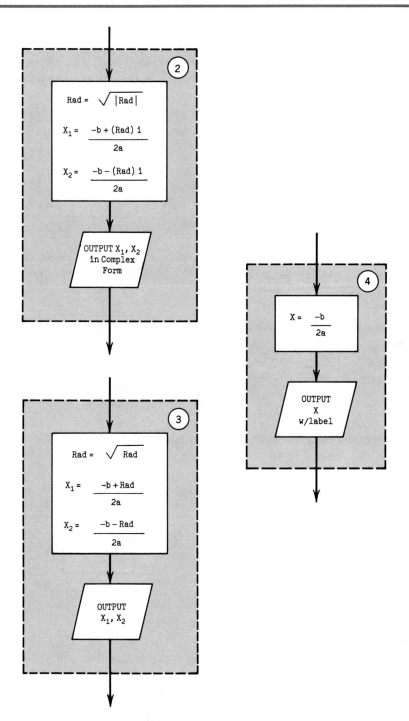

(*b*) Second-level refinement.

(EXAMPLE 2.3 □ Quadratic Equation □ Continued)

value of $b^2 - 4ac$ and use it as the imaginary part of the complex root. If the value of $b^2 - 4ac$ is zero, you have the limiting case where the apex of the parabola represented by the quadratic equation just touches the x axis. For this case there will be only one root; or to be more exact, both roots will be equal to the same value.

Input/Output Design

Input will be the coefficients a, b, and c of the quadratic equation. Output will include the values of the roots and appropriate labels.

Numerical Methods

You will have to ensure that the computer does not try to take the square root of a negative number. No other numerical methods will have to be employed.

Computer Implementation

We will develop both a flowchart and a pseudocode for the quadratic formula implementation. Figure 2.6 shows the flowchart. Sequential and selection control are used. Decomposition and successive refinement are also illustrated. Note the use of the replacement operator

$$Rad = \sqrt{|Rad|} \tag{2.4}$$

which is perfectly acceptable in computer applications. You will find in later chapters that computer variables represent storage locations. The operation of Equation 2.4 takes the absolute value of the present value stored in Rad, finds the square root of the absolute value, and places the resulting value back into storage location Rad. The previous value stored in Rad is now gone. Thus, the equal sign represents a replacement operator, rather than its traditional definition.

The corresponding pseudocode follows. Note the use of indentation to identify the logic hierarchy.

```
INPUT a,b,c (coefficients of quadratic equation)
COMPUTE RAD = (b²-4ac)
START IF (SELECTION STRUCTURE)
IF (RAD < ZERO) THEN
      RAD = SQUAREROOT[ABSOLUTE VALUE(RAD)]
      X₁ = -b/(2a) + RAD/(2a)i
      X₂ = -b/(2a) - RAD/(2a)i
      OUTPUT X₁, X₂ IN COMPLEX FORM
OTHERWISE, IF (RAD > ZERO) THEN
      RAD = SQUAREROOT(RAD)
      X₁ = (-b + RAD)/(2a)
      X₂ = (-b - RAD)/(2a)
      OUTPUT X₁,X₂
OTHERWISE
      X = -b/(2a)
      OUTPUT X
END IF (SELECTION STRUCTURE)
```

EXAMPLE 2.4 □ Particle Trajectory

STATEMENT OF PROBLEM

Any projectile that is traveling near the earth's surface, whether it is an artillery shell or a particle of water, is acted on by the force of gravity. The particle has a velocity parallel with the earth's surface. It also has a velocity perpendicular to the surface. Eventually the particle will impact the earth's surface unless it is acted on by some force in addition to gravity.

    Neglecting the frictional resistance of air, develop a mathematical model that could be used to predict the x and y coordinates of the particle at any time t. Develop a computer algorithm that will use the mathematical model to find an angle or angles that will cause the trajectory to pass through a given point x, y. The particles are initially launched at an angle of θ degrees above the horizontal, as shown in Figure 2.7, and with an initial velocity of V_0. Note that there may be two angles that cause the trajectory to pass through the given point.

MATHEMATICAL DESCRIPTION

Development of general particle trajectory equations is discussed in Appendix G. For this problem, we will consider the equations for the x and y positions for the particle at any time t. They are

$$x = x_0 + V_{0_x} t \tag{2.5}$$
$$y = y_0 + V_{0_y} t - .5gt^2 \tag{2.6}$$

    The modeling of the trajectory of a particle in the earth's atmosphere under the influence of gravity is an application of the parabolic equation to approximate the actual trajectory. It is approximate because the model does not take into account the effects of air resistance on the particle. Any projectile passing through the earth's atmosphere is subject to some amount of air resistance. For

---

FIGURE 2.7   Definition sketch for particle trajectory.

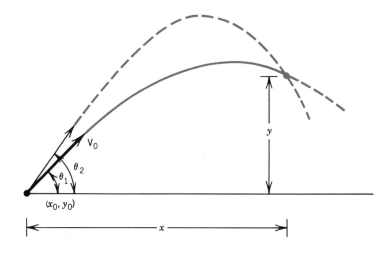

(EXAMPLE 2.4 □ Particle Trajectory □ Continued)

many applications the effects of air resistance are small enough that they can be safely neglected. If air resistance is neglected, that assumption should be stated and justified if necessary.

Equations 2.5 and 2.6 can be combined to develop an expression that can be used to estimate the height y of the trajectory at a distance x from the launch point for a given angle θ and initial velocity V_0. The angle is used to break down the initial velocity into its x and y components. Equation 2.5 can be solved for t and substituted into Equation 2.6, resulting in

$$y = y_0 + V_0 \sin \theta \left(\frac{x - x_0}{V_0 \cos \theta} \right) - 0.5g \left(\frac{x - x_0}{V_0 \cos \theta} \right)^2 \qquad (2.7)$$

Equation 2.7 will be used in subsequent examples to find any angles that make the height of the trajectory y equal to the desired value at the distance x.

ALGORITHM DEVELOPMENT

If you want to find the x and y coordinates of a projectile at any time t, the algorithm will involve breaking the initial velocity into its x and y components and substituting the time t into the equations to find x and y. However, if you want to find an angle θ that will cause the trajectory to be at a particular height at a distance x from the origin, it does not appear that you can rearrange the equations to solve directly for the angle. You must resort to a trial-and-error solution in which you try various angles until you find the one that meets the criteria. You may also find that, for the initial velocity given, the trajectory will not reach the height desired. There may be two angles at which the trajectory meets the criteria for some distance–initial velocity combinations. Your algorithm will have to take all these possibilities into account.

You can start your search at the minimum desired angle and try different values of θ until the height of the trajectory reaches height y for the distance x. Each value of θ will be substituted into Equation 2.7 and the calculated value of y will be compared with the desired value. By looking at the successive values, you can judge whether the solution is approaching or diverging from the desired y value. As the height approaches y, you can make the increment on θ smaller to give a more accurate solution. Once you have found one solution, you have to continue with the solution process to see if there is another solution that fits the conditions of x, y.

Input/Output Design

Input includes the coordinates of the launch point x_0 and y_0, the initial velocity V_0, the acceleration due to gravity, and the coordinates of the desired point on the trajectory x, y. Output will depend on whether one, two, or no angles provide a trajectory that will pass through the point x, y.

Numerical Methods

A method has to be incorporated to initialize the angle θ, increase the angle in set increments through the range desired, and end the iteration.

Computer Implementation

Figure 2.8 shows the decomposition and refinement of the flowchart for this application. We have used a constant angle increment. Therefore, the angle increment has to be small enough so that the height y will not pass beyond the desired y without ever being within the tolerance established for the solution. Also, you should note how the search process is started for the second solution after finding one solution. The first solution is found by starting at the minimum angle and increasing the angle in set increments until either a solution is found, or the angle reaches the maximum value. If a solution is found, the angle is then set equal to the maximum value, and the angle is decreased by the same increment as before until a solution is found. You know that you will find a solution

FIGURE 2.8   Flowchart for particle trajectory.

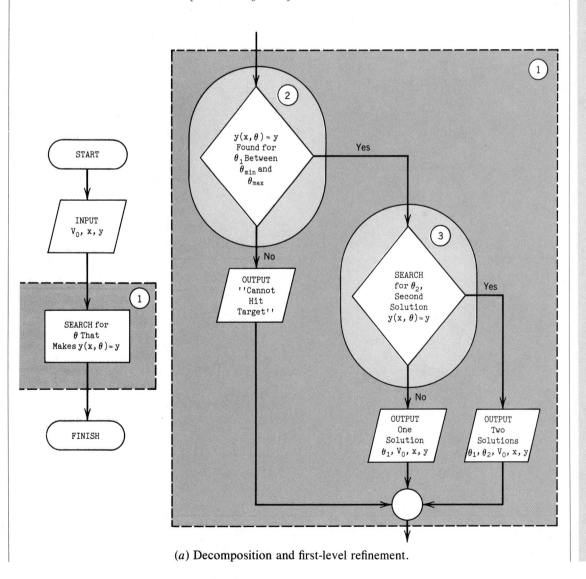

(*a*) Decomposition and first-level refinement.

(EXAMPLE 2.4 ☐ Particle Trajectory ☐ Continued)

FIGURE 2.8 (Continued)

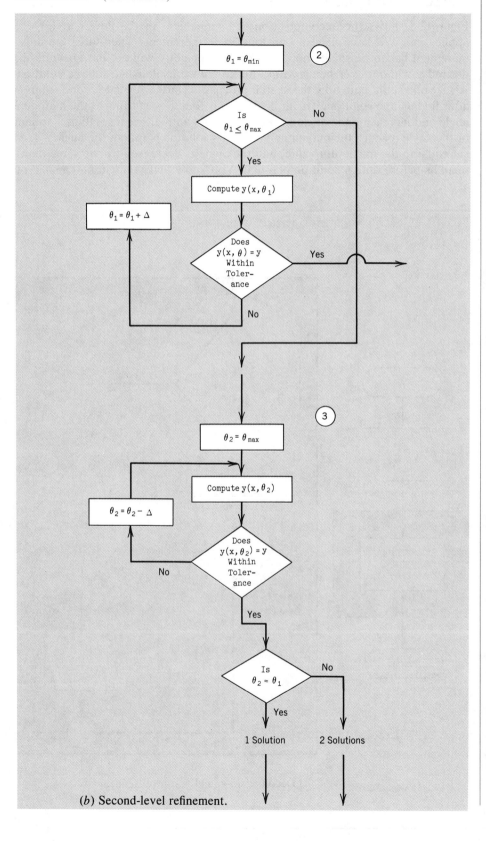

(*b*) Second-level refinement.

this time. Even if a second solution is not found, the angle will eventually reach the value for the first solution. Once you have reached a solution by decreasing the angle, you must check to see if the angle is approximately equal to the angle for the first solution. If it is, you have only one solution. If it is different, you have two solutions. A more sophisticated solution to this problem will be developed in Chapter 4.

2.9 FITTING THE PROPER COMPUTATIONAL TOOL

When you have reached the stage of developing the computer algorithm, it is time to consider what the best computational tool will be for the particular application. In this text you will become familiar with several different computer applications tools, including high-level languages, spreadsheets, equation solvers, and integrated packages. You will also be introduced to database management concepts. While using the text, you may use different computer hardware resources, such as microcomputers, minicomputers, or mainframe computers, and you may use more than one operating system.

When deciding which computer resource or applications package to use, you must consider the form and complexity of the problem. You must also evaluate the convenience of access to the computing resource. The required accuracy of the solution must be taken into account. You must estimate the computational power needed and estimate the probable cost of the application. Costs should include development time, computation time, and user time. In engineering practice the time spent on a problem will equate to money, so it is important to choose the most efficient resource. Above all, you do not want to choose a resource that will take more time to produce results than conventional methods. It may take some experience, and experimenting, before you have a firm grasp on resource selection. Thorough familiarity with the different computer resources will help you. In complex applications, you will probably have to seek outside help to evaluate resources.

As you continue through this text you will encounter discussions about choosing the best computing resource for a particular problem and comparisons among various methods used to solve the same problem. You can also make your own comparisons through the applications exercises at the end of the chapters. You should consider the advantages of using or adapting already existing applications programs over developing a new program. You will have to verify that the model used in the application fits the present problem. If it does, you have saved yourself considerable development time and cost.

SUMMARY

Thorough problem definition is an essential part of any problem-solving process. Computer applications are no exception. Problem definition will lead to the development of a mathematical model to represent the actual processes being studied for the problem. To obtain useful information from a mathematical model, you must have a solution technique. Computer applications will involve writing computer instructions to implement your solution technique. Two tools that help you define the logic of the solution technique are flowcharts and pseudocode. These tools will also help you develop the necessary computer instructions.

You must be precise in your problem definition and development of solution logic. The computer cannot interpret your intent; it can only follow your instructions. Avoid the pitfall of considering the computer to be a thinking machine. If you do not understand the theory of a problem, or the logic of the solution method, do not try to develop a computer application for the problem. You will be destined for frustration and will not achieve useful results. Study the problem and solution techniques until you do understand them.

It is essential to have worked out answers to check computer output against. The time to develop them is during the problem definition and algorithm development stages. Computing answers will help you understand the problem and the solution technique.

REFERENCES

Borse, G. J., *FORTRAN 77 and Numerical Analysis for Engineers*, PWS Publishers, Boston, 1985.

Scanlan, D., "Structured Flowcharts vs. Pseudocode: The Preference for Learning Algorithms With a Graphic Method," *Engineering Education*, American Society for Engineering Education, Wash., DC., December, 1988, pp. 173–177.

EXERCISES

2.1 GENERAL NOTES

The numbers in brackets at the beginning of exercises indicate in which other chapters related exercises can be found. [ALL] indicates that related exercises can be found in each of Chapters 2 through 9. The essential theory for exercises in the text is discussed either in the exercise statement or in one of the appendixes. Appendix A contains answers to selected exercises.

For all exercises in this chapter you should develop a problem definition with all of the following elements:

    Statement of Problem
    Mathematical Description
    Algorithm Development
        Input/Output Design
        Numerical Methods
        Computer Implementation
            Flowchart and/or Pseudocode

Most of the exercises ask you to develop a flowchart. Your instructor may ask you to develop pseudocode also, or ask you to develop pseudocode alone.

It will be necessary for you to develop the mathematical model from the theory presented. Not all equations will be given. You will have to combine and synthesize the mathematical relationships. You will also be expected to perform hand solutions to verify your method and to check your computer results. Your instructor may add one or more of the following options to your exercise assignment:

Input and/or output range checks

Input error trapping

Interactive input and output (the computer asks a question, then accomplishes some task based on the user response)

In many cases you will not be told what to output. It will be up to you to decide what should be output and to carefully design the output for easy interpretation by people using the program.

2.2 MODIFICATION OF EXAMPLES

2.2.1 Example 2.1

Modify the flowchart of Example 2.1 so that either the diameter or the radius can be used as input and so that the computation and output include the diameter, radius, circumference, and area of the circle.

2.2.2 Example 2.2

1. Modify the flowchart of Example 2.2 to incorporate the provisions of the preceding exercise.
2. Modify the flowchart of Example 2.2 so that the beginning and ending radii and the number of radius increments can be input.

2.3 CHEMICAL ENGINEERING

2.3.1 [ALL] Ideal Gas Law

Appendix K discusses the ideal gas law for relating pressure, temperature, and volume for a gas. Two other models, the van der Waals equation and the Redlich–Kwong equation, deal with nonideal gases. Develop a program that will compare the pressures estimated by either or both of the nonideal equations with the pressures given by the ideal gas law. Use a range of volumes at constant temperature. Also, your program should compute the compressibility factors for the van der Waals and Redlich–Kwong equations.

2.3.2 [ALL] Chemical Kinetics

Appendix K discusses first- and second-order kinetics for the decomposition of a compound at a given temperature. Develop a flowchart for a program that will predict the concentration of the compound as a function of time using either first- or second-order kinetics, or both. The user should be able to input the range of times and the number of time increments to be used.

2.3.3 [ALL] Depth of Fluidized-Bed Reactor

Fluidized-bed reactors provide more uniform contact between a fluid and a medium, such as a catalyst, during a reaction process. Appendix K presents the mechanics of fluidized-bed reactors. Develop a flowchart for a program that will estimate the required bed height for a given reactor bed flow rate, unexpanded void fraction, and bed particle size distribution. This exercise is adapted from an example by Borse (1985).

2.3.4 [ALL] Process Design, Gas Separation

Appendix K discusses separation of a gas mixture into its constituents by means of selective absorption of gases in liquids. Develop a flowchart for a program that will estimate the number of plates necessary to reach a desired absorption efficiency for ethane, methane, or propane, and the resulting absorption efficiencies for the other components. The number of absorption plates should be rounded up to the next integer. This exercise is adapted from an example by Borse (1985).

2.3.5 [ALL] Dissolved Oxygen Concentration in Stream

Develop a flowchart for a program that will implement the model for dissolved oxygen concentration in a stream presented in Appendix K and will predict the dissolved oxygen concentration in the river at a number of points downstream from the point of discharge.

2.4 CIVIL ENGINEERING

2.4.1 [ALL] Equilibrium-Truss Analysis

Develop a flowchart for a program that will determine the reactions for a determinate planar truss with one pin and one roller support. Input data will be the coordinates of the two supports, the coordinates of the points of application of up to three point loads, and the magnitude and direction of each of these loads. Refer to Appendix G for the appropriate theory.

2.4.2 [ALL] Open Channel Flow Analysis

1. Appendix G discusses the principles of flow in an open channel. Develop a flowchart for a program that will determine the flow rate in a circular or trapezoidal channel given the necessary dimensional information, the bottom slope, the roughness coefficient, and a specification concerning the system of units to use.
2. Use the information of the previous problem to develop a flowchart for a program that will estimate the depth of flow in a trapezoidal open channel given the flow rate, the other necessary information, and an initial estimate of the depth. Trial and error will have to be used.
3. Develop a flowchart for a program that will estimate a series of flow rate vs. depth values for flow in a trapezoidal open channel.

2.4.3 [ALL] Dam Stability Analysis

Appendix G discusses the principles for estimating hydrostatic forces on submerged surfaces. The cross section of a concrete gravity dam is shown in the figure for this

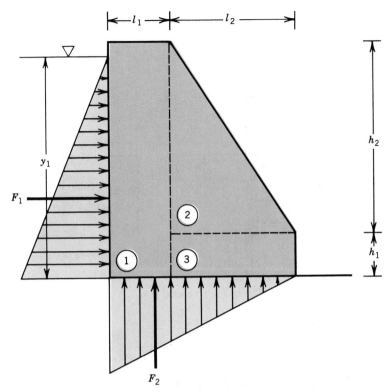

Figure for Exercise 2.4.3

exercise. The dam holds back water of height y_1 behind it. One of the design criteria for a dam of this type is that it has enough weight to prevent its slipping downstream. The water behind the dam produces force F_1. F_1 is the resultant force from the triangular pressure prism shown. Also, if there is any seepage under the dam, there is an uplift force on the dam. To be conservative, it is assumed that the pressure on the upstream corner of the bottom of the dam is the same as the pressure in the water at the same point, or γy_1, and that the pressure decreases linearly until it reaches whatever the pressure is at the downstream bottom corner of the dam. Since there is no water below the dam, the pressure at the downstream corner is zero. Force F_2 is the resultant of this uplift force. It acts through the centroid of the pressure prism on the bottom. The cross section of the dam is homogeneous and made of concrete of specific weight $= 150$ lb/ft³. The dam can be broken up into simple shapes 1, 2, and 3 for analysis. The analysis can be accomplished with an assumed dam length of 1 ft. The frictional force developed between the dam and the foundation must be greater than F_1. Develop the logic of a program that will analyze the stability of a given dam against sliding.

2.4.4 [ALL] Force on Submerged Gate

The figure accompanying this problem shows a gate submerged in a tank of liquid. The gate has to be designed so that it will resist the forces applied to it by the liquid. Although the gate shown is rectangular, Appendix G shows that the methods of analysis are applicable to any common shape. Develop the logic of a program that could be used to find the force and its point of application on either a submerged rectangular, circular, or triangular gate.

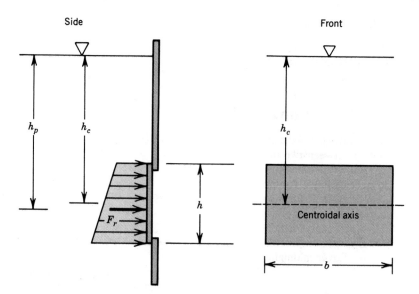

Figure for Exercise 2.4.4

2.4.5 [ALL] Volume of Excavation

Develop a flowchart for a program that could be used to determine the total volume of excavation for a sewer trench. The trench is approximated by a series of rectangular sections, each of length l_i, width b_i, and constant depth d_i.

2.4.6 [4] Classification of Soil Samples

The American Association of State Highway and Transportation Officials provides the following criteria for classifying soils:

Grain Size, mm	Classification
>75	Boulders
2–75	Gravel
0.05–2	Sand
0.002–0.05	Silt
≤0.002	Clay

Develop a flowchart that will classify any number of samples and save the classifications for later use.

2.5 DATA ANALYSIS AND STATISTICS

2.5.1 [ALL] Histogram

Develop a flowchart for a program that will count the frequency with which a series of soil samples falls into each of the classifications given in Exercise 2.4.6.

2.5.2 [ALL] Sorting

Develop a flowchart for a program that will input two values, X and Y, swap the two values (X now contains the value that Y had, and vice versa), and output both the original and new values.

2.5.3 [3,4,5] Time Conversion

1. Develop a flowchart for a program that will determine the elapsed time (number of minutes) between two times on a 24-hr military clock (8:00 A.M. is 0800, 1:00 P.M. is 1300, etc.). The duration will not be more than 24 hr; however, it can extend over midnight from one day to the next.
2. Develop a flowchart for a program that will convert elapsed time in hr : min : s into the equivalent number of seconds.
3. Develop a flowchart for finding the difference in hr : min : s between two times that are also expressed in hr : min : s. The elapsed time will not be more than 12 hr, but it can extend over midnight or noon.

2.5.4 [3,4,5] Date Conversion

1. Develop a flowchart for a program that will convert a Julian date to a calendar date. The Julian date is the sequential day of the year. For example, January 15, 1988 would be 8015, while February 15 of the same year would be 8046. Do not forget that every fourth year is a leap year with 29 days in February, and 366 days in the year.
2. Develop a flowchart for a program that will convert a calendar date to a Julian date.

2.6 ECONOMIC ANALYSIS

2.6.1 [ALL] Economic Formulas

Appendix E discusses economic analysis. Develop a flowchart for a computer application to undertake economic analysis. Given an interest rate i, a number of compounding periods n, a known parameter (P, F, or A), and the parameter you want to solve for (P, F, or A), your program should choose which economic formula to use and then perform the computation. As a check of the procedure, all data will be output at the end of the program. This will be removed later when the procedure is incorporated into a larger program. If the known parameter and the desired parameter are the same, your program should make the output parameter value equal to the input parameter value. For example, if you are doing present worth analysis and the input parameter is the capital cost of the project, the value should not be changed.

2.6.2 [ALL] Nonuniform Series of Payments

Develop a flowchart for the program described in Exercise 4.7.2.

2.6.3 [ALL] Multiple Interest Rates or Number of Compounding Periods

Develop a flowchart for the program described in Exercise 4.7.3.

2.6.4 [ALL] Internal Rate of Return

If you have a given value for a present worth P, and a given value for an annual payment A, you can find the interest rate that will relate these two values through the capital recovery factor CRF. This value for i is called the internal rate of return. Examination of the equation in Appendix E for the capital recovery factor will reveal that you cannot solve explicitly for the internal rate of return. A trial-and-error solution will have to be undertaken. Develop a flowchart for a program that could be used to compute the internal rate of return to a user-specified accuracy.

2.6.5 [ALL] Mortgage Computations

Using the formulas in Appendix E and the information contained in Exercise 3.7.5, develop a flowchart for a program that can be used to determine the size of payments for a given loan at a given interest rate and a range of payback periods.

2.7 ELECTRICAL ENGINEERING

2.7.1 [ALL] Diode Problem

The flow of current through an ideal p-n junction diode is a function of both the applied voltage and the temperature of the diode. The relationship can be represented by

$$I = I_s \left[e^{\left(\frac{QV}{kT} \right)} - 1 \right]$$

in which I = current through the diode, amps; I_s = saturation current, amps; V = voltage across the diode, volts; Q = electron charge, 1.6×10^{-19} coulomb; k = Boltzmann's constant, 1.38×10^{-23} joule/degree K; and T = junction temperature, degrees K. Develop a flowchart for a program to determine the current through the diode between -1.0 V and $+0.3$ V, in increments of 0.1 V. The expected operating temperature range is -20 to $+160°$F. Accomplish your analysis for -20, 70, and 160°.

2.7.2 [ALL] Capacitance and Resistance

Appendix I develops the response functions for a capacitance–resistance circuit subjected to a step input voltage for some time t, and then to no voltage as the circuit is opened. Develop a flowchart for a program that will determine the voltage and current across the capacitor as a function of time.

2.7.3 [ALL] Signal Processing Circuits

Appendix I develops the equations for a resonant electric circuit used as a tuner for radio frequencies. Develop a flowchart for a program that will predict the capacitance required to tune a certain frequency. The necessary resistance and the voltage across the resistance should be computed for a selected Δf as a fraction of the tuned frequency. Your program should be able to compute the parameters for a selected number of frequencies. This exercise is adapted from an example by Borse (1985).

2.7.4 [3,4,5] Parallel and Series Resistance

Develop a flowchart for a program that could determine the equivalent resistance of two or more resistors in either series or parallel. Up to ten resistances should be accommodated. Appendix I presents the theory of parallel and series resistance.

2.7.5 [3,4,5] Parallel and Series Capacitance

Develop a flowchart for a program that could determine the equivalent capacitance of two or more capacitors in either series or parallel. Up to ten capacitances should be accommodated. Appendix I presents the theory of parallel and series capacitance.

2.8 ENGINEERING MATHEMATICS

2.8.1 [ALL] Determinants and Cramer's Rule

Appendix F presents methods for calculating the determinant of a matrix and illustrates the use of determinants to solve systems of linear equations through Cramer's rule. Develop a flowchart for a program that could be used to find the determinant of either a 2×2 or a 3×3 matrix.

2.8.2 [ALL] Vector Cross Product, Unit Vector, Directional Cosines, Vector Components

Develop a flowchart for a program that will calculate the magnitude, components, and unit vector for a vector. Any of the following input data should be accommodated:

   a. Magnitude and any two points on the vector,
   b. Two end-points of the vector,
   c. Magnitude and unit vector, or
   d. Vector components.

Refer to Appendix F for information on vector algebra.

2.8.3 [ALL] Centroid and Moment of Inertia

Appendix G contains information and formulas for finding the centroid and moment of inertia of common shapes used in engineering analysis. Develop a flowchart for a program that will find the centroids and moments of inertia with respect to the x and y axes for a rectangle or a circle.

2.8.4 [ALL] Population Growth

The following model

$$P = P_0(1 + INCR)^t$$

is often used to estimate unconstrained population growth. In this model, $P_0 =$ the initial population, $INCR =$ the fractional increase in the population for each growth period, and $t =$ the number of growth periods. Develop a flowchart for a program that will estimate total population and increase per growth period. The time for the population to double should also be estimated. Use input data of initial population, growth rate, and number of growth periods.

2.8.5 [ALL] Vertical and Horizontal Curves

Geometric design for transportation networks or mechanical components often involves fitting curves between two points. The figure accompanying this exercise shows a circular curve that has been fit through points 1 and 2. The radius and angles θ and α will be

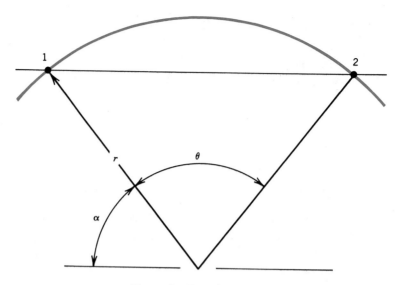

Figure for Exercise 2.8.5

input values. Develop a flowchart for a program that will determine the height of the circular curve above the horizontal line through points 1 and 2 for each degree of curvature. Also, compute the highest point on the curve.

2.8.6 [3,4] Area of Triangle

Develop a flowchart for a program that could find the area of a triangle given the coordinates of its three apexes. Label the apexes of the triangle 1, 2, and 3 so that apex 2 is the middle apex with respect to the x axis. This will simplify the necessary logic of your program.

2.8.7 [4,5,6] Table of Squares and Cubes

Develop a flowchart for the computation of the squares and cubes of values of X between a lower bound and an upper bound with given increments. Output should include a table of values for X, X^2, and X^3.

2.8.8 [4,5] Angles and Sides of a Triangle

Develop a flowchart for the problem described in Exercise 4.9.9.

2.9 INDUSTRIAL ENGINEERING

2.9.1 [ALL] Project Management

Appendix J presents the critical path method for determining the expected duration for a project. Develop a flowchart for a program that will input the expected duration and the start and finish node for each activity of the project using activity on arrow notation. Use the input data to calculate the earliest expected time, the latest possible time, and the slack for each node. You will have to have logical structures that determine which preceding activities lead into a node, and which successor activities are dependent on the node.

2.9.2 [ALL] Quality Control

Parts coming off a production line are checked for several kinds of defects, including:

1. Material defect
2. Bent bracket
3. Out of tolerance dimensions
4. Deficient weld
5. Misaligned drill holes
6. Deficient spring

Some parts may have multiple defects. The order given above is the order in which the parts are checked for the various defects. Develop a flowchart for a program that will count the total number of each type of defect and the total number of defective parts. It should also compute the percentage of the total number of defects for each individual defect and the average number of defects per defective part.

2.9.3 [ALL] Managerial Decision Making

Develop a flowchart for a program that will find the break-even point for a production process. Functions for the total cost and total return will use the number of units produced per year as an independent variable. Appendix J discusses break-even analysis.

2.9.4 [ALL] Queuing (Wait Line) Theory

Appendix J introduces queuing theory and applies it to a problem for finding the optimum number of repairmen to minimize repair costs. Develop a flowchart for a program that will determine the cost and optimum number of repairmen. Input data will include number of machines, maximum number of repairmen, failure rate, repair rate, repairman wages, and loss per machine. This exercise is adapted from an example by Borse (1985).

2.10 MECHANICAL AND AEROSPACE ENGINEERING

2.10.1 [ALL] Shear and Bending Moment

Develop a flowchart for a program that will determine the shear and bending moment as a function of the distance along a beam supported on each end and subjected to a point load, a uniform load, or both a point and a uniform load. Use the free-body technique discussed in Appendix G. Reactions will be input, as will the length of the beam, the magnitude and position of a point load, and the magnitude of a uniform load. The uniform load, if there is one, is assumed to act along the whole length of the beam. The program should use a selected number of increments along the beam.

2.10.2 [ALL] Crank Assembly Analysis

Appendix G develops a mathematical model for the motion of a simple crank and piston assembly. Develop a flowchart for a program that will determine the angle beta and the velocity of the piston for a selected number of angles (angles will go from 0 to 360°) if the crank turns at a constant angular velocity.

2.10.3 [ALL] Friction

A crate and dolly are shown in the figure accompanying this exercise. When a force is applied as shown, four things can happen: the crate can slip on the dolly, the crate can tip on the dolly, the dolly can slip on the platform, or the force may not be large enough to produce any movement. Refer to Appendix G for development of the theory of friction. Develop a flowchart for a program that will determine what will happen for a given applied force.

2.10.4 [ALL] Variation of Atmospheric Pressure with Altitude

Various models have been developed to relate thermodynamic principles to the behavior of the earth's atmosphere. One model that assumes a constant change in temperature with altitude is given here:

$$P = P_0 \left[\frac{(T_0 - \alpha Z)}{T_0} \right]^{\frac{g}{\alpha R}}$$

P is the pressure (pounds per square inch absolute, psia) at altitude Z (feet). P_0 is the pressure and T_0 is the temperature (°R), at ground level. Alpha is the decrease in tem-

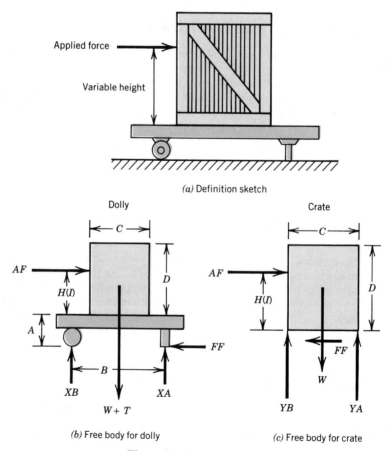

(a) Definition sketch

(b) Free body for dolly

(c) Free body for crate

Figure for Exercise 2.10.3

perature per foot of elevation as you rise into the atmosphere. R is the engineering gas constant and equals 1,715 ft^2/s^2(°R) for air. The local acceleration due to gravity is represented by g. Develop a flowchart for a program that will estimate alpha if the other parameters are given.

2.10.5 [ALL] Aerospace Physics

Appendix G develops the general equations for the path of a particle subjected to a constant acceleration in both the x and y directions. Develop a flowchart for a program that will determine the x and y coordinates, the x and y components of the velocity, and the magnitude and angle with the x axis of the velocity for a number of time increments. The program will start from given initial conditions of position and velocity. Provisions should be made in the program to discontinue the acceleration at some time t, but still continue tracking the particle for an additional number of time steps.

2.10.6 [ALL] Energy Loss in Circular Pipeline

Appendix G develops the theory of energy loss in a circular pipeline. Use that theory for the following problems:

1. Develop a flowchart for a program that would accomplish the following:

   Input data should be D, ε, ν, Q, and L
   If $Re \leq 2{,}000$, then $f = 64/Re$.
   If $Re > 2{,}000$, then one of three options will be executed, at the discretion of the user:
   > Calculate energy loss using both the Chen and Swamee–Jain equations
   > Calculate energy loss using the Chen equation
   > Calculate energy loss using the Swamee–Jain equation

2. Develop a flowchart for a program that will estimate energy loss using the Colebrook equation.

3. Develop a flowchart for a program that will estimate the flow rate in a circular pipe if the energy loss is given. Use the Chen equation to estimate f.

2.10.7 [ALL] Torsion

Appendix G develops a mathematical model for a hollow shaft subjected to a torsional stress. Develop a flowchart for a program that will estimate the required outside diameter for a given inside diameter, torque, maximum allowable angular deformation, elastic shear modulus, length, and maximum allowable shear stress.

2.11 NUMERICAL METHODS

2.11.1 [ALL] Interpolation

Quite often in engineering practice it is necessary to interpolate a value of a dependent variable between two values of the corresponding independent variable. Linear interpolation assumes that the relationship between the independent and dependent variables is linear over the region of interpolation. However, it can produce errors when the function is nonlinear, as is shown in the accompanying figure. If linear interpolation is used to estimate y for $x = 2.5$, you would get approximately 1.1, whereas the figure shows that the correct value is approximately 0.75. Quadratic interpolation is an improved way of accounting for the nonlinearities. It uses three points to generate a second-order

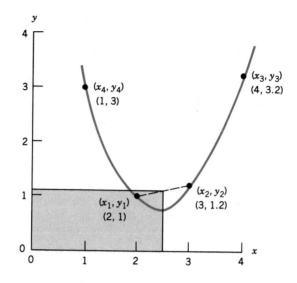

$$\text{Linear: } y(x) = \frac{(x_2 - x)y_1 + (x - x_1)y_2}{x_2 - x_1}$$

$$\text{Quadratic: } y(x) = f_1(x)y_1 + f_2(x)y_2 + f_3(x)y_3$$

$$f_1(x) = \frac{(x - x_2)(x - x_3)}{(x_1 - x_2)(x_1 - x_3)} \qquad f_2(x) = \frac{(x - x_1)(x - x_3)}{(x_2 - x_1)(x_2 - x_3)}$$

$$f_3(x) = \frac{(x - x_1)(x - x_2)}{(x_3 - x_1)(x_3 - x_2)}$$

Figure for Exercise 2.11.1

(parabolic) interpolating function. One method for estimating the quadratic interpolating polynomial is called Lagrange's formula, which is given in the figure. Develop a flow-chart for a program that allows the user to choose either a linear interpolation function or the Lagrange quadratic interpolation function.

2.11.2 [ALL] Solution of Polynomial Equations

Develop flowcharts for programs to solve the problems posed by Exercise 4.12.2.

2.12 PROBABILITY AND SIMULATION

2.12.1 [ALL] Simulation of Two Dice

If two dice are thrown, the value resulting when you add the two numbers on the top faces of the dice can vary from 2 to 12. There is only one way that you can get a 2 or a 12, but in between there may be several combinations that will give a certain value. For example, 3 + 3, 4 + 2, and 5 + 1 all give 6. If you throw the dice many times, the number of occurrences of a certain value divided by the total number of throws should approach the number of theoretical combinations that can add up to a certain value divided by the total number of possible combinations. Develop a flowchart for a program that will simulate the tossing of two dice *n* times, and develop the fractional portion of the total number of throws that produce a given value. Note that you will have to use a random number generator to accomplish the simulation. Assume that there is a random number function available on your computer system.

2.12.2 [ALL] Normal Probability Distribution

Appendix H describes some of the characteristics of the normal probability distribution. Develop a flowchart for a program to estimate the area under the normal curve for a given number of standard deviations from the mean. Since there is no analytical function for the area under the normal probability distribution, you will have to use tabulated values and interpolate between values. Exercise 2.11.1 discusses linear interpolation. You will use tabulated values for every tenth of a standard deviation on one side of the mean for your program. Use Z as defined in Appendix H to determine how many standard deviations a point is from the mean.

CHAPTER THREE

FORTRAN PROGRAMMING: INTRODUCTION

3.1 INTRODUCTION

There are two different but allied processes involved in learning a computer language. The first process is the problem definition/computer algorithm development process described in Chapter 2. You have to properly define the problem and develop its mathematical representation, and you have to specify the solution algorithm and translate that solution algorithm into a computer algorithm. This process is generic in that it does not depend on any particular computer language. The second process involves applying the rules of the computer language to write a computer program to implement the algorithm. You must check the results of the program against known answers. Once it is producing correct answers, you can use the program to generate useful results. You must complete the first process before attempting to develop a computer program, no matter how well you know the programming language. If you do not understand the logic of the problem, do not try to write a program.

When you are writing computer programs, the simplest, well-structured program that works is generally the best. Today's computers are faster and larger than older models. Unless you have an application that requires a great number of computations, you do not have to be overly concerned about producing the most efficient program. Minimizing the computational time of your program and compressing your computer code into as few lines as possible can take considerable effort. Efficient programs are also often harder to interpret by others. You have to weigh the trade-off between the savings to be realized by a more efficient program and the cost to develop that program.

Several concepts have to be introduced before you will be ready to write computer programs. These will be presented in this chapter. By the end of the chapter you will be creating simple, but complete, programs. In succeeding chapters additional concepts are introduced that greatly expand the capabilities of FORTRAN as a problem-solving language. The power of FORTRAN is the sum of its parts. Therefore, it is essential that each concept be understood before going on to the next one. Each new concept builds on and reinforces previous concepts. In the end you should have a clear picture of how the major blocks of the FORTRAN language fit together and how they can be used for implementing well-structured computer algorithms.

3.1.1 What is a Program?

A program is a sequence of instructions, only one of which can be carried out by the computer at any one time. The reason that computers are useful tools is that they can carry out these individual instructions at a very fast rate. Computers can accomplish in the range of one million instructions per second, depending on the type and model of computer that is being used. After completing an instruction, it must be clear what instruction will be performed next. No computer has yet been developed that can think for itself. The programmer must specify the decision rules that govern which instruction will be carried out next.

There are several different types of instructions that might be involved in any FORTRAN program. Input instructions allow the user to enter data into the computer. Output instructions provide the ability to extract data, either the input data or calculated data, from the computer. Data can be generated within a program, so it is possible to have a program without input instructions. However, a program without some form of output instructions does not serve any useful purpose. Storage instructions are actually a special form of input/output instructions. Large volumes of data for input are generally stored on some peripheral device, such as a disk pack. Output can also be stored on peripheral devices for archive storage or later use. Computation instructions describe the mathematical form of the model. Decision instructions provide the rules that determine which instruction will be carried out next. High-level languages, such as FORTRAN, allow writing these instructions in a form easier for the programmer or user to understand. A program will incorporate the instructions in a sequence that accomplishes the desired analysis when the program is executed.

3.1.2 Executing a Program

Executing or running a program requires an additional set of instructions. These instructions translate the high-level language instructions into a form that the computer can use. You will have to compile, link, and run the program. Compilation translates the high-level language into machine language. As you have already learned, well-structured programs can be broken down into modules that perform specific procedures. Many of these modules can be used in more than one application. Computer operating systems make it possible, and preferable in many cases, to compile modules of a program separately. Each compiled module will be stored in a separate file and can be accessed by later applications. Prior to actually running the program, a linker joins these compiled modules with the main program in preparation for the execution stage. During the running stage the program is executed, data files may be accessed, and output is produced. Figure 3.1 shows the sequence of operations involved in compiling and running a program. Note that on the DEC VAX three separate files are involved in the process. The .FOR file is the source program. The compilation stage produces the .OBJ file, which contains the compiled program. When the main program and modules are linked together, the .EXE file is created, which is the file that is run.

Batch processing and time sharing are the two primary methods used for processing FORTRAN programs on computer systems. In batch processing the user prepares the program, data, and processing instructions using the text editor. The processing instructions include commands for compilation and execution (running) and are written using the system job control language. The program (or job)

FIGURE 3.1   Program compilation and execution sequence.

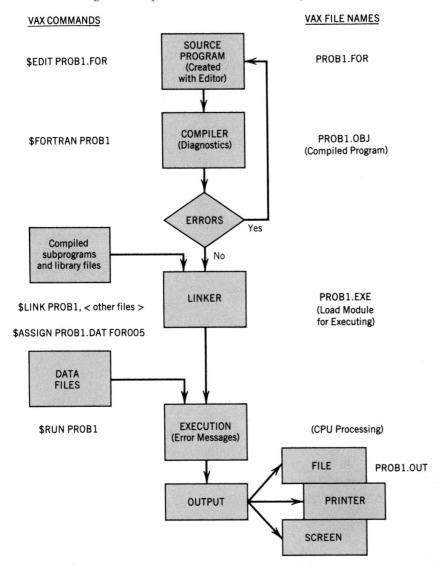

VAX COMMANDS

VAX FILE NAMES

$EDIT PROB1.FOR

SOURCE PROGRAM (Created with Editor)

PROB1.FOR

$FORTRAN PROB1

COMPILER (Diagnostics)

PROB1.OBJ (Compiled Program)

ERRORS — Yes

Compiled subprograms and library files

No

$LINK PROB1, < other files >

LINKER

PROB1.EXE (Load Module for Executing)

$ASSIGN PROB1.DAT FOR005

DATA FILES

$RUN PROB1

EXECUTION (Error Messages)

(CPU Processing)

FILE — PROB1.OUT

OUTPUT

PRINTER

SCREEN

is then submitted to be run. Stored files are generally used to submit program, data, and execution instructions. They can also be submitted in a card deck, but this method is becoming obsolete. Once the program is run, the output is returned, any necessary modifications in the program or data are made, and the cycle is repeated for subsequent executions. As you might deduce, this can be a time-consuming and inefficient process. Therefore, most installations reserve batch processing for programs that require special handling. These might be large programs that involve a lot of computation time or that require use of numerous or specialized peripheral devices. The batch programs can be run when few users are accessing the system. This allows for efficient use of the computing resource and prevents the degradation of service for other users that large programs may cause.

Time sharing, or conversational computing, is the other type of processing. It allows numerous users to share the computer resources simultaneously. The operating system software allows the computer to keep track of who is using the computer at any particular time and what they are doing. The terminal, peripheral device files, or a combination of both can be used for input or output. Job control commands can be entered either through the terminal or through a command file. When you give the proper instructions to execute the command file, it in turn executes the instructions you have placed in the file.

Personal computers use elements of both batch and time sharing to process programs. It is basically a batch process because you have complete control over the computing resource. When you run your program or application, you are submitting it. However, the turnaround time is rapid, and you have ready access to disk files for editing and resubmittal.

3.2 HISTORY OF FORTRAN

FORTRAN, short for FORmula TRANslation language, was developed by IBM in the late 1950s to make it easier for scientific and engineering programmers to write applications programs for the computer. Prior to this time applications had been programmed in assembly language, which was difficult to learn and cumbersome to apply. FORTRAN predates BASIC, PASCAL, and most of the other high-level languages. Since it was developed by IBM, and IBM had traditionally been a leader in the computer field, other computer manufacturers started developing their own FORTRAN compilers to stay competitive. Some incompatibilities developed among the various compilers. In 1961 IBM set about developing an improved FORTRAN, which eliminated most of the machine dependencies. The result of this effort was FORTRAN IV.

In 1966 the American National Standards Institute set up a committee to develop a standard FORTRAN. The committee adopted most of the conventions of FORTRAN IV. This standard FORTRAN became known as FORTRAN 66, and many engineering applications were written using it. A weakness in FORTRAN 66 was that it was not a structured language. Although programs developed with FORTRAN 66 were efficient for computations, they were difficult to interpret and modify by anyone but the original programmer. As corrections and modifications were made to a program, the logic of the program became even more obscure.

In the 1970s, computer applications were developed for many disciplines within the business and scientific communities. Specialized applications led to the development of specialized high-level languages, including ALGOL, COBOL, AND PL/1. The generalized languages BASIC and PASCAL were also gaining adherents during this time. The vast majority of engineering applications, however, still used FORTRAN. Various users of FORTRAN 66 were making improvements to the standard. In 1977 it became necessary to establish another standard, which incorporated the improvements. FORTRAN 77, the result of this standardization, incorporated many of the concepts of structured programming found in PASCAL. FORTRAN 77 has remained the standard to date; however, as with FORTRAN 66, variations have sprung up that implement improvements, mainly in structured programming features. One of the most popular variations is WATFIV. It was developed by the University of Waterloo, Canada, and is used by many universities. By and large, FORTRAN has remained a machine-independent language.

This text will concentrate on using standard FORTRAN 77, with brief discussions of nonstandard features that extend the FORTRAN language.

A new FORTRAN standard is being prepared, and it appears that it will be released in 1990. Among many other changes, the new FORTRAN will include enhanced array operations, dynamic data structures, user-defined data types, pointers, and new regulations for numeric precision.

3.3 STRUCTURED PROGRAMMING IN THE FORTRAN ENVIRONMENT

All computer applications should be developed using structured programming and top-down design. Structured programming is a set of rules that define good programming style and make computer code applications more understandable. Top-down design is the process of decomposition and successive refinement that provide all the details of the solution logic needed to write a program. Both of these concepts have been discussed in a general sense in Chapter 2. This discussion will refine the concepts for applications in the FORTRAN programming environment.

In order to use structured programming techniques, the programmer has to adopt a top-down strategy for applications development. The first step is decomposition of the problem into a series of smaller problems, each representing a distinct logic block, or module. Successive refinement fills in the details of the solution method for each smaller problem. Flowcharts and pseudocode should be used to illustrate the solution logic within each module, as well as the logical relationship among modules.

You can incorporate many modules into one program. However, it is better practice to develop modules as separate subprograms when the procedure will be used more than once in a program, or when there is the possibility that the procedure may be used again in future applications. The subprograms for a particular application are coordinated and controlled by a main program. FORTRAN facilitates the development and saving of these subprograms as separate files. The logic used in the subprograms should be generalized so that the subprograms can be used with a number of applications without modification. Since the subprograms are separate program segments, the user must understand how to transmit data to, and receive data from, the subprogram modules.

FORTRAN syntax for subprograms will be developed in Chapter 5. A flowchart symbol for referencing a separate procedure was given in Figure 2.1. This symbol is applicable to FORTRAN subprograms and will be used in Example 3.1 to depict the use of separate subprogram modules for the development of a statistics application. When you examine the flowchart for the main program and subprogram modules, note the depiction of the variable values that have to be passed to and from the subprogram modules. The programming examples and exercises in Chapters 3 and 4 will not require you to use subprograms. However, you should be aware as you write them that many of your programs, or logic segments from within the programs, would be readily adaptable to separate subprogram procedures.

Using subprogram modules lets you write and test each module separately. This simplifies program correcting, or debugging. Once all options of a module are tested, it can be used in other programs without retesting, as long as the user understands the capabilities and limitations of the module. Modules for large programs can be worked on by several programmers independently without sacrificing generality or understandability. Use of subprograms can also make it easier to add capabilities later on without obscuring program structure and logic.

A good test of the quality of the structure of a program is to ask yourself, "Would I be able to read, understand, and use my program if I were someone else looking at it?"

Example 3.1 will illustrate application of decomposition and successive refinement with subprogram structures for a basic statistical analysis program.

EXAMPLE 3.1 □ Application of Decomposition and Successive Refinement to Statistical Analysis

STATEMENT OF PROBLEM

Develop a structured computer algorithm to compute the following basic statistical parameters: mean, median, and standard deviation of a data set.

MATHEMATICAL DESCRIPTION

Theoretical development of the mathematical model is covered in Appendix H. The necessary equations are as follows:

$$\text{Mean} = \bar{X} = \frac{\sum\limits_{i=1}^{n} X_i}{n} \tag{3.1}$$

If n is odd

$$\text{Median} = X_{(n/2 + .5)}$$

and if n is even   (3.2)

$$\text{Median} = 0.5[X_{(n/2)} + X_{(n/2+1)}]$$

$$\text{Standard deviation} = S = \sqrt{\frac{\sum\limits_{i=1}^{n} X_i^2}{n} - \bar{X}^2} = \sqrt{\frac{\Sigma(X_i - \bar{X})^2}{n}} \tag{3.3}$$

ALGORITHM DEVELOPMENT

Decomposition identifies five program modules as shown in Figure 3.2. The input and output modules can be included in the main program. The modules for computing the mean, median, and standard deviation will be developed as separate modules, since they represent separate procedures and may be useful in future applications. Note that it will be necessary to order the data before you can find the median.

FIGURE 3.2 Decomposition and successive refinement of statistical analysis problem.

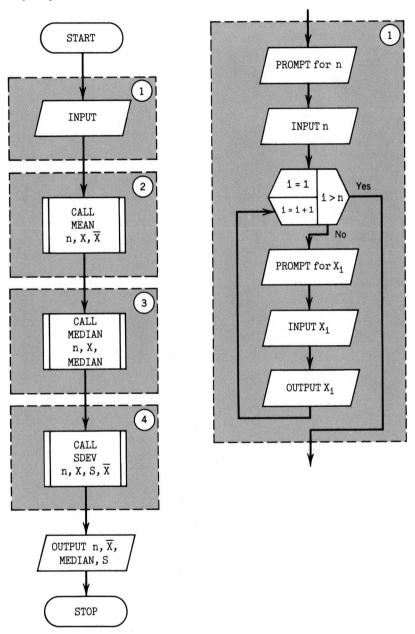

(*a*) Decomposition and refinement of INPUT block.

(EXAMPLE 3.1 □   Application of Decomposition and Successive Refinement to Statistical Analysis □ Continued)

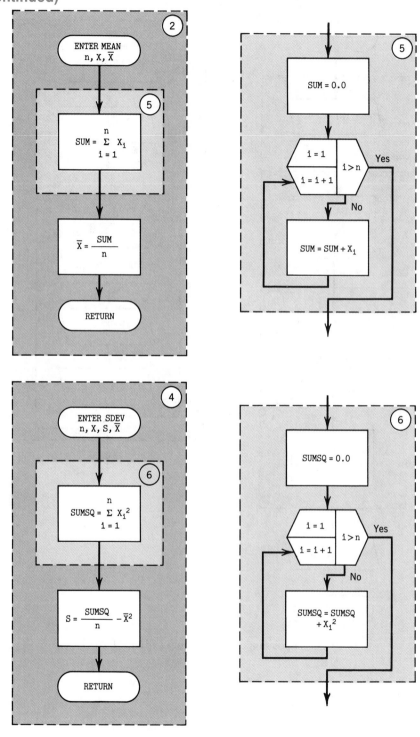

(*b*) Refinement of MEAN and SDEV modules.

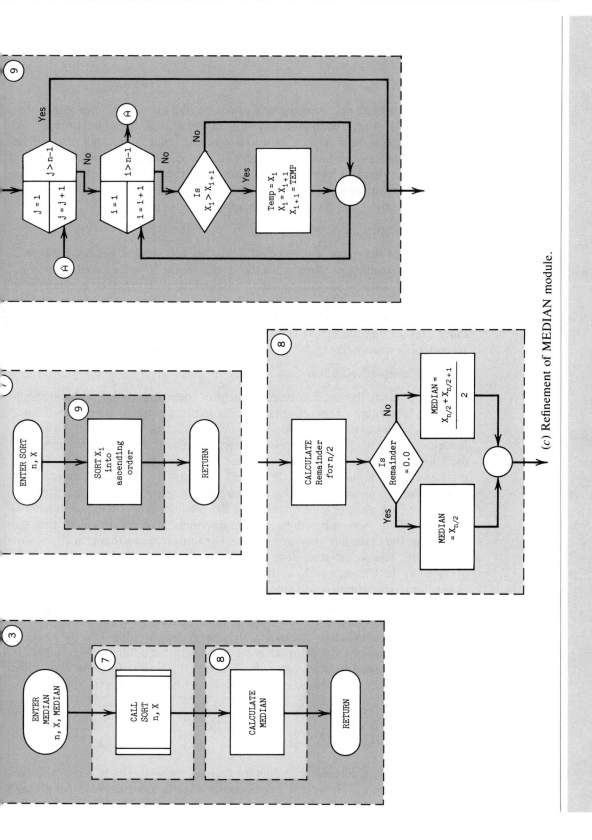

(c) Refinement of MEDIAN module.

(EXAMPLE 3.1   □   Application of Decomposition and Successive Refinement to Statistical Analysis   □ Continued)

Input/Output Design

Some method has to be implemented to control the number of data values that will be loaded into the computer. In this instance, the number of data values will be input, and that value will be used to control the input loop. Input data will be output with appropriate labels right after each value is input. This will display the input data for checking even if input errors cause the program to stop executing before the final results are output. Calculated values for the mean, median, and standard deviation will be output with appropriate labels.

Numerical Methods

Calculation of the mean and standard deviation will involve summing the data values and summing the square of the data values. A sorting routine will be developed to order the data values in ascending order. Either ascending or descending order could be used to determine the median. This sorting routine will be developed as a separate module that can be called by the median module. Later, this sorting routine will be generalized so that it can accommodate either ascending or descending order. It will be used in several applications in the text.

Computer Implementation

Figure 3.2 shows the successive refinement of the decomposed problem into a computer algorithm. The refinement for a particular block is identified by a circled number next to the block and the same circled number next to the refined flowchart segment. Input requires one stage of refinement. The mean and standard deviation modules require two stages of refinement. The first stage defines the steps in the module, while the second stage defines the algorithm for summing the appropriate values. The median module needs three stages of refinement. The first stage defines the steps of the module. A second stage identifies the steps of the sort submodule and the decisions involved in calculating the median for the cases $n =$ even or $n =$ odd. The third stage further refines the sort submodule. The sort routine developed for this application will be described in more detail in Chapter 4. Note that when the sort module is used in future applications, its definition will not have to be refined beyond the reference to it in a flowchart CALL logic block.

The variables that will be sent from the calling module to the called module are shown in the flowchart call block, and also in the entry block to the module. One of the most common errors in using subprogram modules in FORTRAN is the failure to properly pass values back and forth between calling and called modules. Including the variables in the flowchart helps to prevent this.

3.4 PROGRAM ERRORS

Using proper structured programming techniques helps you to develop correct computer programs. However, it does not replace the need to check for program errors. Some errors will almost always be present in newly developed applications. In fact, users and developers are still finding errors in programs that have been used for years. This is why it is important to verify the output of any computer program you use against answers that are known to be correct.

Computer program errors are commonly referred to as "bugs," and the process of correcting errors in a program or application as "debugging." The origin of these terms is interesting, and is related by Chapra and Canale (1986, p. 246). Rear Admiral (RET) Grace Hopper was one of the pioneers in the development of computer languages. In 1945 she was working with one of the earliest electromechanical computers. When the computer stopped working, she and her colleagues opened it up to see what was amiss. They found a moth stuck in one of the relays. Its removal presumably solved the problem, thus the computer had been debugged. Electronic bugs in today's computer applications exhibit many of the characteristics of Admiral Hopper's real bug. They pop up when and where you least expect them, and unless you approach finding them in a reasoned manner, they can be extremely hard to locate. However, once you have located them, they are usually fairly easy to correct.

3.4.1 Types of Errors

Errors can generally be classified into one of four types. Listed in the chronological order in which they are usually encountered, they are system command language errors, errors in FORTRAN syntax, execution errors, and logic errors. System command language errors can occur in the commands your computer system requires to compile and execute a program. These errors are relatively easy to identify and correct because the running of the application stops right where they occur, and a descriptive message is output. Reference to the system documentation or consultation with the system experts should indicate how to fix the problem. FORTRAN syntax errors can be typographical errors or mistakes in the syntax (form) of FORTRAN statements. Syntax errors are identified during the compilation stage. Error messages that give an indication of what the errors are and where they occur are produced. Generally these messages are a big help in locating and correcting syntax errors. Program execution cannot proceed beyond the compilation stage until all critical syntax errors have been corrected. Once the program is compiled and linked with any external modules, it can be run. During the running of the program, execution errors can occur. These generally involve trying to use illegal mathematical operations (taking a log of zero, dividing by zero, etc.), attempting operations that result in numeric values that are outside the capability of the computer to handle, or making input data errors. The program will stop execution wherever execution errors occur, and a descriptive message will be printed out. The final type of errors, logic errors, are the most difficult to locate and correct. They are caused by errors in your approach to solving the problem and can be located and corrected only through the careful detective work that will be described in the next section. No error messages will be printed out by the computer for logic errors. You will have to produce your own error messages. Logic errors could involve your computer algorithm, your numerical algorithm, your basic definition of the mathematical model, or some combination of these. When your computer output does not agree with known answers, you must suspect logic errors. First you can check to make sure that your output instructions provide values for the proper variables. Then you must systematically approach locating and correcting the problem or problems.

3.4.2 Logic Debugging Strategies

Debugging logic errors involves good detective work and hypothesis testing. You have to examine the evidence, make a hypothesis concerning what is causing the error, then test the hypothesis. The evidence will, in most cases, have to be developed by you. The best way to do this is to put intermediate output statements within the program so you can trace variable values throughout the computational process. Strategically placing these output statements and comparing the output to known correct values will allow you to isolate a few lines of computer code in which the error condition is developing. Careful examination of these few lines should give you a good idea of what is wrong and how to fix it. If the fix seems to correct the error, then you must check all alternate logic paths to make sure that they are also correct. You have to make sure that changes in the logic in one area will not introduce errors in other areas. If one fix does not correct the errors, it should be removed before trying the next fix. It is also better to apply one correction at a time. If multiple fixes do not correct the errors, you end up having to take them all back out when one of them may at least partly solve the problem. The extra output statements can be removed when they are no longer needed for debugging.

Programs should be debugged from the top of the logic flow downward. Programs should be developed in stages, with each stage checked before moving on to the next stage. Each module should be checked independently, and then rechecked when it is incorporated into the applications program. Stages might include input, computation stage one, computation stage two, and output. Before you can be sure that a program is error free, multiple options (logic paths) and extreme values have to be tested.

3.5 INTERNAL AND EXTERNAL PROGRAM DOCUMENTATION

Well-structured programs still require some sort of documentation that will show other engineers how to use them. Chapter 9 will discuss program documentation in some detail. The following is a brief introduction.

Internal documentation is included in the program itself. COMMENT statements (covered later in this chapter) can be used to identify and define modules within a program. Comment statements can also be used to define variable meanings and to give general information about the program, such as title and author, or operating instructions. If the program is interactive, the output prompts giving instructions for data entry are a type of internal documentation.

Most short programs need only internal documentation. However, large programs with many options, and programs for which the computer code is not readily available to the user, require some form of external documentation. The external documentation should be sufficiently thorough to allow use of the program without reference to outside experts or other documentation. An outline of what thorough documentation should contain is given below. These external documentation concepts are applicable to uses other than FORTRAN programming.

Elements of External Documentation

Description of the problem to be solved

Capabilities and limitations of the program

Theoretical development on which the program is based

Description of the program structure with a flowchart

Special machine requirements (if required)

    Memory requirements

    Tape and disk requirements

User data input requirements and instructions

Sample problem

    Statement

    Data input

    Program output

Program listing

3.6 DATA TYPES (CONSTANTS AND VARIABLES)

Constants represent fixed values, or groups of characters called character strings, in the computer. Variables represent storage locations for values that can be manipulated by a program.

3.6.1 Constants

Several different types of constants can be defined and used in FORTRAN. Each has its own characteristics and limitations, so each will be discussed separately.

3.6.1.1 Integer Constants

Integer constants represent whole integer values. Since there cannot be a decimal fraction for an integer value, integer constants never have a decimal point associated with them. Examples of correct integer constants are

$$635 \qquad 10 \qquad -110$$

while the value 6.35 would not be a correct integer constant. The number 2,300,000 is not a correct integer constant either, because numbers are not stored in the computer with embedded commas.

The allowable range of values for integer constants varies somewhat depending on the architecture of the computer system being used. The range for 32-bit machines is approximately $\pm2,000,000,000$ (2×10^9).

3.6.1.2 Real Constants

Real constants represent decimal numbers and will always have a decimal point associated with them. Examples of correct real constants are

$$3.1416 \qquad 0.0001 \qquad 3.59E7 \qquad 1.245E-06$$

where 3.59E7 and 1.245E−06 show the way that the computer represents decimal values in scientific notation. The significant figures and the powers of 10 are displayed. For example, 3.59E7 represents 3.59×10^7, or 35,900,000, while 1.24E−06 represents 1.245×10^{-6}, or 0.000001245. Scientific notation is valuable in computer applications as a tool for displaying the magnitude and significant figures for large or small numbers without displaying all the decimal places.

You can enter a computer constant with any number of significant figures. However, computer computations use only as many significant figures as the computer running the application can store. Computers vary in the number of significant figures they can store. Models in use range from 6 to 14 significant figures. The reference manual for your computer will tell you how many significant figures you can use. There are also limitations on how big the scientific notation exponent can be. VAX systems can use values from E+38 to E−38, while the CRAY supercomputer can use values from E+2465 to E−2465.

3.6.1.3 Character Constants

Character constants are made up of character strings enclosed in apostrophes (single quotes).

'John Doe'   'TRIANGLE'   '12345'

are all acceptable character strings. Note that '12345' would not be treated as the number 12345, but the separate characters 1, 2, 3, 4, and 5.

3.6.1.4 Complex Constants

Complex constants represent complex numbers as real and imaginary parts, $a + bi$, in which $i = \sqrt{-1}$. The complex constant (4.5,2.0) represents $4.5 + 2.0i$ in the FORTRAN language. The parentheses are necessary to designate the constant as complex.

3.6.1.5 Double Precision Constants

Double precision constants are useful when using double precision computations. The double precision specification allows the storage and use of more significant figures in order to improve computational accuracy. For example, VAX systems allow storage of seven significant figures under normal precision and 16 significant digits under double precision. The increase slightly more than doubles the number of significant figures because the sign and decimal have to be stored only once. The constants

1.223176549D0   675498928.0123D−6   1.234D7

are all acceptable double precision constants. Note that the double precision notation also carries the scientific notation designator after the D. Therefore,

```
1.223176549D0 = 1.223176549,
674598928.0123D−6 = 674.5989280123,
1.2340000674D7 = 12340000.674, and
1.234D0 = 1.234
```

Although 1.234 has less than seven significant digits, it should still be defined with the double precision modifier if it is to be used in double precision computations. This is because decimal fractions cannot always be represented by even binary numbers in machine notation. In those cases, using double precision increases the accuracy of the constant.

3.6.2 Variables

As indicated previously, variables in FORTRAN refer to storage locations in the computer. When a new piece of data is stored in a location, it replaces whatever data were there previously, and the old data are lost. If it is important to save both pieces of data, two variable locations must be defined, and one piece put into each. Depending on how the variable is defined, numerical values, character strings, or logical constants can be stored in the space allocated to the variable. Numerical and character types are defined in the same manner as they were previously defined for constants. Logical variables store true or false information.

3.6.2.1 Names

Variable names can consist of from one to six letters or numbers. The first character of any variable name must be a letter. FORTRAN-reserved words (any words used to define FORTRAN statement syntax or functions) cannot be used to define variables, since the computer cannot differentiate the variable name from statement syntax. It is best to use descriptive variable names to give readers of your program a better idea of what variables mean without having to refer to a list of variable definitions. This is another element of good structured programming. Examples of valid variable names that are also descriptive are the following:

Variable Name	Meaning
MEAN	Mean (average) value of a set of data
AREA	Area of polygon
DELTA	Increment for independent variable

Examples of invalid variable names and the reasons are the following:

Variable Name	Reason for Invalidity
3RD	Starts with number
I+J	Illegal symbol (+)
DELTAREA	Too many characters

3.6.2.2 Specifying Variable Types

A particular variable can store only one type of data, so FORTRAN has a means of specifying variable types. There are several ways of doing this; however, the best way is to use explicit declaration statements to declare what type a variable is. Declarations include

```
INTEGER,
REAL,
CHARACTER,
COMPLEX,
DOUBLE PRECISION, and
LOGICAL.
```

The general form of the declaration statement is:

```
DECLARATION   variable list
```

in which variable list is the list of variables to be declared of the particular type. For example, the declarations

```
INTEGER I,J,COUNT
REAL AREA,PI,MEAN
```

specify that variables I, J, and COUNT will contain integer values, and variables AREA, PI, and MEAN will contain decimal values.

Declaration statements, by necessity, have to be at the beginning of the program because a variable cannot be used until its type is defined. You do not have to use declaration statements to define your variable types, but by doing so you give any readers of your program explicit information on variable types. This parallels the requirement in PASCAL that all variables and constants be declared and is good structured programming practice.

The second method for specifying integer or real variable types is to use the default typing built into FORTRAN. In the absence of an explicit declaration statement to the contrary, any variable starting with one of the letters I, J, K, L, M, or N is an integer variable. Variables starting with any other letter, and not declared as another type, are treated as real variables. This convention is a holdover from earlier versions of FORTRAN, and you will see it used in many programs. It is better structured programming practice to use declaration statements and avoid the default.

A third way of declaring the type of variables is to use an implicit declaration. The IMPLICIT statement redefines the default specifications described in the preceding paragraph. For example,

```
IMPLICIT INTEGER(A-F), REAL(G-P), COMPLEX(Q-S), CHARACTER(T-V),
         DOUBLE PRECISION(W-Z)
```

would establish that by default variables beginning with A through F are of integer type, G through P real, and so on. As with the original default, explicit declaration will override the implicit default. If you use explicit declaration statements, as previously recommended, the implicit declaration is unnecessary and can be confusing.

The IMPLICIT declaration provides us with a method of checking for typographical or spelling errors in numeric variable name entry. You can use the statement

```
IMPLICIT CHARACTER*1(A-Z)
```

to make the default variable type a character variable. You can override the default for your program variables by explicitly declaring the appropriate variable types. If you misspell any variable names, the program will assume they are character variables because their names will not match any of the explicit declarations. You cannot mix character and arithmetic variables in the same expression. Therefore, the inclusion of the misspelled variables in arithmetic expressions will produce error messages.

3.6.2.3 Integer Variables

Integer variables are used to store integer values. Rules for the range of values allowed are the same as for integer constants. A real value can be placed into an integer variable storage location, but it will become an integer value. The decimal fraction part of the value will be truncated, not rounded, and the decimal point removed. Integer variables are used mainly as counting or indexing mechanisms. For example, an integer variable can be used to store the number of times a program loop has been executed. Integer variables are preferable for this type of application, as will be shown in Chapter 4. It is best to avoid integer variable arithmetic operations involving division because of the truncation of the remainder. Examples of problems this can cause will be illustrated in a subsequent section. The following statement would declare variables I, J, N, and COUNT as integer variables.

```
INTEGER I,J,N,COUNT
```

3.6.2.4 Real Variables

Real variables store decimal values, in either standard or scientific notation. Actually, the computer does not know the difference. It merely stores any value between the exponential limits for the machine. It is only during input and output that a differentiation is made. Rules for storage of significant figures and limits on exponents are the same as described for real constants. If an integer value is stored in a real variable location, it becomes a real value and has a decimal point attached. Examples of variables declared as real are

```
REAL MAX,MIN,SUM,AGADRO
MAX = 1.E10
MIN = -1.0E10
SUM = 0.0
AGADRO = 6.023E23
```

MAX and MIN could represent the upper and lower limit on a range of values, SUM = 0.0 could be the initialization of a variable used to store the sum of several addition operations in a summing loop, and AGADRO is Avogadro's number from chemistry, 6.023×10^{23}. The latter demonstrates the value of being able to represent large values in scientific notation.

3.6.2.5 Character Variables

Character variables define storage locations for character strings. The length of the character string that can be stored in a character variable is defined along with the CHARACTER declaration. For example

```
CHARACTER FNAME*15,MIDIN*2,LNAME*20
```

defines FNAME as a character variable that will store up to 15 letters of a first name, MIDIN will store the middle initial and period, and LNAME will store up to 20 characters of the last name. The declaration

```
CHARACTER*10 SHAPE,LIQTYP
```

defines two character variables that can each store up to 10 characters. SHAPE could denote the type of shape being acted on by a liquid and LIQTYP the liquid type. When you are defining the size of character variables, it is important to carefully determine the maximum size that will be needed. Otherwise, characters may be lost because a character variable cannot store more characters than it is specified to have. The following rules apply if the character string stored in a character variable is shorter, equal to, or longer than the assigned length of the variable:

Character string shorter than variable length. If the length of the character string is *m*, the character string is stored in the first *m* positions of the character variable. The remaining positions are assigned blank characters. Thus

```
CHARACTER NAME*10
NAME = 'John Doe'
```

would actually result in 'John Doe ' being stored in variable NAME.

Character string the same length as assigned variable length.   Whatever is in the character string will be stored in the variable.

Character string longer than variable length.   When a character string is longer than the length of the variable location in which it is to be stored, not all the characters will fit. If the assigned length of the character variable is *n*, the first *n* characters of the string will be stored, and the rest will be truncated. Thus

```
CHARACTER NAME*10
NAME = 'John Q. Public'
```

would result in John Q. Pu being stored in NAME. Note that all blank spaces are counted as characters. When we get to Chapter 6, you will see that a different convention holds for formatted character string input. To avoid any problems or confusion, it is advisable to make sure your assigned character variable length is sufficient to handle any character string that you will want to put into it. There are very few instances when you will purposely want to truncate a character string.

3.6.2.6 Complex Variables

Complex variables store both real and imaginary parts of complex number values.

```
COMPLEX A,B,Z,R
```

would define four complex variables, each of which would contain two values, the real (*a*) and imaginary (*b*) coefficients in the value $a + bi$. Complex values are most frequently associated with electrical engineering and will be illustrated in that context. Other applications in advanced mathematics and calculus are beyond the scope of this text. Complex arithmetic operations are discussed in Appendix I, and an example of their FORTRAN application is given later in this chapter.

3.6.2.7 Double Precision Variables

Double precision variables increase the number of significant figures that can be stored, as described previously for double precision constants. They do not, however, increase the numerical range of values that can be stored. Double precision variables are useful when very accurate results are required, or when many iterative calculations may lead to unacceptable round-off error for single precision variables. Very large dollar values can also be represented accurately to single dollars, or even cents, with double precision variables. If you wanted to use $100,000,200,346.49 accurate to cents,

```
DOUBLE PRECISION DOLLAR
DOLLAR = 100000200346.49D00
```

would store the proper value in location DOLLAR.

3.6.2.8 Logical Variables

Logical variables have two possible values: true or false. Thus, logical constants are either .TRUE. or .FALSE.. The periods before and after the logical constant are required. Logical variables are normally used to test a condition in a program. They are not usually input or output. The statements

```
LOGICAL TEST
TEST = .FALSE.
```

would store the value false in location TEST. Use of logical variables will be discussed in Chapter 4.

3.7 OPERATIONS, EXPRESSIONS, AND COMPUTER-SUPPLIED FUNCTIONS

Expressions are combinations of variables, constants, and operators. FORTRAN evaluates expressions according to a defined set of rules. The evaluated expressions produce numeric, character, or logical values. When you study the assignment statement in the next section, you will see that an equation in FORTRAN is made up of a variable on the left side of the equal sign, and one or more expressions on the right of the equal sign. Thus, an expression is not an equation, but equations are made up of one or more expressions.

Character and logical expressions will be discussed in later sections. The present section will concentrate on the evaluation of numeric expressions.

The following symbols are used to represent the associated mathematical operations:

+	addition
−	subtraction
*	multiplication
/	division
**	exponentiation

Two operators cannot appear directly next to each other in an expression. However, it is acceptable to separate the two operators with parentheses. Thus

        X**−2

is not acceptable, while

        X**(−2)

is acceptable, and would represent X^{-2}.

3.7.1 Order of Operations

Parentheses.   Any part of an expression within parentheses is evaluated before the rest of the expression. If sets of parentheses are nested one inside the other, as in ((A*B)**N), the innermost set is evaluated first, then the next set, continuing to the outside set. Thus, *A* would be multiplied by *B* first, and the result taken to the *N* power. Parentheses are necessary for the proper representation of some expressions. Other times the use of parentheses is optional, but their use may help to identify particular parts of a computational sequence. Unnecessary overuse of parentheses can obscure the meaning of expressions. The expression

        ((3.)*((X−6.)/(Y+4.)))

is correct, but its equivalent

        3.*(X−6.)/(Y+4.)

is easier to read and interpret. In the first representation, you have to pause and figure out which parentheses are paired.

Exponentiation.   Exponentiation is the first mathematical operation to be undertaken. Multiple exponentiation is accomplished from right to left. For example, if $A = 5$, $B = 2$, and $C = 3$,

```
A**B**C = A**(B**C) = 5**8 = 390625
```

B is taken to the C power, and A is taken to the resulting power. Use of parentheses can change the order of operations and result in quite different values. For example

```
(A**B)**C = 25**3 = 15625.
```

Also

```
A**B*A**C = (5**2)*(5**3) = 3125,
```

while

```
A**(B*A)**C = 5**10**3 = 5**1000,
```

which would result in a number too large for most computers to store.

If the exponent is an integer value, exponentiation is accomplished as a series of multiplications. Thus, any arithmetic expression can be raised to either a positive or negative integer exponent. Exponentiation with a real exponent is accomplished with an internal FORTRAN procedure using logs. Since it is impossible to take the log of a negative number, only expressions that result in real values that are greater than or equal to zero can be raised to real powers. If only integer exponents are going to be used, it is better practice to use integer variables or constants to represent the exponents.

Multiplication and Division.   Multiplication and/or division is accomplished after exponentiation from left to right in the expression. For example, the expression

```
6.*2.-3./4.+4.*5.
```

at this stage would evaluate to

```
12. - 0.75 + 20.
```

with 6.*2. being accomplished first, followed by 3./4., followed by 4.*5..

Addition and Subtraction.   Addition and subtraction are accomplished from left to right and are the final computational priorities. This would finish the evaluation of the expression shown above and would result in a value for the expression of 31.25.

It is important to understand the order of operations, because it can have a significant impact on calculated answers. For example, one of the real roots of a quadratic equation can be found by evaluating the expression

$$X = \frac{-B + \sqrt{B^2 - 4AC}}{2A} \tag{3.4}$$

If you write the FORTRAN expression in the form

```
-B + B**2 - 4.*A*C**0.5/2.*A
```

the precedence of operations would give

Priority 1	B^2 $C^{0.5}$
Priority 2	4.*A
Priority 3	4.*A*C$^{0.5}$
Priority 4	4.*A*C$^{0.5}$/2.
Priority 5	2.*A*C$^{0.5}$*A
Priority 6	−B + B^2 − 2.*A^2*C$^{0.5}$

which is not at all what you want to do. Since you want to be sure that you take the square root of $B^2 - 4AC$, enclose that part of the expression in parentheses. Also, the whole numerator has to be divided by 2A. Putting 2A in parentheses will ensure that it is treated as a unit in the evaluation. That is still not enough, however. You also have to isolate the whole numerator so it can be divided by 2A. Do this by enclosing the numerator in parentheses. The proper FORTRAN expression would be

```
(-B + (B**2 - 4*A*C)**0.5)/(2*A)
```

Example 4.6 will modify the above expression, and use it to solve for the real or complex roots of any quadratic equation. Development of the above expression illustrates the power of using parentheses to control the order of operations.

3.7.2 Writing FORTRAN Expressions

The order of operations and use of parentheses must be kept in mind when writing FORTRAN expressions to ensure that they will represent the correct algebraic expressions. Shown below are several algebraic expressions and the corresponding FORTRAN expressions.

Algebraic Expression	*FORTRAN Expression*
$\sqrt{X^2 + Y^2}$	(X**2 + Y**2)**0.5
$h_c + \dfrac{I_c}{h_c A}$	HC + IC/(HC*A)
$\left(\dfrac{M_1 - M_2}{M_1 + M_2}\right) V_{1i}$	(M1 − M2)/(M1 + M2)*V1I
$\sqrt{\dfrac{k}{m}}\,(A^2 - X^2)$	(K/M*(A**2−X**2))**0.5
$\dfrac{\pi G t \rho m r}{R^2}\left(\dfrac{R^2 - r^2}{X^2} + 1\right) dX$	(PI*G*T*RHO*M*R1)/R**2*((R**2−R1**2)/X**2+1.)*DX

Note that in the last expression, PI was used to represent π. A value would have to be placed in PI earlier in the program for it to be used in an expression. FORTRAN does not have a built-in value for π or other commonly used constants. Most of the spreadsheets and other microcomputer applications packages that will be illustrated later in the text do have internal values for π and other constants. Also, in the last expression, the Greek letter ρ was represented by the word RHO, and the small r was represented by R1 to differentiate it from capital R, which was a separate variable. Using a small r would not have worked because

FORTRAN does not differentiate between lower and uppercase letters. R and r would represent the same storage location.

3.7.3 Evaluation of Expressions

In debugging programs it is often necessary to evaluate numerical expressions by hand to find out where an error in translation from algebraic to FORTRAN expression may have occurred. Again, it is essential to follow the order of operations rules closely. For example, the algebraic expression

$$\frac{x \cdot y^n}{x + 15} + y^3 \cdot x \tag{3.5}$$

would evaluate to 24.67 when $x = 3$, $n = 2$, and $y = 2$. Suppose that an intermediate output statement that you placed in the program has indicated that the value of the expression is incorrect. You would then go back to your FORTRAN expression to determine where an error in transcription may have occurred. If your expression in the program was

```
(X+Y)**N/X+15.+Y**3*X
```

the order of computations would be

```
(3.+2.)
(5.)**2      2.**3
25./3.       8.*3.
8.33 + 15. + 24.
       47.33
```

The first error occurred when x was added to y rather than being multiplied by it. This is an obvious error and was probably caused by a typing mistake. However, if you correct this error, the final value of the expression is 51., which is still incorrect. This shows that you cannot assume that the rest of the expression is correct after finding one error. A more careful examination will show that the first set of parentheses was not needed, and actually caused an error. Only y was supposed to be taken to the n power, but the parentheses caused x times y to be taken to the n power. Further examination of the computational sequence reveals that xy^n was being divided by x, not by $x + 15$. as desired. In this case, parentheses are required to combine the elements of the denominator. The corrected FORTRAN expression is

```
X*Y**N/(X+15.)+Y**3*X
```

3.7.4 Mixed Mode Operations

It is best not to mix real and integer variables and constants in the same expression, except as previously described for exponentiation. Expressions containing both real and integer variables and/or constants are said to be of mixed mode. In older versions of FORTRAN mixed mode was not allowed at all and resulted in a syntax error message during compilation. FORTRAN 77 allows mixed mode. This makes the language more flexible and works to the programmer's advantage in certain situations. The reason for avoiding mixed mode whenever possible lies in the rules FORTRAN applies to evaluating mixed mode operations. The rules are as follows:

1. If integer and real quantities are involved in an operation, the integer quantity is converted to a real number prior to the operation, and the result is a real value.
2. If two integer quantities are involved in an operation, the result is an integer value. If the resulting single integer value is used with a real quantity in a subsequent operation, Rule 1 applies for the subsequent operation.

Problems can arise when two integer values are used in division. Since an integer value cannot contain a decimal part, decimal fractions resulting from integer division are truncated. The expression

    (1/3)*Y*Z

would always evaluate to zero, because (1/3) involves two integer constants, and the resulting value would be zero, not 0.3333 as desired. The expression (1./3.) would give the correct value.

Also, if n and m are both integer variables and have the values of 6 and 4 respectively, the expression

    Y**(N/M)

would evaluate to Y^1, not $Y^{1.5}$. In open channel flow analysis, it is often necessary to take a quantity called the hydraulic radius, R_h, to the 2/3 power. An easy mistake to make is to write the expression

    RH**(2/3)

which would compute the hydraulic radius as R_h^0 or 1.0. The expression RH**(2./3.) would give the desired result.

An example of when the rules of mixed mode operations work to your advantage is in statistical computations. Assume that you have entered n data values into the computer and have stored the number of data values, n, in an integer variable location named N. To compute the mean value of the data, Equation 3.1 shows that you need to sum the data values and divide by n. As long as the sum is a real value, you can divide by the integer value N and have the result still be real. This saves you the inconvenience of having to convert N to a real value by other means. FORTRAN does have a method for temporarily converting an integer value to a real value for a particular operation. It involves using a function built into FORTRAN, called an intrinsic function. The text will use this function to help you recognize mixed mode operations and to help you avoid some common problems in your own programming.

3.7.5 Intrinsic Functions

You could write program segments to accomplish mathematical operations such as finding the sine or cosine of an angle, or taking the base 10 log or natural log of a value. Nearly everyone who programs with FORTRAN will want to use these same functions, so it would be inefficient to have these users writing their own procedures. Therefore, most of the commonly used functions are included in the FORTRAN language as intrinsic functions. The functions can be accessed directly from FORTRAN expressions that you write. For example, if you wanted to find the x component of a two-dimensional vector of magnitude V that makes an angle θ with the x axis, you could use

```
VX = V*COS(THETA)
```

As you remember from trigonometry, theta is the argument of the cosine function. In FORTRAN, the arguments of functions can also be FORTRAN expressions. For example, the expression

$$\sqrt{X^2 + Y^2}$$

was previously converted to the FORTRAN expression

```
(X**2+Y**2)**0.5.
```

However, there is also a square root function in FORTRAN, SQRT(expression), which takes the square root of the resulting value when the expression is evaluated. Thus

```
SQRT(X**2+Y**2)
```

would give the same result as the previous expression.

Table 3.1 shows several of the most frequently used intrinsic functions. In FORTRAN 77, these functions are generic in that the type of the value returned by the function is the same as the type of the argument. For example, if X and Y are declared as REAL variables, the SQRT function used above would return a real value, while if X and Y were declared as INTEGER variables, the function would return an integer value.

Additional intrinsic functions are included in Appendix C. One important thing to remember when using trigonometric functions in FORTRAN and most other high-level languages and applications packages is that the value of the argument has to be in terms of radians, not degrees. For inverse trigonometric functions, arcsine, and the like, the value for the function is returned in radians. It is a simple matter to convert degrees to radians in your program using the expression

(no. of radians) = (pi/180)*(no. of degrees)

TABLE 3.1   FORTRAN INTRINSIC FUNCTIONS

Function Name	Function Produces	Function Type	Comments
ABS (X)	Absolute value of X	Same as Argument	
LOG (X)	Natural log of X	Same as Argument	
LOG10 (X)	Log 10 of X	Same as Argument	
ATAN (X)	Arctangent of X	Same as Argument	Radians
COS (X)	Cosine of X	Same as Argument	Radians
EXP (X)	e to the power X	Same as Argument	
INT (X)	Integer value of X	Integer	
MOD (X1,X2)	Remainder of $X1/X2$	Same as Argument	
REAL (I)	Real value of I	Real	
SIN (X)	Sine of X	Same as Argument	Radians
SQRT (X)	Square root of X	Same as Argument	
TAN (X)	Tangent of X	Same as Argument	Radians

The INT and REAL functions are slightly different from the other generic functions, in that they always produce integer and real values, but their arguments can still be of any numeric type. These functions are used to temporarily convert values from one type to the other to avoid mixed mode arithmetic. For example, if variables MEAN and SUM are defined as REAL and N is defined as INTEGER, the equation

```
MEAN = SUM/REAL(N)
```

would calculate the mean value without any mixed mode operations.

Most of the generic functions also have type-specific counterparts that will accept only a certain type of argument and return a value of the same type. These type-specific functions are given in Appendix C. Older FORTRAN versions did not have generic functions, so you will find the type-specific functions used in programs written using these versions. For new programs, it is best to use generic functions. Appendix C will also illustrate which intrinsic functions can be used for double precision and complex operations.

3.7.6 Effects of Significant Figures on Computations

The limited number of significant figures that computers can store and use in computation can cause problems. The cumulative errors caused by the inexact representation of numbers are called round-off errors. Serious numerical round-off errors can develop if many computations have to be accomplished. Even single computations can cause significant round-off error. For example, if your computer can carry seven significant figures, the computational sequence

```
0.4000000 + 1234567. - 1234567.
```

would result in a value of zero, rather than 0.4. When added to 1234567., the decimal four would appear in the eighth significant figure place and would be lost. Thus, you have to be cautious when large and small values will be used together in computations.

Also, the computational sequence

```
10.0/3.0*3.0
```

would result in a value of 9.999999 rather than 10.0, if the computer carries seven significant figures. Errors of this type can cause problems when you are checking to see if one value is equal to another. The computer will check all significant places and will not consider the values equal unless all places are equal.

If round-off errors prove to be a problem, the programmer should use double precision for the arithmetic operations. However, the DOUBLE PRECISION specification increases the amount of computer memory required to run a program. Therefore, it should not be used unnecessarily. Computer induced round-off errors will be discussed more thoroughly in Chapter 9.

Some textbooks refer to computer round-off errors as precision errors, but this is a confusing application of the precision concept. As discussed in Chapter 1, the precision of experimental data is a measure of the reproducibility of results from one test to the next under similar conditions. The precision of a number is defined as the number of significant figures that it has. When a computer is running programs, the cumulative effect of errors caused by the number of significant figures carried is an error in accuracy. The value calculated by the computer is different from the actual value by some amount of error. If the program is run again, the error will be the same. Thus, the error does not satisfy the statistical definition of precision error. The author will refer to these errors as round-off errors, not precision errors.

3.8 ASSIGNMENT STATEMENT

Up to this point you have used the equal sign (=) as an algebraic operator indicating that the numerical expression on the left of the equal sign has the same numerical value as the numerical expression to the right of the equal sign. This definition will now be expanded to reflect its meaning in FORTRAN and other computer languages and applications packages. In computer applications an equation is an assignment statement and the equal sign is a replacement operator. The assignment statement has the form

```
variable = expression
```

which means that the *expression* on the right is evaluated and the result placed in the storage location defined by the *variable* on the left. The = is called a replacement operator because if there is presently anything in the storage location, it will be *replaced* by the new expression evaluation. We have avoided using the word value to describe the evaluation of the expression and the replacement operation because the process is applicable to numerical, character, and logical operations.

The assignment statement

```
A = B + C
```

results in the value of $B + C$ being put into storage location A. If there is presently a value in A, it will be replaced by the new value. For example, if 80. is stored in variable location A, the statement

```
A = 28.
```

would store the value 28 in A and would discard the value 80. However, if the statement is

```
A = A + 28.
```

the present value of 80 in A would be increased by 28, and the resulting value of 108 would be stored in A. This convention provides a convenient way of producing a series of values for an independent variable to be used in calculations. When used in conjunction with the looping mechanisms that will be discussed in Chapter 4, the assignment statement

```
A = A + INC
```

would add the value of the increment, INC, to the present value of A and store the result in A each time through the loop.

The equal sign is a replacement operator, so only one variable can be placed to the left of the equal sign. There can be no operators to the left of the equal sign. Since intrinsic functions represent procedures, not memory locations, they also cannot be on the left of the equal sign.

Assignment transforms the data type of the result of the expression on the right to the type of the variable on the left. For example, if the evaluated expression is of integer type and the variable to the left of the equal sign is of real type, the stored value would be of real type. This provides a convenient method of storing a real equivalent of an integer value, or vice versa, without losing the original integer value. If N is an integer variable and XN is defined as a real variable, the statement

```
XN = N
```

would store the real value of N in XN without changing the type of N. At times you may find this preferable to using the REAL or INT intrinsic functions.

3.8.1 Exponent Underflow and Overflow

Exponent overflow or underflow occurs when the assignment operation tries to place a value resulting from the evaluation of an expression that is either too large or too small for the particular computer system into a variable location. With VAX systems the limit is 10 to the plus or minus 38th power. Thus, if $X = 1.E30$,

```
Y = X**2
```

would result in an overflow error, while

```
Y = 1./X**2
```

would result in an underflow error. This is an execution error and would result in termination of the program at the time that it occurs. Values that would result in an overflow or underflow error on one computer system might not on another.

3.8.2 Use Previously Calculated Variables to Reduce Computations

When you are developing a series of assignment statements to compute values for desired variables, it is helpful to think about the interrelations among the calculations. This can reduce the complexity of expressions and can make a program much easier to read and understand. For example, Figure 3.3 shows a triangle and several equations for determining the moment of inertia about the centroidal axis (I_c), the moment of inertia of the area about an axis parallel to the base, (I_x), the area (A), and the distance to the centroid of the triangle from an axis parallel to the base (y_c).

The second equation in the series is an application of the parallel axis theorem to find I_x, the moment of inertia about an axis parallel to the base. It requires use of the area, the moment of inertia about the centroidal axis, and the distance from the centroidal axis to the parallel axis. You could put all these calculations into the expression. However, it is much better practice to calculate the individual parts in separate assignment statements and use the results in the expression for the parallel axis moment of inertia. The expression for I_x should be put last in order since it uses the results of the other expressions. The following series of

FIGURE 3.3   Moment of inertia for triangle.

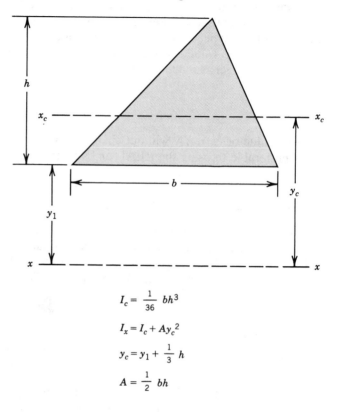

$$I_c = \frac{1}{36}\, bh^3$$

$$I_x = I_c + Ay_c{}^2$$

$$y_c = y_1 + \frac{1}{3}\, h$$

$$A = \frac{1}{2}\, bh$$

FORTRAN assignment and declaration statements would calculate the desired variable values:

```
REAL IC,IX,B,H,A,YC,Y1
IC = (1./36.)*B*H**3
A = 0.5*B*H
YC = Y1 + H/3.
IX = IC + A*YC**2
```

Note the use of the REAL statement to define the types of the variables used. Not only does the sequential development make it easier to interpret what is being calculated, it also provides the intermediate values for use elsewhere in the program.

Preliminary calculation of portions of an expression can also be important in cases where logical decisions have to be made. Take, for example, the calculation of the two roots of a quadratic equation. Equation 3.4 showed the calculation of one root of a quadratic equation. Calculation of the other root would require using

$$X = \frac{-B - \sqrt{B^2 - 4AC}}{2A} \tag{3.6}$$

However, before calculating either of the roots, you have to make sure that $B^2 - 4AC$ is not negative. If the result is negative, the roots are imaginary, and trying to evaluate either Equation 3.4 or 3.6 would result in a FORTRAN execution error. After testing $B^2 - 4AC$, the program should branch to separate computational sequences for real and imaginary roots. Example 4.6 will develop a program for doing this.

3.8.3 Breaking Statements into Smaller Segments

Long statements should be broken into shorter segments to make it easier to follow the logic of the statement and to make it easier to isolate errors when debugging. Parts of a long statement may also be repetitious. These can be separated out as preliminary calculations to make computations more efficient. Many long equations have natural breaks that make it easy to divide them into shorter segments. As a general rule, it is better not to extend a computational statement beyond one line in a FORTRAN program, even though the language allows you to do this. As will be shown in a subsequent section of this chapter, only 66 characters per FORTRAN program line are available for actual statements. It does not take an overly complex equation to exceed this number.

An example of a long equation is given in Equation 3.7.

$$D = \left\{ D_{in} - \frac{k_1 L_{in}}{(k_2 - k_1)(1 + 4k_1 E/u^2)^{1/2}} \right\} \cdot e^{\left\{ \frac{u^2}{2E}[1 - (1 + 4k_2 E/u^2)^{1/2}]t \right\}}$$
$$+ \frac{k_1 L_{in}}{(k_2 - k_1)(1 + 4k_1 E/u^2)^{1/2}} \cdot e^{\left\{ \frac{u^2}{2E}[1 - (1 + 4k_1 E/u^2)^{1/2}]t \right\}} \tag{3.7}$$

This is an equation for estimating the amount of dissolved oxygen removed from a stream through biological oxidation of pollutants. The equation is used for illustrative purposes, so we will not define all the variables. It can be seen, however, that there are several parts of the equation that repeat. For example,

$$E/u^2$$

appears several times, as do

$$(1 + 4k_1 E/u^2)^{1/2}$$

and

$$(k_2 - k_1).$$

Also, a natural breakpoint occurs at the plus sign between the two long portions. This equation is best divided into six separate FORTRAN statements. Variables that will be used in later statements are evaluated first. Assuming that all variables have been declared as REAL, the six statements would be:

```
P = E/U**2
Q = SQRT(1.+4.*K1*P)
R = (K2-K1)
D1 = (DIN-K1*LIN/(R*Q))*EXP(1./(2.*P)*(1.-SQRT(1.+4.*K2*P))T)
D2 = K1*LIN/(R*Q)*EXP(1./(2.*P)*(1.-Q)*T)
D = D1 + D2
```

Note that the slight difference in the exponent for the first portion of the equation prevented its duplication. Also, 1./P was used for U**2/E.

3.9 INTRODUCTORY INPUT AND OUTPUT

Computer programs are designed to manipulate data or produce numerical results for a mathematical model. You have to be able to provide the program with the necessary input data values. You also have to be able to output the appropriate data from the program. Neither input nor output is automatic. You have to give the computer explicit instructions as to how it is to input and output data. READ, PRINT, and WRITE statements give you these capabilities.

List-directed input and output is the simplest method for inputting or outputting data. The name is derived from the set, or list, of variables that will be input or output. This is also called free format or unformatted input and output, because you have little control over the form. List-directed input and output may be adequate for many applications. However, you can also control the form of FORTRAN input or output through FORMAT statements. Formatted input and output will be covered in Chapter 6. By that time you will have mastered the basics of FORTRAN and will be able to appreciate the value of formatting for arranging your input or output in a specified way. Development of list-directed input and output for FORTRAN programs will be discussed in detail in this section. Before proceeding with input and output, we need to discuss how the computer links your program with peripheral devices that can be used for input or output.

3.9.1 Specification of Input and Output Devices

The computer has to be given explicit instructions about where to look for input data and where to send output data. To accomplish this, each peripheral device to be used by a program is assigned a unique unit number. A disk drive can be assigned multiple unit numbers to connect the computer with a particular file on the disk. Most computer systems have default unit designations for frequently used options such as terminal input and output. However, you can override these defaults, and you will have to assign unit numbers for any devices not covered by the defaults. You can instruct the computer, through statements in your FORTRAN program, to use a particular unit in accomplishing the desired input or output operation. It is also possible to direct input and output through use of the operating system command language for your particular computer. Since this is machine specific, it will not be used as a general rule in programs developed in this text. FORTRAN specifications will be used to direct input and output.

Most computer systems reserve unit 5 for input from a terminal, and unit 6 for output to a terminal. If you are using only terminal input and output, you do not have to assign any unit identifiers. Systems also reserve certain numbers for peripheral devices such as plotters and card punches. You should consult your local operating system manuals to see which values are reserved. Any other numbers can be used to refer to the disk storage devices of the system.

Although it is possible to supersede the standard specifications and use units 5 and 6 for something other than terminal input and output, it is better practice not to do so. Units 5 and 6 are readily understood as terminal input and output. Using them for other purposes may be confusing to people studying a program. In some programs it will be necessary to have input from both the terminal and one or more data files, or to have output to both the terminal and output data files. In those cases it will be necessary to use different unit numbers for each type of input and output. Using units 5 and 6 to designate terminal input and output will facilitate understanding of the program by other users.

A convenient way of specifying units for input and output is through use of the PARAMETER statement. It has the form

```
PARAMETER (variable₁=expression₁, variable₂=expression₂,...)
```

and has the effect of setting *variable₁* equal to the evaluation of *expression₁*, *variable₂* equal to the evaluation of *expression₂*, and so on, at the beginning of program execution. To designate units for input and output, the expressions will be integer constants. However, the PARAMETER statement has more general uses, and the expressions can be any valid FORTRAN expressions. The variables will take on the values assigned to them by the PARAMETER statement and can be used to designate units for READ and WRITE statements in the program. The values assigned by the PARAMETER statement cannot be changed elsewhere in the program. A good feature of the PARAMETER statement is that if you need to change the value of all the unit numbers for read or write statements, you can do it for the whole program by changing the value in the PARAMETER statement.

You can also use mnemonic variables to designate commonly used input and output devices. For example, in the programs in the text, we will use the following unit numbers and variable names:

Type	Unit Number	Variable Name
Terminal Input	5	R
Terminal Output	6	W
File Input	4	FR
File Output	7	FW

with R standing for read, and W standing for write. The following statements would define the unit identification variables and set their values:

```
INTEGER R,W,FR,FW
PARAMETER (R=5,W=6,FR=4,FW=7)
```

You will see how these are used when we discuss READ and WRITE statements.

3.9.2 Terminal Input

Input of data directly from the terminal is efficient if the amount of data is minimal or if the data will be different each time the program is run. It can also be efficient for larger amounts of data if you include in your program the option of saving the input data for later use. We will incorporate this option into later examples in the text. If you are going to input significant amounts of data through the terminal, you need to include some type of error trapping in your program. Error trapping, which will be discussed in Chapter 4, can detect certain types of errors in input data from the terminal. You can correct the incorrect data using loop structures, and the program will continue to execute. Otherwise, execution would terminate at the point of the incorrect data entry.

3.9.3 READ Statement for Terminal Input

For list-directed (unformatted) input, the form of the READ statement is

```
READ (R,*) variable list
```

in which R is the variable that contains the unit number for the input, and the *
indicates that the input is unformatted.

Assuming X and Y are real, the statement

```
READ (R,*) X,Y
```

would expect you to type in values for X and Y. The program would pause in its
execution until you did. The normal procedure is to type both values on the same
line, separated by either a comma or a space. Thus, either

```
21.2,33.5
```

or

```
21.2 33.5
```

would assign the value 21.2 to storage location X and 33.5 to storage location Y. If
you typed only one number and then pressed the return or enter key, the program
would wait for you to enter the other value. In the section below on terminal
output you will learn how to send prompts to the terminal. These remind you
when variable values need to be entered.

3.9.4 File Input

File input is generally preferable for large amounts of data. The data file can be
generated using the text editor for your computer system. The file can be edited to
make changes or correct errors. However, you lose the interactive nature of
terminal input and the ability to trap errors. Some of the data files used in the text
will be relatively short and could probably as easily be input through the terminal.
However, they have been developed as data files for illustrative purposes. Other
data files used in the text are fairly long and will give you a good feel for the
efficiency of data file input.

3.9.5 READ Statement for File Input

Each line of the data file is called a record and can be up to 80 characters long.
Whatever you type on a line of input from the terminal also represents a record.
Specific rules hold as to how READ statements treat records. These rules make
the reading of data more flexible and will be useful when we discuss formatted
input in Chapter 6. You also need to understand them if you are to avoid problems
in list-directed input from files.

1. Each execution of a READ statement results in a new input data record
   being accessed. The normal processing sequence is to take the next line
   of data following the last one read in.
2. If the variable list of the READ statement is shorter than the number of
   data values in a record, enough data values, starting from the first one,
   are read from the record to satisfy the READ statement, and the remainder are ignored.

3. If the variable list of the READ statement is longer than the number of data values in the record, the READ statement will take enough values from the next record to satisfy the variable list. If the next record still does not have enough data values to satisfy the variable list, succeeding records will be read until the variable list is satisfied. Any remaining values on the last record accessed will be ignored, as the next READ statement will go to a new input record.
4. If a slash, /, is encountered in the input record, the remainder of the input variable list is ignored. In other words, if the variable list for the READ statement contained four variables and a slash was encountered in the input data record after two values had been read in, the first two variables of the list would have values entered for them and the second two would not. This is not a commonly used feature of FORTRAN, but you should not be surprised if you see it used occasionally.
5. Data files have an end-of-file marker after the last input data record. This end-of-file marker is added by the operating system software after you finish entering your input data records and save your file. If a READ statement tries to read beyond the last input data record, the end-of-file marker will be encountered, and an execution error message will be printed. Execution of the program would terminate at that point. End-of-file errors usually occur because you have not matched up the data file entries with your input variable lists, and the program has been forced to read extra records to satisfy variable lists.

The following example will illustrate the foregoing rules:

All variables are real. The INTEGER variable FR will be used for file input to signify File Read. Assume that the value of FR has previously been set at 4 by a PARAMETER statement. Given the following READ statements:

```
READ (FR,*) A,B,C
READ (FR,*) D,E,F
READ (FR,*) G,H,I
READ (FR,*) A,B,C
```

and the following data file:

```
1.,2.,3.,4.
5. 6.
7.,8.,9.,10.
11.,12.,13.,
14./
```

The following values would be assigned to the variables:

```
A = 14.
B = 2.
C = 3.
D = 5.
E = 6.
F = 7.
G = 11.
H = 12.
I = 13.
```

Note that A would initially be assigned the value 1., but the last READ statement would assign 14. to A, so the 1. would be displaced. B and C would not be changed by the final statement because of the slash in the input record.

There also is an alternate form of the list-directed READ statement

```
READ *,A,B,C
```

that leaves out the unit designation. This form would read only from the default device set through operating system commands, which is usually the terminal. Since this form is not as flexible as the form with the unit number, it will not be used in the examples of the text. You may see it used in some programs.

3.9.6 Input File Specification

In order to read data from a file, or to write data to a file, the program has to know what file to look for. When you save your data file, you assign it a name, such as INPUT.DAT. The .DAT extender defines the file as a data file. The READ statements use a unit identifier to establish where to look for input data. Somehow a link has to be made between the unit number and the data file that you want to have input. This can be accomplished through systems commands that are system specific or through use of the OPEN statement within the FORTRAN program. The latter method will be used throughout the text. The OPEN statement has the form

```
OPEN(UNIT=integer expression,FILE=filename,STATUS=type)
```

in which the *integer expression* for the UNIT specifies the unit identification number, the *filename* is the name assigned to the file stored on the system peripheral, and the *type* is either 'OLD', 'NEW', or 'SCRATCH'. 'OLD' signifies that the file already exists. Input data files will have the status 'OLD'. 'NEW' signifies that the file is an output file and will be saved on the peripheral device identified by the unit number. If the file already exists, an error will result. Therefore, the file has to be deleted after each run of the program, or the status must be changed to 'OLD'. 'SCRATCH' signifies that the file is a temporary file and will not be saved after the program is executed. Scratch files are used for large programs that store intermediate output temporarily to avoid using up computer memory, then read the data back in when it is needed. The filename and the type have to be enclosed in apostrophes since they are character strings.

Some systems also require a REWIND statement before the input file is read. This ensures that the pointer for the input file is placed at the first record of the file. REWIND is automatically accomplished for most systems. However, the REWIND statement can be used within a program to reposition the pointer at the beginning of a file if you desire to use the same data records twice in a program. It is not often that this is necessary, or wise, but the option is there. The REWIND statement has the form

```
REWIND (UNIT=integer expression)
```

in which the *integer expression* again specifies the unit identification number. The OPEN and REWIND statements for the file we described above would be

```
OPEN (UNIT=FR,FILE='INPUT.DAT',TYPE='OLD')
REWIND (UNIT=FR)
```

The filename can also be entered into a variable location and then used in the OPEN statement. For example, if you declared NAME as a character variable, used the terminal READ statement

```
READ (R,*) NAME
```

and entered 'INPUT.DAT' for the character value, the OPEN statement would be written

```
OPEN(UNIT=FR,FILE=NAME,TYPE='OLD')
```

This allows you to enter the name of the file to be accessed as part of the input data for the program execution, rather than having to have it in the program. The filename would have to be entered through a terminal READ, since you have not opened the input data file yet.

There is also a CLOSE command that can be used to disconnect a file from a program. It is optional unless you want to open a second file with the same unit number. All files are automatically closed when program execution terminates. The CLOSE command has the form

```
CLOSE(UNIT=integer expression,STATUS=destination)
```

in which the UNIT parameter is the same as before, and destination is either 'KEEP' if the file is to be saved, or 'DELETE' if the file is to be deleted.

3.9.7 Terminal Output

Terminal output is useful for providing prompts for terminal input or for an initial look at output to assist in program debugging. It has the disadvantage that once a particular part of the output has scrolled off the screen, you cannot go back to that portion without rerunning the program. List-directed terminal output can be accomplished through any of the following statements:

```
PRINT *, expression list,
WRITE *, expression list, or
WRITE (W,*) expression list.
```

The first two forms will produce output to the terminal unless you redirect them through operating system commands. You cannot redirect their output to any other devices through program statements. The third form can be assigned to a particular unit through the OPEN statement, just like the READ (FR,*) statement. Using our convention, we would use WRITE (FW,*) for file output. Execution of any of the output statements causes the output device to go to a new line of output.

The *expression list* can consist of any combination of variables, FORTRAN expressions, and output character strings (called literals), separated by commas. Thus,

```
PRINT *, 'Input a value for TIME.'
```

would print a prompt to the screen requesting that the user input a value for TIME, and the series of statements

```
X=2.
Y=3.
WRITE (W,*) 'X = ',X,'Y = ',Y,'X+Y = ',X+Y
```

would output the following line of data to the terminal:

```
X = 2.0000000 Y = 3.0000000 X + Y = 5.0000000
```

The variables to be output are imbedded between the labels and separated from the labels by commas. Information is output in the order that the computer senses it in moving from left to right across the output expression list. Therefore, the variable value will be output immediately after the label for that variable value. If you want the list-directed output to be on three different lines, you will have to use three separate WRITE statements. You can also use extra WRITE or PRINT statements to insert blank lines between lines of output. The statements

```
X = 2.
Y = 3.
WRITE (W,*) 'X = ',X
WRITE (W,*) ' '
WRITE (W,*) 'Y = ',Y
WRITE (W,*) ' '
WRITE (W,*) 'X+Y = ',X+Y
```

would output the following five lines of data:

```
X = 2.0000000

Y = 3.0000000

X+Y = 5.0000000
```

Note that the list-directed output of real values fills a fixed number of places with zeros added on the right to fill the field. The number of spaces in the field and the space between fields vary among systems. A field of nine and one space between fields will be used in the text for illustrative purposes.

The PRINT statement is useful for differentiating input prompts from normal program output. For normal program output, I prefer to use the general form of WRITE(W,*).

The following series of statements illustrates the input of the coefficients for the quadratic equation

$$X = \frac{-B \pm \sqrt{B^2 - 4AC}}{2A} \tag{3.8}$$

through terminal input with a prompt message and output of the values to check them. Assume that R and W have been assigned the values 5 and 6 through the PARAMETER statement.

```
PRINT *, 'Input the values of A, B, and C'
READ (R,*) A,B,C
WRITE (W,*) 'A = ',A,'B = ',B,'C = ',C
```

The computer screen for the execution of this series of statements would be

```
Input the values of A, B, and C
3.1,-4.2,2.9
A = 3.1000000 B = -4.200000 C = 2.9000000
```

The values entered for *A*, *B*, and *C* are output to make sure they have been entered properly. This is a good practice to follow in any program that you write.

Use of input prompts for terminal input is called conversational, or user friendly, computing. This can be an easy and powerful way to input data to programs and can be used to create data files that can be used during subsequent executions of a program.

As was shown previously, literal output is useful for input prompts and labeling output values. It can also be used to output titles, column headings, and other explanatory information about a program. However, the information output through literals will be the same for each execution of the program. You have to modify the program itself to change the output. Output labels that change from run to run of the program can be accommodated by using character input and output, which will be discussed in a later section of this chapter. When using literal output to define column headings, it is necessary to plan for the spacing of the headings so they will appear over the appropriate column of output. For example, assume that you have five columns of data that you want to output. The first two columns contain X and Y data values for regression analysis. Columns three and four contain values for the x and y values squared. Column five contains values for the cross product of the data values, $X*Y$. For list-directed output, each column of data occupies nine output spaces, with one space between columns. For terminal output 80 spaces are allowed for each line of output. Output going to a printer can use up to 132 spaces per line.

You decide that you want to center the following headings

```
    X    Y    X**2    Y**2    X*Y
    -    -    ----    ----    ---
```

over the middle of the columns of data. For this simple example it would be easy enough to count how many spaces there should be between column headings to accommodate the columns. When numerous columns of varying width are used, it is helpful to have some sort of graphical grid with output spaces numbered from one to the maximum number allowed at the top of the graph sheet.

Figure 3.4 shows such a graph for 80 column output. This is actually a FORTRAN coding form which can assist in program entry. Program entry will be illustrated in Figure 3.5. For output design, you have to be concerned with only the column numbers. The headings can be positioned over the output columns, and the number of spaces required counted from the graph. Using Figure 3.4, it

FIGURE 3.4 FORTRAN coding form used for output design.

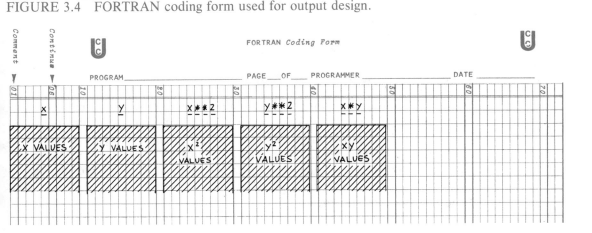

can be seen that the following statements would position the desired headings over the columns:

```
WRITE(W,*) '    X          Y          X**2      Y**2      X*Y'
WRITE(W,*) '    -          -          ----      ----      ---'
```

One additional feature of literal output should be discussed. You may want to output a character string with an imbedded apostrophe. For example, assume that you want to output the words *Kirchoff's Laws* as part of an electrical engineering applications program. If you try to output this as 'Kirchoff's Laws', the computer will be confused by the second apostrophe because it will think it should be marking the end of the character string rather than forming the possessive case. FORTRAN gets around this problem by using the convention that two apostrophes together represent an imbedded apostrophe. Thus the literal 'Kirchoff''s Laws' would be output correctly as Kirchoff's Laws.

3.9.8 File Output

List-directed file output has the same form as list-directed terminal output, with the exception that FW, signifying file write, should be used instead of W for the variable defining the unit number. Thus

```
PARAMETER (FW=7)
OPEN (UNIT=FW,NAME='OUTPUT.DAT',STATUS='NEW')
WRITE (FW,*) 'A = ',A,'B = ',B,'C = ',C
```

would output the coefficients of the quadratic formula, Equation 3.8, to a file named OUTPUT.DAT through unit 7. File output provides a permanent record that can be called back later. It can even be edited, if desired. Many computer systems require that hard copy output be accomplished from a file, unless you are working at a terminal or workstation that allows you to print output as you go along. Also, you can page through file output at your own pace and can return to previously observed sections of the output. With terminal output, there are ways of stopping the scrolling of the output past the screen, but stopping at the desired spot is difficult to control because of the speed of scrolling. Also, once you have passed a certain point with terminal output, the only way to look at a previous section is to rerun the program.

In the next section you will apply the concepts learned in the previous sections to write a simple, but complete, program.

3.10 PROGRAM STRUCTURE

3.10.1 Comment Statements

Comment statements are inserted within a program to provide internal documentation. They can be used to identify logic modules within the program, to provide variable definitions, to describe the purpose of the program, to identify the subject and writer of the program, or to provide any other information that the programmer feels would be helpful to people using the program. Although comment statements are effective communications tools, it is important not to clutter up your programs by overusing them.

A line of a FORTRAN program has 80 possible positions for characters. It is convenient to think of these positions as 80 columns numbered one to 80 from the left. Placing a C or a * in the first column of a FORTRAN statement line identifies that line as a comment statement. Columns 2–80 of a comment statement can contain any characters the programmer wants to include. Comment statements are ignored during the execution of the program, and they will not appear with program output. They will appear when you look at the program listing on the screen or output it to a printer.

3.10.2 Statement Organization (Column Groups)

3.10.2.1 Address Label

Columns one through five of any FORTRAN statement can contain an integer address label from 1 to 99999. Note the dual purpose of column 1: either as a comment identifier or as part of an address label. Address labels are used to identify a certain statement for reference by other statements in the program. The same address label cannot be used for more than one statement within a program, or the computer will be confused when told to refer to that address.

3.10.2.2 Continuation

Column 6 is used to denote continuation of a FORTRAN statement from the previous line. Any character other than a zero and a blank can be used to identify a continuation line. A FORTRAN statement can be continued for several lines in this manner. For computational statements, it is better to separate long statements into multiple shorter statements, as was previously discussed. For other types of statements it is quite acceptable to use continuation lines. Long declaration statements are an example. Also, FORMAT statements, which will be discussed in Chapter 6, are often too long for a single line. Continuation lines cannot have address labels.

3.10.2.3 FORTRAN Statement

Columns 7–72 are reserved for the actual FORTRAN statement. Anything beyond column 72 will be ignored during program compilation and execution, even though there are 80 columns in the input line. Most terminals do not identify the columns as you are typing in statements, so you have to be sure not to go beyond column 72. A good way to keep track of the columns is to write your program out on FORTRAN coding forms that are available at most stationary and computer supply stores. Figure 3.5 shows one such form. Note that the column groupings are identified on the form to assist you. You can also use these forms to organize input data, remembering that all 80 columns are available for data, and that for data input you can disregard the FORTRAN column groupings. It is also important to remember that when you are inputting a FORTRAN file from a terminal, you have to space your cursor over to column 7 before starting a FORTRAN statement that does not have an address label and is not a continuation line. It is, however, permissible to start a FORTRAN statement after column 7. We will use this to our advantage to identify various levels of branching logic starting in Chapter 4.

FIGURE 3.5 FORTRAN coding form.

FORTRAN *Coding Form*

PROGRAM_____ PAGE___OF___ PROGRAMMER_____ DATE_____

```
      PROGRAM CIRCLE
**-----------------------------------------------------------------**
* Program for computing the area of a circle given                  *
*    diameter.                                                      *
* Developed by T. K. Jewell, 10/14/87                               *
**-----------------------------------------------------------------**
      REAL AREA,D,PI
      PT=3.14159
C   Input Diameter
      PRINT *,'Please input value of diameter, in feet.'
      READ (R,*) D
C   Calculate area
      AREA = PI*D**2/4.
C   Output diameter and area
      WRITE (W,*) 'Diameter = ',D,' feet'
      WRITE (W,*) 'Area = ',AREA,' Square feet'
      END
```

3.10.2.4 Statement Identification

Columns 73–80 are reserved for a statement identification or serial number and are ignored in compilation and execution. The identification number is not the same as an address label. Originally, FORTRAN provided these columns for the statement identification number as a means of keeping track of the order of punch cards in a program deck. Not all statements had address labels; and even when they did, the labels did not have to be used in any particular order. If a punch card deck accidentally became shuffled, it would be difficult to put the cards back in order. The serial numbers in columns 73–80 provided a means of reindexing the shuffled cards.

Nearly all processing is now done from files saved on disk, and the editors used to create the files automatically reindex them whenever statements are added or deleted. Although there is no longer any need for the statement identification, FORTRAN still provides for it.

3.10.3 Executable and Nonexecutable Statements

Any statement that causes some processing to take place is called an executable statement. Nonexecutable statements provide information necessary for proper execution of the program and processing of data, but they do not actually cause any processing of information. Some examples of nonexecutable statements are declaration and specification statements; the PROGRAM statement; FORMAT statements that provide information on how to treat data for READ, WRITE, and PRINT statements; and comment statements.

3.10.4 Program Organization

The following section will describe the general order in which types of statements will occur in a FORTRAN program.

3.10.4.1 PROGRAM Statement

The PROGRAM statement is optional. It merely serves to name the program. If used it must be the first statement of the program. It has the form

```
PROGRAM program name
```

No variable within a program can have the same name as the program name.

3.10.4.2 Specification and Declaration Statements

Specification and declaration statements come immediately after the PROGRAM statement. Comment statements can be interspersed with them, or precede them, if desired. Specification and declaration statements include

```
CHARACTER
COMPLEX
DOUBLE PRECISION
INTEGER
LOGICAL
REAL
PARAMETER
INTRINSIC
DIMENSION
COMMON
DATA
```

All but the last three of these have been introduced previously. The DIMEN-SION and COMMON statements provide ways of defining arrays for FORTRAN programs. DIMENSION statements will be discussed in a later section of this chapter. COMMON statements will be discussed in Chapter 7. The DATA statement is an alternate way of assigning values to variables at the start of a program. DATA statements can actually be placed anywhere in the program, but they are usually put at the end of the specification statements. No matter where they are located, DATA statements only assign values to the appropriate variables at the beginning of program execution. Use of the DATA statement will be discussed in more detail in Chapter 6.

Specification and declaration statements have to precede the first executable statement of the program.

3.10.4.3 Statement Functions

Statement functions are user-supplied, one-line functions that can be referenced from the program numerous times, just like intrinsic functions. Rules for developing statement functions will be discussed in the next section. Statement functions must be placed after specification and declaration statements, but before any executable statements.

3.10.4.4 Main Body of Program

The main body of the program can include any of the other types of statements previously discussed, as well as branching or looping statement structures, and statements that instruct the computer to access and execute a separate subprogram.

3.10.4.5 STOP and END Statements

The END statement must always be included as the last statement of a program or subprogram. It instructs the computer that this is the end of the program and all files should be closed. The computer is then ready for the next program. The END statement cannot be given an address label, and so it cannot be branched to directly. However, you can branch to the statement immediately preceding it.

The STOP statement has much the same effect as the END statement. A STOP statement can be located anywhere in a program and terminates execution of the program when it is encountered. The STOP statement is addressable, so it can be used for a branching destination. It is useful for terminating execution of the program when an error condition that will produce erroneous results develops. However, you will have to include logic in the program to detect the error. You

FIGURE 3.6   Order of FORTRAN statements.

Fortran Statements				
PROGRAM				Comments
IMPLICIT		PARAMETER	FORMAT	
INTEGER REAL CHARACTER DOUBLE PRECISION COMPLEX	*Type specifications*			
DIMENSION COMMON EXTERNAL INTRINSIC LOGICAL	*Specification statements*			
STATEMENT FUNCTIONS				

Assignment Statements
DO
CONTINUE
IF
ELSE
ELSE IF
END IF
GO TO
CALL
RETURN
STOP
OPEN
CLOSE
REWIND
READ
WRITE
PRINT
	DATA			

END				
Subprograms SUBROUTINE/FUNCTION				

should have the program print out a message explaining the error before the STOP is executed.

3.10.4.6 Subprograms (Procedures)

Subprograms are separate program modules that can be accessed from within the main program or from another subprogram. Their use will be discussed in Chapter 5. Subprograms for a program can be placed in any order after the main program. However, they can also be separate compiled files that can be accessed by the present program. This is a better practice for any subprograms that will be accessed by more than one program.

Figure 3.6, which is adapted from Borse (1985), shows a schematic of the general order of FORTRAN statements.

Example 3.2 will show the development of a simple, but complete program for determining the area of a circle. Figure 3.5 shows the resulting program placed on a FORTRAN coding form. Note the header information placed at the beginning of the program describing the program, identifying the author, and indicating when the program was written. Also note the use of units when identifying output. Whenever there are units associated with a problem, they should be identified in the output.

This program would be of limited practical use, because each time you wanted to get a new area, you would have to execute the whole program again. In Chapter 4 we will learn how to repeat a procedure such as this as many times as necessary.

EXAMPLE 3.2   □   Computing Area of Circle

STATEMENT OF PROBLEM

Write a program to determine the area of a circle if you are given the diameter.

MATHEMATICAL DESCRIPTION

$$\text{Area} = \frac{\pi D^2}{4} \qquad (3.9)$$

The value for π can be assigned, or it can be calculated to the maximum number of significant figures allowed by the computer as

$$\boxed{\pi = \cos^{-1}(-1.)} \qquad (3.10)$$

ALGORITHM DEVELOPMENT

Input/Output Design

Input the diameter with the appropriate prompt. Output the diameter and area with appropriate labels.

Numerical Methods

No special numerical methods are required.

(EXAMPLE 3.2 □ Computing Area of Circle □ Continued)

Computer Implementation

A pseudocode representation of the program is:

```
START
ASSIGN VALUE TO PI
READ DIAMETER FROM TERMINAL
CALCULATE AREA = PI*D**2/4.
WRITE OUT DIAMETER AND AREA WITH LABELS
END
```

PROGRAM DEVELOPMENT

```
      PROGRAM CIRCLE
******************************************************************
*  Program for computing the area of a circle of given          *
*    diameter.                                                   *
*  Developed by: T. K. Jewell                    10/14/87        *
******************************************************************
      REAL AREA,D,PI
      INTEGER R,W
      PARAMETER (R=5,W=6)
      PI = ACOS(-1.)
C  Input diameter
      PRINT *, 'Please input value of diameter, in feet.
      READ (R,*) D
C  Calculate area
      AREA = PI*D**2/4.
C  Output diameter and area
      WRITE (W,*) 'Diameter = ',D,' feet'
      WRITE (W,*) 'Area = ',AREA,' square feet'
      END
```

PROGRAM TESTING

Use your calculator to verify the area for any diameter.

Now that you have seen how to develop a complete program, several new concepts will be introduced that will help you to develop more sophisticated and more structured programs. These concepts will be used again in later chapters.

3.11 CHARACTER VARIABLE OPERATIONS AND DEFINITIONS

As described in the DATA TYPES section of this chapter, the CHARACTER declaration can be used to designate that selected variable storage locations will contain character information. The number of characters stored in each variable location is governed by the length assigned by the CHARACTER declaration. Several rules were discussed concerning the placement of characters into the character variable if the length of the character string is different from the length of the variable.

Character variables are useful for inputting, operating on, and outputting text material when running FORTRAN programs. This is material that will not be the same each time the program is run. Headings and labels that do not change are best handled by the methods described earlier in this chapter, or by formatted output, which will be covered in Chapter 6.

Two operators that accomplish specific tasks on character strings are the concatenation and the substring operators. The concatenation operator is signified by a double slash, //. It joins two character strings together. For example, the following series of statements

```
CHARACTER FIRST*10,MIDDLE*2,LAST*15,WHOLE*27
FIRST = 'Ace'
MIDDLE = 'B.'
LAST = 'Student'
WHOLE = LAST//FIRST//MIDDLE
```

would result in the character string

```
Student        Ace       B.
```

being stored in variable WHOLE. The extra spaces are inserted because the character strings stored in LAST and FIRST are not as long as the CHARACTER statement allows, so the remaining positions are filled with blanks. The information stored in FIRST, MIDDLE, and LAST would not be affected by the concatenation.

The substring operator, denoted by $(i:j)$, extracts the ith through jth character of the string it is operating on. For example,

```
WHOLE = FIRST(1:1)//MIDDLE//LAST
```

would extract the first initial from the name above and store it, along with the middle initial and last name, in WHOLE. WHOLE would contain the character string

```
AB.Student
```

which is not quite what you would like. You can add characters and spaces by using literals within the concatenation string. The statement

```
WHOLE = FIRST(1:1)//'. '//MIDDLE//' '//LAST
```

would store

```
A. B. Student
```

in WHOLE.

The substring operator can extract any of the characters within a character string. For example, $(3:5)$ would extract the third through fifth characters of the referenced variable. The parameters of the substring operator can also be integer expressions. If $I = 2$ and $J = 5$, the operator $(I:J)$ would extract the second through fifth characters of the referenced character string.

Character variables are useful for inputting and outputting variable program titles or labels. The statements

```
CHARACTER*80 TITLE1,TITLE2
PRINT *, 'Input two line title for program'
READ(R,*) TITLE1
READ(R,*) TITLE2
WRITE(W,*) TITLE1
WRITE(W,*) TITLE2
```

would input and output any two-line title that you desire. Each line of the title could be up to 80 characters long.

3.12 ARRAYS AND SUBSCRIPTED VARIABLES

Subscripted variables, or arrays, are used for reading in, operating on, and printing out a series of values for a variable. They reduce the number of variables used and are essential for undertaking matrix operations with a computer.

An array is an ordered set or collection of data. Vectors and matrices are two common types of arrays that engineers work with. Vectors and vector algebra, as well as matrices and linear algebra, are discussed in Appendix F. An element of an array is signified by a variable with a series of subscripts, the number of subscripts denoting the dimension of the array. For example, in matrix algebra the variable a_{ij} represents the i,j element of the two-dimensional array A. In FORTRAN this would be represented by A(I,J). The FORTRAN language is capable of handling up to seven-dimensional arrays, although most practical problems do not require the use of more than three-dimensional arrays.

Arrays should be used whenever you have multiple values for a variable that you want to use at more than one point in a program. For example, assume that you have 20 pairs of values for X and Y that you want to run a linear regression on. (We will develop a program for regression analysis in Chapter 4. It involves finding the best-fit line through a set of data.) The algorithm requires that you find the sums of the individual data values and the sums of the individual data values squared. Also assume that later in the program you want to construct an X,Y plot of the data. It is not practical to define 40 separate variables to hold the data. You could use two simple variables X and Y and a loop to read in each X,Y pair while doing the regression analysis. However, that data would not be available when you want to develop the plot. Using an array for X and an array for Y will allow you to reserve 20 storage locations for each variable. You can refer to each storage location by its variable name and subscript index. You will be able to use the values stored in the locations any number of times during execution of the program. If you had wanted only to do the regression analysis, you could have treated the X,Y pairs as simple variables. This would result in a simpler program. There is no reason to use arrays when simple variables will do the job.

Several applications of one- and two-dimensional arrays will be illustrated in Chapters 3–6. Chapter 7 will contain a more in-depth discussion of two-dimensional and higher-order arrays.

A vector is a good example of a one-dimensional array. An n-dimensional vector is still only a one-dimensional array, because there is only one subscript that denotes the components of the vector. A matrix is a prime example of a two-dimensional array. An element of the array is a_{ij}, with i indicating the row, and j the column of the matrix.

To use an array or subscripted variable in FORTRAN, sufficient space has to be set aside in computer memory to store all the values associated with the array. Programs can use a subset of the reserved array space, but cannot increase the size of the space during execution. However, all the reserved space will be set aside in computer memory during execution. Therefore, you should be careful to specify the maximum realistic size you will expect to use. Unrealistically large arrays will waste computer memory. Each value of an array is actually stored in a separate storage location. One variable name is used to identify the array, and the individual storage elements of the array are defined by the subscripts associated with the array.

Specification of arrays can be accomplished in one of several ways. If you declare all variables through declaration statements, the best way to define arrays is also through the declaration statements. For example,

```
REAL A(10,10),B(10),C(10,3)
```

would set aside 100 storage locations for A, labeled A(1,1) through A(10,10), 10 storage locations for B, labeled B(1) through B(10), and 30 storage locations for C, labeled C(1,1) through C(10,3). Integer arrays are used less often, but the INTEGER statement could be used to define arrays in the same manner.

A second way to define an array is through use of the DIMENSION statement. The statements

```
REAL A,B,C
DIMENSION A(10,10),B(10),C(10,3)
```

would accomplish the same thing as the single REAL statement above. A third way of defining an array is through use of the COMMON statement, which will be discussed in Chapter 7.

The size parameter of the declaration statement has to be an integer value. It cannot be a variable in the main program. However, its value can be set by a PARAMETER statement. Usually numerous arrays will have the same size, and use of the PARAMETER statement allows changing all the array sizes by changing one statement. If you are using a PARAMETER statement to define the size of an array, the PARAMETER statement must precede the declaration statement for the array. The statements

```
INTEGER N,M
PARAMETER (N=10,M=3)
REAL A(N,N),B(N),C(N,M)
```

would specify the same arrays as before.

The normal declaration of an array requires that the subscript values vary from one to the upper limit. You can also set subscript ranges to any value that you desire. For example,

```
REAL VEL(0:50)
INTEGER YEAR(1960:2000)
```

would define a 51-element, one-dimensional, real array for VEL (velocity) with subscripts from 0 to 50, and a 41-element, one-dimensional, integer array for YEAR with subscripts from 1960 to 2000. Velocity problems frequently start with V_0, and many problems deal with data for a series of years, so the usefulness of this option can easily be seen.

None of the subscript values for a variable can exceed the limits set on that subscript value by the declaration statement. Each of the dimensions of the array should be assigned a size that is a reasonable maximum for the corresponding subscript of the variable. You have to decide what is a reasonable maximum for the particular application. For example, C(10,3) would allow storage of up to three values for each i subscript value. For $i = 1$, these would be C_{11}, C_{12}, and C_{13}. You could store the concentration of a chemical constituent at up to three points in a system for up to 10 times in the array C. You can store less than the specified maximum number of values in an array, but not more. Thus, you could store concentrations for two points at five different times, but not four points for eight times.

Appendix F develops the theory of matrix algebra. We will use matrix operations in later chapters of the text. Matrices can be represented by two-dimensional arrays in FORTRAN. Individual elements a_{ij} of the matrix **A** are represented by $A(I,J)$, in which I and J are integer expressions whose values identify the storage location. We are used to thinking of matrices in terms of i and j representing the element at the intersection of the ith row and the jth column. The computer does not actually store the values in a two-dimensional matrix. However, the FORTRAN language allows us to identify the storage locations by the row–column format and to use the elements of the array just as though they were in matrix form.

The following statements define three one-dimensional arrays representing two-dimensional vectors. The user is asked to input the components for two of these vectors. The program then adds the components of the vectors together and stores the results in the third vector storage locations.

```
REAL VEC1(2),VEC2(2),VECSUM(2)
PRINT *, 'Input x and y for vector 1 followed by same for 2 '
READ (R,*) VEC1(1),VEC1(2),VEC2(1),VEC2(2)
VECSUM(1) = VEC1(1) + VEC2(1)
VECSUM(2) = VEC1(2) + VEC2(2)
```

The size of each array was specified as two, because there are two components in a two-dimensional vector. The three vectors were stored in separate one-dimensional arrays to facilitate identification of the separate vectors. All three could be stored in a single 3×2 array, but then each would be more difficult to identify in a program.

You can see that if you had many elements in the array, it would be inconvenient to have to write a whole series of statements, one for each element. The repetition structures that you will learn in Chapter 4 will give you a way of accomplishing this more efficiently.

3.13 USER-DEFINED FUNCTIONS (STATEMENT FUNCTIONS)

Statement functions allow the programmer to specify an expression that is to be used several times within a program as a function. When the expression is to be used within the program, it is accessed through its function name and specified arguments, just like an intrinsic function. Statement functions are not designed to take the place of intrinsic functions, but to augment them for specific applications programs. In fact, statement functions cannot redefine an intrinsic function. Therefore, you cannot use the name of an intrinsic function as the name of a statement function. You also cannot use the same name for a function statement that you use for a variable in a program.

Statement functions are placed after all specification statements, but before the first executable statement. They have the form

```
F(dummy arguments) = expression using dummy arguments as variables
```

When you want to use the function in an expression anywhere in the program, you reference it by including

```
F(actual arguments)
```

as part of the expression to be evaluated. When the program encounters an expression that contains the function, the values for the actual arguments are substituted into the function for the corresponding dummy arguments. FORTRAN evaluates the function using the substituted values. It then uses the resulting function value in the evaluation of the expression containing the function reference.

Your understanding of dummy and actual arguments and their relation to each other will be increased by examining the following example. In the statement function

```
FX(Y) = Y**2 + Y**3 - 4.*Y**4
```

Y is the dummy argument. Somewhere later in the program the following statement appears:

```
PXY = FX(Z) + 60.*X
```

Z is the actual argument, and its present value is substituted into the function for the dummy argument Y. The function is evaluated, and the resulting value of function $FX(Z)$ is used when $FX(Z) + 60.*X$ is evaluated. If $Z = 2.0$ and $X = 1.0$ in the above example, $FX(Z)$ would evaluate to -52.0, and PXY would be assigned the value 8.0.

Actual arguments can be any valid arithmetic expression. However, the type (mode) of each actual argument must agree in type with the corresponding dummy argument. Corresponding means that the actual and dummy arguments are in the same position in the respective argument list. The type (mode) of the function statement itself is governed by the type of the function name. The type can be specified in any of the same ways as a variable.

The following series of statements illustrates the different possibilities for actual arguments:

```
REAL Y,X,Z,W,A,B
INTEGER N
Y(X,N) = X**N + X - 5.
X = 3.
N = 2
Z = Y(X,N) + 4.
W = Y(5.,N)
A = Y(6.,1)*W
B = Y(W-Z-10.,N-1)
```

Note the correspondence of the types of the actual and dummy arguments. Also, the type of the function has been declared in the REAL statement. The variable locations would contain the following values: $Z = 11.$, $W = 25.$, $A = 175.$, and $B = 3..$

The statement function looks similar to a subscripted variable. It cannot be a subscripted variable, however, because you have not given it a dimension in a REAL, INTEGER, or DIMENSION statement. The compiler assumes that any variable with parentheses after it is a statement function if it has not been defined as an array. This similarity can result in some interesting diagnostic messages. If you forget to declare a subscripted variable as an array, most FORTRAN compilers will consider the reference to the undefined subscripted variable as a reference to an undefined statement function.

Example 3.3 illustrates the application of a statement function to find the length of the hypotenuse of a right triangle by the Pythagorean theorem. The function will be used to find the length of two vectors, and then to find the length of their sum.

EXAMPLE 3.3   □   Summation and Length of Vectors

STATEMENT OF PROBLEM

A program is to be developed that will find the sum of two two-dimensional vectors and the lengths of both the original vectors and the sum vector. Since vector length will have to be found several times, a statement function will be developed for computing the length.

MATHEMATICAL DESCRIPTION

Two vectors are added together by adding the corresponding components.

$$ \mathbf{C} = \mathbf{A} + \mathbf{B} = (c_x, c_y) = (a_x + b_x, a_y + b_y) \qquad (3.11) $$

The length of a two-dimensional vector is found by the Pythagorean theorem.

$$ C = \sqrt{c_x^2 + c_y^2} \qquad (3.12) $$

ALGORITHM DEVELOPMENT

Input/Output Design

Input the x and y components of A and B. Output components of all three vectors and their lengths.

Numerical Methods

No special numerical methods required.

Computer Implementation

Pseudocode Decomposition

```
DECLARATION STATEMENTS
STATEMENT FUNCTION FOR VECTOR LENGTH
INPUT VALUES
COMPUTE SUM OF VECTORS A AND B
COMPUTE LENGTHS OF A, B, AND C
OUTPUT VALUES
```

Pseudocode Refinement

```
DECLARATION STATEMENTS
    ALL VARIABLES USED ARE REAL
STATEMENT FUNCTION FOR VECTOR LENGTH
    LENGTH(x,y)  =    SQRT(x**2+y**2)
```

```
     INPUT VALUES
          READ aₓ,aᵧ
          READ bₓ,bᵧ
     COMPUTE SUM OF VECTORS A AND B
          cₓ=aₓ+bₓ
          cᵧ=aᵧ+bᵧ
     COMPUTE LENGTHS OF A, B, AND C
          REFERENCE LENGTH(x,y) USING x and y COMPONENTS OF EACH VECTOR AS
     ACTUAL ARGUMENTS
     OUTPUT VALUES
          OUTPUT COMPONENTS AND LENGTH OF A
          OUTPUT COMPONENTS AND LENGTH OF B
          OUTPUT COMPONENTS AND LENGTH OF C
```

PROGRAM DEVELOPMENT

```
     REAL LENGTH,AX,AY,BX,BY,CX,CY,LENA,LENB,LENC
     INTEGER R,W
     PARAMETER (R=5,W=6)
     LENGTH(X,Y) = SQRT(X**2+Y**2)
     PRINT *, 'Input X and Y components of vector A'
     READ (R,*) AX,AY
     PRINT *, 'Input X and Y components of vector B'
     READ (R,*) BX,BY
C  Compute sum of vectors A and B (C = A + B)
     CX = AX + BX
     CY = AY + BY
C  Compute length of vectors A, B, and C
     LENA = LENGTH(AX,AY)
     LENB = LENGTH(BX,BY)
     LENC = LENGTH(CX,CY)
     WRITE(W,*) 'AX = ',AX,'AY = ',AY,'Length of vector A = ',LENA
     WRITE(W,*) 'BX = ',BX,'BY = ',BY,'Length of vector B = ',LENB
     WRITE(W,*) 'CX = ',CX,'CY = ',CY,'Length of vector C = ',LENC
     END
```

PROGRAM TESTING

For the vectors $\mathbf{A} = (3,4)$ and $\mathbf{B} = (5,-8)$, $\mathbf{C} = \mathbf{A} + \mathbf{B} = (8,-4)$, length of $\mathbf{A} = 5$, length of $\mathbf{B} = 9.43$, and length of $\mathbf{C} = 8.94$.

Subscripted variables could also have been used for this problem. Simple variables were used purposely for this example to concentrate on the use of statement functions. Exercise 3.2.1 will redo this example using subscripted variables.

3.14 COMPLEX NUMBER INPUT, OUTPUT, AND OPERATIONS

A first course in electrical and electronic circuits generally covers alternating current circuits. Alternating current is represented by a sinusoidal function of time. Appendix I shows how this sinusoidal function can be represented as a phasor for calculations. The phasor represents the magnitude and phase angle of the current, voltage, or impedance in the circuits. The differences in phase angle among quantities used in calculations must be taken into account when analyzing alternating current circuits. Since phasors can also be represented as complex numbers, it is essential to understand complex number operations to analyze these types of circuits.

The FORTRAN language has the capability of inputting, operating on, and outputting complex numbers. Any variable that has been declared as type COMPLEX will be read in as a complex number. Two variables will actually be read in, the first being assigned to the real part and the second to the imaginary part. When using complex variables in expressions, you have to make sure that all appropriate variables are declared as complex, including the variable to the left of the replacement operator. When outputting complex variables, it is good practice to identify the output by putting parentheses around each complex number. Parentheses are included automatically if you use list-directed output.

FORTRAN can also deal with phasors by treating them as one-dimensional arrays of size two. The first element of the array would contain the magnitude, and the second element the angle. Transformations from phasor to complex, and vice versa, can easily be made.

Most of the generic functions have complex counterparts, and there are several intrinsic functions written specifically for complex operations. These can be found in Appendix C. The following functions facilitate working with complex numbers:

Function	Argument	Description
REAL	Complex	Returns the real part of the complex argument
AIMAG	Complex	Returns the imaginary part of the complex argument
CONJG	Complex	Returns complex conjugate of complex argument. The complex conjugate of $(a + bi)$ is $(a - bi)$.
COMPLX	Real	Converts two real arguments, a and b, into the complex number $(a + bi)$.

Example 3.4 gives an example of a brief but complete program for inputting, using, and outputting complex variables. In this example, two alternating current circuits enter a third circuit at a junction. The current in the third circuit is the sum of the two currents coming into the junction. The program will compute this current. Note that electrical engineers prefer to denote the imaginary part of the complex value by the letter j before the imaginary value, rather than i after the value. This helps to avoid confusion with current, denoted by I or i. The latter convention will be employed in this example, and $20 + 30i$ will be expressed as $20 + j30$.

EXAMPLE 3.4 □ Complex Variable Operations for Circuit Analysis

STATEMENT OF PROBLEM

A certain alternating current circuit has a voltage phasor of $130\angle 90°$ volts, and an impedance (Z) of $5\angle 30°$ ohms. It is joined at a junction by a second circuit having a current of $15 + j45$ amps. A third circuit leads off the junction, with the

FIGURE 3.7   Alternating current circuit junction.

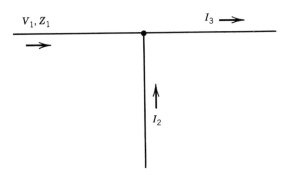

current in this circuit being $I_1 + I_2$. Figure 3.7 shows the junction. Develop a program to input the appropriate data, determine I_3, and output the appropriate values. For output, I_1, I_2, and I_3 should be in complex form.

MATHEMATICAL DESCRIPTION

For the first circuit

$$I_1 = \frac{V_1}{Z_1} = \frac{130\angle90°}{5\angle30°} = \frac{130}{5} \angle 90 - 30° = 26\angle60° \qquad (3.13)$$

I_1 can be converted to rectangular form by

$$\begin{aligned} I_1 &= a + jb = 26\angle60° = M\angle\theta \\ a &= M \cos \theta = 26 \cos 60° = 13 \\ b &= M \sin \theta = 26 \sin 60° = 22.52 \end{aligned} \qquad (3.14)$$

and the current in circuit three computed by

$$I_3 = I_1 + I_2 = 13 + 15 + j(22.52 + 45) = 28 + j67.52$$

ALGORITHM DEVELOPMENT

Input/Output Design

Input voltage and impedance for circuit one in phasor notation. A one-dimensional, two-element, subscripted variable will be used for each phasor. The first element will store M and the second will store θ. The current for circuit two will be input in complex notation. Input values should be printed out for checking. Final output of I_1, I_2, and I_3 will be in complex notation.

Numerical Methods

No special numerical methods are required for this problem.

Computer Implementation

Conversion from phasor to rectangular form can be accomplished as shown in Equation 3.14. The CMPLX function can be used to combine the real and imaginary parts into the complex format. Figure 3.8 gives the flowchart for the program.

(EXAMPLE 3.4 □ Complex Variable Operations for Circuit Analysis □ Continued)

FIGURE 3.8   Flowchart for complex variable alternating current circuit problem.

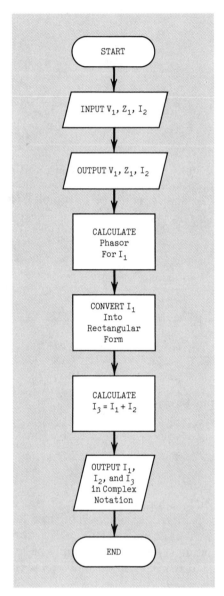

PROGRAM DEVELOPMENT

```
      PROGRAM CIRCUIT
*****************************************************************
*  A program to find the current in an alternating            *
*     current circuit being fed by two other circuits.        *
*  Developed by: T. K. Jewell                    10/13/87      *
*****************************************************************
```

```
C  Vl, Zl and IIl = voltage, impedance, and current in
C  circuit one, in phasor notation.
C  Il, I2, and I3 = current in circuits one, two, and three.
       COMPLEX Il,I2,I3
       REAL Vl(2),Zl(2),IIl(2)
       INTEGER R,W
       PARAMETER (R=5,W=6)
C  Input data values
       PRINT *, 'Input magnitude and phase angle for Vl  '
       READ (R,*) Vl(1),Vl(2)
       PRINT *, 'Input magnitude and phase angle for Zl  '
       READ (R,*) Zl(1),Zl(2)
       PRINT *, 'Input I2 as (a + jb) '
       READ (R,*) I2
C  Output input values
       WRITE (W,*)'Voltage l phasor = ',Vl(1),' at ',Vl(2),' degrees'
       WRITE (W,*)'impedance l phasor = ',Zl(1),' at ',Zl(2),' degrees'
       WRITE (W,*)'Current 2 = ',I2,' amps'
C  Calculate phasor for Il
       IIl(1) = Vl(1)/Zl(1)
       IIl(2) = Vl(2)-Zl(2)
C  Convert phase angle for Il from degrees to radians
       IIl(2) = IIl(2)/57.296
C  Convert Il to rectangular form
       a = IIl(1)*COS(IIl(2))
       b = IIl(1)*SIN(IIl(2))
       Il = CMPLX(a,b)
C  Calculate I3
       I3 = Il + I2
C  Output values
       WRITE (W,*) ' '
       WRITE (W,*) 'OUTPUT VALUES FOR CURRENT'
       WRITE (W,*) ' '
       WRITE (W,*) 'Il = ',Il,' amps'
       WRITE (W,*) 'I2 = ',I2,' amps'
       WRITE (W,*) 'I3 = ',I3,' amps'
       END
```

PROGRAM TESTING

Program output is shown below. This agrees with the calculated values from the mathematical description section above.

```
Input magnitude and phase angle for Vl    130,90

Input magnitude and phase angle for Zl    5,30

Input I2 as (a + jb)  (15,45)

Voltage l phasor =      130.00000000 at         90.00000000 degrees
Impedance l phasor =       5.00000000 at         30.00000000 degrees
Current 2 = (        15.00000000,      45.00000000) amps

OUTPUT VALUES FOR CURRENT

Il = (        13.00009155,       22.51660728) amps
I2 = (        15.00000000,       45.00000000) amps
I3 = (        28.00009155,       67.51660919) amps
```

3.15 DOUBLE PRECISION OPERATIONS

Double precision operations are the same as normal operations, except that you have to remember to declare all appropriate variables as DOUBLE PRECISION and to use double precision functions. Example 3.5 will illustrate the application of double precision in the determination of the area of a circle.

EXAMPLE 3.5 □ Using Double Precision to Find Area of Circle

STATEMENT OF PROBLEM

You are required to find the area of a circle to at least 12 significant figures. Modify the program of Example 3.2 to accomplish this. The problem development is the same as for Example 3.2, and so we will skip down to program development.

PROGRAM DEVELOPMENT

```
      PROGRAM DCIRCLE
**********************************************************************
*  Program for computing the double precision area of a circle *
*     of given diameter.                                       *
*  Developed by: T. K. Jewell                          1/89 *
**********************************************************************
      DOUBLE PRECISION AREA, D, PI
      INTEGER R,W
      PARAMETER (R=5,W=6)
      PI = DACOS(-1.D0)
C  Input diameter
      PRINT *, 'Please input value of diameter, in feet. '
      READ (R,*) D
C  Calculate area
      AREA = PI*D**2/4.D0
C  Output diameter and area
      WRITE (W,*) 'Diameter = ',D,' feet'
      WRITE (W,*) 'Area = ',AREA,' square feet'
      END
```

PROGRAM TESTING

Note the use of the double precision trigonometric function, and the designation of appropriate constants as double precision. The output appears below.

```
Please input value of diameter, in feet.   2

Diameter =      2.00000000000000000 feet
Area =          3.14159265358979312 square feet
```

SUMMARY

In Chapter 3 you have learned many of the essential elements of FORTRAN programming. You have learned how to define and represent various types of constants and variables, the meaning of the replacement operator (=) in

FORTRAN, how to evaluate FORTRAN expressions, and how to input and output data. You were introduced to the rules for putting these elements together to form a program. Character variable operations as well as arrays and subscripted variables were introduced. Two types of functions were discussed: intrinsic functions that are available through the FORTRAN compiler, and statement functions that you can define. You were introduced to using separate subprograms to represent logic modules and to the essential elements of FORTRAN program development and documentation. Three specialized variable types—complex, double precision, and logical—were discussed.

All these concepts will be used in later chapters. In this chapter you have learned sufficient FORTRAN language syntax to write complete programs. However, the programs developed are simple, straight-through programs that do not allow any options or repetition. In succeeding chapters you will add elements to these basic building blocks that will enable you to use FORTRAN as a flexible and powerful tool.

REFERENCES

Borse, G. J., *FORTRAN 77 and Numerical Analysis for Engineers*, PWS Publishers, Boston, 1985, Chapter 9.

Chapra, S. C., and Canale, R. P., *Introduction to Computing for Engineers*, McGraw-Hill, New York, 1986, Chapters 15–17.

Ellis, T. M. R., *A Structured Approach to FORTRAN 77 Programming*, Addison-Wesley, London, 1982, Chapters 1–3.

Etter, D. M., *Structured FORTRAN 77 for Engineers and Scientists*, 2nd. ed., The Benjamin/Cummings Publishing Company, Inc., Menlo Park, CA, 1987, Chapters 1–2.

EXERCISES

3.1 GENERAL NOTES

Problem solutions should be in the form of complete programs, including input and output. Programs should be debugged and running when submitted to your instructor.

All your programs should be as general as possible. This will be especially important when you get to Chapter 5 and start using separate subprograms as program modules for different applications. Any program-specific computer code will have to be included in the main, coordinating program or in subprograms written specifically for the application.

Your instructor will inform you concerning which type of input and output (terminal or file) are to be used for your solutions to programming exercises. The solutions presented in Appendix A and in the *Instructor's Manual* use data file input to save space and to have the data available to instructors. When you are asked to use terminal input, be sure to use appropriate input prompts.

Appropriate output with labels should be generated for each exercise. You should always output your input data for error checking.

Use descriptive variable names. Values for variables that may change from one run of the program to the next should be input rather than set in the program. However, frequently used constants can be defined as variables with the value initialized at the beginning of the program, since these will not change from run to run.

Use spaces within FORTRAN statements to help identify significant portions of the expression logic.

It is essential to have worked out solutions to check your computer results against. The following tips will help you in your debugging of programs.

1. There can be only one variable to the left of the replacement operator (=).
2. Zero cannot be taken to the zero power.
3. The computer and operating system are very seldom at fault when errors occur. Before you run your program, check it and the data files carefully for typographical errors, misalignment, and organization of your data files.
4. Easily misstyped characters include I instead of l (lower case L) and 1 (one), or O (capital O) instead of 0 (zero).
5. Eliminate FORTRAN syntax and execution errors using the error messages provided by the compiler. Check output against known values. If it checks, test other options. If it does not, intermediate output statements can be used to track down the logic problem. Carefully determine which variables should be tested, and where in the program they should be tested. Test logic step by step from the top down. Before making extensive changes to correct errors, store the present version of your program. If your corrections do not work out, you may want to go back to where you were. Make sure you are using a current version of your program when debugging. Check declarations. Check for mixed mode.

3.2 MODIFICATION OF EXAMPLES

3.2.1 Example 3.3

Modify Example 3.3 so that subscripted variables can be used for the components of the vectors **A, B,** and **C.**

3.3 GENERAL SYNTAX AND STATEMENT STRUCTURE

Answers for exercises preceeded by an * are in Appendix A.

3.3.1 FORTRAN Constants

*1. Indicate which of the following constants are incorrect FORTRAN constants and why. For those that are correct, indicate the type of constant represented.

```
3.567     3.*4.      123,456,789     123456789     1.56E6      −2.567E45
43.56*E13    3.0    CONSTANT    'Mary'    '3-dimensional'    1.235E−4.5
```

*2. Write the following as correct FORTRAN constants. Numbers without decimal places should be written as integer constants.

```
John Doe
1.375×10¹²
3,364,185
0.0000000000326
3.567856241894
1234.0×10⁻¹⁵
−3,278
```

3.3.2 Identification of Correct and Incorrect FORTRAN Variables

*1. Indicate which of the following variables are incorrect FORTRAN variables and why. For those that are correct, indicate the type of variable represented. There are no declaration statements in this exercise.

```
A     I     1JUMP     SMAX     MAXT     X-RAY     HUM     LAMB     C*30
MAXIMUM     X12     WRITE
```

2. Indicate which of the following variables are incorrect FORTRAN variables and why. For those that are correct, indicate the type of variable represented. The indicated declarations are in effect.

Declarations
```
REAL A, I
CHARACTER X12*30, MAST*10
INTEGER DULL, INDEX
```

Variables
```
A     X12     IMAX     MINX12     XINDEX     X12*30     DULL     MAST     I
REAL     REALDULL     INDEX     WINDEX
```

3. Indicate which of the following are legal integer variables. Tell what is wrong with those that are not legal. Note the declaration statements in effect.

Declarations
```
REAL I,J,K,L,M,N
INTEGER a,b,c
```

Variables
```
I     L.FOR     II     INTEG     INTEGRATION     HUMP     MAX     YMAX     MAXY
A*30     FORMAT
```

4. Which of the following are not legal real variables, taking into account the given declarations?

Declarations
```
REAL YMIN, I, J
INTEGER RSQ, T10, Y
```

Variables
```
IXRAY     I     T10     NAME#     YMIN     KMAX     M     MX     XRSQ     J
OVER     6IMAX
```

3.3.3 Evaluation of Expressions

1. What values would be assigned to X by the following statements?

```
X = ((3. + 2.)/(4. + (2.**2)+1.5))**.5/2.
X = 3.*(3./(2.**2)+2.)
X = ((4.-4./5.)**(2./3.)+2.**4*((9.+5.)/(4.+2.))*4.)**(-1)
```

2. Evaluate each of the following expressions using the given values. Default variable types are used unless otherwise indicated.

a. $y = \sin(\text{theta})*Vo*(x/(\cos(\text{theta})*Vo)) + .5*a*(x/(\cos(\text{theta})*Vo))**2$

for: theta = 1.; Vo = 100.; x = 200.; a = −32.2

*b. s = (vl**2-vo**2)/(2*a)

   for: $v1 = 50.; vo = 100.; a = -10.0$

c. REAL nl,n2
   q = nl/((n2-nl)/R-nl/p)

   for: $n1 = 1.00; n2 = 1.333; p = 30.0; R = 6.00$

*d. Yc = (q**2/g)**(1./3.)

   for: $q = 2.; g = 32.2$

e. E = y + q**2/(g*y**3)

   for: $q = 100.; g = 32.2; y = 4.$

*f. Z = (A/B+3.+((A+B/2.)**2-((A+B)/7.)**N))

   for: $A = 3.; B = 4.; N = 3$

3. Try to simplify the expression of part a of the preceding problem.
4. Are the following statements correct? If not, what is wrong with them? Default type declarations are in effect.

   a. X = 4.*I/NX

   *b. Y = 4.*Y/-3.

   c. A = A - B + C/D**2

   *d. AN*3. = B + C

3.3.4 Evaluation of Program Sequences

*1. What value would be stored for each variable (q, r, s, t, u, v) at the end of the following short program? Assume that default type declarations are used.

```
a=3.
b=2.
n=3
q=4.
r=q**n
s=r/a
b=b**n
u=b*q
q=q+r
v=b+2.
```

2. What would be stored for each variable $A, B, C, D, E, F,$ and G at the end of the following program?

```
CHARACTER*20 A,B
INTEGER C,D
READ (5,*) A,B,C,E
C=C+2
D=C+3
F=3.*E
G=F*E
END
```

The following input is typed when the program is run:

'Chapter 3','Problem 4',4,3.

3.3.5 Development of Expressions

Express the following algebraic equations as proper FORTRAN assignment statements. Longer equations should be divided up into several smaller statements.

1.
$$Y = \frac{(a + b)^n + 3a^2 + 1.5}{2c + 1}$$

*2. Manning's equation for open channel flow:

$$Q = \frac{1.49}{n} (A)(R_h)^{2/3} S_0^{1/2}$$

3. The equivalent resistance for a certain parallel-series circuit:

$$R_{tot} = \cfrac{1}{\cfrac{1}{R_2 + \left[\cfrac{1}{\frac{1}{R_3} + \frac{1}{R_4}}\right]} + \cfrac{1}{R_5}}$$

*4. The repayment schedule (A dollars/year) for a loan of P dollars, at an interest rate of i, over n years:

$$A = P\left[\frac{(1 + i)^n i}{(1 + i)^n - 1}\right]$$

5. The concentration of dissolved oxygen in a body of water modeled as a completely mixed chemical reactor:

$$D = \frac{t_d D_{in}}{t_d + k_2} + \frac{k_1 t_d L_{in}}{(t_d + k_2)(t_d + k_1)}$$

*6. The concentration of dissolved oxygen in a stream of water leaving a vessel modeled as a plug flow chemical reactor:

$$D = \frac{k_1}{k_2 - k_1} (L_{in})(e^{-k_1 t} - e^{-k_2 t}) + (D_{in})e^{-k_2 t}$$

7. The pressure of a gas at the stagnation point:

$$p = p_0 \left[1 + m^2 \left(\frac{k - 1}{2}\right)\right]^{\left(\frac{k}{k-1}\right)}$$

*8. The van der Waals equation for pressure of a gas:

$$p = \frac{RT}{\frac{V}{n} - b} - a \left(\frac{n}{V}\right)^2$$

9. The electric field at a distance x from the center of a ring-shaped conductor:

$$E = \frac{1}{4\pi\varepsilon_0} \left[\frac{QX}{(X^2 + a^2)^{3/2}}\right]$$

3.3.6 Familiarization with Command Procedures and File Manipulation

The purpose of this exercise is to familiarize you with the operating system and file manipulation for the particular type of computer system (Mainframe, mini-, or micro-) you will be working with. You will write a simple program and make several modifications to this program to practice using various methods of file manipulation and program execution available on your computer. You will be writing a program to find the area and moment of inertia about the centroidal axis for a triangle of base b and height h, when the base is parallel to the x axis. Use $b = 4$ and $h = 6$ to test your program. Your instructor may request that you submit some or all of these files via your computer system electronic mail utility, or through a communications link using file-transfer software. You may also be asked to hand in hard-copy output.

1. Write a program that will ask for data input from your keyboard and output answers to your screen. Prompts should be used for input, and all output must be properly labeled. Name this file PROB1A.FOR.
2. Modify your program so it will accept input from a data file, and output the answers to your screen. Name your program file PROB1B.FOR and your data file PROB1B.DAT.
3. Prepare a command file that will execute your program from part 2. Name this file PROB1B.COM.
4. Prepare a command file (call it PROB1C.COM) that will execute your program from part 2 and save the output in a data file named PROB1C.OUT.
5. Hand in one hard copy of your program file with line numbers and error messages.

3.3.7 Character String Operations

Write a series of FORTRAN statements that will accomplish the following:
1. Specify four character variables, one of length 2, one of length 4, one of length 6, and one of length 10.
2. Assign 'abcd' to the 4-character variable, and 'efghij' to the 6-character variable.
3. Concatenate these two strings to produce 'abcdefghij' for the 10-character variable.
4. Assign the substring 'de' of this 10-character variable to the 2-character variable.
5. Output all these character variables, one under the other.

3.3.8 Input and Output of Variable Text Matter

Write a program that will use a character array to input and output any four lines of text that you desire. Each line should be able to accommodate 80 characters.

3.3.9 Subscripted Variables and Arrays

*1. Develop REAL and INTEGER definitions for the following arrays:

Real variable A to store a 20×40 matrix

Real variable X to store 40 values, starting with X_0

Integer variable YEAR to store data from the years 1900–2000. The year should be the index of the array.

*2. Determine the total storage capacity for each subscripted variable in the following DIMENSION statement.

```
DIMENSION X(30),Y(2,30),Z(0:100),A(2,3,6),I(-10:10)
```

*3. What are the possible ranges for the subscripts in the previous exercise? How many different values can be stored for each variable?

4. Assuming that the defined size of the array is sufficient, which element of each array would be referenced by the following variables if $I = 2$ and $J = 6$?

    A(J,I)   A(I+1,J-1)   X(I+J)   X(2*I-1)   D(1,I+1,I+J+1)

*5. Describe what the following character specification statements accomplish:

```
CHARACTER*8 A(10,20),C(30),D
CHARACTER F(10,15)*6,G*9,H*2
```

3.3.10 Discussion Questions on Evaluating Errors

1. Your output does not indicate any FORTRAN errors, but you do not get any answers output. What might you suspect is wrong, and how would you approach tracking down the problem?

2. What would an error message "DATA-END OF FILE" indicate to you?

3. What would an error message "FLOATING POINT-DIVIDE BY ZERO" indicate to you?

4. Your program runs, but you get nonsensical answers out. What would be your actions to try to diagnose the problem or problems?

5. Why is it advantageous to use separate subprograms for computational procedures?

3.3.11 Finding Errors in Function Statements and References

Indicate any errors in the following function statement and associated statements referencing it. Show how the errors could be corrected.

a. X(A,B,N) = A**2 + B**M/(A*B) + 6.*A**M

.

Z = 4. + X(D,E,3.)/X(A,B,I)

.

W = X(I,C,D)

*b. A(X,Y,N) = X**N + Y**N + X*Y

.

.

F = 3.*A(2,X,Y)**2

3.3.12 List-Directed Input and Output

1. Show what would be output by the following program:

```
INTEGER I,J,K
REAL A,B,C,D,E
READ *,A,B,C,I,J,K
A = A + B
D = A + C
E = D**J
K = I + J*K + K
PRINT *,A,B,C,D,E,K
END
```

if the following values are input: 3.,4.,5.,2,3,2

*2. What will be printed by the following program if it reads the data shown?

```
REAL A,B,C,D,E,F,G,H,I
READ *,A,B,C
READ *,D,E,F
READ *,G,H,I
READ *,A,B,C
READ *,G,H,I
PRINT *,A,B,C,A+B,D,E,F,G,H,I
END
```

```
1.235 4.89 6.5
0.65,9
28 39
1      ,       13
32.5/
9.8/
22.0/45.6/7.7
```

3.4 CHEMICAL ENGINEERING

Note: Appendix K contains additional information on mathematical models for the exercises in this section.

3.4.1 [ALL] Ideal Gas Law

Write a program to compute the pressure using the van der Waals equation and the ideal gas law for a given set of conditions. Also compute the compressibility factor and organize your output so the two values can easily be compared. Check your program using the following data: $R = 0.0826$ atm-ℓ/g mole/°K, $T_c = 405.6$°K, $P_c = 112.5$ atm, molecular weight $= 17$, $W = 732.7$ g, $T = 25$°C, and volume $= 100\ \ell$.

3.4.2 [ALL] Chemical Kinetics

Develop a program to estimate the concentration of a chemical in a mixture at time t using both first- and second-order kinetics to model the reaction. Use function statements for the rate equations. Test your program at 40 minutes if the initial concentration is 0.05 mol/ℓ and the first- and second-order rates are $k_1 = 0.011$ and $k_2 = 0.31$.

3.4.3 [ALL] Depth of Fluidized-Bed Reactor

Develop a program that will estimate the expanded bed height for a fluidized-bed reactor made up of two sizes of particles given the unexpanded-bed height, the unexpanded void fractions, and the fraction and expanded-bed porosity for each of the components. Test your program using an unexpanded height of 5.0 m, and a void fraction for the unexpanded bed of 0.45. Particle size one makes up 0.6 of the bed and has an expanded-bed porosity of 0.37. Particle size two makes up 0.4 of the bed and has an expanded-bed porosity of 0.26.

3.4.4 [ALL] Process Design, Gas Separation

Develop a program that will estimate the absorption efficiency for a particular gas in a packed tower with a given number of plates. Test your program for five plates and methane gas with the following properties:

	Top	Bottom
k	8.0	9.5
Gas flow (mol/s)	50	53
Liquid flow (mol/s)	100	97

3.4.5 [ALL] Dissolved Oxygen Concentration in Stream

Develop a program to estimate the dissolved oxygen in a stream at a given distance downstream from the point of discharge. Use the data given in Exercise 4.4.5 to test your program at a distance of 10,000 ft downstream.

3.5 CIVIL ENGINEERING

Note: Appendix G contains additional information for the exercises in this section.

3.5.1 [ALL] Equilibrium-Truss Analysis

Given the reactions and the angle between the two members of a planar truss that form joint A shown in Figure G.1, write a program that will solve for the forces F_{ab} and F_{ac}. Test your program using a reaction in the y direction of 20 kips, a reaction in the x direction of 5 kips, and an angle of 60°.

3.5.2 [ALL] Open Channel Flow Analysis

Develop a program to estimate the flow-rate in a trapezoidal open channel using Manning's equation and English units. Determine Q when $y = 4$ ft, $n = 0.03$, $S_0 = 0.001$ ft/ft, $b = 4$ ft, and $z = 2$ ft/ft.

3.5.3 [ALL] Dam Stability Analysis

Develop a program to estimate all forces, including the frictional resistance force, acting on the concrete gravity dam of Exercise 2.4.3. Test your program on the following dam:

$l_1 = 40$ ft

$l_2 = 60$ ft

$h_1 = 10$ ft

$h_2 = 90$ ft

$y_1 = 80$ ft

Coefficient of static friction between dam and its foundation = 0.4

Specific weight of water = 62.4 lb/ft^3

Specific weight of concrete = 150 lb/ft^3

3.5.4 [ALL] Force on Submerged Gate

1. Develop a complete program to find the force and its point of application on a vertical rectangular gate of base b and height h. The centroid of the gate is at

depth h_c below the surface of a liquid of known specific weight. Test your program on a gate 4 ft wide by 6 ft high. Its top edge is 10 ft under the surface of water. Water has a specific weight of 62.4 lb/ft^3.

2. Modify the flowchart for Exercise 2.4.4 to incorporate a separate subprogram to compute the force and point of application for each shape.

3.5.5 [ALL] Volume of Excavation

Develop a program to determine the volume of a trench with trapezoidal end sections that are not necessarily the same size. Use t as the top width of an end section, b as the bottom width, and h as the height. The horizontal distance between the end sections is L. Test your program using the following data (all values are in feet):

$$t_1 = 20, \qquad b_1 = 10, \qquad h_1 = 5, \qquad t_2 = 30, \qquad b_2 = 15, \qquad h_2 = 7.5, \qquad L = 500$$

3.6 DATA ANALYSIS AND STATISTICS

3.6.1 [ALL] Histogram

One of the steps in developing a histogram for screen or printer output is to designate character variables to contain the symbols that will be used to form the histogram. Write a short program that can be used to input and output the following symbols: * + ! − #. Each symbol should be stored in a separate character variable location.

3.6.2 [ALL] Sorting

Develop a program to implement the flowchart of Exercise 2.5.2.

3.6.3 [2,4,5] Time Conversion

A character string is used to input the time of day in the format hr : min : s, for the hours, minutes, and seconds of the time. Develop a program that will extract the hour, minute, and second data from this string and store each in a separate variable. Then recombine the hour and minute data into the military time format (for example, 1430 for 2:30 P.M.). Output all the original and derived data. Test your program with the data 14 : 32 : 45.

3.6.4 [2,4,5] Date Conversion

A character string is used to input the day of the year in the standard format mo/dy/yr. Develop a program that will extract the month, day, and year data from this string and store the data in separate variables. Before storing the data for the year, add the characters 19 to the front of it, so 88 would be stored as 1988. Output all input and derived data with appropriate labels. Use the data 06/21/87 to test your program.

3.7 ECONOMIC ANALYSIS

Note: Additional information for the problems of this section can be found in Appendix E.

3.7.1 [ALL] Economic Formulas

1. Develop a program to compute both the present worth P and the future worth F of a uniform series of payments A. The payments will be made over n compounding periods at i percent interest. Test your program for an annual payment of $2,000 over 20 years at 8.5 percent interest.

2. Modify the flowchart of Exercise 2.6.1 to incorporate separate subprograms for computing each of the economic formulas.

3.7.2 [ALL] Nonuniform Series of Payments

Develop a program that could be used to find the present worth of a nonuniform series of payments over a three-year period. Use the data for the first three years of Exercise 4.7.2 to check your program. Use a statement function to compute the present worth for each payment.

3.7.3 [ALL] Multiple Interest Rates or Number of Compounding Periods

Develop a program that will find the present worth of a uniform series of payments for three different interest rates. Test your program with annual payments of $2,000 over 20 years at 7.0, 8.5, and 10 percent.

3.7.4 [ALL] Internal Rate of Return

Develop a program that will compute the difference between a given present worth and a calculated present worth using the present worth factor for a uniform series of payments. Use a given present worth of $100,000, and annual payments of $14,000 for 16 years at 10 percent to check your program.

3.7.5 [ALL] Mortgage Payments

Develop a program that will determine what size monthly payments are required to retire a loan of P dollars in n years at i percent annual interest. Most lending institutions compound the interest on loans monthly, but advertise yearly rates. They divide the yearly rate by 12 when applying the capital recovery factor. Use a loan of $40,000 at 8.5 percent for 25 years to check your program.

3.8 ELECTRICAL ENGINEERING

Note: Additional information for the problems of this section is found in Appendix I.

3.8.1 [ALL] Diode Problem

Develop a program to estimate the current through the junction diode described in Exercise 2.7.1 for a given temperature, saturation current, and applied voltage. Test your program for a temperature of 70°F, an applied voltage of 0.30 V, and a saturation current of 2 μA.

3.8.2 [ALL] Capacitance and Resistance

Develop a program to estimate the voltage and current across the capacitor of the capacitance/resistance circuit of Exercise 2.7.2 at any time t after the switch is closed. Use a voltage of 3 V, a resistance of 10,000 Ω, a capacitance of 2×10^{-6} Farads, and a time of 0.02 s to test your program.

3.8.3 [ALL] Signal Processing Circuits

Develop a program to estimate the capacitance required to tune a desired frequency for a given inductance. Also compute the resistance necessary and the voltage across the resistor for a selectivity of $f_0/25$. Use a tuning frequency of 500,000 Hz, $L = 10^{-5}$ henrys, and $V_0 = 10$ V at 600,000 Hz.

3.8.4 [2,4,5] Parallel and Series Resistance

Develop a program that will find the equivalent resistance for two resistors in series and in parallel. Use resistors of 10 and 20 Ω to test your program.

3.8.5 [2,4,5] Parallel and Series Capacitance

Develop a program that will find the equivalent capacitance for two capacitors in series and in parallel. Use capacitors of 1×10^{-6} and 2×10^{-6} Farads to test your program.

3.9 ENGINEERING MATHEMATICS

3.9.1 [ALL] Determinant and Cramer's Rule

Develop a program that will find the determinant of a 2×2 matrix. Use the matrix

$$\begin{vmatrix} 2 & 3 \\ 1 & 5 \end{vmatrix}$$

to test your program.

3.9.2 [ALL] Vector Cross Product, Unit Vector, Directional Cosines, Vector Components

1. The vector cross product described in Appendix F can be used to determine the moment about the origin produced by a three-dimensional force vector applied at a point r. The moment vector would be given by

$$\mathbf{M} = \mathbf{r} \times \mathbf{F}$$

Write a program that will find the moment vector given the components of $\mathbf{r}$ and $\mathbf{F}$. Also have your program determine the magnitude of all three vectors (use a statement function for this) and the unit vector for the moment. Use subscripted variables for the components of $\mathbf{M}$, $\mathbf{r}$, and $\mathbf{F}$.

2. Develop a program that will input the coordinates of two points in the X-Y plane, (x_1, y_1) and (x_2, y_2), and calculate the following data based on those points:

   a. The length of the straight line joining the two points
   b. The coordinates of the midpoint of the line

c. The slope and intercept of the line

d. The equation for the line that forms the perpendicular bisector with the given line

Refer to Appendix F for additional information. Test your program using the points (20,30) and (50,75).

3.9.3 [ALL] Centroid and Moment of Inertia

1. Write a program to calculate the centroids and moments of inertia about the x and y axes of a rectangle of base b and height h. Test your program for a rectangle 2 ft by 4 ft.

2. Develop a program to compute the moments of inertia for an I beam about both its weak and strong axes (the X-X and Y-Y axes in the figure). Test your program for a beam with the following dimensions: $a = 2$ in., $b = 12$ in., $c = 6$ in., and $d = 1$ in.

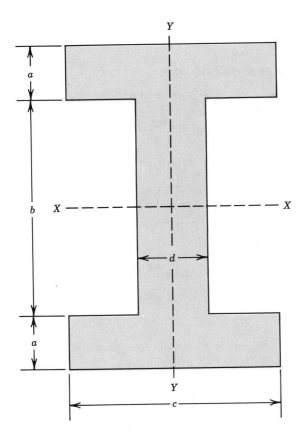

Figure for Exercise 3.9.3

3. Given the base and height of a triangle with the base parallel to, but not necessarily coincident with, the X axis, develop a program to determine:

a. The area of the triangle

b. The moment of inertia of the triangle about its X-X centroidal axis

c. The moment of inertia of the triangle about the X-X axis

d. The distance from the X axis to the centroid of the triangle

Test your program with a triangle of base 10 ft and height 8 ft when the base is 10 ft above the *X-X* axis.

4. Modify the flowchart of Exercise 2.8.3 to incorporate the calculation of the centroid and moment of inertia for each shape in a separate subprogram.

3.9.4 [ALL] Population Growth

Develop a program that will estimate the population *n* years in the future for growth rate *INCR* and a present population of P_0 using both the discrete and continuous growth formulas. Exercise 2.8.4 presented the discrete model. The continuous model is $P = P_0 e^{(INCR)t}$, with the same variable definitions as the discrete model. Test your program for an initial population of 10,000, a growth rate of 0.15, and 10 years.

3.9.5 [ALL] Vertical and Horizontal Curves

Develop a program that will compute the chord and arc length for a vertical curve that has a given radius and a given subtended angle. Test your program with a curve of radius 1000 ft that subtends an angle of 30°.

3.9.6 [2,4] Area of Triangle

Develop a program to estimate the area of a triangle given the coordinates of the three corners in order from left to right, and when the middle corner is higher than either the left or right corners. Use the points (3,4), (4,6), and (6,5) to test your program.

3.9.7 [4,5] Vector Dot Product

Write a statement function that will calculate the dot product of two vectors of dimension 3. The dot product is the sum of the products of corresponding coordinates between the two vectors. Test this statement function by using the vectors $\mathbf{X} = (3,2,-4)$ and $\mathbf{Y} = (-2,1,0)$.

3.9.8 Evaluating a Quadratic Equation

Write a statement function that will evaluate a quadratic equation

$$Y = aX^2 + bX + c$$

for given values of *a*, *b*, *c*, and *X*. Show several calls to this function statement that illustrate the various forms of actual arguments, variables, constants, and expressions.

3.9.9 Area and Sides of Right Triangle

A and *B* are the two legs of a right triangle. Write a program that will accomplish the following tasks:

1. Find the area of two separate right triangles, and add the areas together.
2. Find the sums of the corresponding sides, i.e., *A1* + *A2*.
3. Compute the area of the triangle found in task 2.
4. Compute the ratio of the area from task 3 to the total area from task 1.
5. Output all the data with labels.

Use statement functions to compute the area of a triangle and to add corresponding sides.

3.10 INDUSTRIAL ENGINEERING

Note: Appendix J contains additional information for the exercises of this section.

3.10.1 [ALL] Project Management

Develop a program that will add a given activity duration to a given earliest expected time for a node and use the sum for the earliest expected time for the node on the other end of the activity. Refer to Appendix J for the necessary definitions. Output the node identifiers, the earliest expected time for each node, and the activity duration.

3.10.2 [ALL] Quality Control

Develop a program that will compute the percentage of each type of defect found given the total number of defective parts and the number of times each defect was found. Use the following results for 20 defective parts:

Defect Number	Number of Times Found
1	10
2	2
3	6
4	16
5	3
6	1

3.10.3 [ALL] Managerial Decision Making

Develop a program that will compute the return, total cost, and the difference between return and total cost for a selected number of units. Use returns of $8N$, a fixed cost of $12,000, and a variable cost of $4N$. Run your program several times to try to estimate the break-even point. Confirm your answer by solving the simultaneous equations.

3.10.4 [ALL] Queuing (Wait Line) Theory

Develop a program that can be used to estimate the probability that one, two, or three machines are down when there are 5 machines and 3 repairmen. Use a failure rate of 0.2 machines/hr, and a repair rate of 0.8/hr.

3.11 MECHANICAL AND AEROSPACE ENGINEERING

Note: Appendix G contains additional information for the exercises of this section.

3.11.1 [ALL] Shear and Bending Moment

Develop a program that will estimate the shear and bending moment at any point x along the beam for a beam simply supported at each end and subjected to a uniform load. Test your program for a beam length of 20 ft and a load of 2,000 lb/ft, and for $x = 0$, $x = 10$, and $x = 20$ ft.

3.11.2 [ALL] Crank Assembly Analysis

Develop a program to estimate the angle beta, the angular velocity of the connecting rod, and the velocity of the piston for a given crank angle and crank angular velocity. Test your program for a crank length of 3 in., a connecting rod length of 8 in., a crank speed of 2,000 rpm, and crank angles of 40° and 90°.

3.11.3 [ALL] Friction

Develop a program that will compute the force required for slipping and tipping of the crate and slipping of the dolly for a given height of application of the applied force for the dolly/crate combination shown in Exercise 2.10.3. Use applicable data from Exercise 4.11.3 to test your program.

3.11.4 [ALL] Variation of Atmospheric Pressure with Altitude

Develop a program to solve for the atmospheric pressure at a given altitude using the formulation of Exercise 2.10.4. Test your program by estimating the pressure at 20,000 ft if $R = 1,715$ ft^2/s^2(°R), the lapse rate is 0.00356°R/ft, and the temperature and pressure at sea-level are 59°F and 14.7 psia. Note: °R = °F + 460.

3.11.5 [ALL] Aerospace Physics

Develop a program to determine the x and y coordinates, the x and y components of the velocity, and the magnitude of the velocity for a particle subjected to a constant acceleration in both the x and y directions for any time t. Use $X_0 = 100$ ft, $Y_0 = 50$ ft, $V_0 = 0$, $a_y = -10$ ft/s^2, $a_x = 4$ ft/s^2, and $t = 8$ s to test your program.

3.11.6 [ALL] Energy Loss in Circular Pipeline

1. Develop a program to find the energy loss in a circular pipeline using the Chen equation. Use the following data to check your program. The pipe has a diameter of 8.0 in., and is 1,000 ft long. The pipe is made of cast iron, which has roughness projections of 0.00085 ft. Gasoline is flowing at the rate of 3.0 ft^3/s. It has a dynamic viscosity of 0.47×10^{-5} ft^2/s. Appendix G contains additional information for this problem.
2. Modify the flowchart of Exercise 2.10.6(1.) to incorporate separate subprograms to evaluate the Chen and Swamee–Jain equations.

3.11.7 [ALL] Torsion

Develop a program that will estimate the angular deformation for a hollow cylinder subjected to a given torsional stress. Use $OD = 5$ in., $ID = 2.5$ in., $L = 4$ ft, $G = 1.1 \times 10^7$ lb/in.^{2}, and torque = 16,000 lb-ft to test your program.

3.12 NUMERICAL METHODS

3.12.1 [ALL] Interpolation

1. Write a program to accomplish quadratic interpolation (as described in Exercise 2.11.1) for a function $y(x)$ given three points. Use points 1, 2, and 3 shown in the figure for Exercise 2.11.1 to test your program for $x = 2.5$.

2. Modify the flowchart of Exercise 2.11.1 to incorporate separate subprograms to accomplish linear and Lagrange interpolation.

3.12.2 [ALL] Solution of Polynomial Equation

Develop a program that will evaluate up to a fifth-order polynomial for a given value of the independent variable. The coefficients of the polynomial should be input, and the polynomial should be evaluated in a statement function. Use your program to evaluate $f(x) = x^4 - 3x^2 + 5$ for $x = 2$.

3.12.3 [4,5,7,8] Data Smoothing

You have stored four experimental values of X_i in a one-dimensional array. You want to reduce the effects of random errors on these data by applying a smoothing function. One example of a smoothing function is the equation

$$X_i = \frac{(X_{i-1} + X_i + X_{i+1})}{3}$$

which is applied to all but the first and last data values. Write a program to smooth the second and third data values of a set. Output both the original and the smoothed data.

3.12.4 [5] Rounding-off Values

Write a statement or statements that will round off a value for X in the computer to n decimal places.

3.13 PROBABILITY AND SIMULATION

3.13.1 [ALL] Simulation of Two Dice

Write a program that will simulate a single throw of two dice. Print out the value of each die and the total. Run your program at least three times to demonstrate that you are getting different values each time. Use the random number intrinsic function available on your system.

3.13.2 [ALL] Normal Probability Distribution

Develop a program with a function statement for computing Z, given an estimated value, mean, and standard deviation.

CHAPTER FOUR

FORTRAN PROGRAMMING: CONTROL STRUCTURES

4.1 INTRODUCTION

Selection and repetition control are key elements of programming. These control structures allow programmers to solve problems with branching or looping logic. Without them, we would be limited to programs that execute one instruction after another, with no opportunity for alternate paths or loops. Examples of situations that require some sort of control structure are the successive input and use of a number of values for a variable, and the performance of a program task if or while a certain condition is met.

Most numerical methods algorithms are repetitive and iterate toward more accurate approximations of the correct answer. Therefore, they require control structures for their implementation on computers. The algorithm will have checks to determine if successive approximations are converging toward a solution. If the convergence criteria are met, the iterations cease; if not, iterations continue. Programming these decisions requires more control structures.

4.2 CONTROL STRUCTURES AND STRUCTURED PROGRAMMING

Control structures are particularly helpful in creating well-structured computer programs. On the other hand, poorly designed and implemented control structures can obscure program logic, making it difficult for anyone other than the author to understand or modify the program. The control structures included in FORTRAN 77 have made it much easier for the programmer to implement structured programming.

Sequential control is the default control. It executes the next instruction on completion of the present instruction. This is the type of control used in the programs of Chapter 3. Although sequential control is an essential ingredient of structured programming, it is not sufficient for producing programs that can solve most realistic problems. The selection and repetition control structures add the necessary flexibility to FORTRAN programming and will be the main emphasis in this chapter.

Selection control allows the construction of decision blocks in programs. If the decision block, or conditional statement, is true, then some sequence of statements is executed. If the conditional statement is false, an alternative sequence of statements is executed. These selection control structures can be combined to form almost any logical decision structure desired.

Repetition control repeats a certain sequence of instructions a number of times. The number of times either can be absolute or can be based on some condition being true.

4.3 SELECTION CONTROL

Selection control allows the computer to decide which program steps are to be performed next, based on the outcome of a programmed test. The test might consist of a question such as, "Is X greater than 5.0?" If the value of X is greater than 5.0, one set of program steps is carried out. If the value of X is less than or equal to 5.0, a second set of program steps is carried out. By combining these tests properly, any number of alternative sets of program steps can be accommodated. The next section will show you how to construct the logical tests involved in the question, and succeeding sections will show how to use those logical tests in control structures to select the desired set of program steps.

4.3.1 Relational Operators and Logical Expressions

Relational operators can be used to form logical expressions that evaluate to a condition of either true or false.

4.3.1.1 Relational Operators

Relational operators form the heart of the question to be asked in selecting the next set of statements to be executed. In "Is X equal to 5.0?" the relational operator is "equal to." You are testing X and 5.0 based on what the relational operator specifies.

Relational operators in FORTRAN have the following forms:

Relational Operator	Meaning
.EQ.	Equal to
.NE.	Not equal to
.LT.	Less than
.GT.	Greater than
.LE.	Less than or equal to
.GE.	Greater than or equal to

The periods before and after the relational operator are required in FORTRAN programming.

4.3.1.2 Logical Expressions

Logical expressions consist of two FORTRAN expressions linked by a relational operator.

```
       FORTRAN    Relational    FORTRAN
     (Expression  .Operator.   Expression)
```

Each of the expressions is evaluated, and then they are compared using the relational operator. If the relational operator is satisfied, then the logical expression evaluates to 'true'. If the results of the two expressions do not satisfy the relational operator (the condition of the relational operator is not met), then the logical expression evaluates to 'false'. These logical expressions can be used to form the decision blocks of selection control structures.

Real, integer, character, double precision, and complex expressions, variables, or constants can be compared in logical expressions. However, since each type of data is stored differently, the two FORTRAN expressions must be of the same type for proper comparisons to be made. Even integer and real values are stored differently in the computer, and an expression that evaluates to an integer value should not be compared with an expression that evaluates to a real value in a logical expression.

4.3.1.3 Compound Logical Expressions

Compound logical expressions make multiple comparisons among two or more complete logical expressions, linked by one of the connectors .OR. or .AND.

```
(logical expression.AND.logical expression)
(logical expression.OR.logical expression)
```

and evaluate to either 'true' or 'false' based on the compound comparison. For example, if DIA is the diameter of a pipe, in inches, the compound expression

```
(DIA.GE.6.0.AND.DIA.LE.36.0)
```

would evaluate to true if the diameter of the pipe is between 6 and 36 in., while

```
(DIA.LT.6.0.OR.DIA.GT.36.0)
```

would evaluate to true if the diameter is either less than 6 in. or more than 36 in. With the .AND. connector, the compound logical expression is true only if both of the connected logical expressions are true, and with the .OR. connector, the compound logical expression is true if either or both of the connected logical expressions are true. As with the relational operators, the periods before and after the connector are part of the connector and must be included. Parentheses and/or spaces can be used to reduce the confusion as to what might be a decimal point or part of a relational operator or connector. Adding a zero after the decimal point can also help. For example, if the expression above had been written

```
(DIA.LT.6..OR.DIA.GT.36.)
```

the decimal point and period in succession between 6 and OR are confusing. The original way of writing the expression is a better presentation. If you prefer, you could add parentheses and spaces

```
( (DIA.LT.6.0) .OR. (DIA.GT.36.0) )
```

to further clarify the meaning. More than two logical expressions can be connected together, but you have to be careful of the hierarchy of evaluation, which will be discussed after the next section on negation.

4.3.1.4 Negation

The negation operator, .NOT., changes the result of a logical expression to the opposite case. The expression

```
(.NOT.(DIA.GE.6.0))
```

would evaluate to 'false' if the diameter is greater than or equal to 6.0 in. The negation operator has the effect of changing the logical expression to

$$(DIA.LT.6.0)$$

In most cases the second form would be better, since it produces a more understandable program.

4.3.1.5 Hierarchy

As is the case with arithmetic expressions, logical expressions, or compound logical expressions, have a hierarchy which specifies the order of operations in determining whether the whole expression evaluates to 'true' or 'false'. All arithmetic expressions are evaluated before any logical expressions. Parentheses within logical expressions have the same effect as parentheses in arithmetic expressions. Anything within the parentheses is evaluated first. Precedence for logical operators starts with the relational operators (.EQ., .NE., .LT., .GT., .LE., and .GE.) from left to right. Next in order of precedence is the negation operator .NOT., followed by the connector .AND., which is followed by the connector .OR. at the bottom of the hierarchy. Within the same priority, precedence is from left to right.

Parentheses can be used to control the proper precedence in compound logical expressions. For example, the compound logical expression

```
.NOT.(YMAX.GT.(0.0).AND.YMIN.GT.(0.0).OR.YMAX.LT.(0.0).AND.YMIN.LT.(0.0))
```

was developed to determine if a function goes through zero in an interval. The compound logical expression within the outside parentheses evaluates to 'true' if the function does not pass through zero in the interval. The .NOT. operator inverts this so the total expression will evaluate to 'true' when the function does pass through zero. The outside parentheses are essential. Otherwise the .NOT. operator would only negate the result of the YMAX.GT.(0.0) logical expression, since .NOT. has precedence over .AND.. In the expressions, YMAX is the maximum value of the function in the interval, and YMIN is the minimum value. First the four simple logical expressions are evaluated. Then the first two logical expressions are compared to see if both YMAX and YMIN are positive, and the second two compared to see if both YMAX and YMIN are negative. If either of these compound logical expressions is true, the function does not pass through zero, and the .OR. connector will cause the compound expression to evaluate to 'true'. The .NOT. operator will change this to 'false'.

After studying this expression, you might think that there could be a simpler way to express this logical test. Indeed, one which does not require use of the .NOT. operator is considerably less confusing. The compound logical expression,

```
(YMAX.GT.(0.0).AND.YMIN.LT.(0.0))
```

will evaluate to true if the function does pass through zero. Simplifications such as these produce easily understood and well-structured programs. The best programming practice is to use the least-complicated structures that will get the job done correctly.

4.3.1.6 Danger of .EQ. and .NE. Operators

Use of .EQ. and .NE. relational operators with real expressions can cause problems because of the inexact storage of decimal values in the computer. Unless the results of the expressions are identical out to the last significant digit stored, the computer will treat them as being unequal. For example, if 2.999999 is stored in location X, the logical expression

$$(X.EQ.3.0)$$

would evaluate to 'false' even though for all practical purposes the stored values are the same. If you know how much error you can tolerate in the comparison, you can avoid the problem by constructing the logical test as

$$(ABS(X-3.0).LT.(0.0001))$$

which will evaluate to 'true' if the values are within 0.0001 of each other. The absolute value function is necessary because the error could be on either side of 3.0.

4.3.1.7 Character Variable and Constant Comparisons

Character variables can be compared with character constants or other character variables through use of logical expressions. ASCII codes (Appendix B) are used to make a character-by-character comparison from left to right. Because ASCII codes are used, capital letters are treated differently from lowercase ones. Character strings do not have to be the same length to be compared. FORTRAN adds blanks to the end of the shorter string before making the comparison. For two strings to be equal, all characters in corresponding positions between the strings have to be identical, including case for letters. When ASCII codes are used for comparison, numbers are given first precedence. Their ASCII codes appear before the codes for letters in Table B.1. Capital letters have second precedence, and lowercase letters third precedence. Various symbols are interspersed between each of these major categories. Thus,

Jones < jones

Anderson < Anderson

Addison < Altman

MATTHEW < THOMAS

Matthew < THOMAS

If several character strings that contain only letters of the same case are compared and sorted according to their character 'values', the sorted set of character strings will appear in alphabetical (lexicographic) order. Character numbers (numbers stored as characters, not values) cannot be compared with real or integer numeric values.

4.3.2 Logical IF Statement

The logical IF statement is a powerful control statement. It was the primary selection control structure in older versions of FORTRAN. In FORTRAN 77, it has been somewhat supplanted by the IF THEN ELSE structures that we will study next. However, the logical IF statement still has an important place in structured programming. It has the form

```
IF(logical expression) executable FORTRAN statement
```

in which the *logical expression* can be any of the logic structures that we discussed in the previous section, and the *executable FORTRAN statement* is any executable FORTRAN statement other than a DO statement or another IF statement. If the logical expression evaluates to 'false', control shifts directly to the next statement below the IF. If the logical expression evaluates to 'true', then the statement to the right is executed. Unless the statement to the right transfers control to another part of the program, which is not a good practice in structured programming, program control will transfer to the next statement below the IF after the statement to the right is executed. Typically the statement to the right is a computation or input/output statement.

As an example, the logical IF statement

```
IF(X.GT.XMAX) XMAX=X
```

is designed to test whether the present value of X is greater than the previous maximum value of X, stored in XMAX. If the present value is greater, the statement to the right is executed, and the new maximum value is stored in XMAX.

4.3.3 IF THEN ELSE Structure

The inclusion of the IF THEN ELSE structure in FORTRAN 77 has facilitated the development of structured FORTRAN programs. The IF THEN ELSE structure makes it possible to execute several statements if the logical test is true, or several different statements if the logical test is false. This is a distinct advantage over the single statement executed by the logical IF when the logical expression evaluates to 'true'. There are several different forms of the IF THEN ELSE. Each will be discussed below.

4.3.3.1 IF THEN

The simplest form of the IF THEN ELSE is the block IF, or the IF THEN structure. This form will execute a number of statements if the logical test is true. If the logical test is false, control drops through to the next statement after the end of the structure. The structure has the general form,

```
IF(logical expression) THEN
        .
      Block 1
        .
END IF
```

where *Block 1* represents any series of FORTRAN statements. The flowchart of Figure 4.1 shows the logic of the IF THEN structure.

The series of statements given below can be used to check each element of a one-dimensional array to see if the stored value is larger than XMAX. In addition to storing the new value in XMAX if it is bigger than the present value, the structure will also store the subscript number of the array element that contains the largest value in variable location IMAX. If X(I) is not greater than XMAX, program execution goes directly to the statement after the ENDIF.

```
IF(X(I).GT.XMAX) THEN
      XMAX=X(I)
      IMAX=I
END IF
```

FIGURE 4.1   Flowchart depiction of the IF THEN structure.

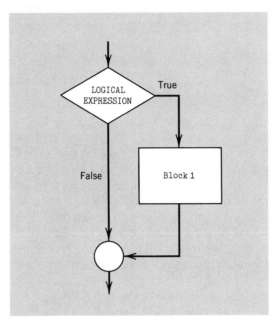

Note that the statements within the IF THEN structure are indented. This assists in identifying the different levels of logic used in a program and is especially important when using some of the more complex structures that can be formed with the IF THEN ELSE. Indentation is another technique that helps to add structure to the program.

Use of an IF THEN structure to sort data will be illustrated in Example 4.10. When you are sorting values in an array into ascending or descending order, it is necessary to switch values in adjacent storage locations of the array. To switch the values you have to temporarily store the first value in an auxiliary location. The second value is then assigned to the first storage location, and the temporarily stored value is assigned to the second storage location. If I is the subscript index and X is the array being sorted, the series of FORTRAN statements would be

```
IF (X(I).GT.X(I+1)) THEN
    TEMP = X(I)
    X(I) = X(I+1)
    X(I+1) = TEMP
END IF
```

This structure would be incorporated into other control structures to accomplish the sorting of values.

4.3.3.2 IF THEN ELSE

The IF THEN ELSE structure allows the choice of two alternate paths after evaluating the logical expression. One path is followed if the logical expression evaluates to 'true', and another path is followed if the logical expression evaluates to 'false'. The structure has the form

```
IF(logical expression) THEN
      .
      Block 1
      .
ELSE
      .
      Block 2
      .
END IF
```

The statements of *Block 1* are executed if the logical expression is true, and the statements of *Block 2* are executed if the logical expression is false. It is an either/or proposition. Both of the sets of statements cannot be executed in the same iteration of the structure. After execution of the selected block, program control normally shifts to the next statement after the END IF statement. You can include statements in the structure that shift control to another part of the program. In structured programming it is best to avoid such shifts in control whenever possible. However, it is quite permissible to call other procedures, such as subprograms or functions, from within the IF THEN ELSE structure. Figure 4.2 gives a flowchart depiction of the IF THEN ELSE structure.

Example 4.1 shows the application of the IF THEN ELSE structure for calculating the energy loss in a circular pipeline based on the type of flow that will exist in the pipe. The type of flow can be predicted by the value of a parameter called the Reynolds number.

FIGURE 4.2   Flowchart depiction of IF THEN ELSE structure.

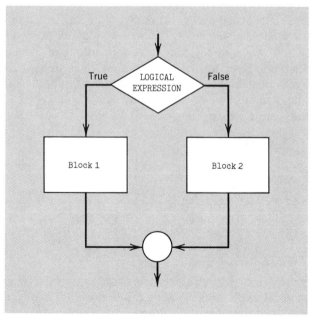

EXAMPLE 4.1 □ Energy Loss for Flow in a Circular Pipe

STATEMENT OF PROBLEM

Section 5.2 of Appendix G develops the theory of energy loss in a circular pipeline. Develop a program that will predict the energy loss for a circular pipe for either laminar or turbulent flow. Use the Chen equation to estimate the friction factor f for turbulent flow.

MATHEMATICAL DESCRIPTION

The formulas for Reynolds number and energy loss are developed as Equations G.24 and G.25 of Appendix G. For laminar flow $f = 64/Re$, and for turbulent flow, f will be computed using the Chen equation.

ALGORITHM DEVELOPMENT

Input/Output Design

Values for the following variables will have to be input: flow rate (Q), diameter (D), length of pipe (L), kinematic viscosity (ν), and roughness projection size (ε). A value for the acceleration due to gravity g will have to be specified or input. Since all the other data are input in English units, it is appropriate to specify the value of g through a FORTRAN statement. Also, π has to be given a value.

Computer Implementation

An IF THEN ELSE structure can be used to select the proper formula for estimating the friction factor f. A pseudocode representation of the program follows:

```
INPUT Q,D,L,E,NU
SPECIFY G AND PI
CALCULATE AREA   AREA = PI*D**2/4.
CALCULATE VELOCITY  VEL = Q/AREA
CALCULATE REYNOLDS NUMBER  RE = VEL*D/NU
IF RE <= 2000 THEN
     CALCULATE F FOR LAMINAR FLOW
     SPECIFY LABEL AS LAMINAR
ELSE
     CALCULATE F FOR TURBULENT FLOW
     SPECIFY LABEL AS TURBULENT
END IF
CALCULATE ENERGY LOSS  HL = F*L/D*VEL**2/(2*G)
OUTPUT INPUT DATA, RE, LABEL, F, AND HL
END
```

Note the use of a label to identify the type of flow. Figure 4.3 gives a flowchart for this program.

(EXAMPLE 4.1   □   Energy Loss for Flow in a Circular Pipe   □   Continued)

FIGURE 4.3   Flowchart for energy loss in circular pipe.

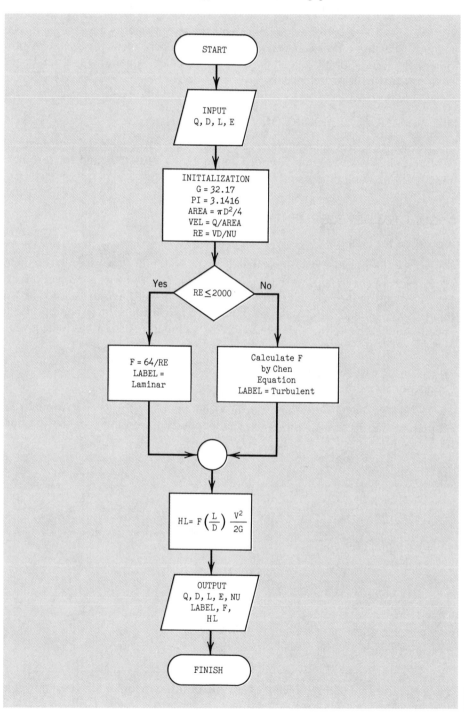

PROGRAM DEVELOPMENT

```
      PROGRAM HEADLOSS
************************************************************************
*     Calculating energy loss in a circular pipe               *
*     Developed by:  T. K. Jewell      November 1987           *
************************************************************************
C
C   D = Diameter of pipe, ft
C   E = Average size of roughness projections, ft
C   AREA = Area of pipe, sq ft
C   VEL = Velocity of flow, ft/s
C   NU = Kinematic viscosity, ft**2/s
C   Q = Volumetric flow rate, ft**3/s
C   L = Length of pipe, ft
C   F = Friction factor
C   HL = Energy loss, feet
C
      REAL D, AREA, VEL, NU, Q, L, E, RE, F, HL, G, PI
      CHARACTER LABEL*9
      INTEGER R,W
      PARAMETER (R=5,W=6)
      PRINT *, 'Input flow rate, cubic feet per second  '
      READ (R,*) Q
      PRINT *, 'Input diameter, feet  '
      READ (R,*) D
      PRINT *, 'Input length of pipe, feet  '
      READ (R,*) L
      PRINT *, 'Input size of roughness projections, feet  '
      READ (R,*) E
      PRINT *, 'Input kinematic viscosity, feet**2/s  '
      READ (R,*) NU
C    Initial calculations and specifications
      G = 32.17
      PI = 3.1416
      AREA = PI*D**2/4.
      VEL = Q/AREA
      RE = VEL*D/NU
C    Calculating f, based on Re
      IF (RE.LE.2000.) THEN
          F = 64./RE
          LABEL = 'LAMINAR'
      ELSE
          A1 = 5.8506/RE**0.8981
          A2 = 1./2.8257*(E/D)**1.1098
          A = A2 + A1
          F1 = E/(3.7065*D) - 5.0452/RE*LOG10(A)
          F = (-2.01*LOG10(F1))**(-2)
          LABEL = 'TURBULENT'
      END IF
      HL = F*L/D*VEL**2/(2.*G)
C  Output appropriate data
      WRITE (W,*)
      WRITE (W,*) 'Flow rate = ',Q,' ft**3/s'
      WRITE (W,*) 'Pipe diameter = ',D,' feet'
      WRITE (W,*) 'Length of pipe = ',L,' feet'
      WRITE (W,*) 'Size of roughness projections = ',E,' feet'
      WRITE (W,*) 'Kinematic viscosity = ',NU,' ft**2/s'
      WRITE (W,*)
      WRITE (W,*) 'Flow is ',LABEL,', the friction factor is ',F
      WRITE (W,*) ' and the energy loss is ',HL,' feet'
      END
```

(EXAMPLE 4.1 □   Energy Loss for Flow in a Circular Pipe □   Continued)

PROGRAM TESTING

Use a diameter of 1.0 ft, a flow rate of 1.57 ft³/s, a length of pipe of 100 ft, a kinematic viscosity of 2.0×10^{-5} ft²/s, and a roughness size of 0.00085 ft to test the program. This input produces turbulent flow with a Reynolds number of approximately 1×10^5. The friction factor f is approximately 0.0215, and the energy loss for a 100-foot section of pipe is 0.134 feet. You can test the laminar flow portion of the program by adjusting the kinematic viscosity enough to bring the Reynolds number below 2,000. For this case, make sure that f is being calculated as $64/Re$.

4.3.3.3 Cascaded IF THEN ELSE

The cascaded IF THEN ELSE is often referred to as the case structure, since it provides a means of choosing among several alternatives, or cases. It has the general form

```
IF(logical expression 1) THEN
       .
       Block 1
       .
ELSE IF(logical expression 2) THEN
       .
       Block 2
       .
ELSE IF(logical expression 3) THEN
       .
       Block 3
       .
ELSE
       .
       Block 4
       .
END IF
```

Any number of blocks can be added, depending on how many alternatives need to be examined. However, only one of the blocks of the structure will be executed. If the first logical expression is true, then *Block 1* is executed. If it is false, the second logical expression is checked. If the second logical expression is true, *Block 2* is executed. The pattern is continued until one of the logical expressions evaluates to 'true' or the final ELSE is reached. The block after the ELSE, in this case *Block 4*, will be executed only if all the logical expressions evaluate to 'false'. Once any of the blocks is executed, control shifts to the statement after the END IF, unless control is shifted elsewhere by the executed block. Figure 4.4 shows a flowchart depiction of the cascaded IF THEN ELSE.

The order of conditions can be important if the choice of alternatives is to be based on a range of values for a particular variable. For example, assume that you want to select your course of action based on the value of X falling within certain ranges of values. If you plan on using .LE. relational operators, the ranges will have to be in ascending order. If you want to execute Block 1 for values of X between minus infinity and 4.0, *Block 2* for values of X greater than 4.0 but less than 8.0, and *Block 3* for any value of X greater than or equal to 8.0, the cascaded

IF THEN ELSE

```
IF(X.LE.4.0) THEN
     Block 1
ELSE IF(X.LT.8.0) THEN
     Block 2
ELSE
     Block 3
END IF
```

FIGURE 4.4 Flowchart depiction of the cascaded IF THEN ELSE structure.

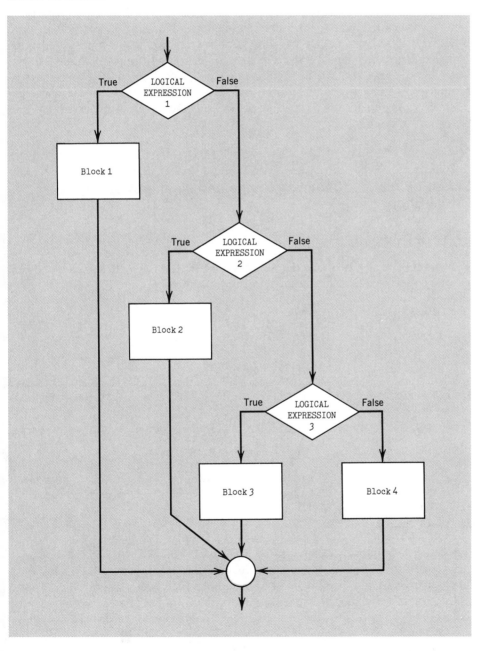

will accomplish the task. However, if the first two logical expressions are reversed, any value of *X* less than 8.0 will make the first logical expression true, and *Block 2* will never be executed. You could use compound logical expressions to more concisely define the range for each option. In the above segment, if the second logical expression is written as

$$(X.GT.(4.0).AND.X.LT.(8.0))$$

it does not matter what order the first two alternatives appear in.

Another use of the cascaded IF THEN ELSE is to choose a computational sequence based on some characteristic value or character string. Example 4.2 develops a program for computing the area of a circle, rectangle, or triangle based on the character string name of the shape. The structure is assumed to be part of a larger program, so no formal output is included. However, intermediate write statements have been included for the purpose of debugging the options. Also, an error message is included if none of the shapes matches the input character string. Note that the first letter of the shape name has been extracted to make the actual comparison. This helps prevent a failure to match the shape with the name because of typing errors. Also, the comparison accommodates either an upper- or a lowercase first letter.

EXAMPLE 4.2 □ Choosing Shape and Finding Area of that Shape

STATEMENT OF PROBLEM

A program is being developed that needs to use the area of a chosen shape. The option shapes are a circle, a triangle, or a rectangle.

MATHEMATICAL DESCRIPTION

Area of Circle

$$\text{Area} = \pi r^2 \tag{4.1}$$

Area of Triangle

$$\text{Area} = \tfrac{1}{2}bh \tag{4.2}$$

Area of Rectangle

$$\text{Area} = bh \tag{4.3}$$

ALGORITHM DEVELOPMENT

Input/Output Design

The shape description will have to be input first. This will require a character variable of length 9 to accommodate a rectangle. When inputting the character data, apostrophes must be placed at each end of the character string. Therefore, it is desirable to include the apostrophes in the prompt. These are embedded apostrophes, so two consecutive apostrophes will produce the desired output.

The cascaded IF THEN ELSE will branch to the appropriate block of statements. If the shape is a circle, you will have to input the radius. If the shape is a triangle or rectangle, you will have to input the base and height. If none of the shapes matches up, an error message will be output. Temporary output statements will be included to debug the options.

Computer Implementation

Figure 4.5 shows the flowchart for this program.

PROGRAM DEVELOPMENT

```
      PROGRAM SHAPES
*****************************************************************
*     Finding area of figure based on shape identification     *
*     Developed by:  T. K. Jewell          November, 1987       *
*****************************************************************
C
C   SHAPE = Shape identification character variable
C   AREA = Area of shape, ft**2
C   RADIUS = Radius of circle, ft
C   BASE = Base of rectangle or triangle, ft
C   HEIGHT = Height of rectangle or triangle, ft.
C
      REAL AREA,RADIUS,BASE,HEIGHT,PI
      CHARACTER SHAPE*9
      INTEGER R
      PARAMETER (R=5)
      PI = 3.1416
C
C  Input shape identifier and branch to appropriate computational
C    sequence.
C
      PRINT *, 'Input shape identifier:  ''CIRCLE'', ''TRIANGLE'','
      PRINT *, ' or ''RECTANGLE''  '
      READ (R,*) SHAPE
      IF (SHAPE(1:1).EQ.'C'.OR.SHAPE(1:1).EQ.'c') THEN
         PRINT *, 'Input radius of circle, feet'
         READ (R,*) RADIUS
         AREA = PI*RADIUS**2
         PRINT *, 'RADIUS and AREA for circle',RADIUS,AREA
      ELSE IF(SHAPE(1:1).EQ.'T'.OR.SHAPE(1:1).EQ.'t') THEN
         PRINT *, 'Input base and height of triangle, feet'
         READ (R,*) BASE,HEIGHT
         AREA = 0.5*BASE*HEIGHT
         PRINT *, 'BASE, HEIGHT, and AREA for triangle',
     1               BASE,HEIGHT,AREA
      ELSE IF(SHAPE(1:1).EQ.'R'.OR.SHAPE(1:1).EQ.'r') THEN
         PRINT *, 'Input base and height of rectangle, feet'
         READ (R,*) BASE,HEIGHT
         AREA = BASE*HEIGHT
         PRINT *, 'BASE, HEIGHT, and AREA for rectangle',
     1               BASE,HEIGHT,AREA
      ELSE
         PRINT *, 'The shape requested is not a circle, ',
     1               'triangle, or rectangle'
      END IF
      END
```

(EXAMPLE 4.2 □ Choosing Shape and Finding Area of that Shape □ Continued)

FIGURE 4.5 Flowchart for choice of shape and area problem.

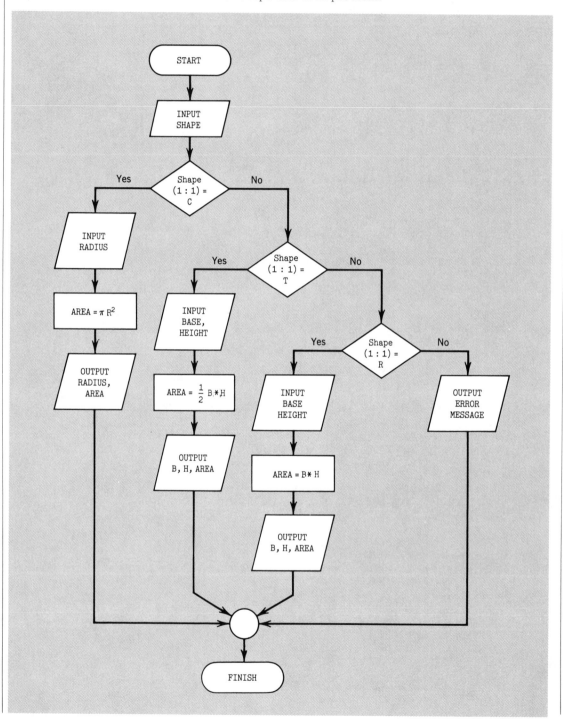

PROGRAM TESTING

Try specifying each shape, as well as a shape that is not programmed. Use a value of one for the input data to check out the computation.

4.3.3.4 Nested IF THEN ELSE

IF THEN ELSE structures can be nested to form detailed logical decision patterns. These structures are not complicated as long as you have a clear picture in your mind of the questions that should be asked and the decisions that the computer should be making. Otherwise, your program logical structure will be muddled, and your program may not run properly. The general form is

```
IF(logical expression 1) THEN
      .
      Block 1
      .
      IF(logical expression 1a) THEN
            .
            Block 1a
            .
      ELSE
            .
            Block 1b
            .
      END IF
      .
ELSE
      .
      Block 2
      .
END IF
```

The nested IF THEN ELSE (*Blocks 1a and 1b*) can be contained anywhere within *Block 1*. Figure 4.6 gives the flowchart depiction of this nested IF THEN ELSE structure.

Almost any combination of nestings is allowed, provided that each nested IF THEN ELSE is contained entirely within the same branch of the next higher IF THEN ELSE structure. In the preceding example, the nested IF THEN ELSE (*Blocks 1a* and *1b*) is contained within the THEN branch of the first-order IF THEN ELSE (*Block 1*). Additional levels of nesting are also permissible, as long as the rule on nesting is observed.

Example 4.3 shows the development of a program that requires a nested IF THEN ELSE. The program will calculate information about a particle trajectory problem. Particle trajectory theory is discussed in Section 4.2 of Appendix G. A stipulation for this application is that the initial angle of the trajectory must be between 30° and 75°. If the angle is within these bounds, the maximum height and distance of the trajectory will be calculated and printed out. If it is not, the program will print out an error message. The program will be set up to use either metric or English units. A nested IF THEN ELSE is required to output the results with the proper units.

FIGURE 4.6   Flowchart depiction of nested IF THEN ELSE structure.

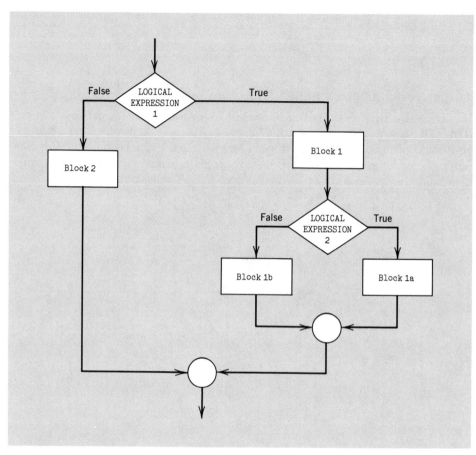

EXAMPLE 4.3 □ **Maximum Height and Distance for Particle Trajectory**

STATEMENT OF PROBLEM

Develop a program that will calculate the maximum height and maximum distance of a trajectory over flat terrain if the initial velocity and trajectory angle are known. Wind resistance can be ignored. Calculate and output the parameters only if the initial trajectory angle is between 30° and 75°. Output an error message if the angle is not within the specified bounds. The user should be able to specify metric or English units.

MATHEMATICAL DESCRIPTION

The equations for particle trajectory in Appendix G, Equations G.15, are rewritten here in FORTRAN form. The time it takes to reach the high point of the trajectory is given by Equation G.15 as

```
TYMAX = VO*SIN(THETA)/G
```

in which $V0$ = initial velocity, THETA = angle of trajectory, and G = local acceleration due to gravity. Using this time, the equation for the height of the trajectory at any point can be rewritten as

```
YMAX = VO*SIN(THETA)*TYMAX-0.5*G*TYMAX**2
```

which gives the maximum height of the trajectory. The maximum distance for the trajectory is

```
XMAX = VO*COS(THETA)*2.0*VO*SIN(THETA)/G
```

ALGORITHM DEVELOPMENT

Input/Output Design

The velocity and trajectory angle will be input through the terminal. The program will also request a value for the local acceleration due to gravity. If the angle is within the specified limits, output will consist of the input data and YMAX and XMAX with appropriate labels. Output for cases with angles beyond the limits will give the angle and an error message.

An IF THEN ELSE structure nested within the computation branch of the program (angle within limits) will output the proper units. Values of G less than 10.0 will be considered to be metric units, and values of G equal to or greater than 10.0 will signify English units.

Computer Implementation

A pseudocode representation of the program is given below. Figure 4.7 shows the flowchart.

```
INPUT VO AND THETA
INPUT G
CONVERT THETA TO RADIANS, CALL THE NEW VALUE ANGLE
IF THETA >= 30 AND <= 75 THEN
     CALCULATE TYMAX, USING ANGLE
     CALCULATE YMAX
     CALCULATE XMAX
     IF G < 10.0 THEN
          OUTPUT VO, THETA, YMAX, and XMAX WITH METRIC UNITS
     ELSE
          OUTPUT VO, THETA, YMAX, and XMAX WITH ENGLISH UNITS
     END IF
ELSE
     OUTPUT THETA WITH ERROR MESSAGE FOR OUT OF RANGE
END IF
END
```

PROGRAM DEVELOPMENT

```
      PROGRAM TRAJECTORY
*******************************************************************
*     Program for calculating maximum height and distance for a   *
*       particle trajectory. Wind resistance is ignored           *
*     Developed by: T. K. Jewell              November 1987        *
*******************************************************************
C
C     Variable definitions
C  VO = initial velocity
C  THETA = angle of trajectory in degrees
```

**(EXAMPLE 4.3 □ Maximum Height and Distance for
Particle Trajectory □ Continued)**

FIGURE 4.7 Flowchart for particle trajectory application.

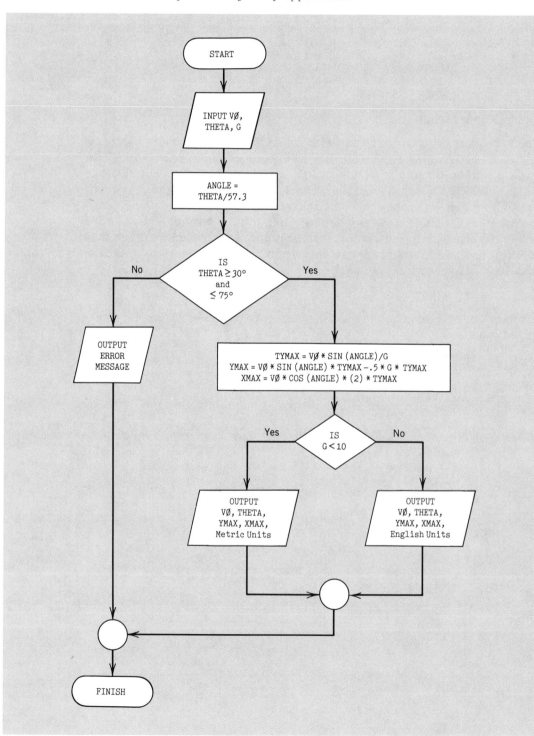

```
C   G = local acceleration due to gravity
C   ANGLE = angle of trajectory in radians
C   TYMAX = time, in seconds, to reach the top of the trajectory
C   YMAX = maximum height of trajectory, feet
C   XMAX = distance from launch to impact, feet
C
        REAL VO,THETA,G,ANGLE,TYMAX,YMAX,XMAX
        INTEGER R,W
        PARAMETER (R=5,W=6)
C
C   Input VO and THETA
        PRINT *, 'Input initial velocity (ft or m/s) and angle',
     1      (degrees) '
        READ (R,*) VO,THETA
C   Input G
        PRINT *, 'Input acceleration due to gravity, m/s**2 or ft/s**2 '
        READ (R,*) G
C   Specifying ANGLE
        ANGLE = THETA/57.3
C   Calculating YMAX and XMAX if THETA is within bounds
        IF (THETA.GE.(30.0).AND.THETA.LT.(75.0)) THEN
            TYMAX = VO*SIN(ANGLE)/G
            YMAX = VO*SIN(ANGLE)*TYMAX-0.5*G*TYMAX**2
            XMAX = VO*COS(ANGLE)*2.0*VO*SIN(ANGLE)/G
C   Output data and results
            WRITE (W,*)
            IF (G.LT.10.0) THEN
C   Output for metric units
                WRITE (W,*) 'For VO = ',VO,' m/s, and THETA = ',THETA,
     1                  ' degrees,'
                WRITE (W,*) 'Maximum height of trajectory =',YMAX,' m,',
     1                  ' and'
                WRITE (W,*) 'distance from launch to impact =',XMAX,
     1                  ' m.'
            ELSE
C   Output for English units
                WRITE (W,*) 'For VO = ',VO,' ft/s, and THETA = ',THETA,
     1                  ' degrees,'
                WRITE (W,*) 'Maximum height of trajectory =',YMAX,' ft,',
     1                  ' and'
                WRITE (W,*) 'distance from launch to impact =',XMAX,
     1                  ' ft.'
            END IF
        ELSE
            WRITE (R,*) 'The specified angle, ',THETA,' degrees,',
     1                  'is not within limits'
        END IF
        END
```

PROGRAM TESTING

For $G = 32.17$ ft/s^2, an initial angle of 45°, and an initial velocity of 100 ft/s, ANGLE = 0.785 radians, TYMAX = 2.20 s, YMAX = 77.7 ft, and XMAX = 310.8 ft. If the answers are not correct, insert WRITE statements to check the values of ANGLE and TYMAX, then carefully check the equation that is giving erroneous results. Also run the program inputting an angle less than 30° and more than 75° to test the angle limits. Use an input value for G of 9.81 m/s^2 to test the metric option.

Example 4.4 illustrates use of a nested IF THEN ELSE within a cascaded IF THEN ELSE to evaluate the roots of a quadratic equation,

$$AX^2 + BX + C = 0 \tag{4.4}$$

using the quadratic formula

$$X = \frac{-B \pm \sqrt{B^2 - 4AC}}{2A} \tag{4.5}$$

This example is an adaptation of the generic flowchart and pseudocode of Example 2.3. Example 2.3 covers development of the problem up through computer algorithm formulation, and should be reviewed. We will use the cascaded IF THEN ELSE to accommodate the three cases that might be encountered when applying the quadratic formula to a quadratic equation. If A is zero, the equation reduces to a linear equation, and there is only one root. If $B^2 - 4AC$ is negative, there is a pair of complex conjugate roots, while a zero or positive $B^2 - 4AC$ produces two real roots. A nested IF THEN ELSE is used within the third case to differentiate between positive and zero values for $B^2 - 4AC$. A positive $B^2 - 4AC$ gives two real and unequal roots. A zero value for $B^2 - 4AC$ indicates that the parabola represented by the quadratic equation just touches the x axis, and the roots are equal. For the complex case, the values are converted to complex variable notation for later use in the program. This algorithm could also be implemented using a four-tiered cascaded IF THEN ELSE.

EXAMPLE 4.4   □   **Quadratic Equation with Real or Complex Roots**

ALGORITHM DEVELOPMENT

Computer Implementation

Example 2.3 developed generic pseudocode for the quadratic equation algorithm. A more FORTRAN specific form is developed for this example. It also incorporates the additional case for A equal to zero and the nested IF THEN ELSE for real roots.

```
SPECIFY VARIABLES
      REAL A,B,C,X1,X2,RAD,IX1
      COMPLEX X1COMP,X2COMP
INPUT VALUES FOR A,B,C
OUTPUT A,B,C FOR ERROR CHECKING
CALCULATE RAD = B**2 - 4.*A*C
IF A.EQ.(0.0) THEN
      OUTPUT 'ONLY ONE ROOT'
      COMPUTE X=-C/B
      OUTPUT X
ELSE IF (RAD.LT.(0.0)) THEN
      CALCULATE COMPLEX ROOTS
      X1 = -B/(2.*A)
      IX1 = SQRT(ABS(RAD))/(2.*A)
      OUTPUT COMPLEX ROOTS
          X1+jIX1
          X1-jIX1
```

```
          CONVERT TO COMPLEX VARIABLES FOR LATER USE IN THE PROGRAM
          X1COMP = COMPLX(X1,IX1)
          X2COMP = CONJG(X1COMP)
ELSE
          CALCULATE REAL ROOTS
          RAD = SQRT(RAD)
          IF(RAD.GT.0.0) THEN
               X1 = (-B + RAD)/(2.*A)
               X2 = (-B - RAD)/(2.*A)
               OUTPUT REAL ROOTS, X1 AND X2
          ELSE
               ROOTS ARE EQUAL
               X1 = -B/(2.*A)
               X2 = X1
               OUTPUT X1 AND X2, AND MESSAGE THAT PARABOLA JUST TOUCHES X AXIS
          END IF
END IF
END
```

PROGRAM DEVELOPMENT

```
      PROGRAM QUADRATIC
*****************************************************************************
*     Program for calculating real or complex roots of a quadratic     *
*        equation                                                       *
*     Developed by: T. K. Jewell                      November 1987     *
*****************************************************************************
C  Variable definitions
C  A, B, and C = coefficients of quadratic eqn: A*X**2+B*X+C=0.0
C  X1 and X2 = real roots of quadratic equation, X1 also acts as
C     real portion of complex root.
C  RAD = B**2 - 4.*A*C
C  IX1 = imaginary portion of complex root
C  X1COMP and X2COMP are complex conjugate roots of the quadratic
C     equation
C
      REAL A,B,C,X1,X2,RAD,IX1
      COMPLEX X1COMP,X2COMP
      INTEGER R,W
      PARAMETER (R=5,W=6)
C  INPUT VALUES FOR A,B,C
      PRINT *, 'Input values for coefficients A, B, and C'
      READ (R,*) A,B,C
C  OUTPUT A,B,C FOR ERROR CHECKING
      WRITE (W,*) 'Input data for checking'
      WRITE (W,*) 'A = ',A,' B = ',B,' C = ',C
      RAD = B**2 - 4.*A*C
C  Start of cascaded IF THEN ELSE
C  Check for A=0.0
      IF (A.EQ.(0.0)) THEN
          WRITE (W,*)  'ONLY ONE ROOT'
          X=-C/B
          WRITE (W,*) 'X = ',X
C  Check for RAD < 0.0 (Imaginary roots)
      ELSE IF(RAD.LT.(0.0)) THEN
C  Calculate complex roots
          X1 = -B/(2.*A)
          IX1 = SQRT(ABS(RAD))/(2.*A)
          WRITE(W,*) 'COMPLEX ROOTS'
          WRITE(W,*) X1,' + j',IX1
          WRITE(W,*) X1,' - j',IX1
```

(EXAMPLE 4.4   □   Quadratic Equation with Real or Complex Roots   □   Continued)

```
C  Convert to complex variables for later use in the program
         X1COMP = CMPLX(X1,IX1)
         X2COMP = CONJG(X1COMP)
      ELSE
         RAD = SQRT(RAD)
C  Start of nested IF THEN ELSE
         IF (RAD.GT.(0.0)) THEN
C  Calculate real roots
            X1 = (-B + RAD)/(2.*A)
            X2 = (-B - RAD)/(2.*A)
C  Output real roots, X1 and X2
            WRITE(W,*) 'REAL ROOTS'
            WRITE(W,*) 'X1 = ',X1,' and X2 = ',X2
         ELSE
C  Calculate and output equal real roots.
            X1 = -B/(2.*A)
            X2 = X1
            WRITE(W,*) 'APEX OF PARABOLA JUST TOUCHES X AXIS'
            WRITE(W,*) 'TWO EQUAL ROOTS, X1 = X2 = ',X1
         END IF
      END IF
      END
```

PROGRAM TESTING

Develop values for A, B, and C that will test all four branches of the program. If the product $4AC$ is larger than B^2, complex roots will result.

4.3.4 Arithmetic IF Statement

The arithmetic IF statement is an older structure that has been replaced by the cascaded IF THEN ELSE for most applications. However, you may see the arithmetic IF used in programs developed before structured programming became the norm, so it is important to be able to recognize it. The arithmetic IF has the form

```
IF(FORTRAN expression) a,b,c
```

in which FORTRAN *expression* = any legitimate FORTRAN expression, and a, b, and c are statement labels.

The FORTRAN expression is evaluated, and if it is less than zero, program control shifts to statement a; if it is equal to zero, control shifts to statement b; and if it is greater than zero, control shifts to statement c. For example,

```
IF(X-6.) 100,200,300
```

will shift control to statement 100 when X is less than 6, will shift control to statement 200 when X equals 6, and will shift control to statement 300 when X is greater than 6. If two of the labels are the same, control will shift to that statement if either of the conditions is met. The arithmetic IF is a three-branch decision structure, as is shown in Figure 4.8.

An alternative form is

```
IF(FORTRAN expression) a,b
```

FIGURE 4.8   Flowchart depiction of arithmetic IF statement.

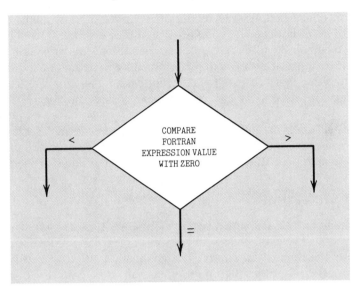

In this form, control shifts to statement *a* if the *FORTRAN expression* does not evaluate to zero, and shifts to *b* if the expression does evaluate to zero.

4.3.5 Computed GO TO Statement

The computed GO TO is another structure that has been superseded in structured programming. It will shift control to one of several statements depending on the value of an integer index expression or variable. The computed GO TO has the form

```
GO TO (n1,n2,n3, . . . , nr),integer expression
```

Values for the *integer expression* can vary from 1 to *r*. If the integer expression equals 1, control shifts to statement n1, if it equals 2, control shifts to statement n2, and so on. If the integer expression has a value less than 1 or larger than *r*, control shifts to the next statement after the computed GO TO. For example,

```
GO TO (100,200,300,400),I
```

would shift control to statement 100 if $I = 1$, to statement 200 if $I = 2$, to statement 300 if $I = 3$, and to statement 400 if $I = 4$. If I is less than 1 or greater than 4, the next statement in the program will be executed. The computed GO TO acts somewhat like the cascaded IF THEN ELSE, except that it does not have the built-in ability to bypass alternative choices en route to the end of the structure.

Both of the described supplementary structures require use of the GO TO statement to complete their logical structure. The GO TO statement will be discussed in the next section on looping structures. Its use should be limited to very specific applications in which it creates conditional loops. Otherwise, use of the GO TO will obscure the structure of a program. Both the arithmetic IF and computed GO TO should be avoided when you are developing structured programs.

4.4 REPETITION CONTROL

Repetition control allows the repetition of a set of statements either a set number of times or until some condition is met. Numerous problems could not be adequately modeled by a FORTRAN program without this provision. For example, repetition control allows you to repeat an entire program or a modular portion of a program a number of times for different cases. You can also use it to keep track of subscript values for arrays or to sum a large number of values for a variable.

First we will look at repetition structures that are normally executed a set number of times, and then at repetition structures that are executed until a logical expression (conditional statement) changes from true to false or vice versa.

4.4.1 DO Statement

The DO statement is an absolute repetition structure. It is best suited for counting-type loops that will be executed a set number of times. The number of times can be a variable that is input by the user. The counter for the loop can be used as a subscript index, or for any other appropriate purpose within the loop.

4.4.1.1 Form

The general form of the DO statement is

```
DO k index=initial,limit.increment

    FORTRAN statements

k   FORTRAN statement
```

The parameter k is a statement label that indicates where the end of the loop is located. The FORTRAN statements between the DO statement and statement k will be repeated a number of times. *Index = initial, limit, increment* are the parameters that control the number of times the loop is executed. *Index* is the index variable for the loop. When the DO statement is encountered in the execution process, *index* is set equal to the value assigned to the parameter *initial*. Before the loop is executed, the value in *index* is compared with the value in *limit*. If *index* is less than or equal to *limit*, the statements within the loop are executed. If *index* is greater than *limit*, control transfers to the next statement after statement k. When the loop is executed and statement k is reached, control shifts back to the DO statement and the value in *increment* is added to the value in *index*. *Index* is again compared with *limit*. It should be noted that *index* will have a value larger than *limit* when the repetition of the loop finishes. Also, if *index* is initially larger than *limit* (*initial > limit*), the loop will not be executed even once.

A flowchart element depicting the DO loop structure is shown in Figure 4.9. This is similar to the generic structure for a loop shown in Figure 2.3a.

Spaces within FORTRAN statements are optional, but with the DO structure it is important to include spaces to avoid confusion for readers of the program. For example,

```
DO100I=1,32,2
```

is actually a DO statement, but the DO100I could be misinterpreted as a variable name if the spaces are not included.

FIGURE 4.9 Flowchart depiction of DO structure.

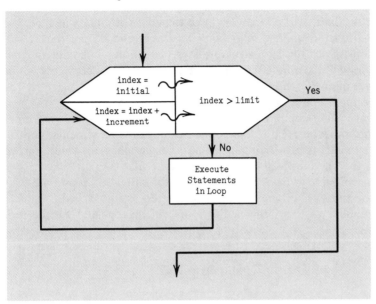

```
DO 100 I=1,32,2
```

is much easier to identify as a DO structure.

It is also permissible, but optional, to put a comma after the loop end statement identification number, which would make the previous DO statement appear as

```
DO 100,I=1,32,2
```

4.4.1.2 CONTINUE Statement

The final statement of the DO loop can be any executable FORTRAN statement except another DO, an IF statement, or any statement that causes transfer of control (e.g., END, RETURN, Arithmetic IF, or GO TO). However, it is much better to use a standard statement to indicate the end of a DO loop. The CONTINUE statement is specifically designed for this purpose. It is an executable statement that can be assigned an address label. If you use the CONTINUE statement to inidcate either the end of a DO loop or a branch destination point for a conditional loop, whenever it is encountered in a program you will know that it is associated with a looping structure. Throughout the text,

```
k   CONTINUE
```

will be used to indicate the end of a DO loop or conditional loop branch point. Thus

```
DO 100 I = 1,100,2
        .
    FORTRAN statements
        .
100   CONTINUE
```

would repeat the statements between the DO and 100 CONTINUE while *I* varies from 1 to 100 in increments of 2.

4.4.1.3 Rules for DO Structures

Index can be either an integer or real variable. However, the rest of the parameters (*initial, limit, and increment*) have to agree with *index* in type.

The parameters *initial, limit,* and *increment* can be constants, variables, or expressions. Once the loop structure starts executing, changing the values of the parameters will have no effect on the loop. However, statements within the loop should not alter the value of the index of the loop.

If *increment* is left out, it is assumed to have a value of 1.

A negative increment (decrement) can be used. If a negative increment is used, then *initial* must be greater than *limit*. For negative increments, control exits the loop when *index* is less than *limit*. If *initial* is greater than *limit*, and *increment* is positive, the DO loop will not be executed at all.

It is permissible to have negative values for either *initial* or *limit*, as long as the rest of the rules governing DO loop parameters are observed.

The number of times that a DO loop will be repeated can be found by the equation

$$Number\ of\ repetitions = INT\left(\frac{limit - initial}{increment}\right) + 1 \qquad (4.6)$$

The INT function is used to denote that the quotient in the brackets is treated as it would be by the INT intrinsic function in FORTRAN. Any decimal portion of the quotient is truncated. For example, if *initial* = 1, *limit* = 20, and *increment* = 3, the loop would be executed

$$INT((20-1)/3)+1 = 7$$

times. It would be executed with values of 1, 4, 7, 10, 13, 16, and 19 for *index*. However, after the seventh repetition, *index* would equal 22, which is greater than 20, so control would transfer to the next statement after statement *k*.

The following are all legal DO structures. If the default options for real and integer variables are observed, and $N = 50$, $M = 25$, $K = 2$, and $Y = 80.0$, each loop would be executed the indicated number of times.

DO Structure	Number of Loop Executions
DO 100 I=1,100	100
DO 100 I=1,100,3	34
DO 100 I=200,1,−1	200
DO 400 I=1,N	50
DO 400 I=M,N	26
DO 400 I=M,N,K	13
DO 400 I=−10,10,1	21
DO 200 X=0.0,10.0,.1	101
DO 200 X=0.0,10.0	11
DO 200 X=50.0,0.0,−1.25	41
DO 200 X=0.0,Y,2.5	33
DO 200 X=Y,0.0,−5.0	17

DO loops can be nested within each other. However, the inner DO loop must be entirely contained within the outer loop. Several levels of nesting can be constructed if the algorithm being implemented calls for them. Unnecessary levels of nesting should be avoided, because they will obscure the program logic. When DO loops are nested, the inner DO loop is finished before control shifts back to the outer DO loop. Each time the outer loop is executed, the inner DO loop will go through its complete cycle of iterations. For example, if the outer loop calls for 20 repetitions, and the inner loop calls for 10 repetitions, the inner loop will actually be executed 200 times before the outer loop is finished. The examples shown below illustrate the proper nesting of DO loops and indicate how many times each loop will be executed. In the first example, when $I = 1$, J will be increased from 1 to 40, then I will be increased to 2 and J increased once more from 1 to 40. Thus the inner loop is executed a total of 20×40 or 800 times. Similar analysis will determine how many times the other loops will be executed.

```
        DO 100 I=1,20          Outer Loop: 20 times
          DO 200 J=1,40
                .              Inner Loop: 800 times
                .
                .
200       CONTINUE
100 CONTINUE

        DO 100 I=31,60         Outer Loop: 30 times
          DO 100 J=1,20
                .              Inner Loop: 600 times
                .
100 CONTINUE

        DO 100 I=1,10          Outer Loop: 10 times
          DO 200 J=1,20        Middle Loop: 200 times
            DO 300 K=1,30
                .              Inner Loop: 6000 times
                .
300         CONTINUE
200       CONTINUE
100 CONTINUE

        DO 100 I=1,10          Outer Loop: 10 times
          DO 200 X=0.0,10.0,.2
                .              Inner Loop: 510 times
                .
200       CONTINUE
100 CONTINUE
```

Note that the same CONTINUE statement can be used as the ending point for more than one nested DO loop. Although this is proper FORTRAN syntax, I prefer not to use it in programs. Using separate ending statements helps to identify the loop nesting.

Shown below are two examples of improper nesting.

```
  +-----DO 100 I=1,10       +---------DO 100 I=1,15
+-|-----DO 200 J=1,10       | +-------DO 200 J=1,10
| |                         | | +-----DO 300 K=2,10
| |                         | | |
| +-100 CONTINUE            | | |
+---200 CONTINUE            | | +-300 CONTINUE
                            +-|---100 CONTINUE
                              +---200 CONTINUE
```

Drawing lines from the DO statement to the ending statement will help you detect improperly nested loops. If any of the lines cross, as in the above examples, the loops are improperly nested.

The same index variable cannot be used for a nested DO loop that is used for the nesting loop. By using the same variable, you would be redefining the index within the outer loop. In the example above, even if the loops were properly nested, the variable I could not have been used for the index in the 200 or 300 loops because it was used for the 100 loop. However, it is permissible to use the same index for several DO loops that are not nested with each other in a program. It is also permissible to transfer out of a DO loop. However, it is much better to use a conditional loop if you want to execute a loop until some condition holds. Branching out of a DO loop would require use of a GO TO. The GO TO statement and conditional loops will be discussed in a subsequent section. You should not transfer into a DO loop, as the DO parameters will not be properly initialized. A DO loop should always be entered through the DO statement.

Several examples will now be presented. These examples develop various applications of DO loops, as well as integrate previously discussed programming concepts and structures into realistic applications. Each example will be preceded by a short explanation of the concepts covered by the program developed.

4.4.1.4 Summation of Elements of Arrays

Example 4.5 uses DO loops to add like elements of two arrays and store the result in a third array. It then sums the squares of the elements of the third array. The one-dimensional arrays used in the example represent n-dimensional vectors.

Note the initialization of the summing variables before entering the summing loop. You could get away with not initializing values for this problem, because the summing sequences are used only once. However, for more complex programs the same summing sequence may be used many times. If the summing variable is not reinitialized each time the summing procedure is used, erroneous results will be generated because the summing variable will add the new values to its previous value.

EXAMPLE 4.5 □ **Summation and Length of**
n-Dimensional Vectors

STATEMENT OF PROBLEM

Develop a program to add two vectors together and find the magnitude (norm) of the resultant. The program should have the capability of working with up to 10-dimensional vectors.

MATHEMATICAL DESCRIPTION

Appendix F develops vector notation and vector algebra. For this program, you will need to use vector addition and the magnitude (norm) of a vector. Vector addition is defined as $c_i = a_i + b_i$, for $i = 1$ to n in $\mathbf{C} = \mathbf{A} + \mathbf{B}$, while the norm, or magnitude, is given by

$$|C| = \sqrt{\sum_{i=1}^{n} c_i^2} \tag{4.7}$$

ALGORITHM DEVELOPMENT

Input/Output Design

Vector arrays **A, B,** and **C** have to be defined as real arrays. The dimension should be as large as the largest vector we want the program to accommodate, which in this case is 10. We need to input the n components of vector arrays **A** and **B.** We can accomplish this by inputting a value for n and using a DO loop with an index that goes from 1 to n to enter a_i and b_i. The values for the index variable can also be used as the subscript index for the vector arrays. Since subscript indexes have to be integer variables, the DO loop index and parameters should be integer quantities.

Output should include n, the components of the vectors **A, B,** and **C,** and the norm of **C.** Any headings for output must be output before the output loop, to prevent the headings from being output n times.

Computer Implementation

All the input, computation, and output of arrays for this program could be done within a single loop. However, in most practical problems requiring arrays, the arrays will be used for several operations. It is better structured programming to split up the different operations into procedures. Each procedure can do a particular task. Procedures can often be used in several programs if they are written in a general manner. Procedures such as these can be saved as separate subprograms and accessed from within the present program as needed. Developing procedures into subprograms will be covered in Chapter 5. For this program, it is sufficient to split up the program into procedures for inputting arrays, computing the sum of two vectors, computing the norm of a vector, and outputting the arrays. This may seem somewhat cumbersome at this time. However, you will see its value when we start doing more complicated programs with procedures that are used several times. The procedures developed for this program are somewhat specific, because you have not yet learned about subprograms. Chapter 5 will show you how they can be generalized. Four separate DO loops will be used for this program, one for each of the procedures described above. Figure 4.10 shows the flowchart for the program.

PROGRAM DEVELOPMENT

```
************************************************************************
*    Program for adding two vectors and finding the norm of the       *
*       resultant                                                     *
*    Developed by:  T. K. Jewell            November, 1987            *
************************************************************************
C
C  VARIABLE DEFINITIONS
C  A and B = n-dimensional vectors to be added together
C  C = n-dimensional vector to store summation of A and B
C  N = dimension of vectors A, B, and C
C  SUMCSQ = summing variable for squared components of vector C
C  NORMC = magnitude (norm) of vector C
C
      REAL A(10),B(10),C(10),SUMCSQ,NORMC
      INTEGER I,N,R,W
      PARAMETER (R=5,W=6)
```

FIGURE 4.10 Flowchart for summation of vectors and finding length.

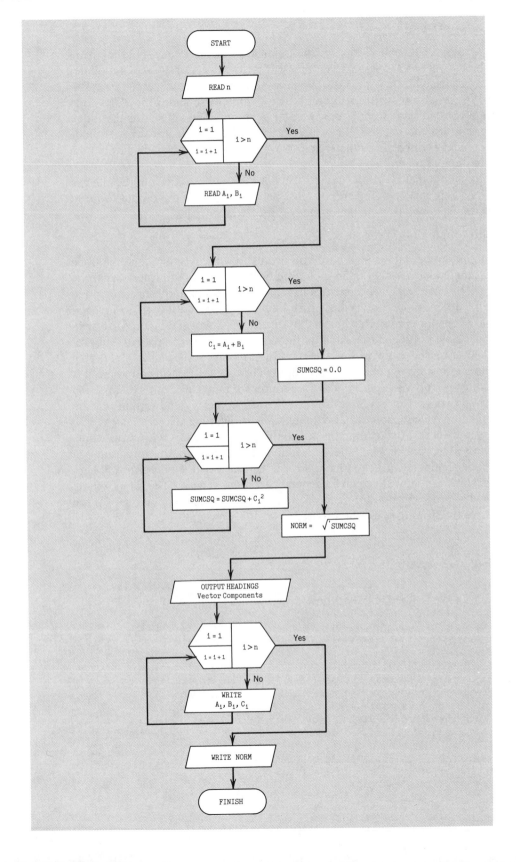

(EXAMPLE 4.5   ◻   Summation and Length of
***n*-Dimensional Vectors   ◻   Continued)**

```
C  Input of vectors A and B
      PRINT *, 'Input the dimension of the vectors '
      READ (R,*) N
      PRINT *, 'The vectors are ',n,' dimensional'
      DO 100, I=1,N
         PRINT *, 'Input component ',I,' of vectors A and B '
         READ (W,*) A(I),B(I)
  100 CONTINUE
C  Calculate C = A + B
      DO 200 I=1,N
         C(I) = A(I) + B(I)
  200 CONTINUE
C  Calculate magnitude (norm) of vector C
      SUMCSQ = 0.0
      DO 300 I=1,N
         SUMCSQ = SUMCSQ + C(I)**2
  300 CONTINUE
      NORMC = SQRT(SUMCSQ)
C    Output results
      WRITE (W,*) 'Components of vectors A, B, and C are:'
      WRITE (W,*)
      WRITE (W,*) '     Vector A        Vector B        Vector C'
      DO 400 I=1,N
         WRITE (W,*) A(I),B(I),C(I)
  400 CONTINUE
      WRITE (W,*)
      WRITE (W,*) 'The magnitude of vector C is: ',NORMC
      END
```

PROGRAM TESTING

For the vectors $\mathbf{A} = (3,6,4)$ and $\mathbf{B} = (1,3,5)$, $\mathbf{A} + \mathbf{B} = (4,9,9)$ and $|\mathbf{A} + \mathbf{B}| = 13.34$. For $\mathbf{A} = (2,1,5,7,9)$ and $\mathbf{B} = (8,-3,18,-6,-10)$, $\mathbf{A} + \mathbf{B} = (10,-2,23,1,-1)$ and $|\mathbf{A} + \mathbf{B}| = 25.20$.

Example 4.6 illustrates the calculation of the mean and standard deviation statistics for several data values without resorting to subscripted variables. Data input for this program will be from a data file. Only the summary statistics will be available for use later in the program. The original data will no longer be available. Note the use of the REAL function to convert the value of N to a real quantity for computation of the statistics.

EXAMPLE 4.6   ◻   Data Analysis: Mean and Standard Deviation

STATEMENT OF PROBLEM

Develop a program that will determine the mean and standard deviation of n data points.

(EXAMPLE 4.6 □   Data Analysis: Mean and Standard Deviation □   Continued)

MATHEMATICAL DESCRIPTION

Appendix H discusses the mean and standard deviation of a data set. The mean value is given by the equation

$$\text{Mean} = \bar{X} = \frac{\sum_{i=1}^{n} X_i}{n} \tag{4.8}$$

FIGURE 4.11   Flowchart for finding mean and standard deviation.

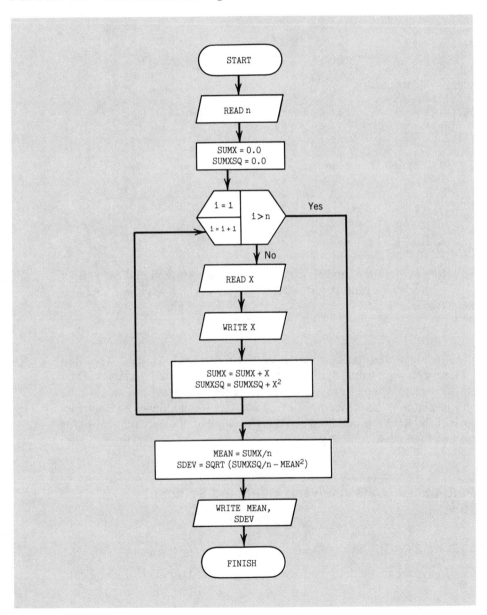

while the standard deviation is given by the equation

$$\text{Standard deviation} = S = \sqrt{\frac{\sum\limits_{i=1}^{n} X_i^2}{n} - \bar{X}^2} \tag{4.9}$$

ALGORITHM DEVELOPMENT

Input/Output Design

Prior to any input, the data file has to be opened. The amount of input will again be controlled by a limit parameter, the value for which will be input prior to start of data input. Since no subscripted variables are being used in this problem, all computations involving data values will have to be accomplished during the input loop. The equations for computing the mean and standard deviation allow us to do this. Input data checks will also have to be accomplished during the input loop. After completion of the loop, the mean and standard deviation can be computed and output.

Since input will be via a data file, we must also design the data file layout. The parameter n will be the first data value input, so it should appear on the first line of the data file. The values for the variable to be analyzed will be read in at the rate of one value per repetition of the loop. According to the rules for READ statements, a new line of data is accessed each time a READ statement is executed. Thus, the values to be analyzed must appear one per line below the value for n. Note that input prompts are not needed when data are coming from a data file.

Computer Implementation

Figure 4.11 shows the flowchart for Example 4.6. All data value computations are accomplished within the single loop, while the mean and standard deviation are computed after termination of the loop.

PROGRAM DEVELOPMENT

```
**********************************************************************
*     Mean and standard deviation of a set of data                  *
*     Developed by:  T. K. Jewell     November 1987                  *
**********************************************************************
C
C  Variable definitions
C  n = number of data values
C  X = individual data value
C  MEAN = mean value of data set
C  SUMX = summation variable for X
C  SUMXSQ = summation variable for X**2
C  SDEV = standard deviation of data set
C
      REAL X,MEAN,SDEV
      INTEGER N,FR,W
      PARAMETER (FR=4,W=6)
C  Input number of data points
      OPEN(UNIT=FR,FILE='EX4_6.DAT',STATUS='OLD')
      READ(FR,*) N
      WRITE(W,*) 'The number of data points is: ',N
```

(EXAMPLE 4.6 □ Data Analysis: Mean and Standard Deviation □ Continued)

```
C  Initialize summing variables
      SUMX = 0.0
      SUMXSQ = 0.0
C  Start of loop for data input and computations
      DO 200 I=1,N
         READ (FR,*) X
         WRITE (W,*) 'Data value no. ',I,' = ',X
         SUMX = SUMX + X
         SUMXSQ = SUMXSQ + X**2
  200 CONTINUE
C  Compute the mean and standard deviation
      MEAN = SUMX/REAL(N)
      SDEV = SQRT(SUMXSQ/REAL(N)-MEAN**2)
      WRITE(W,*)
      WRITE(W,*) 'Mean = ',MEAN,' and Standard Deviation = ',SDEV
      END
```

PROGRAM TESTING

For the program above, you have chosen the name EX4_6.DAT to identify the file name for the input. Therefore, you have to use your editor to create the following file named EX4_6.DAT.

```
10
3.5
4.8
5.65
3.33
6.27
7.5
8
6.778
1.11115
6.97
```

For this set of 10 data points, the mean = 5.391 and standard deviation = 2.074.

4.4.1.5 Search for Minimum and Maximum Values

The next example is the first of several search programs that you will either use or write. You want to determine the maximum and minimum values stored in a one-dimensional array, but you do not need to know which elements of the array contain the values. Search routines are one type of computer-assisted numerical method. We cannot find the minimum or maximum by merely substituting values into equations. The numerical technique for this example is a simple one. You will initialize the values for the maximum and minimum storage variables and successively compare each array value with the present maximum and minimum values. If the array value is less or more, it replaces the value in the appropriate storage location. Proper initialization is an essential part of any type of search routine if correct comparisons and replacements are to be made.

Also introduced in this example is the use of a variable to specify the input data file. This allows you to use any data file without altering and recompiling the program. Input for this problem comes from both the terminal (NAME of data file) and from the data file (all the rest of the data). Therefore, both R and FR appear in the parameter statement.

**EXAMPLE 4.7 □ Searching for Minimum and
Maximum Values in an Array**

STATEMENT OF PROBLEM

Find the minimum and maximum values stored in a one-dimensional array, $X(I)$. The array will have a maximum size of 100. However, only n of the elements of the array will have values stored in them. Therefore, the search must be limited to n elements to avoid including any empty elements.

ALGORITHM DEVELOPMENT

Input/Output Design

Control your input loop by the limit n. A variable NAME will be used to define the input data file. The data file will be set up the same as for Example 4.6.

Numerical Methods

Variables that will store the maximum and minimum values of the array have to be defined. These variables must be given an initial value so that there is something to check against when the loop for determining the maximum and minimum values starts. The values should not be left at zero, because if there are no values in the array that are less than or equal to zero, the minimum value will erroneously stay at zero. For array searches such as this, a convenient initialization value for the minimum and maximum variables is the value of the first element of the array. Once the minimum and maximum values are initialized, the values in the array are successively compared with the minimum and maximum. If a particular value is less than the minimum, it replaces the value in the minimum storage location. Similarly, values larger than the maximum replace the value in the maximum value storage location. After completion of the comparisons, the overall minimum and maximum values will be stored.

Computer Implementation

A separate input loop will be used. The actual search loop will go from 2 to n, because the first value has been used to initialize the minimum and maximum variables. Figure 4.12 shows the flowchart for this program.

PROGRAM DEVELOPMENT

```
**********************************************************************
*    Determining minimum and maximum value of elements of a        *
*       one-dimensional array                                      *
*    Developed by: T. K. Jewell             November 1987          *
**********************************************************************
C
C  Variable definitions
C  X(I) = array to be searched
C  XMAX = maximum value in array X
C  XMIN = minimum value in array X
C  N = size of array X
       REAL X(100),XMAX,XMIN
       INTEGER N,I,R,FR,W
       CHARACTER NAME*14
```

FIGURE 4.12 Flowchart for searching for minimum and maximum values in an array.

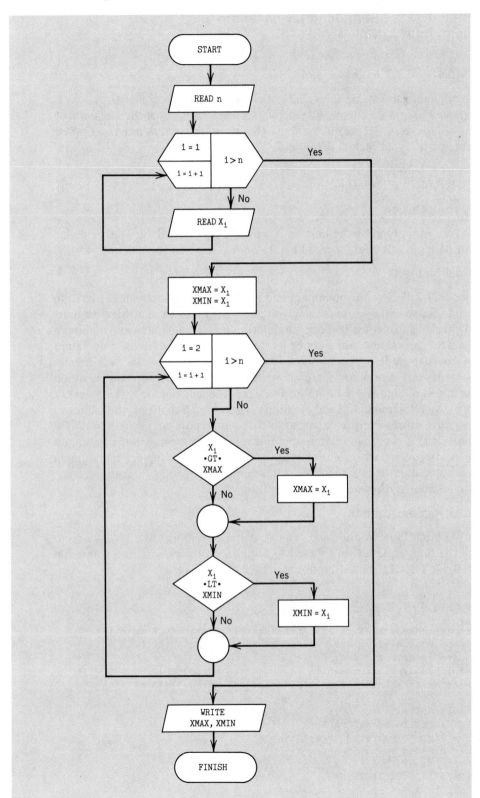

(EXAMPLE 4.7 □ Searching for Minimum and Maximum Values in an Array □ Continued)

```
      PARAMETER (R=5,FR=4,W=6)
      PRINT *, 'Input data file name '
      READ(R,*) NAME
C  Input array to be analyzed
      OPEN(UNIT=FR,FILE=NAME,STATUS='OLD')
      READ(FR,*) N
      DO 100 I=1,N
         READ(FR,*) X(I)
  100 CONTINUE
C  Initialize XMAX and XMIN
      XMAX = X(1)
      XMIN = X(1)
C  Search routine
      DO 200 I=2,N
         IF(X(I).GT.XMAX) XMAX = X(I)
         IF(X(I).LT.XMIN) XMIN = X(I)
  200 CONTINUE
C  Print results
      WRITE(W,*) 'Maximum value in array is ',XMAX
      WRITE(W,*) 'Minimum value in array is ',XMIN
      END
```

PROGRAM TESTING

Use the data file EX4_6.DAT from Example 4.6 to test this program. When you input the file name, you will have to enclose it in single quotes, 'EX4_6.DAT'. You could also develop a data file with all positive or all negative values to verify that the program operates correctly under those conditions.

4.4.1.6 Real Indexes and Parameters for DO

You can use real indexes and parameters for a DO loop to increase a real variable value by an increment a number of times. However, the use of real parameters and indexes may cause a loop to stop at an incorrect point. If you are using a real loop of DO 100 $X=1.0,3.0,0.01$, you would expect the loop to stop after executing the loop with $X=3.0$. However, if computer round-off causes the value of X to be slightly larger than 3.0 when the value should be exactly 3.0, the loop will stop one iteration too soon. This does not always happen, and it can be corrected by changing the *limit* of the DO loop slightly. Example 4.8 will develop a program to calculate a table of the areas of circles using real indexes and parameters for the DO loop.

When you are operating on an array containing real values, you are better off using integer indexes and parameters for the DO loops, because you need to use integer values for the array subscript indexes. Using integer indexes also avoids any problems with the loop not concluding at the proper time due to computer round-off error. If you have real values that need to be increased within the loop, you can initialize the values before starting the loop and add an incremental value each time through the loop. Storing the real values in an array is advantageous because it allows you to use them later in the program. Example 4.9 will modify the program of Example 4.8 to use subscripted variables with integer index and parameters for the DO loop.

EXAMPLE 4.8   □   Areas of Circles

STATEMENT OF PROBLEM

Develop a program that will calculate a table of the areas of circles with radii from 0.5 to 12.0 in. in increments of 0.5 in.

MATHEMATICAL DESCRIPTION

The equation for the area of a circle is area = πr^2.

ALGORITHM DEVELOPMENT

Input/Output Design

Although the problem description does not explicitly call for input data, it would be restrictive to specify the limits for the problem in the FORTRAN statements. As we described earlier, programs should be written in as general a form as possible. This makes it easier to use the program repeatedly, or to modify the procedures of the program for incorporation into other programs. Therefore, *initial, limit,* and *increment* will be treated as input variables. You can verify the correctness of the input data by examining the output table.

Output will be a table of values of radii and areas.

Computer Implementation

Figure 4.13 shows the flowchart for Example 4.8. Note that the output has to be in the same loop as the calculation because of the lack of subscripted variables.

PROGRAM DEVELOPMENT

```
*********************************************************************
*     Program for calculation of areas of circles                  *
*     Developed by:  T. K. Jewell            November 1987          *
*********************************************************************
C  Variable definitions
C  RADIUS = radius of circle (inches)
C  AREA = area of circle (inches**2)
C  INIT = starting value for radius
C  LIMIT = ending value for radius
C  DELT = increment of radius
C
      REAL RADIUS,AREA,INIT,LIMIT,DELT
      INTEGER R,W
      PARAMETER (R=5,W=6)
      PI = 3.1416
      PRINT *,'Input initial, final, and incremental radius values.
      READ (W,*) INIT, FINAL,DELT
C  Output headings for table
      WRITE(W,*) '        RADIUS        AREA'
      WRITE(W,*) '       (inches)     (sq ft)'
C  Computational loop
      DO 100 RADIUS=INIT,FINAL,DELT
         AREA = PI*RADIUS**2
         WRITE(W,*) RADIUS,AREA
  100 CONTINUE
      END
```

FIGURE 4.13   Flowchart for areas of circles.

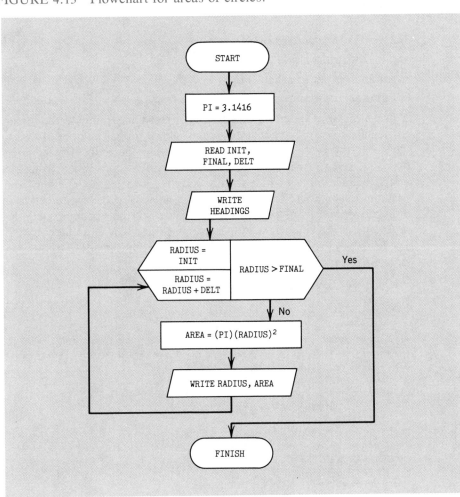

PROGRAM TESTING

Input the given values and check the table for correctness. The area for $r = 0.5$ in. should be 0.785 in.^{2} and for $r = 12.0$ in., the area is 452.4 in.^{2}.

The above example will now be revised to save the values generated in an array. Note that the user must specify the same data as in Example 4.8. The number of iterations for the DO loop is calculated within the program.

EXAMPLE 4.9 □ Areas of Circles Using Subscripted Variables

STATEMENT OF PROBLEM

This problem is the same as Example 4.8, except that subscripted variables will be used to store the values developed.

(EXAMPLE 4.9 □ Areas of Circles Using Subscripted Variables □ Continued)

ALGORITHM DEVELOPMENT

Input/Output Design

Input and output are the same as in Example 4.8.

Computer Implementation

You will have to specify the arrays to be used and compute the upper limit for the DO loop to ensure accomplishment of the proper number of iterations. The value for the radius will be increased within the loop. The generic intrinsic function ANINT will be used to make sure that conversion from a real to an integer value will not cause a truncation error in the upper limit. The ANINT function evaluates to the whole number nearest to the value of the argument. The conversion from real to integer value will be accomplished through the replacement statement. Note the error message if N is greater than 99. This is one less than the upper limit on the array size, because RADIUS($I+1$) is calculated within the loop.

PROGRAM DEVELOPMENT

```
**********************************************************************
*    Program for calculating the areas of circles and             *
*       saving the areas in an array.                             *
*    Developed by:  T. K. Jewell            November 1987          *
**********************************************************************
C  Variable definitions
C  RADIUS = radius of circle (inches)
C  AREA = area of circle (inches**2)
C  INIT = starting value for radius
C  FINAL = ending value for radius
C  DELT = increment of radius
C  N = number of iterations for the loop
      REAL RADIUS(100),AREA(100),INIT,LIMIT,DELT,FINAL
      INTEGER R,W,N,
      PARAMETER (R=5,W=6)
      PI = 3.1416
      PRINT *,'Input initial, final, and incremental radius values.
      READ (R,*) INIT,FINAL,DELT
C  Compute the number of iterations for the loop
      N = ANINT((FINAL-INIT)/DELT)+1
      IF(N.GT.99) PRINT *,'N is larger than allowed'
C  Output headings for table
      WRITE(W,*) '  RADIUS     AREA'
      WRITE(W,*) ' (inches)   (sq ft)'
C  Computational loop
      RADIUS(1)=INIT
      DO 100 I=1,N
      AREA(I) = PI*RADIUS(I)**2
      WRITE(W,*) RADIUS(I),AREA(I)
      RADIUS(I+1) = RADIUS(I) + DELT
  100 CONTINUE
      END
```

PROGRAM TESTING

Use the data listed in Example 4.8 to test this program.

4.4.1.7 Nesting of DO Loops

The rules for nesting DO loops have already been discussed. Nested DO loops are necessary for programs using two-dimensional or higher arrays. Each of the nested loops will control one of the subscript indexes. Input and output of two-dimensional arrays will be developed in Chapter 5. In this section, we will develop two examples that require the use of nested DO loops.

Many applications require the sorting of data into numerical or alphabetical order. Methods for sorting numeric or text data generally require multiple passes through the data. One loop will control the processing during a particular pass. This loop will be nested within another loop that controls the number of passes through the data. Another type of problem that will require nested loops is one in which you want to change the value of one variable a number of times while a second variable is held constant. After the changes in the first variable are completed, the second variable is to be increased and the variation in the first variable repeated. Nested DO loops will be necessary to accomplish this task.

We will now illustrate how to sort numeric data stored in a one-dimensional array into ascending order. Simple modifications to this program could cause it to

FIGURE 4.14   Schematic of bubble sort algorithm for descending order.

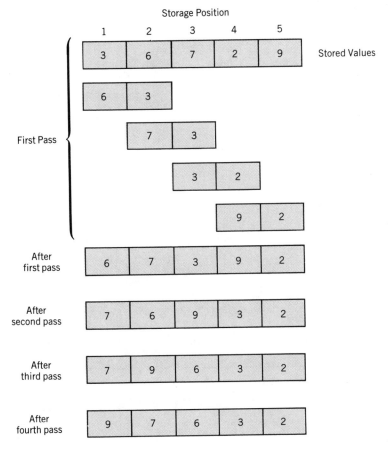

sort in descending order or to sort data alphabetically. The program could also be extended to sort multidimensional arrays. Several methods of sorting data are available. Two methods, the bubble and Shell sort, will be presented here. A program will be developed for the bubble sort. It is the easier of the two to understand and program. The Shell sort is a more efficient algorithm. Development of a program to implement the Shell sort algorithm is included as a chapter-end exercise.

In the bubble sort for ascending order, one element of the array is compared

FIGURE 4.15   Schematic of Shell sort algorithm for descending order.

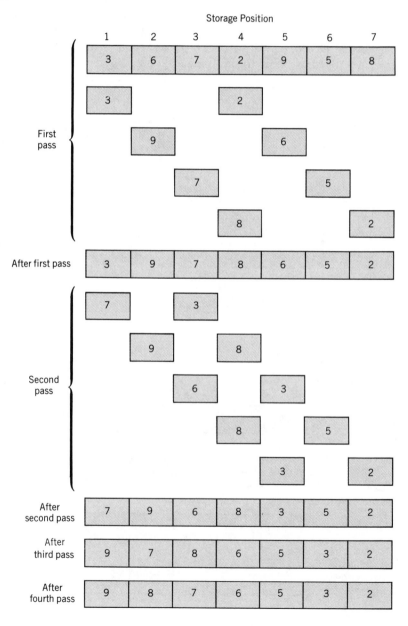

with the next. If the first is larger than the second, the values in the two storage locations are switched so the smaller value will be in the first storage location. This requires the temporary storage of one of the values during the switching process. Next the second and third elements are compared and, if necessary, are switched. The process is repeated until the next to last and last elements for the array are examined. A single pass through the array will move a smaller element only one position upward in the array. Thus, multiple passes are required to be certain that the minimum value rises to the top of the array. In the worst case, where the smallest value is initially in the last element of the array, $n - 1$ passes through an n-element array will be required. The movement of the smaller elements to the top of the array gives rise to the name bubble sort. Figure 4.14 shows a schematic of the process for sorting into descending order.

The Shell sort is named for its developer, Donald Shell. It is a more efficient variation of the bubble sort. Nonadjacent elements of the array are compared in a systematic manner. At the first stage, the first element of the array is compared with the element halfway through the array. If the first is greater than the element halfway through, the values are switched. In an array with 10 elements, an interval of five is used, and the first element would be compared with the sixth. Next, the second would be compared with the seventh, and so forth, until the fifth element is compared with the tenth. This level of detail is repeated until there is a pass with no switches. For an array with an odd number of elements, the interval would be the integer value of the number of elements divided by two. Thus, the interval would be five for an array of either 10 or 11 elements. After a pass with no switches, the difference between the elements compared is halved. For 10 elements, the new interval would be two. Passes are again made through the array until no switches occur. The interval continues to halve until it becomes one. At this point, the sorts become bubble sorts, but the array has already been rearranged in an approximate order. The bubble sort passes are also continued only until there are no switches. Figure 4.15 shows a schematic of the Shell sort algorithm for sorting in descending order. As arrays become larger, the Shell sort exhibits an increasing margin of efficiency over the bubble sort.

Example 4.10 will implement the bubble sort to sort an array into ascending order. Implementation of the Shell sort will be included as an exercise at the end of the chapter, as will the development of a program to accomplish sorts of either ascending or descending order.

EXAMPLE 4.10 □ Sorting Subscripted Array into Ascending Order by Bubble Sort Method

STATEMENT OF PROBLEM

Develop a program to sort data in an array into ascending order using the bubble sort algorithm.

ALGORITHM DEVELOPMENT

Input/Output Design

The size of the array n and the array to be sorted must be input. Note that the input procedure is identical to the input procedure for Example 4.7, demonstrat-

(EXAMPLE 4.10 □ Sorting Subscripted Array into Ascending Order by Bubble Sort Method □ Continued)

FIGURE 4.16 Flowchart for ascending bubble sort of subscripted array.

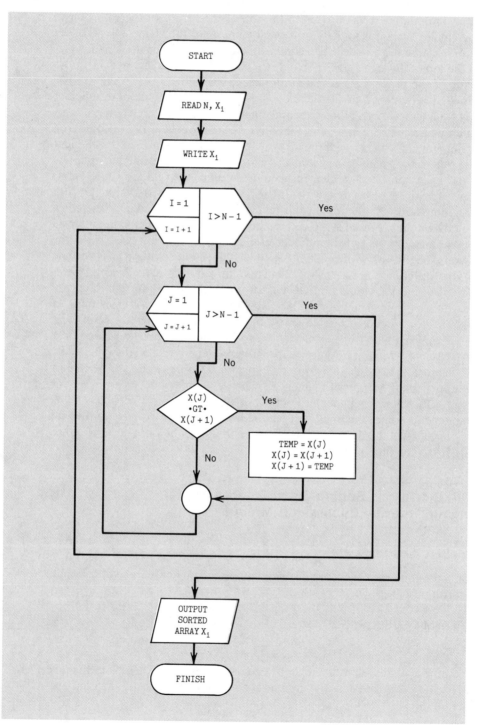

ing the value of being able to store and use procedures as subprograms. Output will include *n*, the unsorted array, and the sorted array. Since the original array is not saved, it will have to be output before the sorting starts. Intermediate output statements can be placed within the loops for sorting if the sort routine does not perform properly.

Computer Implementation

The inner loop has to be accomplished $n - 1$ times to ensure that each adjacent pair of elements is checked. The outer loop also has to be executed $n - 1$ times to ensure that all elements are sorted in the proper order. Figure 4.16 gives a flowchart for the program.

PROGRAM DEVELOPMENT

```
*******************************************************************
*     Program for bubble sort                                    *
*     Developed by:   T. K. Jewell     November 1987             *
*******************************************************************
C
C  Variable definitions
C  X(I) = array elements to be sorted
C  N = size of array
C  TEMP = temporary storage location for the value in X(J) during the
C     swapping procedure
      REAL X(100)
      INTEGER N,R,FR,W
      CHARACTER NAME*12
      PARAMETER (R=5,FR=4,W=6)
      PRINT *, 'Input data file name'
      READ(R,*) NAME
C  Input array to be analyzed
      OPEN(UNIT=FR,FILE=NAME,STATUS='OLD')
      READ(FR,*) N
      WRITE(W,*) 'For n = ',N,' the elements of the array are'
      DO 100 I=1,N
         READ(FR,*) X(I)
         WRITE(W,*) X(I)
  100 CONTINUE
      DO 200 I=1,N-1
         DO 300 J=1,N-1
            IF (X(J).GT.X(J+1)) THEN
               TEMP = X(J)
               X(J) = X(J+1)
               X(J+1) = TEMP
            END IF
  300    CONTINUE
  200 CONTINUE
C  Output sorted array
      WRITE(W,*)
      WRITE(W,*) 'The sorted array is'
      DO 400 I=1,N
         WRITE(W,*) X(I)
  400 CONTINUE
      END
```

PROGRAM TESTING

Use the data file from Example 4.6 to test the program

The second example will show a DO loop with real index and parameters nested within a loop with integer index and parameters. The inner loop modifies the program of Example 4.3 to figure the maximum height and distance of a trajectory for angles from 1° to 89° in increments of 2°. The outer loop controls the number of velocities that will be used.

EXAMPLE 4.11 □ Multiple Particle Trajectories with Varying Angles and Initial Velocities

STATEMENT OF PROBLEM

Modify the program of Example 4.3 to calculate the maximum height and distance for a trajectory with angles from 1° to 89° in increments of 2°, for any number of velocities.

MATHEMATICAL DESCRIPTION

Same as for Example 4.3

ALGORITHM DEVELOPMENT

Input/Output Design

An integer variable will be used to control the number of times the outer loop (the number of velocities) will be executed. The only other input will be the initial velocity during each execution of the outer loop. Output will include headings, the initial velocity, the angle, the time to maximum height, the maximum height, and the maximum distance for each angle.

Computer Implementation

Figure 4.17 shows the flowchart for this program.

PROGRAM DEVELOPMENT

```
*****************************************************************************
*    Program to determine particle trajectory data for varying angles *
*       and initial velocities                                        *
*    Developed by:  T. K. Jewell                         November 1987 *
*****************************************************************************
C  Variable definitions
C  VO = initial velocity
C  THETA = angle of trajectory in degrees
C  G = local acceleration due to gravity
C  ANGLE = angle of trajectory in radians
C  TYMAX = time, in seconds, to reach the top of the trajectory
C  YMAX = maximum height of trajectory, feet
C  XMAX = distance from launch to impact, feet
C  N = number of velocities to be input
       REAL VO,THETA,G,ANGLE,TYMAX,YMAX,XMAX
       INTEGER R,W,N
       PARAMETER (R=5,W=6)
       G = 32.17
C
```

FIGURE 4.17 Flowchart for multiple particle trajectories.

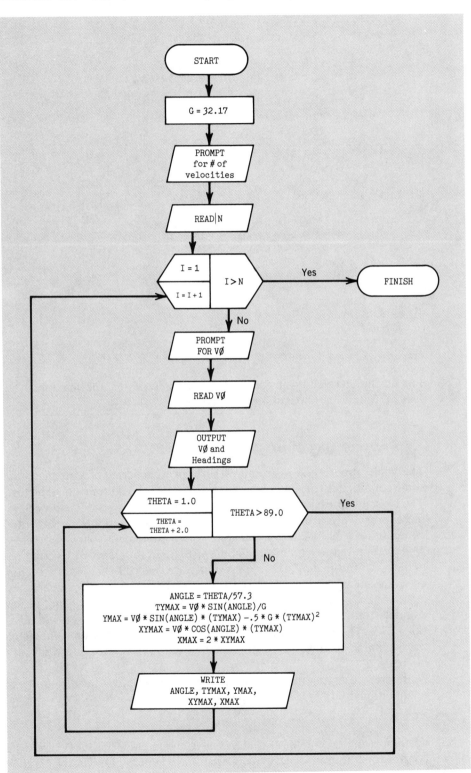

(EXAMPLE 4.11 □ Multiple Particle Trajectories with Varying Angles and Initial Velocities □ Continued)

```
C  Outer loop for different velocities
      PRINT *, 'Input number of velocities to be used  '
      READ(R,*) N
      DO 100 I=1,N
         PRINT *, 'Input initial velocity (ft/s)  '
         READ (R,*) VO
         WRITE(W,*)
         WRITE(W,*) 'For an initial velocity of',VO,' the data are:'
         WRITE(W,*)
         WRITE(W,*) '                              Time to          Maximum',
     1'        Dist to        Maximum'
         WRITE(W,*) '        Angle           Maximum Y        Height  ',
     1'        Max Ht.        Distance'
C  Inner loop for varying angle
         DO 200 THETA = 1.0,89.0,2.0
            ANGLE = THETA/57.3
            TYMAX = VO*SIN(ANGLE)/G
            YMAX = VO*SIN(ANGLE)*TYMAX-0.5*G*TYMAX**2
            XYMAX = VO*COS(ANGLE)*TYMAX
            XMAX = 2.*XYMAX
            WRITE (W,*) THETA,TYMAX,YMAX,XYMAX,XMAX
  200    CONTINUE
  100 CONTINUE
      END
```

PROGRAM TESTING

For an initial velocity of 500 f/s and an angle of 45°, the maximum height is 1,942.6 ft and the maximum distance is 7,771.2 ft.

4.4.2 While Structure

The while structure is a structure that repeats a series of statements as long as some condition is true and exits the loop through the bottom when the test is false. The logical expression is tested at the beginning of each execution of the loop. If the expression is true, statements in the body of the loop are executed. If the expression is false, control transfers to the statement following the loop. The while structure is useful in developing structured programs with conditional loops in them. Unfortunately, it is not presently a standard FORTRAN structure. Some compilers incorporate a DO WHILE or WHILE structure. However, they are not standardized. Since it is not standard FORTRAN, the while will not be used in the examples of this text. As an alternative, a block IF will be used with an unconditional GO TO statement to complete the conditional loop. Comment statements at the beginning and end of the loop will be used to indicate that a while structure is being used.

Before presenting the form of the while structure, we need to understand the unconditional GO TO. It has the form

```
      GO TO n
```

where *n* is any executable and addressable statement. The GO TO causes program control to shift immediately to statement *n*. Prior to FORTRAN 77, the GO TO was used extensively to produce structures somewhat similar to the IF THEN

ELSE. However, the use of many GO TOs made the logic difficult to follow and was one of the primary reasons for implementing the newer structured programming forms. Also, when the GO TO transfers control to a point earlier in the program, an infinite loop can develop if there is no way to escape from the loop. GO TOs should be used only when they are absolutely necessary. The IF THEN approximation of the while structure is a time when use of the GO TO is appropriate.

The form of the IF THEN approximation of the while structure is

```
C WHILE (logical expression) is true, DO n
    m  IF(logical expression) THEN

        FORTRAN statements

    n  CONTINUE
        GO TO m
        END IF
C End of WHILE structure
```

The *n* CONTINUE is used to identity the end of the conditional loop, although it is the GO TO *m* statement that actually accomplishes the transfer of control back up to statement *m*. Figure 4.18 shows a flowchart depiction of the while structure.

You can see that the construction of the logical expression is important. It has to be constructed so that at some time the logical expression will evaluate to 'false', or the loop will be infinite. However, this is also true of the formal while loop.

Example 4.12 will illustrate a problem in which a conditional loop is desirable. In this case, the condition is that at least one switch is made during a particular pass of the bubble sort through an array. When no switches are made, the array is in the proper order.

FIGURE 4.18   Flowchart depiction of While structure.

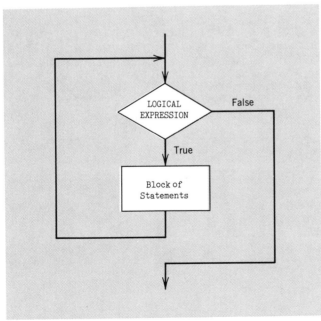

EXAMPLE 4.12 □ Bubble Sort with Stop After Pass with No Switches

STATEMENT OF PROBLEM

The bubble sort will be more efficient if the successive passes stop when the array is sorted. If a pass is made and no switches occur, the array has to be in the proper order. Include a check in your program so that the passes will stop if no switches are made.

ALGORITHM DEVELOPMENT

Input/Output Design

Input and output are the same as for Example 4.10.

Computer Implementation

The inner loop is modified to become a conditional loop. A counting variable is added. This variable will have a value greater than zero if any switches are made during a pass through the array. Each time a switch is made, the variable is increased by one. Before each pass through the array, the counting variable is reinitialized to zero. If there are no switches, it will have a value of zero when the logical expression controlling the outer repetition loop is evaluated. This will cause the logical expression (COUNT.GT.(0)) to evaluate to 'false', which will stop the sorting procedure. Figure 4.19 shows the applicable portion of the flowchart for the while structure.

PROGRAM DEVELOPMENT

```
*********************************************************************
*     Program for bubble sort with stop check for sorted array     *
*     Developed by:  T. K. Jewell                 November 1987     *
*********************************************************************
C
C Variable definitions
C  X(I) = array elements to be sorted
C  N = size of array
      REAL X(100)
      INTEGER N,R,FR,W,COUNT
      CHARACTER NAME*12
      PARAMETER (R=5,FR=4,W=6)
      PRINT *, 'Input data file name'
      READ(R,*) NAME
C  Input array to be analyzed
      OPEN(UNIT=FR,FILE=NAME,STATUS='OLD')
      READ(FR,*) N
      WRITE(W,*) 'For n = ',N,' the elements of the array are'
      DO 100 I=1,N
         READ(FR,*) X(I)
         WRITE(W,*) X(I)
  100 CONTINUE
C
C  Initialize counter for switches for first time through loop
      COUNT = 1
C
```

FIGURE 4.19   Flowchart segments for bubble sort with stop after no switches.

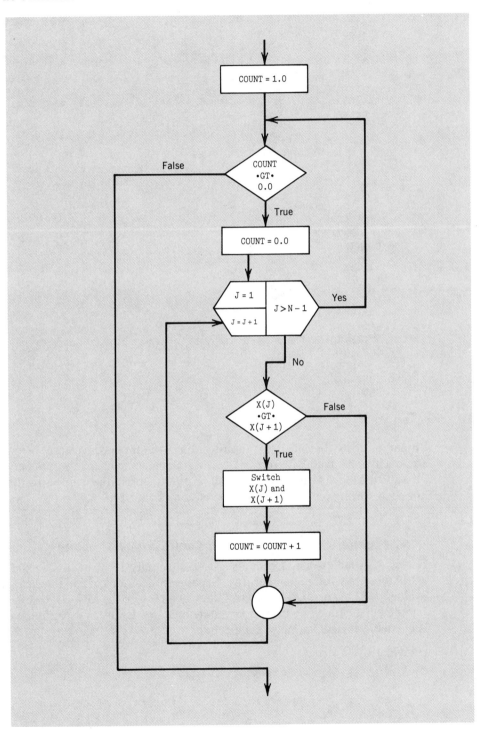

(EXAMPLE 4.12 ☐ Bubble Sort with Stop After Pass with No Switches ☐ Continued)

```
C  WHILE (COUNT.GT.(0)) is true, DO 200
C
   500 IF(COUNT.GT.(0)) THEN
C
C  Initialize counter so it will be zero if no switches are made
         COUNT = 0
C
C  Start inner loop for comparing adjacent elements and switching
C     if necessary
C
         DO 300 J=1,N-1
            IF (X(J).GT.X(J+1)) THEN
               TEMP = X(J)
               X(J) = X(J+1)
               X(J+1) = TEMP
               COUNT = COUNT + 1
            END IF
   300      CONTINUE
   200 CONTINUE
         GO TO 500
         END IF
C  End of WHILE structure
C
C  Output sorted array
         WRITE(W,*)
         WRITE(W,*) 'The sorted array is'
         DO 400 I=1,N
            WRITE(W,*) X(I)
   400 CONTINUE
         END
```

PROGRAM TESTING

Input an array that is almost in order, but will still need at least two switches. You will need to add intermediate write statements to check the number of times the sorting loop is executed. It would be best to output the whole array at the end of each sorting loop to see the progression of the sort.

4.4.2.1 Using Logical Variables to Control While Structures

Logical variables can also be used to control any IF structure, including while structures. As you remember from Section 3.6.2.8, logical variables can store the values .TRUE. or .FALSE.. Thus, they can replace the logical expressions of IF statements. Shown below are key statements of Example 4.12, followed by the equivalent logical variable modifications.

Original while structure

```
      INTEGER N,R,FR,W,COUNT
         .
C  Initialize counter for switches for first time through loop
         COUNT = 1
         .
C  WHILE (COUNT.GT.(0)) is true, DO 200
         .
   500 IF(COUNT.GT.(0)) THEN
         .
```

```
C   Initialize counter so it will be zero if no switches are made
          COUNT = 0
             .
          DO 300 J=1,N-1
             IF (X(J).GT.X(J+1)) THEN
                   .
                   COUNT = COUNT + 1
             END IF
  300 CONTINUE
  200 CONTINUE
          GO TO 500
          END IF
```

Logical variable equivalent

```
          INTEGER N,R,FR,W
          LOGICAL SWITCH
             .
C   Initialize SWITCH for first time through loop
          SWITCH = .TRUE.
             .
C   WHILE (SWITCH) is true, DO 200
             .
  500 IF(SWITCH) THEN
             .
C   Initialize counter so it will be zero if no switches are made
          SWITCH = .FALSE.
             .
          DO 300 J=1,N-1
             IF (X(J).GT.X(J+1)) THEN
                   .
                   SWITCH = .TRUE.
             END IF
  300 CONTINUE
  200 CONTINUE
          GO TO 500
          END IF
```

Logical variables can also be compared with each other by using the .EQV. and .NEQV. relational operators. These operators are evaluated last in the order of precedence, after the .OR. connector. The other connectors (.NOT., .AND., and .OR.) can also be used, but not the normal relational operators (.EQ., etc.). For example, if TEST1 and TEST2 are logical variables that are used to check two different criteria for a program, then the logical expression

$$(TEST1.EQV.TEST2)$$

would evaluate to .TRUE. if TEST1 and TEST 2 are both true or both false, while

$$(TEST1.AND.TEST2)$$

will evaluate to .TRUE. only if both TEST1 and TEST2 are true.

4.5 CONTROLLING THE AMOUNT OF INPUT

The DO loop with a limit of N is one of the most commonly used methods for controlling input. It is generally used for controlling input for arrays and is often used for any type of input that has a predetermined number of elements. This is especially true if the number of elements also has to be used for computations, as would be the case when determining the mean and standard deviation for a set of data values. When you are using a DO loop to control input from a data file, it is important that the upper limit agrees with the number of data records to be input. If the DO loop finishes without reading all the data records, then data that was supposed to be included in the analysis are not input. If the DO loop tries to access more data records than there are in the data file, the READ statement encounters the end of file mark on the input data file. This causes an error condition and stops program execution.

Three other methods of controlling input will now be presented. Each of these can be used to input a varying amount of data without specifying beforehand how many elements there will be. The advantage is that the user does not have to be concerned with the number of input data values. However, if any calculations require a value for the number of data elements, or if subscripted variables are to be used, then some sort of counting or index variable has to be initialized and increased within the loop.

4.5.1 END =

The END = modifies the READ statement. It is actually a special form of the while structure. It has the form

```
READ(FR,*,END=n) variable list
```

and works just like the normal READ when reading data values. However, when the end of file mark is encountered in the input file, the computer transfers control to statement *n* rather than registering an error. This can be combined with a GO TO to form a while loop of the form

```
C  Start of WHILE loop for input
   m        CONTINUE
      READ(FR,*,END=n) variable list
      .
      other FORTRAN statements (optional)
      .
      GO TO m
C  End of WHILE loop for input
   n        CONTINUE
```

The GO TO *m* and *m* CONTINUE statements form the data input loop, and the END=*n* and *n* CONTINUE form the exit from the loop. There cannot be an infinite loop for this structure, because the input data file has to have a finite length. This structure is useful only for data file input, unless you want to input an end of file marker through the keyboard. Figure 4.20 shows a flowchart depiction of the END = structure for input.

4.5.2 Sentinel Variables

A second alternative for controlling input is to use sentinel variable values. One of the variables in the input variable list is chosen as the sentinel variable. A certain value of this variable will indicate that data input is completed. If the variable will also be used in computations, the value chosen as the sentinel must not be a value that would normally be input for the variable. For instance, if the values for the sentinel variable might be expected to range from 1 to 100, you might choose 999

FIGURE 4.20   Flowchart depiction of END = structure.

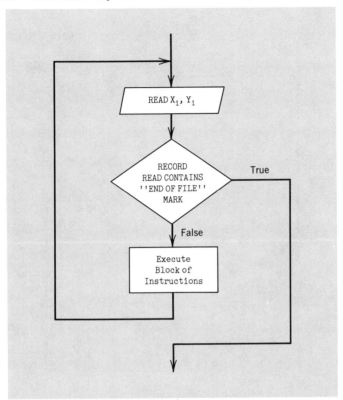

as the value that would signal the end of input data. In other cases, the sentinel variable may be used only to control input. An IF structure is used to read values only until the sentinel value is encountered. To make the choice of the sentinel value more flexible, a sentinel value greater than or equal to, or less than or equal to, may be specified, and the .GE. or .LE. operators used in the IF. For example, any value .GE. 999 could be used to signify the end of input data. When you are using free format data file input, it is important to give values for the other variables in the input list for the final record along with the sentinel value, or the end of file mark will be read and an error indicated. The structure again needs a GO TO to complete the conditional loop and has the general form

```
C   Start of sentinel input loop
    m        READ(FR,*) sentinel variable, variable list
       IF(sentinel variable.NE.sentinel value) THEN
          .
          FORTRAN statements using sentinel variable and variable list
          .
          GO TO m
       END IF
C   End of sentinel input loop
```

Figure 4.21 shows a flowchart depiction of the sentinel input control structure.

Note that this structure works for either file or terminal input. However, you must be sure to tell the user how input control is to be accomplished. For data file

FIGURE 4.21   Flowchart depiction of the sentinel input control structure.

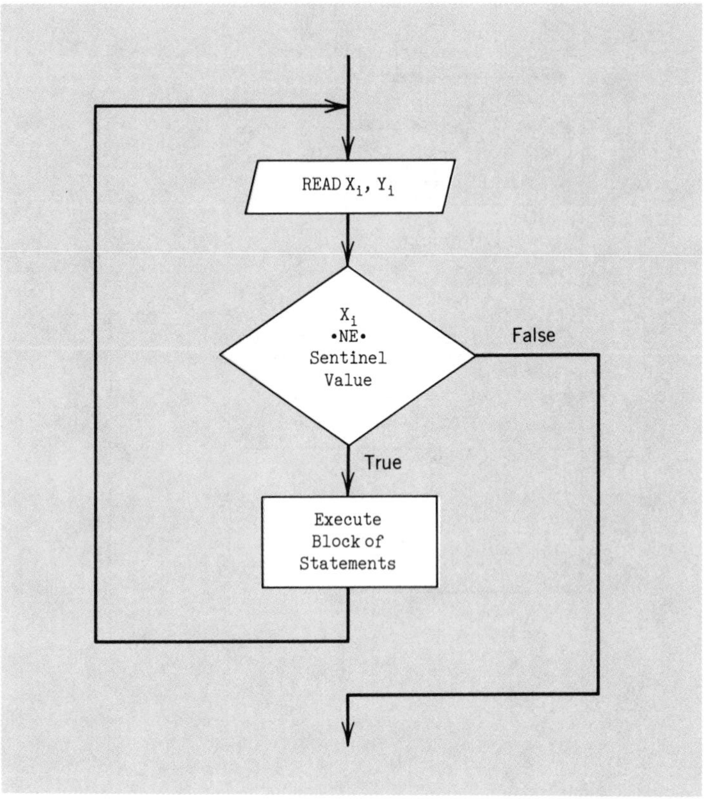

input, the instructions must be part of the input data instructions. For terminal input, input prompts can be used.

With terminal input, sentinel input control can be used to input any number of cases until you are ready to exit the program. For example, the program of Example 4.11 inputs a number of velocities to determine trajectory data. In Example 4.11 the number of repetitions was set by requesting the number from the user prior to starting input of velocities. You might want to set the program up so that you could decide to exit the program after examining the calculated data. $V0$ could be the sentinel variable. A value of zero would be appropriate, because you would not normally input a velocity of zero. Modification of the program of Example 4.11 to accommodate sentinel control will be included as an exercise at the end of the chapter. Example 4.13 will illustrate application of sentinel control in the computation of the norm of an n-dimensional vector.

4.5.3 Inquiry and GO TO

Inquiry and GO TO is another effective way to control input. It is used to control repetition of a program or part of a program. Additional data input is normally a part of this repetition. The decision block is placed at the end of the loop. A prompt is output to the terminal asking the user if another iteration is desired. If

the user answers in the affirmative, control moves back to the start of the repetition loop. If the answer is negative, the program continues with statements after
the inquiry. Inquiry and GO TO has the general form

```
      CHARACTER CHECK*3
         .
C  Start of inquiry and GO TO input control structure
   m        CONTINUE
      READ(*,) variable list
         .
      FORTRAN statements
         .
      PRINT *, 'Do you want to use another case? (Yes or No)'
      PRINT *, 'Be sure to put apostrophes before and after your reply'
      READ(R,*) CHECK
      IF(CHECK(1:1).EQ.'Y'.OR.CHECK(1:1).EQ.'y') GO TO m
C  End of inquiry and GO TO input control structure
```

Figure 4.22 shows a flowchart depiction of the inquiry and GO TO structure. This
method could also have been used effectively to control input to the trajectory
program of Example 4.11.

Example 4.13 is an adaptation of Example 4.5 to illustrate application of both
sentinel variable and inquiry and GO TO control in the same program. The program has been shortened to one that will find the norm of an *n*-dimensional vector.

FIGURE 4.22 Flowchart depiction of the inquiry and GO TO
structure.

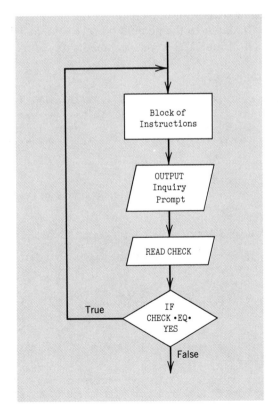

Note that the index for the subscripted variable storing the vector components has to be initialized and increased in the loop since a DO structure is not being used.

EXAMPLE 4.13 □ **Finding the Norm of an *n*-Dimensional Vector with Sentinel and Inquiry Control**

STATEMENT OF PROBLEM

Develop a program that will determine the magnitude of a number of *n*-dimensional vectors. Sentinel control should be used to control the dimension of the vector, and inquiry control used to control the number of vectors that will be examined. The maximum vector component size that is to be expected is 80.

ALGORITHM DEVELOPMENT

Input/Output Design

Output will be modified to include the proper prompts for the sentinel and inquiry control. Input data will no longer include the dimension of the vector. Component values will be input until the sentinel value is encountered. Since any component will not be expected to have a magnitude of more than 80., the sentinel value could be specified as any component value of 100. or larger.

Computer Implementation

Figure 4.23 shows the flowchart for implementing the sentinel and inquiry structures for controlling the input to the adaptation of Example 4.5. Note how the structures of Figures 4.21 and 4.22 were combined.

PROGRAM DEVELOPMENT

```
**********************************************************************
*    Program for finding the norm of an n-dimensional vector       *
*    Developed by:  T. K. Jewell            November, 1987          *
**********************************************************************
C
C  VARIABLE DEFINITIONS
C  C = n-dimensional vector
C  SUMCSQ = summing variable for squared components of vector C
C  NORMC = magnitude (norm) of vector C
C
      REAL C(10),SUMCSQ,NORMC
      INTEGER I,R,W
      CHARACTER LOOP*3
      PARAMETER (R=5,W=6)
C
C  Start of inquiry controlled input loop
  200 CONTINUE
      I=1
C
C  Start of sentinel input loop
  100 CONTINUE
      PRINT *, 'Input component ',I,' of vector C  '
      PRINT *, 'A value >= 100. indicates the end of data   '
```

FIGURE 4.23 Flowchart for sentinel and inquiry control for program to compute the norm of an *n*-dimensional vector.

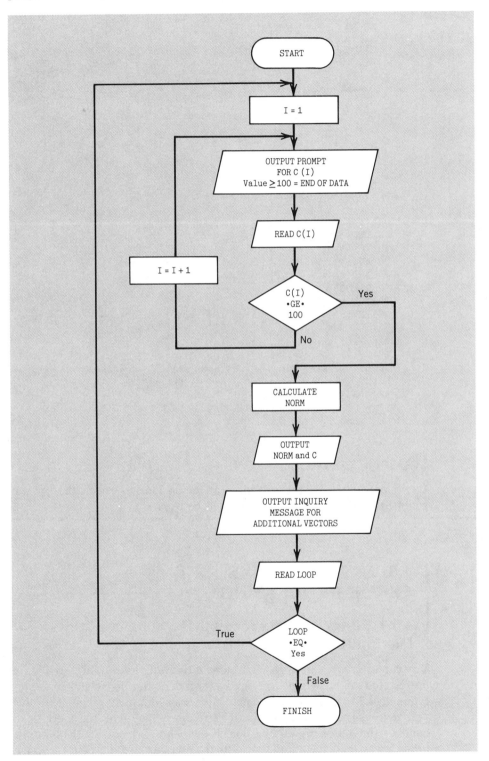

(EXAMPLE 4.13 □ **Finding the Norm of an n-Dimensional Vector with Sentinel and Inquiry Control** □ **Continued)**

```
      READ (W,*) C(I)
      IF (C(I).LT.(100.0)) THEN
         I = I + 1
         GO TO 100
      END IF
C  End of sentinel input loop
C
C  Calculate magnitude (norm) of vector C
      SUMCSQ = 0.0
      N = I - 1
      DO 300 I=1,N
         SUMCSQ = SUMCSQ + C(I)**2
  300 CONTINUE
      NORMC = SQRT(SUMCSQ)
C
C  Output results
      WRITE (W,*) 'Components of vector C are:'
      WRITE (W,*)
      DO 400 I=1,N
         WRITE (W,*) C(I)
  400 CONTINUE
      WRITE (W,*)
      WRITE (W,*) 'The magnitude of vector C is: ',NORMC
C
C  Inquiry for additional input
      PRINT *, 'Do you wish to input another vector? (Yes or No)'
      PRINT *, 'Make sure to put apostrophes on both sides of your'
      PRINT *, ' response
      READ(R,*) LOOP
      IF (LOOP(1:1).EQ.'Y'.OR.LOOP(1:1).EQ.'y') GO TO 200
C  End of inquiry loop
      END
```

PROGRAM TESTING

Use at least two vectors of different dimensions to check the program. The summed vectors of Example 4.5 could be used.

4.6 RANGE CHECKS AND ERROR PROCESSING

4.6.1 Range Checks for Both Terminal and Data File Input

You can use range checks to determine if the values input for variables are within preselected limits. We incorporated a form of input range check in the program of Example 4.3 to make sure the angle input was within the range of 30° to 75°. Now you have learned additional control structures that give you more flexibility for handling the out-of-range condition. If the values are outside the limits, an appropriate error message should be output. You can use a STOP statement to halt execution if it is inadvisable to continue with the program after encountering such

FIGURE 4.24   Flowchart depiction of the range check structure.

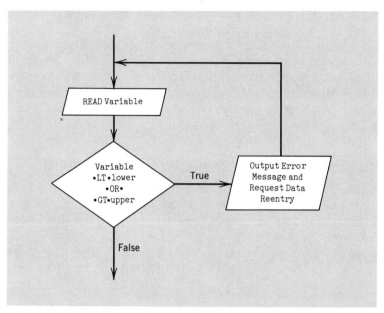

an out-of-range condition. If you are using terminal input, you can loop back up to reinput the out-of-range value. The general form for a range check that requests corrected data is

```
C   Input data with range check
    m       CONTINUE
        READ(R,*) variable
        IF(variable.LT.lower.OR.variable.GT.upper) THEN
            PRINT *, 'The value of variable is out of range.'
            PRINT *, ' Please reinput the value'
            GO TO m
        END IF
C   End of data input with range check
```

The logical expression can be modified to reflect the type of range check you want to make. The parameters *lower* and *upper* would have to be defined either within the program or by data input prior to making the range check. Figure 4.24 shows a flowchart depiction of the input data range check structure.

4.6.2 Error Checking During Program Execution

You can also use range checks to detect out-of-range values for calculated variables. If the value is not in the acceptable range, you should output an error message. You may want to have the program stop execution at that point. At the least, you should warn the user that the computed results may be inaccurate.

Many numerical methods require the repetitive estimation of variable values until some convergence criteria are met. This requires a conditional loop. Computer round-off errors or numerical instabilities in the solution technique may keep a program from meeting the convergence criterion required to terminate the conditional loop. You can detect this type of problem by including a maximum allowable number of iterations in the program logic. You can add a counter and an additional IF structure in the loop. The IF structure should take appropriate action if the value of the counter exceeds the maximum number of times you want to go through the loop. You may want to print an error message and stop the program. Or you may want to exit the conditional loop and have the user correct the condition that caused the problem.

If the divisor in a FORTRAN expression is zero, you will get a divide-by-zero error, and program execution will terminate. If it is possible that in the normal course of computations a divisor may become zero, you should include an IF THEN ELSE structure to avoid evaluating the expression when the divisor is zero. An error message or warning could be printed out on this stem of the IF THEN ELSE.

4.6.3 Input Data Error Trapping for Terminal Input

When you have been inputting a significant amount of data from a terminal, it is frustrating to make a typing or other entry error near the end of your program and find that you have to start the process all over again. With a little extra programming work, you can prevent this problem. The ERR = structure allows you to loop back up and reinput an incorrect entry. The ERR = is similar in operation to the END = structure that is used to control file input. It has the general form

```
C   Start of input with error checks.
    m        CONTINUE
        READ(R,*,ERR=k) variable list
        GO TO n
C
C   Error condition exists
    k        CONTINUE
        PRINT *, 'There is an error in the values input for the last'
        PRINT *, ' variable list.  Please reinput values carefully.'
        GO TO m
C
C   No error condition exists, end of input
    n        CONTINUE
```

Figure 4.25 shows a flowchart depicting the error trapping structure.

4.6.4 Application of Error Trapping and Range Checks

Example 4.14 shows application of both error trapping and input range checks to a particle trajectory problem. The example will apply particle trajectory mechanics to find whether a stream of water from a fire truck will hit a window. The effects of air friction on the stream are ignored. Background for this example was developed in Example 2.4. Examples 4.3 and 4.11 have developed other applications for particle trajectory theory.

The program for Example 4.14 requires use of several of the control structures developed in this chapter. The range check and error trapping structures are included with the input module of the larger program. If the velocity entered is larger than the practical upper limit of 150 ft/s, the program user is asked to reinput the data. Detection of any input data errors also results in the user being asked to reinput the data. Figure 4.27b clearly shows the logic of the range check and error trapping structures combined together. However, in a larger program

FIGURE 4.25   Flowchart depiction of error trapping structure.

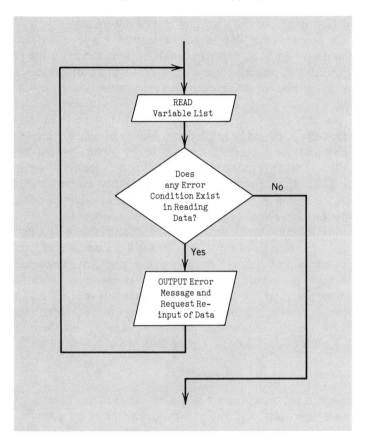

the branching for numerous range checks could be confusing. Use of a subprogram for input and read error testing could greatly reduce the amount of branching required.

Error trapping will not be included with all examples in the text in the interest of saving space. However, it is always a good idea to use it for programs with any significant amount of terminal input. Range checks should also be used whenever they are appropriate.

EXAMPLE 4.14   □   Hitting Window with Water Jet

STATEMENT OF PROBLEM

A fire truck is positioned a distance x away from, and a distance y below, a window. It is necessary to have a stream of water from the deck gun of the truck enter the window. The angle of inclination of the deck gun is θ. Develop a program that will determine what value or values of θ will cause the water to enter the window for a given velocity of exit from the nozzle V_0. The window can be considered to extend 6 in. above and below the center point defined by y.

(EXAMPLE 4.14   □   Hitting Window with Water Jet   □ Continued)

The angle of inclination for the deck gun should not be larger than 75° from the horizontal, nor smaller than 30° from the horizontal. Air resistance on the water jet can be ignored. Figure 4.26 provides a definition sketch for the problem.

     The program should provide the option of evaluating multiple cases. The user should have the option of routing output to the terminal or to both the terminal and an output file.

     If it is impossible to hit the window for a certain value of initial velocity, a message to that effect should be printed out. The output should include the value of the velocity and the maximum distance the trajectory would travel. If the trajectory at least makes it to the wall containing the window, the minimum and maximum height of the trajectory at the wall should be output. The user should then be given the option of exiting or inputting the next values of velocity, x, and y to be evaluated.

     For values of initial velocity that will allow the stream to approach the window from below, it is possible that the stream will continue to rise above the window, then come back down to go through it again. Other conditions will only allow the stream to pass through the window once. For trajectories that will pass through the window, print out whether one or two angles of inclination will cause the stream to hit. For each angle that does hit the window, output V_0, the angle of inclination θ, the maximum height of the trajectory, and the angle that the stream makes with the window as it goes through.

MATHEMATICAL DESCRIPTION

If air resistance is ignored, the jet of water can be considered as a series of individual water particles, and the governing equations for particle mechanics can be used to estimate the trajectory of the water jet.

FIGURE 4.26   Definition sketch for hitting window with water jet.

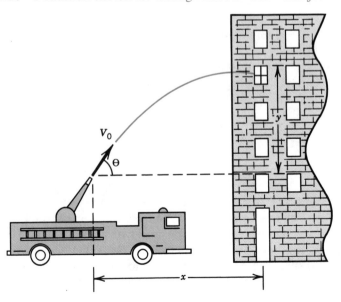

The following equations from Section 4.2 of Appendix G will be needed to develop this program.

$$V_y = V_{0_y} - gt = V_0 \sin \theta - gt$$
$$V_x = V_{0_x} = V_0 \cos \theta \qquad (4.10)$$
$$x = x_0 + V_x t = x_0 + V_0 \cos \theta t \qquad (4.11)$$

$$y = y_0 + V_{0_y} t - \frac{1}{2} gt^2$$

$$= y_0 + V_0 \sin \theta t - \frac{1}{2} gt^2 \qquad (4.12)$$

If x_0 and y_0 are zero, you can combine Equations 4.10 and 4.11 to estimate the height of the trajectory at any distance x.

$$y(x) = x \tan \theta - \frac{1}{2} g \left(\frac{x}{V_0 \cos \theta} \right)^2 \qquad (4.13)$$

Additional information on the development of these equations can be found in Appendix G.

If the water stream does not hit the window, the maximum distance that the stream would travel can be found by determining the horizontal distance that the stream would travel for an angle of inclination of 45°. If the maximum distance is greater than the distance to the wall, the trajectory will hit the wall below the window. For this situation, the distance to the wall should be output as the maximum distance.

ALGORITHM DEVELOPMENT

Input/Output Design

Input will include the distance to and height of the window as well as the initial velocity. The size of the angle increment for the search routine will also be input. The maximum practical velocity for a water jet is about 150 ft/s, so this value will be included as an input data range limit. Error trapping will be used for the input data.

Since the minimum and maximum angles are constant for this problem, they will be specified through a DATA statement. Input and output units for angles will be in degrees. Input will be from the keyboard. The user will have the option of directing output either to the screen or to both the screen and a file. At the end of the main part of the program, an inquiry will be output requesting whether the user wants to enter new data for another execution of the program.

Numerical Methods

A starting point has to be chosen. The minimum angle is a convenient point. Before starting the loop to increase the angle, you can make some checks that will simplify the logic of the rest of the program. First you can compute the maximum horizontal distance that the trajectory will travel for an angle of 45°. If this distance is less than x, the trajectory will never hit the wall, so it cannot possibly hit the window. We will refer to this as case one. Next, you can compute the height of the trajectory at distance x for the minimum and maximum angles allowed by the program. If the trajectory height is above the window for

(EXAMPLE 4.14 ☐ Hitting Window with Water Jet ☐ Continued)

both the minimum and maximum angles, the trajectory cannot possibly pass through the window, but you know that it will hit the wall. Therefore, all you have to compute is the minimum and maximum height of the trajectory at the wall. The distance traveled has to be x. We will refer to this as case two. If the trajectory is above the window for one extreme and below it for the other extreme, the trajectory has to pass through the window only once. There is no need to compute the maximum and minimum heights on the wall for this case. We will refer to this as case three. If the trajectory is at or below the window for both extremes, then the trajectory will pass through the window zero, one, or two times. This will be referred to as case four. It will hit zero times if the height of contact with the wall never reaches the window (case 4a), once if the height reaches the window but never goes above it (case 4b), and twice if it hits the window, rises above the window, then falls back through it as the angle increases (case 4c). You will need to determine the minimum and maximum heights on the wall for case four to accommodate instances when the stream never reaches the window.

For cases three and four, you will have to increase the angle in some orderly manner to search for the angle that causes the stream to pass through the window. There are several ways of approaching this procedure. One is to use a small increment that will steadily approach the window. Another approach is initially to use a fairly large angle increment, check to see when the stream crosses the target height y, and then decrease the angle back toward the target using smaller increments. This procedure can be repeated a number of times, using a smaller increment each time the stream passes over the target until the desired accuracy is achieved. The accuracy is defined as the difference between the estimated height of the trajectory at the wall for angle θ and the height y. The latter method is preferable when very accurate results are required and using a small increment would result in excessive computation time. For this problem you will not be able to adjust the deck gun to more than 0.1° accuracy. Using a constant 0.1° increment will not result in excessive computation time.

The window that you are trying to hit is 6 in. above and below the target elevation y. If your increment is excessively large, there is a danger that the stream may skip from one side of the window to the other for successive computations. Your accuracy check would not be satisfied, so the program would falsely indicate that the stream had not hit the window. An increment of 0.1° should be small enough, but if it is not, you will be able to tell by looking at the minimum and maximum heights on the wall. If the printout says that you never hit the window but the maximum height shows that the stream went above the window, you can rerun the data with a smaller increment.

For cases three and four, there is actually a range of angles that will cause the stream to pass through the window. Our objective is to find the first angle that causes the trajectory to fall within the target height ±6 in. When accomplishing case three analyses, all you have to do is search for the first angle. When case four holds, and the stream does hit the window, you have to have a method of checking for a second angle. You should continue to increase the angle until the trajectory moves above the window. When it does, set a check variable and continue to increase the angle until the trajectory comes back down to the window. You know that it will come back down for case four.

The search method used here is different from the method that was used for the examples of Chapter 2. In Chapter 2 we used a single search routine, and when the trajectory hit the window we jumped to the maximum angle and worked our way back down searching for the second angle. This would work for most cases but would make it more difficult to identify cases where the trajectory hits the window but does not rise above it.

Computer Implementation

Figure 4.27 shows the flowchart for this program. Note the initial decomposition of the problem in Figure 4.27*a*. This part of the program will be developed and debugged before the range and error checks and the logic of each case is added. The logic for the range check and error structure and the cases, shown in Figures 4.27*b* through 4.27*e*, can then be added and debugged individually. A separate flowchart is not included for case one because it requires only the output of a message and the value for XMAX at 45°.

INITIAL PROGRAM DEVELOPMENT

The program for the initial decomposition will be developed and debugged first, then the logic for each case will be added.

```
*****************************************************************************
*  Program for finding the angle that causes a water stream to pass   *
*     through a window a distance x from, and a height y above, the   *
*     start of the stream trajectory                                  *
*  Developed by: T. Jewell  March, 1983.  Modified February 1989.     *
*****************************************************************************
C  Variable Definitions
C  YX = statement function for calculating height of trajectory at X
C  X = distance to target, feet.
C  Y = height of target.  The window is 6 in. above and below the target.
C  XMAX45 = maximum horizontal distance for trajectory at 45 degrees.
C  TXMAX = time to reach XMAX45
C  HTMIN = height of trajectory at wall for minimum angle.
C  HTMAX = height of trajectory at wall for maximum angle.
C  OUT = check variable for output.  1 = output to both screen and file.
C  VO = initial velocity for water stream, ft/s.
C  THETA = angle of trajectory in degrees.
C  ANGLE = angle of trajectory in radians.
C  REPEAT = check variable for entering a new set of data.  Yes = new data.
C  TMIN = minimum angle for search.
C  TMAX = maximum angle for search.
C  DTHETA = angle increment to use in search program.
       REAL YX,X,Y,XMAX45,HTMIN,HTMAX,VO,THETA,ANGLE,TMIN,TMAX,DTHETA,
      1      TXMAX
       INTEGER OUT,R,W,FW
       CHARACTER REPEAT*3,NAME*10
       PARAMETER (R=5,W=6,FW=4)
       DATA TMIN,TMAX/30.,75./
       YX(VO,X,A) = TAN(A)*X - .5*32.17*(X/(VO*COS(A)))**2
C  Select output option
C
       PRINT *, 'Output will be to the screen.  If you also want output'
       PRINT *, '  to a file, type 1, followed by a space, followed by'
       PRINT *, '  the file name for the output file.  For example:'
       PRINT *, ' '
       PRINT *, '1 "TRAJEC.OUT" '
       READ (R,*) OUT,NAME
       PRINT *, 'If you do not want to output to a file, type any other'
```

(EXAMPLE 4.14 □ Hitting Window with Water Jet □ Continued)

```
      PRINT *, 'number, a space, and any letter. For example:'
      PRINT *, '2 "R" '
      IF(OUT.EQ.1) OPEN(UNIT=FW,FILE=NAME,STATUS='NEW')
C  Beginning of major loop for each set of data values.
C
   10 CONTINUE
      PRINT *, 'Input initial velocity (ft/s), distance to wall (ft),'
      PRINT *, ' height of target (ft), and increment for angle '
      PRINT *, ' search (degrees).'
      READ (R,*) VO,X,Y,DTHETA
      WRITE(W,*)
      WRITE(W,*) 'FOR THIS RUN, THE INITIAL VELOCITY IS: ',VO
      WRITE(W,*) 'THE DISTANCE TO THE WALL IS: ',X
      WRITE(W,*) 'THE HEIGHT OF THE TARGET IS: ',Y,' and'
      WRITE(W,*) 'THE INCREMENT FOR THE ANGLE IS: ',DTHETA
      IF(OUT.EQ.1) THEN
      WRITE(FW,*)
      WRITE(FW,*) 'FOR THIS RUN, THE INITIAL VELOCITY IS: ',VO
      WRITE(FW,*) 'THE DISTANCE TO THE WALL IS: ',X
      WRITE(FW,*) 'THE HEIGHT OF THE TARGET IS: ',Y,' and'
      WRITE(FW,*) 'THE INCREMENT FOR THE ANGLE IS: ',DTHETA
C  Compute initial values for determining case.
C
      ANGLE = 45./57.3
      TXMAX = 2.*(VO*SIN(ANGLE)/32.17)
      XMAX45 = VO*COS(ANGLE)*TXMAX
      HTMIN = YX(VO,X,TMIN/57.3)
      HTMAX = YX(VO,X,TMAX/57.3)
C  Beginning of Branching for Different Cases
C
      IF(XMAX45.LT.X) THEN
C  Case 1: Stream never makes it to the wall.
C
      PRINT *, 'CASE 1'
      ELSE IF(HTMIN.GT.Y.AND.HTMAX.GT.Y) THEN
C  Case 2:  Stream hits the wall above the window for all angles.
C
      PRINT *, 'CASE 2'
      ELSE IF ((HTMAX.GT.Y.AND.HTMIN.LT.Y)
     1          .OR.(HTMAX.LT.Y.AND.HTMIN.GT.Y)) THEN
C  Case 3:  Stream hits the window once.
C
      PRINT *, 'CASE 3'
      ELSE
C  Case 4:  Check for stream hitting zero, one, or two times.
C
      PRINT *, 'CASE 4'
      END IF
C  Inquiry for additional input sets
C
      PRINT *, 'Do you want to input an additional set of data?'
      PRINT *, ' Type "Yes" or "No" (either caps or small)'
      READ(R,*) REPEAT
      IF(REPEAT(1:1).EQ.'y'.OR.REPEAT(1:1).EQ.'Y') GO TO 10
      STOP
      END
```

FIGURE 4.27   Flowchart for hitting window with water jet.

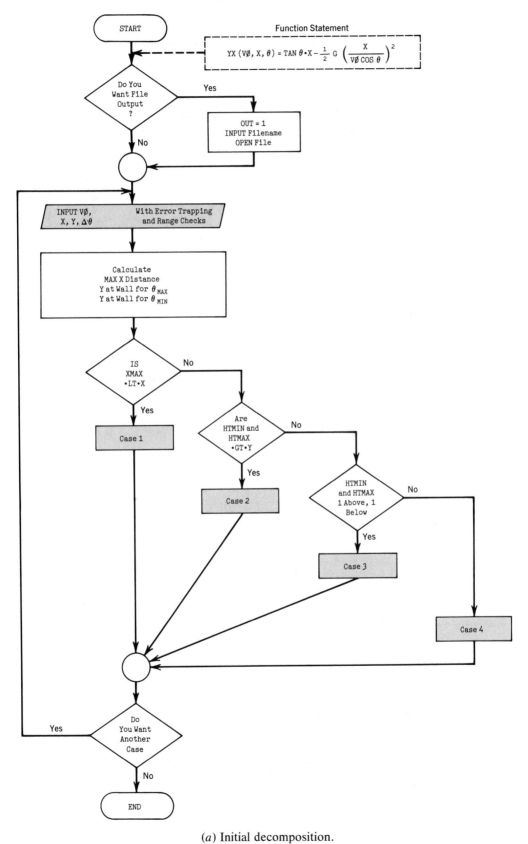

(*a*) Initial decomposition.

(EXAMPLE 4.14 □ Hitting Window with Water Jet □ Continued)

FIGURE 4.27 (Continued)

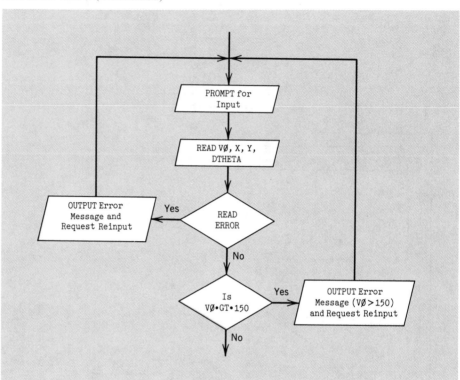

(b) Range checks and error trapping module.

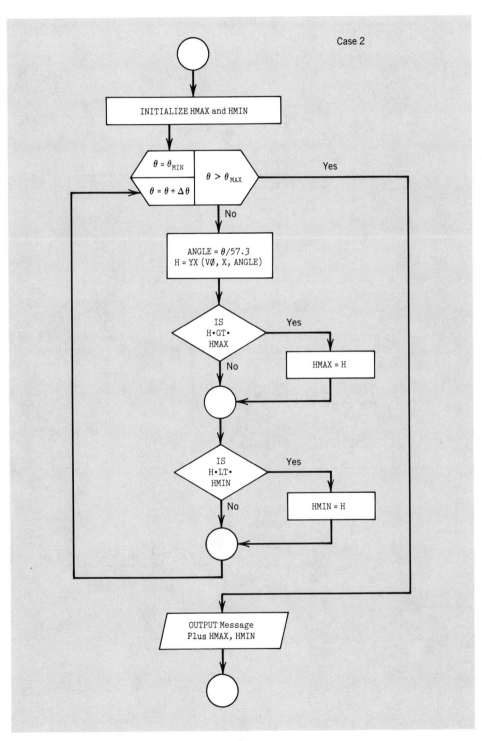

(*c*) Case 2.

(EXAMPLE 4.14 □ Hitting Window with Water Jet □ Continued)

FIGURE 4.27 (Continued)

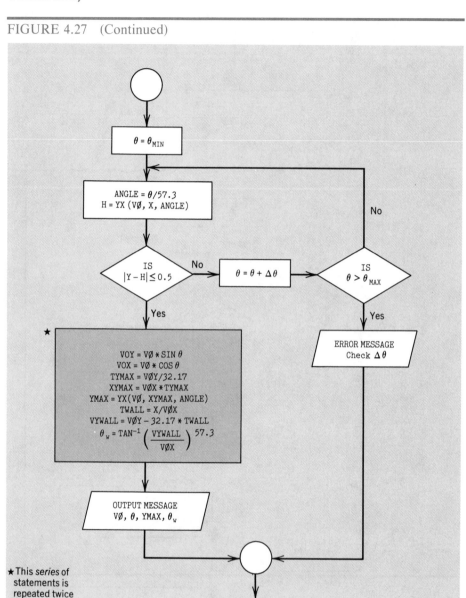

★This *series* of statements is repeated twice in Figure 4.27*e*

(*d*) Case 3.

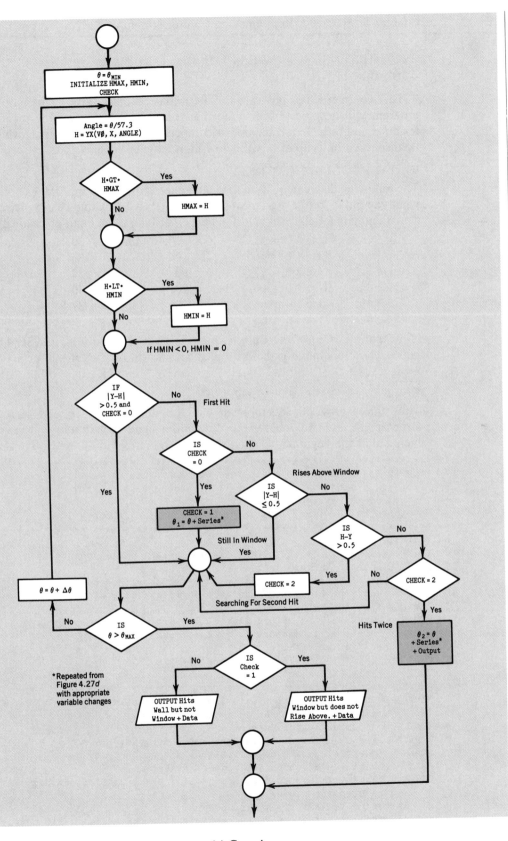

(e) Case 4.

(EXAMPLE 4.14 □ Hitting Window with Water Jet □ Continued)

This completes the program for the initial decomposition. This part of the program will now be tested. Note that the structures for writing to both the terminal and to a file are somewhat cumbersome. Formatted output, which will be discussed in Chapter 6, will allow more efficient dual output.

INITIAL PROGRAM TESTING

The following data can be used to evaluate the initial program. It does not matter what you put in for the increment at this point, but $0.1°$ would be a logical choice. The branching for all cases is tested here. It also tests the input of multiple cases.

Case 1:	Use $V0 = 55$,	$X = 100$,	and $Y = 30$
Case 2:	Use $V0 = 100$,	$X = 100$,	and $Y = 30$
Case 3:	Use $V0 = 77$,	$X = 90$,	and $Y = 20$
Case 4a:	Use $V0 = 66$,	$X = 90$,	and $Y = 40$
Case 4b:	Use $V0 = 66$,	$X = 90$,	and $Y = 37.5$
Case 4c:	Use $V0 = 66$,	$X = 90$,	and $Y = 35$

Note that all three of the Case 4 situations will print out only CASE 4 at this point. The branching for these three subcases will be part of the further refinement.

REFINED PROGRAM DEVELOPMENT

The refined program development will commence with the start of the case structure. A few additional variable definitions and declarations will have to be included at the beginning of the program. The refined program is given below.

```
*****************************************************************
*  Program for finding the angle that causes a water stream to pass  *
*     through a window a distance x from, and a height y above, the   *
*     start of the stream trajectory                                  *
*  Developed by: T. Jewell   March, 1983.  Modified February 1989.    *
*****************************************************************
C  Variable Definitions
C  YX = statement function for calculating height of trajectory at X
C  X = distance to target, feet.
C  Y = height of target.  The window is 6 in. above and below the target.
C  XMAX45 = maximum horizontal distance for trajectory at 45 degrees.
C  TXMAX = time to reach XMAX45
C  HTMIN = height of trajectory at wall for minimum angle.
C  HTMAX = height of trajectory at wall for maximum angle.
C  OUT = check variable for output.  1 = output to both screen and file.
C  VO = initial velocity for water stream, ft/s.
C  THETA = angle of trajectory in degrees.
C  ANGLE = angle of trajectory in radians.
C  REPEAT = check variable for entering a new set of data.  Yes = new data.
C  TMIN = minimum angle for search.
C  TMAX = maximum angle for search.
C  DTHETA = angle increment to use in search program.
C  HMAX = highest elevation on wall hit by stream
C  HMIN = lowest elevation on wall hit by stream
C  CHECK = check variable used in Case 4 to indicate the stream has
C          reached the window
C  H = height of trajectory at distance x
C  THETAW = angle that the stream makes with the horizontal as it passes
C           through the window
C  THETA1 = inclination angle for first hit in Case 4
C  YMAX1 = maximum height of trajectory for first hit in Case 4
```

```
C   THETW1 = angle of trajectory at window for first hit in Case 4
C   THETA2 = inclination angle for second hit in Case 4
C   YMAX2 = maximum height of trajectory for second hit in Case 4
C   THETW2 = angle of trajectory at window for second hit in Case 4
C   VOX = x component of the initial velocity
C   VOY = y component of the initial velocity
C   TYMAX = time to reach the high point in the trajectory
C   XYMAX = x distance to the high point in the trajectory
C   YMAX = y value for the high point in the trajectory
C   TWALL = time for the stream to hit the wall
C   VYWALL = y component of the velocity at the wall
C
        REAL YX,X,Y,XMAX45,HTMIN,HTMAX,VO,THETA,ANGLE,TMIN,TMAX,DTHETA,
     1      TXMAX
        INTEGER OUT,R,W,FW
        REAL HMAX,HMIN,H,THETA1,THETA2,VOX,VOY,TYMAX,XYMAX,YMAX,TWALL,
     1      VYWALL,THETAW,YMAX1,THETW1,YMAX2,THETW2
        INTEGER CHECK
        CHARACTER REPEAT*3,NAME*14
        PARAMETER (R=5,W=6,FW=4)
        DATA TMIN,TMAX/30.,75./
        YX(VO,X,A) = TAN(A)*X - .5*32.17*(X/(VO*COS(A)))**2
C   Select output option
C
        PRINT *, 'Output will be to the screen.  If you also want output'
        PRINT *, ' to a file, type 1, followed by a space, followed by'
        PRINT *, ' the file name for the output file.  For example:'
        PRINT *, ' '
        PRINT *, '1 ''TRAJEC.OUT'''
        PRINT *, ' '
        PRINT *, 'If you do not want output to a file, type any other'
        PRINT *, ' number, a space, and any letter; for example'
        PRINT *,' '
        PRINT *, '2 ''R'' '
        READ (R,*) OUT,NAME
        IF(OUT.EQ.1) OPEN(UNIT=FW,FILE=NAME,STATUS='NEW')
C   Beginning of major loop for each set of data values.
C
    10 CONTINUE
C   Data input with error trapping and range check for velocity.
        PRINT *, 'Input initial velocity (ft/s), distance to wall (ft),'
        PRINT *, ' height of target (ft), and increment for angle '
        PRINT *, ' search (degrees).'
        READ (R,*,ERR=100) VO,X,Y,DTHETA
            GO TO 110
C
C   Error condition exists
C
   100 CONTINUE
        PRINT *, 'Error occurred in input.  Please reinput.'
            GO TO 10
C
C   Error condition does not exist, continue and check VO <= 150 fps.
C
        IF (VO.GT.150.) THEN
            PRINT *, 'VO is greater than 150 ft/s.'
            PRINT *, 'Please enter a reasonable value.'
            GO TO 10
        END IF
        WRITE (W,*)    'FOR THIS RUN, THE INITIAL VELOCITY IS: ',VO
        WRITE (W,*) 'THE DISTANCE TO THE WALL IS:',X
```

(EXAMPLE 4.14 □ **Hitting Window with Water Jet** □ **Continued)**

```
      WRITE (W,*) 'THE HEIGHT OF THE TARGET IS:',Y
      WRITE (W,*) 'THE INCREMENT FOR THE ANGLE IS:',DTHETA
      IF(OUT.EQ.1) THEN
      WRITE (FW,*)    'FOR THIS RUN, THE INITIAL VELOCITY IS: ',VO
      WRITE (FW,*) 'THE DISTANCE TO THE WALL IS:',X
      WRITE (FW,*) 'THE HEIGHT OF THE TARGET IS:',Y
      WRITE (FW,*) 'THE INCREMENT FOR THE ANGLE IS:',DTHETA
      END IF
C  Compute initial values for determining case.
C
      ANGLE = 45./57.3
      TXMAX = 2.*(VO*SIN(ANGLE)/32.17)
      XMAX45 = VO*COS(ANGLE)*TXMAX
      HTMIN = YX(VO,X,TMIN/57.3)
      HTMAX = YX(VO,X,TMAX/57.3)
C
C  Beginning of Branching for Different Cases
C
      IF(XMAX45.LT.X) THEN
C  Case 1: Stream never makes it to the wall.
C
          WRITE (W,*) 'For this set of data, the stream never makes it '
          WRITE (W,*) 'to the wall.  The maximum distance for the '
          WRITE (W,*) 'trajectory is ',XMAX45
          WRITE (W,*)
          IF(OUT.EQ.1) THEN
          WRITE (FW,*) 'For this set of data, the stream never makes it'
          WRITE (FW,*) ' to the wall.  The maximum distance for the '
          WRITE (FW,*) 'trajectory is ',XMAX45
          WRITE (FW,*)
          END IF
C
      ELSE IF(HTMIN.GT.Y.AND.HTMAX.GT.Y) THEN
C  Case 2:  Stream hits the wall above the window for all angles.
C
          HMAX = -1.0E20
          HMIN = 1.0E20
          DO 20 THETA = TMIN,TMAX,DTHETA
              ANGLE = THETA/57.3
              H = YX(VO,X,ANGLE)
              IF(H.GT.HMAX) HMAX = H
              IF(H.LT.HMIN) HMIN = H
   20     CONTINUE
          WRITE (W,*) 'For this set of data, the stream hits the wall'
          WRITE (W,*) 'above the window for all angles.'
          WRITE (W,*) 'The highest it hits is: ',HMAX
          WRITE (W,*) 'The lowest it hits is: ',HMIN
          WRITE (W,*)
          IF (OUT.EQ.1) THEN
          WRITE (FW,*) 'For this set of data, the stream hits the wall'
          WRITE (FW,*) 'above the window for all angles.'
          WRITE (FW,*) 'The highest it hits is: ',HMAX
          WRITE (FW,*) 'The lowest it hits is: ',HMIN
          WRITE (FW,*)
          END IF
C
      ELSE IF ((HTMAX.GT.Y.AND.HTMIN.LT.Y)
     1         .OR.(HTMAX.LT.Y.AND.HTMIN.GT.Y)) THEN
```

```
C   Case 3:   Stream hits the window once.
C
          THETA = TMIN
C         Start of Conditional Loop:   WHILE THETA.LE.TMAX, DO 41
     40   CONTINUE
          ANGLE = THETA/57.3
          H = YX(VO,X,ANGLE)
          IF (ABS(Y-H).LE.0.5) THEN
              VOY = VO*SIN(ANGLE)
              VOX = VO*COS(ANGLE)
              TYMAX = VOY/32.17
              XYMAX = TYMAX*VOX
              YMAX = YX(VO,XYMAX,ANGLE)
              TWALL = X/VOX
              VYWALL = VOY - 32.17*TWALL
              THETAW = ATAN(VYWALL/VOX)*57.3
              WRITE (W,*) 'For this data, the stream hits the window '
              WRITE (W,*) ' once.'
              WRITE (W,*) 'The initial velocity is: ',VO,' ft/s'
              WRITE (W,*) 'The angle of the deck gun is: ',THETA,' Deg.'
              WRITE (W,*) 'The maximum trajectory height is: ',YMAX,' ft'
              WRITE (W,*) 'The angle that the stream makes with the'
              WRITE (W,*) ' window is: ',THETAW,' degrees.'
              WRITE (W,*)
              IF(OUT.EQ.1) THEN
              WRITE (FW,*) 'For this data, the stream hits the window '
              WRITE (FW,*) ' once.'
              WRITE (FW,*) 'The initial velocity is: ',VO,' ft/s'
              WRITE (FW,*) 'The angle of the deck gun is: ',THETA,' Deg.'
              WRITE (FW,*) 'The max. trajectory height is: ',YMAX,' ft'
              WRITE (FW,*) 'The angle that the stream makes with the'
              WRITE (FW,*) ' window is: ',THETAW,' degrees.'
              WRITE (FW,*)
              END IF
          ELSE
              THETA = THETA+DTHETA
              IF (THETA.GT.TMAX) THEN
                  WRITE (W,*) 'The program did not find a trajectory '
                  WRITE (W,*) ' height that fell within the window.'
                  WRITE (W,*) ' Your increment is probably too large.'
                  WRITE (W,*) ' Please check it and reenter your data.'
                  WRITE (W,*)
              ELSE
                  GO TO 40
              END IF
          END IF
C 41      End of Conditional Loop: CONTINUE (END OF WHILE STRUCTURE)
      ELSE
C   Case 4:   Check for stream hitting zero, one, or two times.
C
          THETA = TMIN
          HMAX = -1.0E20
          HMIN = 1.0E20
          CHECK = 0
C         Start of Conditional Loop: WHILE (THETA.LE.TMAX), DO 51
     50   CONTINUE
          ANGLE = THETA/57.3
          H = YX(VO,X,ANGLE)
          IF (H.GT.HMAX) HMAX = H
          IF (H.LT.HMIN) HMIN = H
          IF (HMIN.LT.0.0) HMIN = 0.0
```

(EXAMPLE 4.14 □ **Hitting Window with Water Jet** □ **Continued)**

```
       IF (ABS(Y-H).GT.0.5.AND.CHECK.EQ.0) THEN
C          If the above condition is true, the angle is still increasing,
C          in the search for the first entry into the window.  The next step
C          is to check to make sure that the angle has not reached the upper
C          limit at the end of the case structure for cases 4 a, b, and c.
       ELSE IF (CHECK.EQ.0) THEN
C          This is the first time the stream passes through the window.
           CHECK = 1
           THETA1 = THETA
           VOY = VO*SIN(ANGLE)
           VOX = VO*COS(ANGLE)
           TYMAX = VOY/32.17
           XYMAX = TYMAX*VOX
           YMAX1 = YX(VO,XYMAX,ANGLE)
           TWALL = X/VOX
           VYWALL = VOY - 32.17*TWALL
           THETW1 = ATAN(VYWALL/VOX)*57.3
       ELSE IF (ABS(Y-H).LE.0.5.AND.CHECK.EQ.1) THEN
C          The trajectory is still hitting within the window.  Check to make
C          sure the maximum angle has not been reached, then increase and
C          continue.
       ELSE IF (H-Y.GT.0.5) THEN
           CHECK = 2
C          The trajectory has risen above the window.  We are now searching
C          for the second hit.  Check to make sure that the angle has not
C          reached the upper limit at the end of the case structure for
C          cases 4 a, b, and c.
       ELSE IF (CHECK.EQ.2) THEN
C          The trajectory has now come back down and hit the window for a
C          second time.  Data for the second angle will be computed, data
C          for both angles output, and a GO TO used to skip to the end of
C          Case 4.
           THETA2 = THETA
           VOY = VO*SIN(ANGLE)
           VOX = VO*COS(ANGLE)
           TYMAX = VOY/32.17
           XYMAX = TYMAX*VOX
           YMAX2 = YX(VO,XYMAX,ANGLE)
           TWALL = X/VOX
           VYWALL = VOY - 32.17*TWALL
           THETW2 = ATAN(VYWALL/VOX)*57.3
           WRITE (W,*) 'For an initial velocity of ',VO,' ft/s,'
           WRITE (W,*) ' the trajectory hits the window two times.'
           WRITE (W,*) 'For the first hit'
           WRITE (W,*) 'The angle of the deck gun is: ',THETA1,' Deg.'
           WRITE (W,*) 'The max. trajectory height is: ',YMAX1,' ft'
           WRITE (W,*) 'The angle that the stream makes with the'
           WRITE (W,*) '  window is: ',THETW1,' degrees.'
           WRITE (W,*)
           WRITE (W,*) 'For the second hit'
           WRITE (W,*) 'The angle of the deck gun is: ',THETA2,' Deg.'
           WRITE (W,*) 'The max. trajectory height is: ',YMAX2,' ft'
           WRITE (W,*) 'The angle that the stream makes with the'
           WRITE (W,*) '  window is: ',THETW2,' degrees.'
           WRITE (W,*)
           IF(OUT.EQ.1) THEN
           WRITE (FW,*) 'For an initial velocity of ',VO,' ft/s,'
           WRITE (FW,*) ' the trajectory hits the window two times.'
```

```
             WRITE (FW,*) 'For the first hit'
             WRITE (FW,*) 'The angle of the deck gun is ',THETA1,' Deg.'
             WRITE (FW,*) 'The max. trajectory height is: ',YMAX1,' ft'
             WRITE (FW,*) 'The angle that the stream makes with the'
             WRITE (FW,*) ' window is: ',THETW1,' degrees.'
             WRITE (FW,*)
             WRITE (FW,*) 'For the second hit'
             WRITE (FW,*) 'The angle of the deck gun is ',THETA2,' Deg.'
             WRITE (FW,*) 'The max. trajectory height is: ',YMAX2,' ft'
             WRITE (FW,*) 'The angle that the stream makes with the'
             WRITE (FW,*) ' window is: ',THETW2,' degrees.'
             WRITE (FW,*)
             END IF
C            Skip to the end of Case 4
             GO TO 60
          END IF
          IF (THETA.GE.TMAX) THEN
             IF (CHECK.EQ.1) THEN
                WRITE (W,*) 'For an initial velocity of ',V0,' ft/s,'
                WRITE (W,*) ' the trajectory hits the window once,'
                WRITE (W,*) ' but does not rise above it.'
                WRITE (W,*)
             WRITE (W,*) 'For the first hit'
             WRITE (W,*) 'The angle of the deck gun is: ',THETA1,' Deg.'
             WRITE (W,*) 'The max. trajectory height is: ',YMAX1,' ft'
             WRITE (W,*) 'The angle that the stream makes with the'
             WRITE (W,*) ' window is: ',THETW1,' degrees.'
             WRITE (W,*)
                IF(OUT.EQ.1) THEN
                WRITE (FW,*) 'For an initial velocity of ',V0,' ft/s,'
                WRITE (FW,*) ' the trajectory hits the window once,'
                WRITE (FW,*) ' but does not rise above it.'
                WRITE (FW,*)
             WRITE (FW,*) 'For the first hit'
             WRITE (FW,*) 'The angle of the deck gun is ',THETA1,' Deg.'
             WRITE (FW,*) 'The max. trajectory height is: ',YMAX1,' ft'
             WRITE (FW,*) 'The angle that the stream makes with the'
             WRITE (FW,*) ' window is: ',THETW1,' degrees.'
             WRITE (FW,*)
                END IF
             ELSE IF (HMAX.LT.Y) THEN
                WRITE (W,*) 'For this set of data, the stream hits the'
                WRITE (W,*) ' wall, but does not rise high enough to '
                WRITE (W,*) ' hit the window.'
                WRITE (W,*) 'The highest it hits is: ',HMAX
                WRITE (W,*) 'The lowest it hits is: ',HMIN
                WRITE (W,*)
                IF (OUT.EQ.1) THEN
                WRITE (FW,*) 'For this set of data, the stream hits the'
                WRITE (FW,*) ' wall, but does not rise high enough to '
                WRITE (FW,*) ' hit the window.'
                WRITE (FW,*) 'The highest it hits is: ',HMAX
                WRITE (FW,*) 'The lowest it hits is: ',HMIN
                WRITE (FW,*)
                END IF
             ELSE
                WRITE (W,*) 'The program did not find a trajectory '
                WRITE (W,*) ' height that fell within the window.'
                WRITE (W,*) ' Your increment is probably too large.'
                WRITE (W,*) ' Please check it and reenter your data.'
                WRITE (W,*)
```

(EXAMPLE 4.14 □ Hitting Window with Water Jet □ Continued)

```
             END IF
          ELSE
             THETA = THETA + DTHETA
             GO TO 50
C   51    End of Conditional Loop:  CONTINUE (END OF WHILE STRUCTURE)
          END IF
C         End of Case 4
   60     CONTINUE
       END IF
C
C      End of Cases one through four.
C
C  Inquiry for additional input sets
C
       PRINT *, 'Do you want to input an additional set of data?'
       PRINT *, '  Type ''Yes'' or ''No'' (either caps or small) '
       READ(R,*) REPEAT
       IF(REPEAT(1:1).EQ.'y'.OR.REPEAT(1:1).EQ.'Y') GO TO 10
       STOP
       END
```

REFINED PROGRAM TESTING

The output for the six cases used to test the refined program is shown below. The results can be verified by hand calculations. You should also run the program with intentional data input errors and with an initial velocity greater than 150 ft/s to test the error trapping and range check.

```
FOR THIS RUN, THE INITIAL VELOCITY IS:      55.00000000
THE DISTANCE TO THE WALL IS:    100.00000000
THE HEIGHT OF THE TARGET IS:     30.00000000
THE INCREMENT FOR THE ANGLE IS:      0.10000000
For this set of data, the stream never makes it
 to the wall.  The maximum distance for the
 trajectory is       94.03170780

FOR THIS RUN, THE INITIAL VELOCITY IS:      100.00000000
THE DISTANCE TO THE WALL IS:    100.00000000
THE HEIGHT OF THE TARGET IS:     30.00000000
THE INCREMENT FOR THE ANGLE IS:      0.10000000
For this set of data, the stream hits the wall
 above the window for all angles.
The highest it hits is:      139.33891300
The lowest it hits is:        36.28417970

FOR THIS RUN, THE INITIAL VELOCITY IS:      77.00000000
THE DISTANCE TO THE WALL IS:     90.00000000
THE HEIGHT OF THE TARGET IS:     20.00000000
THE INCREMENT FOR THE ANGLE IS:      0.10000000
For this data, the stream hits the window
 once.
The initial velocity is:      77.00000000 ft/s
The angle of the deck gun is:      74.29936220 Deg.
The max. trajectory height is:      85.39823150 ft
The angle that the stream makes with the
 window is: -72.16811370 degrees.

FOR THIS RUN, THE INITIAL VELOCITY IS:      66.00000000
THE DISTANCE TO THE WALL IS:     90.00000000
```

```
THE HEIGHT OF THE TARGET IS:     40.00000000
THE INCREMENT FOR THE ANGLE IS:      0.10000000
For this set of data, the stream hits the
 wall, but does not rise high enough to
 hit the window.
The highest it hits is:     37.79270170
The lowest it hits is:    0.00000000E-01

FOR THIS RUN, THE INITIAL VELOCITY IS:      66.00000000
THE DISTANCE TO THE WALL IS:     90.00000000
THE HEIGHT OF THE TARGET IS:     37.50000000
THE INCREMENT FOR THE ANGLE IS:      0.10000000
For an initial velocity of      66.00000000 ft/s,
 the trajectory hits the window once,
 but does not rise above it.
For the first hit
The angle of the deck gun is:      53.39968110 Deg.
The max. trajectory height is:     43.63074490 ft
The angle that the stream makes with the
 window is:     -27.61575700 degrees.

FOR THIS RUN, THE INITIAL VELOCITY IS:      66.00000000
THE DISTANCE TO THE WALL IS:     90.00000000
THE HEIGHT OF THE TARGET IS:     35.00000000
THE INCREMENT FOR THE ANGLE IS:      0.10000000
For an initial velocity of      66.00000000 ft/s,
 the trajectory hits the window two times.
For the first hit
The angle of the deck gun is:      49.59973910 Deg.
The max. trajectory height is:     39.25905990 ft
The angle that the stream makes with the
 window is:     -22.15950390 degrees.
For the second hit
The angle of the deck gun is:      60.69956970 Deg.
The max. trajectory height is:     51.48338320 ft
The angle that the stream makes with the
 window is:     -44.79701610 degrees.
```

SUMMARY

The control structures of selection and repetition provide you with the ability to analyze realistic problems with the FORTRAN computer language. You have learned several different structures. Now you have to learn how to apply them to the logical solution structure for problems. These structures automate the steps you would go through. In order to apply them correctly, you have to understand the complete logic of the problem solution.

In applying selection and repetition structures, you should pay particular attention to the structure of your programs. These structures can be used to develop well-structured programs. However, if they are used ineffectively they can obscure the logic of the program. After developing your own programs, check them to find ways that you could improve the structure. This will help you learn to develop better-structured programs the first time through.

One of the most useful tools in debugging programs is to output intermediate results to see what your selection and repetition structures are doing. You should also use literals to flag where in the program you outputted the intermediate results. Outputting the values for variables used in logical expressions and for variables used as indexes or parameters in DO loops will help you understand why structures are not performing as you expect them to.

REFERENCES

Borse, G. J., *FORTRAN 77 and Numerical Analysis for Engineers*, PWS Publishers, Boston, 1985.

Chapra, S. C., and Canale, R. P., *Introduction to Computing for Engineers*, McGraw-Hill, New York, 1986, Chapter 23.

Ellis, T. M. R., *A Structured Approach to FORTRAN 77 Programming*, Addison-Wesley, London, 1982, Chapters 4 and 5.

Etter, D. M., *Structured FORTRAN 77 for Engineers and Scientists*, 2nd. ed., The Benjamin/Cummings Publishing Company, Inc., Menlo Park, CA, 1987, Chapter 3.

Kaufman, C. D., and Janna, W. S., "The Use of the Microcomputer as a Supplement to the Study of Pipe Flow," *Journal*, Vol. 6, No. 3, Computers in Education Division of ASEE, Wash., D.C., July–Sept. 1986.

Press, W. H., Flannery, B. P., Teukolsky, S. A., and Vetterling, W. T., *Numerical Recipes: The Art of Scientific Computing*, Cambridge University Press, Cambridge, 1986, Chapter 5.

Red, W. E., and Mooring, B., *Engineering: Fundamentals of Problem Solving*, Brooks/Cole Engineering Division, Wadsworth Inc., Monterey, CA, 1983.

EXERCISES

4.1 GENERAL NOTES

You should make your programs as general as possible. Some ways you can do this are to input data through variables, to set arrays to some maximum size and input the actual size for the problem, and to input parameters for conditional and absolute loops when appropriate.

All input data should be output.

Do not send output to a file or printer until you are sure you do not have any infinite loops. Infinite loop output to a file will fill up your allocated disk storage space. Infinite loop output to a printer will waste paper if your system does not have length checks for printouts. Do not make hard copies of files unnecessarily. It slows you down and is a waste of paper. Look at your file via the editor or screen review commands. Hard copies can be helpful when you are debugging longer programs or checking the logic of extended portions of the program.

Always complete a hand solution of your problem using the logic sequence that the computer will go through. This will help you in your debugging. It will also help you develop the program logic that you will need.

4.2 MODIFICATION OF EXAMPLES

4.2.1 Example 4.4

1. Modify Example 4.4 so that it uses a four-tiered cascaded IF THEN ELSE structure.

2. Modify the program of Example 4.4 to determine if the parabola opens up or down.
3. Press et al. (1986) have recommended that the following relationships be used to compute the real roots of a quadratic equation to avoid numerical inaccuracies when either of the coefficients A or C (or both) are small.

$$q = -\frac{1}{2}[B + sgn(B)\sqrt{B^2 - 4AC}]$$

$$x_1 + \frac{q}{A} \qquad x_2 = \frac{C}{q}$$

The function sgn(B) means the sign of B. Modify the program of Example 4.4 to use this method for computing the real roots. Compare the output with the output for the original program.

4.2.2 Example 4.5

1. Modify the program of Example 4.5 to also find the magnitude of vectors **A** and **B.**
2. Modify the program of Example 4.5 so that it will provide the option of finding either the sum or difference of the two vectors.

4.2.3 Example 4.6

1. Modify the program for Example 4.6 to stop entry with a zero data value.
2. Modify the program of Example 4.6 to stop entry with the END = clause.

4.2.4 Example 4.7

Modify the program of Example 4.7 so the elements of the array that contain the minimum and maximum values are also identified.

4.2.5 Example 4.10

1. Modify the program of Example 4.10 to copy the array into a second array and then sort the second array. Output both arrays.
2. Modify the program of Example 4.10 so that the array can be sorted into either ascending or descending order, at the option of the user.
3. Modify Example 4.10 so that your program will sort character arrays into ascending or descending alphabetical order.

4.2.6 Example 4.11

Modify the program of Example 4.11 so that it will use sentinel control of input. Use V0 = 0.0 as the sentinel value. Test the modified program with the data of Example 4.11, and then test with V0 = 0.0.

4.2.7 Example 4.14

1. Modify Example 4.14 so that it will output the range of angles that hits the window each time the stream passes through it.
2. Make the necessary modifications in the program of Example 4.14 to accommodate either English or metric units.
3. Add error trapping for all the input of the program for Example 4.14.

4.3 GENERAL SYNTAX AND STATEMENT STRUCTURE

4.3.1 Selection Control

Develop a program to input the diameter of a circular pipe. Include error trapping and range checks to make sure the diameter is between 6 and 36 in. Output the diameter and cross-sectional area of the pipe. Obtain a printout that shows that you have thoroughly checked all the error conditions of the program.

4.3.2 Developing DO Structures

1. The formula

$$Z = \frac{e^{ax} - e^{-ax}}{2} \sin(x + b) + (a) \log \left(\frac{b + x}{2} \right)$$

   is to be evaluated for all combinations of values of x from 0.0 to 2.0 in increments of 0.1, values of a from -1.0 to 1.0 in increments of 0.1, and values of b from 0.5 to 1.5 in increments of 0.2. Develop a set of nested DO loops that will accomplish this. How many values of Z would be generated?

*2. The following two-dimensional array is stored in the computer:

6.	7.	5.	8.	1.	
16.	21.	8.	2.	100.	= **A**
21.	33.	2.	5.	25.	

   Develop a program segment that will search this array for the largest numeric value, then divide each element of the array by that value.

3. Given two polynomial functions $F1(x)$ and $F2(x)$, develop a program segment that will evaluate $F1(x) - F2(x)$ for $x = 0., 0.5, 1.0, \ldots, 10.0$. Output x, $F1(x)$, $F2(x)$, and DIFF without using subscripted variables.

*4. Do the previous problem with subscripted variables

4.3.3 Evaluating the Results of DO Structures

1. What is the final value of Y for the following statements?

   a.
```
      Y = 0.0
      DO 30 X = 1.,5.,.5
      Y = Y + X
   30 CONTINUE
```

   *b.
```
      Y = 0.0
      DO 20 I=1,6
         DO 30 J=1,15
            IF (Y.GT.100.) THEN
               Y = Y/2.
               PRINT *,Y
               STOP
            ELSE
               Y = Y + 1.
            END IF
   30      CONTINUE
   20 CONTINUE
```

2. After the following statements have been executed, what is the value of K?

```
      DO 700 I = 1,100
         X = I - 1
         DO 700 J = 1,10
            Z = J + 4
            IF(X**2 - 2.0*X .EQ. Z) GO TO 600
700      CONTINUE
      GO TO 800
600 K = I + J
800 CONTINUE
```

   *How could the above program segment have been written in a more structured fashion?

3. Identify the sequence of operations in each of the following program segments by specifying the output from the output statements:

   a.
```
      DO 100 I=1,4
         J = I**2
         IF (J.LE.5) THEN
            K = J
         END IF
         PRINT *,I,J,K
100   CONTINUE
```

   *b.
```
      DO 400 I=1,5
         J = I**2
         IF (J.GT.15) THEN
            K = 0
         ELSE
            K = J
         END IF
         PRINT *,I,J,K
400   CONTINUE
```

   c.
```
      DO 700 I=5,2
         PRINT *,I
700   CONTINUE
```

4.3.4 Finding Errors in DO Structures and DO Statements

1. Various forms of the DO statement are given below. Some are correct, but others violate one or more of the basic rules governing DO statements. Default variable types are in effect. Identify violations, and indicate suitable corrective measures.

```
   a. DO 100 III = II,I,3
  *b. DO N 100 = 1,20
   c. DO 20 J = X1,X2,2
  *d. DO 40 I = N,M,N
   e. DO 200 I = 40
  *f. DO 300 A = 1,N4,1
   g. DO 20 X = 10.,0.,-.1
  *h. DO 30 I = 20,1
   i. DO 40 I = 1,N+1,4
  *j. DO I = 1,20
```

2. Indicate what, if anything, is wrong with each of the following sets of DO loops.

a.
```
    DO 20 I = 1,M
        DO 30 J = 1,N
30  I = I-1
20  CONTINUE
```

*b.
```
    DO 30 I = 1,4
        DO 50 J = 1,M
            DO 40 K = 1,N

30  CONTINUE
40              CONTINUE
50      CONTINUE
```

c.
```
    GO TO 300
    DO 400 I = 1,100

300     IF (X.GT.1000.) GO TO 500

400 CONTINUE
500 CONTINUE
```

*d.
```
    DO 100 I = 1,100,-1
        DO 100 K = 3,50,2

100 CONTINUE
```

e.
```
    DO J = 1,20
        PRINT *,J,J**3
10  CONTINUE
```

*f.
```
    DO 10 I = 1,N
        DO 20 J = 1,M
            PRINT *,I,J,I**N,I+J,J**M,M*N
10  CONTINUE
20      CONTINUE
```

g.
```
    READ (5,*) N
    DO 200 J = 1,N
        N = N-2
        IF(N.LT.0) GO TO 300
200 CONTINUE
300 CONTINUE
```

*h.
```
    DO 100 I = 1,100
        DO 100 J = 3,50,2
            DO 100 I = 1,20

100 CONTINUE
```

4.3.5 Evaluating Logical Expressions

*1. Indicate how the following logical expressions would evaluate ('true' or 'false'). For each of the examples, $X = 3.$, $Y = 5.$, and $Z = 2.$.

```
(X - 3.0 .LT. 2.0 .AND. Y + 4.0 .GT. 9.0)

(X .LT. 3. .OR. Y .LT. 4.)

(X*Y .LT. Y**2)

(X .GT. Y .AND X .GT. Z .OR. Y .LT. Z)
```

2. What are the values of the following logical expressions?

   a.   (1.GT.2)
*b.   ((2*2).GE.4)
   c.   ((1+3).LE.4)
*d.   (3.LT.4.AND.7.GE.5)
   e.   ((3.GT.2.OR.(1+2).LT.3).AND.4.GE.3)
*f.   (3.GT.2.AND.(1+2).LT.3.OR.4.LE.3)

4.3.6 Conditional Loops

Develop a conditional loop that will execute while the value of X is greater than zero. X starts out with a value of UPPER and is decreased each time through the loop by the increment DECRE. Within the loop, Z is set equal to $1./X**3$, and the values of X and Z are printed out. Include an escape mechanism to exit the loop if more than 1,000 repetitions take place.

4.4 CHEMICAL ENGINEERING

4.4.1 [ALL] Ideal Gas Law

Develop a program to implement the flowchart for Exercise 2.3.1. Use $R = 0.0826$ atm-ℓ/gmol/°K, $T_c = 405.6$°K, $P_c = 112.5$ atm, $T = 25$°C, MW $= 7$ (ammonia), $W = 732$ grams, and volumes from 90 to 120 ℓ in 2-ℓ increments.

4.4.2 [ALL] Chemical Kinetics

Develop a program to implement the flowchart for Exercise 2.3.2. Use $C_0 = 0.05$ mol/ℓ, $k_1 = 0.011$, and $k_2 = 0.31$. Use times from 0 to 240 min in increments of 10 min.

4.4.3 [ALL] Depth of Fluidized-Bed Reactor

Develop a program to implement the flowchart for Exercise 2.3.3. Use an unexpanded height of 5.0 ft, an unexpanded void fraction of 0.45, reactor flow rate of 0.004 m³/s/m², viscosity $= 8.13 \times 10^{-3}$ N-s/m², fluid density $= 1,000$ kg/m³, particle density $= 2,666$ kg/m³, and particle size and distribution as follows:

Size Fraction	Particle Size (10^{-3} m)
.25	1.0
.15	2.0
.20	4.0
.30	6.0
.10	8.0

4.4.4 [ALL] Process Design, Gas Separation

Develop a program to implement the flowchart for Exercise 2.3.4. You desire to reach a 95 percent absorption efficiency for ethane. The following data should be used:

	Top	Bottom
k Ethane	1.439	1.470
k Methane	8.000	9.500
k Propane	0.480	0.510
Gas Flow (mol/s)	50	53
Liquid Flow (mol/s)	100	97

4.4.5 [ALL] Dissolved Oxygen Concentration in Stream

Exercise 2.3.5 developed a flowchart for a program to estimate the dissolved oxygen concentration in a river downstream from a point source of pollution. The mathematical relationships were developed as a function of the distance below the discharge point. You are now to develop a FORTRAN program to implement that flowchart. The dissolved oxygen concentration in the river and the distance downstream should be stored in separate one-dimensional arrays. Continue the analysis until the pollution has moved 20 miles downstream, or 41 data points have been calculated, whichever occurs first. Test your program with the following data:

Initial concentration of pollutant = 12.4 mg/ℓ

Velocity of river = 0.30 ft/s

Initial dissolved oxygen deficit = 4.1 mg/ℓ

Deoxygenation constant = 0.50 day^{-1}

Temperature of river water = 20°C

Dissolved oxygen saturation concentration = 12.37 mg/ℓ

Diffusivity of oxygen in water at 20°C = 0.81 × 10^{-4} ft^2/s

Depth of river = 4.2 ft

4.5 CIVIL ENGINEERING

4.5.1 [ALL] Equilibrium-Truss Analysis

Develop a program to implement the flowchart of Exercise 2.4.1. Use your program to find the reactions for the truss shown in Figure G.1 of Appendix G, and for the truss shown in the figure accompanying this exercise.

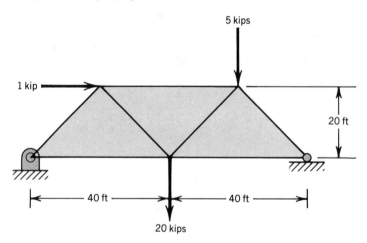

Figure for Exercise 4.5.1

4.5.2 [ALL] Open Channel Flow Analysis

1. [ADVANCED] Develop a program to implement the flowchart of Exercise 2.4.2 (2) to estimate the depth of flow for a trapezoidal channel given the flow rate and

the necessary dimensions. Use your program to determine the depth when $b = 4$ ft, $z = 2$ ft/ft, $n = 0.03$, $S_0 = 0.005$ ft/ft, and $Q = 100$ ft^3/s.

2. Develop a program to implement the flowchart of Exercise 2.4.2 (3) to estimate a series of flow rate vs. depth values for a trapezoidal open channel. Two separate one-dimensional arrays should be used to store the flow rate and depth data. Use your program to estimate the flow rate for depths ranging from 0.0 ft to 10.0 ft in increments of 0.5 ft for a trapezoidal channel with $b = 4$ ft, $n = 0.03$, $S_0 = 0.001$ ft/ft, and $z = 2$ ft/ft.

4.5.3 [ALL] Dam Stability Analysis

Develop a program to test the stability against sliding for the concrete gravity dam of Exercise 2.4.3. Use the data of Exercise 3.5.3 to test the program.

4.5.4 [ALL] Force on Submerged Gate

Develop a program to implement the flowchart of Exercise 2.4.4. Test the program using the data of Exercise 3.5.4. In addition, test it with a circular gate of 6-ft radius and a triangular gate of base 8 ft and height 10 ft. The top of both of these gates is at the same depth as the top of the rectangular gate.

4.5.5 [ALL] Volume of Excavation

1. You are to develop a program to calculate excavation requirements for sewer trenches. The cross-section of the trench will be rectangular, but the width and depth of the trench will change as you progress from one station to the next along the length of the trench. The depth is assumed to vary linearly between stations. The width is constant between stations, but can change at stations. Program input will include trench width, ground surface elevation, and sewer invert (bottom) elevation data for up to 50 distance increments, or stations, along the line. The distance increments are not necessarily equal. Write your program using subscripted variables and a function statement to compute the incremental fill.
   Use the following data to test your program:

Station	Station Distance (ft)	Trench Width STA i to i+1 (ft)	Ground Surface Elevation (ft)	Invert Elevation (ft)
0	0	4	56	50
1	300	4	58	49
2	500	5	54	48
3	800	5	60	47
4	1200	6	62	45
5	1400	—	53	44

2. Modify the program developed above to handle trapezoidal cross sections, and the continuous change of cross section from one station to the next. Test your program with the data given above, assuming that the cross-section widths at stations 0 and 1 are 4 ft, at stations 2 and 3 they are 5 ft, and at stations 4 and 5 they are

6 ft. Also test it assuming that the sides of the trapezoid are sloped at 45° with a bottom width equal to the trench width given above.

4.5.6 [2] Classification of Soil Samples

Write a program to implement the flowchart of Exercise 2.4.6. Use the following data to test your program:

Sample	Grain Size, mm
1	0.001
2	1.2
3	5.1
4	100
5	0.003
6	75
7	2.0

4.5.7 [8] Column Buckling

1. The elastic buckling load of a slender column with both ends pinned is given by

$$P_{cr} = \frac{\pi^2 EI}{L^2}$$

in which P_{cr} = critical load that will produce buckling, lbs or newtons; E = modulus of elasticity, psi or N/m²; I = moment of inertia of the column cross-sectional area about its weak axis, in.⁴ or m⁴; and L = effective length, inches or meters. The effective length for a column pinned at both ends is the length of the column. Develop a program that will predict the critical loading for a 1 × 2-in. solid bar made out of steel that has a modulus of elasticity of 29.0×10^6 psi for lengths from 12 in. to 60 in. in increments of 2 in. Values for length and critical load should be stored in one-dimensional arrays.

2. Two formulas are presented for determining the safe loading L in pounds per square inch on a column with slenderness ratio of S. The slenderness ratio is the effective length divided by the minimum radius of gyration of the cross section. The equations are

$$L = \begin{cases} 16{,}500 - 0.475S^2 & \text{if } S < 100 \\ \dfrac{17{,}900}{2 + (S^2/17{,}900)} & \text{if } S \geq 100 \end{cases}$$

Develop a program that will input a slenderness ratio and then estimate the safe loading. Use $S = 60$ and $S = 150$ to test your program.

4.5.8 [6,8] Addition of Two Hydrographs

Two unsteady flow hydrographs are to be added together where two pipes come together and enter a third, larger pipe. Each of the input hydrographs has 12 values of flow recorded at one-hour intervals. Develop a program that will add the hourly flows for the two hydrographs together and output all three hydrographs. Your program should also determine the average flow rate for each hydrograph and the cumulative volume of flow as a function of time for the summation hydrograph.

Hour	Hydrograph 1	Hydrograph 2
1	30	25
2	20	20
3	10	5
4	10	10
5	20	15
6	40	45
7	60	50
8	70	80
9	40	50
10	40	30
11	50	40
12	20	10

4.6 DATA ANALYSIS AND STATISTICS

4.6.1 [ALL] Histogram

Develop a program to implement the flowchart of Exercise 2.5.1 to count the frequency of occurrence of soil classifications for a series of samples. Test your program using at least 50 grain-size data points.

4.6.2 [ALL] Sorting

1. Develop a program to sort values in an array by the Shell sort method.
2. Add a counter to the bubble and Shell sort programs to evaluate the relative efficiency of each. Test the two programs on several arrays of varying size.

4.6.3 [2,3,5] Time Conversion

Develop programs for one or more of the flowcharts for time conversion developed in Exercise 2.5.3. Input several different times to test your programs.

4.6.4 [2,3,5] Date Conversion

Develop a program for conversion from calendar to Julian date, or vice versa, according to the flowcharts developed for Exercise 2.5.4.

4.6.5 [5,6,7] Sampling and Distribution of Means

A 51-element array contains the numbers from zero to 50. Develop a program that will sample 10 values from this array with replacement, which means that the same value can be drawn more than once for a sample. You will need to use a random number generator for the sampling process. The mean value for the sample is to be determined. The process is to be repeated a number of times, with a new random sample taken for each repetition. A frequency distribution for the mean values is to be calculated and stored in a separate array. Mean values will be rounded to the nearest whole number, with the frequency counts taken for the resulting whole numbers. Run your program for 100, 1,000, and 10,000 repetitions.

4.7 ECONOMIC ANALYSIS

4.7.1 [ALL] Economic Formulas

Develop a program for evaluating economic formulas as described in Exercise 2.6.1. Test your program with representative data that will test all of the options, P to A, P to F, P to P, A to F, A to P, A to A, F to P, F to A, and F to F.

4.7.2 [ALL] Nonuniform Series of Payments

Develop a program that will convert any nonuniform series of payments into either an equivalent present worth or an equivalent future worth. *Hint:* A nonuniform series of payments can be treated as a number of single payments P, with different numbers of compounding periods. Up to 100 payments should be accommodated by your program. Use the following data to test your program and to find both the present worth and future worth at an interest rate of 8 percent:

Year	Payment $
1	2,000
2	1,500
3	3,000
4	1,000
5	1,250
6	1,800
7	2,000
8	2,200

4.7.3 [ALL] Multiple Interest Rates or Number of Compounding Periods

Develop a program to compute economic formulas for multiple interest rates or multiple number of compounding periods. The PWF, SFF, and USPWF economic factors should be computed. For the multiple interest rates, input the starting rate, ending rate, and number of intermediate rates that you want to look at. For the number of compounding periods, input the first number, last number, and the desired increment for compounding periods in between. Test your program for interest rates from 6 to 10 percent, with seven intermediate rates. Also test your program with compounding periods from 10 to 20 in increments of 2.

4.7.4 [ALL] Internal Rate of Return

Develop a program to implement the flowchart of Exercise 2.6.4. Check your program with a present worth of $40,000 and annual payments of $4,000 over 25 years. Check your answer by substituting back into the capital recovery factor.

4.7.5 [ALL] Mortgage Computations

Develop a program to implement the flowchart of Exercise 2.6.5. Test your program for a $10,000 auto loan at 11 percent for 5, 6, 7, or 8 years. Payments will be made monthly. Also test it to find the monthly payments for a home mortgage of $80,000 at 9 percent, for 20, 25, 30, or 35 years.

4.8 ELECTRICAL ENGINEERING

4.8.1 [ALL] Diode Problem

Develop a program to implement the flowchart of Exercise 2.7.1.

4.8.2 [ALL] Capacitance and Resistance

Develop a program to implement the flowchart of Exercise 2.7.2. Use the data of Exercise 3.8.2 and time increments of 0.002 s. The switch is closed for 0.05 s, then opened up for 0.05 s.

4.8.3 [ALL] Signal Processing Circuits

Develop a program to implement the flowchart of Exercise 2.7.3. Use the data of Exercise 3.8.3 and use driving frequencies from 500,000 Hz to 1,600,000 Hz in increments of 100,000 Hz. Test your program for tuning frequencies of 500,000; 1,000,000; and 1,500,000 Hz.

4.8.4 [2,3,5] Parallel and Series Resistance

Develop a program to implement the flowchart of Exercise 2.7.4. Use resistances of 5, 10, and 20 ohms, and find both the equivalent series and parallel resistances.

4.8.5 [2,3,5] Parallel and Series Capacitance

Develop a program to implement the flowchart of Exercise 2.7.5. Use capacitances of 1×10^{-6}, 2×10^{-6}, and 10×10^{-6}, and find both the equivalent series and parallel capacitances.

4.9 ENGINEERING MATHEMATICS

4.9.1 [ALL] Determinant and Cramer's Rule

Develop a program to implement the flowchart of Exercise 2.8.1. Find the determinant for the following two matrices:

$$\begin{vmatrix} 2 & 3 \\ 1 & 5 \end{vmatrix} \qquad \begin{vmatrix} 2 & 3 & 1 \\ 0 & 1 & 2 \\ 4 & 2 & 0 \end{vmatrix}$$

4.9.2 [ALL] Vector Cross Product, Unit Vector, Directional Cosines, Vector Components

Develop a program to implement the flowchart of Exercise 2.8.2. Use the following data to check the different cases:

   a. Magnitude = 50.4, and points = (11,17,30) and (0,0,0)
   b. Two end points = (0,0,0) and (22,34,30)
   c. Magnitude = 50.4 and unit vector = 0.437$\mathbf{i}$ + 0.675$\mathbf{j}$ + 0.595$\mathbf{k}$
   d. Components = 22$\mathbf{i}$ + 34$\mathbf{j}$ + 30$\mathbf{k}$

4.9.3 [ALL] Centroids and Moments of Inertia

Develop a program to implement the flowchart of Exercise 2.8.3. Test your program by finding the area, centroid, and moment of inertia about the centroidal axes of a rectangle of base 20 in. and height 25 in. Do the same for a circle of radius 20 in., but also find its moments of inertia about the x and y axes when the center of the circle is at 100,50.

4.9.4 [ALL] Population Growth

Develop a program to implement the flowchart of Exercise 2.8.4. Use the data for Exercise 3.9.4 and generate estimates for each of the next 20 yr.

4.9.5 [ALL] Vertical and Horizontal Curves

Develop a program to implement the flowchart of Exercise 2.8.5. Test your program using a radius of 500 ft, angle $\theta = 100°$, and angle $\alpha = 40°$.

4.9.6 [2,3] Area of Triangle

Develop a program to implement the flowchart of Exercise 2.8.6. Test your program using the following three sets of data:

Set 1 (3,4), (6,5), (5,6)

Set 2 (3,4), (4,5), (5,6)

Set 3 (3,4), (2,5), (5,6)

4.9.7 [2,5,6] Table of Squares and Cubes

Write a program to implement the flowchart of Exercise 2.8.7. Your output should be X, X^2, and X^3 for values of X from 1 to 10 in increments of 1. Include headings in your output. The lower bound, upper bound, and increment should be input as variables.

4.9.8 [3,5] Vector Dot Product

Develop a program to determine the vector dot product for two n-dimensional vectors. Test your program using vectors $\mathbf{X} = (3,2,-4)$ and $\mathbf{Y} = (-2,1,0)$, and also $\mathbf{X} = (1,3,2,-4)$ and $\mathbf{Y} = (-2,1,0,-5)$.

4.9.9 [2,5] Angles and Sides of a Triangle

Given a triangle with sides a, b, and c; opposite angles A, B, and C; and the following triangle rules:

$$A + B + C = 180°$$
$$c = \sqrt{a^2 + b^2 - 2ab \cos C}$$
$$\frac{a}{\sin A} = \frac{b}{\sin B} = \frac{c}{\sin C}$$

Develop a program that will accomplish the following:

a. If sides a, b, and c are known, compute angles A, B, and C.
b. If two sides and an angle are given, compute the other side and the remaining two angles.

4.9.10 [5] Evaluation of Series

Series approximations for e^x and $\sin(x)$ are given by the equations

$$e^x = 1 + x + \frac{x^2}{2} + \frac{x^3}{3!} + \frac{x^4}{4!} + \cdots$$

$$\sin x = x - \frac{x^3}{3!} + \frac{x^5}{5!} - \frac{x^7}{7!} + \cdots$$

Develop a program that will compare either of these approximations with the corresponding intrinsic function for a given value of the argument x as each term of the approximation is added. Continue until the difference between the approximation and the intrinsic function is less than a selected amount. Output should provide data as each term is added, including the value of the ith term. Output the number of terms that it takes to meet the convergence criterion. Compare the computer results with hand calculator results for the same value of the argument. Use $X = 1$ to check the e^x approximation, and $x = 1$ radian to check the $\sin(x)$ approximation.

4.9.11 [5] Calculating the Area of an Irregular Shape

The area of any polygon defined by n points can be found by applying the algorithm

$$\text{Area} = \frac{1}{2} \left| \sum_{k=0}^{n-1} (x_k y_{k+1} - x_{k+1} y_k) \right|$$

in which x_k and y_k are the x and y coordinates of point k. Set x_0 and y_0 equal to x_n and y_n. Develop a program to implement this algorithm for any area defined by up to 100 points, and test it with the four points (1,4), (3,4), (4,3), and (4,1).

4.9.12 [5,6] Volume and Surface Area of Box, Sphere, or Cone

Develop a program to find the volume and surface area of a box, sphere, or cone. Use a character variable to select the shape to be analyzed. Test all three options.

4.9.13 Determination if Point Is Inside, On, or Outside a Circle of Radius *r*

1. Develop a program to determine whether a given point is inside, on, or outside a circle of radius r. Data should be output under the following headings:

    X   Y   R   INSIDE   ON   OUTSIDE

   A 1 should be output under the appropriate heading to indicate where the point falls, and zeros output under the other two columns. For example, if the point (x, y) is outside the circle, a 1 should be output under OUTSIDE, and zeros under INSIDE and ON. Any number of data sets should be accommodated. Test your program using a circle of radius 5 and the three sets of data, (6,6), (4,3), and (3,3).
2. Modify the previous program to use character variables to indicate the position of the point. The column headings should be

    X   Y   R   POSITION

   and the words INSIDE, ON, or OUTSIDE should be output in the POSITION column.

4.10 INDUSTRIAL ENGINEERING

4.10.1 [ALL] Project Management

Develop a program to implement the flowchart of Exercise 2.9.1. Use the network shown in Figure J.1 of Appendix J to test your program.

4.10.2 [ALL] Quality Control

Develop a program to implement the flowchart of Exercise 2.9.2. Use the following data to test your program:

	Defect					
Part	*1*	*2*	*3*	*4*	*5*	*6*
1	X			X		
2			X	X		
3	X			X		
4		X		X	X	
5			X			
6	X			X		
7				X		
8				X		
9	X		X	X		
10	X					
11					X	
12	X	X		X		
13						X
14			X	X		
15	X			X		
16				X		
17	X		X	X		
18	X			X		
19	X			X	X	
20			X	X		

4.10.3 [ALL] Managerial Decision Making

Develop a program to implement the flowchart of Exercise 2.9.3. Use a total cost function of \$10,000 + \$4P − $\$P^{0.5}$, and a return function of \$8P + $\$1.5P^{1.2}$.

4.10.4 [ALL] Queuing (Wait Line) Theory

Develop a program to implement the flowchart of Exercise 2.9.4. Use your program to find the optimum number of repairmen when 65 machines are available, up to ten repairmen can be hired, the failure rate for the machines is 0.02/hr, the repair rate is 0.4/hr, wages for repairmen are \$16/hr, and the loss per machine per hour is \$300.

4.11 MECHANICAL AND AEROSPACE ENGINEERING

4.11.1 [ALL] Shear and Bending Moment

Develop a program to implement the flowchart of Exercise 2.10.1. Find the shear and bending moment for a 20-ft simply-supported beam subjected to a uniform load of 2,000 lb/ft. Also find the shear and bending moment for the same beam subjected to a point load of 5,000 lb at 8 ft from the left end. Then find the shear and bending moment when both loads are applied simultaneously.

4.11.2 [ALL] Crank Assembly Analysis

Develop a program to implement the flowchart of Exercise 2.10.2. Use the data from Exercise 3.11.2 and angles from 0 to 360° in 10° increments.

4.11.3 [ALL] Friction

Develop a program for the flowchart of Exercise 2.10.3. Output whether nothing would happen, the crate would slip, the crate would tip, or the dolly would slip. Use the following data to check your program:

Height of dolly (A) = 0.25 m

Length of dolly (B) = 1.5 m

Length of crate (C) = 0.6 m

Height of crate (D) = 1.5 m

Weight of dolly (T) = 98 N

Coefficient of friction for crate (U) = 0.35

Coefficient of friction for dolly (V) = 0.50

Weight of crate (W) = 588 N

and the following combinations of applied force and point of application:

Height of Applied Force (m)	Magnitude of Force (N)
0.50	100
0.50	165
0.90	210
1.40	140

4.11.4 [ALL] Variation of Atmospheric Pressure with Altitude

Develop a program to implement the flowchart of Exercise 2.10.4. Test your program for the following data: $T_0 = 59°F$, $p_0 = 14.7$ psia, $Z = 20,000$ ft, and $p = 6.79$ psia.

4.11.5 [ALL] Aerospace Physics

Develop a program to implement the flowchart of Exercise 2.10.5. Use the data of Exercise 3.11.5 and acceleration for 5 s followed by 3 s of constant-velocity motion. Use increments of 0.2 s.

4.11.6 [ALL] Energy Loss in Circular Pipeline

Flowcharts for three applications of energy loss in a circular pipeline were developed in Exercise 2.10.6. Develop a FORTRAN program or programs to implement one or more of these flowcharts. Check your programs with the data of Exercise 3.11.6.

4.11.7 [ALL] Torsion

Develop a program to implement the flowchart of Exercise 2.10.7. Determine the OD for an ID = 2.5 in., a length of 4 ft, $G = 1.1 \times 10^7$ psi, torque = 16,000 lb-ft, maximum angular deformation = 0.5°, and maximum allowable torsional stress = 8,000 lb/in.^{2}.

4.12 NUMERICAL METHODS

4.12.1 [ALL] Interpolation

Develop a program for the problem of Exercise 2.11.1 for linear or quadratic interpolation. Test your program by estimating the linear interpolation for y when $x = 2.5$, and the quadratic interpolation for the same value of x, using points 1, 2, and 3; then using points 1, 2, and 4.

4.12.2 [ALL] Solution of Polynomial Equation

1. Develop a program to search for a root of a third-order or higher polynomial equation within an interval. Use a small interval and check for a change in the sign of the function. Interpolate between the last two values to approximate the root. Use the equation

$$Y = 0.4X^4 - 5.717X^3 + 28.35X^2 - 56.03X + 35.00$$

   to check your program, and search for the smallest positive root in the interval $X = 1$ to $X = 6$.
2. Develop a program to search an interval to see if there is a minimum or maximum value for a third-order or higher polynomial (finding zero of derivative). Use the second derivative to determine if the point is a minimum or maximum. Use a small interval and check for a change in the sign of the derivative. Interpolate between the last two values to approximate the independent variable value that will give zero for the derivative. Find the minimum and maximum points for the equation above in the interval $X = 1$ to $X = 6$.

4.12.3 [3,5,7,8] Data Smoothing

A one-dimensional array named D contains 50 elements of experimental data. You desire to apply a smoothing routine to these data to help reduce the effects of random errors. To do this, you will generate a new array named E, the elements of which will be calculated by

$$E_i = \frac{(D_{i-1} + D_i + D_{i+1})}{3}$$

The first element of E will be the average of the first two elements of D. The last element of E will be the average of the last two elements of D.

4.12.4 [5,6,7,8,9] Saddle Point of a Hyperbolic Paraboloid

The equation

$$z = 2x^2 - 3y^2 + x + y + 8 = f(x,y)$$

represents a model of a hyperbolic paraboloid surface that has been used for some modernistic roof systems. The shape of this surface is represented by the accompanying sketch. Your job is to locate the coordinates of the saddle point of this surface and to determine the height of the roof at that point. You will do this first by writing and running a computer program that will find the saddle point to two decimal place accuracy (both x and y accurate to two decimal places), then you will confirm your answer through the methods of calculus. Input to your program will include the initial point, the initial increment, the final increment, and the coefficients of the equation. Use a function statement for $f(x,y)$.

Preliminary investigations have shown that the saddle point lies somewhere in the region bounded by $x = \pm 1.0$ and $y = \pm 1.0$. A good approach to the computer algorithm would be to

a. Start at point 1,1.
b. Vary y in increments of 0.1 until you find an approximate maximum. You will need to have a test for the maximum point.
c. Keeping the final value of y from (b), vary x in increments of 0.1 until you find an approximate minimum point (test for minimum point).
d. Reduce increment size to 0.01 and repeat steps (b) and (c).

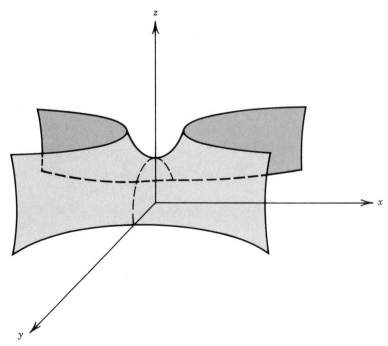

Figure for Exercise 4.12.4

4.12.5 [5,6,7,8,9] Polynomial Surface

Generalize the program of the previous problem to solve any polynomial surface described by two variables. The program should find a local minimum or maximum given any starting point. Direction of travel will be computed. Use the function

$$f(x,y) = 100x - x^2 - 2xy + 200y - 3y^2$$

to test your program.

4.12.6 Real Roots of Cubic Equation

Given a cubic polynomial equation of the form

$$X^3 + a_1 X^2 + a_2 X + a_3 = 0$$

you can solve for three real roots by the following method. First compute

$$Q = \frac{a_1^2 - 3a_2}{9} \qquad R = \frac{2a_1^3 - 9a_1 a_2 + 27a_3}{54}$$

Next check that $Q^3 - R^2 \geq 0.0$. If it is, then there are three real roots that can be found through

$$\theta = \cos^{-1}(R/\sqrt{Q^3})$$

$$X_1 = -2\sqrt{Q} \cos\left(\frac{\theta}{3}\right) - \frac{a_1}{3}$$

$$X_2 = -2\sqrt{Q} \cos\left(\frac{\theta + 2\pi}{3}\right) - \frac{a_1}{3}$$

$$X_3 = -2\sqrt{Q} \cos\left(\frac{\theta + 4\pi}{3}\right) - \frac{a_1}{3}$$

Develop a program to check a cubic equation to see if it has three real roots, and if it does, to solve for those roots. Test your program using the following equations:

$$X^3 - 2X^2 - 5X + 6 = 0 \qquad \text{and} \qquad X^3 + 3X^2 - 3X + 3 = 0$$

4.13 PROBABILITY AND SIMULATION

4.13.1 [ALL] Simulation of Two Dice

Develop a program for the flowchart of Exercise 2.12.1. Use the RND (random number generator) intrinsic function of your computer system. Test your program for 100 and 1,000 tosses of the dice.

4.13.2 [ALL] Normal Probability Distribution

Develop a program to implement the flowchart of Exercise 2.12.2. Test your program for plus and minus one, two, and three standard deviations.

CHAPTER FIVE

FORTRAN PROGRAMMING: SUBPROGRAMS

5.1 INTRODUCTION

SUBROUTINE and FUNCTION subprograms are separate program segments that can work together to form efficient and easily interpreted applications programs. Each subprogram should be a module that implements a specific procedure. A subprogram can be used several times during program execution. Subprograms can be stored in separate files and compiled individually. The compiled subprograms can be linked to the coordinating main program prior to program execution. Subprograms can be saved in subprogram libraries for later use. Most computer systems have a library of general purpose subprograms that can be accessed from any program.

Global variables are variables that are used by both the subprogram and the accessing program. The accessing program can be a main program or another subprogram, but not the subprogram itself. Values for global variables are made available to the subprogram through arguments. The arguments have to follow given rules as to type and order. The rules are similar to the actual and dummy argument rules of function statements. Subprograms can also use local variables. The local variables temporarily store intermediate values for the subprogram computations, but are not needed in other parts of the program. SUBROUTINE and FUNCTION subprograms pass values back to the accessing program in different ways. Each method will be discussed in the appropriate section.

Learning how to use subprograms and pass information back and forth is well worth the effort in terms of adding structure to your programs and making development of more complex programs much easier. They allow modules already programmed by yourself or others to be used in new programs that are being developed. You can append a subprogram to a program and use it without rewriting the subprogram in terms of the variable names used in the current program. This is an important characteristic of subprograms that will be examined in detail in subsequent sections. You can also access a subprogram from several different points in a program. Thus, you will not have to duplicate the code for a particular procedure or use multiple shifts of control that will obscure the logic of the program. Subprograms provide an organizing tool that helps you to write well-structured programs.

Passing of information to and from subprograms is probably the hardest part of subprogram use for beginning programmers to master. The rules for doing this will be covered in detail in a later section. In general, you have to include specifications in both the accessing program segment and the subprogram to make values developed in the accessing program available in the subprogram. The reverse process is necessary to send data back over to the accessing program segment after completion of the subprogram. Figure 5.1 shows a schematic of how subprograms interact with other program segments. Consider the arrows to represent both shifts in execution control and passing of appropriate data. Figure 5.1*a*

FIGURE 5.1   Schematic of subprogram interaction with other program segments.

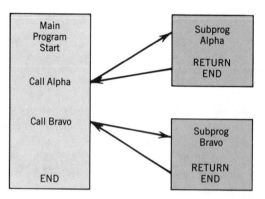

(*a*) Program references to two separate subprograms.

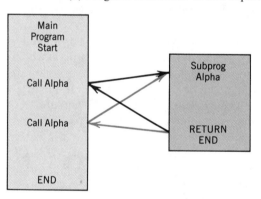

(*b*) Two references to the same subprogram.

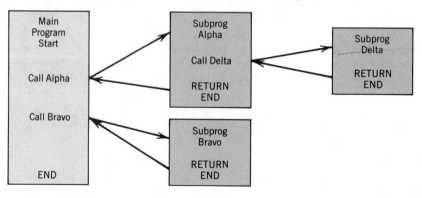

(*c*) Reference to subprogram from within another subprogram.

shows a program that uses two separate subprograms. After each subprogram is executed, control shifts back to the position in the main program from where it was referenced. Figure 5.1*b* shows two references to the same subprogram. Control shifts back to where subprogram Alpha was referenced each time the subprogram is finished. Figure 5.1*c* shows access to a subprogram from within another subprogram. After completion of subprogram Delta, control shifts back to subprogram Alpha, not to the main program. Control shifts back to the main program after completion of subprogram Alpha.

The programmer can specify what arguments will be used in the subprogram and how they will be passed to and from the accessing program. This adds to the versatility of the subprogram structure but also makes it necessary to observe all the rules in designing your subprograms and their related arguments. You will also have all the responsibility of debugging and verifying your own subprograms. Once a subprogram is debugged it can be used in other programs without having to be debugged again.

5.2   GENERAL RULES

Several general rules that apply to both SUBROUTINE and FUNCTION subprograms will now be developed, then applications of each of the structures will be discussed in more detail.

Subprograms are separate program elements. Therefore, statement numbers can be the same in the referencing program and the subprogram. Also, subprograms must have an identifiable beginning and end. The beginning of each subprogram is a statement indicating what type of subprogram it is. These statements will be discussed in the specific sections on each type of subprogram. The end of the subprogram is signified by an END statement, as is the case in normal programs. Some means also has to be included to transfer control back to the accessing program. This is accomplished by a RETURN statement. A RETURN statement returns control to the calling or referencing statement. Once the RETURN is executed, the program continues executing the accessing program from the referencing statement onward. A subprogram may have more than one RETURN statement. However, multiple returns can make the program hard to follow and should be avoided if possible. Normally there will be a RETURN statement just before the END statement of the subprogram.

5.2.1 Subprogram Arguments

When you studied the use of intrinsic and statement functions, you learned the concept of dummy and actual arguments. Arguments provide the function with the values it needs to calculate the value of the function. Subprograms also use arguments to make variable values from the main program available to the subprogram. The argument list of the accessing program is called the actual argument list, whereas the argument list of the subprogram is called the dummy argument list. This relationship is similar to what you observed with statement functions. Actual arguments can be constants, variables, or expressions. Dummy arguments should be variables. The reference for the subprogram in the accessing program makes the computer memory addresses for the actual argument list available to the dummy argument list. The subprogram is then able to use the values in these storage locations during subprogram execution.

With the exception of COMMON blocks, which will be discussed in Chapter 7, the use of argument lists is the only way to make variable values from the main program available to the subprogram. Thus, if a variable does not appear in the argument list, its value (or values in the case of arrays) cannot be used in the subprogram or passed back to the accessing program. This is a critical requirement that has to be observed if you are to master the use of subprograms.

Actual arguments referenced in the accessing program have to agree in mode, number, and order with the dummy arguments used in the subprogram. The names associated with the actual and dummy arguments do not have to agree. This is what allows a subprogram procedure to be used for several purposes in a program. For example, assume that you have developed a subprogram for finding the length of the hypotenuse for a right triangle. The dummy arguments for the subprogram might be

$$(X,Y,H)$$

in which X and Y represent the lengths of the sides of the right triangle, and H represents the calculated hypotenuse. Assume that at some point in the accessing program you want to find the magnitude of a two-dimensional velocity vector from its components in the x and y planes. If VX and VY are the x and y components of the velocity, the actual arguments (VX,VY,VEL) would make the velocity components available to the subprogram and would allow the main program to use the computed value of the velocity as the value stored in VEL. Note that this is a general representation of the actual and dummy arguments. The way that values are made available to the accessing program by FUNCTION subprograms is somewhat different from the way values are returned by SUBROUTINE subprograms. However, the general concept of dummy and actual arguments holds for both types of subprograms.

To continue our example, suppose that in another part of the accessing program you want to find the radius of a circle when the x and y coordinates of a point on the circumference of the circle are known. You can use the same subprogram, but the actual arguments would reflect the use of the variables in the accessing program. The actual arguments might be (X,Y,RADIUS) with X and Y being the coordinates and RADIUS the radius. One subprogram has been used to accomplish two different tasks when each requires the same mathematical procedure.

Some other examples of actual and dummy arguments follow with explanation as to what value would be available to the subprogram for each. All are simple variables, and the default type declaration holds.

Actual argument list   `(X,Y+5.,Y**2,N+1,SIN(X),X+Y,6.,1)`
Dummy argument list   `(A,B,C,L,D,E,F,M)`

Variable A in the subprogram will have the value stored in location X of the accessing program. Five will be added to the value of Y in the accessing program, and the result will be the value of B in the subprogram. Y^2 will be the value of C. Variable L of the subprogram will use the integer value in storage location N of the accessing program plus 1. The sine of X will be the value for variable D of the subprogram. The value stored in location X will be added to the value stored in Y, and the result will be the value of E in the subprogram. Variable F of the subprogram will have the value 6.0 and variable M the integer value 1. All the values for the dummy argument variables will be available to be used in the subprogram.

5.2.2 Passing Arrays to and from Subprograms

If one of the arguments is an array, its dimensions must be specified in both the main program and the subprogram. For example,

```
C  Accessing Program
      REAL A(20,20)
C  Actual argument list
         (A,B,C)
C  Subprogram
C  Dummy argument list
         (A,B,C)
      REAL A(20,20)
```

would make the 20 × 20 array for *A* from the accessing program available to the subprogram, along with the two simple variables, *B* and *C*. Note that only the variable name for the array is used in the argument lists. When this is done, FORTRAN understands that the whole array is to be made available.

You may also use array indexes in actual argument lists. However, this signifies that individual elements of the array are transmitted, not the whole array. For example, in

```
C  Accessing Program
      REAL X(20)
C  First actual argument list
         (X(5),B,C)

C  Second actual argument list
         (X(I),B,C)
C  Subprogram
C  Dummy argument list
         (A,B,C)
```

the first actual argument list would make the fifth element of array *X* available to the subprogram, but as a simple variable value to be used for variable *A* in the subprogram. Note that *A* is not defined as an array in the subprogram, because it is not intended to be used as an array. The second actual argument list would make the *i*th element of the *X* array available to the subprogram. The value of *I* would have to be specified by a DO loop or some other means.

In the main program you have to specify the size of arrays with constants. The sizes of arrays in subprograms can be specified by either constants or variables. This allows you to specify array sizes as large as are actually used in the subprogram. To use variable dimensions in a subprogram, the maximum array sizes have to be specified in the main program. When the array is transferred over to a subprogram, an array size equal to or smaller than the array size specified in the main program can be specified. For example, assume that you had dimensioned a one-dimensional array *B* to be 25 elements in the main program, but had filled up only a section of the array that is *N* elements long. When you transfer the array over to a subprogram you could do the following:

```
      REAL B(25)

      Main program actual arguments (B,N)

      Subprogram dummy arguments (B,N)
      REAL B(N)
```

This would make the first *N* elements of the array *B* available to the subprogram.

Problems can arise if you try to use a subset of a two-dimensional array stored in the accessing program, because the computer does not store two-dimensional arrays in a rectangular format. For example, if the array stored in the main program is dimensioned for 6×4, but you have filled up only a 3×2 subset of this matrix in the accessing program and want to use that 3×2 subset in the subprogram, the actual and dummy arguments

```
REAL A(6,4)
Actual arguments (A,3,2)

Dummy arguments (A,N,M)
REAL A(N,M)
```

would set up a 3×2 matrix in the subprogram but would not make the correct values available to the subprogram. This is because the computer actually stores the elements of the two-dimensional array by columns. You can avoid this problem by not using variable dimensions for two-dimensional arrays. For example

```
REAL A(6,4)
Actual arguments (A,3,2)

Dummy arguments (A,N,M)
REAL A(6,4)
```

would allow you to use a 3×2 subset of the original matrix with no problems. There are certain instances when variable dimensions for two-dimensional arrays are required. One such instance will be illustrated in Chapter 7.

5.2.3 Compiling and Linking Subprograms

If subprograms are included in the same file as the main program, the order within the file does not matter as long as the program and subprograms do not overlap. It is customary to place the subprograms after the main program. Subprograms may also be compiled as separate programs. If so, they must be linked with the main program before execution. Although the process for doing this is similar for different computer systems, the actual commands are system dependent. The commands used by the Digital Equipment Corporation VAX/VMS operating system will be used here to illustrate the procedure. Assume that you have a main program in a file called VECTOR.FOR. The .FOR filename extension indicates a FORTRAN source file. You also have two other files, PYTH.FOR and ADD.FOR, that contain subprograms. You want to access these subprograms from the main program contained in file VECTOR.FOR. If none of the files had been compiled previously, the following series of commands would have to be used:

```
FORTRAN VECTOR
FORTRAN ADD
FORTRAN PYTH
LINK VECTOR,ADD,PYTH
RUN VECTOR
```

The first three statements compile the three programs and would give diagnostics if compile errors were encountered. Each of these commands would create a file that would contain the compiled version of that program element. The fourth statement links the three compiled files together so that the subprograms can be accessed. This step produces a fourth new file, called VECTOR.EXE, that contains the executable version of the linked, compiled files. VECTOR, the main program, has to be the first file listed in the LINK command, followed by the names of the other compiled subprogram elements to be linked, in any order. At the beginning of these commands, you would have had three files; VECTOR.FOR, ADD.FOR, and PYTH.FOR. After the three FORTRAN compile commands, you will have three additional files; VECTOR.OBJ, ADD.OBJ, and PYTH.OBJ. After the LINK command you would add a seventh file, VECTOR.EXE. All these files would appear if you looked at a directory of your disk storage. However, only the .FOR files can be viewed with a text editor. The RUN command actually executes the VECTOR.EXE file. You do not have to include the file extenders (.FOR, .EXE, etc.) with the various commands because the command looks for, and will accept, only a file with the proper extender. Once the LINK command has been used, the VECTOR.EXE file can be RUN any number of times. You have only to compile files during the initial, debugging stage. Once the program is running properly, you do not want to recompile it each time, because the compilation and linking waste computer time. The compiled subprograms can also be used in other programs. Let's say that you now want to compile and execute a file called FORCE.FOR that also will execute the two subprograms ADD and PYTH. The following series of commands will suffice:

```
FORTRAN FORCE
LINK FORCE,ADD,PYTH
RUN FORCE
```

Since ADD and PYTH have already been compiled, you do not need to recompile them. Just link them to the program in which they will be used.

There is an extension of standard FORTRAN 77 that allows you to incorporate stored FORTRAN source files into the compilation without compiling them as separate files and linking them together. The statement

```
INCLUDE 'READ.FOR'
```

would insert all the FORTRAN code of file READ.FOR at the location of the INCLUDE statement. Any FORTRAN statements after the INCLUDE would be moved down to accommodate the insertion. The INCLUDE may well become a part of standard FORTRAN during future revisions.

5.3 SUBROUTINE SUBPROGRAMS

SUBROUTINE subprograms are the more general of the two subprogram structures. References to SUBROUTINES in the accessing program are made through CALL statements. The CALL statement transfers control to the SUBROUTINE and passes values for the actual arguments to the SUBROUTINE. The SUBROUTINE is executed until a RETURN statement is encountered. The RETURN shifts control back to the CALL statement. Program execution then continues with the next statement after the CALL. The SUBROUTINE is executed each time a CALL statement references it.

The actual and dummy arguments for SUBROUTINE subprograms provide a two-way street. Values can be made available to the SUBROUTINE subprogram by the accessing program through the actual arguments. New values can be computed within the subroutine and then made available to the accessing program through the dummy arguments. In fact, values sent over can be changed within the SUBROUTINE, and the new values will be stored in the appropriate actual argument storage locations. For example,

```
        .
      CALL FIRST(X,Y,Z)
        .
      CALL FIRST(A,B,C)
        .
      END
      SUBROUTINE FIRST(Q,R,S)
        .
      RETURN
      END
```

makes values for X, Y, and Z available to variables Q, R, and S of SUBROUTINE FIRST when the first CALL statement is executed. The values of Q, R, and S at the completion of the subroutine are the values that will be stored in locations X, Y, and Z when execution continues after the CALL statement in the accessing program. When the second call is encountered, the values in the accessing program for variables A, B, and C are made available to the subroutine as the values for variables Q, R, and S. After executing SUBROUTINE FIRST for the second time, the values for Q, R, and S are stored in locations A, B, and C of the accessing program. Control then shifts to the statement after the second call. The following brief example will demonstrate how subroutines can change values sent over, as well as develop new values. All variables are REAL.

```
C  Main program
        .
      A = 2.
      B = 3.
      C = 4.
      PRINT *, 'Before CALL to SUBROUTINE CALC',A,B,C,D,E
      CALL CALC(A,B,C,D,E)
      PRINT *, 'After first call to SUBROUTINE CALC',A,B,C,D,E
      CALL CALC(A+C,B+2.,C,D,E)
      PRINT *, 'After second call to SUBROUTINE CALC',A,B,C,D,E
      C = C/10.
      CALL CALC (1.0,2.0,C,E,D)
      PRINT *, 'After third call to SUBROUTINE CALC',A,B,C,D,E
      END
      SUBROUTINE CALC(T,U,V,W,X)
      PRINT *, 'Top of SUBROUTINE CALC',T,U,V,W,X
      V = V**2-T
      W = T + U
      X = V + U + A + B
      PRINT *, 'Bottom of SUBROUTINE CALC',T,U,V,W,X
      RETURN
      END
```

The output would be:

```
Before CALL to SUBROUTINE CALC  2.0  3.0  4.0  0.0  0.0
Top of SUBROUTINE CALC  2.0  3.0  4.0  0.0  0.0
Bottom of SUBROUTINE CALC  2.0  3.0  14.0  5.0  17.0
```

```
After first call to SUBROUTINE CALC   2.0   3.0   14.0   5.0   17.0
Top of SUBROUTINE CALC   16.0   5.0   14.0   5.0   17.0
Bottom of SUBROUTINE CALC    16.0   5.0   180.0   21.0   201.0
After second call to SUBROUTINE CALC   16.0   5.0   180.0   21.0   201.0
Top of SUBROUTINE CALC   1.0   2.0   18.0   201.0   21.0
Bottom of SUBROUTINE CALC   1.0   2.0   323.   3.   325.0
After second call to SUBROUTINE CALC   16.0   5.0   323.0   325.0   3.0
```

The remaining zero places have been left off the output to shorten it. The subroutine contains five dummy arguments. Variables T, U, and V have values when the subroutine is called. The three calculation lines of the subroutine use these values to compute values for W and X, and also to modify the value of V. Thus, T and U are input arguments, and W and X are output arguments. V is both an input and an output argument. It transfers in a value, but that value is changed by the subroutine, and a different value is sent back. The third calculation statement of the subroutine

```
      X = V + U + A + B
```

contains a common subprogram programming error. Variables A and B are used in the calculation, but they have not been defined through the dummy arguments or defined through calculation statements in the subroutine. Therefore, their values are zero. The main program defined values for actual arguments A, B, and C before the first call of CALC. Variables D and E have not been defined, so their value is zero. At the start of the subroutine, T, U, V, W, and X will use the values of A, B, C, D, and E. V, W, and X are modified by the calculation statements, but T and U are not. The values for T, U, V, W, and X at the end of the subroutine are stored in locations A, B, C, D, and E of the calling program. These values are then used for the second call to CALC. However, this time the first two actual arguments are expressions. The resulting values for the expressions are the values for the dummy variables T and U. Storage locations A and B of the calling program keep the values they contained before the second CALL to CALC. In the third call to the subprogram, the first two actual arguments are constants, while the D and E actual arguments have been reversed. Variables T and U of the subroutine will use the constant values from the actual arguments. At the end of the subroutine, the value for W will be stored in location E and the value for X will be stored in D.

The name of a SUBROUTINE subprogram identifies the subroutine. No mode is associated with it since the name of the subroutine is not a storage location. However, the rules for number of characters and allowable characters are the same as the rules for naming variables.

5.4 FUNCTION SUBPROGRAMS

FUNCTION subprograms are similar in many ways to SUBROUTINE subprograms, but they also have some of the properties of statement functions. Many values and arrays can be made available to the FUNCTION subprogram through the actual and dummy arguments, but only one value is returned. This value is returned through the function name rather than through the dummy and actual arguments. The rules for the actual and dummy arguments of the FUNCTION subprogram are the same as the rules for the function statement and the rules for the input arguments of a SUBROUTINE subprogram. Multiple lines of code and subscripted variables may be used in the FUNCTION subprogram, giving it an advantage over the statement function.

The reference to a FUNCTION subprogram in the accessing program looks identical to a reference to a statement function. For example, the reference to $FX(Y)$ in the following statement

```
PXY = FX(Y) + 60.*X
```

could refer to a statement function, a FUNCTION subprogram, or a subscripted variable. The context of the program will determine which structure it refers to. FORTRAN will first check to see if there is a subscripted variable of this form, and then it will look for a statement function or FUNCTION subprogram. If none of the structures is found, an error message will be returned. If $FX(Y)$ is a FUNCTION subprogram, the value of Y will be used in the function and the resulting value of the function will be substituted for $FX(Y)$ when the expression

```
FX(Y) + 60.*X
```

is evaluated.

The mode of a function can be defined explicitly by using the proper declaration in front of FUNCTION. REAL FUNCTION CALC indicates that the FUNCTION subprogram CALC would return a real value. The mode also has to be defined in the accessing program by inclusion of the function name in the declaration statements. If arrays are to be sent over to the FUNCTION subprogram, their dimensions have to be specified in the subprogram through DIMENSION or declaration statements. This is the same as for the SUBROUTINE subprogram.

Prior to execution of the RETURN statement, a statement has to define a value for the function. For example, the statement

```
CALC = FORTRAN expression
```

could appear in REAL FUNCTION CALC to establish the value that will be returned by the function. An example will demonstrate many of these concepts.

We have already developed a program to determine the norm (magnitude) of an n-dimensional vector. This is a procedure that lends itself to a FUNCTION subprogram. Several components of the vector are sent over as a one-dimensional array, and then a DO loop is needed to add up the squared elements of the vector. This could not be accomplished with a statement function, since the statement function could not accommodate the variable number of elements. The norm could be found using a SUBROUTINE subprogram. However, only one value, the norm, would be generated. Therefore, the FUNCTION subprogram is an appropriate choice. Example 5.1 shows the adaptation of Example 4.13 to incorporate a FUNCTION subprogram. Note that the function NORM is declared as a REAL variable in the accessing program. Also, the dimension of the vector in the FUNCTION is a variable dimension (n). This is a proper use of the variable dimension since there is no ambiguity in the storage of one-dimensional arrays. The FUNCTION subprogram can now be used with any program to find the norm of any n-dimensional vector.

EXAMPLE 5.1 □ FUNCTION Subprogram to Find the
Norm of an *n*-Dimensional Vector

STATEMENT OF PROBLEM

Modify the program of Example 4.13 to use a FUNCTION subprogram to find the norm of an *n*-dimensional vector.

ALGORITHM DEVELOPMENT

Input/Output Design

Same as for Example 4.13.

Computer Implementation

Figure 5.2 shows the flowchart for the FUNCTION subprogram. Note that the portion of the program for calculating the norm has been lifted from the main body of the program and placed in REAL FUNCTION NORM. The REAL

FIGURE 5.2 Flowchart for vector norm FUNCTION subprogram.

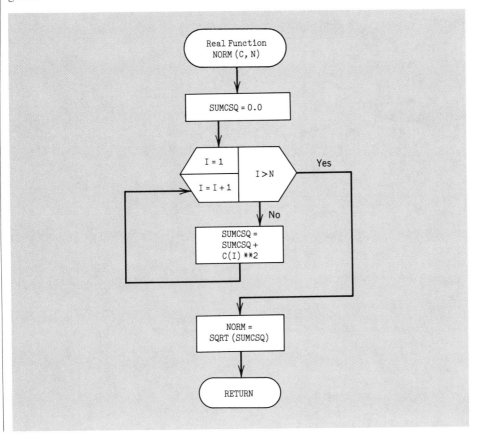

(EXAMPLE 5.1 □ FUNCTION Subprogram to Find the Norm of an *n*-Dimensional Vector □ Continued)

specifies that a real value will be returned. The function reference will be identified in the definition portion of the main program. This will remind you to include the subprogram in the linking process if it is presently stored as a separate file.

PROGRAM DEVELOPMENT

The subprogram is shown appended to the main program for clarity of presentation in the text. If the subprogram is to be used for several programs, it would be better to save it as a separate file.

```
****************************************************************************
*     Program for finding the norm of an n-dimensional vector         *
*        using FUNCTION subprogram                                    *
*     Developed by:  T. K. Jewell            November 1987            *
****************************************************************************
C
C     VARIABLE DEFINITIONS
C     C = n-dimensional vector
C     NORM = FUNCTION subprogram for finding magnitude (norm) of vector C
C
      REAL C(10),SUMCSQ,NORM,NORMC
      INTEGER I,R,W
      CHARACTER LOOP*3
      PARAMETER (R=5,W=6)
C
C  Start of inquiry controlled input loop
  200 CONTINUE
      I=1
C
C  Start of sentinel input loop
  100 CONTINUE
      PRINT *, 'Input component ',I,' of vector C'
      PRINT *, 'A value >= 100 indicates end of input.
      READ (W,*) C(I)
      IF (C(I).LT.(100.0)) THEN
         I = I + 1
         GO TO 100
      END IF
      N = I - 1
C  End of sentinel input loop
C
C  Calculate magnitude (norm) of vector C
      NORMC = NORM(C,N)
C
C  Output results
      WRITE (W,*) 'Components of vector C are:'
      WRITE (W,*)
      DO 400 I=1,N
         WRITE (W,*) C(I)
  400 CONTINUE
      WRITE (W,*)
      WRITE (W,*) 'The magnitude of vector C is: ',NORMC
C
C  Inquiry for additional input
```

```
      PRINT *, 'Do you wish to input another vector? (Yes or No)'
      PRINT *, 'Make sure to put apostrophes on both sides of your'
      PRINT *, ' response'
      READ(R,*) LOOP
      IF(LOOP(1:1).EQ.'Y'.OR.LOOP(1:1).EQ.'y') GO TO 200
C   End of inquiry loop
      END
      REAL FUNCTION NORM(C,N)
C   C = Vector to find the norm of
C   SUMSQ = summing variable for squared components of vector C
C   N = dimension of vector
      REAL C(N),SUMSQ
      INTEGER N
      SUMCSQ = 0.0
      DO 300 I=1,N
         SUMCSQ = SUMCSQ + C(I)**2
  300 CONTINUE
      NORM = SQRT(SUMCSQ)
      RETURN
      END
```

PROGRAM TESTING

Check the program with the same vectors that you used for Example 4.13.

We will now use the FUNCTION subprogram developed in Example 5.1, along with SUBROUTINE subprograms, to input two n-dimensional vectors, add the two n-dimensional vectors together, and find the norm of all three vectors. Input will be from a data file.

A new structure will be introduced that assists in inputting and outputting arrays. If you use a DO loop to increment the subscript index and input or output values, only one value can be input or output per line. FORTRAN contains an alternative form of the DO statement called the implied DO loop. The implied DO loop allows the reading of multiple values for a subscripted variable from the same input record and allows output of multiple values to the same line.

The implied DO loop has the same form for READ, PRINT, or WRITE statements. An implied DO loop for a READ statement will be illustrated here. It has the form

```
      READ(R,*) (A(I),I=1,N)
```

The structure $(A(I),I=1,N)$ has the same effect as a DO loop, except that the variation of the index I does not represent separate executions of the READ statement. Therefore, the READ statement will accept multiple values from the same line of the data file (input data record). If all the values to be input for the array will not fit on one input line (record), the data can be continued on as many subsequent lines as are necessary. The READ statement will continue to go to new input records until it satisfies the input variable list.

If an output line for an implied DO loop has more elements than your computer system will output per line in free format, the remaining values will be output on succeeding lines. You will gain more control over the number of elements output per line when we cover formatted output in Chapter 6.

EXAMPLE 5.2 □ Summation and Length of *n*-Dimensional Vectors

STATEMENT OF PROBLEM

Develop a program to input two vectors, add the two vectors together, and find the magnitude (norm) of all three vectors. The program should have the capability to work with up to 10-dimensional vectors. The magnitude of the vectors will be found using FUNCTION subprogram NORM. Input of the *n*-dimensional vectors and the adding together of the two vectors will be by subprogram. Input for this problem will be through a data file.

MATHEMATICAL DESCRIPTION

Appendix F develops vector notation and vector algebra. For this program, you will need to use vector addition and the magnitude (norm) of a vector. Example 4.5 discusses these in more depth.

ALGORITHM DEVELOPMENT

Input/Output Design

Vector arrays **A, B,** and **C** have to be defined as real arrays. The size of the arrays should be as large as the largest vector we want the program to accommodate, in this case, 10. *N*, the dimension of the vectors to be added, will be input through the main program and passed over to the subprogram to control input. If *N* were input through the subprogram, it would have to be input each time the subprogram is called. The subprogram for inputting the vectors will be called twice. Since multiple values will be input, a SUBROUTINE subprogram will be used. The READ statement of SUBROUTINE VECTOR will use an implied DO loop, so multiple values can be read from the same line, until the input variable list is satisfied. A new input line will have to be started for the second call to SUBROUTINE VECTOR. For the vectors of Example 4.5, three lines of input would suffice: one for *N*, and one for each of the vectors.

Output should include *n*, the components of the vectors **A, B,** and **C,** and the norm of vectors **A, B,** and **C.** Any headings for output must be output before the output loop. If the headings are included in the loop, they will be output *n* times.

Computer Implementation

A SUBROUTINE called ADD will add the two vectors. After addition, the REAL FUNCTION NORM can be called three times to compute the magnitude of each of the vectors. Figure 5.3 gives a flowchart for the program and subprograms. In the flowchart for SUBROUTINE VECTOR, an implied DO loop is represented by a DO structure with dotted outlines. A previous example developed the flowchart for NORM, so it is not repeated in Figure 5.3.

PROGRAM DEVELOPMENT

```
*********************************************************************
*    Program for adding two vectors and finding the norm of the     *
*       resultant                                                    *
*    Developed by:  T. K. Jewell                      November 1987  *
*********************************************************************
```

FIGURE 5.3 Flowchart with subprograms for lengths of two vectors and the length of their sum.

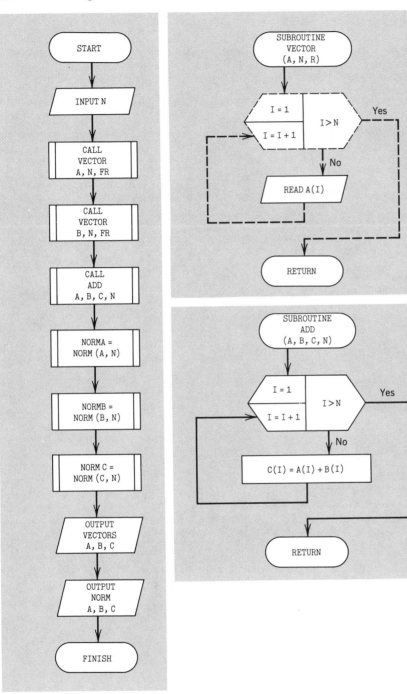

(EXAMPLE 5.2 □ **Summation and Length of**
n-**Dimensional Vectors** □ **Continued)**

```
C
C     VARIABLE DEFINITIONS
C     A and B = n-dimensional vectors to be added together
C     C = n-dimensional vector to store summation of A and B
C     N = dimension of vectors A, B, and C
C     NORMA, NORMB, NORMC = magnitude (norm) of vectors A, B, and C
C
      REAL A(10),B(10),C(10),NORMA,NORMB,NORMC,NORM
      INTEGER I,N,FR,W
      PARAMETER (FR=4,W=6)
C Input of vectors A and B
      OPEN (UNIT=FR,FILE='EX5_2.DAT',STATUS='OLD')
C Input the dimension of the vectors
      READ (FR,*) N
      WRITE (W,*) 'The vectors are ',n,'-dimensional'
C
C Input vectors A and B
      CALL VECTOR (A,N,FR)
      CALL VECTOR (B,N,FR)
C
C Add vectors A and B together to get C
      CALL ADD (A,B,C,N)
C
C Calculate magnitude (norm) of vectors A, B, and C
      NORMA = NORM(A,N)
      NORMB = NORM(B,N)
      NORMC = NORM(C,N)
C
C Output results
      WRITE (W,*)
      WRITE (W,*) 'Components of vectors A, B, and C are:'
      WRITE (W,*)
      WRITE (W,*) '      Vector A        Vector B        Vector C'
      DO 400 I=1,N
         WRITE (W,*) A(I),B(I),C(I)
  400 CONTINUE
      WRITE (W,*)
      WRITE (W,*) 'The magnitude of vector A is: ',NORMA
      WRITE (W,*) 'The magnitude of vector B is: ',NORMB
      WRITE (W,*) 'The magnitude of vector C is: ',NORMC
      END
C
C Subprogram for determining the norm of a vector
      REAL FUNCTION NORM(C,N)
C  C = Vector to find the norm of
C  SUMCSQ = summing variable for squared components of vector C
C  N = dimension of vector
      INTEGER N
      REAL C(N),SUMCSQ
      SUMCSQ = 0.0
      DO 300 I=1,N
         SUMCSQ = SUMCSQ + C(I)**2
  300 CONTINUE
      NORM = SQRT(SUMCSQ)
      RETURN
      END
C
      SUBROUTINE VECTOR(A,N,R)
```

```
      C  Subprogram for inputting n-dimensional vector
      C  R = input unit identifier, either terminal or file
      INTEGER N,R
      REAL A(N)
      READ (R,*) (A(I),I=1,N)
      RETURN
      END
C
      SUBROUTINE ADD(A,B,C,N)
C  Subprogram for adding two n-dimensional vectors together
      INTEGER N
      REAL A(N),B(N),C(N)
C  Calculate C = A + B
      DO 200 I=1,N
         C(I) = A(I) + B(I)
  200 CONTINUE
      RETURN
      END
```

PROGRAM TESTING

Use the vectors of Example 4.5 to check the modified program.

5.5 STATISTICAL SUBROUTINES

In Example 5.3 we will develop three subprograms that can be used to find the mean, median, and standard deviation for a set of data. The subprograms will reference other subprograms wherever appropriate. It will be assumed that the number of data points and the data values have been read in previously. A skeleton main program will be included for the purpose of debugging the subprograms. Example 5.4 will develop a fourth subprogram that can be used to perform a linear regression on two sets of data and to determine the correlation coefficient for the regression. Example 5.5 will use all these subprograms to investigate the relationship between stress–strain data for a steel specimen. This whole section will demonstrate the process of developing a modular program, debugging the various modules as they are developed, then combining the tested modules into a comprehensive program. Other future applications can use these same subprogram modules.

EXAMPLE 5.3 □ Mean, Median, and Standard Deviation

STATEMENT OF PROBLEM

Develop three subprograms that can compute the mean, median, and standard deviation for a set of data. Have the main subprograms reference other subprograms wherever appropriate. Reference to Examples 3.1, 4.6, and 4.10 will be helpful. Write a short main program to test the modules.

MATHEMATICAL DESCRIPTION

Appendix H describes the statistics to be calculated. The mathematical description was developed in Example 3.1.

(EXAMPLE 5.3 □ Mean, Median, and Standard Deviation □ Continued)

ALGORITHM DEVELOPMENT

Input/Output Design

Input and output of data will not be included in these subprograms. It is assumed that input and output are taken care of elsewhere in the application. The proper arrays and variables have to be included in the actual and dummy argument lists to make the desired data available to the subprograms. The subprograms will need to have access to the array containing the data and to a variable defining the size of the data array. The standard deviation module must also have access to the mean value. Variable dimensions can be used in the subprograms. Information to be passed back to the accessing program will be the mean, median, and mode statistics. Since only one value has to be passed back by each subprogram, FUNCTION subprograms are appropriate. The argument lists could be

```
REAL FUNCTION MEAN(A,N)
REAL FUNCTION MEDIAN(A,N)
REAL FUNCTION SDEV(A,N,XBAR)
```

in which A is the data array, N is the size of the array, and $XBAR$ is the variable name for the mean in the main program. The names of the functions are self-explanatory.

Numerical Methods

The summations for the mean and standard deviation are straightforward. They will be developed as FUNCTION subprograms because the same calculations may be required by other programs. The median is more complex, because the data has to be sorted into ascending or descending order before it can be used. The data array will be written into a dummy array for ordering, and the ordered dummy array will be used for finding the median. This leaves the original data in the original order in case you want to use it for other analysis, such as regression. The dummy array can be a local variable to the MEDIAN subprogram.

Computer Implementation

The proper data has to be passed from each subprogram to the subprograms it references. A FUNCTION subprogram will be used to sum the array values and a second FUNCTION subprogram used to sum the squares of the data values. Each of these will require access to the appropriate array and the size of the array. A SUBROUTINE will be adapted from the program of Example 4.10 for sorting the dummy array passed over from the MEDIAN subprogram.

Figure 5.4 shows the six modules included in this program development. It does not include the test main program.

PROGRAM DEVELOPMENT

```
***************************************************************************
*  Test program for statistics modules                                    *
*  Developed by:  T. K. Jewell              November 1987                  *
***************************************************************************
C  X = data array
C  XBAR = mean value
C  XMED = median value
C  XSDEV = standard deviation of X array
C  N = size of the array
```

FIGURE 5.4 Flowchart for mean, median, and standard deviation subprograms.

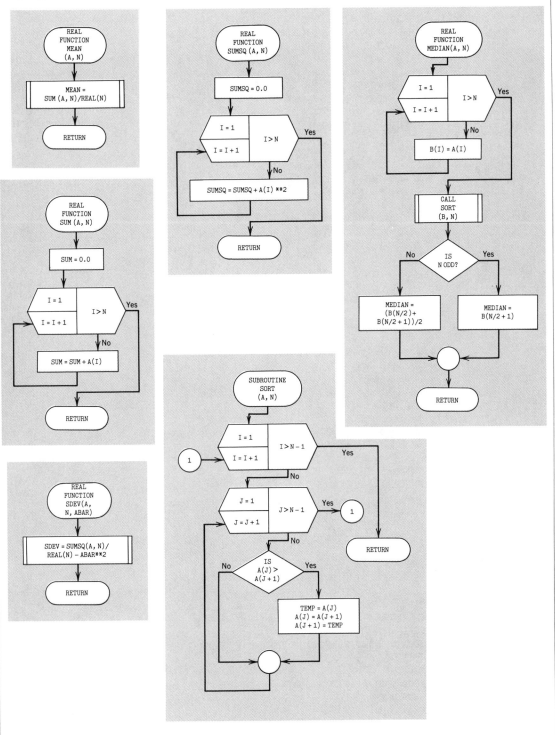

(EXAMPLE 5.3 □ Mean, Median, and Standard Deviation □ Continued)

```
          INTEGER N,R,W
          REAL X(10),XBAR,XMED,XSDEV,MEAN,MEDIAN,SDEV
          PARAMETER (R=5,W=6)
          PRINT *, 'Input N, the number of data points  '
          READ(R,*) N
          PRINT *, 'Input the data values  '
          READ(R,*) (X(I),I=1,N)
          PRINT *, 'For N = ',N,' the data values are:'
          PRINT *, (X(I),I=1,N)
          PRINT *, ' '
C
C   Compute and print the mean
          XBAR = MEAN(X,N)
          PRINT *, 'The mean value is ',XBAR
C
C   Compute and print the median
          XMED = MEDIAN(X,N)
          PRINT *, 'The median value is',XMED
C
C   Compute and print the standard deviation
          XSDEV = SDEV(X,N,XBAR)
          PRINT *, 'The standard deviation is',XSDEV
          END
          REAL FUNCTION MEAN(A,N)
C
C   Module for computing the mean of a set of data
          INTEGER N
          REAL A(N)
          MEAN = SUM(A,N)/REAL(N)
          RETURN
          END
          REAL FUNCTION SUM(A,N)
C
C   Module for summing data elements in a one-dimensional array
          INTEGER N
          REAL A(N)
          SUM = 0.0
          DO 100 I=1,N
             SUM = SUM + A(I)
    100 CONTINUE
          RETURN
          END
          REAL FUNCTION SDEV(A,N,ABAR)
C
C   Module for computing the standard deviation of a set of data
          INTEGER N
          REAL A(N)
          SDEV = SQRT(SUMSQ(A,N)/REAL(N) - ABAR**2)
          RETURN
          END
          REAL FUNCTION SUMSQ(A,N)
          INTEGER N
          REAL A(N)
          SUMSQ = 0.0
          DO 100 I=1,N
             SUMSQ = SUMSQ + A(I)**2
```

```
  100 CONTINUE
      RETURN
      END
      REAL FUNCTION MEDIAN(A,N)
C
C  Module for finding the median value of an unordered set of data
      INTEGER N
      REAL A(N),B(10)
C
C  Copying array A into dummy array B so B can be sorted
      DO 100 I=1,N
         B(I)=A(I)
  100 CONTINUE
C
C  Sorting array B
      CALL SORT(B,N)
C
C  Finding median of ordered set B
      IF (MOD(N,2).NE.0) THEN
         MEDIAN = B(N/2+1)
      ELSE
         MEDIAN = (B(N/2)+B(N/2+1))/2.0
      END IF
      RETURN
      END
      SUBROUTINE SORT(A,N)
C  Sorting in ascending order
      INTEGER N
      REAL A(N),TEMP
      DO 200 I=1,N-1
         DO 300 J=1,N-1
            IF (A(J).GT.A(J+1)) THEN
               TEMP = A(J)
               A(J) = A(J+1)
               A(J+1) = TEMP
            END IF
  300 CONTINUE
  200 CONTINUE
      RETURN
      END
```

PROGRAM TESTING

For the data set 2,4,6,8,10, XBAR = 6, XMED = 6, and XSDEV = 2.83. For the data set 0,2,4,6,8,10, XBAR = 5, XMED = 5, and XSDEV = 3.41.

Example 5.4 will develop a subprogram for determining the regression coefficients and correlation coefficient for a set of X-Y data. A skeleton program will also be included to test the subprogram for a given set of data containing 10 elements.

EXAMPLE 5.4 □ Linear Regression

STATEMENT OF PROBLEM

Develop a SUBROUTINE subprogram that will compute the regression coefficients and the correlation coefficient for the best fit straight line through a set of

(EXAMPLE 5.4   □   Linear Regression   □   Continued)

FIGURE 5.5   Flowchart for linear regression subprogram.

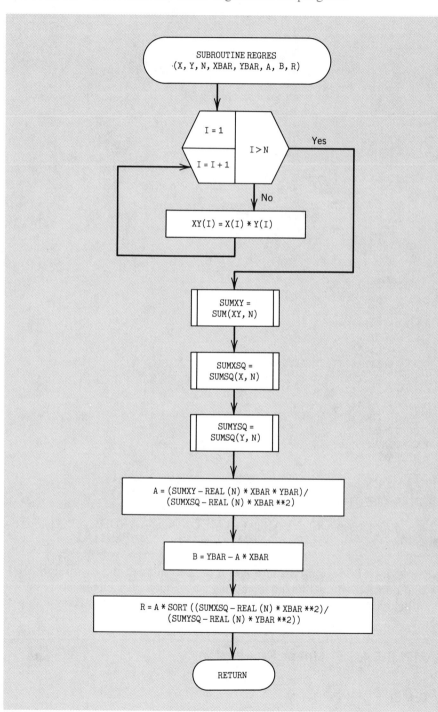

X-Y data. X is the independent variable, and Y is the dependent variable. The X and Y data values and the number of data points have previously been entered. You can assume that the mean values for the X and Y data have already been computed. They can be part of the data passed over to the subprogram. Also, any of the subprograms developed in Example 5.3 can be used for this program.

MATHEMATICAL DESCRIPTION

The equations for computing the regression coefficients, a and b, and the correlation coefficient, r, are

$$a = \frac{n(\Sigma XY) - (\Sigma X)(\Sigma Y)}{n(\Sigma X^2) - (\Sigma X)^2} = \frac{\Sigma XY - n\bar{X} \cdot \bar{Y}}{\Sigma X^2 - n\bar{X}^2} \qquad (5.1)$$

$$b = \bar{Y} - a\bar{X} \qquad (5.2)$$

$$r = a \sqrt{\frac{\Sigma' X^2}{\Sigma' Y^2}} = a \sqrt{\frac{\Sigma X^2 - n\bar{X}^2}{\Sigma Y^2 - n\bar{Y}^2}} \qquad (5.3)$$

Additional explanation of these equations and the summations used in them are included in Appendix H.

DEVELOPMENT OF ALGORITHM

Input/Output Design

The X and Y arrays have to be passed over to the subprogram along with the number of data points. The regression coefficients, a and b, will be passed back, along with the correlation coefficient, r.

Computer Implementation

Figure 5.5 shows the flowchart for this subprogram. Note the use of the functions developed in Example 5.4.

PROGRAM DEVELOPMENT

```
**********************************************************
*  Main program to test regression SUBROUTINE    *
*  Developed by:  T. K. Jewell  November 1987    *
**********************************************************
      REAL X(10),Y(10),XBAR,YBAR,A,B,R,MEAN
      INTEGER N
      OPEN(UNIT=4,FILE='EX5_4.DAT',STATUS='OLD')
      READ (4,*) N
      DO 20 I=1,10
         READ(4,*) X(I),Y(I)
         PRINT *, X(I),Y(I)
   20 CONTINUE
      XBAR = MEAN(X,N)
      YBAR = MEAN(Y,N)
      CALL REGRES(X,Y,10,XBAR,YBAR,A,B,R)
      PRINT *, XBAR,YBAR,A,B,R
      END
      SUBROUTINE REGRES(X,Y,N,XBAR,YBAR,A,B,R)
C
C  Module for calculating regression coefficients and correlation
```

(EXAMPLE 5.4 □ Linear Regression □ Continued)

```
C     coefficient for X-Y data
      INTEGER N
      REAL X(N),Y(N),XBAR,YBAR,A,B,R,SUM,SUMSQ,SUMXY,SUMXSQ,SUMYSQ
C
C  Compute X-Y cross product summation
      SUMXY = 0.0
      DO 100 I=1,N
         SUMXY = SUMXY + X(I)*Y(I)
  100 CONTINUE
C
C  Compute squared sums for regression and correlation
      SUMXSQ = SUMSQ(X,N)
      SUMYSQ = SUMSQ(Y,N)
C
C  Compute B, A, and R
      A = (SUMXY-REAL(N)*XBAR*YBAR)/(SUMXSQ-REAL(N)*XBAR**2)
      B = YBAR-A*XBAR
      R = A*SQRT((SUMXSQ-REAL(N)*XBAR**2)/(SUMYSQ-REAL(N)*YBAR**2))
      RETURN
      END
```

PROGRAM TESTING

The following set of data,

X	Y
0.9	1.3
0.8	2.0
1.3	2.3
1.4	1.5
1.8	2.5
2.0	2.0
2.5	3.0
2.3	1.8
3.1	2.0
3.5	2.6

produces $a = 0.292$, $b = 1.53$, and $r = 0.51$. You can check the calculation process using the intermediate data $\bar{X} = 1.96$, $\bar{Y} = 2.10$, $\Sigma X = 19.6$, $\Sigma X^2 = 45.74$, $\Sigma XY = 43.3$, $\Sigma Y = 21.0$, $\Sigma Y^2 = 46.48$.

Example 5.5 will use the subprograms of Examples 5.3 and 5.4 to develop an applications program to analyze stress–strain data for a steel specimen. In steel design, the strain is the elongation per unit length of a specimen caused by an applied longitudinal stress (force/unit area). Thus

$$\text{Strain}(\varepsilon) \propto \text{Stress}(\sigma) \tag{5.4}$$

The constant of proportionality is $1/E$, where E is defined as the modulus of elasticity. Therefore

$$\boxed{\varepsilon = \frac{\sigma}{E}} \tag{5.5}$$

Stress is the predictor variable (independent variable) for the estimated strain (dependent variable). Numerous tests have confirmed that, within certain limits, the relationship between stress and strain is linear. If the relationship is linear, then linear regression should be a good method for estimating the constant of proportionality, which can be used to calculate E from materials testing data. In English units, stress and the modulus of elasticity are usually reported in units of pounds per square inch (psi), and strain is given as inch per inch (in./in.). An applications program that can be used to evaluate stress–strain data taken in the laboratory is to be developed. You want to check the mean, median, and standard deviation for the data values for each variable. You also want to find the regression coefficients and correlation coefficient for the data set. The results can be used to calculate the modulus of elasticity and to predict the strain for any number of stresses. A sample data set is given at the end of the program for debugging purposes. The data for this example will be further analyzed in later chapters.

EXAMPLE 5.5 □ Statistical Program for Analyzing Stress–Strain Data

PROBLEM STATEMENT

Develop a main program to find the modulus of elasticity for n ($n \leq 50$) pairs of stress–strain data using linear regression analysis. Use the regression equation to predict strain for any number of stress values. Also, determine the mean, median, and standard deviation for both the stress and strain input data.

ALGORITHM DEVELOPMENT

Input/Output Design

Since there may be numerous data pairs, this program will use data file input. You will have to input the name of the data file through the terminal at the start of the program. The first data value input from the file will be N, the number of data pairs. The independent and dependent variable values should be input in pairs, one pair per line. If you input the data in pairs, the order in which you input the pairs does not matter. Output should be grouped by type. First the input data should be output. The second output group should be the basic statistics for both data sets. The third section should give the regression equation and the estimated value for the modulus of elasticity. The final section should show the interactive portion for computing any number of strain values based on given stress values. This portion will also require further input from the terminal.

Numerical Methods

All the numerical portions of this program have already been programmed and tested as subprograms. The present task is to link them into an application through a main program and to debug that main program.

Computer Implementation

Figure 5.6 shows the flowchart for the main program, with references to the various subprograms that will be called. The subprograms have previously been developed and tested, so they are not repeated in the flowchart.

FIGURE 5.6   Flowchart for the main program of the stress–strain
statistical application.

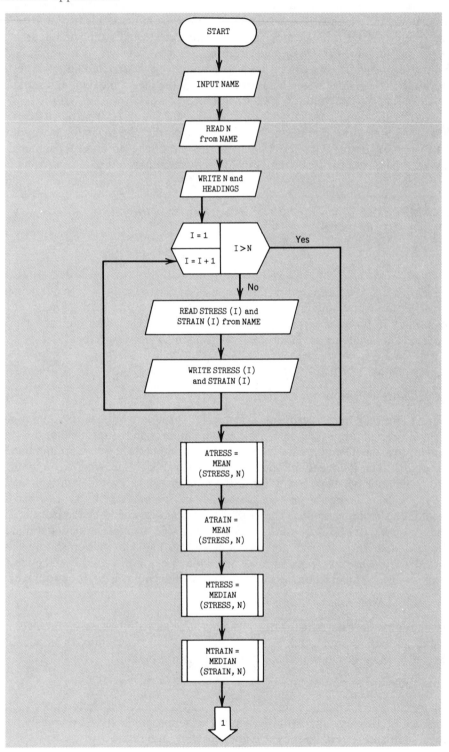

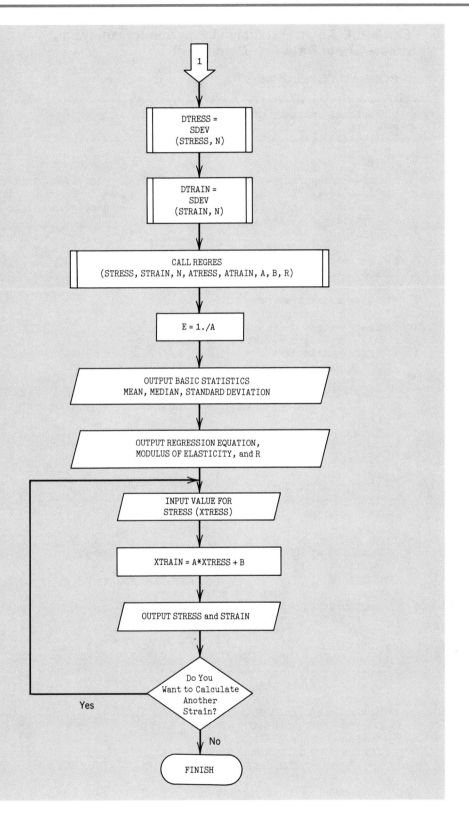

(EXAMPLE 5.5 □ Statistical Program for Analyzing Stress–Strain Data □ Continued)

PROGRAM DEVELOPMENT

```
***************************************************************
*  Statistical analysis of stress strain data                *
*  Developed by: T. K. Jewell      November 1987              *
***************************************************************
C
C  Variable definitions
C  STRESS = array for storing stress data  (psi), independent variable
C  STRAIN = array for storing strain data  (in./in.), dependent variable
C  N = number of data pairs. Limited to <= 50 by dimension specification
C  E = Modulus of elasticity (psi)
C  A = slope of regression line
C  B = intercept of regression line
C  R = correlation coefficient
C  XTRESS = input stress for estimating strain through regression equation
C  XTRAIN = corresponding strain for XTRESS
C  ATRESS = average stress
C  ATRAIN = average strain
C  DTRESS = standard deviation for stress
C  DTRAIN = standard deviation for strain
C  MTRESS = median stress
C  MTRAIN = median strain
C
      REAL STRESS(100),STRAIN(100),E,A,B,R,XTRESS,XTRAIN,MEAN,MEDIAN,
     1     SDEV,SUM,SUMSQ,ATRESS,ATRAIN,DTRESS,DTRAIN,MTRESS,MTRAIN,
     2     CORR
      INTEGER N,R,FR,W
      CHARACTER NAME*15,YES*3
      PARAMETER (R=5,FR=4,W=6)
      PRINT *, 'Input name of data file.'
      PRINT *, 'Include apostrophes before and after.
      READ (R,*) NAME
      OPEN(UNIT=FR,FILE=NAME,STATUS='OLD')
C
C  Input N and n data pairs
      READ (FR,*) N
      WRITE (W,*)
      WRITE (W,*) 'For ',N,' data pairs, the values of stress and'
     1            ' strain are:'
      DO 100 I=1,N
         READ(FR,*) STRESS(I),STRAIN(I)
         WRITE(W,*) STRESS(I),STRAIN(I)
  100 CONTINUE
C
C  Compute statistics and do regression
      ATRESS = MEAN(STRESS,N)
      ATRAIN = MEAN(STRAIN,N)
      MTRESS = MEDIAN(STRESS,N)
      MTRAIN = MEDIAN(STRAIN,N)
      DTRESS = SDEV(STRESS,N,ATRESS)
      DTRAIN = SDEV(STRAIN,N,ATRAIN)
      CALL REGRES(STRESS,STRAIN,N,ATRESS,ATRAIN,A,B,CORR)
C
C  Compute modulus of elasticity
      E = 1./A
C
C  Output basic statistics
```

```
      WRITE(W,*)
      WRITE(W,*) 'The basic statistics for the stress-strain data are:'
      WRITE(W,*)
      WRITE(W,*) 'STRESS'
      WRITE(W,*) 'Mean = ',ATRESS,' psi'
      WRITE(W,*) 'Median = ',MTRESS,' psi'
      WRITE(W,*) 'Standard deviation = ',DTRESS,' psi'
      WRITE(W,*)
      WRITE(W,*) 'STRAIN'
      WRITE(W,*) 'Mean = ',ATRAIN,' in./in.'
      WRITE(W,*) 'Median = ',MTRAIN,' in./in.'
      WRITE(W,*) 'Standard deviation = ',DTRAIN,' in./in.'
      WRITE(W,*)
C
C   Output regression equation
      WRITE(W,*) 'The regression equation is: '
      WRITE(W,*)
      WRITE(W,*) 'STRAIN = ',A,'*STRESS + ',B
      WRITE(W,*)
      WRITE(W,*) 'The modulus of elasticity is ',E,' psi, and'
      WRITE(W,*) 'the correlation coefficient is ',CORR
C
C   Start loop for computing strain
  200 CONTINUE
      PRINT *, 'Input value for stress, psi.  '
      READ(R,*) XTRESS
      XTRAIN = A*XTRESS + B
      WRITE(W,*)
      WRITE(W,*) 'For a stress of ',XTRESS,' the strain is ',XTRAIN,
    1            ' in./in.'
      PRINT *, 'Do you want to input another stress?  (Yes or No)'
      PRINT *, 'Be sure to include apostrophes before and after '
      READ(R,*) YES
      IF(YES(1:1).EQ.'Y'.OR.YES(1:1).EQ.'y') THEN
      GO TO 200
C   End of loop for computing strain
      END IF
      END
```

PROGRAM TESTING

The following stress–strain data can be used to check the model.

Stress (psi)	Strain (in./in.)
20,000	0.0002
16,000	0.0003
22,000	0.0005
26,000	0.0005
30,000	0.0005
28,000	0.0007
36,000	0.0007
32,000	0.0008
38,000	0.0008
38,000	0.0010
44,000	0.0010
38,000	0.0012
44,000	0.0013
42,000	0.0014

(EXAMPLE 5.5 □ Statistical Program for Analyzing Stress–Strain Data □ Continued)

48,000	0.0015
50,000	0.0017
48,000	0.0018
52,000	0.0018
50,000	0.0020
52,000	0.0022

Average stress = 37,700 psi

Average strain = 0.001095 in./in.

Median stress = 38,000 psi

Median strain = 0.0010 in./in.

Standard deviation of stress = 10,910 psi

Standard deviation of strain = 0.0005792 in./in.

Modulus of elasticity, E, = 20.2×10^6 psi

Slope, A, = 4.961×10^{-8}

Intercept, B, = -0.000775

Correlation coefficient, R, = 0.934

For a stress of 53,600 psi, the estimated strain is 0.00188 in./in.

5.6 PARABOLIC ARCH

Many real world phenomena follow a parabolic arch, including the particle ballistic trajectory already discussed and a parabolic support arch for a bridge or other structure. Therefore, it might be advantageous to have a subprogram that can find any number of coordinates in a parabolic arch. The coordinates could be used for graphing or for doing additional computations on force distribution and

FIGURE 5.7   Definition sketch for parabolic arch bridge.

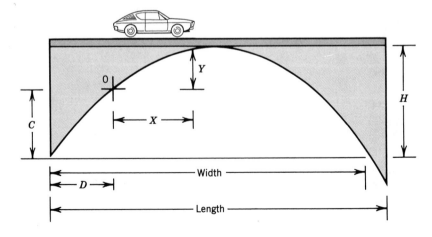

stresses in the arch. The subprogram could be set up with variable dimensions so that any number of points could be found on the arch. Also, the arch could be asymmetrical. The parameters for the parabolic equation could be determined from the maximum trajectory height and distance, H and WIDTH of Figure 5.7, for any parabolic arch, whether it is symmetric or asymmetric. The test data will be for a bridge arch that is slightly asymmetrical.

EXAMPLE 5.6 □ Generating Parabolic Arch Array for Bridge Structure

STATEMENT OF PROBLEM

Develop a subprogram to compute n points of a ballistic, parabolic arch, given the maximum height and distance of the arch, and the initial offset for the first point of the analysis. Figure 5.7 shows the definition sketch for the analysis. Develop a main program to use the subprogram to find 51 points on the arch for an asymmetrical bridge with the initial point of the bridge at the beginning of the arch (C and D equal zero); $H = 100$ ft; WIDTH $= 500$ ft; and a bridge length of 700 ft.

MATHEMATICAL DESCRIPTION

Equation 5.6 gives the mathematical description of the trajectory.

$$Y = A(X + D)^2 + B(X + D) + C$$

$$A = \frac{-H}{(WIDTH/2)^2} \qquad B = -A(WIDTH) \qquad (5.6)$$

Variables are as defined in Figure 5.7.

ALGORITHM DEVELOPMENT

Input/Output Design

Input to the subprogram has to include the initial offset for the beginning of the analysis (D and C of Figure 5.7), the maximum height and distance of the symmetrical arch, the straight line length of the arch, and the number of data points desired. Output will be two arrays, the first containing n values of X; and the second containing the corresponding values of Y, the height of the arch at a distance X. Output of the main program will include all input data and the distance–height arrays.

Computer Implementation

Since there is minimal input data for this program, it can be input efficiently from the terminal. An alternative method of error trapping is employed here. A SUBROUTINE subprogram is appended that is accessed only if an error condition occurs. The subprogram informs you that there is a problem with the last data value entered and that it should be reentered carefully. The subroutine contains its own error trapping loop to catch continued errors. The SUBROUTINE subprogram for developing the arch arrays is straightforward; it is shown in Figure 5.8.

(EXAMPLE 5.6 □ Generating Parabolic Arch Array for Bridge Structure □ Continued)

FIGURE 5.8 Flowchart for parabolic arch bridge program subroutines.

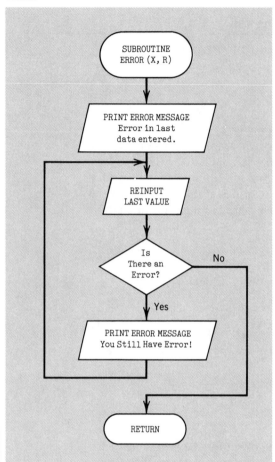
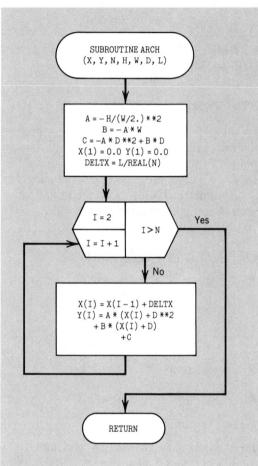

PROGRAM DEVELOPMENT

```
************************************************************************
*  Test program for SUBROUTINE ARCH                                   *
*  Developed by:  T. K. Jewell    December 1987                       *
************************************************************************
C
C  Variable definitions as per subprogram and input prompts
C  Origin is station zero for bridge arch.
       REAL X(201),Y(201),H,WIDTH,D,LENGTH
       INTEGER N,R,W
       PARAMETER (R=5,W=6)
       PRINT *, 'Input number of data points you want to generate'
       READ (R,*) N
       PRINT *, 'Input width of symmetrical parabola'
       READ (R,*,ERR=20) WIDTH
       GO TO 30
    20 CALL ERROR(WIDTH,R)
```

```
   30 CONTINUE
      PRINT *, 'Input maximum height of symmetrical parabola'
      READ (R,*,ERR=40) H
      GO TO 50
   40 CALL ERROR(H,R)
   50 CONTINUE
      PRINT *, 'Input initial X offset (D) from start of parabola'
      READ (R,*,ERR=60) D
      GO TO 70
   60 CALL ERROR(D,R)
   70 CONTINUE
      PRINT *, 'Input length of parabola (span of bridge)'
      READ (R,*,ERR=80) LENGTH
      GO TO 90
   80 CALL ERROR(LENGTH,R)
   90 CONTINUE
C
C  Compute coordinates of arch
      CALL ARCH(X,Y,N,H,WIDTH,D,LENGTH)
C
C  Output arch data
      WRITE(W,*) 'For WIDTH = ',WIDTH,', H = ',H,', D = ',D,
     1           ', and LENGTH =',LENGTH
      WRITE (W,*)
      WRITE(W,*) 'The X and Y coordinates of the arch are:'
      WRITE(W,*) ' POINT        X         Y'
      DO 100 I=1,N
         WRITE(W,*) I,X(I),Y(I)
  100 CONTINUE
      END

      SUBROUTINE ERROR(X,R)
C
C  Subprogram for error trapping
      REAL X
      INTEGER R
      PRINT *, 'You have an error in the last data value entered.'
      PRINT *, 'Please reenter the value carefully'
   20 CONTINUE
      READ (R,*,ERR=50) X
      GO TO 100
   50 CONTINUE
         PRINT *, 'You still have an error condition, please reenter'
         PRINT *, 'the value again, or consult your technical support'
         PRINT *, 'people for help.'
      GO TO 20
  100 CONTINUE
      RETURN
      END

      SUBROUTINE ARCH(X,Y,N,H,W,D,L)
C  Variable definitions
C  H = maximum height for symmetrical parabola of width W
C  W = width to be used for symmetrical parabola.  It is not necessarily the
C      total width of the parabola, but is the width that defines the
C      parameters A and B for the parabolic equation.
C  D = initial offset in the x coordinate of the parabola. This covers the
C      case when the initial point for the simulation is not the initial
C      point for the defined parabola.
C  C = initial offset in the y coordinate corresponding to x offset of D
```

(EXAMPLE 5.6 □ Generating Parabolic Arch Array for Bridge Structure □ Continued)

```
C  L = total width (span) of parabola.
C  N = desired number of points in the array describing the arch.
C  A, B, and C = parameters of parabolic equation
C
      REAL X(N),Y(N),H,W,D,L,A,B,C,DELTX
      INTEGER N
      A = -H/(W/2.)**2
      B = -A*W
      C = -A*D**2 - B*D
      X(1) = 0.0
      Y(1) = 0.0
      DELTX = L/REAL(N-1)
      DO 100 I=2,N
      X(I) = X(I-1) + DELTX
      Y(I) = A*(X(I)+D)**2 + B*(X(I)+D) + C
  100 CONTINUE
      RETURN
      END
```

PROGRAM TESTING

For a maximum height of 100 ft and a symmetrical width of 500 ft, $A = -0.0016$ and $B = 0.8$. If the offset is zero, the Y coordinate for $X = 250$ ft is 100 ft. When $X = 500$, Y is zero. For $X = 700$ ft, Y is -224 ft. If the initial X is offset 100 ft from the beginning of the parabola, $C = -64$ ft. For this case $Y = 36$ ft at $X = 150$ ft. For $X = 400$ ft, $Y = -64$ ft and at $X = 700$ ft, $Y = -448$ ft. Temporary output statements could be inserted in the subprogram to check the values of A, B, and C.

5.7 QUADRATIC EQUATION

Example 4.4 illustrated use of a nested IF THEN ELSE within a cascaded IF THEN ELSE to evaluate the roots of a quadratic equation using the quadratic formula

$$X = \frac{-B \pm \sqrt{B^2 - 4AC}}{2A} \qquad (4.5)$$

A quadratic equation frequently has to be solved as part of another problem. Therefore, it would be useful to have a subprogram to implement the quadratic formula. Example 5.7 will develop a SUBROUTINE subprogram to accomplish this. We will modify the method for calculating complex roots from the program of Example 4.4 to accommodate subroutine data transfer. If complex variables are included in the dummy arguments of the subroutine, then any program using the subroutine would also have to have complex variables defined. Since many applications will not produce complex roots, it would be better to have the subprogram calculate and pass back all roots as real variables. The conversion to complex roots can take place in the accessing program rather than in the subroutine.

EXAMPLE 5.7 □ Quadratic Equation with Real or Complex Roots

STATEMENT OF PROBLEM

Develop a SUBROUTINE subprogram to find the real or complex roots of a quadratic equation using the quadratic formula.

ALGORITHM DEVELOPMENT

Computer Implementation

The cascaded and nested IF THEN ELSE structures used in Example 4.4 can be transferred to the SUBROUTINE with a few minor changes. When you are calling a subprogram several times, it may not be desirable to have a message output each time concerning which case the solution represents. Therefore, the messages will be deleted, and in their place a value will be assigned to a case variable. The value for this variable will be transferred back to the main program. If desired, a message could be output in the main program. Complex variables will not be used in the subprogram. The numeric values of the real and imaginary parts of the roots will be calculated and returned to the main program as REAL variables. In the main program they can be converted to complex variable notation for use later in the program. The following pseudocode shows the portions of the original program that are retained in the SUBROUTINE.

```
SUBROUTINE QUAD(A,B,C,X1,X2,IX1,CASE)
SET X1, X2, AND IX1 = 0.0
CALCULATE RAD = B**2 - 4.*A*C
IF A.EQ.(0.0) THEN
     SET CASE = 1
     COMPUTE X1=-C/B
ELSE IF (RAD.LT.(0.0)) THEN
     SET CASE = 2
     CALCULATE COMPLEX ROOTS
     X1 = -B/(2.*A)
     IX1 = SQRT(ABS(RAD))/(2.*A)
ELSE
     CALCULATE REAL ROOTS
     RAD = SQRT(RAD)
     IF(RAD.GT.0.0) THEN
          SET CASE = 3
          X1 = (-B + RAD)/(2.*A)
          X2 = (-B - RAD)/(2.*A)
     ELSE
          ROOTS ARE EQUAL
          SET CASE = 4
          X1 = -B/(2.*A)
          X2 = X1
     END IF
END IF
RETURN
END
```

(EXAMPLE 5.7 □ Quadratic Equation with Real or Complex Roots □ Continued)

PROGRAM DEVELOPMENT

```
        PROGRAM QUADRATIC
*****************************************************************************
*     Program for testing SUBROUTINE for calculating real               *
*        or complex roots of a quadratic equation                       *
*     Developed by:  T. K. Jewell            December 1987              *
*****************************************************************************
C     Variable definitions
C     A, B, and C = coefficients of quadratic eqn: A*X**2+B*X+C=0.0
C     X1 and X2 = real roots of quadratic equation, X1 also acts as
C        real portion of complex root.
C     IX1 = imaginary portion of complex root
C     X1COMP and X2COMP are complex conjugate roots of the quadratic
C        equation
        REAL A,B,C,X1,X2,IX1
        COMPLEX X1COMP,X2COMP
        INTEGER R,W,CASE
        PARAMETER (R=5,W=6)
C     INPUT VALUES FOR A,B,C
        PRINT *, 'Input values for coefficients A, B, and C'
        READ (R,*) A,B,C
C     OUTPUT A,B,C FOR ERROR CHECKING
        WRITE (W,*) 'Input data for checking'
        WRITE (W,*) 'A = ',A,' B = ',B,' C = ',C
        CALL QUAD(A,B,C,X1,X2,IX1,CASE)
        IF (CASE.EQ.2) THEN
            WRITE(W,*) 'COMPLEX ROOTS'
            WRITE(W,*) X1,' + j',IX1
            WRITE(W,*) X1,' - j',IX1
C     Convert to complex variables for later use in the program
            X1COMP = CMPLX(X1,IX1)
            X2COMP = CONJG(X1COMP)
        ELSE IF (CASE.EQ.3.OR.CASE.EQ.4) THEN
C     Output real roots, X1 and X2
            WRITE(W,*) 'REAL ROOTS'
            WRITE(W,*) 'X1 = ',X1,' and X2 = ',X2
        ELSE
C     There is only one root
            WRITE(W,*) 'SINGLE ROOT, X = ',X1
        END IF
        END
        SUBROUTINE QUAD(A,B,C,X1,X2,IX1,CASE)
        REAL A,B,C,X1,X2,IX1
        INTEGER CASE
        X1 = 0.0
        X2 = 0.0
        IX1 = 0.0
        RAD = B**2-4.*A*C
C     Start of cascaded IF THEN ELSE
C     Check for A=0.0
        IF (A.EQ.(0.0)) THEN
            CASE = 1
            X1=-C/B
C     Check for RAD < 0.0 (Imaginary roots)
        ELSE IF(RAD.LT.(0.0)) THEN
            CASE = 2
```

```
C    Calculate complex roots
        X1 = -B/(2.*A)
        IX1 = SQRT(ABS(RAD))/(2.*A)
     ELSE
C    Calculate real roots
        RAD = SQRT(RAD)
C    Start of nested IF THEN ELSE
        IF (RAD.GT.(0.0)) THEN
           CASE = 3
C    Calculate real roots
           X1 = (-B + RAD)/(2.*A)
           X2 = (-B - RAD)/(2.*A)
        ELSE
C    Calculate equal real roots.
           CASE = 4
           X1 = -B/(2.*A)
           X2 = X1
        END IF
     END IF
     RETURN
     END
```

PROGRAM TESTING

Develop values for A, B, and C that will test all four branches of the program. The same values that you developed for Example 4.4 could be used here.

5.8   INPUT AND OUTPUT OF TWO-DIMENSIONAL ARRAYS

Engineering analysis often involves using matrices or other two-dimensional arrays. We will now develop FORTRAN subprograms for inputting and outputting two-dimensional arrays. The input will be from a data file. These subroutines will allow you to input and output arrays in the normal two-dimensional form. You will find them very useful in developing applications.

Two standard nested DO loops could be used to control input or output of two-dimensional arrays. Input would have the form

```
     DO 100 I=1,N
        DO 200 J=1,M
           READ(R,*) A(I,J)
 200    CONTINUE
 100 CONTINUE
```

If $A(I,J)$ represents a matrix, I would be the row identifier and J would be the column identifier. N is the total number of rows and M the total number of columns. The only problem with the above structure is that the rules for execution of READ statements state that each new execution of the READ will access a new input record. Thus, the input data file for $A(I,J)$ would have to contain one value per line. This would make the input file longer than necessary and would make it hard to check the input values. If a similar structure were used to output an array, it would be output in a column with one element under the other, rather than in the form of a matrix.

Fortunately, the implied DO loop of FORTRAN allows us to read multiple values from the same input data record. We learned about and used the implied DO loop to input a one-dimensional array in Example 5.2. This structure can be extended for input of a two-dimensional array. The form would be

```
      DO 100 I=1,N
         READ(R,*) (A(I,J),J=1,M)
  100 CONTINUE
```

The structure $(A(I,J),J=1,M)$ has the same effect as a nested DO loop, except that the variation of the index J does not represent separate executions of the READ statement. Therefore, the READ statement will accept multiple values from the same line of the data file (input data record). This reads in the matrix a row at a time. If not all the values of the row will fit on one input line (record), the data can be continued on as many subsequent lines as are required to hold all the values for that row of the matrix. The READ statement will continue to go to new input records until it satisfies the input variable list. If the last data for the row take up part of a record, you cannot start the next row in the same input record. When the outer loop index increases to input a new row, the READ is executed again, and a new record will be accessed. Therefore, the next row of the matrix must be started in a fresh record. If you have questions on this, refer back to the rules for input list satisfaction in Chapter 3.

For SUBROUTINE INPUT, the values of N (number of rows) and M (number of columns) will have to be input prior to reading the first row of the array. The data file would include the values of N and M in the first input record, followed by the rows of the array. Each row of the array would have to start on a new record of the data file, no matter where the last element of the previous row was positioned on a record. For example

```
      SUBROUTINE READ(A,N,M,FR)
      REAL A(10,10)
      INTEGER N,M,FR
      READ (FR,*) N,M
      DO 100 I=1,N
         READ(FR,*) (A(I,J),J=1,M)
  100 CONTINUE
      RETURN
      END
```

would input N rows of an $N \times M$ matrix, one row at a time. M elements would be input for each row. Your data file will have to be set up accordingly. Note that variable dimensions are not used in the subprogram because of the problems associated with the way values are actually stored in the computer.

The output subprogram is essentially a duplicate of the input subprogram, with WRITE statements rather than READ statements. For example

```
      SUBROUTINE WRITE(A,N,M,W)
      REAL A(10,10)
      INTEGER N,M,W
      DO 100 I=1,N
         WRITE(W,*) (A(I,J),J=1,M)
  100 CONTINUE
      RETURN
      END
```

would output an $N \times M$ matrix in the usual format.

If an output line of the matrix has more elements than your computer system will output per line in free format, the remaining values will be output on succeeding lines. This makes it difficult to find particular values in the array. Thus, the practical limit on the number of columns in arrays that can be output by this subprogram is the number of values that can be output per line in unformatted output. You will have more control over the number of elements output per line when we cover formatted output in Chapter 6. We will also modify the subroutine in Chapter 7 to output large matrices in blocks that could be pasted together to form the whole matrix. These modifications will allow us to output any size array.

5.9   SYSTEM SUPPLIED AND LIBRARY SUBPROGRAMS

Most computer centers maintain a set of subprograms that can be used with the various languages available on the computers supported. Documentation for these generally includes the name of the subprogram, the type of subprogram (SUB-ROUTINE or FUNCTION), the purpose of the subprogram, a description of the actual arguments that have to be used for the reference to the subprogram, and a brief description of the numerical methods used in the subprogram. The source code for the subprograms may or may not be available. If it is available, you can customize the subprograms for your own uses. If not, you are stuck with using them as published. Some library subprograms will have to be linked with the main program, whereas others can be accessed directly, without linking. Many offer both single and double precision options.

The following is an example of a call to a matrix inversion SUBROUTINE subprogram available on the VAX system. This subprogram is available directly from any program.

```
CALL BAS$MAT_INV(%DESCR(A),%DESCR(AINV))
```

The name of the subprogram is BAS$MAT_INV. A is the matrix to be inverted, and AINV will contain the inverse of A after the subprogram is executed. Source code is not available for this subprogram. It is contained in a compiled form that can be accessed from your program without linking it in the normal manner for subprograms stored in separate files. The interesting thing about this subprogram is that it is actually written in BASIC, but since it is stored in a compiled form, it can be used with FORTRAN programs. Example 7.3 will use BAS$MAT_INV to solve a system of linear equations.

Libraries of subprograms that can be linked with FORTRAN programs are also often available. These will require linking of the subprogram or library prior to executing the accessing program. One popular library is the International Mathematical and Statistical Library (IMSL), (1987). There are three IMSL libraries, one for general applied mathematics, one for statistics, and one for special functions. A subprogram from the IMSL library will accomplish matrix inversion for a truss analysis problem in Example 7.4. Peerless Engineering Services (1986) has developed two volumes of FORTRAN subprograms for use with microcomputer FORTRAN compilers. Subjects usually covered by library subprograms include

Graphics and plotting

Economic analysis

Statistics

Differential equations

Time-series analysis

Matrix operations

Numerical methods

Linear programming

Modeling and simulation

5.10 ADDITIONAL SUBPROGRAM FEATURES

5.10.1 DOUBLE PRECISION Subprograms

DOUBLE PRECISION in FUNCTION and SUBROUTINE subprograms is treated the same way that it is in main programs. All applicable variables have to be declared as DOUBLE PRECISION, and actual arguments that are DOUBLE PRECISION have to correspond with DOUBLE PRECISION dummy arguments. DOUBLE PRECISION placed in front of FUNCTION X(dummy arguments) declares the FUNCTION as DOUBLE PRECISION.

5.10.2 EXTERNAL and INTRINSIC Statements

EXTERNAL and INTRINSIC statements allow function names to be used as arguments in references to subprograms. Then when the corresponding dummy argument is used in the subprogram, it represents the function. The INTRINSIC function allows FORTRAN intrinsic function names to be used as arguments, whereas the EXTERNAL statement accomplishes the same for user-defined functions. The following example will illustrate the form of the statements. SIN and COS are the normal FORTRAN intrinsic functions for sine and cosine, while CTN is a user-defined function for the cotangent.

```
MAIN PROGRAM
      .
EXTERNAL CTN
INTRINSIC SIN,COS
      .
CALL TRIG(ANGLE,SIN,SINE)
      .
CALL TRIG(ANGLE,COS,COSINE)
      .
CALL TRIG(ANGLE,CTN,COTANG)
      .
END

SUBROUTINE TRIG(X,F,Y)
Y = F(X)
RETURN
END
```

```
REAL FUNCTION CTN(X)
CTN = COS(X)/SIN(X)
RETURN
END
```

In the first call to TRIG, F(X) represents SIN(ANGLE) and returns the value to the variable SINE in the calling program. The second does the same for the intrinsic cosine function. The third call uses the FUNCTION CTN(X) subprogram to evaluate the CTN(ANGLE) in SUBROUTINE TRIG.

Examples later in the text will apply these concepts.

5.10.3 ENTRY

It is possible to use the ENTRY statement to enter a subprogram at a point other than the top of the subprogram. However, this is counter to good structured programming procedures. If you need to enter a subprogram at more than one point, then the subprogram should be broken up into multiple subprograms so the logical linkage can be maintained.

5.10.4 SAVE

Values for local variables that are used only in a subprogram are normally lost after the RETURN statement is executed. If for some reason you want to save the values for the next time the subprogram is executed, you can specify the variables to be saved in the subprogram through use of the SAVE specification statement

    SAVE *variable list.*

If the *variable list* is left off, all local variable values are saved.

5.10.5 Alternate RETURN

It is possible to have control return to the accessing program at a point other than the point from which the subprogram was accessed. However, as with the alternate ENTRY, the alternate RETURN is counter to the principles of good structured programming. Its use should be avoided.

SUMMARY

In this chapter you have developed a number of subprograms to implement specific procedures. The examples from the text can become the basis for your own library of subprograms. You will be adding more of your own, and your instructor will be giving you others to use in homework assignments. Through using subprograms, you will gain an appreciation of their value. They make it much easier to develop structured programs. You will also save a great deal of time in the development of applications programs whenever you can use previously developed subprograms.

REFERENCES

Borse, G. J., *FORTRAN 77 and Numerical Analysis for Engineers*, PWS Publishers, Boston, 1985.

Ellis, T. M. R., *A Structured Approach to FORTRAN 77 Programming*, Addison-Wesley, London, 1982, Chapter 10.

Etter, D. M., *Structured FORTRAN 77 for Engineers and Scientists*, 2nd. ed., The Benjamin/Cummings Publishing Company, Inc., Menlo Park, CA, 1987, Chapters 6 and 7.

International Mathematical and Statistical Library (IMSL) Inc., "Math/Library, FORTRAN Subroutines for Mathematical Applications," Version 1.0, Houston, TX, 1987.

Peerless Engineering Service, *Professional FORTRAN Scientific Subroutine Library*, Wiley, New York, 1986.

EXERCISES

5.1 GENERAL NOTES

Develop a main program to test each subprogram or group of subprograms that you develop as part of an exercise.

In general, the same data used to test the corresponding program in Chapter 4 can be used to test the subprograms.

Where appropriate, save your subprograms as separate files, and link them with your main programs during the compilation process. Test subprograms separately before combining them together for a longer applications program. Print statements can be included in either FUNCTION or SUBROUTINE subprograms for debugging.

5.2 MODIFICATION OF EXAMPLES

5.2.1 Example 4.14

Develop a subprogram to accomplish the data input, error checking, and input range checking for Example 4.14.

5.2.2 Example 5.2

Modify the program and subprograms of Example 5.2 so that input can be through the terminal. Include error check loops for the vector components and N.

5.2.3 Example 5.4

1. Add the option of finding the standard error of estimate to the regression subprogram developed for Example 5.4. Refer to Appendix H for definition of the standard error of estimate. The option should be controlled from within the main program. Test the revised program using the data of Example 5.4.

2. Add a subprogram for input of regression data and echo of input. The input subprogram should be accessed from the main program. Any problem identification can be input through the main program.
3. Modify the data input subprogram of part 2 above so that it can be accessed from the regression subprogram. Provisions should be included for a variable number of lines of identification input, up to ten, controlled by a sentinel structure. The output of the regression results should be from the main program.
4. Some phenomena can be modeled by a function of the form

$$y = ae^{bx}$$

which can be linearized by taking the natural logarithms of both sides to yield

$$\ln(y) = \ln(a) + bx$$

Develop a main program to input the barometric pressure vs. altitude data given below and to estimate a and b for the model using altitude as the independent variable. Use the model to estimate the barometric pressure at an altitude of 1,800 m.

Altitude (m)	Barometic Pressure (mm)
0	760
500	714
1000	673
1500	631
2000	594
2500	563

5.2.4 Example 5.5

1. Modify the program of Example 5.5 to include computation of the standard error of estimate. Appendix H gives a definition of the standard error of estimate. Test the revised program using the data of Example 5.5.
2. Modify the program of Example 5.5 so that finding the median and standard deviation is optional. Test both options.

5.3 GENERAL SYNTAX AND STATEMENT STRUCTURE

5.3.1 Calculation of Variable Values After Execution of SUBROUTINE or FUNCTION Subprograms

1. What values would be output by the following program?

```
REAL Y,X
INTEGER N
X = 3.
N = 2
Z = Y((X,N) + 4.
W = Y(5.,N)
A = Y(6.,1) + W
B = Y(W-Z-10.,N-1)
PRINT *,  X,N,Z,W,A,B
END
REAL FUNCTION Y(T,I)
```

```
T1 = T**I
T2 = T - 5.
Y = T1 + T2
RETURN
END
```

*2. What values are printed out by the following program?

```
X = 4.
Y = 3.
CALL RADIUS(X,Y,R)
WRITE (6,*) X,Y,R
END
SUBROUTINE RADIUS (A,B,C)
C = (A**2 + B**2)**0.5
RETURN
END
```

3. What would be the values of X, Y, XVAL, YVAL, and XPRIME after the CALL statement is executed?

```
X = 3.
Y = 5.
CALL POLY(X,Y,XVAL,YVAL,XPRIME)

END
SUBROUTINE POLY(A,B,C,D,E)
C = A**2 + 4.*A - 5.
D = SQRT(B) + 3.
E = 2.*A + 4.
A = C/E
RETURN
END
```

*4. What will be the values of PCT and NWRONG after SUBROUTINE GRADE is referenced by the following CALL statement? Assume NCR = 40

```
CALL GRADE (50,NCR,NWRONG,PCT)

END
SUBROUTINE GRADE (N,NRITE,NRONG,P)
P = NRITE*100/N
NRONG = N - NRITE
RETURN
END
```

5. If the matrix

2.	3.5	−6.	7.3
6.	8.	−1.	3.
8.	−2.	1.5	6.
1.	0.	0.	7.

is stored in a 4 × 4 array named A, what will be the value of T after the following call statement?

```
CALL TRACE (T,A)

END
SUBROUTINE TRACE (T,X)
REAL T,X(4,4)
```

```
      T = 0.0
      DO 10 I=1,4
         T = T + X(I,I)
  10  CONTINUE
      RETURN
      END
```

6. What values are output by the following main programs and subprograms? Indicate what values are printed out where.

a.
```
      B = 5.
      CALL A(2.,B,X)
      PRINT *, X
      CALL A(X,B,Y)
      PRINT *, B,Y
      END
      SUBROUTINE A(G,H,T)
      H = H*2.
      T = G*H - 3.5
      RETURN
      END
```

*b.
```
      I = 5
      CALL B(1,I,J)
      PRINT *, I,J
      CALL B(I,J,K)
      PRINT *, J,K
      END
      SUBROUTINE B(I,J,K)
      J = J/2
      K = I + J + 2
      PRINT *, I
      RETURN
      END
```

c.
```
      2.*4.
      CALL X(C,D)
      CALL X(C+5.,E)
      PRINT *, C,D,E
      END
      SUBROUTINE X(R,S)
      S = R*5.
      R = S/10.
      RETURN
      END
```

7. What values would be output for each of the following sets of FORTRAN statements? Default variable types are in effect.

a.
```
      X = 3.
      Y = 4.
      Z = PYTH(X,Y)
      PRINT *, X,Y,Z
      END
      REAL FUNCTION PYTH(A,B)
      C = A + B - B/A
      A = A/C
      PYTH = C
      RETURN
      END
```

```
*b.    X = 3.
       Y = 4.
       CALL PYTH(X,Y,Z)
       PRINT *,X,Y,Z
       END
       SUBROUTINE PYTH (A,B,C)
       C = A + B - B/A
       A = A/C
       RETURN
       END

 c.    X = 3.
       Y = 4.
       CALL PYTH(X+3.,Y-2.,Z)
       PRINT *,X,Y,Z
       END
       SUBROUTINE PYTH (A,B,C)
       C = A + B - B/A
       A = A/C
       RETURN
       END

*d.    X = 3.
       Y = 4.
       Z = PYTH(X+3.,Y-2.)
       PRINT *,X,Y,Z
       END
       REAL FUNCTION PYTH (A,B)
       C = A + B - B/A
       A = A/C
       PYTH = C
       RETURN
       END
```

8. What values would be output by the following program?

```
       CALL DOUBLE (1.0,4./2.,X)
       PRINT *, X,G
       CALL DOUBLE (X,6.0,G)
       PRINT *, X,G
       CALL DOUBLE (G,G,G)
       PRINT *, X,G
       G = 3.0
       CALL DOUBLE (X**2,G,X)
       PRINT *, X,G
       CALL DOUBLE (G,X,X)
       PRINT *, X,G
       END
       SUBROUTINE DOUBLE (A,B,C)
       C = 2.0*(A+B)
       RETURN
       END
```

5.3.2 Identify and Correct Errors in Calls to SUBROUTINE or FUNCTION Subprograms

*1. All the following calls to SUBROUTINE ACT are incorrect. Why? Make the necessary corrections.

```
REAL X(50),Y(75),W(50)
INTEGER I(75),M
        .
CALL ACT (Z,I,7)
        .
CALL ACT (Z,Y,N,X)
        .
CALL ACT (ACT,I,N,X)
        .
CALL ACT (X,I,M,W)
END
SUBROUTINE ACT (X,J,N,Q)
REAL Q(50),J(75),X
INTEGER J,N
        .
RETURN
END
```

2. Identify any errors in either the referencing statement or the corresponding FUNCTION subprogram definition. Default type declarations are in effect.

```
 a.    Y=5.4-X*S(5,A,B)        REAL FUNCTION S(I,G)
*b.    MEAN(A,B,C)=X*Y         REAL FUNCTION MEAN(X,Y,Z)
 c.    G=ROOT(3*I,K,A)**2      REAL FUNCTION ROOT(N,M,H)
*d.    X=3.5**P-BEAM(L,7.5)    REAL FUNCTION B(M,S)
 e.    B=2.5*SUM(A(I),B(3))    REAL FUNCTION SUM(X,Y)
*f.    PRINT*,X,Y,SUM(X,Y)     REAL FUNCTION SUM(X,Y)
```

5.3.3 Identifying and Correcting General SUBROUTINE and FUNCTION Subprogram Errors

1. Identify and correct errors that occur in the following statements:

```
 a.    READ (5,*) A,B
       X = SUM(A,B)
       Y = SUM(A**2,B**2)
       PRINT *, A,B,X,Y
       END
       SUBROUTINE SUM(T,V)

       RETURN
       END
```

```
*b.    INTEGER K(70),J(70)
       REAL X,Y
       CALL TOTK(K,X)
       CALL TOTK(J,Y)
       END
       SUBROUTINE TOTK(A,B)
       B = 0.
       DO 10 I=1,70
           B = B + A(I)
    10 CONTINUE
       RETURN
       END
```

```
 c.    REAL X(100),D(100)
       CALL XBAR(N,X)
       CALL XBAR(M,K)
```

```
             SUBROUTINE XBAR(N,K)
             REAL XN(100)
                     .
             RETURN
             END
             STOP
             END
```

*d.   CALL (XLEST,A,B)
                .
```
             STOP
             END
             SUBROUTINE XLEST (A,B)
             STOP
             END
             SUBROUTINE XLEST(A,B,X(I))
             Z = A + B
             X(I) = A + Z/B
             A = X(I) + Z*A
             RETURN
             END
```

e.   CALL QUAD (X,A,B,C)
                .
```
             END
             REAL FUNCTION QUAD (X,A,B,C)
             QUAD = A*X**2 + B*X + C
             RETURN
             END
```

*f.   CALL QUAD (X,A,B,C)
                .
```
             END
             SUBROUTINE QUAD(X,A,B,C)
             D = A*X**2 + B*X +C
             RETURN
             END
```

2. Identify any errors in the following FUNCTION subprogram definition statements:

```
   a.    REAL FUNCTION LOAD(Y(100),N,Q,R)
  *b.    REAL FUNCTION STRAIN
   c.    REAL FUNCTION SMALL(X,Y,Z,N,SINE)
  *d.    REAL FUNCTION X(Y,P)
          P = 9.0
          X = P**Y
   e.    REAL FUNCTION MEAN(A,B,C)
          MEAN = (A + B + C)/3.0
  *f.    INTEGER FUNCTION (I,K,M,P)
   g.    REAL FUNCTION A(X,Y**2)
```

5.3.4 Identify and Correct Errors in Passing Arrays to and from Subprograms

Identify any errors in the following program segments, and suggest corrections.

```
a.    REAL X(40),A(50,50),B(40),T,W
      INTEGER POINT(50),I
```

```
       CALL MAT(X,A,T,POINT)
         .
       END
       SUBROUTINE MAT(A,B,C,D)
       REAL A(40),B(50,50),C
       INTEGER D
         .
       RETURN
       END
```

*b.
```
       REAL X(40),A(50,50),B(40),T,W
       INTEGER POINT(50),I
         .
       CALL MAT(X,A,T,I)
         .
       END
       SUBROUTINE MAT(A,B,C,D)
       REAL A(50),B(50,50),C
       INTEGER D
         .
       RETURN
       END
```

c.
```
       REAL X(40),A(50,50),B(40),T,W
       INTEGER POINT(50),I
         .
       CALL MAT(A,X,T,I)
         .
       END
       SUBROUTINE MAT(A,B,C,D)
       REAL A(40),B(50,50),C
       INTEGER D
         .
       RETURN
       END
```

*d.
```
       REAL X(40),A(50,50),B(40),T,W
       INTEGER POINT(50),I
       I = 50
         .
       CALL MAT(A,X,T,I,POINT)
         .
       END
       SUBROUTINE MAT(A,B,C,D,POINT)
       REAL A(I),B(I,I),C
       INTEGER D,POINT(I)
         .
       RETURN
       END
```

5.3.5 Incorrect Programs to Debug

The following program segment produces the output shown.

```
    DO 10 I=1,10
       A = FCN(1.0)
       WRITE (6,*) A,1.0
10  CONTINUE
    END
```

```
REAL FUNCTION FCN(A)
A = A + 1.0
FCN = A
RETURN
END
```

2.000000	2.000000
4.000000	4.000000
8.000000	8.000000
16.00000	16.00000
32.00000	32.00000
64.00000	64.00000
128.0000	128.0000
256.0000	256.0000
512.0000	512.0000
1024.000	1024.000

What is happening to cause this? What modifications could you make to produce all values of 1.000000 in the first column, and all values of 2.000000 in the second column?

5.4 CHEMICAL ENGINEERING

Note: Appendix K contains additional information that relates to the exercises of this section.

5.4.1 [ALL] Ideal Gas Law

Develop separate subprograms for computing the pressure of a gas using the van der Waals equation or the Redleich–Kwong equation. Test your program using the data of Exercise 4.4.1.

5.4.2 [ALL] Chemical Kinetics

Develop a subprogram that will compute the concentration at time t for a chemical involved in a reaction under either first- or second-order kinetics. Test your program using the data of Exercise 4.4.2.

5.4.3 [ALL] Depth of Fluidized-Bed Reactor

Develop a subprogram for finding the void fractions of the constituents of the expanded bed. Test your program using the data of Exercise 4.4.3.

5.4.4 [ALL] Process Design, Gas Separation

Develop a subprogram that will determine the number of plates required to absorb a certain percentage of a given gas, and determine the percentages removed for up to four additional gas constituents. Test your program using the data of Exercise 4.4.4.

5.4.5 [ALL] Dissolved Oxygen Concentration in Stream

Develop a subprogram to determine the dissolved oxygen deficit at a distance X downstream from the point of discharge of a pollutant. Input to the subprogram should be the distance, the average velocity of the river, the diffusivity of oxygen in the water, the oxygen use rate, the initial concentration of pollutant, the initial dissolved oxygen deficit, and the depth of the water. Test your subprogram using the data of Exercise 4.4.5.

5.5 CIVIL ENGINEERING

5.5.1 [ALL] Equilibrium-Truss Analysis

Develop a subprogram to determine the reactions on a planar truss. Use a main program to test the subprogram using the data of Exercise 4.5.1. All input and output should be from the main program.

5.5.2 [ALL] Open Channel Flow Analysis

1. Develop a program to implement the flowchart of Exercise 2.4.2 to estimate the flow rate in either a trapezoidal or circular open channel. Section 5.3 of Appendix G presents the geometric relationships required to solve open-channel flow problems in trapezoidal and circular channels. Use a SUBROUTINE subprogram to estimate the area and hydraulic radius given the type of channel and the appropriate dimensions. Test your program to determine Q for a trapezoidal channel of $y = 4$ ft, $b = 4$ ft, $n = 0.03$, $S_0 = 0.001$ ft/ft, and $z = 2$ ft/ft. Also estimate the flow rate for a circular channel of diameter 15 ft, $n = 0.03$, $S_0 = 0.001$ ft/ft, and $y = 7$ ft.
2. Develop a subprogram for estimating the depth of flow in a trapezoidal open channel when the flow rate and other necessary data are known. Test your subprogram with the data for Exercise 4.5.2.

5.5.3 [ALL] Dam Stability Analysis

Develop a function subprogram to find the force F on a submerged surface, such as the surfaces of the dam of Exercise 2.4.3. Also develop a subroutine to find the weight and centroid of a compound shape, such as the cross section of the dam. Modify the program of Exercise 4.5.3 to use these subprograms. Test the program with the same data and with the addition of a depth of water of 10 ft on the downstream side of the dam. This causes an additional force, F_3, that acts in the opposite direction from force F_1.

5.5.4 [ALL] Force on Submerged Gate

Develop a subroutine subprogram to find the force F and center of pressure h_p for a submerged gate, given the moment of inertia I_c, depth of centroid h_c, area A, and specific weight γ of the liquid exerting the pressure. Also develop separate subroutines to find the area and moment of inertia for a rectangle, a circle, or a triangle, given the appropriate data. You could use the subprogram of Exercise 5.9.3 for this purpose. Modify the program of Exercise 4.5.4 to use these subprograms, and test the program using the data for that exercise.

5.5.5 [ALL] Volume of Excavation

Develop a subprogram to determine the volume of a trench with trapezoidal end sections, using the logic developed in Exercise 3.5.5. Incorporate this subprogram into the program of Exercise 4.5.5 and test the program using the 45° side slope data of that exercise.

5.6 DATA ANALYSIS AND STATISTICS

5.6.1 [ALL] Histogram

Develop a subprogram for determining frequency counts for any set of data. The sub-program should be able to take up to 20 intervals for frequency determination. Variable dimensions should be used so any size data set can be passed over for analysis. Test your program using the data of Exercise 4.6.1.

5.6.2 [ALL] Sorting

1. Develop a subprogram to sort a one-dimensional array in either ascending or descending numerical order using the bubble sort algorithm of Example 4.10.
2. Use the subprogram from part 1 to sort the elements of each row of a two-dimensional array into ascending order.
3. Develop a subprogram to sort a one-dimensional array in either ascending or descending numerical order using the Shell sort algorithm discussed in Chapter 4 and developed as a program in Exercise 4.6.2 (1). If you develop the subprograms for both the bubble sort and Shell sort algorithms, they should use a common data transferal format.
4. Develop a subprogram to sort a one-dimensional character array in either ascending or descending alphabetical order.
5. Develop a subprogram that will move the largest element of a two-dimensional array to the upper left-hand corner of the array, with the next largest values of the array in descending order across the top row. Continue the procedure with the second row, until the smallest element of the array is in the lower right-hand corner of the array.

5.6.3 [2,3,4] Time Conversion

Develop subprograms to perform any of the time conversions of Exercises 4.6.3 and 2.5.3.

5.6.4 [2,3,4] Date Conversion

Develop separate subprograms to convert from calendar to Julian date and from Julian to calendar date.

5.6.5 [4,6,7] Sampling and Distribution of Means

Develop a subprogram to implement the sampling portion of Exercise 4.6.5. Test your modified program in the same manner as you tested the program of Exercise 4.6.5.

5.7 ECONOMIC ANALYSIS

5.7.1 [ALL] Economic Formulas

1. Develop a series of FUNCTION subprograms for evaluation of economic formulas. Write a main program to test each of your formulas.
2. Modify the program of Exercise 4.7.1 into a FUNCTION subprogram that will return the desired value to the main program. Remove all printing output from the subprogram. Test the program with a main program, and include output there. Test the program with the same data you used to test the previous exercise.

5.7.2 [ALL] Nonuniform Series of Payments

1. Modify the program of Exercise 4.7.2 into a SUBROUTINE subprogram that will return both the present worth and the future worth. Test the subprogram with a main program and with the data of Exercise 4.7.2.
2. Modify the program of Exercise 4.7.2 into separate subprograms to determine the present worth and the future worth of a nonuniform series of payments. Select which to use in the main program. Test both options with the data of Exercise 4.7.2.

5.7.3 [ALL] Multiple Interest Rates or Number of Compounding Periods

Incorporate the economic formula FUNCTION subprograms that are needed from Exercise 5.7.1 into the program of Exercise 4.7.3 Test the revised program with the same multiple interest rates and multiple number of compounding periods as in Exercise 4.7.3.

5.7.4 [ALL] Internal Rate of Return

Develop a subprogram to calculate the internal rate of return given the present worth, the annual payment, the number of compounding periods, and an initial estimate of the rate of return. All input and output should be from the main program. Also check the subprogram estimate by substituting back into the capital recovery factor in the main program.

5.7.5 [ALL] Mortgage Computations

1. Develop a subprogram that will determine monthly loan payments for a given loan principal, annual interest rate, and number of years of payback.
2. Develop a subprogram that will determine the amount of principle and the amount of interest paid for each payment period over the life of a loan. It should also compute the total amount of interest paid.

5.8 ELECTRICAL ENGINEERING

5.8.1 [ALL] Diode Problem

Develop a function subprogram to estimate the current through a junction diode, as described in Exercise 2.7.1. Also, develop a separate function subprogram to estimate the voltage required to produce a given current. Test both subprograms with the same data used to test the program of Exercise 4.8.1.

5.8.2 [ALL] Capacitance and Resistance

Develop a subprogram to model the step-response circuit of Exercise 4.8.2. Input and output of data should be from the main program.

5.8.3 [ALL] Signal Processing Circuits

Develop a subprogram to model the tuning circuit of Exercise 4.8.3. Input and output should be from the main program.

5.8.4 [2,3,4] Parallel and Series Resistance

Develop separate FUNCTION subprograms to find equivalent parallel or series resistance.

5.8.5 [2,3,4] Parallel and Series Capacitance

Develop separate FUNCTION subprograms to find equivalent parallel or series capacitance.

5.9 ENGINEERING MATHEMATICS

5.9.1 [ALL] Determinant and Cramer's Rule

Develop separate subprograms for finding the determinant of a 2×2 and a 3×3 matrix.

5.9.2 [ALL] Vector Cross Product, Unit Vector, Directional Cosines, Vector Components

Modify the program of Exercise 4.9.2 into a subprogram. Arguments for the subprogram should include the magnitude, the components of two points on the vector, the components of the unit vector, and the components of the vector. Use a main program to test your subprogram.

5.9.3 [ALL] Centroid and Moment of Inertia

1. Develop a subprogram to determine the area of either a circle or a rectangle.
2. Develop a subprogram that will return the moment of inertia about the centroidal axis for either a rectangle, a circle, or a triangle. Test your function with the data from Exercise 4.9.3.

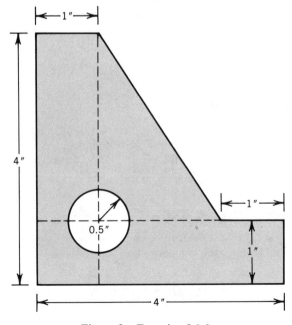

Figure for Exercise 5.9.3

3. Develop a subprogram that will return the distance of the centroid of a compound area from a designated axis, if the shape can be divided up into simple shapes. Test your program using the shape shown. Find the distance to the centroid from the bottom edge of the compound area.

5.9.4 [ALL] Population Growth

Develop subprograms for the continuous and discrete growth models of Exercise 4.9.4.

5.9.5 [ALL] Vertical and Horizontal Curves

Develop a subprogram to analyze the vertical curve data of Exercise 4.9.5. Input and output should be from the main program.

5.9.6 [2,4,6] Table of Squares and Cubes

1. Modify the program of Exercise 4.9.7 so that the values of X, X^2, and X^3 are calculated in a subprogram. Three arrays should be defined in the main program to accommodate up to 50 values for X, X^2, and X^3. Input data will still be the lower bound, upper bound, and increment. The number of values to be computed should be calculated in the main program before calling the subprogram. If the number is greater than 50, an error message is to be output, and a request made for new data. Variable dimensions should be used for the three arrays in the subprogram. Output of all values is to be from the main program.
2. Modify the program of the previous exercise to use a 3×50 two-dimensional array, instead of the three one-dimensional arrays.

5.9.7 [3,4] Vector Dot Product

Develop a FUNCTION subprogram to compute the dot product of two n-dimensional vectors. Test your subprogram using the data of Exercise 4.9.8.

5.9.8 [2,4] Angles and Sides of a Triangle

1. Develop a subprogram for the analysis of Exercise 4.9.9. Input and output should be from the main program.
2. Develop separate subprograms for cases a and b of Exercise 4.9.9.
3. Discuss what you would consider when deciding whether to use no subprograms, one subprogram, or two subprograms in developing a program for this exercise.

5.9.9 [4] Evaluation of Series

Develop FUNCTION subprograms to estimate e^x and $\sin(x)$. Continue to add terms until the final term is less than 0.0001. Use a main program to test these subprograms against the corresponding intrinsic function for the data of Exercise 4.9.10.

5.9.10 [4] Calculating the Area of an Irregular Shape

Develop a subprogram to implement the algorithm of Exercise 4.9.11 to find the area of an irregular shape bounded by a series of points joined by straight lines. Test your subprogram with a main program and the data of Exercise 4.9.11.

5.9.11 [4,6] Volume and Surface Area of Box, Sphere, or Cone

1. Develop a subprogram to determine the volume and surface area of a sphere as a function of either the radius or diameter.
2. Develop two separate FUNCTION subprograms to determine the volume or surface area of a box, a sphere, or a cone, at the option of the user. Test all options of both subprograms with a main program.

5.9.12 Factorials

Write a FUNCTION subprogram that accepts an integer argument and returns as its integer value the factorial of the argument. You may assume that the argument is not less than zero, but be sure that your program produces 1 as the factorial of zero. Test your program by producing factorials up through 10!

5.10 INDUSTRIAL ENGINEERING

5.10.1 [ALL] Project Management

Modify the program of Exercise 4.10.1 into a subprogram that will compute the slack time for each activity of a project if the activity duration, starting node, and ending node are given for each activity. Test your subprogram with the data given in Figure J.1 of Appendix J.

5.10.2 [ALL] Quality Control

Develop a subprogram to find the total number of defects in each category for the data of Exercise 4.10.2.

5.10.3 [ALL] Managerial Decision Making

Develop a subprogram to search for the break-even point for Exercise 4.10.3. Use general functions in the subprogram and make them specific by sending over coefficient and exponent values.

5.10.4 [ALL] Queuing (Wait Line) Theory

Develop a subprogram to determine the probability of n machines being down. Test your program with the data of Exercise 4.10.4.

5.11 MECHANICAL AND AEROSPACE ENGINEERING

5.11.1 [ALL] Shear and Bending Moment

Develop separate subprograms for determining the shear and bending moment for a point load or a uniform load and n selected points along the length of a beam. Refer to Exercise 4.11.1.

5.11.2 [ALL] Crank Assembly Analysis

Develop a subprogram for finding the angle β, the angular velocity of the connecting rod, and the velocity of the piston for a given angle of the crank. Refer to Exercise 4.11.2.

5.11.3 [ALL] Friction

Develop a subprogram to determine whether a crate subjected to a horizontal force at a given height above the bottom will slip or tip. It is in contact with a flat surface with a given coefficient of static friction between the crate and the surface. Incorporate this subprogram into the program of Exercise 4.11.3.

5.11.4 [ALL] Variation of Atmospheric Pressure with Altitude

One method for estimating the pressure in the atmosphere as you rise through it was given in Exercise 2.10.4. This method was based on a constant change in temperature with elevation. Another model is the adiabatic model, which is given by the equation

$$P = P_0 \left[1 - \frac{(k - 1)gZ}{kRT_0} \right]^{\frac{k}{k-1}}$$

in which k is the dimensionless adiabatic constant ($k = 1.4$ for air), and the other variables are as defined in Exercise 2.10.4. Develop a subprogram that will compute the pressure at a given elevation by either of the methods, at the option of the user. The constant lapse rate model is to be used unless the user selects the adiabatic model. Develop a main program to test the subprogram using the data of Exercise 3.11.4. Test both options.

5.11.5 [ALL] Aerospace Physics

Develop a subprogram to determine the position and velocity of a particle at any time t. Test your subprogram with the data of Exercise 4.11.5.

5.11.6 [ALL] Energy Loss in Circular Pipeline

1. Develop a subprogram to estimate the Darcy–Weisbach f as a function of Re using the Colebrook equation. Test your subprogram using a main program with the provisions of Exercises 2.10.6 (b) and 4.11.6.
2. Develop subprograms to implement the flowcharts developed for Exercise 3.11.6 (2). Test both alternatives using the data of Exercise 3.11.6 (1).

5.11.7 [ALL] Torsion

Develop a subprogram for determining the required outside diameter of the shaft of Exercise 4.11.7.

5.12 NUMERICAL METHODS

5.12.1 [ALL] Interpolation

1. Modify the program of Exercise 3.12.1 to include a subprogram that will accomplish quadratic interpolation using the Lagrange interpolating polynomial and a main program to input and output data and call the subprogram. Test your program with the data from Exercise 3.12.1.
2. Modify the program of Exercise 4.12.1 to include a function subprogram that provides the option of using either linear or Lagrange quadratic interpolation functions, at the discretion of the user. A check should be programmed to check that the x value for interpolation is within the range of the given x values. Include input and output in your main program and test both options using the data of Exercise 4.12.1.

5.12.2 [ALL] Solution of Polynomial Equation

Develop a FUNCTION subprogram that will return the estimated slope of a polynomial at any point x, using $f(x-\text{delt}x)$, $f(x)$, and $f(x+\text{delt}x)$. Test your subprogram using the function of Exercise 4.12.2. Compare the computer solution with the analytical derivative for a point x.

5.12.3 [3,4,7,8] Searching Two-Dimensional Array

Develop a subprogram that will search a two-dimensional array to find the largest numeric entry, then divide each element of the array by that value. Use separate subprograms to read in the original array and to output the resulting array. The maximum array size should be 25×25. Use a main program to test the subprogram with at least a 6×6 array.

5.12.4 [3,4,7,8] Data Smoothing

Develop a subprogram to implement the data smoothing function of Exercise 4.12.3. Both the original and smoothed data should be sent back over to the main program.

5.12.5 [3] Rounding-off Values

Develop a FUNCTION subprogram to round off a real value in the computer. Compare the results of your subprogram with the FORTRAN intrinsic function for rounding.

5.12.6 [4,6,7,8,9] Saddle Point of a Hyperbolic Paraboloid

Replace the four separate search patterns used in the program of Exercise 4.12.4 with one general subroutine. Replace the function statement for $f(x,y)$ with a function subprogram.

5.12.7 [4,6,7,8,9] Polynomial Surface

Use the subprogram developed for Exercise 5.12.6 to improve the structure of the program of Exercise 4.12.5. Also, develop subprograms to test optimum points to find out if they are maxima or minima and to determine the direction of travel when the search starts.

5.13 PROBABILITY AND SIMULATION

5.13.1 [ALL] Simulation of Two Dice

Develop a general subroutine that can be used to calculate the frequency distribution for any desired variable. You should allow up to 25 equal increments for summation. The first time the subroutine is called it should calculate the values for each interval and initialize the summing variables for each interval. Values returned by the subroutine should include an array containing the frequency counts for each interval, an array containing the fractional part of the total results that fall in each interval, and an array containing the values for the endpoints of the intervals. Note that the array for the endpoints needs to have one more element than the frequency values for the intervals. Use the subprogram to estimate the frequency distribution for a series of throws of two dice. If your computer system does not have a random number generator available, you can use the subprogram listed in Section 8 of Appendix H. Test your program for 100 and 1,000 tosses of the dice.

5.13.2 [ALL] Normal Probability Distribution

Develop a subprogram for estimating the area under the normal curve according to the procedure used in Exercise 4.13.2.

CHAPTER SIX

FORTRAN PROGRAMMING: ADVANCED INPUT/OUTPUT

6.1 INTRODUCTION

In Chapters 1 through 5 you have learned all the basic structures necessary for developing the logic of FORTRAN applications programs. You have begun to develop fairly complex applications by combining separate procedures into well-structured programs. In this chapter you will learn how to use formatting to help you organize and control your input and output. You will find that formatted output is an indispensable tool. Formatted input can help you organize your data files, but you may not find its use necessary, especially for terminal input.

Formatted output allows you to organize your output for easier reading and understanding. It also permits the control of the number of decimal places and/or number of significant figures that are output. Formatted input requires the placing of input data values within certain column groups. This allows you to design input forms for inputting large volumes of data with minimum input errors.

Many of the rules for formatted input and output are the same as those for list directed input and output. However, you must understand the differences to properly apply formatting. We will first cover the general rules that apply to both formatted input and output. Once you have learned these rules, we will study formatted output, followed by formatted input.

6.2 GENERAL STRUCTURE

You have been using READ and WRITE statements of the form

```
READ(R,*) input variable list
WRITE(W,*) output variable list
```

In these statements, the * indicates unformatted input or output. For formatted input or output, the * is replaced by a statement address label. The address label links the READ or WRITE statement with the appropriate FORMAT statement. FORMAT statements are nonexecutable, but addressable, statements that contain information on how data are to be input or output. For example

```
    READ(R,100) input variable list
    WRITE(W,200) output variable list
200 FORMAT (output format list)
100 FORMAT (input format list)
```

will use the instructions in the *input format list* of FORMAT statement 100 to READ the *input variable list*, and will WRITE the *output variable list* according to the *output format list* instructions of FORMAT statement 200. The input and output format lists have to be enclosed in parentheses.

FORMAT statements follow the general rules for FORTRAN statements. The first five columns are used for the statement identification label. The statement identification label is mandatory for FORMAT statements. Otherwise there would be no linkage between the FORMAT statement and the input or output statements it is meant to modify. Column 6 is used to identify a continuation of a FORMAT statement from the previous line. Continuation will be used frequently for FORMAT statements since they are often lengthy. Columns 7 through 72 can contain format instructions. As with all FORTRAN statements, any format specifications that appear in Columns 73 through 80 will be ignored. Placing specifications in these columns may cause errors during compilation or undesirable output during execution.

Input data records can use 80 columns for either formatted or unformatted input. The READ and FORMAT statements that use the input record will govern how and where data are placed in the record.

FORMAT statements can be placed anywhere in the program. Some programmers prefer to group all their FORMAT statements at the beginning or end of the program. Others like to keep the FORMAT statements near the READ or WRITE statements they will be used with. However, the same FORMAT statement can be referenced by several READ or WRITE statements. It is not always possible to place the FORMAT statement near them all. FORMAT statements that reference READ or WRITE statements in a loop do not have to be in the loop. A READ or WRITE statement cannot reference a FORMAT statement that is contained in a different program module. For example, a READ statement in the main program could not use a FORMAT statement that is contained in a SUBROUTINE subprogram. The following sections will cover the specific instructions that can be incorporated into the input or output format specification lists.

6.2.1 Field Type Specifiers

Field specifiers define what type of data will be input or output for a particular variable, and how those data will be input or output. A field refers to the number of input or output data record columns that will be used by an input or output FORMAT. Decimal points, signs, or other specification information will take up some of the columns of the field, making these columns unavailable for numeric data. The four most commonly used field specifiers are for real, scientific notation, integer, and character data. These four specifiers will be used throughout the text. Other specifications that are used less often will also be defined. Some of the specifiers have identical definitions for input and output. Others appear the same, but have slightly different operational definitions for input and output. It is important to understand these differences before studying their application to input and output situations.

Several format specifications of the same or different types can be put-in the same format list. However, you have to be careful to match the type of field specifier with the corresponding variable type. Otherwise, you may encounter data type mismatch errors during program execution. The format list can also contain other formatting specifications and literal alphanumeric specifications. You can use literals to output labels or headings, as you did previously with unformatted output. Each element of the FORMAT list is separated from the next element by a comma.

6.2.1.1 I Field Specification

The I field specification is for the input and output of integer data. Its general form is

$$rIn$$

in which r is the number of times the specification is repeated, and n is the field width. The form and meaning of the I specification is identical for input or output.

The repeater r in the format specification saves you from having to type in each format specification when the same specification is to be used for several successive fields in the same FORMAT statement. Thus,

$$3I5$$

is the equivalent of

$$I5,I5,I5$$

When inputting values with an integer field specification, you have to be careful where you place the value in the field. If the value is not right-justified in the field, the computer will add zeros at the end to fill the field. If the field width is six, and the first four places are filled with 1234, while the last two places are blank, the value read into the computer would be 123400, not 1234.

If you attempt to input or output a decimal value with the integer format, an error will result that will terminate execution of the program.

6.2.1.2 F Field Specifier

The F field specifier is for input and output of real (decimal) data. Integer data can be handled by this specifier, but it will be treated as real data. The general form of the field is

$$rFn.m$$

in which r is the number of times the format will be repeated and n is the total field width for the specification. Decimal points and negative signs will take up available spaces in the field. The parameter m has different meanings depending on whether it is used for input or output. For input, m indicates the number of places the decimal point will be moved to the left in the input data if no decimal place is included in the input data for the field. If a decimal point is included in the input data field, the m part of the specification is overridden. The F specification will also accept input data in scientific notation form. For output, m indicates the number of decimal places which will be output, regardless of how many the value actually has. For input, m has the capability of changing the value that the computer will store, but it cannot change a value that will be output. For example, if the following data are right-justified (moved as far as possible to the right in the field) in the first ten columns of an input record,

$$123456$$

the specification F10.3 for a READ statement would input the value 123.456, while a specification of F10.4 would input 12.3456. For output, if the value stored in the computer is 12.3456, the specification F10.4 would output the value 12.3456, while the specification F10.2 would output the value 12.35, which is the same value, only to two decimal places rather than four. Note that the value in the final place is rounded according to the normal rules for rounding. If the digit stored in the next place beyond the number being output is less than five, the value is rounded down and the last digit of the output stays the same as the value stored. If the digit stored in the next place is greater than or equal to five, the value is rounded up, and the last digit output is increased by one. It should be emphasized again, however, that this does not change the value stored in the computer; it only modifies the value that appears in the output. If the number of positions to be output, including decimal points and negative signs, is greater than the assigned field width, n asterisks will be output. To correct this problem, you can either increase the field width or decrease the number of decimal places, or both.

Parentheses can be used to group format specifications together. Everything within the parentheses can be repeated by a repeater outside the left parenthesis. The format specification

$$3(F10.2,F5.3)$$

is equivalent to

$$F10.2,F5.3,F10.2,F5.3,F10.2,F5.3$$

Other types of fields can also be included within the parentheses.

6.2.1.3 E Field Specification

The E field specification is for input or output of data in scientific notation (powers of 10). For example, Avogadro's number, 6.023×10^{23} would require a field width of at least 25 if it is handled with an F field specification. With the E field, only enough spaces to accommodate the significant figures and the sign, plus the necessary information to define the power of 10, need to be assigned. Also, the E field accommodates the input or output of a wide range of values, from very large values, to very small decimal fractions. The E field has the general format

$$rEn.m$$

in which r is the number of repetitions of the field specification and n is the total field width. The parameter m governs placement of the decimal point on input, whereas for output it determines the number of significant figures (not decimal places) that will be included. The use of the E specification for input is not necessary, because scientific notation values can be input much more easily through the F specification. The E specification is very useful, however, for output. You can use it to output very large or very small values without worrying about exceeding an F specification field.

In output, part of the E field is taken up by the powers of 10, the decimal place, the sign, and a leading zero. If the value 1,234,500,000 is stored in the computer, and it is output with an E10.4 specification, the output would appear as 0.1235E+10. The output takes up exactly 10 places: four for the significant figures, one for the decimal point, one for the standard leading zero, and four (E+10) for the power of 10. E+10 indicates that the value given is multiplied by 10^{10}. Note the rounding in the last significant figure place. If the value -0.00000000012345 is stored in the computer, and you attempt to output it with the same format, E10.4, you would see ********** in your output. That is because the output would have to include the negative sign, $-0.1235E-09$, which takes up 11 positions. Since the field width is only 10, asterisks are printed. A specification of E11.4 would accommodate the minus sign in front of the value. Therefore, the field width for E specifications should always be at least seven larger than the number of significant digits desired. With any of the format specifications it is allowable to have the field width larger than the number of places that have to be output. In that case, blanks are added to the beginning of the field.

6.2.1.4 A Field Specification

The A field is for inputting and outputting character data. It has the form

$$rAn$$

where, again, r is the number of repetitions of the field specification, and n is the total field width. The A format is useful for working with character arrays. For example, the specification 80A1 could be used to input 80 character variable values, each one character long. When inputting and outputting character data with the A field specification, it is best to use the same length for the field on input and output. That way there is no chance of losing characters in the transfer process.

If you use an A specification without including a field width, the field width defaults to the length of the character variable being input or output.

6.2.2 Alignment Specifiers

The following specifications are useful in aligning and positioning data and headings in output. They are also applicable to input, but are used much less often for that purpose. The X specifier skips spaces within the input or output record, while the T specifier tabs to a particular position in the input or output record. The / specifier in a format specification list causes the next field specifier to look for data on a new line for input, or to go to the next output line for output.

6.2.2.1 X Specification

The X specification can modify the position of the fields in the input or output record. The specification

$$nX$$

tells the computer to skip n spaces before executing the next field specifier. The X specification is used mainly to space output fields on a line so the values are more easily understood. It can also be used in input to skip over values or blank spaces in an input record.

6.2.2.2 Tab Specification

The T specification allows you to move to a particular position in an input or output record. The specification

$$Tn$$

tells the computer to start the next field specifier starting in position n of the record. There are also left and right tab specifiers. TL5 would move left five spaces in the record before executing the next FORMAT instruction. TR5 would tab right five spaces, and is equivalent to a 5X specification.

6.2.2.3 / Specification

When a / is encountered in a format specification list, the next format specification will skip to the next input or output record. The slash is not often used in input, but it can be used to skip portions of a record or whole records. In output, the slash is a convenient way to insert lines between sections of output or to move to a new line without having to execute a new WRITE statement. The specification list

$$(F10.3,//,F10.3)$$

would print two values, according to the F10.3 specification, underneath each other with one blank line in between.

6.2.3 Other FORMAT Specifications

The Ln specification is for input or output of the logical values true or false. On input, if the first nonblank characters in the input data field are either T or .T, the value will be interpreted as .TRUE., no matter what characters fill the rest of the input field. Similarly, F or .F will be input as .FALSE.. On output, $n-1$ blanks will be output followed by either T or F, depending on the stored value.

The G$n.m$ specification can be used for either input or output. On input it has the same effect as the F$n.m$ specification. On output, it provides the flexibility of outputting values in either the F$n.m$ or E$n.m$ formats, depending on the value assigned to the variable being output by the specification. The G specification can be used when you would prefer to have the easier to read F$n.m$ specification but you are not sure that all values to be output will fit within the specification. The G$n.m$ specification is not used all the time because it is somewhat distracting to have the output switch between F and E formats within a series of output values.

6.2.4 Repetition

In Chapter 3 we discussed what happens when an input variable list calls for more data than is contained in an input record, or vice versa. The rules of FORTRAN try to input all the data called for by the input variable list. If the output list exceeds the number of values that can be put on one line, FORTRAN continues output on subsequent lines until the output variable list is satisfied. Similar rules apply to formatted input and output.

If the output or input variable list exceeds the number of FORMAT specifications contained in the referenced FORMAT statement, the computer will again take steps to input or output all the values requested by the input or output variable list. READ statements will repeat the FORMAT, but will start the repetition with the next input data record. WRITE statements will output what is called for by the present repetition of the FORMAT specification list and start a repetition of the FORMAT specification list on a new line of output. Repetitions will be continued until the output variable list is satisfied.

When FORMAT specification lists are repeated, there are rules that govern from what point within the list the repetition starts. If there are no nested parentheses in the specification list, the whole specification list is repeated. If the specification list does contain nested parentheses, the specification list is repeated from the left parenthesis that corresponds to the right-most nested right parenthesis. If the left parenthesis is preceded by a repetition counter, then the repetition count will hold. If you want the whole specification repeated, you should avoid using nested parentheses in the specification.

For both input and output, if the variable list is shorter than the FORMAT specification list, the remainder of the FORMAT specification list is ignored.

Repetition of FORMAT specification lists is very helpful when you are inputting or outputting two-dimensional arrays. FORMAT statements can be used in conjunction with implied DO loops to control the form of array values for input or output and to input and output the array one line at a time.

6.2.5 Additional Concepts

Several READ or WRITE statements can use the same FORMAT statement. However, the same FORMAT should not be used for both READ and WRITE statements. This is because of the printer carriage control characters that can be used with formatted output. Carriage control will be discussed in the next section. Also, most output formats will contain spaces, slashes, or alphanumeric data that would not be correct to use for an input FORMAT. Use of the same FORMAT statement with several READ or WRITE statements makes it unnecessary to duplicate input or output format specifications that are identical in form. The example given below shows the use of the same FORMAT statement for two READ statements, and two different FORMAT statements for WRITE statements.

```
    READ(R,100) input variable list
    WRITE(W,200) output variable list
    READ(R,100) input variable list
    WRITE(W,300) output variable list
100 FORMAT(input format list)
200 FORMAT(output format list)
300 FORMAT(output format list)
```

Specific applications of output FORMAT statements will now be discussed.

6.3 FORMATTED OUTPUT

6.3.1 Introduction

Either WRITE or PRINT statements can have formats associated with them. However, since PRINT statements do not have the versatility of WRITE statements, the use of FORMAT statements with WRITE statements will be emphasized. Examples of PRINT formats will be used for checking some programs. The rules and application of the FORMAT statements are the same for either type of output statement.

The maximum number of characters per line of output can be either 80 or 132, depending on the type of output device you are using. Terminal output is limited to 80 columns, unless you use the compressed character mode, which is hard to read on the screen. Most printers attached to mainframe and minicomputers are set up to handle up to 132 columns, as are some wide-carriage printers for microcomputers. Other microcomputer printers are limited to 80 standard-width columns. In the interest of generality, the formats used in the text will be limited to 80 columns.

6.3.2 Special Features of Output FORMATS

The nH format is for imbedding labels, headings, or other character information in output FORMAT lists. It is also called the Hollerith specification. The specification 20H means that the next 20 characters after the H are to be considered character data for output when the FORMAT statement is referenced. The specification

```
          19HThe Mean value is: ,F10.4
```

is equivalent to

```
          'The Mean value is: ',F10.4
```

I prefer the second form with apostrophes as delimiters because you do not have to worry about counting the number of spaces occupied by the literal. For example, let's say that you miscounted the number of characters in the above Hollerith as

```
          22HThe Mean value is: ,F10.4,
```

When the FORMAT statement is referenced, 22 characters would be considered character data, which would include ,F1, the first part of the F10.4 specification. An error would result because the computer would not know what to do with the 0.4 portion of the specification.

   The first column or position of a formatted output record is used for printer carriage control and does not appear in the output. Output formats should always contain a specification for this first position. This is one of the primary reasons that the same FORMAT statements should not be used for both input and output. Depending on what is in the first position of the output record, different types of carriage control will be implemented. The characters used and their meanings are

Character	Meaning
Blank	Space vertically one line. This is the standard line feed before starting a new input line.
0	Space vertically two lines before printing
—	Space vertically three lines before printing
1	Go to top of next page before starting printing
+	Suppress line feed. This allows overprinting of the last line printed.

Any other character is treated the same as a blank. Some terminals do not use carriage control and will output the whole format specification, including the carriage control. Terminals also do not usually have page control, so a 1 in column one would be treated as a blank.

   Thus, the WRITE and FORMAT statements

```
     WRITE(W,20) X,Y,Z
  20 FORMAT(1X,3F10.3)
```

would print the values for X, Y, and Z according to the F10.3 specification on the next line of output, while the statements

```
      WRITE(W,30) X,Y,Z
   30 FORMAT(1H1,3F10.3)
```

would print the value at the top of the next page. The 1X skips over column 1, which is the equivalent of a blank (normal line feed). The 1H1 is a Hollerith specifier that puts a 1 in column 1, which indicates carriage control shifts to the top of the next page. Being able to shift control to the top of the next page is a good way of organizing tables so they are not split by the bottom of a page.

   If you do not accommodate carriage control in the first column of the FORMAT output specification, the computer still takes whatever is in the first column as carriage control. Thus, if $X = 853.4678$

```
      WRITE(W,40) X
   40 FORMAT('X = ',F10.3)
```

would output

```
=     853.468
```

because the X appears in the first position of the output specification, and is treated as the carriage control. Also, if the value in the computer is 12345.67, the statements

```
      WRITE(W,50) X
   50 FORMAT(F8.2)
```

would output

```
2345.67
```

at the start of the next page, because the digit 1 in the first output position signifies a shift of control to the top of the next page.

   Continuation of FORMAT statements beyond the first line is quite common. This is a case where the advice to keep statements short is not necessarily the best advice. For example, a portion of the output for the program of Example 5.5 (regression subprogram) was written as

```
      WRITE(W,*) 'The regression equation is:'
      WRITE(W,*)
      WRITE(W,*) 'STRAIN =        ',A,'*STRESS + ',B
      WRITE(W,*)
      WRITE(W,*) 'The modulus of elasticity is ',E,' psi, and'
      WRITE(W,*) 'the correlation coefficient is ',R
```

while the same output could be achieved by the following formatted output:

```
      WRITE(W,100) A,B,E,R
  100 FORMAT(1X,'The regression equation is:',//,
     1 1X,'STRAIN = ',F6.4,'*STRESS   + ',F8.5,//,
     2 1X,'The modulus of elasticity is ',E10.3,' psi, and',/,
     3 1X,'the correlation coefficient is ',F5.3)
```

Note how the carriage control has to be included with each new line of output, not each new line of the FORMAT. The three continuation lines are numbered sequentially to show at a glance how many continuations there are. It is important to close any literal strings before continuing to the next line of the FORMAT statement. For example, assume that you split the second and third continuation lines of the above format as follows

```
2          1X,'The modulus of elasticity is ',E10.3,' psi,
3          and',/, 1X,'the correlation coefficient is ',F5.3)
```

The output would insert 20 blanks between psi, and the next word, *and*. The literal is kept open so all the blanks would be considered as part of the literal.

If a colon is included in an output format specification list, the output produced by the FORMAT statement will be terminated at the colon if there are no more variables in the output list. The output will be terminated anyway when the next field specification is encountered, but the colon can prevent outputting unwanted text. If the output list was only A and B for the format of the previous regression example, the output would look like

```
The regression equation is:

STRAIN = x.xxxx*STRESS + xx.xxxxx

The modulus of elasticity is
```

because the output would not terminate until the E10.3 specification was encountered with no corresponding output variable. However, the FORMAT

```
      WRITE(W,100) A,B
  100 FORMAT(1X,'The regression equation is:',//,
     1        1X,'STRAIN = ',F6.4,'*STRESS   + ',F8.5,://,
     2        1X,'The modulus of elasticity is ',E10.3,' psi, and',/,
     3        1X,'the correlation coefficient is ',F5.3)
```

would produce the output

```
The regression equation is:

STRAIN = x.xxxx*STRESS + xx.xxxxx
```

The S, SP, and SS specifications placed in a FORMAT specification list serve to determine if plus signs will be displayed in front of positive output values, as minus signs are output in front of negative values. After the finish of the FORMAT statement, the setting goes back to the default, which is usually not to include plus signs. Any positive values output after an SP specification is encountered will have the + sign included. The SS specifier turns off putting + signs in front of the output values, and the S specifier goes back to whatever the default is, which is usually not to include + signs.

The T specification causes transfer to the indicated column of the output record. T38 would cause the next output field to start in column 38, regardless of whether or not the present position is before or after 38. Thus, this specifier can be used to skip spaces in the output record, but it can also be used to backspace and double strike portions of the output record.

6.3.3 Interpretation of Output FORMAT Specifications

To help you get a better idea of what different FORMAT specifications will produce in the way of output, we will look at a number of representative examples. Assume that the following values are stored in the computer. The default type settings have been used. All variables beginning with the letters I, J, K, L, M, or N are integer variables.

```
PI = 3.141593
E = 30×10⁶
AVOGAD = 6.023×10²³
RADIAN = 57.2958
MILE = 5280
EBASE = 2.718282
TOLER = 1×10⁻¹⁰
XMIN = −50.×10³⁰
```

Each WRITE and FORMAT statement is followed by the corresponding line or lines of output. The following line of numbers will be included with the examples to represent columns on your output.

```
1234567890123456789012345678901234567890123456789012345678901234567890
```

It will help to show you which columns the various elements of output would appear in. For example, the statements given below would produce the output shown.

```
      WRITE(W,20) RADIAN,EBASE,MILE,AVOGAD,TOLER
   20 FORMAT(1X,F8.3,3X,F6.4,3X,I4,3X,2E11.4)
1234567890123456789012345678901234567890123456789012345678901234567890
   57.300   2.7183   5280   0.6023E+24 0.1000E-09
```

A second example shows the importance of including spaces in your output formats, either through use of the X or T*n* specifiers, or by making sure the output field is larger than the number of positions to be filled with numbers and characters.

```
      WRITE(W,20) RADIAN,EBASE,MILE,AVOGAD,TOLER
   20 FORMAT(1X,F8.3,F6.4,I4,2E10.4)
1234567890123456789012345678901234567890123456789012345678901234567890
   57.3002.718352800.6023E+240.1000E-09
```

Note in the next example that any value can be output with an E specification, even an integer value. However, overuse of the E specification can make the output difficult to read.

```
      WRITE(W,30) RADIAN,EBASE,MILE,AVOGAD,TOLER
   30 FORMAT(1X,5E12.3)
1234567890123456789012345678901234567890123456789012345678901234567890
   0.573E+02   0.272E+01   0.528E+04   0.602E+24   0.100E-09
```

The following example shows how the value of π will be output for various format specifications.

```
      WRITE(W,40) PI,PI,PI,PI,PI,PI,PI
   40 FORMAT(1X,F6.0,F6.1,F6.2,F6.3,F6.4,F6.5)
1234567890123456789012345678901234567890123456789012345678901234567890
   3.   3.1 3.14 3.1423.1416******
```

The example below has an output variable list that is longer than the format specification list. The format list is repeated, but on a new line of output.

```
      WRITE(W,50) E,AVOGAD,TOLER,XMIN
   50 FORMAT(1X,2E13.5)
1234567890123456789012345678901234567890123456789012345678901234567890
   0.30000E+08   0.60230E+24
   0.10000E-09 -0.50000E+32
```

In the next example, the output variable list is shorter than the format specification list, so the remainder of the format specifications are ignored. When the second WRITE statement is executed, the FORMAT is repeated from the beginning, and the E9.3 is not large enough to accommodate the minus sign associated with XMIN. The first two FORMAT specifications violate the rule of *n* being at least *m*+7.

```
      WRITE(W,60) E,AVOGAD
      WRITE(W,60) TOLER,XMIN
   60 FORMAT(1X,2(E9.3,3X),2(E12.3))
12345678901234567890123456789012345678901234567890123456789012345678901234567890
0.300E+08   0.602E+24
0.100E-09   *********
```

In the final example below, the value of E is much too large to fit into an F10.6 specification. Two zeros are added to the end of RADIAN because there are only four decimal places stored. The value for TOLER is not zero, even though it appears to be zero in the output. The output is rounded in the last display position which, in this case, is still zero.

```
      WRITE(W,70) PI,E,RADIAN,TOLER
   70 FORMAT(1X,4(F10.6,3X))
12345678901234567890123456789012345678901234567890123456789012345678901234567890
   3.14159   **********   57.295800    0.000000
```

6.3.4 Output Design

Formatted output gives you much more control over the design of output than list-directed output does. To make your output effective, you should consider several things in your output design. First of all, think about the reader. Your output should be organized so that it is clear which data are input data and which are calculated data. Lists of values should be organized in tables whenever it is possible to do so. Headings and labels should be clear in their definition of the meaning of the output. At the same time, try to avoid putting too much information in the output. That can confuse the reader and obscure the important information. If part of the output is more important than other portions, be sure to identify it as such. Output that is not essential, but which may be of interest to some users, can be made optional.

You also have to think about the magnitude and number of significant figures of numerical data to be output. If you know the values for a variable are going to range from 0.01000 to 0.00010 with two significant figures, you would not want to use an F10.3 format because you would likely lose the significant figures on output. The six places to the left of the decimal point would take up space in the output but would contain no useful information. An F8.5 would work well. Five decimal places would capture all the significant figures and would give you an extra place at the beginning of the field for spacing. An E9.2 specification would also work and would always output the two significant figures. If you know that a value is going to fall between 80,000 and 120,000, with 4 significant figures, an F10.3 would still not be a good choice, even though the values would fit within that field width specification. The output of three decimal places that will never contain significant figures is misleading as to the accuracy of the results. An F8.0 or an E11.4 specification would be better.

Look on designing your output as you would look on any writing assignment. Put yourself in the place of the reader and ask yourself if you would be able to understand the output if you were just being introduced to it. The output should be clear to both programmers and nonprogrammers.

Once you have decided how to organize your output, you must transfer that organization onto a page layout. Coding sheets or any other type of paper organized in an 80- or 132-column format can be used to lay out your output. Once it is laid out, the columnar sheet can be used to determine spacing and FORMAT field sizes. A simple example will now be developed to illustrate the process. This example also shows that it is permissible to use formatted output even when the input is list directed.

EXAMPLE 6.1 □ Difference Between Two Polynomial Functions

STATEMENT OF PROBLEM

Develop a program to determine the difference between two polynomial functions that use the same independent variable for a selected number of values between two limits. The independent variable values, polynomial values, and differences are to be stored in arrays. The maximum number of values that will be needed is 101. The polynomial functions will be of up to order 5, and the coefficients for the polynomials will be input. Values of the polynomial coefficients will vary between −9.99 and 9.99, with no more than two decimal places. Values for the independent variable are not expected to exceed ±1,000. Output is to include all input values, plus a table of the values for the various arrays.

MATHEMATICAL DESCRIPTION

The form of the polynomials will be

$$Y = C_1X^5 + C_2X^4 + C_3X^3 + C_4X^2 + C_5X + C_6 \qquad (6.1)$$

ALGORITHM DEVELOPMENT

Input/Output Design

Separate subscripted variables will be used to store the coefficients of the two polynomial functions. Values have to be included for all the coefficients of the polynomial. A zero indicates that the corresponding element of the polynomial is not used. The minimum value, maximum value, and number of increments of the independent variable will also have to be input.

The output for the program should include the input data and a table of values for the independent variable, polynomial, and difference arrays. The input data portion should include problem identification, descriptions of the polynomials to be evaluated, and the independent variable information. The output table should include appropriate headings centered over the columns of values. Figure 6.1 shows the output design developed on FORTRAN coding forms. This output design will be used to develop the output FORMAT specifications. F FORMAT specifications can be used for the output of the polynomial coefficients and the X

(EXAMPLE 6.1 □ Difference Between Two Polynomial Functions □ Continued)

FIGURE 6.1   Output design for Example 6.1 using coding forms.

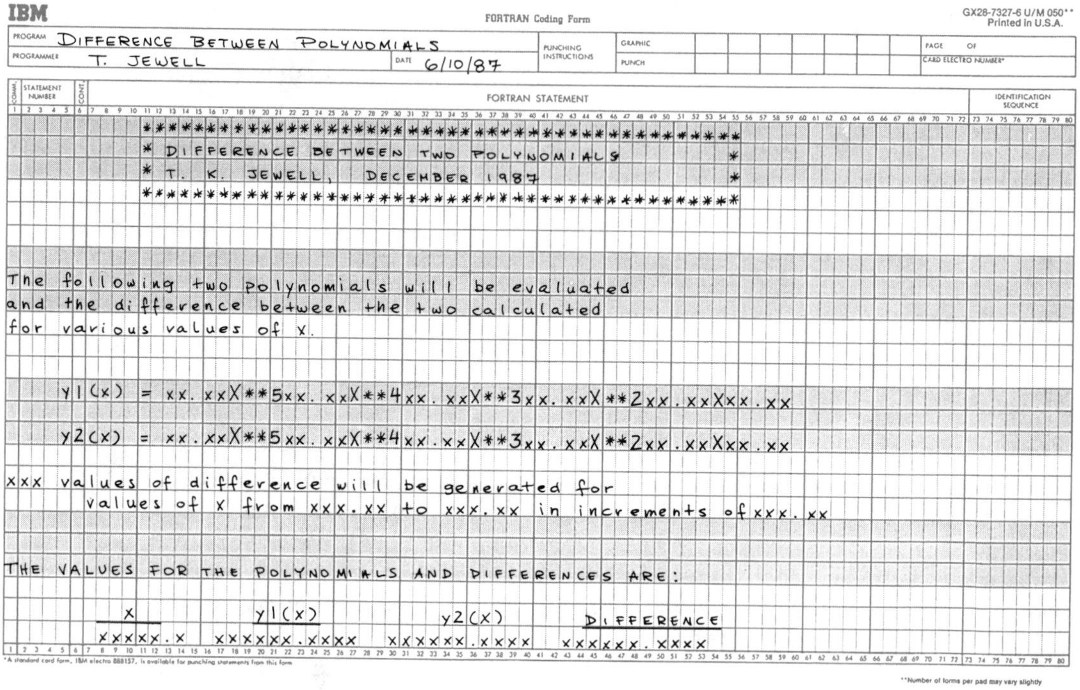

```
***********************************************
 *  DIFFERENCE BETWEEN TWO POLYNOMIALS   *
 * T. K. JEWELL, DECEMBER 1987           *
***********************************************


The following two polynomials will be evaluated
and the difference between the two calculated
for various values of x.

  Y1(x) = xx.xxX**5xx.xxX**4xx.xxX**3xx.xxX**2xx.xxXxx.xx

  Y2(x) = xx.xxX**5xx.xxX**4xx.xxX**3xx.xxX**2xx.xxXxx.xx

xxx values of difference will be generated for
    values of x from xxx.xx to xxx.xx in increments of xxx.xx


THE VALUES FOR THE POLYNOMIALS AND DIFFERENCES ARE:

       x          Y1(x)          Y2(x)         DIFFERENCE
    xxxxx.x    xxxxxxx.xxxx    xxxxxx.xxxx    xxxxxx.xxxx
```

limits. The magnitudes of the polynomials and differences can be very large, on the order of 10^{15}. Therefore, it would be advisable to use E format specifications for that output.

Computer Implementation

We will use a FUNCTION subprogram to evaluate the polynomials. The applicable array for the polynomial coefficients will be made available to the subprogram when it is referenced. The following is a pseudocode representation of the program.

```
SPECIFY ARRAYS, X,Y1,Y2,C1,C2,DIFF
INPUT COEFFICIENTS FOR FIRST POLYNOMIAL (IMPLIED DO)
INPUT COEFFICIENTS FOR SECOND POLYNOMIAL (IMPLIED DO)
INPUT XSTART,XEND,N
CALCULATE DELTX = (XEND-XSTART)/N
SUBSTITUTE X(1) = XSTART
FOR I=1 TO N, DO 100
    Y1(I)=POLY(C1,X(I))
    Y2(I)=POLY(C2,X(I))
    DIFF(I) = Y1(I)-Y2(I)
    X(I+1)=X(I)+DELTX
100 CONTINUE
OUTPUT PROGRAM IDENTIFICATION
```

```
OUTPUT POLYNOMIAL DEFINITIONS
OUTPUT INDEPENDENT VARIABLE INFORMATION
OUTPUT TABLE HEADINGS
OUTPUT TABLE VALUES USING LOOP FOR I=1,N
END
```

PROGRAM DEVELOPMENT

```
*************************************************************************
*  Program for developing table for difference between two          *
*    polynomials illustrating formatted output                      *
*  Developed by:  T. K. Jewell              December 1987           *
*************************************************************************
C  Definition of variables
C  POLY = FUNCTION program to evaluate up to fifth order polynomial
C  C1 and C2 = arrays to store coefficients for two polynomial equations
C  X = independent variable
C  Y1 and Y2 = values of the two polynomials evaluated for X
C  DIFF = difference between Y1 and Y2
C  XSTART = starting value for the independent variable
C  XEND = ending value for the independent variable
C  N = number in intervals (not to exceed 100)
C  DELTX = size of X increment
C
      REAL POLY,C1(6),C2(6),Y1(101),Y2(101),X(102),DIFF(101)
      INTEGER N,FR,W
      PARAMETER(FR=4,W=6)
      OPEN (UNIT=FR,FILE='EX6_1.DAT',STATUS='OLD')
C  Note that X is dimensioned for 102 to allow for last increment if
C    101 values of X are to be used in the array.
C
C  Input of polynomial coefficients
      READ(FR,*) (C1(I),I=1,6)
      READ(FR,*) (C2(I),I=1,6)
C
C  Input of starting and ending values for X, and the number of intervals
      READ(FR,*) XSTART,XEND,N
      DELTX = (XEND-XSTART)/REAL(N)
      X(1) = XSTART
C
C  Loop for calculating data
      DO 100 I=1,N+1
         Y1(I) = POLY(C1,X(I))
         Y2(I) = POLY(C2,X(I))
         DIFF(I) = Y1(I)-Y2(I)
         X(I+1) = X(I) + DELTX
  100 CONTINUE
C
C  Output of heading and input data
      WRITE(W,200)
  200 FORMAT(11X,'***********************************************',/
     1       11X,'* DIFFERENCE BETWEEN TWO POLYNOMIALS        *',/
     2       11X,'* T. K. JEWELL,   DECEMBER 1987             *',/
     3       11X,'***********************************************',///)
      WRITE(W,210)
  210 FORMAT(1X,'The following two polynomials will be evaluated',/
     1       1X,'  and the difference between the two calculated',/
     2       1X,'  for various values of X.'//)
      WRITE(W,220) (C1(J),J=1,6)
      WRITE(W,230) (C2(J),J=1,6)
```

(EXAMPLE 6.1 □ Difference Between Two Polynomial Functions □ Continued)

```
  220 FORMAT(5X,'Y1(X) = ',SP,F5.2,'X**5',F5.2,'X**4',F5.2,'X**3',
     1                    F5.2,'X**2',F5.2,'X',F5.2,/)
  230 FORMAT(5X,'Y2(X) = ',SP,F5.2,'X**5',F5.2,'X**4',F5.2,'X**3',
     1                    F5.2,'X**2',F5.2,'X',F5.2,/)
      WRITE(W,240) N+1,XSTART,XEND,DELTX
  240 FORMAT(1X,I3,' values of difference will be generated for',/
     1      4X,   '    values of X from ',F6.2,' to ',F6.2,
     2                ' in increments of',F6.3,//)
C
C  Output of table
      WRITE(W,250)
  250 FORMAT(1X,'THE VALUES FOR THE POLYNOMIALS AND DIFFERENCES ARE:',
     1  //, 6X,'     X          Y1(X)          Y2(X)        DIFFERENCE')
      WRITE(W,260)
  260 FORMAT(7X,'   ──────       ──────        ──────       ──────────')
      DO 110 I=1,N+1
          WRITE(W,270) X(I),Y1(I),Y2(I),DIFF(I)
  270     FORMAT(8X,F7.1,3(2X,F11.4))
  110 CONTINUE
      END
C
      REAL FUNCTION POLY(C,X)
      REAL C(6),X
      POLY1 = C(1)*X**5 + C(2)*X**4 + C(3)*X**3
      POLY2 = C(4)*X**2 + C(5)*X + C(6)
      POLY = POLY1 + POLY2
      RETURN
      END
```

PROGRAM TESTING

For the following data file input

```
5.0,-3.5,1.0,-5.5,4.0,6.5
0.0,4.0,3.0,-6.0,3.0,5.0
-1.5,1.0,25
```

the output should look like

```
        **************************************************
        * DIFFERENCE BETWEEN TWO POLYNOMIALS             *
        * T. K. JEWELL,   DECEMBER 1987                  *
        **************************************************
```

```
The following two polynomials will be evaluated
  and the difference between the two calculated
  for various values of X.

  Y1(X) = +5.00X**5-3.50X**4+1.00X**3-5.50X**2+4.00X+6.50
  Y2(X) = +0.00X**5+4.00X**4+3.00X**3-6.00X**2+3.00X+5.00

 26 values of difference will be generated for
     values of X from  -1.50 to   1.00 in increments of 0.100
```

THE VALUES FOR THE POLYNOMIALS AND DIFFERENCES ARE:

X	Y1(X)	Y2(X)	DIFFERENCE
−1.5	−70.9375	−2.8750	−68.0625
−1.4	−52.9608	−3.8256	−49.1352
−1.3	−38.7530	−4.2066	−34.5464
−1.2	−27.6472	−4.1296	−23.5176
−1.1	−19.0629	−3.6966	−15.3663
−1.0	−12.5000	−3.0000	−9.5000
−0.9	−7.5328	−2.1226	−5.4102
−0.8	−3.8040	−1.1376	−2.6664
.	.	.	.
.	.	.	.
0.3	7.2158	5.4734	1.7424
0.4	7.2456	5.5344	1.7112
0.5	7.1875	5.6250	1.5625
0.6	7.0712	5.8064	1.2648
0.7	6.9480	6.1494	0.7986
0.8	6.8968	6.7344	0.1624
0.9	7.0301	7.6514	−0.6213
1.0	7.5000	9.0000	−1.5000

6.4 FORMATTED INPUT

Formatted input is not as essential as formatted output. Nonetheless, it is a useful organizational tool for input data files. Instructions can be given to a data entry person as to the columns that certain data have to be placed in. People who have little or no knowledge of FORTRAN can enter the data. A formatted input data file can be proofread and edited much more readily than a free format (list directed) input file.

6.4.1 Special Features of Input FORMATS

If a decimal place is entered into the field of an F specification, the m of the FORMAT (F$n.m$) will be ignored, and the value entered in the field will be input to the computer. No movement of the decimal place will occur. I prefer to use this override for almost all real value input. Unless all the input for which the FORMAT specification is developed is going to have the same number of decimal places, it is easier to assign a general FORMAT, say F8.0, then input the decimal place in the desired place in the input field. As long as the whole value is contained within the defined field, the proper value will be input to the computer.

Values in scientific notation can easily be entered through F specifications. You just have to make sure that the field width is large enough for the significant figures, sign, decimal point, and E±NN notation.

The BN and BZ specifiers determine how nonleading blanks are interpreted in input fields. Normally nonleading blanks are treated as zeros, which can cause the input of erroneous values when the desired value is not right justified in the input field. The BN specifier placed in a FORMAT specification list overrides the default and declares that all nonleading blanks will be treated as null characters. In our previous example in which 1234 was not right justified in its integer input field, the value was entered as 123400. Under the BN specifier, it would have been entered as 1234. Once the BN specifier has been invoked, it applies to the remainder of the specification list. If a whole input field is blank, the computer will still interpret the value as zero. The BZ specifier reinstitutes the default treatment of nonleading blanks as zeros.

6.4.2 Input FORMAT Examples

Several examples of input FORMATS and their effects on the values input will now be developed and discussed. A segment of an input data file will be presented, followed by READ and FORMAT statements, followed by a listing of the values that would be stored in the computer for the input variable list or lists. The first line of the data file in each case is a line of column numbers to help in locating input fields. Unless otherwise noted, the default variable types are used.

DATA FILE

```
12345678901234567890123456789012345678901234567890123456789012345678901234567890
    395   6785 34761      22
   167     218  238 165461
  1  65     616  214    6523
 12345       6111    238  543

        READ(FR,60) X,Y,Z,A
     60 FORMAT(2F6.1)
        READ(FR,70) I,J,K
     70 FORMAT(3I5)
```

$$X = 39.5, \quad Y = 678.5, \quad Z = 167.0, \quad A = 21.8,$$
$$I = 1006, \quad J = 50006, \quad K = 16002$$

The FORMAT specification list for the first READ statement is shorter than the input variable list, so the FORMAT specification list is repeated. A new line of data is accessed, even though there are still data remaining on the first line. Note that the decimal place is moved one place to the left for each value input. The FORMAT specification list for the second READ statement calls for three integer fields of five columns each. The nonleading blanks are input as zeros, and the values shown are input.

DATA FILE

```
12345678901234567890123456789012345678901234567890123456789012345678901234567890
    395   6785 34761      22
   167     218  238 165461
  1  65     616  214    6523

        READ(FR,80) A,B,C,D
     80 FORMAT(3F8.2)
```

$$A = 395.00, \quad B = 678503.47, \quad C = 610000.22, \quad D = 1670.00$$

The input variable list again exceeds the FORMAT specification list. The first three values are read in on the first repetition of the FORMAT list, and the fourth on the second repetition. In this case, the only value read off the second line of the data file is the value for D. The field width is larger than in the first example, so more columns are included in the field for each variable. The decimal point is moved two places to the left. The third line of the data file is not used.

DATA FILE

```
12345678901234567890123456789012345678901234567890123456789012345678901234567890
    395   6785 34761      22
   167     218   238 165461

        READ(FR,90) I,J,K
     90 FORMAT(3I6,3I8)
        READ(FR,90) L,M,N
```

$I = 395,$ $J = 6785,$ $K = 34761,$ $L = 1670,$ $M = 218,$ $N = 2380$

In this case, the FORMAT specification list exceeds the input variable list, so the first three specifications are used to input I, J, and K, and the second three are ignored. The second READ statement repeats the format, and the first three specifications are used again.

For the following examples, it is desired to input the following values for the variables indicated:

```
PI = 3.141593
EBASE = 2.718
RADIAN = 57.296
AVOGAD = 6.023x10²³
```

For the following data file lines and read statements, the values indicated will actually be input.

DATA FILE

```
12345678901234567890123456789012345678901234567890123456789012345678901234567890
   3141593       2718       57296  60231023

        READ(FR,100) PI,EBASE,RADIAN,AVOGAD
    100 FORMAT(4F10.6)
```

$PI = 3.141593,$ $EBASE = 0.002718,$ $RADIAN = 0.057296,$
$$AVOGAD = 60.231023$$

The decimal place is moved six places to the left for each field. The input value for PI is correct, but all the others are in error. The significant figures were right justified without accounting for the decimal place adjustment of the FORMAT specification. The data in the field for AVOGAD is not even close to the form necessary to input scientific notation values.

DATA FILE

```
12345678901234567890123456789012345678901234567890123456789012345678901234567890
   3141593   2718        57296      6.023E23

        READ(FR,100) PI,EBASE,RADIAN,AVOGAD
    100 FORMAT(4F10.6)
```

$PI = 3.141593,$ $EBASE = 2.718,$ $RADIAN = 57.296,$
$$AVOGAD = 6.023 \times 10^{23}$$

The same format is used, but the data in the file have been moved so that when the decimal point is adjusted, the proper values will be input. Also, the value for AVOGAD is input in proper scientific notation form.

DATA FILE

```
12345678901234567890123456789012345678901234567890123456789012345678901234567890
   3141593      2718     57296 6.023E+23

        READ(FR,100) PI,EBASE,RADIAN,AVOGAD
   100 FORMAT(F10.6,F10.3,F10.3,F10.0)
```

$$PI = 3.141593, \quad EBASE = 2.718, \quad RADIAN = 57.296,$$
$$AVOGAD = 6.023 \times 10^{23}$$

In this case, the data file is the same as the first example, but the FORMAT specifications are modified so that the decimal place is moved the proper number of places. The value for AVOGAD has been changed by the addition of the + sign in the E+23. This is also correct scientific notation representation.

DATA FILE

```
12345678901234567890123456789012345678901234567890123456789012345678901234567890
3.141593  2.718      57.296      6.023E23

        READ(FR,100) PI,EBASE,RADIAN,AVOGAD
   100 FORMAT(4F10.6)
```

$$PI = 3.141593, \quad EBASE = 2.718, \quad RADIAN = 57.296,$$
$$AVOGAD = 6.023 \times 10^{23}$$

The values are left justified in the field, but since the decimal places are included in the field, the values are read in correctly.

DATA FILE

```
12345678901234567890123456789012345678901234567890123456789012345678901234567890
   3141593     2718      57296      6.023E23

        READ(FR,100) PI,EBASE,RADIAN,AVOGAD
   100 FORMAT(F10.6,F10.3,F10.3,F10.0)
```

$$PI = 3.141593, \quad EBASE = 2718.000, \quad RADIAN = 57296.000,$$
$$AVOGAD = 6.023 \times 10^{23}$$

In this example, the values for EBASE and RADIAN are not right justified, while the FORMAT specifications are set up to have them right justified. Therefore, zeros are added for the nonleading blanks, and the instructions of the specifications are carried out.

DATA FILE

```
12345678901234567890123456789012345678901234567890123456789012345678901234567890
    3141593       2718      57296     6.023E23

        READ(FR,100) PI,EBASE,RADIAN,AVOGAD
   100 FORMAT(F10.6,F10.3,F10.3,F10.0)
```

$$PI = 0.314159, \quad EBASE = 3000002.718, \quad RADIAN = 57.296,$$
$$AVOGAD = 6.023 \times 10^{2}$$

In this example, the values for PI and AVOGAD have been slipped one place to the right, so the final digit of each is not in the correct field. Thus, the only value that is read in properly is RADIAN. The 3 from the end of the value for PI is in the field for EBASE, so it is included in the value read in for EBASE. Zeros are added for all the nonleading blanks. The second digit of the power of 10 for AVOGAD is out of its field, so the value is multiplied by only 10^{2}, not 10^{23}.

After studying these examples, you should be ready to approach input design with more understanding.

6.4.3 Input Design

For formatted input to be effective, the placement and structure of input data must minimize chances of improper coding and data entry. Input instructions must be clear and complete. Coding sheets or specially prepared forms should be used. The columns for each piece of input data should be clearly identified.

You have to design the organization of your data file and the FORMATS to be used. The organization of your data file has to follow the organization of your input data in your program. Each time a READ statement is executed, you will have to have a new line in the input data file. Also, if you have any READ statements that will use FORMAT repetition, you will have to accommodate the multiple lines of input that the FORMAT repetition will require. If one READ statement is in the main program, and the next READ statement executed is in a subprogram, the information will be read from the next line of the data file, unless other instructions to open a new file have been given in the program.

In designing the input data FORMAT specifications, it is necessary to take into account what type of data is to be input and what magnitude the data will have. When inputting large volumes of data, for example a two-dimensional array, it is important to use the whole input record and to organize the input so that it will look like a table when the data file is examined.

Example 6.1 used list-directed input to enter the necessary information for two polynomial equations, plus the starting value, ending value, and number of increments for the independent variable X. That input will now be modified to use formatted input as an illustration of input design. Only three lines will be needed in the data file. The first record will contain values for the first polynomial, the second record will contain values for the second polynomial, and the third record will contain the data for the independent variable. Six values will be input for each polynomial. The range of values for the polynomial coefficients is -9.99 to 9.99. A maximum of five places are needed if the decimal place is included in the input field. However, to separate the input data values in the data file it is better to make the field a little larger than it needs to be. Since only six values have to be input, you can spread the input out and use a field width of 10 for each value. This is an easy size to remember when you are typing in the data file and will provide ample separation. Since no more than two decimal places will be needed, an F10.2 specification would be appropriate. A decimal place can be included in the input field if you want to override the .2 specification. Three values have to be input for the independent variable. XSTART and XEND will be real values, and the number of increments, N, will be an integer value. Values for XSTART and XEND will not exceed $\pm 1,000$, so an F10.0 would be an appropriate specification. No more than three places will be needed for N, so an I5 would provide sufficient separation, and would go well with the previous field widths of 10. The program statements would be modified as follows:

```
C
C   Input of polynomial coefficients
      READ(FR,20) (C1(I),I=1,6)
      READ(FR,20) (C2(I),I=1,6)
   20 FORMAT(6F10.2)
```

```
C
C  Input of starting and ending values for X, and the number of intervals
     READ(FR,30) XSTART,XEND,N
  30 FORMAT(2F10.0,I5)
C
C  End of input
```

The input instructions could have the form:

Input for Difference Between Polynomials Program

RECORD 1 (5F10.2) Right justify data with two decimal places.
Coefficients for first polynomial

Columns	Variable Name	Description
1–10	C1(1)	Coefficient for $X**5$
11–20	C1(2)	Coefficient for $X**4$
21–30	C1(3)	Coefficient for $X**3$
31–40	C1(4)	Coefficient for $X**2$
41–50	C1(5)	Coefficient for X
51–60	C1(6)	Constant coefficient

RECORD 2 Coefficients for second polynomial. Format and description same as for RECORD 1. Variable names are C2(I).

RECORD 3 (2F10.0,I5) Right justify data. Put decimal point in F fields if need be. Do not put decimal point in I field.
Data for independent variable.

Columns	Variable Name	Description
1–10	XSTART	Initial value for independent variable
11–20	XEND	Final value for independent variable
21–25	N	Number of increments (maximum of 100)

If necessary, information could also be included about the limitations on the coefficient and variable values. The next section will develop a set of array input instructions that could be used for any array input through a standard subprogram.

6.5 INPUT AND OUTPUT OF ARRAYS

Arrays are used extensively in engineering problem solving. You are familiar with using subscripted variables in the algebraic representation of mathematical models. A FORTRAN program to assist in the solution of such a model would require the use of arrays to represent the subscripted variables. Arrays are also essential for undertaking matrix operations, another common tool in engineering analysis. Formatting the input and output of arrays can help to add structure to their usage.

.5.1 SUBROUTINE Subprogram for Input and)utput of Arrays

1 Section 8 of Chapter 5, subroutines were developed to input and output two-imensional arrays by list-directed input. We will now modify these subroutines) use formatted input. Examples of how list repetition affects the input and utput of arrays will also be developed.

Formatted input of arrays should provide the option of using the whole input :cord (80 columns) if need be. Smaller arrays can still use just part of the input :cord. A general format specification that will allow use of 80 columns of input hould be assigned to the array READ statement. The number of input fields that an be fit into the input record is a function of the number of significant figures in 1e input values. A specification of 10F8.m will allow for input of sufficient signifi-ant figures for most applications. If scientific notation values have to be input, 1e field width can be adjusted accordingly, to 8F10.m or even to 16F5.m. The size f the m specifier can be assigned according to how many decimal places you want) input or can be set to zero with the understanding that a decimal place will be 1cluded in the input field. The input format specification used in the sample 1broutine is 10F8.0, which allows up to 10 input values per input record. The 1put loop is set up to read one row of the array at a time. The implied DO loop has *limit* of M, the number of columns in the array. If M is greater than 10, the 'ORMAT specification list will be repeated and a new line of data accessed. Any ize row can be read in this way, 10 values at a time. After using as many input ata records as it requires to complete the first row of the array, the computer will :ad in the next row of the array commencing with the next input data record. The)O 100 loop controls the row index. If the array is 10×25, it will require a total of 0 input records to input the array, three for each row of the array. The first record 'ould have the first 10 values of a row, the second record the next 10 values of the articular row, and the third record the last 5 values of the row. This provides a 1ore structured approach to array input than list-directed input.

EXAMPLE OF SUBROUTINE FOR FORMATTED INPUT OF ARRAYS

```
      SUBROUTINE READ(A,N,M,FR)
      REAL A(10,10)
      INTEGER N,M,FR
      DO 100 I=1,N
         READ(FR,200) (A(I,J),J=1,M)
100   CONTINUE
200   FORMAT(10F8.0)
      RETURN
      END
```

The output subprogram is similar in form to the input subprogram, but with VRITE statements rather than READ statements, and an output FORMAT with arriage control. Note also that on output, the appropriate number of decimal 1laces should be output. The subprogram given below will effectively handle only 1rray output that contains no more than the number of columns that the output ormat allows to be output on a single line. In this case it is 10. If the output 'ariable list of the implied DO calls for more output than the FORMAT specifica-ion allows, the FORMAT will be repeated, but on a new line of output. Thus, the :leventh element of the first row would appear under the first element of the first row, the twenty-first under the eleventh, and so on. This makes it hard to interpret the array output. We will modify this subprogram in Chapter 7 to incorporate output of the rows and columns in blocks. The blocks will maintain the integrity of the array and can be pasted together to form the whole array.

EXAMPLE OF SUBROUTINE FOR OUTPUT OF ARRAYS

```
      SUBROUTINE WRITE(A,N,M,W)
      REAL A(10,10)
      INTEGER N,M,W
      DO 100 I=1,N
         WRITE(W,200) (A(I,J),J=1,M)
  100 CONTINUE
  200 FORMAT(1X,10F8.2)
      RETURN
      END
```

6.5.2 Interaction of FORMAT and Variable Lists in Array Input and Output

Study each of the following sets of input statements until you understand how many input records will be accessed for each. The objective is to input an 8x10 matrix into a two-dimensional array. You can assume that the dimensions for each array are sufficient to accept the values input. The first example does not use an implied DO loop. Therefore, input would require 80 input records, 10 for each row.

```
      DO 100 I=1,8
         DO 200 J=1,10
            READ(FR,10) A(I,J)
  200 CONTINUE
  100 CONTINUE
   10 FORMAT(F10.0)
```

The second example would still require 80 input records. All that has been changed is the input FORMAT specification list. It now allows reading of up to 10 values per input record, but the READ statement still calls for reading only one value on each execution of the DO 200 loop.

```
      DO 100 I=1,8
         DO 200 J=1,10
            READ(FR,10) A(I,J)
  200    CONTINUE
  100 CONTINUE
   10 FORMAT(10F8.0)
```

The next example form would input the same array as the last two examples but would access only eight input records. The implied DO loop and the format specification both allow reading of 10 values per input record.

```
      DO 100 I=1,8
         READ(FR,10) (A(I,J),J=1,10)
  100 CONTINUE
   10 FORMAT(10F8.0)
```

The fourth example would require 16 input records to read in the whole 8 × 10 array. The FORMAT specification list allows only five values to be read in per repetition of the FORMAT. This is not a good use of a general FORMAT specification.

```
      DO 100 I=1,8
          READ(FR,10) (A(I,J),J=1,10)
  100 CONTINUE
   10 FORMAT(5F8.0)
```

The following example would read in a different array, as the implied DO loop calls for reading only five values per line. The array read in would be 8 × 5. The second five FORMAT specifications in the 10 FORMAT would be ignored.

```
      DO 100 I=1,8
          READ(FR,10) (A(I,J),J=1,5)
  100 CONTINUE
   10 FORMAT(10F8.0)
```

The last input example would read in an 8 × 18 array, using 16 input records. The first row of the data file would contain the first 10 values for the first row of the array. The second line of the data file would contain the last eight values for the first row of the array. The third line of the data file would contain the first 10 values of the second row of the array, and so forth, until the last eight values of the eighth row would be read from the sixteenth data record. When setting up your data file, you have to be sure that your data file structure agrees with the way the computer is going to access the data. If you set up your data file for the foregoing example with nine values in each line, the array read in is not going to be correct. The first nine elements of each row would have the proper values, the tenth array element of each row would contain zero, and the eleventh through eighteenth elements would each contain the value that should have been in the previous element.

```
      DO 100 I=1,8
          READ(FR,10) (A(I,J),J=1,18)
  100 CONTINUE
   10 FORMAT(10F8.0)
```

The following examples will help you understand the interactions between output variable lists and FORMAT specification lists when you are outputting arrays. The first example would output all the values in a single column. The first 10 elements of the column would be the first row of the array, the second 10 elements the second row, and so on. A total of 80 lines would be output.

```
      DO 100 I=1,8
          DO 200 J=1,10
              WRITE(W,20) A(I,J)
  200     CONTINUE
  100 CONTINUE
   20 FORMAT(1X,F8.2)
```

The second example will still use 80 lines of output. The FORMAT specification list allows output of up to 10 values per repetition, but the nested DO 200 loop executes the WRITE statement for a single $A(I,J)$ element each time through the loop. Therefore, only one of the 10F8.2 specifications is used, and the rest are ignored.

```
      DO 100 I=1,8
         DO 200 J=1,10
            WRITE(W,20) A(I,J)
200      CONTINUE
100 CONTINUE
 20 FORMAT(1X,10F8.2)
```

The values would still be printed out in a single column for the next example. The implied DO loop will allow output of 10 values per execution of the WRITE statement, but there is only one FORMAT specification in the specification list. Therefore, the FORMAT has to be repeated 10 times for each execution of the WRITE, and each repetition outputs to a new line. Eighty output lines will be needed.

```
      DO 100 I=1,8
         WRITE(W,200) (A(I,J),J=1,10)
100 CONTINUE
200 FORMAT(1X,F8.2)
```

The next example would output the array in the normal matrix format. It would output eight rows of 10 elements each. Eight output records would be needed.

```
      DO 100 I=1,8
         WRITE(W,20) (A(I,J),J=1,10)
100 CONTINUE
 20 FORMAT(1X,10F8.2)
```

Similarly, the following example would output an 8 × 5 array in the proper format. The second five FORMAT specifications are ignored. Eight lines of output would result.

```
      DO 100 I=1,8
         WRITE(W,20) (A(I,J),J=1,5)
100 CONTINUE
 20 FORMAT(1X,10F8.2)
```

The next example will produce 16 lines of output. The first line of output would contain the first 10 elements of the first row of the array, the second line of output would contain the last 4 elements of the first row of the array, the third row of the output would contain the first 10 elements of the second row of the array, and so on.

```
      DO 100 I=1,8
         WRITE(W,20) (A(I,J),J=1,14)
100 CONTINUE
 20 FORMAT(1X,10F8.2)
```

Similarly, the following example will produce 24 lines of output, three for each line of the array. The FORMAT specification list allows only five values per line. This is not good utilization of the output format.

```
      DO 100 I=1,8
         WRITE(W,20) (A(I,J),J=1,14)
100 CONTINUE
 20 FORMAT(1X,5F8.2)
```

When you are designing output FORMATs for arrays, be sure to organize the output structures so the array will be output in a form that is easily understood and interpreted.

6.5.3 The A FORMAT Specification and Character Arrays

Character arrays are commonly used for database management or screen graphics applications. The A FORMAT specification makes it easier to input and output characters to and from these arrays. Character variables and the A FORMAT are also useful for inputting and outputting variable text material for an applications program. Character arrays provide a versatile way of storing the text material.

6.5.3.1 Rules for Formatted Character Input and Output

If the A specification is used for input of character data, and the assigned field width (n) is smaller than the character length assigned to the character variable (m), then the characters in the input field are stored (left justified) in the first n positions of the character variable, and the rest of the positions ($m-n$) of the character variable are filled with blank characters. This is the same procedure as for list-directed input or assignment of character strings to character variables.

If the assigned field width (n) for the A specification is larger than the length assigned to the character variable (m), then the rightmost m characters of the input string will be stored in the character variable location. This is the opposite of what will happen if a larger character string is stored in a character variable through an assignment statement or through list-directed input. In those cases, the leftmost characters will be stored in the character variable location. To avoid any confusion, it is better to make your input formats the same length as the size of the character variable.

On output, if the length of the character variable (m) is smaller than the field width assigned by the A specification (n), then the m characters of the character variable are right justified in the output field, and blanks are inserted in the leading positions.

If the length of the character variable (m) is larger than the field width assigned by the A specification (n), then the leftmost n characters of the character variable are output. Again, to avoid any confusion and unexpected output, it is best to make all three lengths—the character variable size, the input specification, and the output specification—the same.

There is an additional option for both input and output. You can use the A specification but omit the assigned field width n. In this case, the computer assumes the length is the same as the length of the character variable to be input or output. This is the same as the procedure for handling list-directed input and output of character variables.

The following example will show what happens when the lengths of the character variables, input specifications, and output specifications are not all the same.

```
      CHARACTER*4 A,B,C,D,E
      INTEGER FR,W
      PARAMETER (FR=4,W=6)
      OPEN (UNIT=FR,FILE='EX6.DAT',STATUS='OLD')
      A = 'ABCDEF'
      READ (FR,*) B
      READ (FR,20) C
   20 FORMAT(A4)
      READ (FR,30) D
   30 FORMAT(A2)
      READ (FR,35) E
```

```
   35 FORMAT(A6)
      WRITE(W,40) A,B,C,D,E
   40 FORMAT(1X,'The values in the computer are:',/,1X,5(3X,A4))
      WRITE(W,*) 'List directed output'
      WRITE(W,*) A,B,C,D,E
      WRITE(W,50) A,B,C,D,E
   50 FORMAT(1X,'FORMAT length not specified',/,1X,5(3X,A))
      WRITE(W,60) A,B,C,D,E
   60 FORMAT(1X,'FORMAT specification shorter than variable length',/,
     1        1X,5(3X,A3))
      WRITE(W,70) A,B,C,D,E
   70 FORMAT(1X,'FORMAT specification longer than variable length',/,
     1        1X,5(3X,A6))
      END
```

DATA FILE

```
'ABCDEF'
ABCDEF
ABCDEF
ABCDEF
PROGRAM OUTPUT
The values in the computer are:
   ABCD    ABCD    ABCD    AB      CDEF
List directed output
ABCDABCDABCDAB CDEF
Format length not specified
   ABCD    ABCD    ABCD    AB.     CDEF
Format specification shorter than variable length
   ABC    ABC    ABC    AB    CDE
Format specification longer than variable length
    ABCD        ABCD        ABCD        AB        CDEF
```

The first line of the program declared that all character variables would be of length 4. Note that on input, the assignment statement and list-directed READ statement took the leftmost characters in the string, the A2 took the first two characters, and the A6 took the last four characters to put into the character variable of length four. On output the leftmost characters are output when the A specification is smaller than the character variable length, and leading blanks are inserted when the A specification is larger than the character variable length.

We will now develop two examples of formatted array input and output.

6.5.4 Examples of Formatted Array Input and Output

Example 6.2 will develop a program to input and output data for a truss-analysis problem. The data will include problem identification information and the matrices necessary to solve for the forces in the members of the truss. Example 6.3 will develop a program to input and output character grids.

The method of joints and method of sections are two methods for estimating the forces in the members of a truss. The method of sections is preferable for finding the forces in a few of the members. The method of joints provides a way of determining the force in each member of a truss. Most engineering texts present these concepts in a format that lends itself to calculator solutions of textbook problems. Section 3.1 of Appendix G extends the method of joints to produce a system of equations, Equations G.4, for solving for the member forces. That system of equations can be represented in matrix format. Example 6.2 will provide the means of inputting and outputting the matrices. Solution techniques for these matrices will be developed in later chapters.

It would also be useful to be able to input and output identifying information for a particular problem. You could use a character array to store information to identify the truss being analyzed, the particular loading conditions, or any other pertinent identifying data. This information would help you identify printouts for different analyses. Once the information is input, it can be output any number of times during program execution.

Character arrays are also used for generating printer graphics displays. A two-dimensional character variable array with a length of one for each element of the array can be thought of as a two-dimensional character grid. Each cell in the grid could contain any character that you want to assign to it. You could develop programs for plotting X-Y functions using the grid. Example 6.3 will develop a program that will input and output simple letter shapes in a 5×5 character grid. Only four letters will be used for illustration, but all letters could be input into the array called ALPHA. The array ALPHA could then be used to develop large letter titles, or any other combinations of the letters desired.

EXAMPLE 6.2 □ Input and Output of Arrays for Truss Analysis

STATEMENT OF PROBLEM

Develop a program to input and output the coefficient matrix and right-hand-side vector for a truss analysis problem involving n members, with n being less than or equal to 25. Also provide the option of inputting and outputting up to 10 lines of problem identification information.

MATHEMATICAL DESCRIPTION

The theory for the method of joints is presented in Section 3.1 of Appendix G. In the appendix the theory is generalized to show how you can produce a set of simultaneous linear equations for the summations of forces at the joints. The forces in the members are the variables of these equations. You can represent the set of equations in matrix notation and can use matrix operations to solve for the forces in the members. This approach can be used for any size problem. With each additional member, the dimensions of the coefficient matrix ($\mathbf{A}_{n \times n}$), solution vector ($\mathbf{F}_n$), and right-hand-side vector ($\mathbf{B}_n$) will each increase by one. For our example we will analyze the truss shown in Figure 6.2.

The summations of forces at the various joints of the truss of Figure 6.2 can be represented by Equation G.5 as

$$\boxed{\mathbf{A} \cdot \mathbf{F} = \mathbf{B}} \tag{6.2}$$

or by Equation G.6 as

$$
\begin{bmatrix}
1 & -0.4472 & 0 & 0 & 0 & 0 & 0 \\
0 & 0.8944 & 0 & 0 & 0 & 0 & 0 \\
0 & 0.4472 & 0.4472 & -1 & 0 & 0 & 0 \\
0 & 0.8944 & -0.8944 & 0 & 0 & 0 & 0 \\
-1 & 0 & -0.4472 & 0 & 0.4472 & 1 & 0 \\
0 & 0 & 0.8944 & 0 & 0.8944 & 0 & 0 \\
0 & 0 & 0 & 1 & -0.4472 & 0 & -0.4472
\end{bmatrix}
\begin{bmatrix}
FAC \\ FAB \\ FBC \\ FBD \\ FCD \\ FCE \\ FDE
\end{bmatrix}
=
\begin{bmatrix}
0 \\ 5 \\ 0 \\ 0 \\ 0 \\ 10 \\ 0
\end{bmatrix}
\tag{G.6}
$$

(EXAMPLE 6.2 ☐ Input and Output of Arrays for Truss Analysis ☐ Continued)

FIGURE 6.2   Definition sketch for truss analysis.

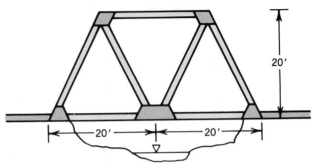

(*a*) Pictorial model of truss.

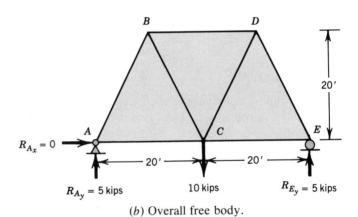

(*b*) Overall free body.

In this matrix equation, **F** is the vector of member forces F_1 through F_7 for the truss. The coefficients in the **A** matrix are the directional sines and cosines of the members of the truss. We will assume that the coefficient matrix has been calculated externally and needs to be input to the computer. The right-hand-side vector **B** will also be input.

ALGORITHM DEVELOPMENT

Input/Output Design

We will modify the subroutines for array input and output discussed in Section 8 of Chapter 5. Since all the arrays are either $n \times n$ or $n \times 1$, you can input *n* once at the beginning of the program. Therefore, we will remove the statement for inputting *n* and *m* from the subroutine. The right-hand-side vector will be input as an $n \times 1$ two-dimensional array so it will be compatible with later matrix multiplication operations. The present problem involves only seven members, so there is no problem with the matrix rows extending to multiple lines of output.

A subscripted variable will be used for the identification information. This will make the identification information available anywhere in the program that you want to output it. Output will be a mirror image of input. To allow for the use of the whole output line, a character variable length of 80 and an A80 specification should be used. However, if you are outputting to a terminal, the 1X for carriage control at the beginning of the line is included in the 80 columns. To avoid insertion of a blank line after each line, an A79 specification and a character length of 79 will be substituted. The character variable itself will be a subscripted variable of dimension 10.

The following instructions should be incorporated into the user instructions for the program:

[RECORD #1] NUMBER OF LINES OF PROBLEM IDENTIFICATION IN-FORMATION
Columns 1–5 (I5) NUMBER Right justify value in field.

[RECORDS #2 through NUMBER+1] PROBLEM IDENTIFICATION IN-FORMATION
Columns 1–79 (A79) IDENT(I),I=1,NUMBER

Up to 10 lines of problem identification information can be put in. RECORD #1 specifies the number of lines to be used. Typical information might include:

JOB ID #

CLIENT NAME

TRUSS LOCATION

TYPE OF TRUSS

LOADING CONDITIONS

ANY OTHER SPECIAL OR UNUSUAL CONDITIONS

Additional instructions could be included for input of the matrix and the right-hand-side vector.

Computer Implementation

A short main program has to be written to read in the variable identification data and the dimensions of the problem. It will then call the input subprogram, write out the appropriate information, and call the output subprogram.

PROGRAM DEVELOPMENT

```
      PROGRAM MATRIX
****************************************************************
*  Input and output of coefficient matrix and right-hand     *
*     side vector for truss analysis                         *
*  Developed by:  T. K. Jewell          December 1987        *
****************************************************************
C
C  Variable definitions
C  A = coefficient matrix (directional cosines)
C  B = right-hand side vector (forces and reactions)
C  N = number of members in truss
C
      REAL A(25,25),B(25,1)
      INTEGER N,FR,R,W,NUMBER
      CHARACTER NAME*14,IDENT(10)*79
```

(EXAMPLE 6.2 □ Input and Output of Arrays for Truss Analysis □ Continued)

```
      PARAMETER (FR=4,R=5,W=6)
      PRINT *, 'Please Input File Name'
      PRINT *, 'This is formatted input, so you do not include'
      PRINT *, 'apostrophes before and after the name.  '
      READ (R,10) NAME
   10 FORMAT (A14)
      OPEN (UNIT=FR,FILE=NAME,STATUS='OLD')
C
C  Input and output problem identification information.
C  NUMBER = number of lines of information desired (up to ten).
      READ(FR,15) NUMBER
   15 FORMAT(I5)
      READ (FR,14) (IDENT(I),I=1,NUMBER)
   14 FORMAT(A79)
C
C  Input of number of members
      READ(FR,20) N
   20 FORMAT(I5)
C
C  Input coefficient matrix.
      CALL READ(A,N,N,FR)
C
C  Input right-hand side vector.
      CALL READ(B,N,1,FR)
C
C  Output identification information and data
      WRITE(W,30)
   30 FORMAT(11X,'MATRIX ANALYSIS OF TRUSSES',//,
     1        1X,'General problem information is:',/)
      WRITE(W,34) (IDENT(I),I=1,NUMBER)
   34 FORMAT(1X,A79)
C
C  Output data
      WRITE(W,32) N
   32 FORMAT(//,1X,'The number of members is: ',I2,///,
     1           1X,'DIRECTIONAL COSINE COEFFICIENT MATRIX')
C
      CALL WRITE(A,N,N,W)
C
      WRITE(W,40)
   40 FORMAT(/,1X,'RIGHT-HAND SIDE VECTOR',/)
C
      CALL WRITE(B,N,1,W)
      END
      SUBROUTINE READ(A,N,M,FR)
      REAL A(10,10)
      INTEGER N,M,FR
      DO 100 I=1,N
         READ(FR,200) (A(I,J),J=1,M)
  100 CONTINUE
  200 FORMAT(10F8.0)
      RETURN
      END
C
      SUBROUTINE WRITE(A,N,M,W)
      REAL A(10,10)
```

```
      INTEGER N,M,W
      DO 100 I=1,N
         WRITE(W,200) (A(I,J),J=1,M)
  100 CONTINUE
  200 FORMAT(1X,10F8.3)
      RETURN
      END
```

PROGRAM TESTING

The input data file, set up according to the structure of the main program and the subprograms, is given below.

DATA FILE

```
12345678901234567890123456789012345678901234567890123456789012345678901234567890
    5
JOB # 456
CLIENT NAME:  City of New York
TRUSS LOCATION:  Central Park South Bridge
TYPE OF TRUSS:  Simple
LOADING CONDITIONS:  One load at midspan
    7
    1.0 -0.4472      0        0        0       0        0
      0  0.8944      0        0        0       0        0
      0  0.4472  0.4472    -1.0        0       0        0
      0  0.8944 -0.8944      0        0       0        0
   -1.0       0 -0.4472      0   0.4472     1.0        0
      0       0  0.8944      0   0.8944       0        0
      0       0       0    1.0  -0.4472       0 -0.4472
      0
    5.0
      0
      0
      0
   10.0
      0
```

The output is shown below. Note that the right-hand-side vector is input and output as a column vector, with one value per line. This is the form it needs to be in for later computations.

```
            MATRIX ANALYSIS OF TRUSSES
General problem information is:

JOB # 456
CLIENT NAME:  City of New York
TRUSS LOCATION:  Central Park South Bridge
TYPE OF TRUSS:  Simple
LOADING CONDITIONS:  One load at midspan
The number of members is:   7

DIRECTIONAL COSINE COEFFICIENT MATRIX
   1.000  -0.447   0.000   0.000   0.000   0.000   0.000
   0.000   0.894   0.000   0.000   0.000   0.000   0.000
   0.000   0.447   0.447  -1.000   0.000   0.000   0.000
   0.000   0.894  -0.894   0.000   0.000   0.000   0.000
  -1.000   0.000  -0.447   0.000   0.447   1.000   0.000
   0.000   0.000   0.894   0.000   0.894   0.000   0.000
   0.000   0.000   0.000   1.000  -0.447   0.000  -0.447
```

(EXAMPLE 6.2 □ Input and Output of Arrays for Truss Analysis □ Continued)

```
RIGHT-HAND-SIDE VECTOR
     0.000
     5.000
     0.000
     0.000
     0.000
    10.000
     0.000
```

EXAMPLE 6.3 □ Input and Output of Character Grids

STATEMENT OF PROBLEM

Develop a program that will input and output enlarged characters of the alphabet. Each enlarged character will be made up of the desired letter stored at the appropriate spots in a 5×5 array. Use the letters A, B, C, and D for illustration.

ALGORITHM DEVELOPMENT

Input/Output Design

Each letter will require a 5×5 grid. There are 26 letters in the alphabet. To store the letters in an array called ALPHA will require a three-dimensional subscripted variable. One subscript will identify which letter of the alphabet is stored; the other two will identify the cell within the 5×5 grid. The letters A, B, C, and D will be input and output to show the proper operation of the program.

Computer Implementation

Each letter will be contained in the data file in the same form as it will be output. Input will be via an implied DO loop for the elements of a row within the letter grid, nested within a DO loop to control the row of the grid, in turn nested within a DO loop to control the letter of the alphabet. Output will involve the same nesting.

PROGRAM DEVELOPMENT

```
****************************************************
*  Letters of the alphabet                         *
*  Developed by:  T. K. Jewell  December 1987       *
****************************************************
C
C  ALPHA = array for storing the letters of the alphabet, each in a 5x5
C          array.
C
      CHARACTER*1 ALPHA(26,5,5)
      INTEGER FR,W
      PARAMETER (FR=3,W=6)
      OPEN (UNIT=FR,FILE='EX6_4.DAT',STATUS='OLD')
C
```

```
C  Input of four sample letters
      DO 100 K=1,4
         DO 200 I=1,5
            READ(FR,10) (ALPHA(K,I,J),J=1,5)
  200    CONTINUE
  100 CONTINUE
   10 FORMAT(5A1)
C
C  Output four sample letters
      DO 300 K=1,4
         DO 400 I=1,5
            WRITE(W,12) (ALPHA(K,I,J),J=1,5)
  400    CONTINUE
         WRITE (W,14)
  300 CONTINUE
   12 FORMAT(1X,5A1)
   14 FORMAT(//)
      END
```

PROGRAM TESTING

Your input data file should look like the following:

DATA FILE

```
  A
 A A
 AAA
A   A
A   A
BBBB
B   B
BBBB
B   B
BBBB
 CCC
C   C
C
C   C
 CCC
DDDD
D   D
D   D
D   D
DDDD
```

and the output would look the same except that there would be two lines between each letter.

6.6 DATA STATEMENT

There are often values that are used every time a program is executed, such as
π or the number of degrees in a radian. It would be inefficient to input these values
each time the program is run. They could be assigned by assignment statements.
However, the DATA statement provides an alternative that can initialize the
values for several variables prior to the start of execution of the program.

6.6.1 DATA Statement Syntax

The DATA statement has the general form

```
DATA variable list/value list/,variable list/value list/
```

Each *variable list* is a list of one or more variable names, array names, array element names, or implied DO lists, separated by commas. Each *value list* is a list of values corresponding in type and number with the appropriate *variable list*. A number of *variable list/value list/* combinations can be included in the same DATA statement. The comma separating each *variable list/value list/* combination can be left out. It is generally easier to interpret the intended values if only one *variable list/value list/* is used with a DATA statement. Examples of the various types of variable lists and their corresponding value lists will be discussed. It is important to remember that the number of values in the */value list/* has to match the number of variables in the *variable list* exactly. If it does not, a fatal error will result. The data values also have to match in type with their corresponding variables.

The DATA statement is a specification statement, so it is nonexecutable. DATA statements must be placed after variable type declaration statements, because the types of the variables in the *variable list* have to be defined. They should be placed before any executable statements. DATA statements do not have to be placed before any executable statements, but if they are placed later in the program, their function of initially assigning values prior to execution will be obscured.

Since the DATA statement initializes values only at the beginning of a program, it cannot be used to reinitialize variables later in the program. Therefore, a variable for finding the summation of a series of values could not be reinitialized with a data statement. However, unlike the PARAMETER statement, there is no restriction on changing a value that has been initialized by a DATA statement. Data statements should not be used to initialize values within subprograms if these values will be modified by the subprogram. The values will be initialized for the first execution, but will not be reinitialized for subsequent executions of the subprogram.

6.6.2 DATA Statement Examples

The following examples will illustrate the use of the DATA statement to initialize variable values.

6.6.2.1 Initialization of PI and RADIAN

The following DATA statement initializes values for the commonly used constants for π and the conversion factor from radians to degrees.

```
REAL PI,RADIAN
DATA PI,RADIAN/3.14159,57.2958/
```

6.6.2.2 Initialization of Array Values for Diameter and Area of Circles

Five combinations of REAL and DATA statements are shown below. Each initializes the values in two arrays. The elements of array D are assigned the diameters for circles of 6, 8, 10, 12, 15, and 18 in. The elements of array AREA are assigned the corresponding circle areas, in square inches. Note that an implied DO loop controls the array indexes when the values are assigned. In DATA statements you have to be careful to keep the data values in the same order as the variable list when you are initializing array values. In the first example, the index is increased after a value of D and a value of AREA are initialized. Therefore, the diameter and area data have to alternate in the data list.

```
REAL D(6),AREA(6)
DATA (D(I),AREA(I),I=1,6)/6.,28.274,8.,50.265,10.,78.540,12.,
1   113.097,15.,176.715,18.,254.469/
```

In the second example, two separate implied DO loops are used. The first implied DO will be completed before the second is started. Therefore, the *value list* contains the complete set of diameter data followed by the area data.

```
DATA (D(I),I=1,6),(AREA(I),I=1,6)/6.,8.,10.,12.,15.,18.,28.274,
1   50.265,78.540,113.097 176.715,254.469/
```

If all the elements of the arrays are to be read in, then the implied DO can be left off the variable list, and the DATA statement would be

```
DATA D,AREA/6.,8.,10.,12.,15.,18.,28.274,50.265,78.540,113.097,
1   176.715,254.469/
```

and the same order of data as in the previous example would be correct. However, if the dimension of the array is larger than the number of values that will be initialized, the implied loop must be used. If it is not, the *variable list* will exceed the *value list* when the whole array is initialized. Thus,

```
REAL D(12),AREA(12)
DATA (D(I),I=1,6),(AREA(I),I=1,6)/6.,8.,10.,12.,15.,18.,28.274,
1   50.265,78.540,113.097,176.715,254.469/
```

would correctly initialize the first six values of D and AREA, but would do nothing to initialize the last six values of each variable. However, the statements

```
REAL D(12),AREA(12)
DATA D,AREA/6.,8.,10.,12.,15.,18.,28.274,50.265,78.540,113.097,
1   176.715,254.469/
```

would cause an error. The computer would try to initialize 12 values for D and 12 values for AREA, but there are only 12 values in the *value list*.

6.6.2.3 Initialization of Character Variables

DATA statements can also initialize character variables. The rules for assignment apply if the length of the character string initialized through the DATA statement does not agree with the declared length of the character variable. The following example assigns characters to variables for use in developing screen plots.

```
CHARACTER*1 MINUS,PLUS,ASTER
DATA MINUS,PLUS,ASTER/'-','+','*'/
```

If the same value or character is to be assigned to several variables, you do not have to repeat the value for each variable. You can use a repetition operator to repeat the value. In the following example, you want to initialize a 5 × 5 character array to contain all X characters.

```
CHARACTER*1 A(5,5)
DATA A/25*'X'/
```

All 25 of the cells of the character array will initially contain the letter X. When a two-dimensional array is assigned values by a DATA statement without use of implied loops, as in the above example, the first subscript is varied first. Thus, the first value would be assigned to A(1,1), the second to A(2,1), and so on. This order would be important only if the array elements are to be initialized to different values.

6.7 EMBEDDED FORMATS

Embedded formats get their name from the fact that they are embedded within the READ or WRITE statement that they modify. Because of this characteristic, an embedded format cannot be used by more than one READ or WRITE statement. However, the embedded format also provides a means of inputting format specifications through character variables. The variable can be used in any number of WRITE statements. After discussing the general provisions and structure of embedded and variable formats, we will apply them to the subprograms for input and output of arrays, and to the program of Example 4.3.

The embedded format will replace the normal reference to a FORMAT statement in the READ or WRITE statement. A normal FORMAT would be

```
    READ(FR,100) X,Y,Z
100 FORMAT(3F10.0)
```

while the equivalent embedded format would be

```
    READ(FR,'(3F10.0)') X,Y,Z
```

The format specification is actually a character string, delimited by ' ', which is interpreted by the computer as a format because it is in place of the normal FORMAT statement reference.

Because the format is represented by a character string, the output of literals and embedded apostrophes has to be treated differently than it has been up to this point. The character string format makes the normal literal delimiting apostrophe an embedded apostrophe. By the rule established in Chapter 3, the literal output delimiting apostrophes can be represented by double apostrophes, '', within the character string delimiting apostrophes. However, that does not allow us to represent any apostrophes embedded within the literal by double apostrophes as we did in Chapter 3. As you might suspect, increasing the number of consecutive apostrophes will be taken as an apostrophe embedded within the literal, which is embedded within the character string for the FORMAT. FORTRAN will interpret four consecutive apostrophes as an apostrophe for this case. An example will illustrate this.

Suppose you want to output the current I for an electrical engineering problem with the following label:

```
By Kirchhoff's Laws, the current is:  xxxxxx.xx
```

You can assume that the variable I has been declared as real. With a normal FORMAT, WRITE combination, the lines of code would be

```
    WRITE(W,100) I
100 FORMAT(6X,'By Kirchhoff''s Laws, the current is: ',F9.2)
```

while with an embedded FORMAT the WRITE statement would be

```
WRITE(W,'(6X,''By Kirchhoff''''s Laws, the current is: '',F9.2)') I
```

Since you have seen that embedded FORMATS can be represented by constant character strings, it should not come as too much of a surprise that embedded FORMATS can also be represented by character variables, allowing you to input the FORMAT. For example, the statements

```
CHARACTER*20 FORM
FORM = '(3F8.0)'
       .
       .
READ(FR,FORM) X,Y,Z
```

would read the values for X, Y, and Z according to the FORMAT specifications stored in character variable FORM. This particular example would allow only a FORMAT specification that took up no more than 20 spaces.

6.8 VARIABLE FORMAT SPECIFICATIONS

The embedded FORMAT can also be represented by a character expression. All the rules of character operations defined in Chapter 3 could be incorporated into the expression. For example, the WRITE statement,

```
WRITE(W,'('//FORM1//FORM2//')') X,Y,Z
```

would concatenate the character strings contained in variables FORM1 and FORM2, with the left and right parentheses, and use the resulting combined character string as the format. FORM1 might contain the label information, and FORM2 the FORMAT specifications. This would allow you to use the labels or the specifications in other embedded formats. Using character expressions allows you to customize output formats for different conditions in the program.

Section 5.1 of this chapter presented SUBROUTINE subprograms for input and output of arrays. Example 6.4 will now modify these subprograms to accept variable formats by using character variables to represent the FORMAT specifications. If you need to change the FORMAT specifications for different applications of the subprograms, this is a good way to do it.

Example 6.4 will allow you to input both the input and output formats. It will also modify the input format for output. Once you study this example, I think you will agree that the best method is to input both formats.

EXAMPLE 6.4 □ Variable FORMAT Specifications for Array Input and Output

PROBLEM STATEMENT

Modify the SUBROUTINE subprograms developed in Section 5.1 of this chapter to accept variable format specifications, and develop a main program to allow testing of the variable formats.

(EXAMPLE 6.4 □   Variable FORMAT Specifications for Array Input and Output □   Continued)

ALGORITHM DEVELOPMENT

Input/Output Design

The variable formats will be input through the main program and sent over to the subprogram. Thus, you will not have to input the formats each time a subprogram using variable formats is called. You will also be able to use the formats at other places in the program to form new formats through character operations. The input format will be used first to output the array to show the values stored. Conversion of the input format to an output format will require use of character operations in a character expression. These will add the 1X specification at the beginning of the format and three spaces between each output field. If the input format is 10F8.4, the corresponding output format would be 1X,10(F8.4,3X). Generating the output format will require separating the 10 and F8.4 through character string operations. For this to work properly, you always have to use two places for the repetition count, even if the actual repetition is less than 10. A note will be placed in the program to alert users to this requirement. Following the output using the modified input format, the user-supplied output format will be used to output the array a second time.

Computer Implementation

The following pseudocode representation shows how the main program should be set up:

```
INPUT, INPUT FORMAT
INPUT, OUTPUT FORMAT
INPUT DIMENSIONS
INPUT ARRAY
OUTPUT ARRAY USING INPUT FORMAT
OUTPUT ARRAY USING OUTPUT FORMAT
END
```

PROGRAM DEVELOPMENT

```
****************************************************
*  Program for testing variable FORMATs            *
*  Developed by:  T. K. Jewell    December 1987    *
****************************************************
C  INFORM = input FORMAT specification
C  INFOPR = input FORMAT modified for output
C  OUTFOR = output FORMAT specification
C  N = # of rows of array    M = # of columns of array
C
       REAL A(10,10)
       INTEGER N,M,R,FR,W
       CHARACTER INFORM*20,INFOR1*22,INFOPR*40,OUTFOR*20,OUTFO1*25,
      1NAME*14
       PARAMETER (FR=3,R=5,W=6)
       PRINT *, 'Please Input File Name  '
       READ (R,'(A14)') NAME
       OPEN (UNIT=FR,FILE=NAME,STATUS='OLD')
C
```

```
C  Read input and output FORMAT specifications.  For this test program only,
C    make sure that the repeater for the input FORMAT takes up the first two
C    positions of the input. If you want to use less than 10 for the
C    repeater, then leave the first position blank.  This is so the expression
C    for INFOPR works properly.
       READ (FR,'(A20)') INFORM,OUTFOR
       INFOPR = '(1X,'//INFORM(1:2)//'('//INFORM(3:20)//',3X))'
       INFOR1 = '('//INFORM//')'
       OUTFO1 = '(1X,'//OUTFOR//')'
C
C  Input the dimensions of the array
       READ (FR,'(2I5)') N,M
C  Input the array
       CALL READ (A,N,M,FR,INFOR1)
C
C  Output according to input FORMAT
       WRITE (W,*) 'Values input to the computer:'
       CALL WRITE(A,N,M,W,INFOPR)
C
C  Output according to the variable FORMAT
       WRITE(W,*) 'Values output by the variable output format:'
       CALL WRITE (A,N,M,W,OUTFO1)
       END
C
       SUBROUTINE READ(A,N,M,FR,FORM)
       REAL A(10,10)
       INTEGER N,M,FR
       CHARACTER FORM*40
       DO 100 I=1,N
          READ(FR,FORM) (A(I,J),J=1,M)
  100 CONTINUE
       RETURN
       END
C
       SUBROUTINE WRITE(A,N,M,W,FORM)
       REAL A(10,10)
       INTEGER N,M,W
       CHARACTER FORM*40
       DO 100 I=1,N
          WRITE(W,FORM) (A(I,J),J=1,M)
  100 CONTINUE
       RETURN
       END
```

PROGRAM TESTING

Use a sample data file and input and output formats to make sure that the values
are being entered and output properly by the variable formats. The following
data file will produce the output shown. You can adjust the data file formats and
observe the changes in the output.

DATA FILE

```
10F8.4
10(F8.2,3X)
    3    5
 1111111 1111111 1111111 1111111 1111111
 2222222 2222222 2222222 2222222 2222222
 3333333 3333333 3333333 3333333 3333333
```

(EXAMPLE 6.4 ☐ Variable FORMAT Specifications for Array Input and Output ☐ Continued)

PROGRAM OUTPUT

```
Values input to the computer:

111.1111    111.1111    111.1111    111.1111    111.1111
222.2222    222.2222    222.2222    222.2222    222.2222
333.3333    333.3333    333.3333    333.3333    333.3333

Values output by the variable output format:

111.11      111.11      111.11      111.11      111.11
222.22      222.22      222.22      222.22      222.22
333.33      333.33      333.33      333.33      333.33
```

We will now use variable embedded formats to streamline a previously developed program. The program of Example 4.3 gave the option of using either metric or English units for a particle trajectory problem. Although the computations were the same for both sets of units, a separate set of output statements had to be used to output the proper labels for the units. Example 6.5 will incorporate variable formats to eliminate the need for separate output sections.

EXAMPLE 6.5 ☐ Variable Output Formats to Accommodate Metric or English Units

STATEMENT OF PROBLEM

Incorporate embedded formats into the program of Example 4.3 to accommodate the output of either metric or English units, depending on the value of G, the acceleration due to gravity.

ALGORITHM DEVELOPMENT

Input/Output Design

Input is the same as for Example 4.3. A nested IF THEN ELSE is still needed, but only to establish the variable units portion of the output format.

Computer Implementation

Figure 6.3 shows the modified flowchart of Figure 4.7.

PROGRAM DEVELOPMENT

Only the portion of the program that has to be modified will be presented here.

```
      CHARACTER UNITS*4
      IF (G.LT.10.0) THEN
         UNITS = '  m.'
      ELSE
         UNITS = ' ft.'
      END IF
      WRITE (W,'(''For V0 = '',F6.1,''''//UNITS//'/s, and THETA = '',
```

```
1              F6.2,'' degrees,'')') V0,THETA
  WRITE (W,'(''Maximum height of trajectory = '',F6.1,'''//UNITS//
1         ', and'')') YMAX
  WRITE (W,'(''distance from launch to impact = '',F6.1,'''//UNITS//
1         '')') XMAX
```

FIGURE 6.3   Flowchart for variable format treatment of English or
metric units.

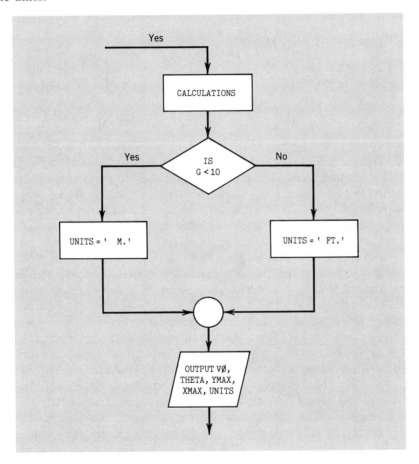

PROGRAM TESTING

Test your program with the data of Example 4.3 to check that the variable
formats are working properly.

6.9 COMBINED INPUT/OUTPUT FROM/TO TERMINAL/FILE

We have already seen how to input information from both the terminal and a
file at the same time. We have also learned that file output has several distinct
advantages over terminal or screen output. It provides a permanent record that
you can edit or read at your leisure. You can also move up or down in the file at

will. You cannot do this with terminal output, because it scrolls from beginning to end. It is frequently convenient to route some output to the terminal and some output to a file. Other times you may want to route the same information to both a terminal and a file. We will now develop different forms of file output, including the simultaneous output to the terminal and a file, as well as several variations in file handling that will provide you with additional options for input and output of information.

6.9.1 Review of File Handling Procedures

To open more than one file for access by a program, you have to include an OPEN statement for each file. As you have learned, the form of the OPEN statement is

```
OPEN (UNIT=integer expression,FILE=character
         expression,STATUS=character expression)
```

Normally the *integer expression* of the UNIT = specification must evaluate to a different integer number for each file to be accessed. They would be the same only if you wanted to write data to the same file that you had previously read, or if you wanted to read data stored in a file earlier in the execution of the program. We have developed a nomenclature for differentiating between file and terminal input and output. If you have more than one file for input, you could assign units to each using integer variables FR1, FR2, and so on. A similar series could be used for output to multiple files. The *character expression* of the FILE = specification has to contain the name of the file to be opened, including the applicable file extension, and any necessary information on the disk drive and directory that contains the file. The *character expression* of the STATUS = specification will contain the character strings 'NEW', 'OLD', or 'SCRATCH'. An OLD file is one that you have already created. READ statements will input values from an OLD file. NEW files will be created by sending information to them through WRITE statements. If you already have a file with the same name as the one used in the OPEN statement, you will not be able to write to that file with a NEW status. This is to protect you from inadvertently overwriting information already in the file. SCRATCH files are temporary files that are used only for storage of data during the execution of the program. SCRATCH files are not saved after the program is finished.

It is also possible to use the ERR = specification with the OPEN statement. This provides you with a method of error recovery if errors are found in any of the parameters of the OPEN statement. It works the same way as the ERR = specification used with READ statements.

If you write information to a file and then want to read it back into the program during the same execution, you have to use the REWIND command to position the disk read heads at the beginning of the file. The REWIND command has the form

```
REWIND integer expression
```

in which *integer expression* designates the unit to be rewound. Thus

```
REWIND FW1
```

would rewind the unit corresponding to the value of FW1.

Another file handling command is the CLOSE command. It is generally not required because files will automatically be closed when the program execution terminates. However, if you ever wanted to use the same unit to access more than one file during execution, you could close the first file that had been opened under that unit designation and open a new file. You could not access these files simultaneously. The CLOSE statement has the general form

```
CLOSE (UNIT = integer expression)
```

in which the value of the *integer expression* designates the unit to be closed.

6.9.2 Input/Output Data Handling Variations

You have many options available to you to accomplish input and output of computer data. They will be described in the following two sections, then several of them will be applied in three variations of the truss analysis problem of Example 6.2. You have also used some of these options in previous examples and exercises.

6.9.2.1 Input Data

Computer programs need to have input data before they can produce useful results. Data that will be the same for repeated executions of the program can be introduced through PARAMETER or DATA statements, or through the use of assignment statements. The PARAMETER statement can initialize only values that will remain constant throughout the program execution. DATA statements can initialize values, but those values can later be changed through assignment statements. Data that may vary from execution to execution should be input through READ statements. In this section we will study the options available for READ input.

You are probably better off using input from the terminal for programs with minimal data input. This will allow you to easily change the input data from one execution to the next. However, you will also have to reinput the data each time you run the program. You should use error trapping to avoid having to start over if you make a mistake. When using straight terminal input, you need to ask yourself if it will take less time to enter the data each time than it would take to modify and save an input data file.

You can use terminal input for programs with large amounts of data if you offer the option of saving the input data in a file. Error trapping is essential for such applications. The program should also be able to retrieve saved data for input. You may want to give the user the option of editing the saved file. The edited file could either replace the old file or be saved under a new name.

Your second option is to input all data from data files. Input for programs that will be run in the batch mode has to be from data files. You can input data from several different files, as long as you open each one and keep your unit specifications straight in your READ statements. If you do not, you will be trying to read data from the wrong file. The decision as to whether or not to use formatted input depends mainly on the intended user. If the person is not familiar with programming languages, then formatted input is probably preferable. It allows you to give explicit instructions as to where data should be placed when input records are being typed. Formatted input files are also easier to edit. Users who are familiar with programming may prefer the flexibility of unformatted input. You will see both formatted and unformatted input used in the remaining examples of the text. You can form your own opinions as to when each is preferable. The primary consideration should be the clarity of the methods of data input to the user.

A third option is to use a combination of file and terminal input. You have seen examples of this already. We have used prompts and terminal input to define the name of an input file. The program then used this name in the OPEN statement and read data from the named file. We have used responses to inquiries to select a particular branch of a logical structure. You can also use a combination of terminal and file input for numerical input. Assume for example that you have a large structure that you want to subject to various loading conditions. Information on the geometry and properties of the structure could be input from a file. The loading data could be input through the terminal. This would give you the flexibility of changing the loading conditions without having to modify the input file.

6.9.2.2 Output of Results

Output can be directed to a terminal, to a file, or to a combination of the two. You can also output to several files if you desire to do so. As with input from several files, you have to be careful about where you are sending particular elements of output. In some text examples we have output prompts to the terminal and program results to a file. The advantages of file output have been discussed previously. The disadvantage is that you do not see the output generated as the program progresses. After execution you have to go to the output file to see the results. This disadvantage can be overcome by having both types of output, at least for key parts of the results. In Example 4.14 we offered the option of outputting to the terminal or to both the terminal and a file.

You can also temporarily output data to a file. In larger programs it may be advantageous to save intermediate results, load a different analysis module into the computer, and then continue with the execution. The saved data can then be read back in and used. Or you may want to save computer memory by using an array for more than one purpose. If you save the values in the array before using it for the second purpose, you have the option of returning to the first use of the array later in the program. You will have to use the REWIND statement to move the file pointer to the beginning of the file before reinputting the temporarily-saved data.

6.9.3 Combined Input/Output Examples

We will now develop several examples incorporating different input/output options. The first, Example 6.6, will incorporate both file and terminal input for a truss analysis program and will give the user the option of outputting all data to the terminal or to a file. Example 6.7 will input both the coefficient matrix and the right-hand-side vector through the terminal, with error trapping, and provide the option of saving both sets of data in a file for later retrieval as input. Output for checking will be directed either to the terminal or to a file, as in Example 6.6. Example 6.8 will allow the editing of any of the elements of the coefficient matrix. Example 6.9 will combine all these elements into a comprehensive program for input, output, and editing of data for the truss analysis problem.

EXAMPLE 6.6 □ Further Development of Input and Output for Truss Analysis

STATEMENT OF PROBLEM

Modify the program of Example 6.2 to accommodate the following input and output:

a. The coefficient matrix will be input from a data file.
b. The right-hand-side vector will be input from the terminal with error trapping.
c. The user will be provided with the option of outputting results either to a file or to the terminal.

ALGORITHM DEVELOPMENT

Input/Output Design

Input of the coefficient array can be the same as it was for Example 6.2 and will use the same subprogram. The right-hand-side vector will be input one element at a time, with error trapping. The value of N, input for the coefficient matrix, will control the number of input values for the right-hand side. A general subprogram for interactive input of an array will be developed from the array input subprogram for file input. When you review this subprogram, note that a formatted PRINT statement is used. One set of output statements will be used. These are the same as the output statements of Example 6.2. Output will be directed to the terminal or to a file by selection of a value for the WRITE statement unit number.

Computer Implementation

Figure 6.4 shows a flowchart for the input and output accomplished for this example.

PROGRAM DEVELOPMENT

```
      PROGRAM MATRIX
**********************************************************
*   TRUSS ANALYSIS                                       *
*   Input of coefficient matrix from file                *
*   Input of right-hand-side vector from terminal        *
*   Output to either file or terminal                    *
*   Developed by:  T. K. Jewell      December 1988       *
**********************************************************
C
C  Variable definitions
C  A = coefficient matrix (directional cosines)
C  B = right-hand-side vector (forces and reactions)
C  N = number of members in truss
C
      REAL A(25,25),B(25,1)
      INTEGER N,FR,R,W,WR
      CHARACTER*14 NAMEIN,NAMOUT
      PARAMETER (FR=3,R=5,W=6)
```

(EXAMPLE 6.6 □ Further Development of Input and Output for Truss Analysis □ Continued)

FIGURE 6.4   Flowchart for further development of truss analysis input and output.

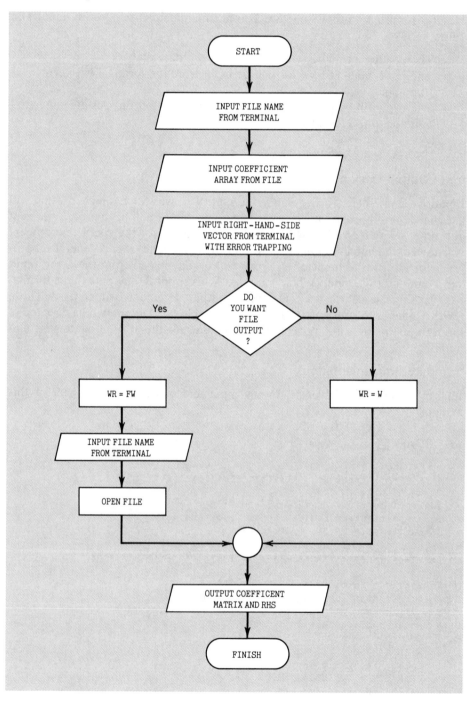

```
      PRINT *, 'Please Input File Name'
      READ (R,10) NAMEIN
   10 FORMAT (A14)
      OPEN (UNIT=FR,FILE=NAMEIN,STATUS='OLD')
C
C  Input of number of members
      READ(FR,20) N
   20 FORMAT(I5)
C
C  Input coefficient matrix.
      CALL READ(A,N,N,FR)
C
C  Input right-hand-side vector from terminal with error trapping.
      PRINT *, 'Input the right-hand-side vector, one term at a time'
      PRINT *, 'NOTE: Since this is a vector, the second subscript',
     1          ' will always be one.'
      CALL TREAD(B,N,1,R)
C
C  Selection of file or terminal output
C
      PRINT *, 'Select file or terminal output'
      PRINT *, '4 = file output, any other number = terminal output'
      READ (R,*) WR
      IF (WR.EQ.4) THEN
         PRINT *, 'Input file name for output, for example, TRUSS.OUT '
         READ (R,'(A14)') NAMOUT
         OPEN (UNIT=4,FILE=NAMOUT,STATUS='NEW')
      ELSE
         WR = W
      END IF
C
C  Output of coefficient matrix and right-hand side
      WRITE(WR,30) N
   30 FORMAT(11X,'MATRIX ANALYSIS OF TRUSSES',//,
     1        1X,'  The number of members is:  ',I2,///,
     2        1X,'DIRECTIONAL COSINE COEFFICIENT MATRIX')
C
      CALL WRITE(A,N,N,WR)
C
      WRITE(WR,40)
   40 FORMAT(1X,'RIGHT-HAND-SIDE VECTOR')
C
      CALL WRITE(B,N,1,WR)
      END
      SUBROUTINE READ(A,N,M,FR)
      REAL A(10,10)
      INTEGER N,M,FR
      DO 100 I=1,N
         READ(FR,200)  (A(I,J),J=1,M)
  100 CONTINUE
  200 FORMAT(10F8.0)
      RETURN
      END
C
      SUBROUTINE TREAD(A,N,M,R)
      REAL A(10,10)
      INTEGER N,M,R
      DO 100 I=1,N
         DO 200 J=1,M
            GO TO 20
```

(EXAMPLE 6.6 ☐ Further Development of Input and Output for Truss Analysis ☐ Continued)

```
      10       CONTINUE
               PRINT *, 'YOU HAVE MADE AN ERROR IN YOUR LAST INPUT VALUE'
               PRINT *, 'PLEASE CHECK AND REINPUT THIS VALUE'
      20       CONTINUE
               PRINT 30,I,J
      30       FORMAT(1X,'Input element (',I2,',',I2,')  ')
               READ(R,*,ERR=10) A(I,J)
     200     CONTINUE
     100 CONTINUE
         RETURN
         END
C
         SUBROUTINE WRITE(A,N,M,W)
         REAL A(10,10)
         INTEGER N,M,W
         DO 100 I=1,N
            WRITE(W,200) (A(I,J),J=1,M)
     100 CONTINUE
     200 FORMAT(1X,10F8.3)
         RETURN
         END
```

PROGRAM TESTING

The input data file is given below.

DATA FILE

```
12345678901234567890123456789012345678901234567890123456789012345678901234567890
    7
   1.0 -0.4472        0        0        0        0        0
     0   0.8944        0        0        0        0        0
     0   0.4472   0.4472     -1.0        0        0        0
     0   0.8944  -0.8944        0        0        0        0
  -1.0        0  -0.4472        0   0.4472      1.0        0
     0        0   0.8944        0   0.8944        0        0
     0        0        0      1.0  -0.4472        0  -0.4472
```

PROGRAM TESTING

The interactive portion of the input would be as follows:

```
Please Input File Name
TRUSS.DAT
Input the right-hand-side vector, one term at a time
NOTE: Since this is a vector, the first subscript will always be one.
Input element (1,1)
0
Input element (2,1)
%.)
YOU HAVE MADE AN ERROR IN YOUR LAST INPUT VALUE
PLEASE CHECK AND REINPUT THIS VALUE
Input element (2,1)
5.0
Input element (3,1)
0
Input element (4,1)
```

```
0
Input element (5,1)
0
Input element (6,1)
10.0
Input element (7,1)
0
Select file or terminal output
4 = file output, any other number = terminal output
4
Input file name for output, for example, TRUSS.OUT
TRUSS.OUT
```

The output portion would be the same as Example 6.2.

EXAMPLE 6.7 □ Input from Terminal with Error Trapping for Truss Analysis

STATEMENT OF PROBLEM

Modify the program of Example 6.6 to accomplish the following input and output tasks:

    a. Input the coefficient array from the terminal.
    b. Input the right-hand-side vector from the terminal.
    c. Give the user the option of combining the coefficient array and right-hand-side vector into a file to be saved for later input.
    d. Output the input data to a file or to the terminal to check values.

ALGORITHM DEVELOPMENT

Input/Output Design

The first item of input must be the number of equations for the problem. The number of equations defines the size of the coefficient matrix and the right-hand-side vector. Subprogram TREAD can then be used to input either the coefficient matrix or the right-hand-side vector. Output for review will be the same as in Example 6.6. The data output to be saved in a file will not have the label information of the review output. It will also be output without formatting to make it easier to preserve significant figures. If you use formatted output, you would have to adjust the format to the magnitude of the coefficients and right-hand-side values.

Computer Implementation

The subprograms from Example 6.6 that can be used without modification will not be repeated here. The subprogram for writing out an array is modified to provide unformatted output and is named TWRITE. The following pseudocode outlines the changes in the main program.

```
INPUT NUMBER OF MEMBERS, N
OUTPUT PROMPTS FOR COEFFICIENT MATRIX INPUT
INPUT COEFFICIENT MATRIX
   CALL TREAD(A,N,N,R)
```

(EXAMPLE 6.7 □ Input from Terminal with Error Trapping for Truss Analysis □ Continued)

```
OUTPUT PROMPTS FOR RIGHT-HAND-SIDE VECTOR INPUT
INPUT RIGHT-HAND-SIDE VECTOR
    CALL TREAD(B,N,1,R)
DO YOU WANT TO SAVE YOUR INPUT FOR FUTURE USE?
    YES:  OUTPUT COEFFICIENT MATRIX FOLLOWED BY RIGHT-HAND-SIDE VECTOR
    NO:   CONTINUE
IS OUTPUT TO BE TO A FILE OR THE TERMINAL?
    FILE: NAME, OPEN, WRITE LABELS AND DATA
    WRITE LABELS AND DATA
END
```

PROGRAM DEVELOPMENT

```
      PROGRAM MATRIX
*******************************************************************
*   TRUSS ANALYSIS                                                *
*   Input of coefficient matrix from terminal                     *
*   Input of right-hand-side vector from terminal                 *
*   Optional output of input data to file for future use          *
*   Output to either file or terminal                             *
*   Developed by:  T. K. Jewell            December 1988           *
*******************************************************************
C
C  Variable definitions
C  A = coefficient matrix (directional cosines)
C  B = right-hand-side vector (forces and reactions)
C  N = number of members in truss
C
      REAL A(25,25),B(25,1)
      INTEGER N,R,W,WR
      CHARACTER NAMEIN*14,NAMOUT*14,SAVE*3
      PARAMETER (R=5,W=6)
C
      PRINT *, 'Input the number of members for the truss  '
      READ(R,*) N
C
C  Input coefficient matrix, with error trapping.
      PRINT *, 'You now need to enter the coefficient matrix,'
      PRINT *, 'one element at a time. Input each row, from'
      PRINT *, 'left to right, before going on the next row  '
      CALL TREAD(A,N,N,R)
C
C  Input right-hand-side vector from terminal with error trapping.
      PRINT *, 'Input the right-hand-side vector, one term at a time'
      PRINT *, 'NOTE: Since this is a vector, the second subscript',
     1         ' will always be one.'
      CALL TREAD(B,N,1,R)
C
C  Saving input file for later use
      PRINT *, 'Do you want to save your input for later use?'
      PRINT *, 'Input Yes or No  '
      READ (R,'(A3)') SAVE
      IF (SAVE(1:1).EQ.'Y'.OR.SAVE(1:1).EQ.'y') THEN
         PRINT *, 'Input file name for data file. It must be '
         PRINT *, 'different from the file name for the output file'
         PRINT *, 'An example would be:  TRUSS.DAT '
         READ (R,'(A14)') NAMEIN
```

```
            WR = 3
            OPEN (UNIT=WR,FILE=NAMEIN,STATUS='NEW')
            WRITE (WR,*) N
            CALL TWRITE(A,N,N,WR)
            CALL TWRITE(B,N,1,WR)
         END IF
C  Selection of file or terminal output
C
         PRINT *, 'Select file or terminal output'
         PRINT *, '4 = file output, any other number = terminal output   '
         READ (R,*) WR
         IF (WR.EQ.4) THEN
            PRINT *, 'Input file name for output, for example, TRUSS.OUT   '
            READ (R,'(A14)') NAMOUT
            OPEN (UNIT=4,FILE=NAMOUT,STATUS='NEW')
         ELSE
            WR = W
         END IF
C
C  Output of coefficient matrix and right-hand-side vector
         WRITE(WR,30) N
      30 FORMAT(11X,'MATRIX ANALYSIS OF TRUSSES',//,
        1         1X,' The number of members is: '   ,I2,///,
        2         1X,'DIRECTIONAL COSINE COEFFICIENT MATRIX')
C
         CALL WRITE(A,N,N,WR)
C
         WRITE(WR,40)
      40 FORMAT(1X,'RIGHT-HAND-SIDE VECTOR')
C
         CALL WRITE(B,N,1,WR)
         END
C
C  Subprogram for saving input file
         SUBROUTINE TWRITE(A,N,M,W)
         REAL A(10,10)
         INTEGER N,M,W
         DO 100 I=1,N
            WRITE(W,*) (A(I,J),J=1,M)
     100 CONTINUE
         RETURN
         END
```

PROGRAM TESTING

Enter the coefficient matrix and right-hand-side vector of Example 6.2. Make
some input errors on purpose to test the error trapping. Test both options for
saving the input data.

EXAMPLE 6.8 □ Editing of Input Array for Truss Analysis

STATEMENT OF PROBLEM

Modify and combine parts of Examples 6.6 and 6.7 to accomplish the following
with the coefficient matrix of the truss analysis problem:

(EXAMPLE 6.8 □ Continued)

FIGURE 6.5 Flowchart for editing of input arrays for truss analysis.

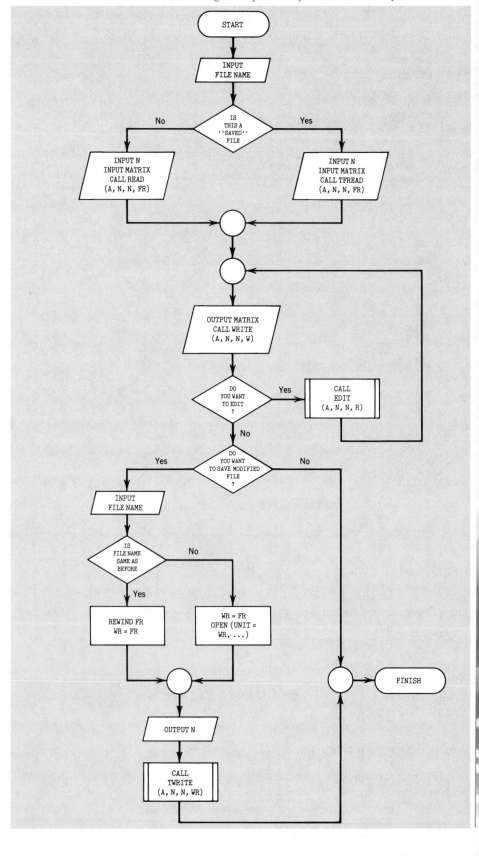

a. Input the coefficient matrix either from a formatted file or from an unformatted saved file. Output the array to the screen for checking.
b. Edit as many elements of the coefficient matrix as you desire.
c. Output the array to the screen again for checking.
d. Edit the array again if necessary.
e. Save the modified array with the option of calling it a new name.

ALGORITHM DEVELOPMENT

Input/Output Design

Input will be a combination of the formatted input of Example 6.6 and a modification of the routine TREAD of Example 6.7. Output for checking can use subprogram WRITE of Example 6.6. Output for the saved file can use subprogram TWRITE of Example 6.7. Subprograms that are used from previous programs will not be repeated in the program listing.

Computer Implementation

Figure 6.5 shows a flowchart for the editing routine of Example 6.8.

PROGRAM DEVELOPMENT

```
      PROGRAM MATRIX
****************************************************************
*  TRUSS ANALYSIS                                             *
*  Input of coefficient matrix from formatted file or         *
*    from unformatted saved file                              *
*  Editing of Coefficient Matrix                              *
*  Developed by:  T. K. Jewell          December 1988         *
****************************************************************
C
C  Variable definitions
C  A = coefficient matrix (directional cosines)
C  N = number of members in truss
C
      REAL A(25,25)
      INTEGER N,FR,FW,R,W,WR
      CHARACTER NAMEIN*14,NAMOUT*14,SAVE*5,CHECK*3
      PARAMETER (FR=3,FW=4,R=5,W=6)
      PRINT *, 'Please Input Name of Input File  '
      READ (R,10) NAMEIN
   10 FORMAT (A14)
      OPEN (UNIT=FR,FILE=NAMEIN,STATUS='OLD')
C
      PRINT *, 'Is this a formatted input or saved file?'
      PRINT *, 'Input SAVED if it is a saved file, otherwise'
      PRINT *, '  it will be considered to be a formatted file.  '
      READ (R,'(A5)') SAVE
      IF (SAVE.NE.'SAVED') THEN
C Input of number of members
         READ(FR,20) N
   20    FORMAT(I5)
C Input coefficient matrix.
         CALL READ(A,N,N,FR)
      ELSE
         READ (FR,*) N
         CALL TFREAD(A,N,N,FR)
      END IF
```

(EXAMPLE 6.8 □ **Editing of Input Array for Truss Analysis** □ **Continued)**

```
C
C  Start of loop for editing
  100 CONTINUE
C
C  Output of coefficient matrix
      WRITE(W,30) N
   30 FORMAT(//,11X,'MATRIX ANALYSIS OF TRUSSES',//,
     1           1X,'  The number of members is:   ',I2,///,
     2           1X,'PRESENT DIRECTIONAL COSINE COEFFICIENT MATRIX',/)
C
      CALL WRITE(A,N,N,W)
C
      PRINT *, 'Do you want to edit the coefficient matrix?'
      PRINT *, 'Please enter Yes or No '
      READ (R,'(A3)') CHECK
      IF (CHECK(1:1).EQ.'Y'.OR.CHECK(1:1).EQ.'y') THEN
         CALL EDIT (A,N,N,R)
         GO TO 100
      END IF
C
      PRINT *, 'Do you want to save the modified file?'
      PRINT *, 'Please enter Yes or No'
      PRINT *, 'NOTE: The modified file will be an unformatted SAVED'
      PRINT *, '      file.  If you save it under the same name as the'
      PRINT *, '      input file, the input file will be lost.  '
      READ (R,'(A3)') CHECK
      IF (CHECK(1:1).EQ.'Y'.OR.CHECK(1:1).EQ.'y') THEN
         PRINT *, 'Input file name, such as TRUSS.DAT  '
         READ (R,'(A14)') NAMOUT
         IF (NAMOUT.EQ.NAMEIN) THEN
            REWIND FR
            WR = FR
         ELSE
            WR = FW
            OPEN (UNIT=WR,FILE=NAMOUT,STATUS='NEW')
         END IF
         WRITE (WR,*) N
         CALL TWRITE(A,N,N,WR)
      END IF
      STOP
      END
      SUBROUTINE TFREAD(A,N,M,R)
      REAL A(10,10)
      INTEGER N,M,R
      DO 100 I=1,N
         READ(R,*) (A(I,J),J=1,M)
  100 CONTINUE
      RETURN
      END
      SUBROUTINE EDIT(A,N,M,R)
      REAL A(10,10)
      INTEGER N,M,R
C
C  Start of editing loop
  100 CONTINUE
      PRINT *, 'Which element of the array do you want to edit?'
```

```
      PRINT *, 'Any subscript out of the range or N and M will'
      PRINT *, '  terminate the editing.  '
      READ (R,*) I,J
      IF (I.LE.N.AND.J.LE.M) THEN
          PRINT 30,I,J
 30 FORMAT(1X,'INPUT NEW VALUE FOR A(',I2,',',I2,')  ')
          GO TO 300
200       CONTINUE
          PRINT *, 'You goofed! Please reenter last value  '
300       CONTINUE
          READ(R,*,ERR=200) A(I,J)
          GO TO 100
      END IF
      PRINT 40
 40 FORMAT (//,'END OF EDITING',//)
      RETURN
      END
```

PROGRAM TESTING

Change several elements in different rows of the matrix to test the program.
Output the modified file, then call it back in for a subsequent execution of the
program to make sure the modified file is being saved properly.

EXAMPLE 6.9 □ Combination of Different Aspects of Truss Analysis

STATEMENT OF PROBLEM

The programs of the previous examples can be modified into subprograms to
accomplish input, output, and editing options for the coefficient matrix and right-
hand-side vector for truss analysis. Each of the options should include the sub-
options shown in the outline. For this problem, we will develop a main program
to access any of these options through a menu structure. Development of the
subprograms is included as a chapter-end exercise.

INPUT OPTIONS SUBPROGRAM

a. All input from a data file developed with the text editor.
b. All input from a file saved after a previous execution of the program.
c. Input of the coefficient matrix from a file and input of the right-hand-side
   vector from the terminal, with the option to save both to a file for future
   input.
d. All input from the terminal, with the option to save to a file for future input.
e. Option to include variable identification information.

OUTPUT OPTIONS SUBPROGRAM (Including labels)

a. Terminal output.
b. File output.
c. Both terminal and file output.

(EXAMPLE 6.9 □ Combination of Different Aspects of Truss Analysis □ Continued)

EDITING OPTIONS SUBPROGRAM

   a. Editing of coefficient matrix with output for review.
   b. Editing of right-hand-side vector with output for review.
   c. Saving edited file under old file name or new file name.

FIGURE 6.6 Flowchart for combined truss analysis program.

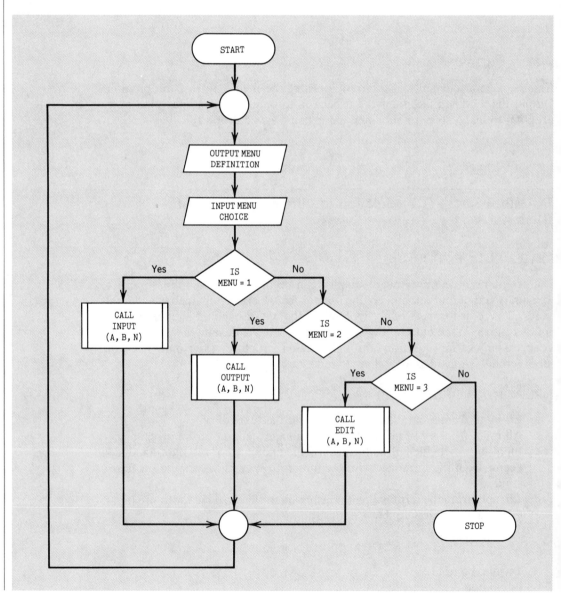

ALGORITHM DEVELOPMENT

Input/Output Design

The choice of which option to use will be determined by a menu, which can be implemented by a case structure. Each choice will be associated with a particular number, as shown in the following list:

> 1 = INPUT
> 2 = OUTPUT
> 3 = EDITING
> 4 = EXIT (Anything other than 1, 2, or 3 will cause exit)

Once you choose an option, input and output are governed by the procedures developed in Examples 6.2, 6.6, 6.7, and 6.8. Some modifications of these procedures are necessary to make them into subprograms and to include all the options listed in the STATEMENT OF PROBLEM for this example.

Computer Implementation

The program will contain a loop structure that allows you to choose any number of options, one after the other. You need to have a way of exiting the loop that is fail-safe. Therefore, any entry other than 1, 2, or 3 will cause program execution to terminate. The input, output, and edit subprograms can still reference all the other subprograms developed. Figure 6.6 shows a flowchart for the main program.

PROGRAM DEVELOPMENT AND TESTING

The main program with menu structure is shown below. Development and testing of the subprograms is left up to the student.

```
******************************************************************
*  Menu program for input, output, and editing of       *
*    Truss Analysis Program                              *
*  Developed by:  T. K. Jewell      February 1989        *
******************************************************************
       REAL A(10,10),B(10,1)
       INTEGER N,R,W,MENU
       PARAMETER (R=5,W=6)
   10  WRITE (W,100)
  100  FORMAT(//,15X,'TRUSS ANALYSIS INPUT/OUTPUT MENU',//,
      1             5X,'Select MENU option from following list:',//,
      2             5X,'1 = INPUT',/,
      3             5X,'2 = OUTPUT',/,
      4             5X,'3 = EDIT',/,
      5             5X,'Anything Else = EXIT',/)
       READ(R,*) MENU
       IF (MENU.EQ.1) THEN
          CALL INPUT(A,B,N)
          GO TO 10
       ELSE IF (MENU.EQ.2) THEN
          CALL OUTPUT(A,B,N)
          GO TO 10
       ELSE IF (MENU.EQ.3) THEN
          CALL EDIT(A,B,N)
          GO TO 10
```

(EXAMPLE 6.9 ☐ Combination of Different Aspects of Truss Analysis ☐ Continued)

```
END IF
STOP
END
```

PROGRAM TESTING

You can use dummy subroutines to test the operation of the menu structure.

6.10 DOUBLE PRECISION INPUT AND OUTPUT

Problems that require a high degree of accuracy or involve a large number of repetitive calculations may require DOUBLE PRECISION variable types and double precision operations. Some library subprograms require the use of DOUBLE PRECISION, while others offer the option. If a variable has been declared as DOUBLE PRECISION, an F specification can be used to input a value for the variable. Scientific notation or double precision notation can be used with the input value. For example, under the specification F14.0, all three of the following entries in the field would result in the same value being stored in the computer.

```
398.67898245
.39867898245E3
398.67898245D0
```

You can output double precision values through either F or E specifications, or through use of the special D specification,

$$aDn.m$$

which works the same as the E specification, but puts a D in place of the E. If a D16.7 is used to output the value shown above, the output would be

```
0.3986790D+03
```

with three blanks placed at the beginning to fill up the field width of 16.

SUMMARY

In Chapter 6 you have learned about formatted input and output and the DATA statement. Formatted input will add structure that assists in data entry and editing. Formatted output can help you organize output so it is easier to read and interpret. The DATA statement is a useful method for initializing variable values at the beginning of program execution. Formatted and unformatted input and output can be combined by the programmer to add flexibility to applications programs.

REFERENCES

Borse, G. J., *FORTRAN 77 and Numerical Analysis for Engineers*, PWS Publishers, Boston, 1985.

Ellis, T. M. R., *A Structured Approach to FORTRAN 77 Programming*, Addison-Wesley, London, 1982, Chapter 6.

Etter, D. M., *Structured FORTRAN 77 for Engineers and Scientists*, 2nd. ed., The Benjamin/Cummings Publishing Company, Inc., Menlo Park, CA, 1987, Chapters 4, 8, and 9.

EXERCISES

6.1 GENERAL NOTES

Except where noted, output for all exercises should be formatted. Be sure to output units with your results.

Whenever possible, incorporate previously developed subprograms into your exercise solutions. You should modify previously developed subprograms to include formatted input or output, as appropriate.

You may find it useful to use FORTRAN coding forms to assist in your design of output formats. Make sure that the order of each output or input variable list corresponds to the order of specifications in the referenced FORMAT statement. Be sure not to extend long FORMAT statements beyond column 72 of the statement record. Be sure to allocate sufficient field width to your formats to accommodate the number of places to be output.

6.2 MODIFICATION OF EXAMPLES

6.2.1 Example 6.6

1. Modify the program of Example 6.6 for unformatted input of the array. Use comments in your program to document how to develop the data file.
2. Add error trapping for the file name input and the OPEN statement for Example 6.6.
3. Add variable formats to the program of Example 6.6. Input the variable formats through the main program, and pass them over to the subprograms.

6.2.2 Example 6.7

Modify the program of Example 6.7 so that input can be either from the screen or from a saved file. You can assume that the saved file will be unformatted.

6.2.3 Example 6.8

Develop the editing procedure of Example 6.8 into a subprogram, and modify the main program to accommodate editing of either the coefficient matrix or the right-hand-side vector.

6.2.4 Example 6.9

Develop the subprograms necessary to implement the program of Example 6.9. Test the program.

6.3 GENERAL SYNTAX AND STATEMENT STRUCTURE

6.3.1 Development of Input and Output Format Specifications

*1. Develop a set of READ and FORMAT statements and the corresponding data records that would store the following values in the indicated variable locations:

$$X = 65.739 \quad Y = 0.3578 \quad Z = 3.00345 \quad I = 30$$
$$J = 650 \quad K = 1000 \quad A = 4.5E\text{-}05 \quad C = 656.789$$

2. The following data are stored in a file as indicated:

```
0800    45.0    29.92
1000    49.5    29.95
1200    50.6    29.96
1400    54.3    29.98
1600    55.4    29.94
```

The first column of data is the time of day, in military format. The second column is the temperature in degrees Fahrenheit. The third column is the barometric pressure in inches of mercury. Develop a program that will read these values into the computer and output them in a tabular form with appropriate headings. Place a solid horizontal line between the headings and the columns of values, and a second horizontal line under the bottom of the columns. The average temperature and barometric pressure should appear below the appropriate columns.

6.3.2 Determination of Values Read in for Given FORMAT Specification

1. Four lines of a data file are shown below. It is assumed that all columns to the right of those given are blank. The letter b indicates a blank space.

```
bbb395bb6785b34761bbbb22
bb167bbbb218bb238b165461
b1bb65bbb616bb214bbb6523
12345bbbbb6111bbb238bb543
```

What values would be read in for the variables given in each of the following READ statements? Each set of READ statements starts at the beginning of the data file.

```
a.    READ(5,60) X,Y,Z,A
   60 FORMAT(2F6.1)
      READ(5,70) I,J,K
   70 FORMAT(3I5)

b.    READ(5,80) A,B,C,D
   80 FORMAT(3F8.2)

c.    READ(5,90) I,J,K
   90 FORMAT(10I6)
      READ(5,90) L,M,N
```

*2. Given the following data records and input statements, state what values would be stored for the various variables. Each set of READ statements starts at the beginning of the data set.

```
bbb385bb66589bbb66781bbb2233
blbb3bb4bbbb5
b6.789bbb5E06bbbbb4.39E-02
12345678998765432111123456789
```

a.   READ (5,100) A,B,C,D
   100 FORMAT(F6.2,F7.1,F8.3,F7.5)
        READ (5,100) E,F

b.   READ (5,200) A,B,C,D
   200 FORMAT(F6.3)

c.   READ (5,300) I,J,K
   300 FORMAT(2I4,/,I2)

d.   READ (5,400) A,B
   400 FORMAT(F6.4)
        READ (5,500) C,D,E
   500 FORMAT(3F9.0)
        READ (5,500) F,G

e.   READ (5,600) A
        READ (5,600) B
        READ (5,600) C
        READ (5,600) D
   600 FORMAT(2F10.1)

f.   READ (5,200) I,J,K,L
   200 FORMAT(I6)

3. Given the following lines of data, what values would be read in for the indicated variables using the combination of READ and FORMAT statements shown?

```
bbbb678bbb1234bbbbb38bbbbbbb4
bb235bb8900bbb-3567b6004
bb1.E23bbb3.56E10
```

a.   READ(5,45) A,B,C
   45 FORMAT(F10.2)

b.   READ(5,50) A,B,C,I,J,K
   50 FORMAT(3F8.4,3I8)

c.   READ(5,60) I,J,K
   60 FORMAT(2I5)
        READ(5,70) X,Y
   70 FORMAT(2F9.3)

*4. What values would be assigned to each variable using the data file given below and the following statements?

```
bbbb5bbbb3
b3.53b6.98b4.00b5.55
b4.98b6.00b5.15b4.30
b3.19b6.50b1.12b1.56
b2.15b6.38b8.98b9.98
b1.00b0.98b6.15b0.69
501bb635bb138bb469bb
```

```
      REAL A(10,10)
      READ (5,100) M,N
100   FORMAT (2I5)
      DO 50 I=1,N
50    READ (5,200) (A(I,J),J=1,M)
200   FORMAT (3F5.3)
```

6.3.3 Form of Output for Given FORMAT Specification

1. The value 345.8756 has been assigned to variable X. What would appear in the output record using each of the following FORMAT statements? Use b to indicate blank spaces.

```
WRITE (6,30) X
```

a. 30 FORMAT(1X,3F10.7)

b. 30 FORMAT(1X,F6.2)

c. 30 FORMAT(1X,F7.5)

d. 30 FORMAT(1X,E13.5)

e. 30 FORMAT(1X,F10.4)

*2. The following values are in the computer for the indicated variables: X=3.56789, Y=245.2837, Z=9987776., I=711, J=12, K=50. Indicate what the output lines would look like for the following combinations of WRITE and FORMAT statements. Indicate blank spaces by using b.

a.
```
      WRITE(6,100) X,Y,Z
100   FORMAT(1X,3(F10.2,3X))
```

b.
```
      WRITE(6,200) X,Y,Z,I,J,K
200   FORMAT(1X,3F8.2,//,1X,3I5)
```

c.
```
      WRITE(6,300) X,Y,Z
300   FORMAT(1X,E11.4)
```

d.
```
      WRITE(6,200) X,Y,I,J,Z,X
200   FORMAT(1X,2F8.2,/,1X,2I5)
```

3. Given the following input data record:

bbbb7986bbbb3.597bbbbb67bbbb98bbbbbbbbb45

a. What values would be assigned to X, Y, I, J, and K by the following statements?

```
      READ(5,100) X,Y,I,J,K
100   FORMAT(F8.2,F10.2,4X,I3,I5,7X,I3)
```

b. What would be the appearance of the output record if the values read in are output according to the following FORMAT specification?

```
      WRITE(6,200) X,Y,I
200   FORMAT(1X,F10.3,F8.2,I8)
```

6.3.4 Determination of Number of Lines of Input or Output for Given FORMAT–Variable List Combinations

How many lines of data would be read in or printed out for the following combinations of READ or WRITE and FORMAT statements?

a.   `DO 20 I=1,8`
   `20 READ(5,100) (A(I,J),J=1,10)`

   `100 FORMAT(10F8.0)`
   `100 FORMAT(8F8.0)`
   `100 FORMAT(4F8.0)`
   `100 FORMAT(15F5.0)`
   `100 FORMAT(5F16.0)`

*b.   `READ(5,100) (N(I),I=1,40),M,L,K`
   `100 FORMAT(14I5)`

c.   `DO 30 I=1,8`
   `30 WRITE(6,200) (A(I,J),J=1,10)`

   `200 FORMAT(1X,5F10.0)`
   `200 FORMAT(1X,8F8.0)`
   `200 FORMAT(1X,10F8.0)`
   `200 FORMAT(1X,15F5.0)`
   `200 FORMAT(1X,F5.0)`

*d.   `DO 50 I=1,30,2`
   `50 WRITE(6,200) X(I)`
   `200 FORMAT(1X,12F10.4)`

6.3.5 Input and Output of Arrays and File Handling

1. Develop a program to input a matrix of dimension up to 10×10 into an array using formatted input. Output the array to an unformatted output file. Also output the array to the terminal, with labels, using formatted output. Reinitialize all the array elements to zero. Output the reinitialized array to confirm that it is all zeros. Rewind the output file and read the matrix values back into the array, and again output the array to the terminal to confirm that the matrix has been read back in properly.
2. Develop a program to read in an $n \times m$ matrix and output its transpose without altering the original matrix. Demonstrate that the original matrix has not been transposed.

6.3.6 Data Statements

1. Develop a data statement that will input values for the area of circles of one, two, three, and four foot radius and store the values in variable locations A1, A2, A3, and A4.
*2. What values would be assigned to each element of the following array by the given data statement?

   `REAL B(2,4)`
   `DATA B/2.5,3.,4.2,5.,6.6,7.5,8.1,9./`

*3. Write a single data statement that will initialize all values of a 20×30 array, $X(I,J)$, to 10.0; set $A(1) = 6.5$, $A(2) = 8.2$, and $A(3) = 9.2$ if A is dimensioned for up to 10 values; and set $L = 30$ and $Q = 22.63$.

4. What values or character strings would be assigned to each of the indicated variables by the following data statement?

```
CHARACTER B*4,N(3)*1
REAL X(10),A
DATA N,X,A,B/1H*,'=',1H',10*0.0,10.589,'TIME'/
```

5. What would be output by the following statements?

```
CHARACTER NAME(10)*3
DATA NAME/3HMAC,3HBUR,3HDON,3HGER,3HALD,3HKIN,3H'S ,3HG  ,2*3H  /
WRITE(6,100) (NAME(I),I=1,10,2)
100 FORMAT(1X,10A3)
```

6.3.7 Design of Output

Develop FORMAT, WRITE, and DO statements that would provide the following output. Assume that all variables have additional significant figures associated with them in the computer (with the exception of the member number). Variables AREA, Ic, and FORCE are dimensioned to 10, but only four values will be printed out. The member number is not dimensioned.

MEMBER NUMBER	CROSS-SECTIONAL AREA (IN**2)	MOMENT OF INERTIA (IN**4)	FORCE (LBS)
1	1.35	100.4	3000.
2	2.48	500.6	-4090.
3	3.68	368.1	2500.
4	4.21	426.5	-4600.

*6.3.8 Identification of Errors in Formatted Input and Output Specifications

Identify any errors in the following FORMAT specifications and make corrections:

a. Input

```
(F6.2F9.3,/)
(3X,F6.2,A5,I6)
('Input length and width',F6.2,F6.4)
(25F5.1)
```

b. Output

```
(4X,Maximum Length = ,F8.3, ft.)
(3I4,F6.2)
(5F7.2,3F(8.2),I5)
```

c.
```
    READ(5,100) (A(I),I=1,25)
100 FORMAT(25F4.0)
```

6.3.9 Changing Selected Characters Within FORMAT Statements

Some older compilers used * as a delimiter for literal output in FORMAT statements. The standard is now ', and most compilers will not accept the *. Assume that you have an older program that uses the * as a delimiter. Develop a program that will search through your FORTRAN program file and change all * in FORMAT statements to ', but will not change * in other parts of the program. You will also have to be able to identify continuation lines for FORMAT statements.

6.4 CHEMICAL ENGINEERING

6.4.1 [ALL] Ideal Gas Law

Develop formatted output for Exercise 5.4.1 or 4.4.1. Include the results of all three models in one table.

6.4.2 [ALL] Chemical Kinetics

Develop formatted output for Exercise 5.4.2 or 4.4.2.

6.4.3 [ALL] Depth of Fluidized-Bed Reactor

Develop formatted output for Exercise 5.4.3 or 4.4.3.

6.4.4 [ALL] Process Design, Gas Separation

Develop formatted output for Exercise 5.4.4 or 4.4.4.

6.4.5 [ALL] Dissolved Oxygen Concentration in Stream

Add formatted output to the programs of Exercises 4.4.5 and 5.4.5. Include a table that gives distance downstream, time of travel, dissolved oxygen deficit, and dissolved oxygen concentration in the stream.

6.5 CIVIL ENGINEERING

6.5.1 [ALL] Equilibrium-Truss Analysis

1. Use formatted output for the data and reactions of Exercise 5.5.1.
2. Modify the program of Exercise 5.5.1 so that the reactions will be stored in a vector with the same form as the right-hand-side vector of Example 6.2. In other words, each element of the vector would correspond to either an X or Y reaction at a particular joint. Any joints that do not have reactions associated with them would have entries of zero. Develop a second vector that would contain the components of the applied loads in the same order as the reaction vector. Your program should then add these two vectors together to produce the right-hand-side vector. Use the data of Exercise 4.5.1 to check your program. Your program

should generate the same right-hand-side vector that was input for Example 6.2 and the other truss analysis examples. Use formatted output to output all three vectors.

6.5.2 [ALL] Open Channel Flow Analysis

1. Develop a program that will combine the trial-and-error solution for a depth of flow in a trapezoidal open channel (Exercise 4.5.2) and the choice of shape to use (Exercise 5.5.2). Use the data of Exercise 5.5.2 to test your program. Use an initial estimate of depth somewhat different from the known depth to test the convergence ability of the algorithm.
2. Engineers often estimate the flow rate in a complex cross-sectional area by dividing the area into simpler shapes, estimating the flow rate through each using Manning's equation, and adding the results together to get an estimate of the total flow. This method is used to estimate the flow rate in rivers when the depth is larger in the main channel but some of the flow is going through the floodplain adjacent to the channel. Develop a program that will accommodate up to five different subareas. Use the subprograms of Exercise 5.5.2 where appropriate. Output should be a table giving velocity, area, and flow rate through each subarea, and the totals for area and flow rate. Use your program to estimate the flow rate through a center rectangular area that is 500 ft wide by 20 ft deep, a left overbank section 600 ft wide by 1 ft deep, and a right overbank section 400 ft wide and 2 ft deep. Use the actual wetted perimeter in computing the hydraulic radius for each section.

6.5.3 [ALL] Dam Stability Analysis

Develop formatted output for the dam stability problem of Exercise 5.5.3. Add analysis of tipping to your program. For the dam to be stable against tipping, the normal force F_n must act somewhere within the bottom portion of the dam. In order to prevent excessive stress in the dam, it is customary to limit the location of the normal force to somewhere in the middle third of the bottom of the dam. You can find the magnitude of F_n from the friction stability analysis and use moments to find its point of application. It is convenient to take moments around one of the bottom corners of the dam. Your output should identify whether the dam is stable against slipping and/or stable against tipping. If it is stable against tipping, you must also identify whether the normal force falls within the desired middle third of the bottom.

6.5.4 [ALL] Force on Submerged Gate

Add formatted output to Exercise 5.5.4. Also develop a table of values of h_c, F, and h_p vs. h, the height of water above the bottom of the gate. You will have to add a new section to your program to account for the situation where the depth of water is less than the height of the gate. Test your program with the gate of Exercise 3.5.4, using depths of water of zero to 20 ft in increments of 0.5 ft.

6.5.5 [ALL] Volume of Excavation

Develop tabular, formatted output for the program of Exercise 5.5.5. One table should contain information on the cross sections at the stations, to include the station, the top width, the top elevation, the bottom width, and the bottom elevation. A second table should contain data on the incremental volumes of the trench, to include the first station of the section, the second station of the section, the length of the section, and the incremental volume of the section. Total values for length and volume should be included at the bottom of the table, under the appropriate columns.

6.5.6 [4,8] Addition of Two Hydrographs

Use formatted output to make the output of Exercise 4.5.8 more readable. Include variable labels in the program.

6.5.7 [7,8] Estimation of Flow in a Channel of Irregular Cross Section

Exercise 8.4.6 in the *Instructors Manual* describes the method used to estimate flow in channels with irregular cross-section when sufficient information can be gathered at one point in the channel. Develop a FORTRAN program to estimate the flow for the data given in Exercise 8.4.6. Use formatted output to produce a table similar to the table given in Exercise 8.4.6.

6.6 DATA ANALYSIS AND STATISTICS

6.6.1 [ALL] Histogram

Develop formatted output for the program of Exercise 5.6.1

6.6.2 [ALL] Sorting

Develop formatted output for any of the programs of Exercise 5.6.2.

6.6.3 [4,5,7] Sampling and Distribution of Means

Develop formatted output for Exercise 5.6.5.

6.7 ECONOMIC ANALYSIS

6.7.1 [ALL] Economic Formulas

1. Develop a program that will accept a series of costs and returns of different types (P, A, or F). The program should use these data to compute the net present worth, NPW. NPW is the difference between the present worth of returns and the present worth of costs. The costs may be treated as negative returns. Use the FUNCTION subprogram of Exercise 5.7.1 to accomplish the economic computations. Use formatted output.
2. Modify the Program of Exercise 6.7.1 (1) so that either net present worth or net uniform annual return can be used to evaluate the economic feasibility of a project.

6.7.2 [ALL] Nonuniform Series of Payments

Develop formatted output for the programs of Exercise 5.7.2

6.7.3 [ALL] Multiple Interest Rates or Number of Compounding Periods

Develop formatted output for the program of Exercise 4.7.3.

6.7.4 [ALL] Internal Rate of Return

Develop formatted output for Exercise 5.7.4.

6.7.5 [ALL] Mortgage Payments

Develop variable identification input for Exercise 5.7.5 that will identify the type of loan and repayment schedule. Use formatted output. Develop a table that gives the payment number, amount paid on principal, amount paid in interest, and remaining principal. Also output summary statistics on the total amount repaid and the total amount of interest paid.

6.8 ELECTRICAL ENGINEERING

6.8.1 [ALL] Diode Problem

Develop an interactive design program that can be used to estimate either the current, the applied voltage, or the operating temperature of a junction diode. Use the subprograms of Exercise 5.8.1 in your program, and use formatted output.

6.8.2 [ALL] Capacitance and Resistance

Develop formatted output for Exercise 5.8.2.

6.8.3 [ALL] Signal Processing Circuits

Develop formatted output for Exercise 5.8.3.

6.9 ENGINEERING MATHEMATICS

6.9.1 [ALL] Determinant and Cramer's Rule

Develop formatted output for Exercise 4.9.1.

6.9.2 [ALL] Vector Cross Product

Develop formatted output for the program of Exercise 5.9.2.

6.9.3 [ALL] Centroid and Moment of Inertia

1. Develop a subroutine subprogram that will find a centroidal axis for a compound area and the moment of inertia of the compound area about the centroidal axis. The subroutine should call the functions developed in Exercise 5.9.3. Test your program by finding the vertical centroidal axis and moment of inertia for the area used in Exercise 5.9.3. Use formatted output.

2. Combine and modify the programs of Exercises 4.9.3 and 5.9.3 to develop a program that will determine the distance to the centroid and the moment of inertia of a compound area with respect to an axis that is parallel to the centroidal axes of the subshapes of the compound area.

6.9.4 [ALL] Population Growth

Develop formatted output for Exercise 4.9.4.

6.9.5 [ALL] Vertical and Horizontal Curves

Develop formatted output for Exercise 5.9.5.

6.9.6 [2,4,5] Table of Squares and Cubes

Develop formatted output for the programs of Exercise 5.9.6.

6.9.7 [4,5] Volume and Surface Area of Box, Sphere, or Cone

Develop formatted output for the programs of Exercise 5.9.11.

6.10 INDUSTRIAL ENGINEERING

6.10.1 [ALL] Project Management

Develop a table of data for the activities of a project. Each row of the table should correspond to a particular activity and should include the beginning and ending node of the activity; the activity duration; the EST, EFT, LST, and LFT; the slack; and an indicator to show whether or not the activity is on the critical path. An * should be used to identify critical path activities.

6.10.2 [ALL] Quality Control

Develop formatted output for Exercise 4.10.2 or 5.10.2. Develop a compact table to output the defect matrix.

6.10.3 [ALL] Managerial Decision Making

Develop formatted output for Exercise 5.10.3.

6.10.4 [ALL] Queuing (Wait Line) Theory

Develop formatted output for Exercise 4.10.4 or 5.10.4.

6.11 MECHANICAL AND AEROSPACE ENGINEERING

6.11.1 [ALL] Shear and Bending Moment

Develop formatted output for Exercise 4.11.1. Output tables giving the individual shear and bending moment contributions of a point and a uniform load, as well as the total shear and bending moment, as a function of the distance along the beam.

6.11.2 [ALL] Crank Assembly Analysis

Develop formatted tabular output for Exercise 5.11.2.

6.11.3 [ALL] Friction

Incorporate formatted output into the programs of Exercises 4.11.3 and 5.11.3.

6.11.4 [ALL] Variation of Atmospheric Pressure with Altitude

Modify the program of Exercise 5.11.4 to calculate the atmospheric pressure at any elevation by both methods. Use formatted output to develop a table for elevations from zero to 50,000 ft in 5,000-ft increments. Include columns in your table for the elevation, the temperature assuming a constant lapse rate, and the atmospheric pressures using both methods.

6.11.5 [ALL] Aerospace Physics

1. Develop formatted output for Exercise 5.11.5.
2. Section 4.3 of Appendix G presents mathematical models that describe the trajectories of rockets as they burn fuel. Two types of rockets, constant thrust and constant acceleration, are modeled. Develop a program with formatted output to compare the trajectory position and velocity for rockets of each type with the same initial conditions. The launch angle is 60°, exhaust velocity is 4,000 m/s, original mass is 50,000 kg, fuel to total mass ratio is 0.55, and the burn rate is 750 kg/s. Model the trajectory until time of burnout.

6.11.6 [ALL] Energy Loss in Circular Pipeline

Develop a program that will compare the results of the three equations for estimating the friction factor for energy loss in a circular pipeline against a given friction factor from the Moody diagram. The Moody diagram is the graphical relationship for estimating f as a function of Reynolds number and ε/D. The Chen, Swamee–Jain, and Colebrook equations are all approximations for the Moody diagram. Your program should compute the percent error between the Moody diagram f value and the estimated f value using each equation. Use the subprograms of Exercise 5.11.6 in developing your program. Formatted output should be used. Test your program for $f_{Moody} = 0.020$, $Re = 2.5 \times 10^5$, and $\varepsilon/D = 0.0008$; and also $f_{Moody} = 0.032$, $Re = 1.4 \times 10^4$, and $\varepsilon/D = 0.002$.

6.11.7 [ALL] Torsion

Develop formatted output for Exercise 5.11.7.

6.12 NUMERICAL METHODS

6.12.1 [ALL] Interpolation

Linear and quadratic interpolation have been used in previous exercises to estimate a value for a dependent variable y for a given value of the independent variable x between two known values for x and y. Higher order polynomial functions can also be used to generate interpolation estimates. Generally the higher order polynomials will produce more accurate interpolation estimates for functions that vary significantly from linear. One method of higher order interpolation is called the divided difference method; its application to develop a cubic interpolation polynomial is illustrated below. Note that the general function for generating the divided differences $[f_i(x_j)]$ could be applied to any order interpolating polynomial. This function is used to develop the divided difference formulas given for the problem. Develop a subprogram that implements the divided difference method to interpolate between two values of x and y using a cubic polynomial. Test your subprogram using the data for points 1 through 4 of Exercise 2.11.1. Also estimate Y for $X = 3.5$ for the following data:

X	-2	1	3	4
Y	-15	3	25	57

Divided Difference Scheme for Cubic Interpolation

$$y(x) = f_0(x_1) + (x - x_1)f_1(x_2) + (x - x_1)(x - x_2)f_2(x_3) + (x - x_1)(x - x_2)(x - x_3)f_3(x_4)$$

in which $f_0(x_1) = y_1$ and in general

$$f_i(x_j) = \frac{f_{i-1}(x_j) - f_{i-1}(x_i)}{x_j - x_i}$$

The following implementation of the general formula can be used for cubic interpolation:

$$f_0(x_2) = y_2 \qquad f_0(x_3) = y_3 \qquad f_0(x_4) = y_4$$

$$f_1(x_2) = \frac{f_0(x_2) - f_0(x_1)}{x_2 - x_1}$$

$$f_2(x_3) = \frac{f_1(x_3) - f_1(x_2)}{x_3 - x_2} \qquad f_1(x_3) = \frac{f_0(x_3) - f_0(x_1)}{x_3 - x_1} \qquad f_1(x_4) = \frac{f_0(x_4) - f_0(x_1)}{x_4 - x_1}$$

$$f_3(x_4) = \frac{f_2(x_4) - f_2(x_3)}{x_4 - x_3} \qquad f_2(x_4) = \frac{f_1(x_4) - f_1(x_2)}{x_4 - x_2}$$

6.12.2 [ALL] Solution of Polynomial Equation

Develop formatted output for Exercise 5.12.2.

6.12.3 [4,5,7,8,9] Saddle Point of a Hyperbolic Paraboloid

Modify the program of Exercise 5.12.6 to make your output look similar to the following example. Question marks indicate where the computer should supply values.

```
Your Name
FINDING THE SADDLE POINT OF A HYPERBOLIC PARABOLOID
The equation is Z = ?X**2 - ?Y**2 + ?X + ?Y + ?
The initial point is (??.??)
The initial increment is ???. The final increment is ???.

Preliminary Searches
---------------------

        X               Y           f(X,Y)
     -------          -------        ------

      ?.??             ?.??           ?.??
      ?.??             ?.??           ?.??
      ?.??             ?.??           ?.??
      ?.??             ?.??           ?.??
      ?.??             ?.??           ?.??
```

```
Solution for Saddle Point with accuracy of ??

    X = ?????
    Y = ?????
    Z = ?????
```

6.12.4 [4,5,7,8,9] Polynomial Surface

Modify the program of Exercise 5.12.7 to include formatted output similar to the type of output used for Exercise 6.12.3. Include variable problem identification, and output whether a minimum or maximum point is found.

6.13 PROBABILITY AND SIMULATION

6.13.1 [ALL] Simulation of Two Dice

Develop a general subroutine to compare any distribution found through simulation with the corresponding theoretical frequency distribution. A maximum of 25 frequency intervals can be assumed. Any labels should be input and output using character variables. The theoretical distribution should be input. Fractional distributions should be used. Use your program from Exercise 5.13.1 to test your subroutine.

6.13.2 [ALL] Normal Probability Distribution

Develop a formatted table of probabilities for a mean value of 5; and standard deviations of 1, 2, and 3, with x varying from zero to 10 in increments of 1. Use a data statement to input the given probabilities used in the previous problems.

CHAPTER SEVEN

FORTRAN PROGRAMMING: WORKING WITH ARRAYS AND MATRICES

7.1 INTRODUCTION

In FORTRAN you can use arrays of up to seven dimensions. Figure 7.1 gives a schematic representation of the storage structure for arrays of up to four dimensions. Most problems that you will undertake will not require variables with more than three subscripts. Therefore, they can be represented by arrays of three or fewer dimensions. If you have to use higher order arrays you can follow the same procedures as for three-dimensional arrays, but you will have to keep track of a larger number of subscript indexes. You will also have to use more nested DO loops to increment the indexes.

You have studied several examples that used one-, two-, or three-dimensional arrays. A one-dimensional array can store the components of a vector. Examples have applied vector operations, such as addition and finding the length of a vector, to these arrays. You have also seen programs for sorting data stored in a one-dimensional array into a desired order. Other programs, such as the parabolic bridge arch application, have stored program data in one-dimensional arrays. You have seen several examples that input and output two-dimensional arrays. An example in Chapter 6 used a three-dimensional array to store character grids for the letters of the alphabet.

This chapter introduces you to more in-depth applications of two-dimensional arrays, including matrix operations. You will also develop additional applications of three-dimensional arrays.

7.2 APPLICATIONS OF TWO-, THREE-, AND HIGHER-DIMENSIONAL ARRAYS

This section contains general descriptions of the use of multidimensional arrays in engineering. It will be followed by development of specific examples of multidimensional array and matrix applications.

Many engineering problems require multidimensional arrays to properly represent the variables and variable operations of the problem. You can determine the required dimension of the array from the number of subscripts used in the mathematical depiction of the problem. For example, if x_i is an element of X, and a_{ij} is an

FIGURE 7.1   Representation of one-, two-, three-, and four-dimensional arrays.

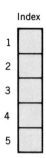

(*a*) One-dimensional array, DIMENSION A(5).

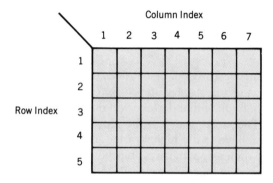

(*b*) Two-dimensional array, DIMENSION A(5,7).

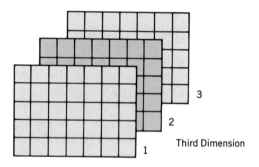

(*c*) Three-dimensional array, DIMENSION A(5,7,3).

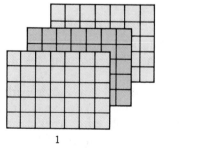
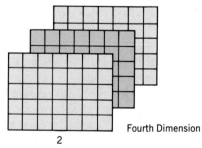

(*d*) Four-dimensional array, DIMENSION A(5,7,3,2).

element of A, then

$$x_i = \sum_{j=1}^{n} a_{ij}$$ (7.1)

would require a one-dimensional array for X and a two-dimensional array for A. If you are following the usual convention, x_i represents the summation of the values in the ith row of A.

7.2.1 Database Management

Engineers have to deal with databases almost every day of their careers. Any large body of data from which you extract particular data to assist in an analysis or design problem represents a database. Examples of databases are the steel design tables of the *Manual of Steel Construction*, the periodic table of the elements, and the CRC *Handbook of Chemistry and Physics*. Engineers may also generate their own databases through field or laboratory data acquisition programs. Meteorological data, water quantity and quality data, and chemical process data are often gathered and analyzed as part of engineering studies.

A database must be organized in some logical format that allows the user to quickly and easily find the desired data values. A common method for accomplishing this is to index the database so that key data values can be used to locate the target data quickly. For example, the steel tables are indexed by the size and weight per foot of the various members represented. A water quality database might be indexed by year, day, and data gathering station.

Operations on databases may include sorting the different entries of the database according to numerical or alphabetical order of an index variable, searching a database for particular data values, preparing summary reports of the data, or using the data values in the database to compute values for additional parameters. All these operations require that a particular data value can be identified and accessed in an efficient manner. When you are dealing with databases in the FORTRAN environment, the quickest and easiest way of organizing and accessing data is through multidimensional arrays. Example 7.1 will illustrate database sorting and searching using FORTRAN programs.

7.2.2 Matrices

By definition, a matrix is a rectangular, two-dimensional array of numbers arranged in rows and columns. Engineers use matrices to represent systems of equations. A rectangular array is used to represent the coefficient matrix of the system of equations. Vectors are a special type of matrix, with only one row or column. Column vectors are used to represent the unknowns and the right-hand-side vector of systems of equations. The complexity of the system of equations depicted can vary from a system of linear equations to a system of nonlinear partial differential equations. Solution of linear equations will require the application of linear algebra operations. Finite difference or finite element techniques may have to be applied to find a solution for a system of nonlinear partial differential equations. In this text, the word matrix will be used to define an array representing systems of linear equations, whereas other two-dimensional problem representations will be referred to as two-dimensional arrays. The solution of linear equations using matrix representation and linear algebra will be examined in some detail in various chapters of the text. Appendix F contains the necessary background for understanding linear algebra and matrix notation.

Example 7.2 will develop a subprogram for accomplishing matrix multiplication. This is one of the linear algebra operations necessary to solve a system of equations. Another key linear algebra operation, the inversion of the coefficient matrix, will be illustrated in Example 7.3. Then Example 7.4 will use both matrix inversion and matrix multiplication to solve for the forces in a determinate planar truss.

7.2.3 Other Multidimensional Arrays in Engineering Applications

There are many engineering problems that use variables that vary with one or more discrete indexes. Each index relates the value of the variable to a corresponding value of another variable. For example, flow versus time in a fluid system can be represented by one-dimensional arrays. FLOW(I) would store the ith value of FLOW, which would correspond with the ith value of time, TIME(I). If it is necessary to store values of flow vs. time and location, a two-dimensional array would be necessary. FLOW(I,J) would store the value of FLOW corresponding to the ith value of TIME and the jth value of location. The location may be a physical location within the system, such as a distance from a reference point. In that case you would also have an array to store the distances. In other instances, the location may be an index identifying a particular point in the system, such as a point where several pipes come together. For these cases it may not be necessary to have another array for storing the relative distances.

Let us assume that the flow situation depicted in the previous paragraph is in a chemical plant. The flow at any point in time and location may contain several chemical constituents. The total flow is the sum of the constituents. A three-dimensional array, FLOW(I,J,K), could be used to store the values of flow rate for the ith time increment, jth location, and kth constituent. A separate two-dimensional array could be used to store the total flow, TOTFLO(I,J), if it is necessary to have the total in memory.

If you want to use an array to store meteorological data that varies with location, year, and month, a three-dimensional array would be required. If the data were also time-dependent, a fourth dimension would be required. The array could be represented as

$$A(I,J,K,L)$$

with I = location index, J = year index, K = month index, and L = time step index. If the location were divided up into state and city, there would be a need for a fifth dimension. However, as the number of dimensions increases, you would have to be conscious of the number of storage locations that would be required for the array. In the above example, if you gathered data for 50 states, six representative cities in each state, 10 years of data, 12 months in the year, and 50 time steps within each month, the array to store the data would take up 1,800,000 storage locations. This would overtax all but the largest computers.

It is possible to store values for more than one variable in the same array, but the practice should be avoided. For example, the three-dimensional array $D(I,J,K)$, with I = position index, J = time step index, and K = flow or pollutant indicator index, could be used to hold both flow rate and quality data for a storm water pollution study. However, this would be confusing since the variable D is not descriptive of either the flow or pollutant data. The only way to know which data are in which position would be to keep track of the subscript identifiers. It would be better to use two or more separate two-dimensional arrays with descriptive variable names. Separating the arrays would not take up any more storage space in the computer.

Examples 7.5 through 7.7 illustrate several operations with two- and three-dimensional arrays, including addition of corresponding members of two or more arrays, addition across columns or down rows of an array, or arithmetic operations using corresponding elements of two or more arrays. You will readily recognize the differences in these operations from the linear algebra operations for matrices and vectors.

7.2.4 Graphics

Two-dimensional character arrays can be used to generate plots or other graphical output. Example 6.3 showed the input and output of block alphabet letters in 5×5 character arrays. This concept can be extended to a plotting grid of any convenient size. An 80×80 character array, with each element of the array one character long, provides an 80×80 plotting grid. One character can be placed in each plotting position. Different characters can be used to represent borders, axes, and function plots. Placement of the characters for the functions can be controlled by proportional expressions. These will determine which plotting point should represent a given function value.

Character plots will not have high resolution because of the limitation on the number of character spaces on the screen or printed page. However, they are useful for examining data or functional relationships on the screen or printed output device without having to transfer the data to a high-resolution plotting environment. They are also useful for teaching the concepts of computer graphics. Since the main thrust of this text is toward engineering problem solving, and since we will see that there are excellent modules for graphing engineering data in many of the generic applications packages, we will not take the time to develop X-Y plotting routines in FORTRAN.

7.3 DATABASE MANAGEMENT

Database management can develop into a career field for some engineers. It involves the storage, retrieval, and manipulation of large quantities of data. Most of us are frequent recipients of the results of database management. For example, if you are receiving mail that you did not request, you can be certain that someone has accessed your name and address from a customized database. Your name and academic records are also kept in a database that is maintained by your college or university. The database management system used by the school administration allows easy access to the data and generation of your grade reports and transcripts.

Most of you took the Scholastic Aptitude Test (SAT) in high school and answered some survey-type questions. Afterwards, colleges began to send you information that was customized to your interests, your geographic location, and how well you did on the exam. This was no accident. Your responses and test results were entered into a database that was used by the colleges you received information from. From the huge list of students who took the test nationwide, a college chooses to send literature to those students who are most likely to be interested in their programs. Searching and extracting data from the database can be accomplished at a small fraction of the cost of sending literature to all the students who took the exam.

In this text you will be introduced to the concepts of database management without going into the specifics of specialized database management programs. In this chapter you will learn two database concepts. You will learn how to sort information in a database according to set criteria. You will also search for and extract desired data from a database.

FORTRAN is not designed to be a database management language. There are languages and applications packages that are designed specifically for this purpose. However, in your FORTRAN programming, you will use many of the concepts of database management to manipulate information in arrays. For example, you may want to sort the information stored in one or more arrays based on the data stored in a particular set of variable locations. The sorting may be in numerical or alphabetical order. You may also use database management techniques to search for data and retrieve it from a file.

In Chapter 8 you will use the database capabilities of the software applications programs LOTUS 1–2-3 and ENABLE. Learning some database management concepts now will help you when you study these applications. It will also increase your ability to choose the proper software tool for specific applications.

Database files are made up of a series of records, each record representing a group of related data. For example, your name, address, and telephone number might make up a record for a telephone company database. The database file would be made up of the records for all of the customers of the telephone company. Another database of the telephone company might contain data on long distance calls. However, something in the long distance call file has to relate the calls to the customer. The logical choice is the telephone number. A database management system that would use these databases would be called a relational database system. The telephone number would be the data value that would relate the long distance calls to the name and address of the appropriate customer.

Database records are somewhat different from FORTRAN data records. As you know, a normal FORTRAN record can consist of up to 80 characters. Some FORTRAN records may contain all numerical data, while others may contain all character data. The organization of the records within the file is governed by the organization of the READ and FORMAT statements. In database files, the records can be much larger than 80 characters. The maximum size of a record is dependent on the database management system being used. For example, in the database module of ENABLE, a record can contain up to 254 fields, and each field can contain up to 254 characters. Therefore, a record can contain up to 64,516 characters. Also, the order of the records within the file does not matter in relational database management systems.

FORTRAN arrays do not lend themselves to storage of database records, because you cannot store both character and numerical data in the same array. It is possible to approximate a database structure by using multiple arrays and storing the data of one record on corresponding lines of the various arrays. However, you have to be sure to sort all rows of corresponding arrays for any sort operation, or the database records will be scrambled.

7.3.1 Sorting Data in Related Arrays

You will often want to sort the records of a database into a specific order. Example 4.12 developed a program to sort a one-dimensional array by the bubble sort method, with a stop when a pass is made with no switches. Example 7.1 will generalize this program to read and sort database records for production and defect rates at an industrial plant. Three database files will be used. The first file will contain the first name, middle initial, and last name of each employee, a unique numeric identification number for each employee, and the number of years of service of the employee. The identification number will be the variable that relates data in the three files. The second data file will contain the employee identification number, plus the monthly production rate for that employee for the year. The third data file will contain the employee identification number, plus the number of defective products reported for the employee each month for the year. The company does not contemplate having more than 25 employees in the particular division making this product. Personnel has reserved employee identification numbers 1001 through 1025 for this division.

7.3.2 Searching Databases

You will not always want to work with a whole database. Many times you will just want to extract information from a database. For example, for the data of the previous example, you might want to know how many workers produced more than a certain number of defective products for the year. You also might want to know the names of these employees. Some other types of data that you could extract from the database might be the production/defect history of a certain employee, or a productivity index for the employee. The productivity index might be the ratio of production to defective items, or the number of months a certain employee produced fewer than a certain number of defective parts. All these operations would require searching the database for the desired data and extracting the data for further processing. The data might be used to generate a report or to calculate additional needed information. The search could be conducted using a number of database files, relating them through a common variable. The search could also be accomplished using a sorted database.

The latter part of Example 7.1 will search the sorted industrial production database to extract information about a certain employee and to compute productivity information for that employee. The procedure could easily be extended to include a group of employees in the search or a group of employee identification numbers.

EXAMPLE 7.1 □ **Industrial Production Database**

STATEMENT OF PROBLEM

Develop a program that will sort the data of the employee database into alphabetical order. After sorting, provide the option of searching for the record of a particular employee. Extract the production and defect data for that employee, and calculate his or her total annual production, total annual number of defects, and the production/defect ratio.

(EXAMPLE 7.1 □ Industrial Production Database □ Continued)

ALGORITHM DEVELOPMENT

Input/Output Design

Although the maximum number of employees is 25, there may be fewer than that actually employed, so the number of employees should be input. The lowest ID number possible for this group should also be input. This will allow you to use the program for any group of employees. Data from the name, ID number, and years of service file can be input. The data will be stored in three separate arrays. After these three arrays are sorted into the proper order, the production and defect data can be input and stored in the same order as the alphabetized data. This will be accomplished using a pointer variable based on the new order of the identification numbers. Note that the order of data in the data files is not affected by the sorting.

The only additional information required for the search will be an inquiry input for the name of the target employee. You should include the option of searching for more than one employee as well as an error trapping option in case the user types an incorrect name.

Numerical Methods

Only the name array needs to be sorted. The other data will be rearranged based on the new order of the names. The bubble sort program of Example 4.12 will be modified into a subprogram that can be used to sort a one-dimensional character array. A second subprogram will be developed to rearrange the order of data in a numerical array to correspond with the alphabetized array. This will be used to rearrange the order of the data in the employee identification number and years of service arrays. This subprogram will be for rearranging integer data. Similar subprograms that would rearrange real or character data could be developed. The original order of the name array has to be preserved in a temporary array for the rearrangement subprogram to work. This can be accomplished in the main program before the sorting subprogram is called. Finally, the data from the production and defect files will be read in and placed in the proper order according to the employee identification numbers.

Computer Implementation

Figure 7.2 shows a flowchart for the main program and the subprogram for rearranging the numeric array according to alphabetical order. The first part of the program is straightforward. Data for names, identification numbers, and years of service are read into three separate arrays. Before the name array is sorted, a dummy array called TNAME is created. This contains the name data in the original order. The flowcharts of Figures 4.16 and 4.19 show the procedure for the bubble sort algorithm. For this application it has been modified into a subroutine named SORT that will sort character data into ascending order. Similar subroutines could be developed to sort real or integer data.

After the NAME array is sorted, the SWITCH subprogram is called twice to rearrange the ID and YEARS arrays into the same order. It accomplishes this by comparing the position of data in the TNAME and NAME arrays. First it stores the data to be arranged in an array called TEMP. Note that even though TEMP is a local array used only in the subprogram, it had to be declared in the main

FIGURE 7.2   Flowchart for database sorting and searching.

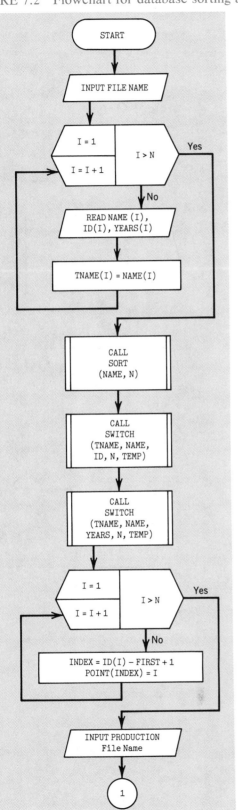
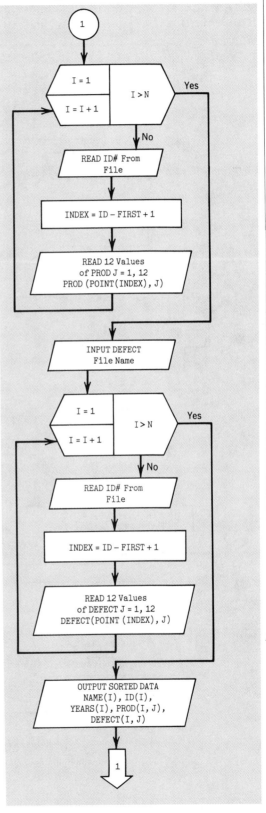

(a)

**(EXAMPLE 7.1   ☐   Industrial Production Database   ☐
Continued)**

FIGURE 7.2   (Continued)

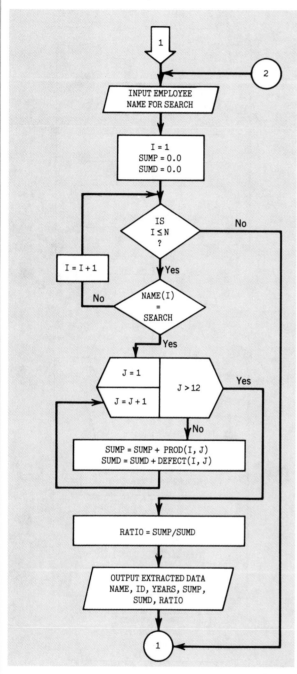
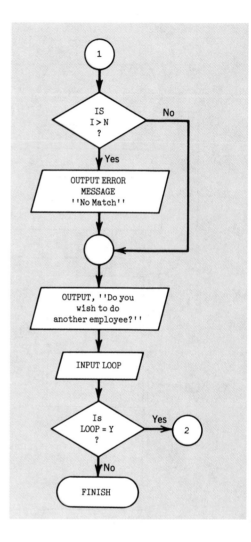

(*a*) Main program.

FIGURE 7.2 (Continued)

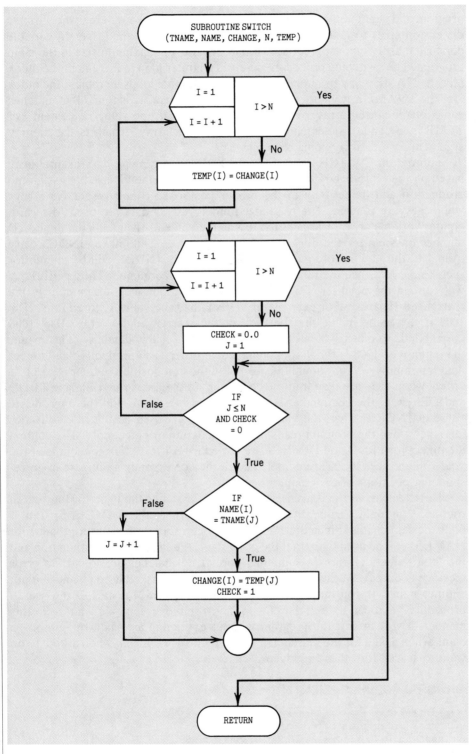

(*b*) Subroutine for switching values in an array.

(EXAMPLE 7.1   □   Industrial Production Database   □ Continued)

program and sent over with the arguments. This is because we are using variable dimension sizes to make the subprogram general. You cannot have a variable dimension array in a subprogram without declaring a maximum size in the main program. The switching procedure involves an outer loop to increment the index *i* for the NAME array from one to *n*. Inner conditional loops increment the index *j* for the TNAME array until the name in the original array is found that matches name$_i$ of the sorted array. When the match is found, the value in element *j* of TEMP is assigned to element *i* of the CHANGE array, which is the dummy argument for the array being rearranged.

Subprogram SWITCH shows one method of rearranging data in one array based on the rearranged order in another. Another method is used to assign the production and defect data to the proper positions in their respective arrays when they are read in. The rearranged employee ID data are used to form a pointer variable array. Variable INDEX is the value of ID$_i$ minus the lowest ID number possible for the division, FIRST, plus one. This makes INDEX vary from 1 to the number of employees in the division. INDEX will be unique for each employee since the identification number is also unique. Thus POINT$_{index}$ will contain the position of a particular employee identification number in the rearranged ID array. For example, if POINT$_8$ had the value 1 stored in it, ID# 1008 is the value in the first position of the rearranged ID array. ID# 1008 corresponds to index = 8 because FIRST = 1001 for this database. This means that employee 1008 is the first in alphabetical order. The production and defect data for employee 1008 should thus be put in the first position of their respective arrays when they are read into the computer. In the loops for reading in PROD and DEFECT, the ID number is read first and is used to find the appropriate value of INDEX. Then the value in POINT$_{index}$ is used as the row subscript identifier when the program reads in the production or defect data. The column identifiers for PROD and DEFECT are varied from 1 to 12 for the months of the year. When these loops are completed, all the corresponding data are in corresponding positions of the computer arrays.

The latter part of the flowchart of Figure 7.2*a* shows the logic for the search routine. You input a name, then the program searches the NAME array until it finds the same name. Since the database has been sorted, the subscript value for NAME(I) can be used to extract the other data. Finding the total production and the total defects requires a summing loop. If no matches are found, an error message is output. At the end of the program you are asked if you want to input another name. If you input *yes*, control shifts back to the start of the search procedure. You can extract data for as many employees as you want to in this manner. You can exit the program without extracting any data by entering a blank when asked for the search name, and then answering *no* when asked if you want to search for other employees.

PROGRAM DEVELOPMENT

```
****************************************************************
*  Industrial Production Data, Database Management            *
*  Developed by:  T. K. Jewell        January 1988            *
****************************************************************
C
C  Variable definitions
```

```
C  NAME = array for employees' names
C  TNAME = array for storage of original NAME array after NAME is sorted
C  ID = array for employee identification numbers
C  YEARS = array for employee years of service
C  PROD = array for employee production data
C  DEFECT = array for employee defect data
C  N = number of employees
C  FIRST = lowest employee number possible for this group
C  SIZE = maximum number of employees allowed in the database
C  POINT = pointer array to place production and defect data in the proper
C          position to agree with the alphabetized names
C  IDPT = employee ID # input to point to position for PROD and DEFECT data
C  CHECK = 0 before search match is made, 1 afterwards
C  SUMP = summing variable for employee production data
C  SUMD = summing variable for employee defect data
C  RATIO = total production/total defects for an employee
       INTEGER N,FIRST,FR,R,W,SIZE,INDEX,IDPT
       PARAMETER (SIZE=25,FR=4,R=5,W=6)
       REAL PROD(SIZE,12),DEFECT(SIZE,12),SUMP,SUMD,RATIO
       INTEGER ID(SIZE),YEARS(SIZE),POINT(SIZE),TEMP(SIZE)
       CHARACTER*30 NAME(SIZE),TNAME(SIZE),SEARCH
       CHARACTER  FILEID*14,LOOP*1
       PRINT *, 'Input employee data file name  '
       READ(R,5) FILEID
     5 FORMAT (A14)
C  Input employee database arrays and copy NAME into TNAME
       OPEN(UNIT=FR,FILE=FILEID,STATUS='OLD')
       READ(FR,'(2I5)') N,FIRST
       DO 100 I=1,N
          READ(FR,110) NAME(I),ID(I),YEARS(I)
          TNAME(I) = NAME(I)
   100 CONTINUE
   110 FORMAT(A30,2X,I4,2X,I2)
       CLOSE (UNIT=FR)
C
C  Sort the NAME array into alphabetical order
       CALL SORT(NAME,N)
C
C  Rearrange ID and YEARS to agree with the alphabetical order of NAME
       CALL SWITCH(TNAME,NAME,ID,N,TEMP)
       CALL SWITCH(TNAME,NAME,YEARS,N,TEMP)
C
C  Read in production and defect databases in the proper order
C  Generate pointing array
       DO 200 I=1,N
          INDEX = ID(I) - FIRST + 1
          POINT(INDEX) = I
   200 CONTINUE
C
C  Read in employee ID number followed by production and defect data
       PRINT *, 'Input production data file name  '
       READ(R,5) FILEID
C  Input production database arrays
       OPEN(UNIT=FR,FILE=FILEID,STATUS='OLD')
       DO 300 I=1,N
          READ (FR,'(I5)') IDPT
          INDEX = IDPT - FIRST + 1
          READ (FR,310) (PROD(POINT(INDEX),J),J=1,12)
   310    FORMAT(12F6.0)
   300 CONTINUE
       CLOSE (UNIT=FR)
```

**(EXAMPLE 7.1   □   Industrial Production Database   □
Continued)**

```
C
      PRINT *, 'Input defect data file name  '
      READ(R,5) FILEID
C  Input defect database arrays
      OPEN(UNIT=FR,FILE=FILEID,STATUS='OLD')
      DO 400 I=1,N
         READ (FR,'(I5)') IDPT
         INDEX = IDPT - FIRST + 1
         READ (FR,310) (DEFECT(POINT(INDEX),J),J=1,12)
  400 CONTINUE
C
C  Output of sorted data
      WRITE (W,500) N,FIRST
  500 FORMAT(1X, 'INDUSTRIAL PRODUCTION DATABASE',///,
     1      1X,'The number of employees is ',I5,/,
     2      1X,'The lowest employee ID number for this group is',I5)
      WRITE (W,510)
  510 FORMAT(///,1X,'SORTED EMPLOYEE NAMES',9X,'ID NUMBERS,',
     1          1X,'YEARS OF SERVICE',/)
      DO 600 I=1,N
      WRITE(W,520) NAME(I),ID(I),YEARS(I)
  520 FORMAT(1X,A30,I5,6X,I5)
  600 CONTINUE
      WRITE(W,530)
  530 FORMAT(//1X,'SORTED PRODUCTION DATA',/)
      WRITE(W,540)
  540 FORMAT(1X,'  ID   JAN.  FEB.  MAR.  APR.  MAY.  JUN.  JUL.  AUG.',
     1                ' SEP.  OCT.  NOV.   DEC.')
      DO 610 I=1,N
         WRITE(W,550) ID(I),(PROD(I,J),J=1,12)
  550    FORMAT(1X,I5,12F6.0)
  610 CONTINUE
      WRITE(W,560)
  560 FORMAT(//,1X,'SORTED DEFECT DATA',/)
      WRITE(W,540)
      DO 620 I=1,N
         WRITE(W,550) ID(I),(DEFECT(I,J),J=1,12)
  620 CONTINUE
      WRITE(W,541)
  541 FORMAT (//)
C
C  Top of loop for employee searches
C  Continuation is by inquiry at the end of the program
 1000 CONTINUE
C
C  Input employee name for search
      PRINT *, 'Input employee name for search'
      PRINT *, 'Example: Doe, John Q.'
      PRINT *
      READ (R,'(A30)') SEARCH
C
C  Start of search
      I = 1
      SUMP = 0.0
      SUMD = 0.0
C  WHILE (I.LE.N) is true, DO 900
 1100 IF (I.LE.N) THEN
```

```
C
C  Check for match of values between NAME(I) and SEARCH
         IF (NAME(I).EQ.SEARCH) THEN
            DO 950 J=1,12
               SUMP = SUMP + PROD(I,J)
               SUMD = SUMD + DEFECT(I,J)
 950        CONTINUE
            RATIO = SUMP/SUMD
C  Output extracted data
            WRITE(W,1200) NAME(I),ID(I),YEARS(I),SUMP,SUMD,RATIO
 1200       FORMAT(1X,'EMPLOYEE NAME:  ',A30,/
     1        1X,'IDENTIFICATION NUMBER:  ',I5,/
     2        1X,'YEARS OF SERVICE:  ',I5,/
     3        1X,'TOTAL PRODUCTION FOR YEAR:  ',F6.0,/
     4        1X,'TOTAL DEFECTS FOR YEAR:  ',F6.0,/
     5        1X,'RATIO OF PRODUCTION TO DEFECTS:  ',F6.1,/)
         ELSE
            I = I+1
            GO TO 1100
         END IF
      END IF
C 900 CONTINUE
C
      IF (I.GT.N) WRITE(W,1210)
 1210    FORMAT(1X,'Your input name did not match any names in the ',
     1           'database.'/1X,'Please check your input format and ',
     2           'typing.  Reinput if you desire.'/)
      PRINT *, 'DO YOU WISH TO INPUT ANOTHER EMPLOYEE NAME? (Y or N)  '
      READ (5,'(A1)') LOOP
      IF (LOOP.EQ.'Y'.OR.LOOP.EQ.'y') GO TO 1000
C  End of inquiry loop for extracting employee data
      END
C
      SUBROUTINE SORT(X,N)
**********************************************************************
*    Subprogram for bubble sort with stop check for sorted array    *
*    Developed by:  T. K. Jewell               January 1988         *
**********************************************************************
C
C  Variable definitions
C    X(I) = array elements to be sorted
C    N = size of array
C    COUNT = flag variable to indicate whether or not switches have been made
C    TEMP = temporary storage location for data being switched
C
      INTEGER N,COUNT
      CHARACTER*30 X(N),TEMP
C
C  Initialize counter for switches for first time through loop
      COUNT = 1
C
C  WHILE (COUNT.GT.(0)) is true, DO 200
C
  500 IF (COUNT.GT.(0)) THEN
C
C  Initialize counter so it will be zero if no switches are made
         COUNT = 0
C
C  Start inner loop for comparing adjacent elements and switching
C    if necessary
```

(EXAMPLE 7.1 □ Industrial Production Database □ Continued)

```
C
          DO 300 J=1,N-1
              IF (X(J).GT.X(J+1)) THEN
                  TEMP = X(J)
                  X(J) = X(J+1)
                  X(J+1) = TEMP
                  COUNT = COUNT + 1
              END IF
  300     CONTINUE
  200 CONTINUE
      GO TO 500
      END IF
C  End of WHILE structure
C
      RETURN
      END
C
      SUBROUTINE SWITCH(TNAME,NAME,CHANGE,N,TEMP)
**********************************************************************
*  SUBROUTINE for rearranging array CHANGE to correspond with the   *
*    alphabetized array NAME                                         *
*  Developed by:  T. K. Jewell                      January 1988     *
**********************************************************************
C
C TNAME = array containing original order of NAME
C NAME = alphabetized array
C CHANGE = array to be rearranged in order corresponding to alphabetized
C          order of NAME
C N = size of arrays
C CHECK = 0 before switch, 1 after switch
C
      INTEGER N,CHANGE(N),TEMP(N),CHECK
      CHARACTER*30 TNAME(N),NAME(N)
C
C Copy CHANGE into TEMP for switching
      DO 100 I=1,N
          TEMP(I) = CHANGE(I)
  100 CONTINUE
C
C Outer loop for placing the correct value in each element of CHANGE
      DO 200 I=1,N
C
C Inner loop for matching an element of NAME with the element of TNAME
C    that contains the same value, and putting the value of the element of
C    TEMP corresponding to the element of TNAME into the element of CHANGE
C    corresponding with the element of NAME.
C
      CHECK = 0
      J = 1
C WHILE (J.LE.N) and (CHECK.EQ.(0)) are true, DO 300
  600     IF (J.LE.N.AND.CHECK.EQ.(0)) THEN
C
C Check for match of values between NAME(I) and TNAME(J)
              IF (NAME(I).EQ.TNAME(J)) THEN
                  CHANGE(I) = TEMP(J)
                  CHECK = 1
              ELSE
```

```
          J = J+1
        END IF
      GO TO 600
      END IF
300    CONTINUE
200 CONTINUE
    RETURN
    END
```

PROGRAM TESTING

Three sample data files are given below. Note that the order of data in the three files has purposely been scrambled. This will show that the program will properly sort the data using the employee identification number.

Data File EX7_1A.DAT

```
 9 1001
Dodd, Pamela B.              1003    8
Jones, John T.              1004    9
Reynolds, Keith B.          1005   15
Johnson, James S.           1001    2
Thomas, Brenda A.           1002   10
Thomas, Elizabeth R.        1006    7
Smith, Rebecca S.           1007    5
Carmen, Joan                1008   13
Flint, Thomas K.            1009    6
```

Data File EX7_1B.DAT

```
1008
   132   115   132   135   135   167   131   121   158   129   141    70
1009
   100   110   108   115   109   111   112   115   118   108   109   106
1001
   155   170   190   188   175   166    88   175   165   158   173   180
1002
   130   132   110   129   108   142   111   112   115   175   124   126
1003
   145   153   115   135   100   134   120   133   130   109   154   153
1004
   143   133   127   140    99   131   132   128   110   127   136   143
1005
   125   140   122   167   190   122   137   165   127   143   132    33
1006
   160   155   153   182   145    78   160   170   165   180   190   200
1007
   130   125   135   123   157   133   128   143   154   154   142    88
```

Data File EX7_1C.DAT

```
1001
    50    45    62    25    35    44    40    21    36    43    36    28
1008
    31    23    26    25    28    25    21    22    28     7    21    38
1002
    20    25    35    36    38    21    18    22    41    77    22    31
1003
    24    24    36    31    25    32    33    24    39    21    18    27
1004
    25    22    34    26    26    27    34    31    35    19    16    24
```

(EXAMPLE 7.1 □ Industrial Production Database □ Continued)

1007											
33	24	28	24	37	22	32	26	22	22	12	33
1005											
26	33	33	27	29	35	37	28	29	18	18	1
1006											
10	15	13	16	14	9	18	19	20	18	16	32
1009											
5	4	6	3	5	7	4	3	6	4	5	7

These three data files will give the output shown below. Data for two employees was extracted, then an erroneous name was entered, and then the program was terminated.

```
Input employee data file name   EX7_1A.DAT

Input production data file name   EX7_1B.DAT

Input defect data file name   EX7_1C.DAT

INDUSTRIAL PRODUCTION DATABASE

The number of employees is      9
The lowest employee ID number for this group is 1001
```

```
SORTED EMPLOYEE NAMES          ID NUMBERS, YEARS OF SERVICE

Carmen, Joan                   1008          13
Dodd, Pamela B.                1003           8
Flint, Thomas K.               1009           6
Johnson, James S.              1001           2
Jones, John T.                 1004           9
Reynolds, Keith B.             1005          15
Smith, Rebecca S.              1007           5
Thomas, Brenda A.              1002          10
Thomas, Elizabeth R.           1006           7
```

```
SORTED PRODUCTION DATA
```

ID	JAN.	FEB.	MAR.	APR.	MAY.	JUN.	JUL.	AUG.	SEP.	OCT.	NOV.	DEC.
1008	132.	115.	132.	135.	135.	167.	131.	121.	158.	129.	141.	70.
1003	145.	153.	115.	135.	100.	134.	120.	133.	130.	109.	154.	153.
1009	100.	110.	108.	115.	109.	111.	112.	115.	118.	108.	109.	106.
1001	155.	170.	190.	188.	175.	166.	88.	175.	165.	158.	173.	180.
1004	143.	133.	127.	140.	99.	131.	132.	128.	110.	127.	136.	143.
1005	125.	140.	122.	167.	190.	122.	137.	165.	127.	143.	132.	33.
1007	130.	125.	135.	123.	157.	133.	128.	143.	154.	154.	142.	88.
1002	130.	132.	110.	129.	108.	142.	111.	112.	115.	175.	124.	126.
1006	160.	155.	153.	182.	145.	78.	160.	170.	165.	180.	190.	200.

```
SORTED DEFECT DATA
```

ID	JAN.	FEB.	MAR.	APR.	MAY.	JUN.	JUL.	AUG.	SEP.	OCT.	NOV.	DEC.
1008	31.	23.	26.	25.	28.	25.	21.	22.	28.	7.	21.	38.
1003	24.	24.	36.	31.	25.	32.	33.	24.	39.	21.	18.	27.
1009	5.	4.	6.	3.	5.	7.	4.	3.	6.	4.	5.	7.
1001	50.	45.	62.	25.	35.	44.	40.	21.	36.	43.	36.	28.
1004	25.	22.	34.	26.	26.	27.	34.	31.	35.	19.	16.	24.
1005	26.	33.	33.	27.	29.	35.	37.	28.	29.	18.	18.	1.
1007	33.	24.	28.	24.	37.	22.	32.	26.	22.	22.	12.	33.
1002	20.	25.	35.	36.	38.	21.	18.	22.	41.	77.	22.	31.
1006	10.	15.	13.	16.	14.	9.	18.	19.	20.	18.	16.	32.

```
Input employee name for search
Example: Doe, John Q.
Johnson, James S.

 EMPLOYEE NAME:  Johnson, James S.
 IDENTIFICATION NUMBER:   1001
 YEARS OF SERVICE:       2
 TOTAL PRODUCTION FOR YEAR:   1983.
 TOTAL DEFECTS FOR YEAR:    465.
 RATIO OF PRODUCTION TO DEFECTS:     4.3

DO YOU WISH TO INPUT ANOTHER EMPLOYEE NAME? (Y or N)  y

Input employee name for search
Example: Doe, John Q.
Carmen, Joan

 EMPLOYEE NAME:  Carmen, Joan
 IDENTIFICATION NUMBER:   1008
 YEARS OF SERVICE:      13
 TOTAL PRODUCTION FOR YEAR:   1566.
 TOTAL DEFECTS FOR YEAR:    295.
 RATIO OF PRODUCTION TO DEFECTS:     5.3

DO YOU WISH TO INPUT ANOTHER EMPLOYEE NAME? (Y or N)  y

Input employee name for search
Example: Doe, John Q.
Doe, Whose E.

 Your input name did not match any names in the database.
 Please check your input format and typing.  Reinput if you desire.

DO YOU WISH TO INPUT ANOTHER EMPLOYEE NAME? (Y or N)  n
```

7.4 ARRAY AND MATRIX OPERATIONS

7.4.1 Matrix Multiplication

Matrix multiplication forms a product of two matrices or arrays. Matrix algebra and the procedures for matrix multiplication are discussed in Appendix F. Matrix multiplication can be accomplished with two rectangular arrays, as long as the number of columns of the first matrix is equal to the number of rows of the second. If this is the case, the matrices are said to be compatible. The logic of this restriction is evident when you examine the general equation for matrix multiplication

$$\mathbf{A} \cdot \mathbf{B} = \mathbf{C} \qquad c_{ik} = \sum_{j=1}^{n} a_{ij} b_{jk} \qquad (7.2)$$

in which i varies from one to the number of rows in $\mathbf{A}$, and k varies from one to the number of columns in $\mathbf{B}$. As j varies, the summation moves across the ith row of the **A** matrix and down the kth column of the **B** matrix. Therefore, if the number of columns of the first is not equal to the number of rows of the second, the a_{ij} and b_{jk} elements will not match up properly as j is increased. If the matrices are compatible, the resulting product matrix will have the number of rows of the **A** matrix and the number of columns of the **B** matrix. If **A** is 5×4 and **B** is 4×3, then **A** $\times$ **B** is defined and would produce a 5×3 matrix. **B** $\times$ **A** would not be defined for this case.

Equation 7.2 lends itself very nicely to computerization as a subroutine subprogram. To make the subprogram general, both of the matrices to be multiplied, as well as the product matrix, should be two-dimensional. This presents no problem since a one-dimensional vector can be represented as an $N \times 1$ two-dimensional array. $B(7)$ and $B(7,1)$ would represent the same number of storage locations. The second notation would identify a seven-element column vector that could be a right-hand-side vector for a linear algebra problem. We will now develop a subroutine for matrix (array) multiplication in Example 7.2. A short main program will be written to test the subroutine.

EXAMPLE 7.2 □ SUBROUTINE for Matrix Multiplication

STATEMENT OF PROBLEM

Develop a SUBROUTINE for matrix multiplication. Use a main program and the two-dimensional array input and output subprograms developed previously to test the subprogram.

MATHEMATICAL DESCRIPTION

Equation 7.2 gives the mathematical formula for matrix multiplication.

ALGORITHM DEVELOPMENT

Input/Output Design

The dimensions and values for arrays **A** and **B** will be input from a data file using SUBROUTINE READ. All arrays will be output using SUBROUTINE WRITE. Variable formats will be used, as developed in Example 6.4. A statement has been added to SUBROUTINE READ to input the dimensions of the matrix. Both subroutines have also been made more general by passing the maximum array size through the arguments. The revised subroutines are included with the example. Subsequent examples will use the revised READ and WRITE subroutines.

Numerical Methods

The product array **C** will be initialized to zero each time the SUBROUTINE is called.

Computer Implementation

The main program is straightforward. Figure 7.3 shows the flowchart for the matrix multiplication SUBROUTINE.

FIGURE 7.3   Flowchart for matrix multiplication SUBROUTINE.

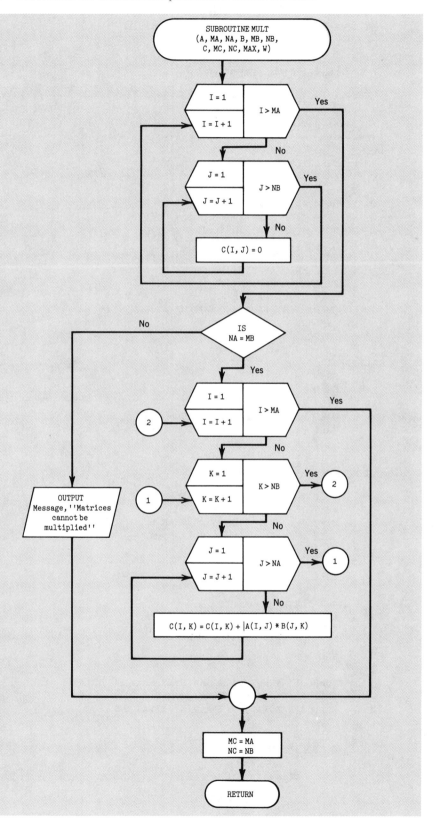

(EXAMPLE 7.2 □ SUBROUTINE for Matrix Multiplication □ Continued)

PROGRAM DEVELOPMENT

```
************************************************************************
*  Main program to check matrix multiplication SUBROUTINE            *
*  Developed by:  T. K. Jewell                  January 1988         *
************************************************************************
C
C  VARIABLE DEFINITIONS
C  A and B = arrays to be multiplied together
C  C = product array
C  MA,NA = # of rows and columns of A
C  MB,NB = # of rows and columns of B
C  MC,NC = # of rows and columns of C
C  MAX = maximum size for arrays
C  INFORM and OUTFOR = input and output format specifications
       INTEGER FR,R,W,MA,NA,MB,NB,MC,NC,MAX
       PARAMETER(FR=4,R=5,W=6,MAX=10)
       REAL A(MAX,MAX),B(MAX,MAX),C(MAX,MAX)
       CHARACTER INFORM*40,OUTFOR*40,NAME*14
       INFORM = '(16F5.0)'
       OUTFOR = '(1X,10F8.2)'
C
C  Input arrays
       PRINT *, 'Input File Name  '
       READ(R,'(A14)') NAME
       OPEN (UNIT=FR,FILE=NAME,STATUS='OLD')
       CALL READ(A,MA,NA,FR,INFORM,MAX)
       CALL READ(B,MB,NB,FR,INFORM,MAX)
C
C  Output A and B to check them
       WRITE(W,10)
   10 FORMAT(1X,'THE ARRAY FOR A IS',//)
       CALL WRITE(A,MA,NA,W,OUTFOR,MAX)
       WRITE(W,12)
   12 FORMAT(///,1X,'THE ARRAY FOR B IS',//)
       CALL WRITE(B,MB,NB,W,OUTFOR,MAX)
C
C  Multiply AxB
       CALL MULT(A,MA,NA,B,MB,NB,C,MC,NC,MAX)
C
C  Output the product array, C
       WRITE(W,14)
   14 FORMAT(///,1X,'THE PRODUCT ARRAY, C, IS',//)
       CALL WRITE(C,MC,NC,W,OUTFOR,MAX)
       END
C
       SUBROUTINE MULT(A,MA,NA,B,MB,NB,C,MC,NC,MAX)
       INTEGER MA,NA,MB,NB,MC,NC,MAX,W
       REAL A(MAX,MAX),B(MAX,MAX),C(MAX,MAX)
C
C  Initialize the C array to zero
       DO 50 I=1,MA
          DO 60 J=1,NB
             C(I,J) = 0.0
```

```
   60     CONTINUE
   50 CONTINUE
C
C  Check for multiplication compatibility between the arrays
      IF (NA.EQ.MB) THEN
         DO 100 I=1,MA
            DO 200 K=1,NB
               DO 300 J=1,NA
               C(I,K) = C(I,K) + A(I,J)*B(J,K)
  300          CONTINUE
  200       CONTINUE
  100    CONTINUE
      ELSE
         PRINT 10
   10    FORMAT(/,1X,'THE MATRICES CANNOT BE MULTIPLIED TOGETHER,',
     1                'NA NOT EQUAL TO MB')
      END IF
      MC = MA
      NC = NB
      RETURN
      END
      SUBROUTINE READ(A,M,N,FR,FORM,MAX)
      INTEGER M,N,FR,MAX
      REAL A(MAX,MAX)
      CHARACTER FORM*40
      READ (FR,'(2I5)') M,N
      DO 100 I=1,M
         READ(FR,FORM) (A(I,J),J=1,N)
  100 CONTINUE
      RETURN
      END
      SUBROUTINE WRITE(A,M,N,W,FORM,MAX)
      INTEGER M,N,W,MAX
      REAL A(MAX,MAX)
      CHARACTER FORM*40
      DO 100 I=1,M
         WRITE (W,FORM) (A(I,J),J=1,N)
  100 CONTINUE
      RETURN
      END
```

PROGRAM TESTING

Use the following matrices to test out the SUBROUTINE:

$$\mathbf{A} \cdot \mathbf{B} = \mathbf{C}$$

$$\mathbf{A} = \begin{bmatrix} 4.0 & 3.5 & 1.8 & 2.6 \\ 3.9 & 1.6 & 1.0 & 4.5 \\ 6.9 & 7.0 & 1.2 & 8.0 \\ 7.0 & 8.0 & 9.0 & 5.0 \\ 10.0 & 8.5 & 9.5 & 6.0 \end{bmatrix}$$

$$\mathbf{B} = \begin{bmatrix} 10.0 & 11.0 & 7.0 \\ 6.5 & 3.4 & 4.8 \\ 5.1 & 6.0 & 3.0 \\ 7.1 & 4.2 & 1.3 \end{bmatrix}$$

(EXAMPLE 7.2 □ SUBROUTINE for Matrix Multiplication □ Continued)

$$\mathbf{C} = \begin{bmatrix} 90.4 & 77.6 & 53.6 \\ 86.5 & 73.2 & 43.8 \\ 177.4 & 140.5 & 95.9 \\ 206.0 & 182.2 & 122.4 \\ 246.3 & 221.1 & 147.1 \end{bmatrix}$$

7.4.2 Matrix Inverse

Section 5.1 of Appendix F presents the rules for developing the inverse of a matrix and describes the properties of the inverse. You can use the inverse of the coefficient matrix to solve for the unknowns in a system of linear equations. The solution vector **X** is found by premultiplying the right-hand-side vector **B** by the inverse of the coefficient matrix $\mathbf{A}^{-1}$.

$$\mathbf{X} = \mathbf{A}^{-1} \cdot \mathbf{B} \tag{7.3}$$

There are algorithms for solving systems of equations that do not require finding the inverse. However, if the right-hand-side vector is changed, the whole system has to be solved again. If the inverse of the coefficient array is known, any number of solutions for different right-hand-side vectors can be found just by using matrix multiplication.

Most computer systems have library subprograms available to accomplish many linear algebra operations, including matrix inversion. Two representative library subprograms for matrix inversion will be illustrated here. Both of these are available on the VAX minicomputer system that I use. The first is a subprogram that does not have to be linked to the main program during compilation to be referenced during program execution. The second example will use a SUBROUTINE from the International Mathematical and Statistical Library, IMSL, described in Chapter 5. The math library of IMSL contains SUBROUTINE LINRG for computing the inverse of a matrix in single precision and SUBROUTINE DLINRG for finding the inverse in double precision. To use the IMSL library, you have to link it to the main program along with any other subprogram files. After the library is linked, you can access any of the library subprograms.

One of the chapter-end exercises in the numerical methods section of Chapter 5 assigns the development of a matrix inversion subprogram. In the unlikely event that your computer system does not have a library matrix inversion subprogram, the solution file for that exercise could be linked to your programs when it is necessary to perform matrix inversion.

7.4.2.1 SUBROUTINE for Matrix Inversion

Example 7.3 demonstrates the procedure for using an inversion routine, called BAS$MAT_INV, which is resident on the VAX 11/780 system that I use. This subroutine does not have to be linked to be used. However, you do need to access it from a SUBROUTINE subprogram that converts the dimensions of the array to be inverted from its maximum size, the DIMENSION limit in the main program, to the actual size of the matrix to be inverted. Unless the matrix to be inverted is the same size as the DIMENSION limit, the BAS$MAT_INV subprogram will see a column or columns of all zero values in the matrix as stored in the computer and will not be able to invert it. This is because the computer does not store a partially filled array in the two-dimensional format that we visualize as a matrix.

EXAMPLE 7.3 □ SUBROUTINE for Matrix Inversion Using Resident Inversion Routine BAS$MAT_INV

STATEMENT OF PROBLEM

Write a main program and subroutine that can be used in conjunction with the library subprogram BAS$MAT_INV to find the inverse of an N × N coefficient matrix. First you will have to convert the N × N coefficient matrix, which is stored in a larger array of dimension MAX × MAX, into an array of dimension N × N.

ALGORITHM DEVELOPMENT

Input/Output Design

Input will be via the READ subprogram. Output will be via the WRITE subprogram.

Computer Implementation

Figure 7.4 shows a flowchart for using the inversion routine BAS$MAT_INV.

PROGRAM DEVELOPMENT

```
****************************************************************
*  Main program for driving matrix inversion subprogram    *
*  Developed by:  T. K. Jewell            January 1988     *
****************************************************************
C  A = two-dimensional matrix to be inverted
C  AINV = inverse of the matrix
C  D and E = dummy matrices to ensure that the matrix sent over to
C            BAS$MAT_INV is NxN
C  N = dimension of matrix
      REAL A(10,10),AINV(10,10),D(10,10),E(10,10)
      INTEGER N,R,FR,W
      CHARACTER NAME*14,INFORM*40,OUTFOR*40
      PARAMETER (R=5,FR=4,W=6)
      INFORM = '(16F5.0)'
      OUTFOR = '(1X,10F8.2)'
      PRINT *,'Input name of input data file  '
      READ (R,'(A14)') NAME
      OPEN(UNIT=FR,FILE=NAME,STATUS='OLD')
      CALL READ(A,N,N,FR,INFORM,10)
      WRITE(W,100)
  100 FORMAT(1X,'The matrix to be inverted is:',/)
      CALL WRITE(A,N,N,W,OUTFOR,10)
C  Call to matrix inversion subprogram
      CALL MATINV(A,AINV,N,D,E)
      WRITE(W,110)
```

(EXAMPLE 7.3 □ SUBROUTINE for Matrix Inversion Using Resident Inversion Routine BAS$MAT_INV □ Continued)

FIGURE 7.4 Flowchart for using inversion routine BAS$MAT_INV.

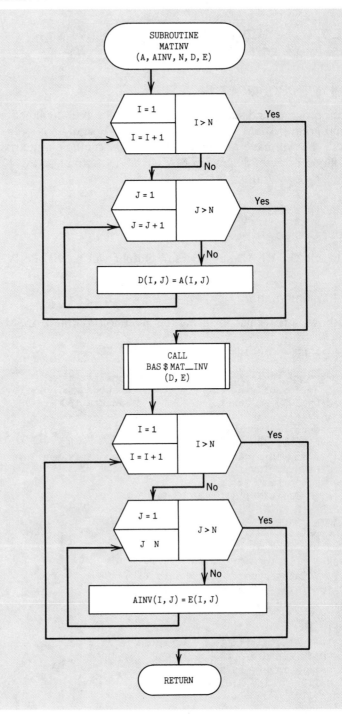

```
  110 FORMAT(1X,'The inverted matrix is:',/)
      CALL WRITE(AINV,N,N,W,OUTFOR,10)
      END
      SUBROUTINE MATINV(A,AINV,N,D,E)
************************************************************
*   SUBROUTINE for converting coefficient matrix to NxN *
*     and finding the inverse                           *
*   Developed by:  T. K. Jewell          January 1988 *
************************************************************
      INTEGER N
      REAL A(10,10),AINV(10,10),D(N,N),E(N,N)
C  Convert A(10,10) into D(N,N)
      DO 10 I=1,N
        DO 20 J=1,N
          D(I,J) = A(I,J)
   20    CONTINUE
   10 CONTINUE
C  CALL matrix inversion subroutine
C  D is the matrix to be inverted, E is the inverse
      CALL BAS$MATINV(%DESCR(D),%DESCR(E))
C  Convert E(N,N) into AINV(10,10)
      DO 30 I=1,N
        DO 40 J=1,N
          AINV(I,J) = E(I,J)
   40    CONTINUE
   30 CONTINUE
      RETURN
      END
```

PROGRAM TESTING

For the matrix

$$\begin{bmatrix} 2. & 3. & 1. \\ 0. & 1. & 2. \\ 4. & 2. & 0. \end{bmatrix}$$

the inverse is

$$\begin{bmatrix} -0.33 & 0.17 & 0.42 \\ 0.67 & -0.33 & -0.33 \\ -0.33 & 0.67 & 0.17 \end{bmatrix}$$

7.4.3 Solution of Systems of Equations

The inverse of a coefficient matrix will now be used to solve a practical problem. The coefficient matrix and right-hand-side vector for a truss analysis problem were input in Example 6.2. The coefficient matrix represents the directional cosines of the truss members, while the right-hand-side vector shows the loads and the reactions. The inverse of the coefficient matrix multiplied by the loading/reaction vector will give the member force vector. The IMSL subprogram LINRG will be used to find the matrix inverse.

EXAMPLE 7.4 □ **Solution of Equilibrium Equations for Determinate Planar Truss**

STATEMENT OF PROBLEM

Develop a program to solve for the forces in up to a 25 member planar truss. Input of the directional cosine matrix and loading/reaction vector will be from a file. Use the IMSL SUBROUTINE LINRG to find the inverse, and the SUBROUTINE MULT of Example 7.2 for matrix multiplication. Test your program with the data of Example 6.2.

MATHEMATICAL DESCRIPTION

Appendices F and G give the mathematical background for this problem. Figure 6.2 shows the schematic and free body diagrams for the truss under analysis.

ALGORITHM DEVELOPMENT

Input/Output Design

The program of Example 6.2 will be used to input the coefficient array and right-hand-side vector of loads and reactions. Note that the solution vector and right-hand side have to be two-dimensional ($N \times 1$) to conform to the convention for the matrix multiplication subprogram. Output will include the solution vector with appropriate labels.

Numerical Methods

Although the code is not available for the IMSL SUBROUTINE LINRG and the solution technique is only briefly described in the documentation, the procedure has been used by so many people that it can safely be considered to be correct. Nonetheless, we can check its accuracy because the solution to the truss problem given has been previously calculated by hand.

Computer Implementation

Figure 7.5 gives a flowchart for the solution of the linear equations of the truss analysis problem. Assume that the main program is stored in a file named TRUSS.FOR, the compiled array input SUBROUTINE READ is stored in a file named READ.OBJ, and the compiled array multiplication SUBROUTINE MULT is stored in a file named MULT.OBJ. The following set of commands would have to be used to run the program under VAX/VMS

```
$FORTRAN TRUSS
$LINK TRUSS,IMSL/LIB,READ,MULT
$RUN TRUSS
```

The instructions in the IMSL documentation (IMSL 1987) for the arguments of matrix inversion procedure LINRG are

```
CALL LINRG(N,A,LDA,AINV,LDAINV)
```

in which

　　N = dimension of coefficient matrix to be inverted

　　A = N × N coefficient matrix to be inverted

FIGURE 7.5 Flowchart for solution of truss analysis linear equations.

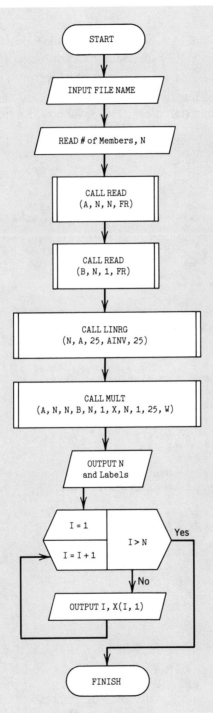

(EXAMPLE 7.4 □ Solution of Equilibrium Equations for Determinate Planar Truss □ Continued)

LDA = leading dimension of A, as specified in the DIMENSION or declaration statement of the main program (upper limit on size of array)

AINV = N × N matrix containing the inverse of A

LDAINV = leading dimension of AINV, as specified in the DIMENSION or declaration statement of the main program (upper limit on size of array)

From these instructions you can deduce that this inversion subprogram takes care of the problem of a partly filled array internally.

PROGRAM DEVELOPMENT

```
      PROGRAM TRUSS
***********************************************************************
*   Input of coefficient matrix and right-hand-side vector for truss *
*      analysis, and solution of the problem by inverse and matrix   *
*      multiplication                                                *
*   Developed by:  T. K. Jewell                      December 1987 *
***********************************************************************
C
C   Variable definitions
C   A = coefficient matrix (directional cosines)
C   B = right-hand-side vector (applied forces and reactions)
C   N = number of members in truss
C   F = solution vector, force in each member of the truss
      REAL A(25,25),B(25,1),F(25,1),AINV(25,25)
      INTEGER N,FR,R,W
      CHARACTER NAME*14,INFORM*40,OUTFOR*40
      PARAMETER (FR=4,R=5,W=6)
      INFORM = '(10F8.0)'
      OUTFOR = '(1X,10F8.2)'
      PRINT *, 'INPUT FILE NAME  '
      READ(R,'(A14)') NAME
      OPEN(UNIT=FR,FILE=NAME,STATUS='OLD')
C
C   Input dimensions and coefficient matrix.
      CALL READ(A,N,N,FR,INFORM,25)
C
C   Input right-hand-side vector.
      CALL READ(B,N,1,FR,INFORM,25)
C
C   Compute inverse of coefficient matrix
      CALL LINRG(N,A,25,AINV,25)
C
C   Multiply AINVxB
      CALL MULT(AINV,N,N,B,N,1,F,N,1,25)
C
C   Output results (Forces in members)
      WRITE(W,30) N
   30 FORMAT(11X,'MATRIX ANALYSIS OF TRUSSES',//,
     1          1X,'  The number of members is:  ',I2,///)
C
      WRITE(W,40)
   40 FORMAT(1X,'VECTOR OF MEMBER FORCES IS',//
     1          3X,'MEMBER    FORCE',
     2          3X,'NUMBER    POUNDS',/)
```

```
      DO 100 I=1,N
         WRITE(W,'1X,I6,F10.2) I,F(I,1)
 100  CONTINUE
      END
```

PROGRAM TESTING

The input data file, set up according to the structure of the main program and the subprograms, is given below.

DATA FILE

```
12345678901234567890123456789012345678901234567890123456789012345678901234567890
    7    7
   1.0 -0.4472       0        0        0        0        0
     0  0.8944       0        0        0        0        0
     0  0.4472  0.4472    -1.0        0        0        0
     0  0.8944 -0.8944       0        0        0        0
  -1.0       0 -0.4472       0   0.4472      1.0        0
     0       0  0.8944       0   0.8944        0        0
     0       0       0     1.0  -0.4472        0  -0.4472
    7    1
     0
   5.0
     0
     0
     0
  10.0
     0
```

For the truss and loading of Example 6.2, the solution is

$$F_1 = F_{AC} = 2.50$$
$$F_2 = F_{AB} = 5.59$$
$$F_3 = F_{BC} = 5.59$$
$$F_4 = F_{BD} = 5.00$$
$$F_5 = F_{CD} = 5.59$$
$$F_6 = F_{CE} = 2.50$$
$$F_7 = F_{DE} = 5.59$$

7.4.4 Array and Matrix Addition and Subtraction

You have used vector (one-dimensional array) addition in previous chapters. Now this procedure will be generalized for arrays of two dimensions or higher.

The first example given below will involve addition of several elements of one array to calculate the corresponding element of a second array. Specifically, corresponding elements of three rows of the first matrix will be added together to find the corresponding element of a specified row in the second matrix.

The second example will illustrate matrix or multidimensional array addition, which follows the same procedure as vector addition. Corresponding elements of the two arrays are added together, so the arrays will have to be the same size. The resulting array will be the same size as the original two. There is no requirement that the arrays be any particular shape, as long as the two arrays are the same size and shape.

The third example will illustrate addition across rows or down columns of an array. These are common procedures for extracting information from an array.

7.4.4.1 Array Addition for Hydrograph Routing

Many engineering problems involve the unsteady flow of liquids or gasses through conduits or open channels. Unsteady flow is flow that changes with time at a certain point in a fluid system. A plot of unsteady flow vs. time in a system carrying water is called a hydrograph. Since water is an incompressible liquid (as opposed to gasses, which compress rather easily) the volumetric flow rate is the appropriate flow parameter to use for representing flow. Volumetric flow rate is normally represented by the letter Q and has units of cubic feet per second, cfs, or cubic meters per second, m^3/s. Figure 7.6a shows a definition sketch of a hydrograph. Figure 7.6b shows how the same hydrograph can be represented by a series of discrete segments of time period Δt. The volume under the curve represents the

FIGURE 7.6   Definition sketch for hydrograph addition.

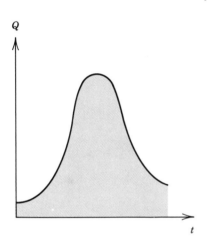

(a) Hydrograph definition sketch.

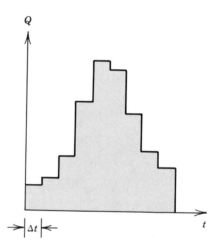

(b) Approximate hydrograph for routing.

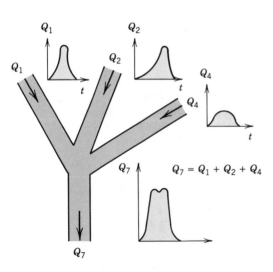

(c) Addition of hydrographs.

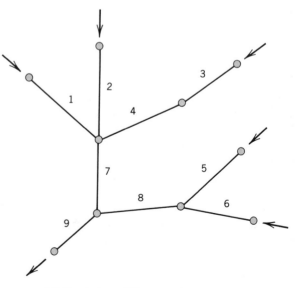

(d) Depiction of flow routing network.

otal volume of flow. It would be approximately the same for either depiction. Theoretically, if Δt becomes infinitely small, the approximate hydrograph of Figure 7.6b would approach the continuous hydrograph, and the volumes would be identical. For most practical applications the hydrograph curve does not follow a smooth mathematical function. Therefore, the discrete representation is a convenient way of depicting the hydrograph for computer processing. Each discrete low rate can be stored in an element of a subscripted array. When two or more conduits come together, the hydrographs for each can be added together to find he combined hydrograph. Figure 7.6c depicts this process for three hydrographs epresented by discrete segments. Array addition would have to be used to add hese hydrographs together.

Example 7.5 will develop a program to add hydrographs together. The resulting nput hydrograph will be stored in a different array from the output hydrographs being added together. This simulates what might be happening in a larger program where a whole network of conduits is linked together in a tree structure, as depicted in Figure 7.6d. The input hydrograph for a conduit is routed through that conduit by a routing algorithm. At the downstream end of the conduit it becomes the output hydrograph for that conduit. When several conduits join at a junction, their output hydrographs can be combined to form the input hydrograph for the next downstream conduit. Thus for the system shown in Figure 7.6d,

$$Q_{7_{in}} = Q_{1_{out}} + Q_{2_{out}} + Q_{4_{out}} \qquad (7.4)$$

Keep in mind that the program you are writing is only a part of the program to handle the whole system. The total program would include an algorithm for routing the flow through a conduit and algorithms for determining which conduits enter and leave each junction of the system. This would allow you to model the flow at all points in the system over a series of time steps.

EXAMPLE 7.5 □ Hydrograph Routing

STATEMENT OF PROBLEM

Develop a program to add output hydrographs together for conduits coming into a junction to form the input hydrograph for the next downstream conduit. Refer to Figure 7.6 for a definition sketch of hydrograph routing. Use separate input and output arrays. Each array should be able to hold input or output hydrographs for up to 10 conduits. Up to 25 discrete time intervals should be allowed for the hydrographs. Your program will have to select which output hydrographs it will add together at a particular joint and which input hydrograph will receive the results.

MATHEMATICAL DESCRIPTION

The input flow rate to a conduit is the sum of the output flow rates for all conduits leading into that conduit. The identification numbers for the output and input conduits will not necessarily be in numerical sequence. For example, in Figure 7.6c the three output conduits are labeled 1, 2, and 4, while the input conduit these lead into is labeled 7. The intervening labels could be used for conduits in

(EXAMPLE 7.5   □   Hydrograph Routing   □   Continued)

other parts of the system. For time interval j, the input flowrate for conduit 7 would be

$$QIN_{7j} = QOUT_{1j} + QOUT_{2j} + QOUT_{4j} \tag{7.5}$$

Your program will need to be able to differentiate the proper indexes for the output and input conduits.

ALGORITHM DEVELOPMENT

Input/Output Design

One 10×25 array (QOUT) will be used to store the output hydrographs, and another 10×25 array (QIN) will be used to store the input hydrographs. The rows will represent the conduits and the columns the time steps for the hydrographs. For this problem, only three rows of the output array and one row of the input array will hold hydrographs. In a complete system model, the other rows would be occupied by output or input hydrographs for other parts of the system. The number of output conduits (hydrographs), the identification number for each, the identification number for the input conduit, and the number of discrete time steps in the problem have to be input first. Then the output hydrographs can be read in. All data input will be by list-directed input. All of both arrays will be output to show that the hydrographs are in the proper positions within the arrays.

Computer Implementation

A one-dimensional array will be used to store the identification numbers for the output conduits. A separate variable will be used to store the identification number for the input conduit. Conduit identification numbers have to be between 1 and the maximum number of conduits allowed; in this case 10. The elements of the output conduit identification array and the input conduit identification variable will be used as pointers to show what row of the hydrograph array a particular set of data should occupy. For the example of Figure 7.6c, POINT(1) = 1, POINT(2) = 2, and POINT(3) = 4. This means that the third hydrograph to be added will be in the fourth row of array QOUT. The subscripted variable POINT will be used to denote the subscript index for the array QOUT. For example, QOUT(POINT(3),4) would denote QOUT(4,4). Use of an integer array as a subscript for another array is acceptable since a particular element of the array POINT just represents a value.

Figure 7.7 shows a flowchart for the program.

PROGRAM DEVELOPMENT

```
******************************************************
*  Program for hydrograph addition                   *
*  Developed by: T. K. Jewell     January 1988       *
******************************************************
C
C VARIABLE DEFINITIONS
C  NOUT = number of output hydrographs
C  POINT = array to store identification numbers of output conduits
C  IDIN = identification number for input conduit
```

FIGURE 7.7 Flowchart for hydrograph addition program.

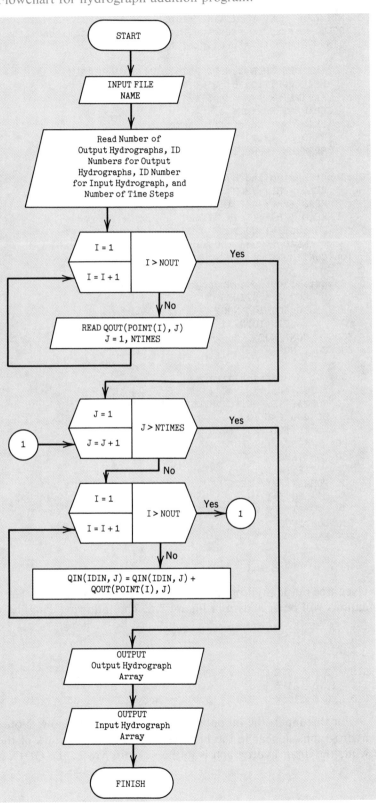

(EXAMPLE 7.5 □ Hydrograph Routing □ Continued)

```
C  NTIMES = number of time steps
C  QOUT = array to store output hydrographs for conduits
C  QIN = array to store input hydrographs
C
       INTEGER FR,R,W,NOUT,POINT(10),IDIN,NTIMES
       REAL QOUT(10,25),QIN(10,25)
       PARAMETER (FR=4,R=5,W=6)
       CHARACTER NAME*14
       PRINT *, 'Input file name  '
       READ(R,'(A14)') NAME
       OPEN(UNIT=FR,FILE=NAME,STATUS='OLD')
C
C  Input system and Hydrograph data
       READ (FR,*) NOUT
       READ (FR,*) (POINT(I),I=1,NOUT)
       READ (FR,*) IDIN,NTIMES
       DO 100 I=1,NOUT
           READ (FR,*) (QOUT(POINT(I),J),J=1,NTIMES)
  100 CONTINUE
C
C  Compute the input Hydrograph
       DO 200 J=1,NTIMES
           DO 300 I=1,NOUT
               QIN(IDIN,J) = QIN(IDIN,J) + QOUT(POINT(I),J)
  300      CONTINUE
  200 CONTINUE
C
C  Output Hydrograph arrays
       WRITE (W,10)
   10 FORMAT(1X,'OUTPUT HYDROGRAPH ARRAY IS',//)
       DO 400 I=1,10
           WRITE(W,20) (QOUT(I,J),J=1,NTIMES)
   20      FORMAT(1X,10F8.1)
  400 CONTINUE
       WRITE (W,30)
   30 FORMAT(////,1X,'INPUT HYDROGRAPH ARRAY IS',//)
       DO 500 I=1,10
           WRITE(W,20) (QIN(I,J),J=1,NTIMES)
  500 CONTINUE
       END
```

PROGRAM TESTING

Use three output hydrographs and 10 time steps for the test. The conduit identifications will be as shown in Figure 7.6. The following data file will be used:

```
3
1 2 4
7 10
0.0   3.5   6.8   10.4   20.0   22.5   30.0   31.0   15.0   5.5
1.0  10.5  21.6   33.2   20.1   10.4    5.1    2.2    1.0   0.1
0.5  20.5  10.4    9.0    5.6    4.5   25.0   35.4   10.0   4.3
```

For this input, the output will appear as shown below. Note that the output hydrographs appear in the first, second, and fourth rows of the QOUT array, while the input hydrograph is in the seventh row of the QUIN array.

OUTPUT HYDROGRAPH ARRAY IS

0.0	3.5	6.8	10.4	20.0	22.5	30.0	31.0	15.0	5.5
1.0	10.5	21.6	33.2	20.1	10.4	5.1	2.2	1.0	0.1
0.0	0.0	0.0	0.0	0.0	0.0	0.0	0.0	0.0	0.0
0.5	20.5	10.4	9.0	5.6	4.5	25.0	35.4	10.0	4.3
0.0	0.0	0.0	0.0	0.0	0.0	0.0	0.0	0.0	0.0
0.0	0.0	0.0	0.0	0.0	0.0	0.0	0.0	0.0	0.0
0.0	0.0	0.0	0.0	0.0	0.0	0.0	0.0	0.0	0.0
0.0	0.0	0.0	0.0	0.0	0.0	0.0	0.0	0.0	0.0
0.0	0.0	0.0	0.0	0.0	0.0	0.0	0.0	0.0	0.0
0.0	0.0	0.0	0.0	0.0	0.0	0.0	0.0	0.0	0.0

INPUT HYDROGRAPH ARRAY IS

0.0	0.0	0.0	0.0	0.0	0.0	0.0	0.0	0.0	0.0
0.0	0.0	0.0	0.0	0.0	0.0	0.0	0.0	0.0	0.0
0.0	0.0	0.0	0.0	0.0	0.0	0.0	0.0	0.0	0.0
0.0	0.0	0.0	0.0	0.0	0.0	0.0	0.0	0.0	0.0
0.0	0.0	0.0	0.0	0.0	0.0	0.0	0.0	0.0	0.0
0.0	0.0	0.0	0.0	0.0	0.0	0.0	0.0	0.0	0.0
1.5	34.5	38.8	52.6	45.7	37.4	60.1	68.6	26.0	9.9
0.0	0.0	0.0	0.0	0.0	0.0	0.0	0.0	0.0	0.0
0.0	0.0	0.0	0.0	0.0	0.0	0.0	0.0	0.0	0.0
0.0	0.0	0.0	0.0	0.0	0.0	0.0	0.0	0.0	0.0

7.4.4.2 Array Addition for Transportation Cost Problem

In the previous example you added several elements of one array together to find the value to be stored in a certain element of a second array. The next example will illustrate the application of array addition to the cost data for a transportation problem. In array addition, you add together corresponding elements of two or more arrays to produce the elements of a new array.

A product is to be shipped from one of several warehouses to one of several retail stores. A given number of units of the product is available at each of the warehouses, and each of the retail stores needs a certain number of units. The end objective is to determine how the product should be shipped from the warehouses to the retail stores to minimize the total cost of meeting the demand of the stores for the product. In order to do this, it is necessary to know the cost of shipping one unit from warehouse i to retail store j. Example 7.6 illustrates how these costs might be developed. Three warehouses will be shipping to four retail stores. Data are available on the distance from each warehouse to each store, the average number of stops encountered along the route due to traffic control, and the value of a traffic index that indicates the density of traffic along the route. Costs will go up as the distance increases, the number of stops increases, and the traffic index shows high traffic densities. The example will assume that you have developed formulas to convert each of these types of data into costs. The total cost of shipment from warehouse i to retail store j would be the sum of the three individual costs. In effect, what you will be doing is converting each of the data arrays into a cost array, and then adding the three arrays together to get the total cost.

EXAMPLE 7.6 □ Transportation Cost Problem

STATEMENT OF PROBLEM

Develop a subprogram to add two matrices together. Use this subprogram to determine the total cost of shipping one unit of product from warehouse i to retail store j. The program should be capable of accepting up to 10 warehouses and 10 retail stores. For this application, three warehouses and four stores will be used. Figure 7.8 shows a definition sketch for the problem. Distance, number of stops, and traffic index data are available between each origin and destination.

MATHEMATICAL DESCRIPTION

The following equations are used to convert the distance, number of stops, and traffic index data into cost data:

$$DCOST = \$0.50*MILES + \$10.00$$
$$SCOST = 0.833*STOPS$$
$$ICOST = 10.0*I^{1.2}$$

ALGORITHM DEVELOPMENT

Input/Output Design

The raw data will be stored in arrays dimensioned for up to 10×10, with the rows representing warehouses and the columns representing retail stores. Each data array will be converted to costs using the appropriate equation. All arrays will be output for checking using SUBROUTINE WRITE, which was developed in Chapter 5.

Computer Implementation

Once the individual arrays have been converted to costs, they can be added together to get the total cost of shipping one unit from warehouse i to retail store j. Figure 7.9 shows a flowchart for the program and the matrix addition subprogram. Note that the elements of DCOST and SCOST are first added together to form COST, then the elements of COST are added to the elements of ICOST to get the final values for the elements of COST.

PROGRAM DEVELOPMENT

```
************************************************************
*  Transportation costs by array addition          *
*  Developed by: T. K. Jewell     January 1988      *
************************************************************
C   MILES = array for distance between warehouse i and store j
C   DCOST = cost for traveling from warehouse i to store j
C   STOPS = average number of stops between warehouse i and store j
C   SCOST = cost due to stops
C   INDEX = traffic index for route between warehouse i and store j
C   ICOST = cost due to traffic index
C   COST = array for total transportation cost from warehouse i to store j
C   M = number of warehouses
C   N = number of stores
        REAL MILES(10,10),DCOST(10,10),STOPS(10,10),SCOST(10,10),
```

```
1      INDEX(10,10),ICOST(10,10), COST(10,10)
       INTEGER N,M,FR,R,W
       CHARACTER NAME*14,OUTFOR*40
       PARAMETER (FR=4,R=5,W=6)
       OUTFOR = '(1X,10F8.2)'
       PRINT *, 'Input File Name  '
       READ (R,'(A14)') NAME
       OPEN(UNIT=FR,FILE=NAME,STATUS='OLD')
```

FIGURE 7.8   Definition sketch for transportation problem.

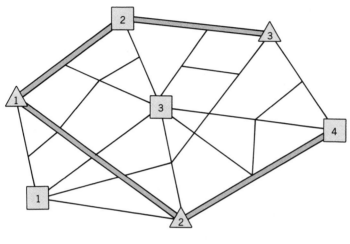

(*a*) Map.

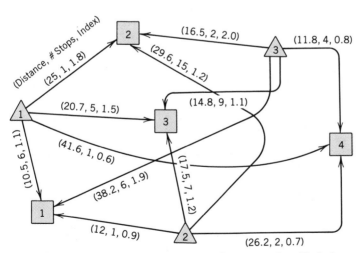

(*b*) Schematic with distance, number of stops, and traffic index.

▢ Destinations

△ Origins

(EXAMPLE 7.6 ☐ Transportation Cost Problem ☐ Continued)

```
C
C  Input distance, number of stops, and traffic index data
       READ (FR,*) M,N
       DO 100 I=1,M
          READ(FR,*) (MILES(I,J),J=1,N)
  100 CONTINUE
       DO 200 I=1,M
          READ(FR,*) (STOPS(I,J),J=1,N)
  200 CONTINUE
       DO 300 I=1,M
          READ(FR,*) (INDEX(I,J),J=1,N)
  300 CONTINUE
C
C  Convert each array to cost
       DO 400 I=1,M
          DO 500 J=1,N
             DCOST(I,J) = 0.50*MILES(I,J) + 10.
             SCOST(I,J) = 0.833*STOPS(I,J)
             ICOST(I,J) = 10.0*INDEX(I,J)**1.2
  500     CONTINUE
  400 CONTINUE
C
C  Add the arrays together
       CALL ADD(DCOST,SCOST,COST,M,N)
       CALL ADD(ICOST,COST,COST,M,N)
C
C  Output all arrays for checking
       WRITE(W,10)
  10 FORMAT(IX,'DISTANCE DATA FROM WAREHOUSE i TO STORE j',//)
       CALL WRITE(MILES,M,N,W,OUTFOR,10)
       WRITE(W,11)
  11 FORMAT(//,1X,'COST DATA FOR DISTANCE TRAVELED FROM i TO j',//)
       CALL WRITE(DCOST,M,N,W,OUTFOR,10)
       WRITE(W,12)
  12 FORMAT(//,1X,'DATA FOR NUMBER OF STOPS BETWEEN i AND j',//)
       CALL WRITE(STOPS,M,N,W,OUTFOR,10)
       WRITE(W,13)
  13 FORMAT(//,1X,'COSTS FOR STOPS BETWEEN i AND j',//)
       CALL WRITE(SCOST,M,N,W,OUTFOR,10)
       WRITE(W,14)
  14 FORMAT(//,1X,'TRAFFIC INDEX FOR ROUTE BETWEEN i AND j',//)
       CALL WRITE(INDEX,M,N,W,OUTFOR,10)
       WRITE(W,15)
  15 FORMAT(//,1X,'COSTS DUE TO TRAFFIC BETWEEN i AND j',//)
       CALL WRITE(ICOST,M,N,W,OUTFOR,10)
       WRITE(W,16)
  16 FORMAT(//,1X,'TOTAL COST OF TRANSPORTATION BETWEEN i AND j',//)
       CALL WRITE(COST,M,N,W,OUTFOR,10)
       END
       SUBROUTINE ADD(A,B,C,M,N)
**************************************************
*  Matrix addition subprogram                    *
*  Developed by:  T. K. Jewell   January 1988    *
**************************************************
       REAL A(10,10),B(10,10),C(10,10)
       INTEGER M,N
```

FIGURE 7.9 Flowchart for transportation problem.

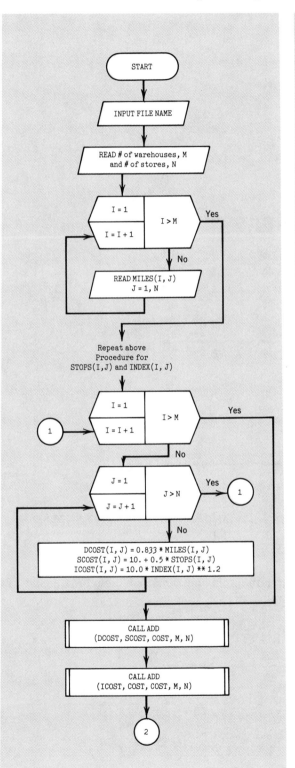
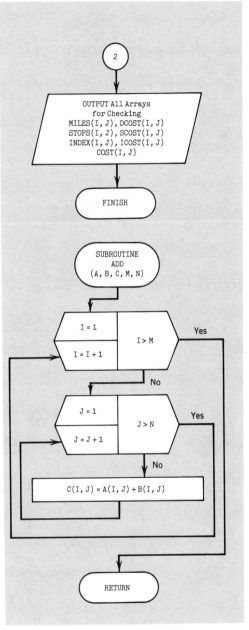

(EXAMPLE 7.6 ☐ Transportation Cost Problem ☐ Continued)

```
      DO 100 I=1,M
        DO 200 J=1,N
          C(I,J) = A(I,J) + B(I,J)
  200   CONTINUE
  100 CONTINUE
      RETURN
      END
```

PROGRAM TESTING

Use the following data:

DISTANCE

Warehouse/Store	1	2	3	4
1	10.5	25.0	20.7	41.6
2	12.0	29.6	17.5	26.2
3	38.2	16.5	14.8	11.8

AVERAGE NUMBER OF STOPS

Warehouse/Store	1	2	3	4
1	6	1	5	1
2	1	15	7	2
3	6	2	9	4

TRAFFIC INDEX

Warehouse/Store	1	2	3	4
1	1.1	1.8	1.5	0.6
2	0.9	1.2	1.2	0.7
3	1.9	2.0	1.1	0.8

The total cost matrix would give the following dollar amounts for transporting one unit from warehouse i to retail store j:

TOTAL COST OF TRANSPORTATION
BETWEEN i AND j ($)

Warehouse/Store	1	2	3	4
1	31.46	43.58	40.78	37.05
2	25.65	49.74	37.03	31.28
3	55.70	42.89	36.11	26.88

7.4.4.3 Summation Across Rows and Down Columns of Array

It is often necessary to add data across rows or down columns of an array. This is actually another form of database management. In Example 7.7, a three-dimensional array is being used to record the number of paying flights in a five-year period for a particular charter airliner from airport i to airport j. The third dimension of the array is used to identify the airliner. It is desired to know the total number of flights the airliner made to and from each city. Therefore, each city can be either an origin or a destination. Note that not all the totals into and out of a particular city are equal, because some flights were ferry flights that did not produce revenue. This program could easily be revised (using elements of the program of Example 7.6, and the array addition subprogram developed for that example) so that several arrays could be added together to give the total number of flights into or out of a particular city for all of the planes of the charter fleet. The same procedure could be used to find the total number of units of several products shipped from each warehouse and the total number of units shipped to each store. Figure 7.10 shows a general definition sketch for trip data.

FIGURE 7.10   Definition sketch for trip data.

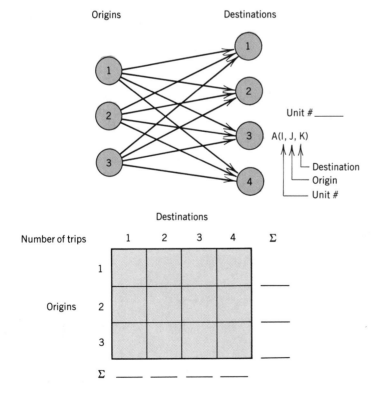

EXAMPLE 7.7 □ **Summation Across Rows and Down Columns of Array to Find Total Number of Trips**

STATEMENT OF PROBLEM

Develop a subprogram that will add across the rows and down the columns of any two-dimensional, $N \times M$ array and store the sums of each in one-dimensional arrays. Use this subprogram to add the data in a three-dimensional flight data array that stores the flight history between selected cities for up to 10 aircraft.

MATHEMATICAL DESCRIPTION

The summations involved in the subprogram are

$$SUMROW_j = \sum_{k=1}^{n} a_{jk} \qquad SUMCOL_k = \sum_{j=1}^{m} a_{jk} \qquad (7.6)$$

ALGORITHM DEVELOPMENT

Input/Output Design

The number of flights from location j to location k for aircraft i can be depicted as FLIGHT(I,J,K). Only one airliner will be used in this example. The aircraft identification number and the number of cities will have to be input first, and then the flight data array. The output of the flight array will be omitted to conserve space. The total flight arrays will be output with appropriate labels. They will be output in a columnar format showing trips from and trips to each city.

Computer Implementation

The total number of trips from location j will be stored in the variable SUMROW(J) and the total number of trips to location k will be stored in SUMCOL(K). Figure 7.11 shows the flowchart for the summation across rows and columns subprogram, as well as the main program to use this subprogram to find the total number of trips to and from each city for the charter airliner. The three-dimensional array of flight data will be converted into a two-dimensional dummy array before calling the row–column summation subprogram. This is because the subprogram for summing is more general if it is developed for a two-dimensional array.

PROGRAM DEVELOPMENT

```
********************************************************************
*  Program for determining total charter flights out of and     *
*    into a particular city                                      *
*  Developed by:  T. K. Jewell                  January 1988     *
********************************************************************
C
C  FLIGHT = three-dimensional array holding flight data from city j to
C           city k for aircraft i
C  SUMROW = array for summation of flights from city j
C  SUMCOL = array for summation of flights to city k
C  DUMMY = dummy two-dimensional array to be sent to summation SUBROUTINE
C  N = number of cities
```

FIGURE 7.11   Flowchart for summation across rows and down columns.

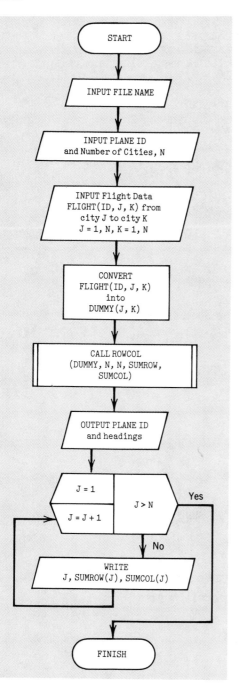

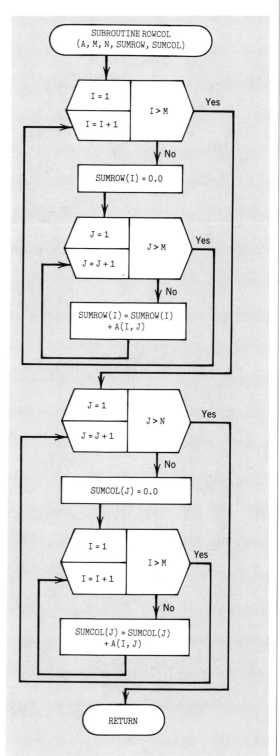

(EXAMPLE 7.7 □ Summation Across Rows and Down Columns of Array to Find Total Number of Trips □ Continued)

```
C  ID = identification number for plane (from 1 to NPLANE)
      INTEGER FR,R,W,N,ID,NPLANE,NCITY
      PARAMETER (FR=4,R=5,W=6,NPLANE=10,NCITY=10)
      REAL FLIGHT(NPLANE,NCITY,NCITY),SUMROW(NCITY),SUMCOL(NCITY),
     1      DUMMY(NCITY,NCITY)
      CHARACTER NAME*14
      PRINT *, 'Input file name for data input  '
      READ (R,'(A14)') NAME
      OPEN (UNIT=FR,FILE=NAME,STATUS='OLD')
      READ(FR,*) ID,N
      DO 100 J=1,N
          READ(FR,*) (FLIGHT(ID,J,K),K=1,N)
  100 CONTINUE
C
C  Convert FLIGHT(ID,J,K) into DUMMY(J,K)
      DO 200 J=1,N
          DO 300 K=1,N
              DUMMY(J,K)=FLIGHT(ID,J,K)
  300     CONTINUE
  200 CONTINUE
C
C  Accomplish summing across rows and columns
      CALL ROWCOL(DUMMY,N,N,SUMROW,SUMCOL)
C
C  Output results
      WRITE(W,10) ID
   10 FORMAT(1X,'TOTAL NUMBER OF FLIGHTS FOR PLANE',I2,//,
     1       1X,' CITY    FROM      TO'/)
      DO 500 J=1,N
          WRITE(W,13) J,SUMROW(J),SUMCOL(J)
  500 CONTINUE
   13 FORMAT(1X,I4,2X,F8.0,F7.0)
      END
C
      SUBROUTINE ROWCOL(A,M,N,SUMROW,SUMCOL)
      INTEGER M,N,MSIZE,NSIZE
      PARAMETER (MSIZE = 10,NSIZE = 10)
      REAL A(MSIZE,NSIZE),SUMROW(M),SUMCOL(N)
C
C  Sum across rows
      DO 100 I=1,M
          SUMROW(I) = 0.0
          DO 200 J=1,N
             SUMROW(I) = SUMROW(I) + A(I,J)
  200     CONTINUE
  100 CONTINUE
C
C  Sum down columns
      DO 300 J=1,N
          SUMCOL(J) = 0.0
          DO 400 I=1,M
              SUMCOL(J) = SUMCOL(J) + A(I,J)
  400     CONTINUE
  300 CONTINUE
      RETURN
      END
```

PROGRAM TESTING

Use the flight data for the six cities shown to test the program.

From City	To City 1	2	3	4	5	6
1	0	10	45	2	21	85
2	14	0	38	31	2	22
3	40	44	0	9	15	19
4	4	26	6	0	2	1
5	18	4	16	1	0	10
6	42	20	16	0	8	0

The output will appear as shown below.

```
TOTAL NUMBER OF FLIGHTS FOR PLANE 1
   CITY    FROM     TO
    1       163.    118.
    2       107.    104.
    3       127.    121.
    4        39.     43.
    5        49.     48.
    6        86.    137.
```

7.5 OUTPUT OF LARGE ARRAYS IN BLOCK FORMAT

Up to this point, we have not tried to output an array with rows longer than one ine of screen output would accommodate. However, many practical problems nvolve much larger arrays. A SUBROUTINE that will output large arrays in a block format will now be developed. These blocks can be connected together to orm a large array.

EXAMPLE 7.8   □   Block Output of Large Array

STATEMENT OF PROBLEM

Develop a SUBROUTINE subprogram that will output a large array in block format. The first block should contain row identifiers, and all columns should be labeled with their indexes.

ALGORITHM DEVELOPMENT

Input/Output Design

The array used to check the SUBROUTINE will be generated from within the main program. The width of the columns, WIDTH, and the number of decimal

(EXAMPLE 7.8 □ Block Output of Large Array □ Continued)

places, DECI, will be input using free format, so different combinations can be tried. It will be assumed that the desired column width will vary from 5 to 10 output columns and that no more than four decimal places will be required. Only F field specifiers will be used, although the subprogram could be modified to accommodate other specifiers. The output format will be a function of the width of the columns and the number of decimal places.

Computer Implementation

If the array to be output is $M \times N$, each block will output M rows and a number of columns that will depend on the required column width. The first block will generally have one less column of array output than subsequent blocks because of the output of row identifiers. A variable format will be used by the SUBROUTINE to accommodate variable column sizes. The last block will have only as many columns as are necessary to finish off the array. For each block, the initial value for the column identifier is indexed to the block number. For example, the initial value for the first block would be 1, for the second block 11, for the third block 21, and so on. The final value for the column index will be the initial value plus nine, or N, whichever is less. Figure 7.12 shows the flowchart for block output.

PROGRAM DEVELOPMENT

```
C  Main program for checking block output of arrays
      INTEGER W,M1,N1
      PARAMETER (W=6,M1=15,N1=25)
      INTEGER WIDTH,DECI,M,N
      REAL A(M1,N1)
C  Generate array to check
      DO 100 I=1,15
         DO 200 J=1,25
            A(I,J) = REAL(I) + 0.1*REAL(J)
  200    CONTINUE
  100 CONTINUE
      M=15
      N=25
      PRINT *, 'Input column width and number of decimal places  '
      READ *, WIDTH,DECI
      CALL WRITEB(A,M,N,W,WIDTH,DECI,M1,N1)
      END
      SUBROUTINE WRITEB(A,M,N,W,WIDTH,DECI,M1,N1)
**************************************************************
*  SUBROUTINE for outputting large arrays in block format   *
*  Developed by:  T. K. Jewell              January 1988     *
**************************************************************
C
C  VARIABLE DEFINITIONS
C  A = array to be output
C  M = # of rows in array
C  M1 = maximum number of rows for dimension
C  N = # of columns in array
C  N1 = maximum number of columns for dimension
C  WIDTH = desired width of output columns
C  DECI = number of decimal places for output
```

```
C  FORM = variable FORMAT for output
C  F1 = array to store character data for first block repeater
C  F2 = array to store character data for subsequent block repeater
C  F3 = array to store character data for column width
C  F4 = array to store character data for # of decimal places
C  COLUMN = array to store column identification numbers
C  INIT = initial value for column index for subsequent blocks after the first
C  NUM = number of columns output per block
C  FINAL = final value for column index for subsequent blocks after the first
       INTEGER M1,N1,M,N,WIDTH,DECI,W,NUM,COLUMN(13),INIT,FINAL
       CHARACTER FORM*40,F1*2(5:10),F2*2(5:10),F3*2(5:10),F4*2(0:4)
       REAL A(M1,N1)
       DATA (F1(J),J=5,10)/'12','10','9','8','7','6'/
       DATA (F2(J),J=5,10)/'13','11','10','8','8','7'/
       DATA (F3(J),J=5,10)/'5','6','7','8','9','10'/
       DATA (F4(J),J=0,4)/'0','1','2','3','4'/
C
C  Output first block
       NUM = 75/(WIDTH + 1)
       IF(NUM.GT.N) NUM = N
       DO 50 J=1,NUM
          COLUMN(J) = J
   50 CONTINUE
       FORM='(3X,'//F1(WIDTH)//'(1X,I'//F3(WIDTH)//'))'
       WRITE(W,FORM) (COLUMN(J),J=1,NUM)
       FORM='(I3,2X,'//F1(WIDTH)//'(1X,F'//F3(WIDTH)//'.'//F4(DECI)//'))'
       DO 100 I=1,M
          WRITE(W,FORM) I, (A(I,J),J=1,NUM)
  100 CONTINUE
C
C  Output subsequent blocks
       FINAL = NUM
       NUM = 80/(WIDTH + 1)
C
C  WHILE (FINAL.LT.N) is true DO 600
  700 CONTINUE
       IF (FINAL.LT.N) THEN
C  Skip to next page for start of new block
          WRITE(W,'(1H1)')
          INIT = FINAL + 1
          FINAL = FINAL + NUM
          IF (FINAL.GT.N) THEN
             FINAL = N
             NUM = N - INIT + 1
          END IF
          FORM='(//,'//F2(WIDTH)//'(1X,I'//F3(WIDTH)//'))'
          DO 200 J=1,NUM
             COLUMN(J) = INIT - 1 + J
  200     CONTINUE
          WRITE (W,FORM) (COLUMN(J),J=1,NUM)
          FORM='('//F2(WIDTH)//'(1X,F'//F3(WIDTH)//'.'//F4(DECI)//'))'
          DO 300 I=1,M
             WRITE(W,FORM) (A(I,J),J=INIT,FINAL)
  300     CONTINUE
C 600 CONTINUE (End of WHILE loop)
          GO TO 700
       END IF
       RETURN
       END
```

(EXAMPLE 7.8   □   Continued)

FIGURE 7.12   Flowchart for block output of large arrays.

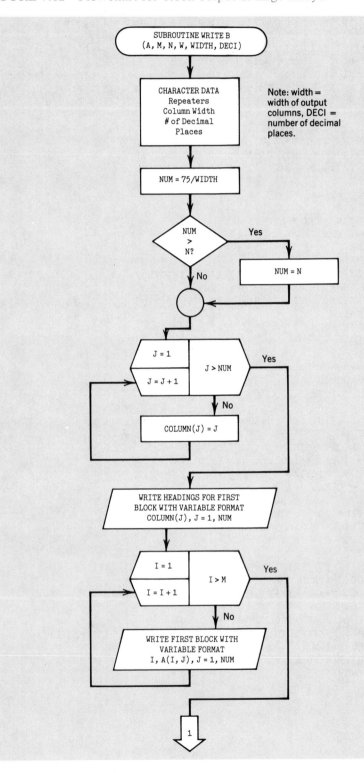

FIGURE 7.12 (Continued)

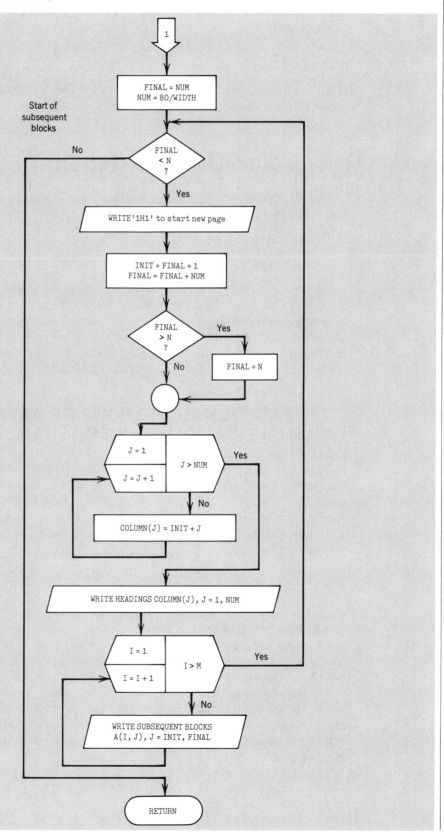

(EXAMPLE 7.8 □ **Block Output of Large Array** □ Continued)

PROGRAM TESTING

Use various column widths and number of decimal places to test out the algorithm. Be sure to check the limits.

7.6 EQUIVALENCE STATEMENT

The EQUIVALENCE statement causes two variables or arrays to use the same memory locations. The statements

```
REAL A(10,10),B(10,10)
INTEGER I,J
EQUIVALENCE (A,B),(I,J)
```

cause the values for the arrays A and B to be stored in the same set of variable locations. This is usually done if A and B are to be used in different parts of the program, so it is permissible to lose the values of one array when the other is used. The values for I and J will be stored in a different common storage location. Use of the EQUIVALENCE statement can conserve memory use for large programs but is confusing and not recommended for structured programming.

7.7 ADDITIONAL METHODS OF DATA TRANSFER AMONG SUBPROGRAMS

Thus far, data transfer to and from SUBROUTINE subprograms, as well as data transfer to FUNCTION subprograms, has been via the dummy and actual arguments. Arrays have had to be declared in both the main program and the subprogram by using DIMENSION or declaration statements. In this section some other methods for transferring data among main programs and subprograms will be discussed.

7.7.1 COMMON Blocks

Common blocks work much like their name implies. The COMMON statement sets aside a block of storage locations. This block of storage locations can be accessed by a main program or any subprograms that are used in an application. You gain access to the storage block through the name of the common block. The positions of the storage locations in the common block are fixed. You can refer to the storage locations using any variable you choose. The value that will be assigned to that variable will depend on the position of the variable in the reference to the common block. When you use common blocks you have to observe the rules for their usage closely. Failure to do so can result in the inadvertent scrambling of data.

COMMON statements can be used to define both arrays and simple variables.

If the type of the variables is to be specified, declaration statements also have to be used. In other words,

```
REAL A,B,X
INTEGER M,N
COMMON A(10,10),B(10),X(10),M,N
```

would define a 10 × 10 real array A, and real arrays B and X , as well as simple integer variables M and N as part of a common block. This same common block could be used in a subprogram to access the values stored. If your common variables do not agree in type and dimension for all subprograms, your subprogram variables may be assigned incorrect values. The names do not have to agree.

The COMMON statement establishes a common storage region in the computer. Data storage methods dictate that the same common region cannot be used for storage of both numeric and character data. Named common, to be discussed in a subsequent section, allows us to define multiple common regions. Some of these can store numeric data, and others can store character data. When common blocks are being used to store variable values that can be accessed from subprograms, those variables are not included in the dummy and actual arguments of the CALL statement or function reference. Variables that are not in the common block are still included in the dummy and actual arguments.

Blank and named common are the two types of common blocks that will be discussed. The use of BLOCK DATA subprograms will also be discussed because they require the use of common blocks.

7.7.1.1 Blank COMMON

Blank common is a group of COMMON statements that are not named in any way to differentiate the parts of the common block. Thus, the entire common block has to appear in every subprogram that accesses common. DATA statements cannot be used to initialize values in blank common.

The following program segments illustrate the use of blank common blocks to assist in reading in student names, identification numbers, and grades for a class. All the data will be sorted into alphabetical order by name. The program will then determine the mean and standard deviation of the grade data. Note that the whole COMMON block appears in each subprogram. Since the common block cannot contain both character and numeric data, the NAME array will not be placed in common. The array NAME and the simple variable N, since they are not in the common block, are included in the actual and dummy arguments. If all the variables that are going to be used in the subprogram are included in the common block, the dummy and actual arguments can be left off. This will be illustrated in the example for named common in the next section.

```
REAL GRADE,GBAR,SDEV
INTEGER ID,N
CHARACTER*30 NAME(50)
COMMON ID(50),GRADE(50),GBAR,SDEV
    .
PRINT *, 'Input number of students'
READ (R,*) N
    .
CALL READ(N,NAME)
    .
```

```
        CALL ALPH(N,NAME)

        CALL CALC(N)
          .
        END
        SUBROUTINE READ(N,NAME)
        REAL GRADE,GBAR,SDEV
        INTEGER ID,N
        CHARACTER*30 NAME(50)
        COMMON ID(50),GRADE(50),GBAR,SDEV
          .
C       Read in n values of NAME, ID, and GRADE
          .
        RETURN
        END
        SUBROUTINE ALPH(N,NAME)
        REAL GRADE,GBAR,SDEV
        INTEGER ID,N
        CHARACTER*30 NAME(50)
        COMMON ID(50),GRADE(50),GBAR,SDEV
          .
C       Rearrange all data in alphabetical order
          .
        RETURN
        END
        SUBROUTINE CALC(N)
        REAL GRADE,GBAR,SDEV
        INTEGER ID,N
        COMMON ID(50),GRADE(50),GBAR,SDEV
          .
C       Compute mean (GBAR) and standard deviation (SDEV)
          .
        RETURN
        END
```

7.7.1.2 Named Common

Named common divides the common block up into named sections. Individual named sections can be accessed by a subprogram, which eliminates the need to specify the entire common block in a subprogram.

The following program segments illustrate the use of named common blocks to assist in the same processing of student grades accomplished with blank common. Note the separate named common block used for the character data of NAME. STUDENT is the named COMMON for NAME, while IDENT is a separate named COMMON for ID, even though they appear on the same line. The program uses a total of four named common blocks. Only the named common block needed for a particular subprogram will be referenced in that subprogram.

```
        REAL GRADE,GBAR,SDEV
        INTEGER ID,N
        CHARACTER*30 NAME

        COMMON/STUDENT/NAME(50)/IDENT/ID(50)
        COMMON/SCORE/GRADE(50),N
        COMMON/STAT/GBAR,SDEV

          .
        PRINT *, 'Input number of students'
        READ (R,*) N
```

```
CALL READ
    .
CALL ALPH
    .
CALL CALC
    .

END

SUBROUTINE READ

REAL GRADE
INTEGER ID,N
CHARACTER*30 NAME

COMMON/STUDENT/NAME(50)/IDENT/ID(50)
COMMON/SCORE/GRADE(50),N

    .
C     Read in n values of NAME, ID, and GRADE
    .
RETURN
END

SUBROUTINE ALPH

REAL GRADE
INTEGER ID,N
CHARACTER*30 NAME

COMMON/STUDENT/NAME(50)/IDENT/ID(50)
COMMON/SCORE/GRADE(50),N

    .
C     Rearrange all data in alphabetical order
    .
RETURN
END

SUBROUTINE CALC

REAL GRADE,GBAR,SDEV
INTEGER N

COMMON/SCORE/GRADE(50),N
COMMON/STAT/GBAR,SDEV

    .
C     Compute mean (GBAR) and standard deviation (SDEV)
    .
RETURN
END
```

It is permissible to use a combination of blank and named common in the same program. None of the programs that you will be developing in this course will require the use of COMMON blocks, but you should be aware of their usage in case you encounter them in programs you are working with. It is advisable to avoid using common blocks in your programming when you can. Their overuse can make it difficult to produce general subprograms.

7.7.2 BLOCK DATA

A BLOCK DATA subprogram can take the place of long DATA blocks in program segments. BLOCK DATA is used in conjunction with named COMMON. This allows a program to access the data from the main program, or from any subprograms, when the appropriate named COMMON is included in the program segment. The program generally searches the data of the BLOCK DATA subprogram and extracts the information it needs. An appropriate use of BLOCK DATA would be the storage of data on section properties for various shapes of structural steel members. The program would extract the data for a particular section based on input data.

BLOCK DATA takes up computer memory. However, it has the advantage of being immediately available when it is needed during program execution. The alternative to BLOCK DATA would be to use file input, which would require a search for each item of information. The file would have to be rewound before another search could be made. This would slow down program execution.

Example 7.9 shows an example of a BLOCK DATA subprogram for storing solar heat gain data. Some of the important data for estimating the portion of home heating that can be accomplished by the sun are the total amount of heat gain available from the sun and the number of degree days for a given month of the heating season. Other supporting data are also given in the example. Note how the different types of data are put in separate named COMMON blocks, so only the needed parts have to be referenced when they are used in subprograms. These data are used with permission of UTS/Building Design and Construction TK Solver Pack/Universal Technical Systems, Inc., 1983.

EXAMPLE 7.9 □ **Solar Heat Gain**

STATEMENT OF PROBLEM

Develop a BLOCK DATA subprogram for storing solar heat gain data.

ALGORITHM DEVELOPMENT

Computer Implementation

The data to be given here can be divided into three main groups: city data, degree day data, and the solar heat gain factor for different latitudes. City data includes name, latitude, and percentage of sunshine. There are 26 cities represented. Degree day data are given for the months of January, February, March, April, May, October, November, December, and the whole heating season for each city. Solar heat gain factors are given for five latitudes and each of the degree day reporting periods. The three major subdivisions will be used for labeling the COMMON blocks.

PROGRAM DEVELOPMENT

```
BLOCK DATA
CHARACTER*12 CITIES(26)
INTEGER LATIT(26)
REAL PERC(26),JAN(26),FEB(26),MAR(26),APR(26),MAY(26),
```

```
1      OCT(26),NOV(26),DEC(26),SEASON(26),GAIN24(9),GAIN32(9),
2      GAIN40(9),GAIN48(9),GAIN56(9)
 COMMON /CITY/CITIES/LOCAT/LATIT,PERC
 COMMON /DEGDAY/JAN,FEB,MAR,APR,MAY,OCT,NOV,DEC,SEASON
 COMMON /GAIN/GAIN24,GAIN32,GAIN40,GAIN48/GAIN56
 DATA /CITIES/'Albuquerque','Atlanta','Boise','Boston',
1      'Buffalo','Charleston','Chicago','Dallas','Denver',
2      'Jacksonville','LasVegas','LosAngeles','Louisville',
3      'Minnepolis','Montreal','Nashville','NewOrleans','NewYork',
4      'Phoenix','StLouis','SanFrancisco','Seattle','Toronto',
5      'Vancouver','Washington','Winnipeg'/
 DATA/LATIT,PERC/35,34,44,42,43,33,42,33,40,30,36,34,38,45,46,36,
1      30,41,33,39,38,48,44,49,39,50,77,61,67,60,52,65,57,65,70,61,
2      85,73,57,58,43,58,59,59,86,59,67,49,44,38,58,50/
 DATA /JAN,FEB,MAR,APR,MAY,OCT,NOV,DEC,SEASON/930,636,1113,1088,
1      1256,443,1150,601,1004,332,688,310,930,1631,1510,778,344,
2      986,474,977,443,738,1233,862,871,2008,703,518,854,972,1145,
3      367,1000,440,851,246,487,230,818,1380,1328,644,241,885,328,
4      801,336,599,1119,723,762,1719,595,428,722,846,1039,273,868,
5      319,800,174,335,202,682,1166,1138,512,177,760,217,651,319,
6      577,1013,676,626,1465,288,147,438,513,645,42,489,90,492,21,
7      111,123,315,621,657,189,24,408,75,270,279,396,616,501,288,
8      813,81,25,245,208,329,0,226,6,254,0,6,68,105,288,288,40,0,
9      118,0,87,239,242,298,310,74,405,229,124,415,316,440,34,279,
*      62,366,12,78,31,248,505,496,158,12,233,22,202,118,329,439,
1      456,217,683,642,417,792,603,777,210,705,321,714,144,387,132,
2      609,1014,864,495,165,540,234,576,231,543,760,657,519,1251,
3      868,648,1017,983,1156,425,1051,524,905,310,617,229,890,1454,
4      1355,732,291,902,415,884,388,657,1111,787,834,1757,4348,
5      2961,5809,5634,7062,1794,5882,2363,5524,1239,2709,1349,4660,
6      8382,7899,3578,1254,4871,1765,4484,3001,4424,6827,5515,4224,
7      10679/
 DATA /GAIN24,GAIN32,GAIN40,GAIN48,GAIN56/52080,39144,28396,
1      14640,11594,41850,49500,53816,290620,52886,43680,36518,
2      21600,15500,46686,50280,52824,319474,50406,45976,43028,
3      29280,22196,49042,47880,48050,335858,43462,45528,47554,
4      36300,29946,48298,41160,38192,330440,30194,41272,49600,
5      41940,37448,43524,28500,21266,293744/
 END
```

PROGRAM TESTING

These data will be used in this chapter for chapter-end exercises and for examples and exercises in later chapters.

SUMMARY

Multidimensional arrays can be used to represent many realistic engineering problems. Two-dimensional arrays that depict matrices are the type that are probably the most familiar to you. The structure of FORTRAN makes it easy to work with the rows and columns of matrices and to develop procedures for implementing matrix operations. In addition to matrices, any problem for which a dependent variable is dependent on more than one independent variable is a good candidate for applying multidimensional arrays. Each index of the subscripted variable representing the dependent variable would correspond to a different independent variable. When you are working with arrays, you must have a clear picture of how you are organizing your data in the array, as well as how you will access and work with it. Subscripted algebraic variables are a big help in this. Use arrays where necessary, but do not overuse them. If a program can be written effectively using simple variables, the forced inclusion of subscripted variables could well obscure the meaning of the program. When you are using subscripted variables, try to use the minimum number of dimensions possible.

In this chapter you have become more familiar with matrix operations and use of other two- and three-dimensional arrays. You have also learned some new tools, among them the block output of large arrays, the COMMON statement and common blocks, and the BLOCK DATA subprogram. This completes your formal instruction on the syntax of the FORTRAN language. In the remaining chapters of the text you will see some additional FORTRAN applications. You will also be introduced to several generic microcomputer packages that are superior to FORTRAN for some applications. You will learn how to decide which package, including FORTRAN, is appropriate for a particular application.

REFERENCES

Borse, G. J., *FORTRAN 77 and Numerical Analysis for Engineers*, PWS Publishers, Boston, 1985.

Ellis, T. M. R., *A Structured Approach to FORTRAN 77 Programming*, Addison-Wesley, London, 1982, Chapters 9 and 11.

International Mathematical and Statistical Library (IMSL) Inc., "Math/Library, FORTRAN Subroutines for Mathematical Applications," Version 1.0, Houston, TX 1987.

James, M. L., Smith, G. M., and Wolford, J. C., *Applied Numerical Methods for Digital Computation*, Harper & Row, New York, 1985, Example 8-1, pp. 591–596, used with permission.

Nyhoff, L., and Leestma, S., *FORTRAN 77 for Engineer's and Scientists*, Macmillan, New York, 1985, Chapters 7, 10, and 12.

Page, R., Didday, R., and Alpert, E., *FORTRAN 77 for Humans*, 2nd ed., West Publishing Co., St. Paul, MN, 1983, Chapters 8 and 14.

Red, E. W., and Mooring, B., *Engineering: Fundamentals of Problem Solving*, Brooks/Cole Engineering Division, Wadsworth, Inc., Monterey, CA, 1983.

EXERCISES

7.1 GENERAL NOTES

All output for programs in this chapter should be formatted.

7.2 MODIFICATION OF EXAMPLES

7.2.1 Example 7.1

1. Write a program to extract the data and make the calculations of Example 7.1 using the three database files rather than the sorted arrays.
2. Modify the program of Example 7.1 to save the sorted database in a file.
3. Modify the program of Example 7.1 to extract the same information and compute the same data, but to do it for all employees whose last name starts with a letter included between two input letters. Modify the program to use the data for employees whose employee identification number falls between two input limits.

7.2.2 Examples 7.2 and 7.3

1. Develop a program to read in an $n \times m$ matrix, determine its transpose in a subprogram, and store the transpose in a separate array. Output the matrix and its transpose. Multiply the matrix by its transpose, and then the transpose by the original matrix to verify the sizes of the resulting matrices. Output both product matrices. Use the following matrix to test your program:

$$
\mathbf{A} = \begin{matrix}
2.0 & 3.5 & 4.3 \\
1.6 & 4.8 & 3.98 \\
0.4 & 6.8 & 1.0 \\
25. & 12. & 6.7
\end{matrix}
$$

2. Develop a program to multiply the matrix **A** for part 1 above and the matrix **B** given below in both orders to demonstrate that matrix multiplication is, in general, not commutative.

$$
\mathbf{B} = \begin{matrix}
1 & 2 & 3 & 4 \\
2 & 3 & 4 & 5 \\
3 & 4 & 5 & 6
\end{matrix}
$$

3. Try to multiply the matrices **A** and **B** of Example 7.2 in the reverse order to show that they are not compatible.
4. Combine the programs and subprograms of Examples 7.2 and 7.3 to find the solution of the set of equations

$$
\begin{aligned}
2X + 3Y + Z &= 4 \\
Y + 2Z &= 6 \\
4X + 2Y &= 12
\end{aligned}
$$

You will need to substitute a call to the matrix inversion subprogram available on your computer system for the BAS$MAT_INV call in SUBROUTINE MATINV.
5. Verify that the inversion routine used in the previous exercise is correct by multiplying the original matrix by the inverse to see if the identity matrix results.
6. Do the previous problem using the alternate matrix inversion routine.
7. Verify that multiplication of a matrix by its inverse is commutative by doing the multiplication in the opposite order.

7.2.3 Example 7.5

Modify the program of Example 7.5 so that several input hydrographs can be generated, each with different upstream conduits feeding in.

7.2.4 Example 7.7

1. Modify the program of Example 7.7 to find the total number of flights to and from the six cities, and the total number of flights between any two cities, for five charter airplanes. Output the totals for each plane, as well as the grand totals.

2. Modify and combine the programs of Examples 7.6 and 7.7 to determine the total number of units of four different products that are shipped from warehouse i to store j in a one-year period, as well as the total units of products that are shipped between any warehouse and any store.

7.3 GENERAL SYNTAX AND STATEMENT STRUCTURE

7.3.1 Character Arrays

1. Describe what the following character specification statements accomplish:

```
CHARACTER*8 A(10,20), C(30), D
CHARACTER F(10,15)*6, G*9, H*2
```

*2. Write a FORTRAN program that will develop the following pattern in a 15 × 15 character array and output the result.

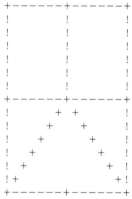

3. Write a FORTRAN program that will produce the following pattern in an 11 × 11 character array and output the pattern:

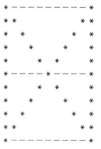

7.3.2 COMMON Blocks

Identify and correct errors in the following program segments. Default variable types are in effect, except where noted.

a.   COMMON ZX,AI
       .
       CALL THT
       .
       END
       SUBROUTINE THT
       COMMON I,J
       .
       RETURN
       END

*b.   REAL M(5)
       COMMON A,B,M
       .
       CALL MTMT
       .
       END
       SUBROUTINE MTMT
       COMMON X,Y,M
       .
       MTM=1
       RETURN
       END

c.   COMMON X,A,P,T
       COMMON G(5),H(5)
       .
       CALL DIM(T)
       .
       END
       SUBROUTINE DIM(A)
       COMMON TX,TP,TA
       COMMON TG(5),TH(5)
       .
       RETURN
       END

*d.   REAL X(10),Y(10),A(10,10)
       COMMON B(30),C(20),Y(10)
       .
       END

e.   REAL X(10),Y(10)
   100 COMMON N,Z
       DO 200 I=10
       .
   200 CONTINUE
       .
       GO TO 100
       .
       END

7.4 CHEMICAL ENGINEERING

7.4.1 [ALL] Ideal Gas Law

Modify the program of Exercise 4.4.1 or 5.4.1 to produce tables showing the variation
of pressure with both temperature and volume using each of the three models: ideal,
van der Waals, and Redlich–Kwong. The tables should be output with appropriate la-

bels along the edges identifying the temperatures and volumes. Store the pressure data in a three-dimensional array with i(row) for temperature, j(column) for volume, and k(depth) for the model (1 = ideal, 2 = van der Waals, 3 = Redlich–Kwong). Use an $11 \times 11 \times 3$ array. Vary the volume from 100 to 120 liters in 2-liter increments, and the temperature from 25 to 45°C in increments of 2 degrees. Other data should be the same as Exercise 4.4.1.

7.4.2 [ALL] Chemical Kinetics

Modify the program of Exercise 4.4.2 or 5.4.2 to show how concentration varies with both time and initial concentration for reactions that follow first or second order kinetics. Appropriate labels should be included with the output. Store the data in an $11 \times 11 \times 2$ array. The i(row) subscript should be for initial concentrations, the j(column) subscript for time, and the k(depth) subscript for the kinetics order. Compare your results with what you would expect from the models. Increment time from 0 to 200 min in 20-min increments. Vary the initial concentration from 0.05 mol/ℓ to 0.10 mol/ℓ in increments of 0.005.

7.4.3 [ALL] Depth of Fluidized-Bed Reactor

Modify Exercise 5.4.3 so that you can develop a table that shows the variation of expanded bed height with varying particle density and reactor bed flow rate. Use particle densities from 2,000 to 3,000 kg/m^3 in increments of 100 kg/m^3, and bed flow rates of 0.002 to 0.010 m/s in increments of 0.001. Store the values for the table in a two-dimensional array.

7.4.4 [ALL] Process Design, Gas Separation

Modify the program of Exercise 4.4.4 or 5.4.4 to develop a table that shows the absorption efficiency of up to five constituents when each of the constituents in turn is subjected to a certain percent absorption. The constituent along the diagonal of the table will be the design constituent, and the remainder of the row will show the removal efficiencies of the other constituents. Data along the diagonal will show the number of plates required to achieve the given absorption efficiency for the selected constituent. Use a two-dimensional array to store the data for the table. Test your program with the data of Exercise 4.4.4.

7.4.5 [ALL] Dissolved Oxygen Concentration in Stream

Modify the program of Exercise 4.4.5 so that you can produce a three-dimensional array containing concentration as a function of distance downstream when the coefficients of deoxygenation and reoxygenation are subject to variation. Rows of the array should vary with the distance downstream, columns with the deoxygenation coefficient, and the depth of the array with the reoxygenation coefficient. Use an $41 \times 8 \times 4$ array. You should be able to output any or all of the layers of the array. This will produce a table of concentration as a function of distance downstream and deoxygenation coefficient for a given reoxygenation coefficient. Test your program for the data of Exercise 4.4.5, with deoxygenation coefficients varying from 0.1 to 0.8 day^{-1}, in increments of 0.1, and the reoxygenation coefficient varying from 0.4 to 0.7 day^{-1}, in increments of 0.1.

7.5 CIVIL ENGINEERING

7.5.1 [ALL] Equilibrium-Truss Analysis

Develop separate subprograms to input the data for Exercise 6.5.1 from either a data file or from the terminal. Include the option of saving the terminal input data to a file. Combine your program with the program of Example 7.4 to determine the forces in the members of the truss. Provide an option in your program for modification of the loading and recomputation of the forces without reinverting the coefficient matrix.

7.5.2 [ALL] Open Channel Flow Analysis

1. Develop a program that will store flow rate values for a series of depth and bottom width values for a trapezoidal open channel. Each row of the array will represent a different depth, and each column will represent a different bottom width. Use $n = 0.03$, $S_0 = 0.001$ ft/ft, and $z = 2$ ft/ft. Develop the array for depths from 0.0 to 10.0 ft in increments of 0.5 ft, and bottom widths from 4.0 to 10.0 ft in increments of 1.0 ft.
2. Develop a search routine that can be used to estimate the depth of flow for a given bottom width and flow rate using the tabular data from Part 1 of this exercise. If there is no depth that satisfies the conditions, an error message should be output and new input requested. Use linear interpolation to refine the value of depth when the flow rate falls between two values of flow rate in the table. Test your program with both values that will fall in the table and outside it.

7.5.3 [ALL] Dam Stability Analysis

Modify the program of Exercise 6.5.3 so that you can produce two tables of output. Use a two-dimensional array to store the interior data for each table. Axis information should be in separate arrays. The first table will contain values of the location of the normal force for various values of water height h and dam bottom width b. The second table will contain character values that will indicate whether or not each h-b combination meets all stability criteria. You can assume that the height of the dam is 20 ft higher than the height of water for all cases. To test your program, use water heights varying from 30 to 130 ft and bottom widths varying from 80 to 200 ft. The top width is set at 40 ft for all cases and the downstream cutoff height is 10 ft.

7.5.4 [ALL] Force on Submerged Gate

The circular gate system shown in the accompanying figure is designed to open automatically when the water in the tank reaches a certain height. The gate will begin to open when the moment caused by the weight is less than the moment caused by the force of the water on the gate. The critical point is where the two moments are exactly equal. Use the concepts developed in Exercise 6.5.4 and the subprogram developed in Exercise 5.5.4. Develop a program to estimate what weight is required for a given height a of the pivot point above the centroid of the gate and a given distance b of the weight moment arm. Have the program develop a table of weights as a function of h_c and b. To test your program, use a value of a of 5 ft and a value of d_0 of 2 ft. Vary h_c from 2 to 20 ft in increments of 0.5 ft, and b from 1 to 10 ft in increments of 0.5 ft.

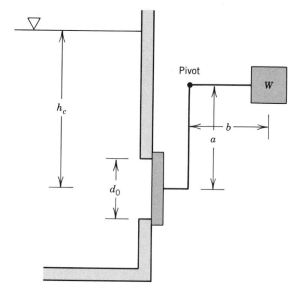

Figure for Exercise 7.5.4

7.5.5 [ALL] Volume of Excavation

Generalize your program of Exercise 6.5.5 to accept cross sections that are quadrilateral but not necessarily trapezoidal. Quadrilateral cross sections are what you will encounter when estimating the amount of cut or fill required for a highway project. Input for each cross section should be the coordinates of the corners of the quadrilateral. Store the data for the cross sections in a 51 × 8 array to accommodate up to 51 cross-sections and four pairs of corner coordinates.

7.5.6 [6,8] Estimation of Flow in a Channel of Irregular Cross Section

Modify the program of Exercise 6.5.7 so that it can solve for a number of flow rates for measurements made at the same cross section at different times. Use a three-dimensional array to store the input geometry and velocity data. The third dimension should be for the different sets of data.

7.6 DATA ANALYSIS AND STATISTICS

7.6.1 [ALL] Histogram

Develop a subprogram that will produce a horizontal histogram for up to 25 data groups. Include axis values, labels, and appropriate title information. The bars of the histogram should be made up of a series of Xs, with the number of Xs proportional to the value being displayed. Use the subprogram to display the data developed for Exercise 6.6.1.

7.6.2 [ALL] Sorting

Develop a program that will sort a two-dimensional array into descending order so that the largest value of the array is in the leftmost position in the first row, the second largest value in the second position of the first row, and so on, until the smallest value s in the lower right-hand corner of the array. Provide the option of accomplishing the sort in ascending order. Both the original and sorted array should be saved and output. Up to an 11 × 11 array should be accommodated. Test your program with a 6 × 6 random array.

7.6.3 [4,5,6] Sampling and Distribution of Means

Use the program of Exercise 7.6.1 to output a histogram of the distribution of means from the sample space of Exercise 6.6.3.

7.7 ECONOMIC ANALYSIS

7.7.1 [ALL] Economic Formulas

Develop a program that will produce a table of values for any of the economic formulas of Appendix E, for a specified number of compounding periods and a range of interest rates. The choice of the formula to be used should be input. Use the FUNCTION subprograms of Exercise 5.7.1 to generate the formula values. Formula values should be stored in a two-dimensional array. Each row of the array will contain the selected formula values for either a whole or a tenth of a percent increment in the interest rate from the previous row. Thus, the number of rows used in the array will be dependent on the range of interest rates that the user chooses, and the increment chosen (either whole or one-tenth percent). The array will contain 10 columns. Each column will contain values for the economic formula in increments one-tenth as large as the increment between rows. If the interest rate for the row is 10.1 percent, the first column would contain the formula for 10.10 percent, the second the value for 10.11 percent, and so forth, with the last column containing the value for 10.19 percent. Thus, you could look up the formula value accurate to two decimal places. If increments of 1 were used for the rows, the columns would show the values to one decimal place. When you output your array, include appropriate labels for the rows and columns.

7.7.2 [ALL] Nonuniform Series of Payments

A factory uses n machines for accomplishing the same task. A two-dimensional array contains data on the return earned by each machine for each of the past m years. Develop a program that will determine which of the machines has earned the most in terms of present worth. Machines should be rank ordered in descending order. Use the data in Table 7.1 to test your program:

TABLE 7.1

		Year					
		1	*2*	*3*	*4*	*5*	*6*
	1	$5,000	$2,000	$4,000	$6,000	$8,000	$9,000
	2	$6,000	$5,500	$7,500	$6,000	$6,800	$4,000
Machine	3	$4,000	$4,500	$5,500	$6,500	$7,500	$8,500
	4	$7,000	$6,500	$7,500	$3,000	$3,500	$5,500
	5	$5,500	$5,500	$6,000	$5,500	$6,000	$1,000

7.7.3 [ALL] Multiple Interest Rates or Number of Compounding Periods

Develop a program to store values for any of the economic factors in a two-dimensional array. The number of compounding periods will vary with the rows and the interest rate will vary with the columns. The upper limit, lower limit, and increment for the number of compounding periods and the interest rates should be input variables. The array should have the capability of being up to 25 × 25. Output should be appropriately labeled. Test your program with 2 to 20 compounding periods in increments of 2, and interest rates from 8 to 12 percent in 0.5 percent increments.

7.7.4 [ALL] Internal Rate of Return

Develop a program to compute the internal rate of return for a range of compounding periods and annual payments for a given present worth. Store the results in an 11 × 11 two-dimensional array. Output the array with appropriate labels. Test your program for a present worth of $20,000, annual payments from $3,000 to $5,000 in $500 increments, and number of compounding periods from 8 to 12 in increments of one.

7.7.5 [ALL] Mortgage Computations

1. Develop a program that will store a loan amortization table (values in the table will be the required monthly payments) for a given interest rate, a selected range of number of payments, and a selected range of principals. Store the data in an 11 × 11 two-dimensional array. Test your program for 48, 60, 72, 84, and 96 payments for loans of $10,000, $12,000, and $15,000 at an interest rate of 10 percent.
2. Accomplish the same analysis, except for a range of interest rates and a range of repayment periods. Use interest rates of 9, 10, 11, and 12 percent for 48, 60, 72, 84, and 96 payments for a $10,000 loan.

7.8 ELECTRICAL ENGINEERING

7.8.1 [ALL] Diode Problem

Develop a table of current values for the diode problem of Exercise 4.8.1 using formatted output. The table should have 11 rows and 6 columns. Each row will represent a different value of voltage between two specified limits. The top row will be the smallest voltage and the bottom row will be the highest voltage. Voltages for the intermediate rows will be developed using equal increments between the limits. The columns will represent increasing temperature, in equal increments, between two specified temperatures. The voltages and temperatures should also be printed out with the table as row and column labels.

7.8.2 [ALL] Capacitance and Resistance

Using Kirchoff's laws, develop a program that could be used to estimate the current through each branch of the circuit shown in Figure I.3 of Appendix I. Initially use $I_{1,out} = 10$ mA, $I_{3,out} = 10$ mA, and all resistances equal to 1,000 Ω to test your program. Estimate voltage drop across each resistor as well as the current. Also test your program with $I_{1,out} = 40$ mA, $I_{3,out} = 10$ mA.

7.8.3 [ALL] Signal Processing Circuits

Modify the program of Exercise 6.8.3 to include a two-dimensional array to store the required voltage across the resistor for the tuning circuit for a range of half-widths and a range of tuning frequencies. You should be able to accommodate up to 20 frequencies and up to 6 half-widths. Use the frequencies selected for Exercise 4.8.3 and half-widths of .1, .2, .4, and .8 times f_0. Output the results with appropriate labels.

7.9 ENGINEERING MATHEMATICS

7.9.1 [ALL] Determinant and Cramer's Rule

Develop a program that will solve either a 2×2 or 3×3 system of linear equations by Cramer's rule and determinants. Use the subprograms developed in Exercise 5.9.1. Test your program by solving the following set of linear equations:

$$
\begin{aligned}
2X_1 + 3X_2 + X_3 &= 4 \\
X_2 + 2X_3 &= 6 \\
4X_1 + 2X_2 \phantom{{}+ 2X_3} &= 12
\end{aligned}
$$

7.9.2 [ALL] Vector Cross Product, Unit Vector, Directional Cosines, Vector Components

Develop a program that can be used to find the moment of a three-dimensional force vector applied at a specified point on a three-dimensional moment arm. Use the subprograms of Exercise 5.9.2 to assist in developing the necessary data for the force and moment vectors. Use your program to find the moment of the system used to illustrate the vector cross product in Section 1.7 of Appendix F. Output should include magnitude, components, and unit vector for the moment vector.

7.9.3 [ALL] Centroid and Moment of Inertia

Modify your program of Exercise 6.9.3 so that you can produce and output two tables of moments of inertia, one about the x-x axis, and one about the y-y axis, for I beams of varying web height w and varying flange width b. The variables are shown in the figure for this exercise. Flange thickness t_1 and web thickness t_2 will be constant for the table. Test your program with web and flange thicknesses of 0.5 in. Vary w from 6 to 14 in., in 1-in. increments. Vary b from 4 in. to 10 in., in increments of 0.5 in.

7.9.4 [ALL] Population Growth

Another model for population growth is called the saturation model. The growth rate slows down as the population increases. The saturation model has the form

$$ P = P_0 + a \left(\frac{t}{b + t} \right) $$

Develop a program that can be used to investigate the behavior of this model as a function of time for various values of the parameter b. Store the data in a two-dimensional array with rows for different values of time and columns for different values of b. Compare the values for this model with the exponential growth model used in previous population growth exercises. Vary b from 2 to 10 in increments of 1, and vary time from 1 to 20 yr to test your program. Use INCR = 0.05 and $a = 0.5$.

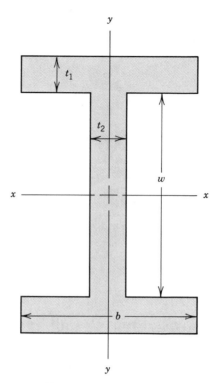

Figure for Exercise 7.9.3

7.9.5 [ALL] Vertical and Horizontal Curves

Develop a program that can be used to help fit a circular curve of radius of curvature r between two lines that approach each other so that each end of the curve is tangent to one of the lines. One method of accomplishing this is by specifying one point on the first line as the first end point of the curve, calculating the coordinates of selected points on the curve, and comparing them with the second line until the curve approaches the second line tangentially. You can determine the required angle of curvature for the circular arc by the angle the two lines make with each other. In order to be tangent to both lines, the point of curvature has to be at the point of intersection of perpendicular lines drawn through the point of tangency. This is illustrated in the figure for Exercise 2.8.5. Use a two-dimensional array to store several sets of curvature data for different initial points. Use the rows for the variation of the angle, and the columns for different starting points. Output the points of tangency and the center of curvature. Use your program to fit a curve of radius 4.0 between the lines $X = 2$ and $Y = 0.167X + 8.5$.

7.10 INDUSTRIAL ENGINEERING

7.10.1 [ALL] Project Management

If each activity of a project requires a certain weekly level of application of a resource, develop a subprogram for your program of Exercise 6.10.1 that will determine the total requirements for the resource each week of the project. Test your subprogram with the network of Figure J.1, and the following data:

Activity	Number of Trucks
1–2	3
3–4	2
2–4	3
2–5	5
4–5	4
4–7	5
6–8	2
7–9	7

7.10.2 [ALL] Quality Control

Modify the program of Exercise 6.10.2 to use a two-dimensional array to store the defect data. Each row should represent a particular defective part, and each column a particular defect. Add the capability of determining the fraction of each defect found if only the first defect uncovered was logged. Compare these statistics with the previous fractions.

7.10.3 [ALL] Managerial Decision Making

Modify the program of Exercise 6.10.3 so that you can test the reaction of the break-even point to a range of values for the fixed costs and the linear coefficient of the return function. Store the data in a two-dimensional array, and output it with appropriate labels. Up to 11 values for the fixed cost and up to 6 values for the linear return should be accommodated. Test your program with fixed costs of $8,000, $10,000, $12,000, $15,000, and $20,000 and linear returns of $6P$, $8P$, $10P$, and $12P$.

7.10.4 [ALL] Queuing (Wait Line) Theory

Modify the program of Exercise 6.10.4 so that the sensitivity of the results can be tested for variations in the reliability of machines and the quality of the work done by repairmen. Store the results in a 10 × 10 array. Machine reliabilities to be tested are failure rates from 0.01 to 0.04 per hour in increments of 0.005, with three different levels of repairman proficiency. The first type of repairman can repair 0.6 units/hr, but charges $25/hr. The second can repair 0.4 units/hr and charges $16/hr. The third can only repair 0.2 units/hr at a rate of $12/hr. Output the results with appropriate labels.

7.11 MECHANICAL AND AEROSPACE ENGINEERING

7.11.1 [ALL] Shear and Bending Moment

Modify the program of Exercise 6.11.1 so that the maximum bending moment can be found when one or both of the supports are moved toward the center of the span, leaving a cantilever end. Store the results in an 11 × 11 two-dimensional array, with the rows representing different locations for the left reaction and the columns representing different locations for the right reaction. Test your program with the data of Exercise 4.11.1 and with the left reaction at 0, 2, 4, 6, and 8 ft and the right reaction at 14, 16, 18, and 20 ft.

7.11.2 [ALL] Crank Assembly Analysis

Modify the Program of Exercise 6.11.2 to produce two-dimensional arrays for each of the computed values based on variations in the angle and length of the crank. Use the angle data of Exercise 4.11.2 and other data of Exercise 3.11.2, with crank lengths of 2.5, 3, 3.5, and 4 in.

7.11.3 [ALL] Friction

Modify the program of Exercise 6.11.3 so that an array of character data will be generated that shows what type of action takes place for application of forces of varying magnitude at various heights on the crate. The entries in the array should be from among the following choices:

NO MOVEMENT

DOLLY SLIPS

CRATE SLIPS

CRATE TIPS

Test your program for heights of 0, .25, .5, .75, 1.0, 1.25, and 1.5 ft and forces of 50, 100, 150, 200, and 250 lb. Other data should be the same as for previous friction exercises. Output the results with appropriate labels.

7.11.4 [ALL] Variation of Atmospheric Pressure with Altitude

Modify the program of Exercise 4.11.4 so that an 11×11 two-dimensional array will store sensitivity data for the constant lapse rate model with variation in the initial temperature T_0 and the lapse rate α. Use temperatures from 40 to 100° Fahrenheit in increments of 10° and lapse rates of -0.003, -0.00325, -0.0035, -0.00375, and -0.004°R/ft. Output the results with appropriate labels.

7.11.5 [ALL] Aerospace Physics

Modify the program of Exercise 6.11.5 to store trajectory data for either a constant thrust or a constant acceleration rocket with variation in the burnout time. After burnout the trajectory should continue until the trajectory impacts a surface level with the point of launch. Model the burnout characteristics for original masses of 30,000, 40,000, 50,000, and 60,000 kg. Other data should be the same as Exercise 6.11.5.

7.11.6 [ALL] Energy Loss in Circular Pipeline

Develop a program that can be used to find the required pipe diameter to carry a given flow rate with a given energy loss. The user should have the option of using any of the formulations for f. Use the data of Example 4.1 to test your program.

7.11.7 [ALL] Torsion

Modify the program of Exercise 6.11.7 to produce a two-dimensional array containing angular deformation data for ranges of inside and outside diameters. Use ID = 2.0, 2.25, 2.5, 2.75, and 3.0 in. and OD = 4.0, 4.5, 5.0, 5.5, and 6.0 in. to test your program. Output the data with appropriate labels.

7.11.8 [8] Heat Loss and Solar Gain

Example 7.9 presents a BLOCK DATA subprogram containing solar heating data. This is the same data used for the TK SOLVER model illustrated in Exercise 8.10.8. Exercise 8.10.8 also presents equations that can be used to predict the solar heat gain through south-facing windows. The BLOCK DATA subprogram contains much of the background data that is needed for 26 selected North American cities. Develop a program that can extract the necessary information from the BLOCK DATA for a given city and use it to determine the solar heat gain using the mathematical model of Exercise 8.10.8. When searching for the GAIN data in COMMON block GAIN, data are given for only the five selected latitudes, whereas the actual latitude of the city is given in the LATIT data. Your program should extract the GAIN data for the lower latitude of the GAIN block interval into which the actual latitude of the city falls. For example, the actual latitude of St. Louis is 39°, so the GAIN32 data would be used. Test your program using the data of Exercise 8.10.8.

7.12 NUMERICAL METHODS

7.12.1 [ALL] Interpolation

Modify the subprogram of Exercise 6.12.1 so that you also have the option of determining the coefficients of the cubic interpolating polynomial. Develop estimates of the function over the interval of points used to generate the interpolating polynomial in Exercise 6.12.1 and compare them with the actual polynomial values.

7.12.2 [ALL] Solution of Polynomial Equation

Modify the program of Exercise 4.12.2 so that when the sign of the function changes, the increment is divided by 10 and the search routine is reversed to search for the root with the smaller increment. Repeat the division and reversal process until the increment used is less than or equal to some predetermined value. The final estimate of the root should be found using this value for the increment. Use the data of Exercise 4.12.2, an initial increment of 0.1, and a final increment of 0.001 to test your program. Output sufficient data so that you can observe the reversal and convergence process.

7.12.3 [3,4,5,8] Data Smoothing

A two-dimensional array $D(10,10)$ is filled with velocity data taken in a grid pattern across a wind tunnel with a hot-wire anemometer. You want to apply a smoothing routine to these data to help reduce the effects of random errors. To do this, you will generate a new array E, the elements of which will be calculated by

$$E_{i,j} = \frac{(D_{i,j-1} + D_{i-1,j} + D_{i,j} + D_{i+1,j} + D_{i,j+1})}{5}$$

The formulas for the boundary and corner elements of E should be modified to account for the lack of data outside the array boundaries. Develop a subprogram that will accomplish this smoothing. Also develop a main program to test the subprogram.

7.12.4 [4,5,6,8,9] Saddle Point of a Hyperbolic Paraboloid

Add a second subprogram to the program of Exercise 6.12.3 that will calculate a grid of values of Z between selected ranges of x and y values. An 11×11 array of values should be generated and printed out, along with appropriate marginal values and labels. The purpose of this display is to determine the approximate shape of the function and approximate locations of critical points. Modify your program to allow you to input the starting point of the search, or to redo the grid with new limits.

7.12.5 [4,5,6,8,9] Polynomial Surface

Accomplish the analysis of Exercise 7.12.4 for any polynomial surface.

7.13 PROBABILITY AND SIMULATION

7.13.1 [ALL] Simulation of Two Dice

Use the subprogram of Exercise 7.6.1 to plot a histogram for the frequency counts found in Exercise 6.13.1.

7.13.2 [ALL] Normal Probability Distribution

Combine the subprogram of Exercise 5.13.2 with the subprograms of Example 5.3 to determine the mean value and the percentage of rods that are expected to fall within a given tolerance. Use your program to determine the percentage of rods that can be expected to fall between 6.795 and 6.805 in. if a sample of 20 rods produces the following lengths:

6.81	6.80	6.79	6.80
6.82	6.80	6.78	6.80
6.81	6.83	6.79	6.77
6.82	6.80	6.78	6.80
6.81	6.81	6.79	6.87

CHAPTER EIGHT

APPLICATIONS PACKAGES: EQUATION SOLVERS, SPREADSHEETS, AND INTEGRATED PACKAGES

8.1 INTRODUCTION

The availability of microcomputers for engineers has increased significantly over the past few years. The memory capacity and other capabilities of these machines has also improved considerably. Many microcomputers now have efficient FORTRAN compilers, so FORTRAN applications are being developed for, or transferred over to, microcomputers. The engineering community also has an increasing interest in generic applications programs. These programs are referred to as generic because they were not developed for a particular discipline but rather to analyze a particular spectrum of problems. For example, equation solvers are useful in finding solutions to systems of linear and nonlinear equations. Spreadsheets work efficiently with tabular data. Generic software packages are extremely powerful applications-oriented programs that can be effective tools in engineering practice. In this chapter you will be introduced to three types of generic software: equation solvers, spreadsheets, and integrated packages.

Generic applications software usually have menu driven commands to facilitate operations such as file loading and manipulation. For example, when you are working with many files, it is difficult to remember all the file names exactly. An option is usually available in the main menu of the program to display a list of the applications files stored on the data disk. From this list you are able to choose the desired working file. Many of the other commands for the package will also be available through menu choices.

Most generic applications packages allow you to produce clear, concise reports through formatting options in the program. They should also be able to import and export data to and from other types of applications packages. Most packages include output device drivers for a variety of printers and plotters. More advanced features of generic packages provide the means of developing applications similar to structured high-level languages.

Generic software packages open up the world of computers to practitioners who have had neither the time nor the inclination to learn a programming language. They are much easier to learn than a high-level language and are intuitive in their application. However, as with any other type of tool, generic software packages have their limitations and rules of usage that must be understood if they are to be used effectively.

Generic software packages provide an excellent introduction to the power of computers for engineering students and practitioners. These tools can often replace writing programs to solve engineering problems. These generic packages are also valuable as educational tools. You must still know the applicable theory and be able to formulate the problem. Generic software packages make it much easier to work through the necessary computations. They allow you to concentrate on the concepts and the investigation of alternatives. The structure of the package allows you to easily and quickly change input data. You can pose many "What if" questions and have the program help you answer them.

Some practitioners and educators feel that these powerful applications-oriented software packages will eliminate the need to teach a programming language at the undergraduate level. Although this assessment may be premature, there is little doubt that these packages are powerful tools that will help you solve appropriate engineering problems. You will be able to solve the problems more efficiently than you can by other means. You can learn the basic features of generic applications packages rapidly and pick up the more sophisticated features as you work on applications. At the upper end of sophistication, generic software packages actually evolve into structured programming languages.

There are numerous individual products for each of these types of generic software. We will illustrate one product for each of the types of application. TK SOLVER, by Universal Technical Systems, Inc., will be used to illustrate equation solvers. The featured spreadsheet will be LOTUS 1-2-3 by Lotus Development Corporation. ENABLE, by Enable Software Inc., will show the capabilities of integrated packages. Each of the featured products has a student or educational version that provides all the capabilities students will need at a reasonable price. We will not try to make you an expert in the use of particular products. What we will do is show you the characteristics of each type of generic software and illustrate how efficiently these packages can solve many engineering problems.

I have found that a good way to become familiar with generic software packages is to develop applications for several engineering problems you have solved previously. You can develop these applications after reading the manufacturer's reference manuals. The examples in the manufacturer's manuals will act as reference material for your own development. However, they usually will not be engineering oriented. The answers to the previously solved engineering problems will act as checks for the output of the packages. Once you have practiced with applications, you can read additional literature to increase your understanding of a package. The learning is a cyclical process of research followed by practice.

8.2 EQUATION SOLVERS

Equation solvers represent a step toward having the computer solve problems without your giving it specific instructions about the method of solution. When you use one of these solvers, you are explaining the problem to the computer and giving it your input values. Then the computer develops the solution without your having to specify all the procedural steps. However, equation solvers do not turn the computer into a thinking machine. Their programmers included all the decision logic that the machine must go through in the software.

8.2.1 General Concepts

Equation solvers accept a set of rules that represent the mathematical model for a problem. The model can be either linear or nonlinear. After you specify values for selected variables, the program tries to solve for the remaining variables. It checks for equation consistency and over- or underspecification of variable values. Equation solvers can also solve problems that require iteration.

Equation solvers will generally have facilities for units conversion. Units conversion allows you to input values for variables in convenient units. The program then converts these units to consistent units for computations. After a solution is reached, the program converts values back to the convenient units for display and output. For example, manufacturers generally list pipe diameters in inches, but formulas for solving fluid flow problems require the use of the diameter in feet. Units conversion allows the input and display of pipe diameters in inches, but converts the values to units of feet for computations.

One limitation of equation solvers is their inability to find solutions to all problems. Direct solutions where values are substituted for the unknowns and the equations are solved pose no problem. When iterative solutions are required, solution convergence is somewhat sensitive to initial estimates for the unknown variable values. Some model/initial value combinations may not converge to a solution. Anyone experienced in numerical analysis knows that this is a problem with many numerical approximation techniques under certain conditions. Systems of nonlinear equations present more problems than systems of linear equations.

8.2.2 TK SOLVER PLUS

TK SOLVER, produced and distributed by Universal Technical Systems, Inc., is an equation processor. A mathematical model consisting of a set of related equations is input. Equations can be linear, nonlinear, or a combination of both. Each equation forms a rule that TK SOLVER will use in its processing. Standard mathematical functions are available for use in the rules. TK SOLVER identifies and creates a table for the variables used in the rules. You enter values for the known variables into the variable table through rules or through lists containing multiple values. Then you instruct the program to solve the set of rules for the remaining variables. You can easily change the output variables without having to alter the rules of the model. For example, if you have $y = f(x)$, you can solve for y by specifying x; or you can solve for x by specifying y.

TK SOLVER will generate values for the output variables after you have properly specified the equation rules and input variable values. If output variables are single-valued, the values will appear in the variable table. TK SOLVER uses lists to store multiple values for variables. These lists are similar to subscripted variables in high-level languages. The list values are available for later computations in the same application, for output in tables or plots, and for saving for later applications.

If the set of equation rules is not independent, or variable values are under- or overspecified, error messages help to pinpoint where the problems are. TK SOLVER will solve for as many unknown values as it can before producing the error messages. If TK SOLVER cannot solve the set of equations directly, it can iterate from an initial estimate toward the convergent solution.

TK SOLVER is a declarative or rule-based programming language. You define the relationships among the variables and set up the constraints. Then you let TK SOLVER proceed with the solution without specifying the exact steps that it will take. Thus, TK SOLVER is an effective engineering knowledge management tool. It combines facilities for creating and storing mathematical models and related data with a processor for generating solutions to particular problems in your discipline. Engineers can use TK SOLVER to create and maintain expert systems for other engineers. The latest version of TK SOLVER includes a library of applications that could be used independently or be incorporated into user applications.

TK SOLVER was originally developed for IBM PC microcomputers and compatibles. Now versions are available for the VAX family of minicomputers, for the Macintosh microcomputer, and for Sun, IBM AIX/RT, and HP/UX workstations. Universal Technical Systems, Inc. continues to develop additional applications for TK SOLVER. A major project was the programming of computer applications to accompany *Roark's Formulas for Stress and Strain*, (Young 1989). The resulting publication is titled *TK Solver System: Roark & Young Application* (1989). The application incorporates the several thousand equations presented in Young's book into a menu structure that makes it easy to access the equations needed for a particular problem. Users can change the equations or add their own if they desire. A similar application is *Exploring Physics with TK Solver Plus*, by Mathews (1989). It supplements *Physics* or *Physics: Classical and Modern*, by Gettys, Keller, and Skove. This application contains 50 models covering 43 chapters of the text. The models can be used for classroom demonstrations or for student problem solving or term projects.

There are two books that are very useful to anyone using TK SOLVER. The book by Konopasek and Jayaraman (1984) does an excellent job of explaining the inner workings of Release 1 of TK SOLVER. Wright (1984) presents many engineering applications of TK SOLVER in his book.

8.2.2.1 Capabilities, Features, and Limitations

Two strengths of TK SOLVER are the ease of developing models using its basic features and its power as a solver of everyday problems. These problems might take from 15 to 30 minutes to solve with paper and calculator. You can solve them in a couple of minutes with a microcomputer and TK SOLVER. TK SOLVER really produces time savings in solving iterative, trial-and-error type problems, such as solving for a root of a polynomial equation. It also eliminates the tedium of doing repetitive computations for a looping algorithm. Repetitive computations are prime sources of errors when you are doing problems by hand.

With TK SOLVER, you can concentrate on the appropriate equations and mathematical structure rather than on your programming technique. This is one of the major advantages of TK SOLVER over a high-level programming language. TK SOLVER will not obscure your understanding of the theory behind problem solutions. You will have to analyze the solution process every step of the way, asking questions such as:

Are the derived equations complete?

What are the known and unknown variables?

Are units compatible within and among equations?

Does the solution make physical sense?

The model development and solution processes actually enhance your understanding of the theory.

TK SOLVER is easy to teach and learn. You can learn the basic rules and syntax of TK SOLVER through a couple of hours of instruction. After you have done a few sample problems, you will realize how much time TK SOLVER will save you. You will continue to use it in solving homework and other problems. As the need arises, you can work into more complex applications. The more advanced features of TK SOLVER make the package as comprehensive, complex, and powerful as a programming language. It will manage complete engineering knowledge bases efficiently, as in the computerization of *Roark's Formulas for Stress and Strain* (Young 1989). However, the package still has the ease of usage and the power of the basic application building blocks that we will study. These are the RULE, VARIABLE, UNITS, LIST, and PLOT sheets.

TK SOLVER is menu driven, either by cursor movement or by typing the first letter of commands. Extensive context-sensitive HELP facilities are included within the program. The program has all the usual intrinsic mathematical functions, along with several unique functions that add to the power of TK SOLVER. Conversion of units of measurement is included. TK SOLVER contains the ability to produce graphs and tables and has an interactive table solver that acts much like the recalculation feature of spreadsheets. Computer memory and acceptable solution time are the only limitations on the size of models that TK SOLVER can process. TK SOLVER checks for independence of the set of equations used and overspecification or underspecification of variable values. Several formats are available for file storage and retrieval.

TK SOLVER, as now formulated, cannot do symbolic calculus. It also does not have built-in functions for accomplishing matrix operations. However, it does come with application programs for applying the well-known algorithms for numerical integration and matrix operations. Solutions to complex iterative models on some microcomputers may take an excessive amount of time. Some models may produce solutions that diverge or that oscillate unacceptably. As you become more familiar with numerical approximation techniques you will find that this can be a problem with any solution algorithm under adverse conditions. Rearrangement or modification of model equations may eliminate such problems. TK SOLVER can find real or imaginary number solutions for most sets of algebraic equations.

8.2.2.2 Sheet Structure

Sheets and subsheets divide the structure of TK SOLVER into modules. Each sheet or subsheet has a particular function in running the package. Figure 8.1 shows a schematic of the sheet structure and the interaction among the sheets and subsheets. Simple applications developed by the user will normally use at least the GLOBAL, VARIABLE, and RULE sheets. The other sheets and subsheets can be accessed according to the needs of the application. You can also develop advanced applications that do not need to use the VARIABLE and RULE sheets. A general introduction to each sheet follows. The examples will illustrate use of subsheets.

When you are developing or working with models, you can bring a different sheet into the active window by pressing = and then selecting the appropriate sheet. You can select a sheet by typing the first letter of the sheet name or by moving the cursor in the selection box to the name and pressing enter.

FIGURE 8.1 Schematic of sheet structure. *(Reprinted from UTS/ TK Solver plus Reference Manual/Universal Technical Systems, Inc./1983)*

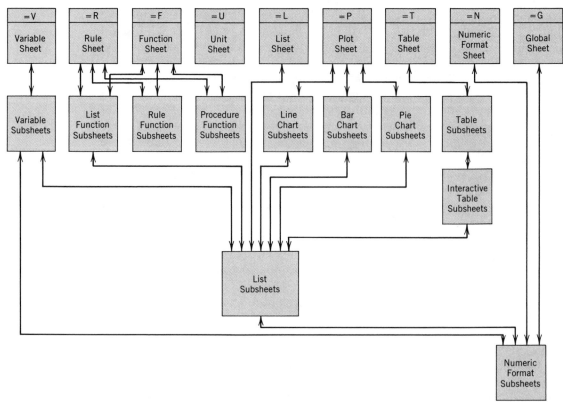

RULE Sheet: The RULE sheet contains the actual equations of the model and references to functions (rule, list, and procedure). The sheet also contains a status column that shows whether or not a rule has been satisfied or shows an error condition for a particular equation.

VARIABLE Sheet: The VARIABLE sheet contains variable definition information, display units, and input data. It also will show output values for the solution variables. A status column is used to designate an input value as an initial guess for iterative solutions or to define the variable as a list variable with multiple values. Each defined variable also has a variable subsheet that contains additional optional information about the variable.

GLOBAL Sheet: The GLOBAL sheet provides general operating parameters such as maximum number of iterations and page formatting instructions.

FORMAT Sheet: The FORMAT sheet allows specification of numeric conventions such as number of decimal places, monetary format, and the like.

UNIT Sheet: The UNIT sheet provides definitions for unit conversions to be accomplished during model solution. This allows input and displayed units to be different from the units used for calculations.

LIST Sheet: The LIST sheet allows the assignment of multiple values to variables for input, calculations, or output. The LIST sheet defines the names of lists.

It also displays the number of values entered into the list, the display units for the variable, and any desired comments. The subsheets will contain the actual values for the individual elements of the list. There is one subsheet for each list variable.

FUNCTION Sheet: The FUNCTION sheet allows the user to define custom-designed functions.

PLOT Sheet: The PLOT sheet allows the definition of any number of plots. The sheet also defines the type of plot, the graphics display mode, the output device, and the plot title. Line charts, pie charts, and bar charts can be generated and sent to a variety of printers and plotters or can be displayed on the screen. A separate plot subsheet defines additional information for each plot.

TABLE Sheet: The TABLE sheet allows the user to produce tables of values for list variables.

8.2.2.3 Commands

As with all software packages, TK SOLVER incorporates certain commands to allow the user to accomplish input, editing, and output of both models and data. This section will define several of the TK SOLVER commands that the examples will use. You should consult the TK SOLVER *Reference Manual* for further information.

Typing a / when you are not in the process of entering information into a particular position on one of the sheets will bring up a window on the screen containing command prompts. A command is selected by typing the first letter of the command or by moving the cursor to the desired command and pressing **ENTER.** This will execute the command, request further information, or bring up a list of subcommand word prompts. The applications examples will illustrate the use of the commands below the initial definition stage. The commands that will be used are:

/B BLANK

Erases the contents of a field or fields in a column.

/C COPY

Copies selected row(s) to the position of the cursor.

/D DELETE

Deletes selected row(s) from the sheet on the screen containing the cursor.

/E EXAMINE

A debugging tool which evaluates a variable or an expression.

/I INSERT

Inserts a selected number of blank rows starting at the present cursor position.

/L LIST

Provides for operations using variables with multiple (list) values.

/M MOVE

Moves selected row(s) to a new position on the screen shown by the cursor.

/P PRINT

Prints a specified block of rows from the sheet containing the cursor.

/Q QUIT

Quits TK SOLVER and returns to the operating system prompt.

/R RESET

Blanks out all the sheets, or selected sheets, so you can load a new model into TK SOLVER. If you do not RESET before loading a new model, the new model will be appended to the model already in memory.

/S STORAGE

Allows the user to save, call back, or delete a model using either a floppy or hard disk drive. If a drive other than the one containing the TK SOLVER system disk will store the data, the user must identify the disk drive in the file specification. For example, B:FLOW would refer to a file named FLOW that will either be loaded from or saved to disk drive B. Typing **/SS B:FLOW** would save the file FLOW.TK to the disk in drive B. Typing **/SL B:FLOW** would load it back in. The /SS refers to Storage Save and the /SL refers to Storage Load. If you do not remember the file name, you can type **/SL B:** and press the **ENTER** key to view a menu of all the TK files stored on the disk in drive B. To load a particular model, all you have to do is move the cursor to that file name and press **ENTER.**

/W WINDOW

Allows the user to select any sheet to fill the screen completely or to divide the screen into two windows. The two windows can contain any selected sheets or subsheets.

! SOLVE

Begins solution of the model defined by the various sheets. TK SOLVER will undertake either a direct or iterative solution depending on specifications contained in the variable sheet. The applications examples will illustrate both direct and iterative solutions.

8.2.2.4 Function Keys

The IBM keyboard contains several function keys, labeled **F1** through **F12.** Programmers can take advantage of these keys to incorporate several keystrokes into a single function key. TK SOLVER uses the following function key commands:

F1	Invokes HELP utility
F2	Cancels entry
F3	Loads model (requests model name)
F4	Saves model
F5	Invokes editor. Allows editing of the entry in the position of the cursor. Selecting EDIT will produce a subcursor within the main cursor. You can move this subcursor back and forth with the arrow keys to position it for deletions or insertions. Changes are entered by pressing the **ENTER** or **RETURN** keys.
F6	Causes active window to fill whole screen
F7	Displays plot from anywhere in TK (PLOT sheet and subsheet must already be defined)
F8	Displays table output from anywhere in TK (TABLE sheet and subsheet must already be defined)
F9	Solves model from anywhere within TK. Pressing ! works only if you are in the VARIABLE or RULE sheets.
F10	Invokes LIST SOLVE

8.2.2.5 Functions

TK SOLVER contains a total of 71 built-in functions that aid in model building. Functions range from functions that return values for π and e (natural log base) through trigonometric, logarithmic, hyperbolic, mathematical, Boolean, and complex functions. TK SOLVER contains many of the same special functions found in popular spreadsheets. It also includes several functions unique to TK SOLVER. Some of these functions deal with lists of values. This section will define mathematical functions used in the applications examples. Section 8.2.2.2.13 will define functions for complex number operations. The TK SOLVER documentation provides definitions for the remaining functions.

$\text{ABS}(x)$ = the absolute value of the argument x.

$\text{COUNT}('X)$ = the number of nonblank elements in the list named X.

$\text{DOT}('X,'Y)$ = the sum of the products of each element of the first list, $'X$, multiplied by the corresponding element in the second list, $'Y$. If the same list name is used for both arguments, the result is the sum of the squared values of each element of the list.

$\text{E}()$ = the approximate value of e, the natural log base. The function has to contain parentheses even though there is no argument. This structure is used in several functions.

$\text{EXP}(x)$ = the value e (2.718281828) raised to the power of the argument x.

$\text{LN}(x)$ = the natural logarithm of the argument x.

$\text{LOG}(x)$ = the base 10 logarithm of the argument x.

$\text{PI}()$ = an approximation for the value of π to 15 decimal places.

$\text{SGN}(x)$ = -1 if x evaluates to <0, 0 if x evaluates to 0, and 1 if x evaluates to >0.

$\text{SIN}(x)$ = sine of the angle x, where x is an expression evaluated in radians. Other standard trigonometric functions include $\text{COS}(x)$ and $\text{TAN}(x)$ and the inverse functions $\text{ACOS}(x)$, $\text{ASIN}(x)$, and $\text{ATAN}(x)$. The inverse functions return the value of the angle in radians. Similar trigonometric functions are available that accept the argument in degrees, for example $\text{SIND}(x)$.

$\text{SQRT}(x)$ = positive square root of the argument x.

$\text{SUM}('X)$ = sum of the values in the list named X.

8.2.2.6 Operations

Model Formulation and Entry After turning on your computer, loading the operating system, and inserting the TK SOLVER program disk, you can load the program by typing **TK.** A successful load will result in the screen display shown in Figure 8.2a, which will contain blank VARIABLE and RULE sheets. The RULE sheet will contain the actual equations of the model. The VARIABLE sheet will display the variables of the model along with associated values, units, and comments. The use of the leftmost (St and S) columns of each of these sheets will be explained as they are used in the applications. When you press the / key, the pop-up menu shown in Figure 8.2b will appear. This menu provides access to the commands of TK SOLVER.

The two portions of the screen shown in Figure 8.2 are windows. Pressing the ; key will cause the cursor to move from one window to the other. You can make changes or additions to the window containing the cursor.

FIGURE 8.2   TK SOLVER screen.

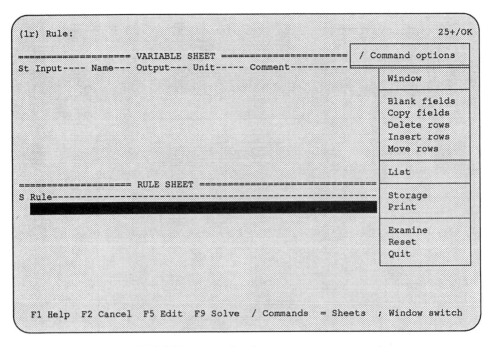

(*a*) Initial screen after start-up.

(*b*) Initial screen showing pop-up menu.

To get from one sheet to another, the user types =. A pop-up menu will display the available sheets. You can go to a sheet by typing the first letter of the sheet desired or by moving the cursor to the desired sheet name and pressing **ENTER**. This sheet will then replace the sheet in the window of the cursor. The applications will illustrate use of the various sheets. All sheets except the RULE, GLOBAL, and UNITS sheets have subsheets that further define the sheet contents. You can move to or from the subsheets by using the > and < keys. Going from a sheet to a subsheet is called diving in TK SOLVER.

You formulate and enter a TK SOLVER model as you would write the equations on a piece of paper for a manual solution. After you initially load TK SOLVER into computer memory, the cursor will appear in the first position of the RULE sheet. You may now input the rules, or equations, of your model. TK SOLVER allows rules of up to 200 characters. It is preferable to divide long rules into two or more shorter rules. After you have finished typing in the first rule, press the **RETURN** or **ENTER** key to enter the rule into the model. The VARIABLE sheet will display all variables of the first equation. Pressing the down cursor movement key will move the cursor to the next empty row in the RULE sheet for entry of another rule. As you enter succeeding equations (rules) by pressing **RETURN** or **ENTER** after typing the equation, any new variables will appear in the VARIABLE sheet. You must remember that TK SOLVER variables are sensitive to case. Therefore, BASE does not represent the same storage location as Base. This is not true for FORTRAN, so you will have to be careful not to mix cases when using TK.

Variable Specification   After completing the equation entry you can move the cursor to the VARIABLE sheet window and add any necessary units and comments. If a direct solution is possible, place the constant values for input variables in the INPUT column of the VARIABLE sheet. The input fields also act as calculators. You could type **2/3** in a field. When you press **ENTER,** TK SOLVER will do the arithmetic operation and display 0.6666667 in the field. For iterative solutions, some of the values in the INPUT column will be initial estimates of variable values. These values will change during iteration. A later section in this chapter will discuss iterative solutions.

Solving   You can start model solution by typing **!** or pressing the **F9** key. During direct solutions TK SOLVER internally rearranges the equations and substitutes values for variables. Output variable values appear in the OUTPUT column of the VARIABLE SHEET. If TK SOLVER cannot solve for all unknown variables, a > will appear in the STATUS column at the appropriate row on either the VARIABLE or RULE sheet. Moving the cursor to the > will display the error message in the message portion of the screen. At this point you could use the context-sensitive help function of TK SOLVER to display additional information about the error condition. Pressing the **F1** function key on the IBM PC keyboard will invoke the HELP function.

Formatting of Output   TK SOLVER incorporates flexible formatting including decimal fixing, zero or blank padding, and right- or left-field justification. You also have the option of indicating negative values by parentheses. You can print any number of sheets to a single text file. This text file can be imported to word processing programs for customizing or for incorporation into longer reports. We will not try to teach you the rules for formatting but will illustrate formatting in later examples. The files for these models are available on your instructor's disk. You may want to borrow it and see how the format specifications are set up.

The GLOBAL sheet contains a field for a global format. You can enter one of the defined formats in this field. All values that are not assigned a specific format will be displayed in this format. However, you can override the global format by specifying a particular format on the VARIABLE subsheet for a variable.

8.2.2.7 Conditional Rules

TK SOLVER supports IF THEN ELSE structures in both the RULE sheet and in PROCEDURE functions. The general form of the structure is

IF *<logical expression>* THEN *<expression 1>* ELSE *<expression 2>*

in which *<logical expression>* is a conditional statement similar to the logical expressions of FORTRAN IF THEN ELSE statements. The relational operators are symbolic rather than text. The relational operators for TK SOLVER, along with their FORTRAN equivalents, are as follows:

Meaning	*TK SOLVER*	*FORTRAN*
Equal	=	.EQ.
Less than	<	.LT.
Greater than	>	.GT.
Not equal	<> or ><	.NE.
Less than or equal	<= or =<	.LE.
Greater than or equal	>= or =>	.GE.

Evaluation of the logical expression to true or false follows the same rules as FORTRAN. In the THEN ELSE portion of the structure, *<expression 1>* and *<expression 2>* are each TK SOLVER rules, with equal signs. These rules cannot be conditional. If the logical expression is true, *<expression 1>* is used in the model. If the logical expression is false, *<expression 2>* is used in the model.

The ELSE portion of the structure is optional. If it is left off, you have an IF THEN structure . An IF THEN ELSE structure might be too long to fit on one line of the screen. Two shorter IF THEN structures could accomplish the same purpose. The only difference is that one of the IF THEN structures would not be used.

Example 8.1 develops some of the details of model entry and solution. It also uses logical structures. The model predicts energy loss in a circular pipe. This is a problem similar to Example 4.1. Section 5.2 of Appendix G contains background theory for the problem. You have to calculate the friction factor differently for laminar and turbulent flow. Therefore, you have to set up some type of logic structure to use the correct equation based on the value of the Reynolds number. Applying the IF THEN ELSE structure of TK SOLVER allows us to do this. We will use two IF THEN structures because of the length of the expression for the Chen equation. If you use an IF THEN ELSE structure, the resulting rule would be longer than the normal terminal or printer display width.

EXAMPLE 8.1 □ **Choice of Energy Loss Equation for Pipe Flow**

STATEMENT OF PROBLEM

Develop a TK SOLVER model that will compute the energy loss for a circular pipe. Given data are the volumetric flow rate through the pipe; the length, diameter, and size of roughness projections of the pipe; and the kinematic viscosity of the fluid flowing. Use the Chen equation to predict the Darcy–Weisbach friction factor for turbulent flow. Refer to Example 4.1 and Appendix G for additional information.

MATHEMATICAL DESCRIPTION

The mathematical development of the problem parallels the development of Example 4.1.

ALGORITHM DEVELOPMENT

Input/Output Design

Input and output design for the basic TK SOLVER model are straightforward. Variables are automatically added to the VARIABLE sheet as they are entered into the rules. You can add variable definition comments to the VARIABLE sheet. Values placed in the INPUT column of the VARIABLE sheet define input variables. All other variables will be considered to be output variables. If you make a typing error when entering rules, you may enter an erroneous variable into the VARIABLE sheet. You can remove it by placing the cursor over the erroneous variable on the VARIABLE sheet, and using the /D command to delete that row of the sheet.

Input data will include the volumetric flow rate, diameter of the pipe, length of the pipe, size of pipe roughness projections, kinematic viscosity of the fluid flowing, and the local acceleration due to gravity. Output will include velocity, Reynolds number, friction factor, and energy loss.

Computer Implementation

You can use a separate expression to calculate the coefficient A of the Chen equation. This will reduce the overall size of the Chen expression. The logic structures for choosing the correct expression for calculating f will be

```
A = 1/2.8257*(epsilon/D)^1.1098 + 5.8506/Re^0.8981
IF Re <= 2000 THEN f=64/Re
IF Re > 2000 THEN f = (-2.01*log10(epsilon/(3.7065*D)
                -5.0452/Re*LOG10(A)))^(-2)
```

The solution process will use one of the conditional statements.

Figure 8.3 shows a schematic of the TK SOLVER implementation of the frictional energy loss model for circular pipe.

PROGRAM DEVELOPMENT

You can develop the model according to the flowchart of Figure 8.3 and the mathematical development of Example 4.1. Figure 8.4a shows the RULE sheet

(EXAMPLE 8.1 □ Choice of Energy Loss Equation for Pipe Flow □ Continued)

and the input VARIABLE sheet. Note that the order of the variables in the VARIABLE sheet has been rearranged to place all the input variables together. The MOVE command was used to do this. Comments have been added to define the variables. Values have been placed in the input column for the input vari-

FIGURE 8.3   Flowchart for energy loss computations.

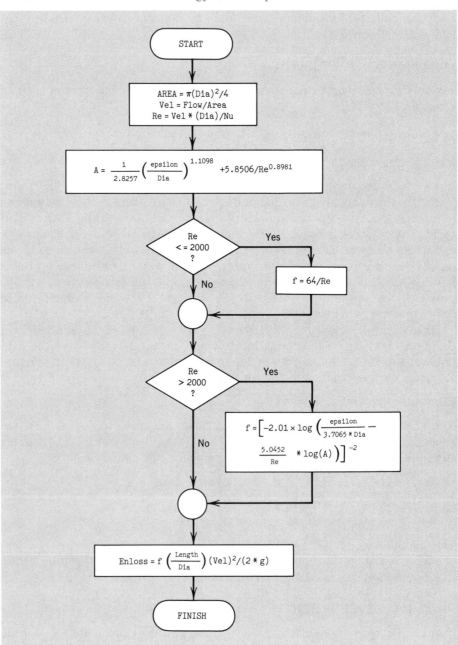

ables. The input units of the diameter are inches, whereas the diameter has to be in feet for computations. Therefore, unit conversion will be necessary.

Figure 8.4b shows the UNIT sheet and the VARIABLE subsheet necessary to define and accomplish the units conversion for Dia, the diameter. You can get to the subsheet for variable Dia by placing the cursor over the variable name and pressing >. After entering the necessary information in the subsheet, you can return to the VARIABLE sheet by pressing <.

PROGRAM TESTING

When you solve this model, an asterisk will remain in the status column of one of the IF THEN structures. If you move the cursor to that cell, the message will indicate "ignored" for the status of the rule.

FIGURE 8.4   Input RULE and VARIABLE sheets for energy loss.

```
=================== VARIABLE SHEET ==============================================
St Input---- Name--- Output--- Unit----- Comment-----------------------------
   .56       Dia                inches    Actual inside diameter of pipe (in)
   6         Length                       Length of pipe (feet)
   .001      Nu                           Kinematic viscosity
   .00000 5  epsilon                      Size of pipe roughness (ft)
   32.17     g                            Acceleration due to gravity (ft/sec^2)
   .015      Flow                         Flowrate in Pipe (ft^3/sec)
             Area                         Cross-sectional area of pipe (ft^2)
             Vel                          Velocity of flow (ft/sec)
             Re                           Reynolds number
             A                            Coefficient for Chen equation
             f                            Calculated friction factor
             Enloss                       Energy loss (ft)
=================== RULE SHEET ==================================================
S Rule------------------------------------------------------------------------
* Area=PI()*Dia^2/4
* Vel=Flow/Area
* Re=Vel*Dia/Nu
* A=1/2.8257*(epsilon/Dia)^1.1098+5.8506/Re^0.8981
* IF Re <= 2000 THEN f=64/Re
* IF Re>2000 THEN f=(-2.01*LOG10(epsilon/(3.7065*Dia)-5.0452/Re*LOG10(A)))^(-2)
* Enloss = f*(Length/Dia)*Vel^2/(2*g)
```

(a) Input model.

```
=================== VARIABLE: Dia ==============================================
Status:
First Guess:
Associated List:
Input Value:              .56
Output Value:
Numeric Format:
Display Unit:             inches
Calculation Unit:        feet
Comment:                 Actual inside diameter of pipe (in)

=================== UNIT SHEET =================================================
From----- To------- Multiply By-- Add Offset--- Comment----------------------
inches    feet       .083333333333
```

(b) Units conversion.

(EXAMPLE 8.1 □ Choice of Energy Loss Equation for Pipe Flow □ Continued)

FIGURE 8.5   Output VARIABLE and RULE sheets for energy loss.

```
================== VARIABLE SHEET =============================================
St Input---- Name--- Output--- Unit----- Comment------------------------------
   .56       Dia               inches    Actual inside diameter of pipe (in)
   6         Length                      Length of pipe (feet)
   .001      Nu                          Kinematic viscosity
   .000005   epsilon                     Size of pipe roughness (ft)
   32.17     g                           Acceleration due to gravity (ft/sec^2)
   .015      Flow                        Flowrate in Pipe (ft^3/sec)
             Area      .00171042         Cross-sectional area of pipe (ft^2)
             Vel       8.7697622         Velocity of flow (ft/sec)
             Re        409.25557         Reynolds number
             A         .02639956         Coefficient for Chen equation
             f         .1563815          Calculated friction factor
             Enloss    24.03391          Energy loss (ft)

=================== RULE SHEET ================================================
S Rule-----------------------------------------------------------------------
  Area=PI()*Dia^2/4
  Vel=Flow/Area
  Re=Vel*Dia/Nu
  A=1/2.8257*(epsilon/Dia)^1.1098+5.8506/Re^0.8981
  IF Re <= 2000 THEN f=64/Re
* IF Re > 2000 THEN f = (-2.01*LOG(epsilon/(3.7065*Dia)-5.0452/Re*LOG(A)))^(-2)
  Enloss = f*(Length/Dia)*Vel^2/(2*g)
```

(*a*) Laminar flow.

```
================== VARIABLE SHEET =============================================
St Input---- Name--- Output--- Unit----- Comment------------------------------
   .56       Dia               inches    Actual inside diameter of pipe (in)
   6         Length                      Length of pipe (feet)
   .00001059 Nu                          Kinematic viscosity
   .000005   epsilon                     Size of pipe roughness (ft)
   32.17     g                           Acceleration due to gravity (ft/sec^2)
   .015      Flow                        Flowrate in Pipe (ft^3/sec)
             Area      .00171042         Cross-sectional area of pipe (ft^2)
             Vel       8.7697622         Velocity of flow (ft/sec)
             Re        38645.474         Reynolds number
             A         .00045804         Coefficient for Chen equation
             f         .02228504         Calculated friction factor
             Enloss    3.4249364         Energy loss (ft)

=================== RULE SHEET ================================================
S Rule-----------------------------------------------------------------------
  Area=PI()*Dia^2/4
  Vel=Flow/Area
  Re=Vel*Dia/Nu
  A=1/2.8257*(epsilon/Dia)^1.1098+5.8506/Re^0.8981
* IF Re <= 2000 THEN f=64/Re
  IF Re > 2000 THEN f = (-2.01*LOG(epsilon/(3.7065*Dia)-5.0452/Re*LOG(A)))^(-2)
  Enloss = f*(Length/Dia)*Vel^2/(2*g)
```

(*b*) Turbulent flow.

Figures 8.5*a* and 8.5*b* show solutions for different kinematic viscosities. One produces laminar flow, and the other produces turbulent flow. The output values are the same as the values of Example 4.1.

8.2.2.8 Iterative Solutions

The iterative solver is a valuable tool for solving complex systems of equations that cannot be solved without using numerical approximation procedures. You also have to invoke the iterative solver whenever a required solution variable appears more than once in an equation, even if you could clear the variable through use of algebra. This is a quirk of TK SOLVER that users have to remember.

When you assign values to input variables, any values that are first approximations for an iterative solution are identified by a G (Guess) in the status column. This will automatically turn on the iterative capabilities of TK SOLVER. Unless a variable has the Guess attribute, TK SOLVER will not try to approximate it. A modified Newton–Raphson procedure is used to find iterative solutions.

The TK SOLVER model of Example 8.2 will use the iterative capabilities of TK SOLVER to solve the problem of hitting a target at a distance X from the origin and a height H above it. This is the same problem solved with a FORTRAN program in Example 4.14.

This is an excellent problem to compare with a high-level language program that can accomplish the same end. It is not a difficult problem to program in FORTRAN, but it takes a fair amount of logic and code. The solution must meet several conditions. Repeating the problem using TK SOLVER shows the equation solver's versatility. After entering the model, you can easily solve for any of the variables as unknowns without having to do any program modification. You could also develop any number of related problems from this model. For example, you could plot the trajectory. Or you could determine an error bound for the trajectory at the wall, assuming that the launch angle was subject to some amount of error.

EXAMPLE 8.2 □ Particle Trajectory Model to Hit Target

STATEMENT OF PROBLEM

Develop a TK SOLVER model to solve the problem of hitting a window at a distance X from, and height H above, the initial launch point. You should follow the specifications of Example 4.14. Section 4.2 of Appendix G also gives more information on particle trajectory problems.

MATHEMATICAL DESCRIPTION

The mathematical description is presented in Example 4.14 and Section 4.2 of Appendix G.

(EXAMPLE 8.2 □ Particle Trajectory Model to Hit Target □ Continued)

FIGURE 8.6 Flowchart for TK SOLVER solution of trial-and-error particle trajectory problem.

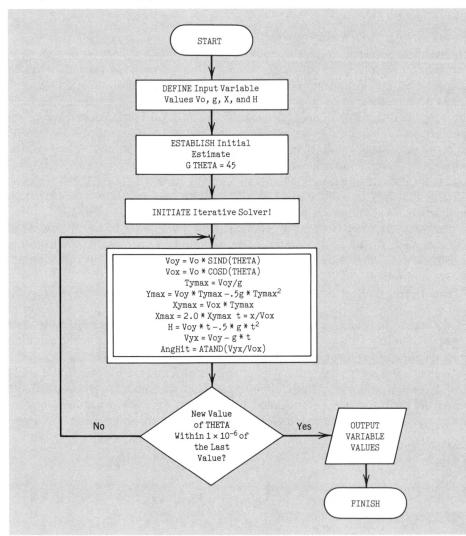

ALGORITHM DEVELOPMENT

Input/Output Design

In this example you will specify values of H and X, and let TK SOLVER find values of t and THETA that satisfy the conditions. Vo, g, X, and H are input variables. To initiate the iterative solver, a G is placed in the status column for THETA, and an initial estimate for the angle is placed in the INPUT field. This indicates that values of THETA will be adjusted during the iteration process. There is no need to guess t because TK SOLVER can solve for it after estimating a value for THETA. The initial value assigned to THETA will influence which solution the model will converge to if there are two angles that will hit the window. The section on model testing will discuss this in more detail.

Computer Implementation

Figure 8.6 shows a flowchart for the TK SOLVER solution of the trial-and-error trajectory problem. TK SOLVER can solve for the angle THETA without using any IF structures. The FORTRAN program for this problem in Example 4.14 required several IF structures. Note that the model uses the trigonometric functions that take arguments in degrees.

PROGRAM DEVELOPMENT

The rules shown in the computation block of Figure 8.6 are entered into the RULE sheet. The input variables are moved to the top of the VARIABLE sheet, as is shown in Figure 8.7. Figure 8.7 also shows that the initial estimate for THETA is 45°. The Vo, X, and H data are those that gave two angles for Example 4.14.

FIGURE 8.7   Input VARIABLE and RULE sheets for trajectory model.

```
=================== VARIABLE SHEET ===========================================
St Input---- Name--- Output--- Unit----- Comment------------------------------
   66        Vo                          Initial velocity, ft/sec
   32.17     g                           Acceleration due to gravity, ft/sec^2
   90        X                           Horizontal distance along trajectory
   37        H                           Height of trajectory at distance X, ft
             Voy                         Y component of initial velocity
             Vox                         X component of initial velocity
             Tymax                       Time to max height of trajectory, sec
             Ymax                        Maximum height of trajectory, ft
             Xymax                       X distance to max height of trajectory
             Xmax                        Maximum length of flat trajectory, ft
             Vyx                         Y velocity at distance X, ft/sec
             AngHit                      Angle of entry into window, degrees
 G 45        THETA                       Angle of inclination, degrees
             t                           Time to reach distance X, sec
=================== RULE SHEET ===============================================
S Rule------------------------------------------------------------------------
* Voy = Vo*SIND(THETA)
* Vox = Vo*COSD(THETA)
* Tymax = Voy/g
* Ymax = Voy*Tymax-0.5*g*Tymax^2
* Xymax = Vox*Tymax
* Xmax = 2.0*Xymax
* t = X/Vox
* H = Voy*t - 0.5*g*t^2
* Vyx = Voy - g*t
* AngHit = ATAND(Vyx/Vox)
```

(EXAMPLE 8.2   □   Particle Trajectory Model to Hit Target   □   Continued)

FIGURE 8.8   Output VARIABLE sheets for trajectory model.

```
==================== VARIABLE SHEET ==========================================
St Input---- Name--- Output--- Unit----- Comment-----------------------------
   66        Vo                           Initial velocity, ft/sec
   32.17     g                            Acceleration due to gravity, ft/sec^2
   90        X                            Horizontal distance along trajectory
   37        H                            Height of trajectory at distance X, ft
             Voy      52.918704           Y component of initial velocity
             Vox      39.441232           X component of initial velocity
             Tymax    1.6449706           Time to max height of trajectory, sec
             Ymax     43.524856           Maximum height of trajectory, ft
             Xymax    64.879667           X distance to max height of trajectory
             Xmax     129.75933           Maximum length of flat trajectory, ft
             Vyx      -20.48925           Y velocity at distance X, ft/sec
             AngHit   -27.45134           Angle of entry into window, degrees
             THETA    53.302196           Angle of inclination, degrees
             t        2.281876            Time to reach distance X, sec
```

(a) Output VARIABLE sheet for initial estimate of THETA = 45°.

```
==================== VARIABLE SHEET ==========================================
St Input---- Name--- Output--- Unit----- Comment-----------------------------
   66        Vo                           Initial velocity, ft/sec
   32.17     g                            Acceleration due to gravity, ft/sec^2
   90        X                            Horizontal distance along trajectory
   37        H                            Height of trajectory at distance X, ft
             Voy      52.918704           Y component of initial velocity
             Vox      39.441232           X component of initial velocity
             Tymax    1.6449706           Time to max height of trajectory, sec
             Ymax     43.524856           Maximum height of trajectory, ft
             Xymax    64.879667           X distance to max height of trajectory
             Xmax     129.75933           Maximum length of flat trajectory, ft
             Vyx      -20.48925           Y velocity at distance X, ft/sec
             AngHit   -27.45134           Angle of entry into window, degrees
 G 75        THETA                        Angle of inclination, degrees
             t        2.281876            Time to reach distance X, sec
```

(b) Modified input VARIABLE sheet for finding second solution.

```
==================== VARIABLE SHEET ==========================================
St Input---- Name--- Output--- Unit----- Comment-----------------------------
   66        Vo                           Initial velocity, ft/sec
   32.17     g                            Acceleration due to gravity, ft/sec^2
   90        X                            Horizontal distance along trajectory
   37        H                            Height of trajectory at distance X, ft
             Voy      56.600263           Y component of initial velocity
             Vox      33.947168           X component of initial velocity
             Tymax    1.7594113           Time to max height of trajectory, sec
             Ymax     49.791572           Maximum height of trajectory, ft
             Xymax    59.727033           X distance to max height of trajectory
             Xmax     119.45407           Maximum length of flat trajectory, ft
             Vyx      -28.68815           Y velocity at distance X, ft/sec
             AngHit   -40.20056           Angle of entry into window, degrees
             THETA    59.045913           Angle of inclination, degrees
             t        2.6511784           Time to reach distance X, sec
```

(c) Output VARIABLE sheet for second solution.

PROGRAM TESTING

Figure 8.8a shows the output variable sheet for the solution of the model input of Figure 8.7. You will recognize this as one of the solutions for case 2 of Example 4.14. The next task is to find the second solution. This can be accomplished by proceeding from the initial solution. You can move the cursor to the status field of the THETA row of the VARIABLE sheet and type **G**. The current value for THETA (53.302196) will move from the output to the input field. This value becomes the initial estimate for a new solution. If you solve at this point, you will repeat the first solution. However, you can change the input value. Move the cursor to the input field for THETA and enter a new initial estimate. An initial estimate to 75°, as shown in Figure 8.8b, will converge to the second angle, as shown in Figure 8.8c. It would be helpful to have a quick method of determining if you need to look for two solutions. You will learn how to gain this information in the sections of this chapter covering TABLE and PLOT sheets.

If you input the data for case 1 of Example 4.14, which is a trajectory that does not hit the window, TK will go through its maximum number of iterations without converging. The default number of iterations is 10, set on the GLOBAL sheet. If you watch the convergence criterion value displayed at the top of the screen with each successive iteration, you will see that the convergence criterion value is not decreasing. This shows that the variable values are not converging toward a solution. For the TK SOLVER trajectory model, lack of convergence means that the stream does not hit the window.

You can also use the model of Example 8.2 to solve for any of the other variables by specifying the proper input variable values. For example, you could solve for the height of the trajectory H at any distance X by specifying Vo, X, and THETA.

The model of Example 4.14 required that the angle of launch be between 30° and 75°. The TK SOLVER model of Example 8.2 could solve for angles outside these limits. It is up to the user to determine that a particular solution does not meet the limiting criteria. You could put comments in the variable sheet to remind users of this requirement. It is also possible to include IF THEN ELSE structures in the model to check for an out-of-limits angle. The following rules

```
IF THETA < 30 THEN Voy = 0.0 ELSE Voy = Vy*SIN(ANGLE)
IF THETA > 75 THEN Voy = 0.0
```

would set the initial velocity in the y direction to zero if the angle is outside the limits. Any variables dependent on Voy would evaluate to zero. This would be the flag that the angle was out of limits. The model would produce the correct value of Voy if the angle were not out of limits.

8.2.2.9 LIST Sheet

So far, the variables that we have used in TK SOLVER models have been single valued. The LIST sheet and subsheets allow manipulation and storage of multiple values for a variable or variables. LIST sheets treat variables much the same as FORTRAN treats subscripted variables. List variables are also essential to the production of TK SOLVER tables and plots. Later sections will illustrate the development of tables and plots.

The LIST sheet identifies which variables will have multiple values associated with them. There are several ways to define list variables. You can define them in the LIST sheet. Entering a list name in the appropriate field in the TABLE, PLOT, or LIST FUNCTION subsheet will define a list variable. Another way is to type **L** in the status column for a variable in the VARIABLE sheet. However, you do not have to include list variables in the VARIABLE sheet to use them in other parts of the TK SOLVER program.

The number of elements in a list is dependent on how many values you place in it. There is no requirement to set a maximum number of values as you had to do with FORTRAN. Also, variables that contain temporary values during the computational process do not have to be list variables. These correspond to the simple variables in FORTRAN that are used in computations but are not saved for later reference.

Each variable identified on the LIST sheet has an associated subsheet that will actually contain the values for the variable. You can enter input values in the LIST subsheets for the input variables. You can also enter values through the interactive table subsheet or import them from other TK SOLVER models, from spreadsheets, and from other types of software. TK SOLVER will place values in output lists during the solution process. Lists will be saved with the model. They also can be saved separately for transfer to other TK SOLVER models or to spreadsheets or other software.

Example 5.4 developed a FORTRAN subprogram for accomplishing regression analysis. Example 5.5 used the subprogram for analyzing stress-strain data. You can also develop a TK SOLVER model that will perform regression analysis on any set of *X-Y* data. You could use the model by itself or append it to other models for reducing or analyzing raw data in the laboratory.

Example 8.3 develops a general TK SOLVER regression model, and then uses it to analyze the stress-strain data of Example 5.5. The example will develop two forms of the model. The first form is similar to a linear regression model first presented by Wiggins (1986). It uses the list capabilities of TK SOLVER and several of the built-in TK functions that accomplish tasks that require considerable programming in FORTRAN. The model will then be modified to present it in a more generic form for use with other TK SOLVER models.

EXAMPLE 8.3   □   Linear Regression

STATEMENT OF PROBLEM

Develop a linear regression model for TK SOLVER, and use the model to analyze the stress-strain data of Example 5.5. Modify the model to accommodate its incorporation into other TK SOLVER applications.

MATHEMATICAL DESCRIPTION

Example 5.4 presented the mathematical model for regression analysis.

ALGORITHM DEVELOPMENT

Input/Output Design

Variables will be given generic names, *X* for the independent variable, and *Y* for the dependent variable. Input data are paired *X-Y* values. The comments for the lists in the first form of the model could contain the definitions for the independent and dependent variables. However, the variable names themselves are not descriptive. The second form of the model allows inclusion of variable definitions in the VARIABLE sheet. You can also use descriptive variable names for the second form.

The independent and dependent variables are the only required lists. Special TK SOLVER functions can access these X and Y data to develop the regression parameters. This example requires entry of stress and strain data directly into the list subsheets for the independent and dependent variables. Example 8.4 will show how you can input data through an interactive table subsheet.

Output for this example will include the number of data values, the mean values for the independent and dependent variables, the slope and intercept of the regression line, and the correlation coefficient. Applications exercises will extend the model to include prediction of strain for a particular stress and the development of a scatterplot of the data.

Computer Implementation

The COUNT function of TK SOLVER will determine N, the number of data points, from the number of X-Y pairs in the LIST sheets. The SUM function will find the summation of the individual data values. The DOT function will compute the required cross product and squared data value summations. For example:

$$\mathrm{DOT}('X,'Y) = \sum_{i=1}^{n} X_i Y_i \qquad \mathrm{DOT}('X,'X) = \sum_{i=1}^{n} X_i^2 \qquad (8.1)$$

Equation 8.1 references the names of the lists through the symbolic names, 'X and 'Y. This is one way that TK SOLVER refers to lists of values in function arguments. A second method will be shown in the revised model.

PROGRAM DEVELOPMENT

The order of entry of the various parts of the model does not matter, unless you want to save the general model as a template without data. In that case you should develop the RULE sheet, rearrange and define the variables as desired in the VARIABLE sheet, and define the variables that will have lists associated with them. Do not, however, enter any values in the subsheets for X and Y.

Figure 8.9a shows the input VARIABLE and RULE sheets for the basic model. Note the use of the built-in COUNT, SUM, and DOT functions. The functions will automatically access the lists during the solution process. Therefore, inclusion of X and Y in the VARIABLE sheet is optional. You also do not need to invoke the LIST SOLVE command, since the model is not carrying out a series of solutions for the list entries. In Example 8.4 you will have to use the LIST SOLVE features and will be able to see the difference in the use of lists.

Figure 8.9c shows the LIST sheet and the associated subsheets for X and Y. Note that you have to be sure to put the X and Y values in the proper order in the two lists. You can begin the data entry process by bringing the LIST sheet up in the active window. Place the cursor over the name of the variable list that you want to enter and type $>$ to dive to that subsheet. After entering the n values for that variable, typing $<$ will return you to the LIST sheet. You can then move the cursor to the second input list variable and repeat the procedure.

(EXAMPLE 8.3 □ Linear Regression □ Continued)

PROGRAM TESTING

Solution of the model will give the VARIABLE sheet shown in Figure 8.9b. Note that the estimates for a, b, and r are the same as in Example 5.5.

MODIFICATION FOR TEMPLATE

Minor changes in this model will form a template that can easily be incorporated into any TK SOLVER model. The functions are given general arguments that will accept specific lists from the model. The VARIABLE sheet can link the function DOT(X,Y) with specific lists. This makes it necessary to include X and Y in the VARIABLE sheet. Figure 8.10a shows the VARIABLE, RULE, and LIST sheets for the modified model. The entry 'STRESS in the input field of variable X links X with the list STRESS. When a function sees X as an argument, it uses the values from STRESS. Similarly, functions use values from STRAIN when Y is the argument. The names of the lists are descriptive of the data stored in them. We did not have to retype all the data into the new lists. All we had to do was to rename the old lists. Figure 8.10b shows that the solved

---

FIGURE 8.9 RULE, VARIABLE, and LIST sheets for TK
SOLVER regression analysis.

```
=================== VARIABLE SHEET ========================================
St Input---- Name--- Output--- Unit----- Comment--------------------------------
             N                            Number of X-Y pairs of data
             Xbar                         Mean value for independent variable
             Ybar                         Mean value for dependent variable
             a                            Slope of regression line
             b                            Intercept of regression line
             r                            Correlation Coefficient
=================== RULE SHEET ============================================
S Rule---------------------------------------------------------------------
* N = COUNT('X)
* Xbar = SUM('X)/N
* Ybar = SUM('Y)/N
* a = (DOT('X,'Y)-N*Xbar*Ybar)/(DOT('X,'X)-N*Xbar^2)
* b = Ybar - a*Xbar
* r = a*SQRT((DOT('X,'X)-N*Xbar^2)/(DOT('Y,'Y)-N*Ybar^2))
```

(a) Input VARIABLE and RULE sheets.

```
=================== VARIABLE SHEET ========================================
St Input---- Name--- Output--- Unit----- Comment--------------------------------
             N        20                  Number of X-Y pairs of data
             Xbar     37700               Mean value for independent variable
             Ybar     .001095             Mean value for dependent variable
             a        4.9605E-8           Slope of regression line
             b        -.0007751           Intercept of regression line
             r        .93390417           Correlation Coefficient
```

(b) Output VARIABLE sheet.

FIGURE 8.9 (Continued)

```
=================== LIST SHEET ===============================================
Name----- Elements-- Unit----- Comment-------------------------------------
X         20                   Independent variable values
Y         20                   Dependent variable values
=================== LIST: X ==================================================
Comment:                       Independent variable values
Numeric Format:
Display Unit:
Calculation Unit:
Element-- Value--------------
1          20000
2          16000
3          22000
4          26000
5          30000
6          28000
7          36000
8          32000
9          38000
10         38000
11         44000
12         38000
13         44000
14         42000
15         48000
16         50000
17         48000
18         52000
19         50000
20         52000
=================== LIST: Y ==================================================
Comment:                       Dependent variable values
Numeric Format:
Display Unit:
Calculation Unit:
Element-- Value--------------
1          .0002
2          .0003
3          .0005
4          .0005
5          .0005
6          .0007
7          .0007
8          .0008
9          .0008
10         .001
11         .001
12         .0012
13         .0013
14         .0014
15         .0015
16         .0017
17         .0018
18         .0018
19         .002
20         .0022
```

(*c*) LIST sheet and subsheets.

(EXAMPLE 8.3 □ Linear Regression □ Continued)

model gives the same results as before. You could append this model to any model as long as the main model does not duplicate the variable names used for the regression analysis. Example 8.5 illustrates development of a rule function for regression that eliminates this requirement.

FIGURE 8.10   TK SOLVER regression template.

```
=================== VARIABLE SHEET ==========================================
St Input---- Name--- Output--- Unit----- Comment-----------------------------
             N                            Number of X-Y pairs of data
             Xbar                         Mean value for independent variable
             Ybar                         Mean value for dependent variable
             a                            Slope of regression line
             b                            Intercept of regression line
             r                            Correlation Coefficient
    'STRESS  X                            Independent variable
    'STRAIN  Y                            Dependent variable
=================== RULE SHEET ===============================================
S Rule----------------------------------------------------------------------
* N = COUNT(X)
* Xbar = SUM(X)/N
* Ybar = SUM(Y)/N
* a = (DOT(X,Y)-N*Xbar*Ybar)/(DOT(X,X)-N*Xbar^2)
* b = Ybar - a*Xbar
* r = a*SQRT((DOT(X,X)-N*Xbar^2)/(DOT(Y,Y)-N*Ybar^2))
=================== LIST SHEET ===============================================
Name----- Elements-- Unit----- Comment--------------------------------------
STRESS    20                   Stress (Independent variable values)
STRAIN    20                   Strain (Dependent variable values)
```

(*a*) RULE sheet, LIST sheet, and input VARIABLE sheet.

```
=================== VARIABLE SHEET ==========================================
St Input---- Name--- Output--- Unit----- Comment-----------------------------
             N        20                  Number of X-Y pairs of data
             Xbar     37700               Mean value for independent variable
             Ybar     .001095             Mean value for dependent variable
             a        4.9605E-8           Slope of regression line
             b        -.0007751           Intercept of regression line
             r        .93390417           Correlation Coefficient
    'STRESS  X                            Independent variable
    'STRAIN  Y                            Dependent variable
```

(*b*) Output VARIABLE sheet.

8.2.2.10 LIST SOLVE

The LIST SOLVE command allows you to execute the rules of a model several times to generate multiple output values in lists. After defining the appropriate lists and typing the letter **L** in the STATUS field for each list variable in the VARIABLE sheet, you have to specify which of the lists are input. This is done by moving the cursor to the STATUS field for the input variable list and typing **I.** The letter I will not appear in the field. However, if you try to solve the model without typing the I, you will get an error. TK SOLVER will use the iterative solver if some of the input lists are guesses. Otherwise it will begin a direct solution. First estimates for iteration can be provided through the First Guess field in the associated variable subsheet, or by placing initial values in the guess list elements. The LIST SOLVE uses the input list values to produce corresponding output values and places those values in the output lists in the proper positions. Example 8.4 will illustrate use of LIST SOLVE.

8.2.2.11   TABLE and PLOT Sheets

The previous section discussed the use of lists to store multiple values for one or more variables. Displaying input and output data is difficult using just the lists, however. You must either look at several pages of hard copy to examine corresponding values or switch between several LIST subsheets. The TABLE sheet brings together these lists and presents them in a tabular format for review. TK SOLVER provides a limited amount of formatting for tables. You can add column headings, separators for columns and groups of rows, and numeric formats.

Example 8.4 develops several lists for the particle trajectory problem. These lists will help in determining if there are any angles that will hit the window for a particular velocity of discharge. The lists will also show how many times the trajectory will hit the window. In this case, LIST SOLVE will generate the output list values. LIST SOLVE is necessary because a separate solution has to generate the data for each element of the lists. Also, the LIST FILL command will be used to easily and accurately generate lists of related values.

For this example, angle will be the independent variable. The input list for the angle will have a lower limit of 30°, an upper limit of 75°, and an increment of 1° between angles. The VARIABLE sheet will specify the initial velocity and distance to the window. The solution will fill lists for H, t, Ymax, and Xmax. The table will include a selected range from the lists. This allows inclusion of only the most pertinent parts of the lists in the table.

After developing the table, we will use the interactive table as a tool to further refine the estimates for the correct angle to hit the window. We will also use the interactive table as a tool for data entry.

There is an interactive TABLE subsheet for each table defined. You can get to it by diving from anywhere within the TABLE subsheet for that table. The interactive subsheet looks similar to the output table, but has some useful additional capabilities. After solving, you can change the contents of a table cell for an input variable. Pressing the **F10** key while in the interactive table begins the list solver, which will solve for the related table entries. Changing the values in the interactive TABLE subsheet also changes the corresponding values in the lists. Solving the interactive table changes the appropriate output list values. The values will appear in the appropriate positions in the table as the solution progresses. This feature of the interactive table is very similar to the recalculation capabilities of spreadsheets, which the next chapter will discuss. The interactive table allows you to change input easily and to see the effects on the output immediately. Example 8.4 will illustrate use of the interactive solving capabilities of TK SOLVER.

You can also use the interactive table as a convenient mechanism to input list values. Once you have defined the lists through the TABLE subsheet, you can dive to the interactive table to enter values for the input lists. You can see all your input data without having to switch between LIST subsheets. In fact, you do not have to go to the LIST subsheets at all. LIST FILL can generate the input values for the interactive table in the same manner as illustrated for lists. When you have finished entering data, pressing **F10** will LIST SOLVE and display the results in the interactive table. The same values will appear in the table defined by the TABLE subsheet.

TK SOLVER offers flexible options for generation and output of plots. Several different formats and screen resolutions are available for line graphs, bar charts, and pie charts. These are controlled through the PLOT sheet.

The possible resolution of screen graphics images varies depending on the graphics capability of your computer. TK SOLVER supports all of the most popular graphics boards for IBM and IBM compatibles. It also has output drivers for popular plotters and laser and dot matrix graphics printers. You cannot save graphics images, but you can save the PLOT sheet and subsheets. The driver programs can use these independently of TK SOLVER to produce graphs.

The PLOT sheet defines all the plots that are to be generated for a model. Each named plot will have a PLOT subsheet. As with the TABLE subsheets, the subsheet defines the particular attributes of the PLOT. The definitions on the PLOT sheet include the name of the plot, the plot type (LINE, BAR, or PIE), the display options, the output device, and the title for the plot. Placing the cursor in the DISPLAY OPTIONS or the OUTPUT DEVICE fields displays the possible options in the prompt line at the top of the screen. An option is chosen by typing the corresponding number. Example 8.4 illustrates the information placed in the PLOT sheet and subsheet for line graphs.

Line graphs are X-Y plots. Various options are available for setting axes, plot limits, and labels. You can display linear or logarithmic scales, a zero axis, and a grid. One X-axis list must be specified, but multiple Y-axis lists can be used. You can specify the first and last elements of the Y-axis list. Example 8.4 uses line graphs to help estimate the angles for which a certain trajectory would hit a window. Both tables and plots require that you have previously defined the lists to be displayed.

EXAMPLE 8.4   □   Table and Plots for Particle Trajectory Problem

STATEMENT OF PROBLEM

Modify the TK SOLVER model of Example 8.2 so that it will produce a set of lists to estimate which angle or angles of a trajectory will hit the window. Use angles between the limits of 30° and 75° in increments of 1°. The VARIABLE sheet will specify the distance to the window X and initial velocity Vo. Dependent variable lists will include the height of the trajectory at distance X (H), the time of travel (t), the maximum height of the trajectory (Ymax), and the maximum level distance of the trajectory (Xmax). Use the interactive TABLE subsheet and solver to show refinement of solution estimates. Illustrate an alternate method of data entry through the interactive table.

Develop a PLOT sheet and subsheets to help estimate the angles for which a water jet will hit the window. Initially plot H vs. initial angle for the full limits of 30° to 75°. Then refine the angle plotted to increase the accuracy of the estimate.

Use the following data to illustrate the table and plot output. The initial velocity of the jet is 66 ft/s. The window is 37 ft above the ground and 90 ft away from the point of launch. These data will illustrate how the table can identify multiple solutions.

MATHEMATICAL DESCRIPTION

The mathematical model will be the same as in Example 8.2, with the exception that H is now a dependent variable rather than a specified value. Therefore, you will not have to use the iterative solver.

ALGORITHM DEVELOPMENT

Input/Output Design

Values for Vo, X, and g will remain as input in the VARIABLE sheet for the model. Define variables THETA, t, H, Ymax, and Xmax as lists by typing the letter **L** in the STATUS column. THETA will be the independent variable, and the others will be dependent variables. Figure 8.11a shows the modified VARI-ABLE sheet. Initially we will develop the output table without formatting and then will revise it to show how formatting helps in the presentation of data.

The only additional input for the plot will be the generation of a straight line at 37 ft to define the height of the window. This line will be created with a list named Window, which will contain the constant value 37.

Computer Implementation

The following series of steps will have to be undertaken to produce the desired table:

```
MODIFY VARIABLE SHEET TO ACCOMMODATE LISTS FOR SELECTED VARIABLES
BRING LIST SHEET INTO ACTIVE WINDOW
DIVE TO THETA SUBSHEET
ENTER VALUES FOR THETA
BRING VARIABLE SHEET INTO ACTIVE WINDOW
SPECIFY THETA AS AN INPUT LIST
INVOKE LIST SOLVE
BRING TABLE SHEET INTO ACTIVE WINDOW
SPECIFY NAME AND TITLE FOR TABLE
DIVE TO TABLE SUBSHEET
SET ATTRIBUTES OF TABLE
SOLVE (PRINT) TABLE
FORMAT TABLE AND RESOLVE
DIVE TO INTERACTIVE TABLE SUBSHEETS
MODIFY INPUT
INVOKE INTERACTIVE SOLVER
```

The following steps will generate the desired plots:

```
BRING PLOT SHEET INTO ACTIVE WINDOW
DEFINE THE NAME (Window), TYPE (Line), DISPLAY OPTION (4. VGA),
    OUTPUT DEVICE (4. HPLASJET), AND TITLE OF PLOT
DIVE TO PLOT SUBSHEET
SET PLOT PARAMETERS
PLOT (F7)
OUTPUT PLOT (Press letter O)
RETURN TO PLOT SUBSHEET (Press ENTER, or any letter)
REDEFINE PARAMETERS TO REFINE PLOT
PLOT
OUTPUT
```

(EXAMPLE 8.4   □   Table and Plots for Particle Trajectory Problem   □   Continued)

PROGRAM DEVELOPMENT

Development of the model will be in the order specified above. Reference will be made to the appropriate portions of Figure 8.11. The INPUT/OUTPUT DE-SIGN section has accomplished the first step.

You can bring the LIST sheet into the active window by typing =L. The variables will already appear in the LIST sheet because they have been defined through the VARIABLE sheet. If you want to change the order of their appear-

---

FIGURE 8.11   Output table for particle trajectory problem.

```
=================== VARIABLE SHEET =========================================
St Input---- Name--- Output--- Unit----- Comment---------------------------
      66        Vo                        Initial velocity, ft/sec
      32.17     g                         Acceleration due to gravity, ft/sec^2
      90        X                         Horizontal distance along trajectory
L               H                         Height of trajectory at distance X, ft
                RAD                       # of degrees in one radian
                ANGLE                     Angle of inclination, radians
                Voy                       Y component of initial velocity
                Vox                       X component of initial velocity
                Tymax                     Time to max height of trajectory, sec
L               Ymax                      Maximum height of trajectory, ft
                Xymax                     X distance to max height of trajectory
L               Xmax                      Maximum length of flat trajectory, ft
                Vyx                       Y velocity at distance X, ft/sec
                AngHit                    Angle of entry into window, degrees
L  45           THETA                     Angle of inclination, degrees
L               t                         Time to reach distance X, sec
=================== RULE SHEET =============================================
S Rule---------------------------------------------------------------------
* RAD = 180/pi()
* ANGLE = THETA/RAD
* Voy = Vo*SIN(ANGLE)
* Vox = Vo*COS(ANGLE)
* Tymax = Voy/g
* Ymax = Voy*Tymax-0.5*g*Tymax^2
* Xymax = Vox*Tymax
* Xmax = 2.0*Xymax
* t = X/Vox
* H = Voy*t - 0.5*g*t^2
* Vyx = Voy - g*t
* AngHit = RAD*ATAN(Vyx/Vox)
=================== LIST SHEET =============================================
Name----- Elements-- Unit----- Comment------------------------------------
THETA      46                   Initial Angle for Trajectory
t                               Time to reach distance X, seconds
H                               Height of trajectory at distance X
Xmax                            Maximum level distance of trajectory, feet
Ymax                            Maximum height of trajectory
```

(*a*) VARIABLE, RULE, and LIST sheets.

FIGURE 8.11 (Continued)

```
==================== LIST: THETA =================================================
Comment:                      Initial Angle for Trajectory
Numeric Format:
Display Unit:
Calculation Unit:
Element-- Value--------------
1         30
2         31
3         32
4         33
5         34
6         35
7         36
8         37
9         38
10        39
11        40
12        41
13        42
14        43
15        44
16        45
17        46
18        47
19        48
20        49
21        50
22        51
23        52
24        53
25        54
26        55
27        56
28        57
29        58
30        59
31        60
32        61
33        62
34        63
35        64
36        65
37        66
38        67
39        68
40        69
41        70
42        71
43        72
44        73
45        74
46        75
```

(*b*) LIST subsheet for THETA.

(EXAMPLE 8.4 □ Table and Plots for Particle Trajectory Problem □ Continued)

FIGURE 8.11 (Continued)

```
=================== LIST SHEET =================================================
Name----- Elements-- Unit----- Comment----------------------------------------
THETA     46                   Initial Angle for Trajectory
t         46                   Time to reach distance X, seconds
H         46                   Height of trajectory at distance X
Xmax      46                   Maximum level distance of trajectory, feet
Ymax      46                   Maximum height of trajectory
=================== TABLE SHEET ================================================
Name---------- Title----------------------------------------------------------
Trajectory     Trajectory Data for Hitting Window
=================== TABLE: Trajectory =========================================
Screen or Printer:         Printer
Title:                     Trajectory Data for Hitting Window
Vertical or Horizontal:    Vertical
Row Separator:
Column Separator:
First Element:             20
Last Element:              35
List----- Numeric Format-- Width-- Heading------------------------------------
THETA                      10      Angle
H                          10      Height
t                          10      Time
Ymax                       10      Max Height
Xmax                       12      Max Distance
```

(c) LIST sheet after solution, TABLE sheet, and trajectory TABLE subsheet.

Angle	Height	Time	Max Height	Max Distance
	Trajectory Data for Hitting Window			
49	34.0415457	2.07852694	38.5626207	134.08789889
50	34.8670494	2.12144158	39.7296508	133.34854125
51	35.6185488	2.16683963	40.8895192	132.44671896
52	36.2843449	2.21491261	42.0408127	131.38353073
53	36.8507472	2.26587292	43.1821287	130.16027192
54	37.3016967	2.31995675	44.3120767	128.77843286
55	37.6183065	2.37742745	45.42928	127.23969711
56	37.7782988	2.43857952	46.5323774	125.5459394
57	37.7553122	2.50374335	47.620025	123.6992233
58	37.5180444	2.57329079	48.6908977	121.70179875
59	37.0291846	2.64764185	49.7436908	119.5560993
60	36.2440768	2.72727273	50.7771215	117.26473916
61	35.1090344	2.81272546	51.789931	114.83051
62	33.5592002	2.90461973	52.780885	112.25637756
63	31.5158097	3.00366718	53.7487765	109.54547801
64	28.8826625	3.11068914	54.6924261	106.70111417

(d) Basic TABLE output.

FIGURE 8.11 (Continued)

```
                    Trajectory Data for Hitting Window
        --------------------------------------------------------------------
        | Angle     | Height     | Time      | Max Height | Max Distance |
        --------------------------------------------------------------------
        | 49        |    34.04   |    2.08   |    38.56   |    134.09    |
        | 50        |    34.87   |    2.12   |    39.73   |    133.35    |
        | 51        |    35.62   |    2.17   |    40.89   |    132.45    |
        | 52        |    36.28   |    2.21   |    42.04   |    131.38    |
        | 53        |    36.85   |    2.27   |    43.18   |    130.16    |
        | 54        |    37.3    |    2.32   |    44.31   |    128.78    |
        | 55        |    37.62   |    2.38   |    45.43   |    127.24    |
        | 56        |    37.78   |    2.44   |    46.53   |    125.55    |
        | 57        |    37.76   |    2.5     |    47.62   |    123.7     |
        | 58        |    37.52   |    2.57   |    48.69   |    121.7     |
        | 59        |    37.03   |    2.65   |    49.74   |    119.56    |
        | 60        |    36.24   |    2.73   |    50.78   |    117.26    |
        | 61        |    35.11   |    2.81   |    51.79   |    114.83    |
        | 62        |    33.56   |    2.9     |    52.78   |    112.26    |
        | 63        |    31.52   |    3       |    53.75   |    109.55    |
        | 64        |    28.88   |    3.11   |    54.69   |    106.7     |
        --------------------------------------------------------------------
```

(*e*) Table with formatting (TABLE subsheet appears in Figure 8.12*a*).

```
(t) Title: Trajectory Data for Hitting Window                      25+/F9

=================== TABLE: Trajectory ==================================
Title:          Trajectory Data for Hitting Window
Element Angle----- Height---- Time------ Max Height Max Distance -------
1        49         34.04     2.08       38.56      134.09
2        50         34.87     2.12       39.73      133.35
3        51         35.62     2.17       40.89      132.45
4        52         36.28     2.21       42.04      131.38
5        53         36.85     2.27       43.18      130.16
6        54         37.3      2.32       44.31      128.78
7        55         37.62     2.38       45.43      127.24
8        56         37.78     2.44       46.53      125.55
9        57         37.76     2.5        47.62      123.7
10       58         37.52     2.57       48.69      121.7
11       59         37.03     2.65       49.74      119.56
12       60         36.24     2.73       50.78      117.26
13       61         35.11     2.81       51.79      114.83
14       62         33.56     2.9        52.78      112.26
15       63         31.52     3          53.75      109.55
16       64         28.88     3.11       54.69      106.7


 F1 Help  F2 Cancel  F5 Edit  F9 Solve  / Commands  = Sheets  ; Window switch
```

(*f*) Interactive TABLE subsheet.

(EXAMPLE 8.4   □   Table and Plots for Particle Trajectory Problem   □   Continued)

ance in the LIST sheet, you can use the /M command. To dive to the THETA subsheet, place the cursor on THETA and type >.

You could enter values for THETA individually, but that would be time consuming and prone to error. If you place the cursor in the first value position in the LIST subsheet for the variable and type an exclamation point, **!,** you will invoke the LIST FILL command. The pop-up menu shows there are four options for filling a list. Selecting ADD STEP will add a specified step increment to each successive value until a specified upper limit is reached. LINEAR SPACING will assign intermediate values so there is equal spacing between all values between given first and last values. MULTIPLY BY STEP will multiply each successive value by a specified multiplier until the specified upper limit is reached. GEOMETRIC SPACING will assign values between the given first and last values so consecutive values differ by a constant factor. We will use the ADD STEP command for this example. The initial angle will be 30°, and THETA will be increased by 1.0° until the final value of 75° is reached. Figure 8.11*b* shows the resulting LIST subsheet for THETA. The LIST sheet of Figure 8.11*a* shows that THETA now has 46 elements.

After defining the list named Window, you will want to enter the height of the window as the value for each list element. You can use LINEAR SPACING LIST FILL with 46 elements and first and last values of 37. This will produce a horizontal line to define the height of the window for your plots.

You can bring the VARIABLE sheet into the active window by typing **=V.** THETA is specified as an input list by moving the cursor to the STATUS column for THETA and typing **I,** for input. The I will not appear in the space, but 45 will appear in the input cell for THETA. This was the last value used as a guess when the model was used for Example 8.3. TK SOLVER remembers the last value for a variable and brings that value into the INPUT cell whenever a G or I is typed in the STATUS column. TK SOLVER will not use this value in the solution. Instead, it will use the values from the associated input list. After the values for the list are exhausted, 45 will again appear in the INPUT cell. To invoke the LIST SOLVE command, type /L and **ENTER,** and observe what happens to the VARIABLE sheet. Intermediate values for the variables will flash on the screen as TK SOLVER uses successive input values. Although it is not required for this example, the LIST SOLVER can also perform iterative solutions when one or more of the input lists contains initial guesses for the output values.

Figure 8.11*c* shows the LIST sheet after solution of the model. Note that all the lists now have 46 elements. If you dive to any of the subsheets for the various lists, you could view those values. However, we will bring the TABLE sheet into the active window by typing **=T.** We will define the table named Trajectory as shown in Figure 8.11*c*. You can define as many tables as you need. From the TABLE sheet you can dive to the TABLE subsheet for Trajectory. Figure 8.11*c* also shows this subsheet. Once in the subsheet, you can choose whether you want the output to go to the screen or the printer. You can enter the title you want for the table and specify whether you want the columns aligned vertically or horizontally (vertical is the normal orientation). The title could be entered either here or in the TABLE sheet. You also have to define the elements to be included in the table. A prior examination of the LIST subsheet for H showed that the elements of the list that were of interest were between elements 20 and 35. We will specify these as the first and last elements of the table. The order of the list names in the subsheet determines the order in which the columns will appear in the table. The default column width is 10 characters. The heading for *Xmax* is accommodated by increasing the column width to 12. The last two characters would not appear if the column width were left at 10. Pressing the **F8** key or the **!** will cause the table to print. Figure 8.11*d* shows the table output.

The table of Figure 8.11*d* contains extra decimal places that are not significant and does not have any type of column separators. You can define twod in the NUMERIC format sheet as a numeric format with two decimal places and centering justification. Entering the twod specification in the NUMERIC FORMAT column of the Trajectory TABLE subsheet will format the table. Entering a − in the ROW SEPARATOR definition cell, and a | in the COLUMN SEPARATOR definition cell will provide row and column separation. Figure 8.11*e* shows the formatted table. The modified TABLE subsheet is shown in Figure 8.12*a*.

From the TABLE subsheet for Trajectory, you can dive to the interactive TABLE subsheet by typing >. The interactive TABLE subsheet for Trajectory shows a table that is similar to the printed table, but without the separator lines. It also includes an ELEMENT column that gives the row number of the table produced by the TABLE subsheet. These are not the element numbers of the original list. You can change values for THETA within the Angle column and press **F10** for LIST SOLVE to change the output accordingly. In Figure 8.11*f*, THETA values of 53.1, 53.2, and 53.3 were substituted for the values 54, 55, and 56. These values will more closely estimate the angle that causes the trajectory to reach the window. When the table is interactively solved, it shows that an angle of 53.3° produces a height of exactly 37 ft. The table also shows that the second angle is about 59°.

An alternate way of developing this model and entering the data will now be presented. We are still using the rule sheet of Example 8.3. You can specify which variables will be associated with lists, and specify THETA as the input list, in the VARIABLE sheet. Then you do not have to go the LIST sheet at all. The VARIABLE sheet will act as a control center. You can go directly to the TABLE sheet and create the Trajectory table. You can specify the variables for the table, and their attributes, through the TABLE subsheet for Trajectory. This time you do not have to specify a range for the elements. You will enter the elements of the lists through the interactive table. Figure 8.12 shows the resulting VARIABLE, TABLE, and TABLE subsheets.

When you dive to the interactive table, it will contain the variable labels, but no data. Move the cursor to the Angle column and enter the input angles that you want to use. LIST FILL can be used as before. Figure 8.12*b* shows the interactive table with 16 values for the angle entered. Inputting the angles generates the element numbers. If there were more input lists, you could repeat the procedure for those lists. You can remain in the interactive subsheet and LIST SOLVE by pressing **F10.** The output values will appear in the interactive table as the solution progresses. Figure 8.12*c* shows the interactive table after completion of the LIST SOLVE.

(EXAMPLE 8.4 ☐   Table and Plots for Particle Trajectory Problem   ☐   Continued)

FIGURE 8.12   List entry through interactive table.

```
==================== VARIABLE SHEET ==========================================
St Input---- Name--- Output--- Unit----- Comment----------------------------
   66        Vo                          Initial velocity, ft/sec
   32.17     g                           Acceleration due to gravity, ft/sec^2
   90        X                           Horizontal distance along trajectory
L            H                           Height of trajectory at distance X, ft
             RAD                         # of degrees in one radian
             ANGLE                       Angle of inclination, radians
             Voy                         Y component of initial velocity
             Vox                         X component of initial velocity
             Tymax                       Time to max height of trajectory, sec
L            Ymax                        Maximum height of trajectory, ft
             Xymax                       X distance to max height of trajectory
L            Xmax                        Maximum length of flat trajectory, ft
             Vyx                         Y velocity at distance X, ft/sec
             AngHit                      Angle of entry into window, degrees
L            THETA                       Angle of inclination, degrees
L            t                           Time to reach distance X, sec

==================== TABLE SHEET =============================================
Name---------- Title---------------------------------------------------------
Trajectory     Trajectory Data for Hitting Window

==================== TABLE: Trajectory ======================================
Screen or Printer:          Printer
Title:                      Trajectory Data for Hitting Window
Vertical or Horizontal:     Vertical
Row Separator:              -
Column Separator:           |
First Element:              1
Last Element:
List----- Numeric Format-- Width-- Heading----------------------------------
THETA                      10      Angle
H         twod             10      Height
t         twod             10      Time
Ymax      twod             10      Max Height
Xmax      twod             12      Max Distance
```

(a) VARIABLE, TABLE, and TABLE subsheets.

FIGURE 8.12 (Continued)

```
(1,1) Angle: 49                                                        25+/F9

==================== TABLE: Trajectory ======================================
Title:      Trajectory Data for Hitting Window
Element Angle----- Height---- Time------ Max Height Max Distance -------
1       49
2       50
3       51
4       52
5       53
6       54
7       55
8       56
9       57
10      58
11      59
12      60
13      61
14      62
15      63
16      64




  F1 Help  F2 Cancel  F5 Edit  F9 Solve  / Commands  = Sheets  ; Window switch
```

(*b*) Interactive table with input list.

```
(1,1) Angle: 49                                                        25+/F9

==================== TABLE: Trajectory ======================================
Title:      Trajectory Data for Hitting Window
Element Angle----- Height---- Time------ Max Height Max Distance -------
1       49        34.04      2.08       38.56      134.09
2       50        34.87      2.12       39.73      133.35
3       51        35.62      2.17       40.89      132.45
4       52        36.28      2.21       42.04      131.38
5       53        36.85      2.27       43.18      130.16
6       54        37.3       2.32       44.31      128.78
7       55        37.62      2.38       45.43      127.24
8       56        37.78      2.44       46.53      125.55
9       57        37.76      2.5        47.62      123.7
10      58        37.52      2.57       48.69      121.7
11      59        37.03      2.65       49.74      119.56
12      60        36.24      2.73       50.78      117.26
13      61        35.11      2.81       51.79      114.83
14      62        33.56      2.9        52.78      112.26
15      63        31.52      3          53.75      109.55
16      64        28.88      3.11       54.69      106.7



  F1 Help  F2 Cancel  F5 Edit  F9 Solve  / Commands  = Sheets  ; Window switch
```

(*c*) Interactive table after LIST SOLVE.

(EXAMPLE 8.4 □ Table and Plots for Particle Trajectory Problem □ Continued)

Now you are ready to produce the trajectory plot. Use **=P** to bring the PLOT sheet into the active window. Figure 8.13*a* shows the PLOT sheet with the proper entries. Next you can use > to dive into the PLOT subsheet for the plot named Window. Figure 8.13*a* shows the settings for the parameters of the plot.

FIGURE 8.13   Line graphs for TK SOLVER.

```
=================== PLOT SHEET =================================================
Name-------- Plot Type-- Display Option- Output Device-- Title-----------------
window       Line chart  4.VGA            4.HPLASJET      Height at X vs. THETA
=================== LINE CHART: window =========================================
Display Scale:              Yes
Display Zero Axes:          None
Display Grid:               Yes
Line Chart Scaling:         Linear
Title:                      Height at X vs. THETA
X-Axis Label:               Initial Angle of Trajectory (degrees)
Y-Axis Label:               Trajectory Height
X-Axis Minimum, Maximum:    30,75
Y-Axis Minimum, Maximum:
X-Axis List:                THETA
Y-Axis--- Style------ Character-- Symbol Count-- First-- Last---
H          Lines        *             0             1      46
```

(*a*) PLOT sheet and subsheet, height vs. angle, 30° to 75°.

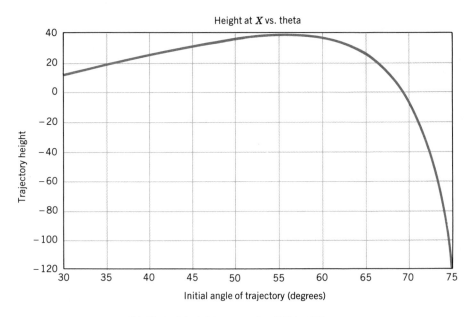

(*b*) Plot of height vs. angle, 30° to 75°.

FIGURE 8.13 (Continued)

```
==================== LINE CHART: window ==========================================
Display Scale:              Yes
Display Zero Axes:          None
Display Grid:               Yes
Line Chart Scaling:         Linear
Title:                      Height at X vs. THETA
X-Axis Label:               Initial Angle of Trajectory (degrees)
Y-Axis Label:               Trajectory Height
X-Axis Minimum, Maximum:    50,60
Y-Axis Minimum, Maximum:
X-Axis List:                THETA
Y-Axis--- Style------ Character-- Symbol Count-- First-- Last---
H          Curves      *           0              1       46
Window     Lines       -           0              1       46
```

(c) Height vs. angle, 50° to 60°.

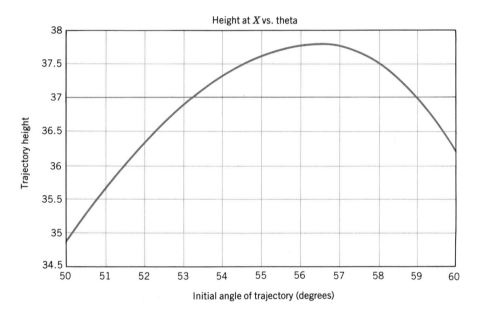

(d) Plot using curve smoothing option.

The PLOT subsheet heading shows what type of plot it represents, as well as the name of the plot. The rest of the entries are largely self-explanatory. The minimum and maximum X-axis values could have been left out, because they are the same as the minimum and maximum values of the list for THETA. The character * is the default character and will be used for plots made in text mode to show the data points.

Pressing the **F7** key will display the resulting plot. Pressing the letter **O** will send the graphics image to the selected output device. Figure 8.13*b* shows the resulting plot. Pressing the **ENTER** key will return you to the PLOT subsheet.

(EXAMPLE 8.4   □   Table and Plots for Particle Trajectory Problem   □   Continued)

Examination of the plot of Figure 8.13*b* shows that the solutions fall between 50° and 60°. Therefore, you can refine the estimates by limiting the plotting area to those angles. In Figure 8.13*c* the *X* axis minimum and maximum are 50 and 60. Also, the second *Y* axis list has been added (Window). This is the list that contains the constant height of the window. Figure 8.13*d* shows the plot for the refined range using the curve smoothing option. The plot is output on an *X-Y* plotter.

PROGRAM TESTING

Output for this example agrees with the output already checked for the previous particle trajectory examples.

8.2.2.12 User Supplied Functions

Three types of functions are possible with TK SOLVER PLUS: LIST, RULE, and PROCEDURE functions. Each type has its own rules for arguments and syntax. It is these functions that define TK SOLVER as a true structured programming language. They help set TK SOLVER apart from other equation solvers and increase its versatility for the serious user in science and engineering. Also, these functions allow the development of standard procedures that can be used in different programs without having to change the variable names for the procedure. These user defined functions are similar in purpose to subprograms in FORTRAN. This text will provide you with a short introduction to the capabilities of these functions. It will be worth your time to continue to study all the nuances of using functions with TK SOLVER. To aid in this, Universal Technical Systems Inc. has developed an excellent tutorial that explores the rules and syntax of TK SOLVER functions.

The FUNCTION sheet defines the various functions that a TK SOLVER model will use. Each defined function will also have a FUNCTION subsheet that will define the parameters of the particular function. The subsheet for rule and procedure functions will also contain the rules of the function. The subsheet for a list function will contain the names of the independent and dependent variable lists.

Three examples will illustrate the use of functions. Key elements of the development of the models will be presented. The instructor's software supplement for the text contains the complete models. You can load the models and run them to gain an appreciation for how functions operate with TK SOLVER. In a later section we will use the TK SOLVER library model for numerical integration using Simpson's rule. This model also gives a good illustration of using functions.

LIST Functions   LIST functions express the relationship between two lists of variable values. One list represents the independent variable, and the other list the dependent variable. A value for the independent variable is sent over to the LIST function as an argument. The corresponding value from the dependent variable list is sent back as the function value. The corresponding value of the dependent variable is determined through table mapping, step mapping, linear interpolation, or cubic interpolation. Subsequent paragraphs will discuss each type of mapping and interpolation.

Table mapping is the only type of mapping or interpolation that requires a one-to-one correspondence between the elements of the independent and dependent variable. The function will accept only values of the independent variable that correspond exactly to one of the values in the independent variable list. The function value will be the corresponding value of the dependent variable list.

Step mapping searches the independent variable list. It locates the interval of the list into which the value of the independent variable argument falls. It then finds the corresponding pair of values from the dependent variable list and returns the first of these values as the value of the function. Any value in the interval will return the same value for the function, which gives rise to the name "step mapping."

Interpolation provides a more accurate way of estimating the dependent variable value when the value of the independent variable argument falls between two values of the independent variable list. TK SOLVER gives you the option of using linear or cubic interpolation. Linear interpolation finds the value of the dependent variable by taking a linear proportion between two values of the dependent variable list. The value of the independent variable argument is used to determine the linear proportion. Linear interpolation works well for functions that are approximately linear, but loses accuracy for nonlinear functions. Cubic interpolation is usually more accurate for interpolation of nonlinear functions. It develops a cubic interpolation function based on four corresponding values of the independent and dependent variable lists. The cubic interpolation function uses the independent variable value to estimate the dependent variable value. The LIST function returns this value. Additional information on the procedures for cubic interpolation can be found in Exercise 6.12.1.

RULE Functions   A RULE function is similar in many ways to a function subprogram in FORTRAN. It expresses a functional relationship between specified variables using dummy and actual arguments. Multiple rules can accomplish the calculations of the function. However, there are several significant differences. The RULE function can return either one or two values. This permits transfer of complex numbers. A maximum of 20 dummy and actual argument variables can be used. Any or all of the arguments can be lists. RULE functions can be accessed through a reference in an expression using the function name, or they can be accessed through a separate CALL statement like a FORTRAN subroutine. If you use the CALL statement, the arguments can also pass values back to the calling statement. TK SOLVER rule functions are not restricted as to what variable values are specified when referencing the function. Thus, the output value of the function may be known, and the function can be used to calculate one of the arguments. TK SOLVER treats the rules of a RULE function the same way as it treats rules in the RULE sheet.

PROCEDURE Functions   PROCEDURE functions are similar to subprograms in high-level languages. They are a series of assignment statements that accept input values, operate on those input values, and produce a corresponding set of output values. The procedure returns the values to the calling program. Note the use of the words "assignment statements" rather than rules. That is because TK SOLVER treats the statements of procedure functions like assignment statements in high-level languages. Version 1 of TK SOLVER PLUS required the use of the operators := for assignment statements and = for rules. Version 1.1 has made this distinction optional. Statements using the = sign are treated as rules when included in the RULE sheet or in RULE functions. PROCEDURE functions treat them as assignment statements.

CALL statements, much like those in BASIC and FORTRAN programs, can access procedure functions. They can also be accessed through references similarly to RULE functions. However when you access PROCEDURE functions through a CALL statement, you can have up to 20 result variables. When they are accessed through references, two is the maximum number of result variables. Both methods of access can have up to 20 parameter and argument variables. A PROCEDURE function can access any other TK SOLVER function. It can also reference itself through recursion.

Argument and result variables in PROCEDURE functions are local to the procedure. Therefore the variable names do not have to agree with the names in the argument list of the referencing statement. The position of variables within the argument and result variable lists must agree in position with the corresponding variables in the referencing statement. Values are passed back and forth by the position, not the name. The parameter variables act much like a common block in FORTRAN. However, the variable names have to be the same as they are in other parts of the model. The order of the variables in the Parameter Variables list does not matter. Also, TK SOLVER does not type variables according to REAL and INTEGER, so there is no need to match types.

PROCEDURE functions also can contain selection and control structures similar to those in FORTRAN and BASIC. Among these are the IF THEN ELSE, FOR NEXT, and GO TO structures.

The IF THEN ELSE structure is similar to the conditional rules discussed in Section 8.2.2.7 of this chapter. It also has some additional capabilities when used in the procedure functions. The basic form

    IF <logical expression> THEN <expression 1> ELSE <expression 2>

can have the form

    IF <logical expression> THEN GO TO <label> ELSE <expression 2>

or

    IF <logical expression> THEN RETURN ELSE <expression 2>

The second and third expressions could also have been written without the ELSE clauses at the end. The GO TO structure shifts control to the labeled statement in the procedure function. It will also search for a specified character string. RETURN has the same connotation that it does in high-level languages; it returns control to the location of the CALL statement. The GO TO and RETURN statements can also be separate rules in the PROCEDURE function.

The FOR NEXT structure is the same as the structure of the same name in BASIC. It is similar to the DO loop in FORTRAN. It allows the modeler to specify the repetition of a certain series of rules a selected number of times.

Example 8.5 develops a general RULE function to accomplish the regression analysis of Example 8.3.

Example 8.6 uses all three types of functions in the same model. The list function will employ cubic interpolation. The problem is the routing of an un-

teady flow hydrograph through a storage basin that has sloped sides and a circu-
ar hole, or orifice, in the bottom. The orifice allows water to flow out of the
torage basin. This type of temporary storage basin might be used in a chemical
•lant, in a pumping station, or in flood control and storm water management
eservoirs. The problem is to estimate what the outflow hydrograph is going to be
or a given input hydrograph. The amount stored in the reservoir and the rate of
•utflow are both dependent on the depth within the reservoir. The depth is in turn
, function of the amount in storage at any particular time. Therefore, you have to
ise some type of tabular or iterative algorithm to solve the problem.

   You can apply numerical integration to estimate the value for the integral of a
unction without using the analytical integral. Chapter 9 discusses numerical inte-
·ration in more detail. Example 8.7 illustrates one of the applications library
nodels provided with the TK SOLVER package.·It is called SIMPSON.TK, and
mplements Simpson's rule for numerically integrating a function. SIMPSON.TK
ises both RULE and PROCEDURE functions. The model will be combined with
 short driving model to find the moment of inertia for a triangle about a specified
xis.

EXAMPLE 8.5 □ RULE Function for Regression Analysis

STATEMENT OF PROBLEM

Develop a RULE function to accomplish the regression analysis of Example 8.3.
Make it as general as possible.

MATHEMATICAL DESCRIPTION

Example 8.3 developed the rules for the mathematical description of regression
analysis. Example 5.4 presented the mathematical model.

ALGORITHM DEVELOPMENT

Input/Output Design

You can develop the RULE function directly from the model of Example 8.3.
The VARIABLE sheet of the revised model will use descriptive definitions to
make it clearer what data are being analyzed. The model of Example 8.3 already
includes the input lists. In practice, the input lists for regression will be part of
the model into which the regression RULE function is merged. Output of values
will be to the VARIABLE sheet of the calling model.

Computer Implementation

The following sequence of steps will be needed to create the regression RULE
function:

```
LOAD MODEL OF EXAMPLE 8.3
GO TO FUNCTION SHEET
DEFINE REGRESSION RULE FUNCTION (Named REGRESS)
DIVE TO REGRESS FUNCTION SUBSHEET
DEFINE ALL VARIABLES AS ARGUMENT VARIABLES (X,Y,N,Xbar,Ybar,a,b,r)
```

(EXAMPLE 8.5 □ RULE Function for Regression Analysis □ Continued)

```
USE THE /COPY COMMAND TO COPY THE RULES FROM THE RULE SHEET TO THE
    REGRESS FUNCTION SUBSHEET
MOVE TO THE RULE SHEET AND DELETE THE RULES
ADD FUNCTION CALL TO RULE SHEET
MOVE TO VARIABLE SHEET AND MODIFY TO TEST THE MODEL
TEST MODEL
SAVE RULE FUNCTION USING /STORAGE FUNCTION OPTION
```

Since the function is calculating three parameters (a, b, and r), it is better to include all of them as argument variables. This bypasses the restrictions placed on result variables in RULE functions. The RULE function can have a maximum of two result variables. If it has two they have to be related, such as the real and imaginary components of a complex number. The only restriction on the argument variables is that there cannot be more than 20 of them. Argument variables can pass values back and forth just like the dummy and actual arguments of subroutine subprograms in FORTRAN. To use the RULE FUNCTION in this form you must use a CALL statement in the RULE sheet, rather than invoke the rule function through a function reference. Example 8.6 will illustrate the latter option.

Using the /STORAGE FUNCTION option saves the FUNCTION sheet and the RULE FUNCTION subsheet in a separate file. You can later merge the function into any TK SOLVER application.

PROGRAM DEVELOPMENT

Figure 8.14a shows the FUNCTION sheet and the REGRESS RULE FUNCTION subsheet developed according to the implementation instructions above. The variable names are in their general form. Figure 8.14b shows the VARIABLE and RULE sheets used to test the RULE function. All the old variables were deleted and new ones were defined. The variable and list names and comments are descriptive of the actual data. Values for the argument variables are passed back and forth by position. Therefore, the CALL statement can use the descriptive names.

PROGRAM TESTING

The VARIABLE sheet of Figure 8.14 shows the solved model. You can see that the answers are the same as those produced in Example 8.3. Again, it was not necessary to invoke LIST SOLVE. In applications, the VARIABLE and RULE sheets would normally contain other rules and variables to accomplish additional analysis. For example, you might want to calculate estimated values for the strain or calculate other statistics such as the standard error of estimate.

Figure 8.14c shows an alternate form of the VARIABLE and RULE sheets that gives the same results. In this form you do not have to refer to the dependent and independent variables in the VARIABLE sheet. The arguments of the CALL statement link them to the calling model. I prefer to use the form shown in Figure 8.14b because it presents all the variable definitions in one place, the VARIABLE sheet. The second form shows that the input variable lists are independent of the variable definitions on the VARIABLE sheet.

FIGURE 8.14   Regression rule function.

```
================== FUNCTION SHEET =========================================
Name---------- Type----- Arguments-- Comment----------------------------------
REGRESS         Rule      8;0        Rule Function for Regression

================== RULE FUNCTION: REGRESS =================================
Comment:                Rule Function for Regression
Parameter Variables:
Argument Variables:   X,Y,N,Xbar,Ybar,a,b,r
Result Variables:
S Rule----------------------------------------------------------------------
  N = COUNT(X)
  Xbar = SUM(X)/N
  Ybar = SUM(Y)/N
  a = (DOT(X,Y)-N*Xbar*Ybar)/(DOT(X,X)-N*Xbar^2)
  b = Ybar - a*Xbar
  r = a*SQRT((DOT(X,X)-N*Xbar^2)/(DOT(Y,Y)-N*Ybar^2))
```

(*a*) FUNCTION sheet and subsheet.

```
================== VARIABLE SHEET =========================================
St Input---- Name--- Output--- Unit----- Comment----------------------------
             N       20                  Number of specimens
             Mstress 37700               Average stress
             Mstrain .001095             Average strain
             a       4.9605E-8           Slope of regression line
             b       -.0007751           Intercept of regression line
             r       .93390417           Correlation Coefficient
  'STRESS    Stress                      Measured values for stress
  'STRAIN    Strain                      Measured values for strain

================== RULE SHEET ============================================
S Rule----------------------------------------------------------------------
  CALL REGRESS(Stress,Strain,N,Mstress,Mstrain,a,b,r)
```

(*b*) VARIABLE and RULE sheets for testing.

```
================== VARIABLE SHEET =========================================
St Input---- Name--- Output--- Unit----- Comment----------------------------
             N       20                  Number of specimens
             Mstress 37700               Average stress
             Mstrain .001095             Average strain
             a       4.9605E-8           Slope of regression line
             b       -.0007751           Intercept of regression line
             r       .93390417           Correlation Coefficient

================== RULE SHEET ============================================
S Rule----------------------------------------------------------------------
  CALL REGRESS('STRESS,'STRAIN,N,Mstress,Mstrain,a,b,r)
```

(*c*) Alternate form of VARIABLE and RULE sheets.

EXAMPLE 8.6 □ **Reservoir Routing Illustration of LIST, RULE, and PROCEDURE Functions**

STATEMENT OF PROBLEM

Illustrate a TK SOLVER model for estimating the outflow hydrograph from a storage reservoir. Input data will include the inflow hydrograph, the geometry of the basin, and the type of outflow control device. Figure 8.15 gives a definition sketch for this problem.

MATHEMATICAL DESCRIPTION

A common method for modeling an unsteady flow hydrograph through a reservoir is the Storage-Indication Working Curve Method (or Modified Puls Method). The outflow is assumed to be a function of the depth of water in the reservoir. This is a valid assumption as long as the volume of the reservoir is large in comparison with the inflow. You have to develop relationships for the elevation vs. outflow and for the elevation vs. volume in storage. If these two functions are known, you can define the $2S/t + O$ curve. In this expression, S is the volume in storage, t is the time-step length used for the routing simulation, and O is the outflow rate. The complete inflow hydrograph, as well as the initial outflow, must be known. The principle of conservation of mass states that the rate of change of the volume in storage is the difference between the inflow and outflow at any time. This leads to the following expression for determining $2S/t + O$ for a time step:

$$\left[\frac{2S}{t} + O\right]_i = I_{i-1} + I_i + \left[\frac{2S}{t} + O\right]_{i-1} - 2O_{i-1} \tag{8.2}$$

If the analysis is being done by hand, you can determine the outflow O for the time step from a plot of $2S/t + O$ vs. O. You can repeat the procedure to

FIGURE 8.15   Definition sketch for unsteady flow routing through a reservoir.

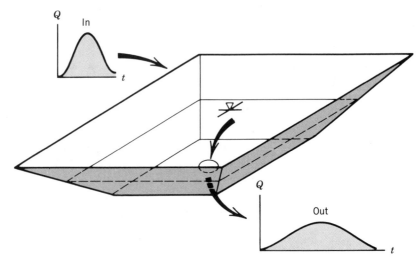

determine $2S/t + O$ and O for subsequent time steps. A computer program can use an interpolation routine to determine O for a given $2S/t + O$.

ALGORITHM DEVELOPMENT

Input/Output Design

Input will consist of the list of values for the input hydrograph, as well as the initial storage and outflow values. The model will produce a table containing data on the inflow, time, $2S/t + O$, depth in the storage basin, volume in storage, and outflow. A plot will show the inflow and routed outflow on the same graph.

Numerical Methods

A LIST function named Stor will estimate the outflow for a given value of $2S/t + O$ using cubic interpolation. The data of this function are dependent on the geometry of the basin and the outlet structure. The data will not change for different input hydrographs for the same basin. Therefore, there is no need to produce the data for each solution of the model. A separate TK SOLVER model generated the lists for the function. The present model will merge these lists into the function. A PROCEDURE function in the reservoir routing model could also have generated these data.

As with most estimation techniques, the shorter the time step is, the more accurate the results will be. An infinitely small time step approaches the differential equation solution as a limit.

Computer Implementation

As described above, LIST function Stor estimates the outflow, OUT. The argument is the computed value of $2S/t + O$ (STORFUNC) from Equation 8.2. The lists for the function have 61 elements to increase accuracy.

The routing procedure has to use a PROCEDURE function to accommodate the references to values computed during the previous time step. Computation of the $2S/t + O$ value for the current time step requires these backwards references. The model also incorporates a loop structure with an index for the time steps. In TK SOLVER, these operations can be done only with a PROCEDURE function.

Separate RULE functions compute the depth in the storage reservoir and the volume in storage. This keeps the model as general as possible. If you want to change either the size or geometry of the basin or outlet structure, you can change the RULE functions. You would also have to recompute the $2S/t + O$ vs. O data for function Stor.

A pseudocode representation of the unsteady flow routing model follows:

```
GENERATE STORAGE FUNCTION (2S/t + O) AND O LISTS
DEFINE FUNCTION STOR WITH SF (2S/t + O) AS THE INDEPENDENT VARIABLE, AND
      OUTFLOW (Out) AS THE DEPENDENT VARIABLE
SAVE FUNCTION SHEET AND Stor SUBSHEET USING /SF
RESET ALL SHEETS
LOAD FUNCTION SHEET AND Stor SUBSHEET
BRING FUNCTION SHEET INTO THE ACTIVE WINDOW
DEFINE ADDITIONAL FUNCTIONS Resroute (PROCEDURE), Depth (RULE), AND Vol (RULE)
DIVE TO SUBSHEET FOR Vol RULE FUNCTION
      ENTER ARGUMENT AND RESULT VARIABLES
      DEFINE GEOMETRIC DATA
```

**(EXAMPLE 8.6 □   Reservoir Routing Illustration of
LIST, RULE, and PROCEDURE Functions □
Continued)**

```
     DEFINE RULE FOR COMPUTING STORAGE VOLUME AS A FUNCTION OF DEPTH
RETURN TO FUNCTION SHEET AND DIVE TO SUBSHEET FOR Depth PROCEDURE
     ENTER ARGUMENT AND RESULT VARIABLES
     ENTER RULE FOR AREA OF CIRCLE
     DEFINE RULE FOR COMPUTING DEPTH AS A FUNCTION OF OUTFLOW
RETURN TO FUNCTION SHEET AND DIVE TO SUBSHEET FOR Resroute PROCEDURE
     INITIALIZE VARIABLES FOR FIRST TIME STEP
          Time = 0.0
          IN[1] = inzero
          y[1] = yzero
          OUT[1] = outzero
     CALL Vol FOR INITIAL CONDITIONS
     CALCULATE STORAGE FUNCTION FOR INITIAL CONDITIONS
     FOR i = 2 TO n
          CALCULATE TIME
          CALCULATE STORAGE FUNCTION i FROM EQUATION 8.2
          IF STORAGE FUNCTION <= 0.0, SET EQUAL TO ZERO
          CALCULATE OUT USING FUNCTION Stor
          CALCULATE DEPTH IN RESERVOIR USING FUNCTION Depth
          CALCULATE VOLUME IN STORAGE USING FUNCTION Vol
     NEXT i
BRING RULE SHEET INTO ACTIVE WINDOW
     INPUT CALL FOR Resroute PROCEDURE FUNCTION
BRING VARIABLE SHEET INTO ACTIVE WINDOW
     ASSIGN INPUT VALUES FOR yzero, outzero, n, inzero, and t
     DEFINE VARIABLES IN, OUT, Storage, y, STORFUNC, and Time (For information
          only)
BRING TABLE SHEET INTO ACTIVE WINDOW
     DEFINE TABLE Reservoir
DIVE TO Reservoir SUBSHEET
     DEFINE PARAMETERS FOR TABLE
     DEFINE LISTS FOR TABLE
          IN, Time, STORFUNC, y, Storage, and OUT
DIVE TO INTERACTIVE TABLE
     ENTER INPUT HYDROGRAPH
     SOLVE MODEL
BRING PLOT SHEET INTO ACTIVE WINDOW
     DEFINE PARAMETERS FOR PLOT
```

PROGRAM DEVELOPMENT

Figure 8.16 shows the development of the model. The Resroute PROCEDURE function subsheet illustrates some additional features of PROCEDURE functions. Note the definition of Parameter Variables. Values for these variables are automatically brought over from the VARIABLE sheet. The notation STORFUNC[i] refers to the ith element of the list variable STORFUNC. The effect is similar to a subscripted variable in FORTRAN, except that you do not have to define an upper limit on the dimension size. The assignment operator := is used in PROCEDURE function Resroute. If you are using Release 1.1 of TK SOLVER, you could replace := with =.

The PROCEDURE function calls function Vol two different ways to illustrate the difference. The first reference uses a CALL statement. This passes the value for yzero over as the argument variable value. The CALL returns the value for

FIGURE 8.16   Input sheets for unsteady flow routing.

```
==================== LIST FUNCTION: Stor =====================================
Comment:                 Storage Function (2S/t+0) generated by separate model
Domain List:             SF
Mapping:                 Cubic
Range List:              Out
Element-- Domain------------- Range--------------
1         0                   0
2         21.52               14.7
3         34.47               20.79
4         46.02               25.47
5         56.88               29.4
6         67.29               32.88
7         77.41               36.01
8         87.31               38.9
9         97.03               41.58
10        106.63              44.11
11        116.12              46.49
12        125.53              48.76
13        134.87              50.93
14        144.15              53.01
15        153.39              55.01
16        162.59              56.94
17        171.75              58.81
18        180.9               60.62
19        190.02              62.38
20        199.13              64.09
 .          .                   .

 .          .                   .
40        381.95              91.82
41        391.23              92.99
42        400.54              94.14
43        409.87              95.28
44        419.22              96.41
45        428.6               97.52
46        438                 98.63
47        447.42              99.72
48        456.86              100.79
49        466.33              101.86
50        475.83              102.92
51        485.35              103.96
52        494.9               105
53        504.47              106.02
54        514.07              107.03
55        523.7               108.04
56        533.35              109.04
57        543.04              110.02
58        552.75              111
59        562.49              111.97
60        572.26              112.93
61        582.06              113.88
```

(*a*) LIST function Stor imported from previous TK model.

(EXAMPLE 8.6 □ Reservoir Routing Illustration of LIST, RULE, and PROCEDURE Functions □ Continued)

FIGURE 8.16 (Continued)

```
=================== FUNCTION SHEET ========================================
Name----------- Type----- Arguments-- Comment------------------------------
Stor            List       1;1         Storage Function (2S/t+0) generated by se
Resroute        Procedure  6;0         Unsteady Flow Routing Through a Reservoir
Vol             Rule       1;1         Storage vs. depth for a basin with trapez
Depth           Rule       1;1         Depth as a function of outflow

=================== RULE FUNCTION: Vol ====================================
Comment:                Storage vs. depth for a basin with trapezoidal X-section
Parameter Variables:
Argument Variables:   y
Result Variables:     Storage
S Rule---------------------------------------------------------------------
  B1 = 175                ''First bottom dimension, ft
  B2 = 175                ''Second bottom dimension, ft
  SS = 2                  ''Side slop of basin, ft/ft
  Storage=((B1+2*SS*y)*(B2+2*SS*y)+B1*B2)/2*y

=================== RULE FUNCTION: Depth ==================================
Comment:                Depth as a function of outflow
Parameter Variables:
Argument Variables:   Outflow
Result Variables:     y
S Rule---------------------------------------------------------------------
  D = 3                    ''Diameter of orifice, ft
  Cd = 0.82                ''Discharge coefficient for orifice
  g = 32.17                ''Acceleration due to gravity, ft/sec^2
  Area = pi()*D^2/4        ''Area of orifice, ft^2
  y = (1/(2.*g))*Outflow^2/Cd^2/Area^2

=================== PROCEDURE FUNCTION: Resroute ==========================
Comment:                Unsteady Flow Routing Through a Reservoir
Parameter Variables:  t,yzero,outzero,n,inzero
Input Variables:      IN,OUT,Storage,y,STORFUNC,Time
Output Variables:
S Statement----------------------------------------------------------------
  Time[1]:=0
  IN[1]:=inzero
  y[1]:=yzero
  OUT[1]:=outzero
  Call Vol(yzero;Storage[1])
  STORFUNC[1]:=Stor(outzero)
  for i := 2 to n
      Time[i]:=Time[i-1]+t/60
      STORFUNC[i]:=IN[i-1]+IN[i]+STORFUNC[i-1]-2*OUT[i-1]
      IF (STORFUNC[i]  0.0) THEN STORFUNC[i]:= 0.0
      OUT[i]:=Stor(STORFUNC[i])
      y[i]:=Depth(OUT[i])
      Storage[i]:=Vol(y[i])
  next i
```

(b) FUNCTION sheet, and Vol, Depth, and Resroute subsheets.

FIGURE 8.16 (Continued)

```
=================== TABLE: Reservoir =======================================
Screen or Printer:          Screen
Title:                      UNSTEADY FLOW ROUTING THROUGH A RESERVOIR
Vertical or Horizontal:     Vertical
Row Separator:              -
Column Separator:           |
First Element:              1
Last Element:               11
List----- Numeric Format-- Width-- Heading----------------------------------
IN                          10      Inflow
Time                        10      Time
STORFUNC                    10      2S/t+0
y                           10      Depth
Storage                     10      Storage
OUT                         10      Outflow

=================== VARIABLE SHEET =========================================
St Input---- Name--- Output--- Unit----- Comment---------------------------
   0         yzero                        Depth of water in basin at time zero
   0         outzero                      Outflow at time zero, ft^3/sec
   11        n                            Number of time steps
   0         inzero                       Inflow at time zero
   900       t                            Time step length, sec
             IN                           Inflow hydrograph, ft^3/sec
             OUT                          Outflow hydrograph, ft^3/sec
             Storage                      Volume in storage, ft^3
             y                            Depth of water in basin, ft
             STORFUN                      2*Storage/t + OUT  [2S/t + 0]
             Time                         Elapsed time in simulation, minutes

=================== RULE SHEET =============================================
S Rule---------------------------------------------------------------------
* CALL Resroute('IN,'OUT,'Storage,'y,'STORFUNC,'Time)
```

(c) TABLE subsheet, VARIABLE, and RULE sheets.

```
=================== TABLE: Reservoir =======================================
Title:      UNSTEADY FLOW ROUTING THROUGH A RESERVOIR
Element Inflow---- Time------ 2S/t+0---- Depth----- Storage--- Outflow---
1        0
2        2
3        39
4        129
5        168
6        108
7        60
8        21
9        5
10       0
11       0
```

(d) Interactive subsheet for reservoir showing input hydrograph.

volume in storage and stores that value in Storage[1] as the result variable. The second reference to Vol uses the function name and argument variable in an assignment statement. The value for y[i] is sent over as the value of the argument variable. The function value is the value of the result variable. The assignment operator stores the value in Storage[i].

(EXAMPLE 8.6 ☐ **Reservoir Routing Illustration of LIST, RULE, and PROCEDURE Functions** ☐ **Continued)**

FIGURE 8.17   Table and plot for unsteady flow routing.

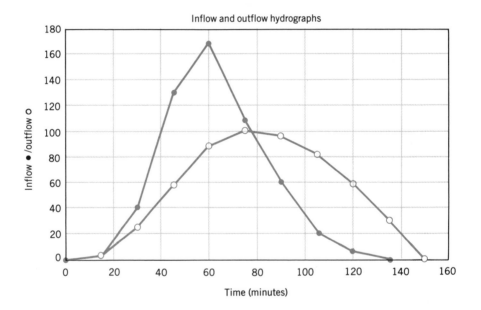

UNSTEADY FLOW ROUTING THROUGH A RESERVOIR

Inflow	Time	2S/t+O	Depth	Storage	Outflow
0	0	0	0	0	0
2	15	2	0	30.44	1.47
39	30	40.07	.25	7629.16	23.14
129	45	161.79	1.49	47255.45	56.78
168	60	345.23	3.5	116199.93	87.01
108	75	447.21	4.6	156382.48	99.69
60	90	415.82	4.26	143920.26	96
21	105	304.82	3.06	100560.39	81.35
5	120	168.11	1.56	49517.7	58.07
0	135	56.96	.4	12388.33	29.44
0	150	0	0	0	0

Note that the VARIABLE sheet did not have to define the variables IN, OUT, Storage, y, STORFUNC, and Time to solve the model. They were put there for reference and documentation.

PROGRAM TESTING

Figure 8.17 shows the Plot and Table for the output of the routing model. Hand computations will verify that the output agrees with the mathematical development. Also, the plots look reasonable, since the peak output is lower in magnitude and occurs later than the peak input.

EXAMPLE 8.7 □ **Integration to Find Moment of Inertia About a Specified Axis**

STATEMENT OF PROBLEM

Use the TK SOLVER library model SIMPSON.TK to find the moment of inertia about the base of a triangle of base 6 and height 9. Section 1.3 of Appendix G discusses finding moments of inertia for areas. Example 9.4 will use numerical integration to estimate the moment of inertia for a triangle. It provides a definition sketch for this example.

MATHEMATICAL DESCRIPTION

Equation 8.3 shows the integral necessary to find the desired moment of inertia.

$$I_x = \int_0^h y^2 \left(b - \left(\frac{b}{h}\right) y\right) dy \qquad (8.3)$$

ALGORITHM DEVELOPMENT

Input/Output Design

Values of b and h have to be input.

FIGURE 8.18   TK SOLVER model for finding moment of inertia.

```
=================== FUNCTION SHEET ==========================================
Name---------- Type----- Arguments-- Comment----------------------------------
simpson         Procedure 4;1        Definite integral, Simpson rule
simp2           Procedure 3;1        Definite integral, Simpson rule with prec
simp3           Rule      3;1        Definite integral, main part of recursive
simp3r          Procedure 7;1        Definite integral, recursive part
SIMPSON_doc     Rule      0;0
Ix              Rule      1;1        Moment of Inertia

=================== RULE FUNCTION: Ix ======================================
Comment:             Moment of Inertia
Parameter Variables: b,h
Argument Variables:  y
Result Variables:    dI
S Rule---------------------------------------------------------------------
  dI = y^2*(b-(b/h)*y)

=================== VARIABLE SHEET =========================================
St Input---- Name--- Output--- Unit----- Comment------------------------------
             b       6                   Base of triangle
             h       9                   Height of triangle
             Ix      364.5               Moment of inertia about base

=================== RULE SHEET ============================================
S Rule---------------------------------------------------------------------
  b = 6.0
  h = 9.0
  Ix = simpson('Ix,0,h,100)
```

(a) User supplied information.

**(EXAMPLE 8.7 □   Integration to Find Moment of
Inertia About a Specified Axis □   Continued)**

FIGURE 8.18   (Continued)

```
==================== PROCEDURE FUNCTION: simpson =============================
Comment:                Definite integral, Simpson rule
Parameter Variables:
Input Variables:        fun,a,b,n
Output Variables:       value
S Statement-------------------------------------------------------------------
  '' Notation:  fun     name of rule, procedure or built-in function
  ''                    defining the integrand
  ''            a,b     lower and upper limit
  ''            n       number of integration steps (must be even number)
  ''            value   value of definite integral
  if mod(n,2)<>0 then error:= odd_number_of_intervals
  h:= (b-a)/n
  x:= a
  k:= 1
  value:= 0
  for i=2 to n
      x:= x + h
      k:= 3 - k
      value:= value + k*apply(fun,x)
  next i
  value:= (2*value+apply(fun,a)+apply(fun,b))*h/3

==================== RULE FUNCTION: SIMPSON_doc =============================
Comment:
Parameter Variables:
Argument Variables:
Result Variables:
S Rule-----------------------------------------------------------------------
  '' SIMPSON INTEGRATORS
  '' This collection of tools is used for numerical integration using Simpson's
  '' Rule with optional precision control.  Extensive descriptions below.
  '' HOW TO USE:
  '' 1. Enter the function you wish to integrate as a procedure or rule
  ''    function with 1 Argument, 1 Result, & any number of Parameter Variables.
  '' 2. Enter any rule or statement of the form RESULT = simpson('fun,LL,UL,n)
  ''    where 'fun is the name of your rule function, LL & UL are the limits
  ''    and n is the number of integration steps (n MUST BE AN EVEN INTEGER!).
  ''    In general, as n increases, precision increases & speed decreases.
  '' 2b. If you are using one of the two modified Simpson integrators, DO NOT
  ''     include n.  Specifically,  RESULT = simp2('fun,LL,UL) or
  ''     RESULT = simp3('fun,LL,UL)  is the proper syntax in this case.
  '' TK BONUS: Using any of these 3 tools, you can input a value for RESULT
  '' and backsolve (iteratively) for the corresponding value of the function.

  '' Additional notes on the various tools included in this file:
  '' 1. simpson  is the standard method of numerical integration using second
  ''    degree polynomial approximation of the integrand.
  '' 2. simp2 uses Simpson's rule for evaluating definite integrals.  It keeps
  ''    doubling the number of intervals until the difference between the
  ''    current and previous result falls below a defined tolerance limit
  ''    (see the conditional statement in the procedure, it is set to 1e-8;
  ''     also, an absolute or relative tolerance may be used).
  '' 3. simp3 is part of a recursive version of a procedure that evaluates the
```

(b) Key elements of SIMPSON model.

FIGURE 8.18   (Continued)

```
'  '    definite integral with required absolute or relative tolerance (set to
'  '    1e-8 in the last rule).  simp3r is the other part.
'  '    simp3r refers to itself, repeatedly refining the value of the integral
'  '    until the difference between the last two results falls below the
'  '    required absolute tolerance (the local variable eps).
```

(*b*) Key elements of SIMPSON model (continued).

PROGRAM DEVELOPMENT

Figure 8.18 shows the development of the TK SOLVER model using the library model SIMPSON.TK. This model accomplishes numerical integration using Simpson's rule. Figure 8.18*a* shows the parts that you have to add to the library model to solve the problem at hand. You must define the rule function Ix, which contains the function to be integrated. You also have to make the appropriate entries in the RULE sheet to specify the dimensions and to call the procedure function for applying Simpson's rule. The VARIABLE sheet shows the output values after solution. Figure 8.18*b* shows key parts of the library model.

PROGRAM TESTING

Using the formula $I_c = bh^3/36$ and the parallel axis theorem, $I_x = I_c + y^2A$, a value of 3 in. for y gives an I_x of 364.5 in.^{4}. This is the same value provided by the TK SOLVER model.

8.2.2.13 Additional TK SOLVER Features

TK SOLVER supports complex constants, variables, operations, and functions. TK SOLVER PLUS includes an extensive set of library models that can be imported for use in applications models or modified as required for particular applications. There are several sets of prepared models, or TK Solverpacks, for specific disciplines. TK SOLVER can also exchange data with popular spreadsheets and other types of applications software.

Complex Number Support   TK SOLVER represents complex numbers by pairs of numeric expressions separated by a comma. Parentheses enclose the complex expression. Thus, (*a*,*b*) would represent $a+jb$ and (1,2) would represent $1+j2$. The complex number representations can be used as operands in arithmetic operations or as arguments of complex functions. The complex functions include

$$\text{POWER}((c,d),e) \quad = (c+jd)^e$$
PTOR(*arg1*,*arg2*)   = polar-rectangular conversion using radians
                          arg1 = magnitude, *arg2* = phase angle
PTORD(*arg1*,*arg2*) = polar-rectangular conversion using degrees
RTOP(*arg1*,*arg2*)   = rectangular-polar conversion using radians
                          arg1 = real portion, *arg2* = imaginary portion
RTOPD(*arg1*,*arg2*) = rectangular-polar conversion using degrees

Figure 8.19 shows a TK SOLVER solution of the current addition problem of Example 3.4. This illustrates the complex number operations of TK SOLVER.

Library of Models   An extensive set of library models is supplied with the TK SOLVER PLUS package. The models can directly assist in the solution of prob-

FIGURE 8.19   Complex variable operations.

```
==================== VARIABLE SHEET ===========================================
St Input---- Name--- Output--- Unit----- Comment----------------------------------
             iil1                amps        current 1, phasor notation
             iil2                amps        current 1, phase angle (degrees)
     130     vl1                 volts       voltage 1, phasor notation
     90      vl2                 volts       voltage 1, phase angle (degrees)
     5       zl1                 ohms        impedance 1, phasor notation
     30      zl2                 ohms        impedance 1, phase angle (degrees)
             ilr                 amps        current 1, real portion
             ilj                 amps        current 1, imaginary portion
     15      i2r                 amps        current 2, real portion
     45      i2j                 amps        current 2, imaginary portion

                                             THE ANSWERS ARE:
             i3r                 amps        current 3, real portion
             i3j                 amps        current 3, imaginary portion
==================== RULE SHEET ===============================================
S Rule-------------------------------------------------------------------------
* iil1 = vl1/zl1
* iil2 = vl2-zl2
* (ilr,ilj) = PTORD(iil1,iil2)
* (i3r,i3j) = (ilr,ilj) + (i2r,i2j)
```

(*a*) Input VARIABLE and RULE sheets.

```
==================== VARIABLE SHEET ===========================================
St Input---- Name--- Output--- Unit----- Comment----------------------------------
             iil1    26          amps        current 1, phasor notation
             iil2    60          amps        current 1, phase angle (degrees)
     130     vl1                 volts       voltage 1, phasor notation
     90      vl2                 volts       voltage 1, phase angle (degrees)
     5       zl1                 ohms        impedance 1, phasor notation
     30      zl2                 ohms        impedance 1, phase angle (degrees)
             ilr     13          amps        current 1, real portion
             ilj     22.51666    amps        current 1, imaginary portion
     15      i2r                 amps        current 2, real portion
     45      i2j                 amps        current 2, imaginary portion

                                             THE ANSWERS ARE:
             i3r     28          amps        current 3, real portion
             i3j     67.51666    amps        current 3, imaginary portion
```

(*b*) Output VARIABLE sheet.

lems, or comprehensive applications can use them as modules. The following is a
list of the general categories for these models.

Algebraic equations

Matrix operations

Differentiation and integration

Differential equations

Complex functions and variables

Special functions and miscellaneous

Financial functions and models

Statistical models and procedures

TK Solverpacks   Several TK Solverpacks provide sets of models developed for specific purposes. These include a floppy disk containing the model files and a manual describing each of the models and giving input data instructions. Subject areas for the Solverpacks include

Mechanical engineering

Financial analysis

Building design and construction

Introductory science

Engineering calculations

Electrical engineering

Units conversion

Stress-strain calculations

File Formats   TK SOLVER can exchange data with LOTUS 1-2-3 spread-sheets that have the file extender .WKS. TK SOLVER models will accept lists of values from LOTUS files. You can also transfer TK SOLVER lists into LOTUS 1-2-3 spreadsheets. TK SOLVER can create or import files containing list values in the .DIF format used by other spreadsheets or in standard ASCII format.

8.2.2.14 TK SOLVER Summary

TK SOLVER is consistently reviewed as one of the most powerful equation solvers on the market. As an equation solver, TK SOLVER is both easy to use and versatile. However, it is much more than an equation solver. It has developed into a highly structured programming language that can aid in the solution of many sophisticated engineering problems. TK SOLVER can also provide a sub-stantial time savings when compared with conventional high-level programming languages. It automatically accomplishes many of the program organizational chores.

8.2.3 Equation Solver Summary

You have received an introduction to the equation solver TK SOLVER. Through this introduction you should have gained an appreciation of equation solvers as engineering tools. Programs such as these bring you closer to being able to tell the computer what you want it to do, rather than having to supply the computer with step-by-step instructions as to how to accomplish a task. They are especially valuable as solvers of trial-and-error and nonlinear problems that are difficult or tedious to solve by hand. You can develop high-level language programs for these problems, but their development takes a significant amount of time and effort. Equation solver applications for these problems can typically be developed, de-bugged, and run in a matter of minutes. Furthermore, any time you have a similar problem in the future, the application file will act as a template for the new solution. However, you also need to remember that there are many problems for which a high-level program is still preferable. These include complex problems with many iterations and options, problems with large amounts of data manipula-tion, and problems that require customized output.

8.3 SPREADSHEETS

8.3.1 Introduction

The use of spreadsheet programs by engineers is a relatively recent development. The first spreadsheet, VisiCalc, was developed in 1978. It was designed as an electronic worksheet to aid in the processing of financial and accounting tables and in making financial projections. Business applications of VisiCalc blossomed quickly as people saw how much it could reduce the drudgery of working with tables of figures. They also recognized that spreadsheets could substantially reduce error rates. Since then, numerous powerful spreadsheet programs have been developed.

There are many spreadsheets available. The core concepts of all of them are similar; however, they differ in the details of implementation. They also differ considerably in the extra features, such as graphics and database management, that are supported. We will use LOTUS 1-2-3, by Lotus Development, to illustrate spreadsheet concepts. We will also use the spreadsheet portion of the integrated package ENABLE. LOTUS 1-2-3 has been the best-selling spreadsheet for several years and has spawned many lookalike or act-alike clones, some of which have some features superior to those of 1-2-3. Once you consider all the features that you need in a spreadsheet, the one that you decide to use is basically a matter of personal preference. Versions of spreadsheets are available for all types of computers, from mainframes to microcomputers. They are most often associated with personal computers, largely because they are the type of tool that works best in the interactive environment of microcomputers. Applications in this text will assume that you are using a microcomputer with a hard disk.

Tutorials are available for most of the popular spreadsheets. Some are provided by the software developers, whereas others have been developed by independent firms. These may be useful to first-time users. Numerous sources of preprogrammed spreadsheets (or templates) are appearing. These templates may help you produce results sooner and may act as a training vehicle. However, you would still be responsible for ensuring that the spreadsheet is giving correct answers and that it is applicable to the particular problem at hand. An understanding of the structure and programming of spreadsheets is essential if you are to be able to accomplish this effectively.

8.3.2 General Concepts and Characteristics

Spreadsheets have been referred to as personal productivity tools. The basic component of a spreadsheet is an electronic worksheet consisting of a two dimensional table with row and column labels that appears on the computer screen. Each element of the table defined by a row and column intersection is called a cell. The cells actually represent storage locations. Tabular numeric values or alphabetic character data can be inserted into the cells. Formulas can be superimposed on the numeric cells. These formulas gather values from other referenced numeric cells and operate on them through mathematical operators and functions to produce a value for the resident cell of the formula. The numeric value is what shows on the screen. Formulas can be viewed and edited by using various commands that are available for this purpose. Portions of the spreadsheet can be moved around, and groups of rows and/or columns can be added or deleted to adjust the spreadsheet for various conditions. Properly entered cell references will automatically be adjusted for the change in the spreadsheet layout. Standard mathematical functions, such as trigonometric functions and exponentiation, are available. Spreadsheets also have special functions to accomplish tasks such as summing rows and columns of values. Most spreadsheets feature logical functions and built-in economic formulas. You can also sort rows or columns into ascending or descending numeric or alphabetic order based on values or character strings in a selected row or column. Windows that split the spreadsheet into different work areas can be defined so that distant portions of the spreadsheet can be viewed simultaneously on the screen. Various degrees of output formatting are available, including fixing the number of decimal places, centering within columns, and designating dollars and cents or percentage data. Most newer spreadsheets allow for variable column width. Some of the newest spreadsheets integrate the spreadsheet operation with other popular microcomputer operations, such as database management, graphics, and word processing.

Spreadsheets are generally referred to as nonprocedural approaches, in that the development of applications does not require writing a high-level program. However, there are many similarities to programming in the development of the logic of problem solution and the formatting of output. In fact, macros can actually approach structured programs in their sophistication. Macros are user-defined series of keystrokes that can be executed with a single command. They can be used to develop procedures similar to subprograms.

Spreadsheets are powerful tools for solving routine engineering problems. They are ideally suited for problems that can be displayed in a tabular format. When values are changed in a spreadsheet, the spreadsheet will recalculate values for cells whose formulas are dependent on the changed cells. This facilitates asking "What if" questions, or testing alternate designs. Spreadsheets can also help you understand the processes being modeled by estimating the dynamic response of the system to changing input. You can use spreadsheets to test the sensitivity of a model to input variables.

Spreadsheets are not the answer for all applications. For example, multivariate relationships are difficult to model with spreadsheets, and recalculation can be slow for complex tabular problems with iteration. Part of the challenge of being computer literate is knowing when a particular computational tool is the appropriate choice for a problem. After studying the spreadsheet examples in this chapter, you will have a better idea of the types of problems that lend themselves to spreadsheet analysis.

The cells of spreadsheets can store character data, numeric data, or formulas. They perform the storage location definition function of variables in high-level languages. Character data stored in cells can be used for labels and can also be operated on using character operators and functions. In spreadsheets, there is no differentiation between real and integer storage locations for numeric data. However, integer values will be displayed without a decimal point. Formats can be assigned to the cells to establish how the data will be displayed. The formulas assigned to cells act like assignment statements. They evaluate mathematical expressions using values stored in referenced cells and replace the old value in the host cell with a new value. Formulas can contain operators, functions, and references to other cells. Most spreadsheets also allow conditional structures. These evaluate a logical expression and compute the value for a cell using one formula if the logical expression is 'true'. If the logical expression is 'false', a second formula computes the value for the cell. If a formula for a cell computes a new value, the new value will replace whatever value was originally in the cell.

Formulas refer to other cells to gather values needed to make computations. The references to these cells can be either absolute or relative. Absolute cell references always refer to the same cell, no matter where the formula might be copied to within the spreadsheet. Relative cells, on the other hand, refer to a cell that is a certain number of cells away from the formula cell. If the formula is copied to a cell somewhere else in the spreadsheet, the reference will be to a cell the same relative distance from the new location. Cells can be named and referred to by the name rather than their location within the cell grid network. Naming of cells produces formulas that look similar to expressions in a high-level language. Named, absolute, and relative cell references are all useful in developing spreadsheet applications, and each will be illustrated in the chapter examples.

Spreadsheets are generally menu driven. This helps simplify applications development. Most spreadsheets will also allow the merging of worksheets from different applications files and may allow the referencing of other stored worksheets to import data.

8.3.3 Spreadsheet Applications Development

The development of spreadsheet applications parallels the process for developing high-level language applications. Because the cells of spreadsheets can contain both data and formulas, the input and output design and computational logic design are more integrated than with high-level languages. The character and numeric data for cells are what you normally see on the screen. If there is a formula associated with the cell the cursor is over, it is displayed somewhere on the screen. The other formulas stay in the background. When you print out the spreadsheet you will normally print only the character and numeric data. However, most spreadsheets give you the option of displaying or printing out all formulas. You might want to look ahead to Example 8.8 to view an example of a spreadsheet. You will see the display of character and numeric data for the stress-strain analysis of Examples 5.5 and 8.3.

You should designate one portion of your spreadsheet for input. Labels and input data instructions can be included in the spreadsheet. Input is automatically echoed since it appears in the spreadsheet. Cells other than the input data cells can be protected so that their contents cannot be inadvertently changed by users who are not familiar with spreadsheet structures. Both input and output cells can be formatted for display. As is the case in FORTRAN, formatting does not change stored values.

You need to consider the page layout of your spreadsheet. If you want the spreadsheet to appear on successive pages when it is printed out, you will need to organize your spreadsheet accordingly. Each output page can contain a certain number of columns, depending on the column widths you have assigned. You should place output information in the leftmost columns of the spreadsheet that can be output as a page. You can use the cells to the right of the output section for intermediate calculations that do not need to be output. If you want to output a wider spreadsheet, you can paste the output pages together. The spreadsheet will automatically be output in a block format. If you are using a dot matrix printer with continuous feed paper, there are accessory programs available that will shift the output 90 degrees.

It is more difficult to develop repetitive structures for spreadsheets than it is for high-level languages or equation solvers. It can be done using macros and/or circular references. An example of a circular reference would be three cells with formulas in all three. The formula for the first cell would refer to the second cell, the formula for the second cell would refer to the third cell, and the formula for the third cell would refer to the first cell. The spreadsheet would have to iterate to approach values that would satisfy all the formulas.

Most spreadsheets automatically recalculate values as you change or add values or formulas to cells. This is one of the strengths of the spreadsheet application. However, recalculation can slow down spreadsheet development if it is performed each time you expand a spreadsheet to a new cell. You may also want to enter or change more than one input data value before recalculating. Therefore, spreadsheets give you the option of shutting off automatic recalculation during development and data entry. You can still perform recalculation by issuing the appropriate spreadsheet command.

After you develop a spreadsheet, it can be used to solve many similar problems by expanding or contracting the input data area. The data for the originally developed spreadsheet can be left in and modified for new applications, or the input data can be removed, saving just the skeleton spreadsheet. Whichever method is used, you have to provide instructions to the user on how to adjust the problem size and how to input data. Spreadsheets designed to be used for multiple applications are generally called templates.

The methodology for developing spreadsheet applications will be covered in detail in the examples that follow.

8.3.4 LOTUS 1-2-3

8.3.4.1 Introduction

LOTUS 1-2-3 has been the best-selling spreadsheet package for several years. Many spreadsheets that have similar features and command structures have been developed. Since 1-2-3 is such a widely used program, there is a considerable amount of user support available. This includes the Lotus PROMPT support program of the LOTUS Development Corporation, *LOTUS* magazine published by LOTUS Publishing Company (a subsidiary of LOTUS Development), the World of Lotus Bulletin Board, and the *123 User's Journal* published by a third party. There have been many books published on LOTUS applications and how to use the program. Several tutorials are available. Some of these are on cassette or video tapes. Numerous short courses and seminars on business and engineering applications are held each year. Lotus also markets training courses and materials to assist instructors.

There is also a student edition of LOTUS 1-2-3 (O'Leary 1989), which is the same as the commercial Release 2.01 version in many respects. However, it will accept only spreadsheets up to 64 columns by 256 rows. Spreadsheets saved using the commercial version can be loaded into the student version if they are smaller than the size limitation. Spreadsheets saved with the student version can be loaded into the commercial version. The student edition has the capability of transferring files to and from other software packages. The examples in the text will use commands and syntax that are compatible with Release 2.01 and the student version of LOTUS 1-2-3. A new student edition of Release 2.2 (O'Leary 1990) does away with the size limitation.

FIGURE 8.20 Initial LOTUS 1-2-3 screen.

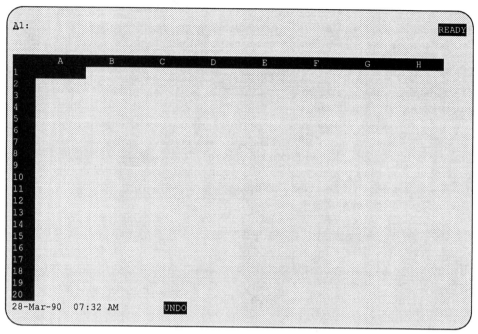

(*a*) Initial screen.

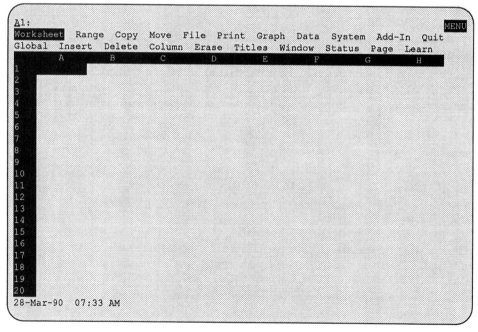

(*b*) Initial screen with command menu.

After you have accomplished the necessary system commands, you can load the LOTUS 1-2-3 program by typing 123. Figure 8.20a shows the initial LOTUS 1-2-3 screen. Note that the rows are identified by numbers, and the columns by letters. After the first run through the alphabet, the columns are identified by AA, AB, and so forth. Intersections of rows and columns form the cells of the spreadsheet. The cell in the upper left-hand corner of the spreadsheet would be labeled A1, while the cell in the third row and second column would be B3. The cursor initially appears in cell A1. It can be moved around the spreadsheet by using the cursor arrow keys of the microcomputer keyboard. If the cursor is moved beyond the edge of the portion of the spreadsheet on the screen, the spreadsheet will scroll to include the new position of the cursor. The command menu for 1-2-3 does not appear on the screen until you type /. The screen will then appear as in Figure 8.20b. While the command menu is on the screen, you cannot enter anything into the spreadsheet proper. The use of the commands will be described in Section 8.3.4.3.

8.3.4.2 Capabilities and Limitations

LOTUS 1-2-3 has extensive graphics and database capabilities. Graphical display of data can be in the form of bar, pie, or X-Y plots. Database commands include sorting, query, matrix, and regression operations. Many of these will be illustrated in the examples to follow. A number of the commands of LOTUS 1-2-3 can be executed from the function keys on your computer. The references given here will be for the IBM keyboard. The actual keys used on other keyboards may be different, but the end result will be the same.

The latest release of LOTUS 1-2-3, Release 3.0, has several new features of interest to users. A three-dimensional feature allows you to stack up to 256 sheets on top of each other in memory. Formulas that refer to cells and ranges in multiple worksheets can be written. Links can also be made to spreadsheets stored on disk. You can now print graphics from the screen, rather than having to go to a separate print graph program. Graphics output has been improved to what most users consider to be presentation quality. Graphs and data can be displayed together and graphs are automatically updated as the data change. Minimal recalculation is available in Release 3.0. This means that the program recalculates only cells that are dependent on modified cells. This speeds up recalculation for larger spreadsheets. Macros can be recorded and stored in libraries. There is an undo function that reverses unwanted commands.

Release 3.0 also supports disk-based database handling. A driver is available that will allow you to manipulate files in the Ashton-Tate dBase format of dBase II, dBase III Plus, and others, using LOTUS 1-2-3 data commands.

There is also a second new version of LOTUS 1-2-3. This is Release 2.2. It incorporates file linking, enhanced graphics, an undo facility, a macro recorder, minimal recalculation, and additional screen drivers. Release 2.2 was designed to run on any personal computer that the earlier version, Release 2.01, would run on. Release 3.0, on the other hand, will run only on machines with at least one megabyte of memory. Many users believe that Release 2.2 will become the version of choice.

A limitation of LOTUS 1-2-3 is that the alignment of numeric entries within cells cannot be changed. They are always right-justified. Labels can be either left-, center-, or right-justified.

8.3.4.3 Commands

The command menu for LOTUS 1-2-3 is brought up on the screen by typing /. Any LOTUS 1-2-3 commands can be accessed by typing the first letter of the command or moving the cursor to the command and pressing **[enter]**. For presentation clarity, the whole command title will be typed out in the text. Most of the commands have one or more levels of subordinate menus, as will be seen in the description of the commands. After you have selected the desired options, you can execute the command by pressing **[enter]**. Wherever you are in the menu structure, you can press the **[esc]** key and back out to the next higher level without executing a command. If you press **[esc]** enough times, you will return to the READY condition with no menu displayed.

Commands used in the text will be defined here. There are others that are defined in the 1-2-3 documentation. Indentation of commands indicates the hierarchical structure of the command menus. The word ''worksheet'' used in the commands, and in other places with reference to LOTUS 1-2-3, is synonymous with ''spreadsheet.'' A range refers to a block of cells. For example, A1..B2 would represent the block of cells A1, A2, B1, and B2.

/COPY: Copy a range of the worksheet into another range.

/DATA

> SORT: Sorting rows of the spreadsheet according to the contents of selected key columns.
>> DATA-RANGE: Portion of worksheet to be sorted.
>> PRIMARY-KEY: Column of range that will be used for primary sort.
>> SECONDARY-KEY: Column of range that will be used for secondary sort in case of ties in primary sort.
>> GO: Execute sort.
>> QUIT: Returns to READY mode.

> QUERY: Searching for and acting on selected records of the database portion of a worksheet.
>> INPUT: Range of worksheet that will be searched in the query.
>> CRITERION: Criteria that you establish for selecting records from the worksheet database. Criteria can be absolute or conditional.
>> OUTPUT: Range of worksheet into which extracted records will be copied.
>> FIND: Finds and highlights records that meet your criteria.
>> EXTRACT: Copies records that match your criteria into a specified portion of the worksheet.
>> QUIT: Returns to READY mode.

> DISTRIBUTION: Creates a frequency distribution of the values in a selected range.

> MATRIX
>> INVERT: Computes the inverse of a matrix contained in a specified range of the worksheet, and puts the inverse in a separate section of the worksheet.
>> MULTIPLY: Multiplies one matrix (identified as a range in the worksheet) by a second matrix (range of worksheet), and puts the product matrix in a specified range of the worksheet.

REGRESSION

    X-RANGE: Worksheet range of independent variable.

    Y-RANGE: Worksheet range of dependent variable.

    OUTPUT-RANGE: Worksheet range that will contain the computer regression parameters.

    INTERCEPT: Intercept can be calculated or set to zero.

    GO: Executes regression analysis.

    QUIT: Returns to READY mode.

/FILE: Saving, retrieving, combining, and extracting portions of worksheet files.

/GRAPH

    TYPE

        LINE: X-Y plot for independent variables with equal increments, such as data values for each month of the year.

        BAR: Dependent variable values are represented by bars of varying length. Multiple bars are used for multiple dependent variables.

        XY: X-Y plot for general X-Y data.

        STACKED BAR: Multiple bars placed one on top of another.

        PIE: Portion of data total shown as slice of pie.

    X A B C D E F: Defines range of independent, and up to six dependent (A-F) variables.

    VIEW: Displays the current graph on the screen. You can also view the current graph from the READY mode by pressing the **F10** key.

    SAVE: Stores the current graph in a graph file. This is necessary before you use the PRINTGRAPH program to print or plot the graph.

    OPTIONS

        LEGEND: Adds a legend below the graph to indicate what each symbol, color, or crosshatching represents in the graph.

        FORMAT: Allows you to display data points as symbols, draw lines between each data point, or both.

        TITLES: Allows you to place titles on the graph and the X and Y axes.

        GRID: Adds or removes grid lines on your graph display.

        SCALE: Allows you to specify the range of values for the X and Y axis, or to let 1-2-3 determine the range from the data.

        QUIT: Returns to GRAPH menu.

    QUIT: Returns to READY mode.

/MOVE: Moves a range of cell entries from one part of the worksheet to another.

/PRINT: Prints all or a portion of a worksheet to either a printer or file.

/QUIT: Concludes 1-2-3 session and returns to the operating system prompt. 1-2-3 does not automatically save your worksheet, so you must do that before using QUIT.

/RANGE: Manipulates a range of cells.

    FORMAT: Sets the numeric format for a range of cells.

    LABEL: Aligns existing labels in a range of cells.

    ERASE: Removes the contents of cells in a selected range.

/SYSTEM: Allows you to use operating system commands without permanently leaving 1-2-3.

/WORKSHEET

GLOBAL: Commands that affect the whole worksheet unless overridden by commands for a range of cells.

FORMAT: Sets default numeric format.

LABEL-PREFIX: Sets the alignment of labels for the whole worksheet.

COLUMN-WIDTH: Sets the width for all columns except those set using the /WORKSHEET COLUMN COLUMN-WIDTH command.

RECALCULATION: Controls worksheet recalculation options.

INSERT: Used to insert one or more blank rows or columns in the worksheet.

DELETE: Used to delete one or more rows or columns in the worksheet. All cell entries in the affected rows or columns are lost.

COLUMN: Used to set the width of an individual column, or to hide the column from the display.

ERASE: Removes the current worksheet and gives you a blank worksheet. If you want to keep the old worksheet, you must save it before using /WORKSHEET ERASE.

WINDOW: Divides the worksheet into two horizontal or vertical windows. Allows you to view and work with two different portions of the worksheet at one time.

8.3.4.4 Function Key Commands

Several special commands are available through the keyboard function keys. For the IBM keyboard, the function keys, associated commands, and descriptions are as follows:

F1	HELP	Invokes the HELP facility
F2	EDIT	Places highlighted entry on the control panel for editing
F3	NAME	Produces a menu of the current range names
F4	ABS	Cycles a cell address through relative, absolute, and mixed references
F5	GOTO	Moves the cell pointer to the cell you specify
F6	WINDOW	Switches the cursor between the two windows when there is a split screen
F7	QUERY	Repeats the last /DATA QUERY command
F8	TABLE	Repeats the last /DATA TABLE command specified
F9	CALC	Recalculates worksheet formulas
F10	GRAPH	Displays the graph most recently specified

8.3.4.5 Functions

Most standard mathematical functions and many special functions are available for use with LOTUS 1-2-3 spreadsheets. The functions of LOTUS 1-2-3 are all preceded by the @ symbol. For example, @SUM(range) is the function for summing a range of cells. The functions that will be used in the examples and exercises of this chapter are as follows:

@ABS(*expression*)	Absolute value of the *expression*
@ATAN(*expression*)	Arc tangent. Returns the radian angle of the tangent value given by the *expression*.

@AVG(*list*)	Average value of the *list* of cells. A *list* can contain one or more numbers, numeric formulas, references to ranges that contain numbers or numeric formulas, or any combination of numbers, formulas, and references to ranges. The formula @AVG(B1..B3,D1,1.0) would return the average of the values in cells B1, B2, B3, and D1, and 1.0.
@COS(*expression*)	Returns the cosine of the radian angle value of the *expression*
@COUNT(*list*)	Returns the number of nonblank cells in the *list* of ranges
@EXP(*expression*)	Constant *e* (2.71823) to the power of the value of the *expression*
@LN(*expression*)	Computes the natural log of the value of the *expression*
@LOG(*expression*)	Computes the log base 10 of the value of the *expression*
@MAX(*list*)	Maximum value in a *list* of cells
@MIN(*list*)	Minimum value in a *list* of cells
@PI	Value of π to 18 significant figures
@SQRT(*expression*)	Return the square root of the value of the *expression*
@STDS(*list*)	Returns the sample standard deviation of the values in a *list* of cells
@SUM(*list*)	Returns the sum of the values in a *list* of cells
@SIN(*expression*)	Returns the sine of the radian angle value of the *expression*
@TAN(*expression*)	Returns the tangent of the radian angle value of the *expression*

8.3.4.6 Economic Functions

The following economic functions are available in LOTUS 1-2-3. These are all functions for the economic formulas found in Appendix E. Payments are assumed to be made at the end of the compounding period. The function arguments can be values or cell references.

@PV(*payments,interest,term*)	The definitions of the parameters are *present-value* or PV = present value, *future-value* or FV = future value, *interest* = interest rate per compounding period, *term* or TERM = number of compounding periods, *payments* or PMT = payment or return per compounding period, RATE = interest rate necessary for a *present-value* to reach a *future-value* in a given *term*.
@FV(*payments,interest,term*)	
@TERM(*payments,interest,future-value*)	
@PMT(*present-value,interest,term*)	
@RATE(*future-value,present-value,term*)	
@NPV(*interest,range*)	Returns the net present value for a nonuniform series of future cash flow values contained in a *range* of cells using the interest per compounding period of *interest*.

| @IRR(*guess*,*range*) | Returns the internal rate of return for a series of cash flows defined by *range*. *Guess* = initial estimate of interest rate. |

8.3.4.7 Formulas

Formulas can be constructed for cells so that values can be brought in from other cells, combined and operated on, and the resulting value displayed in the resident cell. This value can in turn be referenced by other cells. The structure of a formula is similar to the structure of an expression in FORTRAN. Parentheses can be used to group portions of a formula. As is the case with most computer languages, any parts of the formula in parentheses will be evaluated first, from the innermost set outward, before the rest of the formula is evaluated. The order of operations is similar to that for FORTRAN. Any of the functions previously described can be included in the formula, along with the mathematical operators used in the BASIC computer language:

+	Addition
−	Subtraction
*	Multiplication
/	Division
^	Exponentiation

When you enter data into cells for LOTUS 1-2-3 you have to differentiate between character, formula, and numeric entries. Any cell entry in 1-2-3 that starts with one of the characters

$$0\ 1\ 2\ 3\ 4\ 5\ 6\ 7\ 8\ 9\ .\ +\ -\ (\ @\ \#\ \$$$

is considered to be a value or formula cell. Any other first character marks the cell as a label or text cell. If you want to start a label with one of the above characters you can enter ', ", or ^ as the first character. The character ' indicates left justification of the label, " indicates right justification, and ^ indicates centering. The justification character of the label will not appear in the spreadsheet when the label is entered. However, it will appear when you edit the label and can be changed through the EDIT command. For example,

$$^1\text{-}2\text{-}3$$

would center the label 1-2-3 in the cell into which it is entered. Label justification can also be changed through the /RANGE LABEL command. Left justification is the default. LOTUS 1-2-3 does not provide the option of justifying numeric values in cells. They are always right-justified.

Formulas for LOTUS 1-2-3 can be printed using the /PRINT, PRINTER, OPTIONS, OTHER, CELL-FORMULAS series of commands. Formulas are printed out one under another, with the cell identifier along the left border. There is no option for printing out row and column identifiers with a normal spreadsheet printout.

8.3.4.8 Relative and Absolute Cell References

References to other cells in a formula can be absolute, relative, or by name. Naming cells will be illustrated in Example 8.9. Absolute cell references refer to a cell by its row and column identifier. For example, $B$3 refers to cell B3. The $ is what identifies the reference as absolute. When an absolute reference is entered into a formula, it will refer to the same cell no matter where in the spreadsheet the formula might be moved or copied. Relative cell references, on the other hand, refer to cells by their position with respect to the formula cell. For example, if the formula C9+1 appears in cell C10, the formula will add one to the value in cell C9 and store that value in cell C10. If you copy this formula to cell D20, it will appear as D19+1. One will be added to the value in D19 and placed in D20. The reference is always to a cell one cell above the formula cell. Absence of the $ makes the reference relative. Relative cell references facilitate the rapid duplication of formulas for rows or columns of the spreadsheet. It is also possible to make one part of the cell identifier relative, and the other part absolute. Thus B$10 is absolute with respect to row 10 but relative with respect to column B.

8.3.4.9 Graphics

LOTUS 1-2-3 has the capability of producing line, bar, stacked bar, pie, and XY plots. The line plot is a specialized X-Y plot that has specified interval labels along the X axis defined by the contents of a range of cells. These interval labels need not be numerical. They could be the months of the year or days of the week. The corresponding Y value is placed on the graph above the interval label. This Y value could, for example, be an employee's production data during January, February, and so on. The XY plot is a general X-Y plot in which values for one or more dependent variables are placed in a continuous XY grid in the correct position for the corresponding independent variable value. The data for the stress-strain measurements of Example 8.8 will be used to illustrate the XY graphing capabilities of LOTUS 1-2-3. The use of graphics to help analyze a problem will also be illustrated.

8.3.4.10 Model Formulation and Entry

Before you begin to build a spreadsheet, you must have a thorough knowledge of the theory and mathematical expressions of the process or procedure to be modeled. If you approach a problem by preparing a draft of the tabular or model format on a sheet of paper, you can save yourself a great deal of time during the model entry stage. Explanatory material and labels can be entered either before or after the main body of the spreadsheet model is entered. I prefer entering at least the basic information of the title and the row and column labels prior to formulating the model. This information will help you remember where to place the different data and formulas that you will be adding during the computation development process. Input data values can be entered directly into the designated rows or columns. Formulas for a column of a table can be generated quickly by entering the formula in the top cell of the table using relative references and then using the COPY command to duplicate it in the remaining cells of the column. The same procedure can be used to duplicate formulas for multiple cells in a row. The COPY command can also be used to copy all the formats, formulas, characters, and values from a column, row, or range of cells to another portion of the spreadsheet. The EDIT command can then be used to make necessary modifications in the duplicated cell entries.

After you have entered the model, you must have known solution values to compare with the spreadsheet values. If the known values of the output do not agree with the spreadsheet, then cell references and formulas have to be checked and corrected as necessary. Debugging is facilitated by the fact that much of the intermediate output is immediately available in the spreadsheet.

We will discuss enough commands for you to be able to enter and use the spreadsheets of this chapter. There are many more commands and alternate methods of using some of the commands available in LOTUS 1-2-3. You can increase your expertise by studying these in the reference manual.

Spreadsheets are ideal tools for assisting with the analysis of problems that can be arrayed in a tabular format. Finding the best fit straight line through a set of data points using regression analysis is one such problem. Example 8.8 develops a LOTUS 1-2-3 spreadsheet for analyzing the stress-strain data of Examples 5.5 and 8.3. Graphical analysis will also be used to assist in the analysis.

EXAMPLE 8.8 □ Regression Analysis for Stress-Strain Data

STATEMENT OF PROBLEM

Develop a LOTUS 1-2-3 spreadsheet to find the slope, intercept, correlation coefficient, and coefficient of determination for the stress–strain data of Example 5.5. Stress should be the independent variable, strain the dependent variable. Make the main body of the spreadsheet as general as possible. After entering and debugging the general portion of the spreadsheet, add the calculation of the modulus of elasticity for the stress–strain data. Use the graphics capabilities of LOTUS 1-2-3 to further analyze the data. If necessary, remove any unrepresentative data.

MATHEMATICAL DESCRIPTION

Section 6 of Appendix H develops the mathematical model for regression analysis. It has been used previously in Examples 5.5 and 8.3. The modulus of elasticity for the stress–strain problem is the inverse of the slope of the regression line.

ALGORITHM DEVELOPMENT

Input/Output Design

Before you start to develop the spreadsheet, you must have a thorough knowledge of the theory and solution techniques to be used. It is also important to plan the layout of the spreadsheet input and output. The spreadsheet should be as compact as possible. Whenever feasible, it is best to place the output portion of the spreadsheet in the first eight columns. This is the number of columns normally printed out on a page. Intermediate calculations that do not need to be printed out can be accomplished in columns to the right of eight.

All references to the independent and dependent variables will be in terms of the generic variables X and Y in the main part of the spreadsheet. Space will be set aside in the initial portion of the spreadsheet for up to four lines of variable definition or any other information you want to include. Also, space will be allocated at the end of the spreadsheet for any specific computations you want to include. For this application we will add the computation of the modulus of elasticity.

Computer Implementation

The /DATA REGRESSION command of LOTUS 1-2-3 will be used to generate the slope, intercept, and coefficient of determination for the data. X is the independent variable, and Y is the dependent variable. The standard output will be expanded to include the correlation coefficient, Y estimate for a given X, and modulus of elasticity.

PROGRAM DEVELOPMENT

Figure 8.21 shows the spreadsheet for the linear regression. Note that row and column numbers are included in Figures 8.21a and b for reference purposes. Normally these would not be included in a printout for a report. A detailed description of the entry of the various elements of the spreadsheet follows.

FIGURE 8.21   LOTUS 1-2-3 regression model.

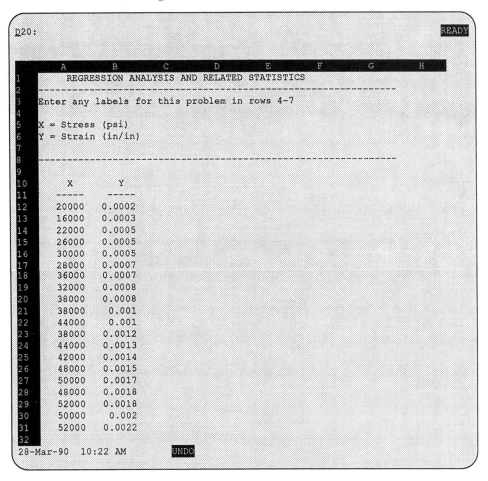

(a) Label and input portion of spreadsheet.

**(EXAMPLE 8.8 □ Regression Analysis for
Stress-Strain Data □ Continued)**

FIGURE 8.21   (Continued)

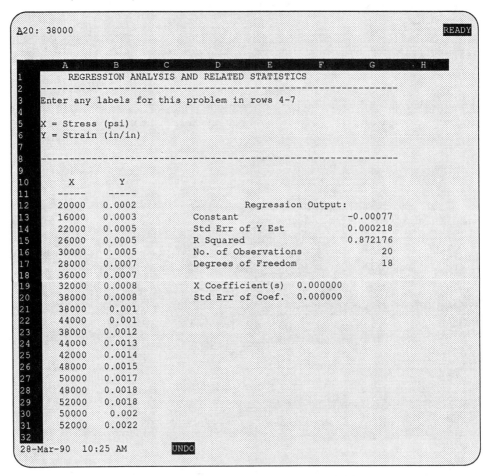

(*b*) Spreadsheet with /DATA REGRESSION block.

FIGURE 8.21   LOTUS 1-2-3 regression model.

```
              REGRESSION ANALYSIS AND RELATED STATISTICS
      -------------------------------------------------------------
      Enter any labels for this problem in rows 4-7
      X = Stress (psi)
      Y = Strain (in/in)
      -------------------------------------------------------------

          X        Y
        -----    -----
        20000   0.0002                    Regression Output:
        16000   0.0003        Constant                    -0.00077
        22000   0.0005        Std Err of Y Est            0.000218
        26000   0.0005        R Squared                   0.872176
        30000   0.0005        No. of Observations               20
        28000   0.0007        Degrees of Freedom                18
        36000   0.0007
        32000   0.0008        X Coefficient(s)   4.96E-08
        38000   0.0008        Std Err of Coef.   4.48E-09
        38000    0.001
        44000    0.001        Correlation Coefficient        0.934
        38000   0.0012
        44000   0.0013        For a value of X =    53600
        42000   0.0014        The estimated Y = 0.001883
        48000   0.0015
        50000   0.0017
        48000   0.0018
        52000   0.0018
        50000    0.002
        52000   0.0022
```

(*c*) Completed spreadsheet.

```
A1:  '     REGRESSION ANALYSIS AND RELATED STATISTICS
A2:  '---------
B2:  '---------
C2:  '---------
D2:  '---------
E2:  '---------
F2:  '---------
G2:  '---------
A3:  'Enter any labels for this problem in rows 4-7
A5:  'X = Stress (psi)
A6:  'Y = Strain (in/in)
A8:  '---------
B8:  '---------
C8:  '---------
D8:  '---------
E8:  '---------
F8:  '---------
G8:  '---------
A10: '       X
B10: '       Y
A11: '    -----
B11: '    -----
A12: 20000
B12: 0.0002
E12: 'Regression Output:
```

(*d*) Formula printout.

(EXAMPLE 8.8 □ Regression Analysis for Stress-Strain Data □ Continued)

FIGURE 8.21 (Continued)

```
A13: 16000
B13: 0.0003
D13: 'Constant
G13: -0.0007750988
A14: 22000
B14: 0.0005
D14: 'Std Err of Y Est
G14: 0.0002182797
A15: 26000
B15: 0.0005
D15: 'R Squared
G15: 0.872176994
A16: 30000
B16: 0.0005
D16: 'No. of Observations
G16: 20
A17: 28000
B17: 0.0007
D17: 'Degrees of Freedom
G17: 18
A18: 36000
B18: 0.0007
A19: 32000
B19: 0.0008
D19: 'X Coefficient(s)
F19: (S2) 0.0000000496
A20: 38000
B20: 0.0008
D20: 'Std Err of Coef.
F20: (S2) 0.0000000045
A21: 38000
B21: 0.001
A22: 44000
B22: 0.001
D22: 'Correlation Coefficient
G22: (F3) @SQRT(G15)
A23: 38000
B23: 0.0012
A24: 44000
B24: 0.0013
D24: 'For a value of X =
F24: 53600
A25: 42000
B25: 0.0014
D25: 'The estimated Y =
F25: +F19*F24+G13
A26: 48000
B26: 0.0015
A27: 50000
B27: 0.0017
A28: 48000
B28: 0.0018
A29: 52000
```

(*d*) Formula printout (continued).

FIGURE 8.21 (Continued)

```
B29: 0.0018
A30: 50000
B30: 0.002
A31: 52000
B31: 0.0022
```

(*d*) Formula printout (continued).

Columns one and two of the spreadsheet can be used to contain the values for *X* and *Y*. You should enter this part of the spreadsheet before developing any of the statistical formulas. The designations of data will intentionally be left general so that once the spreadsheet is developed, it can easily be adapted for other problems. Space is left to insert labels for a particular problem. After the headings have been entered, the *X* and *Y* data values can be entered in columns one and two.

A more detailed explanation of the entry of the elements of the spreadsheet will now be accomplished. To illustrate the commands used, the whole command identifier will be spelled out. When you are selecting commands from the LOTUS 1-2-3 menu, you can either move the cursor to the command and press **[enter]** or type the first letter of the command. You do not have to type the whole command.

The applicable portions of Figure 8.21 should be referenced as the development of the spreadsheet is described. The title of the spreadsheet can be entered in cell A1. When a label is too long to be displayed in a single cell, the remainder will automatically carry over into succeeding cells. The default column width with LOTUS 1-2-3 is nine characters wide.

To start the underline of the second row, move the cursor to A2, and enter '---------. The /COPY command can be used to complete the underline. When using this command, you specify first the range from which you want to copy and then the range that you want to copy to. With the cursor still in A2, type /**COPY**. FROM: A2..A2 will appear as the first submenu prompt. This is your prompt to designate the range from which to copy. Since you want to copy the data in A2, you just have to press **[enter]** to designate A2 as the copy from range. If the cursor had been somewhere else, you could move it to A2 and press **[enter]**. Now the prompt TO: A2..A2 will appear. You want to copy the data into cells B2 through G2, so move the cursor to B2, and B2..B2 will appear in the prompt. Press **[.]** to anchor the first part of the range at B2. Now move the cursor to G2. As you do this, the range will change and will be highlighted in reverse video on the screen. When the cursor is in G2, the prompt will read TO: B2..G2. By pressing **[enter],** you will complete the command and copy the underline as shown in Figure 8.21*a*.

The second underline can be generated in the following manner. Move the cursor to A2, and type /**COPY**. Anchor the FROM range at A2 by pressing **[.],** and complete the range by moving the cursor to G2 and pressing **[enter].** In response to the TO: prompt, move the cursor to A8 and press **[enter].** You do not have to specify the whole target range. LOTUS 1-2-3 assumes that you want to copy the FROM: range into an equivalent range starting at A8. The other labels can be entered as shown in Figure 8.21*a*.

The *X* and *Y* data can be entered in columns A and B. The headings over the *X* column (X and -----) were entered by using spaces to center them over the

(EXAMPLE 8.8 □   Regression Analysis for Stress-Strain Data   □   Continued)

numerical entries. The headings for the Y column were copied from the X column, and the EDIT function was used to change X to Y. To use the EDIT function, move the cursor to C10 and press the **F2** function key. The arrow keys will move the cursor back and forth in the field. The **[backspace]** key will delete characters to the left of the cursor. By typing characters you will insert them.

We can now generate the regression statistics by using the /DATA REGRESSION command of LOTUS 1-2-3. When you look up this command in the LOTUS reference manual, you will find that the standard output produced by the command takes up a block four columns by nine rows. Since the extra columns are not needed for X^2, Y^2, and $X*Y$, this block can be placed next to the X and Y data. When you type /**DATA REGRESSION,** you are asked for the X (independent variable) range. Enter the range **A12..A31.** The regression routine of LOTUS 1-2-3 will accomplish multiple linear regression with up to 16 independent variables. Since we are using linear regression with one independent variable, there is only one column in the range for X. The range for the dependent variable, Y, data has to be entered for the Y-RANGE prompt. This would be B12..B31. The OUTPUT-RANGE specifies where the standard output block will be placed. All you have to specify is the cell for the upper left-hand corner of the output. For our example, D12 was specified as this cell. The INTERCEPT subcommand allows you to specify how the Y intercept for the regression line will be computed. You can select either COMPUTE or ZERO. ZERO will force the intercept to be zero, while COMPUTE will allow the intercept to be computed based on the data. Even though we know that the strain should be zero for zero stress, we want to compute the intercept to check for possible errors in the data, so COMPUTE is chosen. Finally, the GO subcommand, which executes the routine, is selected.

The standard regression output block is shown in Figure 8.21b. The constant is the Y intercept, and the X coefficient is the slope. It appears that the X coefficient is zero, but that is only because the slope is a very small number. If you place the cursor over the cell containing this value, cell F19, the value 0.0000000496 will appear in the control panel portion at the top of the screen. We can reformat this cell, along with the standard error of coefficient cell, so that the true values will be displayed. The following series of commands will do this:

```
/RANGE
     FORMAT
          SCIENTIFIC
               DECIMAL PLACES: 2
          RANGE:  F19..F20 [enter]
```

The values will be displayed in scientific notation, as is shown in Figure 8.21c.

The standard error of coefficient can be used to develop a confidence interval for the slope of the regression line. Other elements of the standard output include the standard error of estimate (Equation H.7 of Appendix H), the coefficient of determination (R Squared), the number of observations, and the degrees of freedom. The degrees of freedom is the number of independent data values after estimation of the regression coefficients and is used in confidence interval testing of the regression results. Confidence interval testing is beyond the scope of this text, but a description of it can be found in standard statistical analysis texts.

Spreadsheet computation of the standard error of estimate has been included as an exercise at the end of this chapter.

We can use the results provided in the output block to generate the additional data that we desire, including the correlation coefficient and estimated value of Y for a specified value of X. Enter the label Correlation Coefficient in cell D22, move the cursor to cell G22, and enter the formula @**SQRT(G15)** to calculate the correlation coefficient as the square root of the coefficient of determination. The relative cell reference G15 can be generated by moving the cursor to that cell after @**SQRT(** has been typed, and can be finalized by closing the parentheses. The value in this cell is to be displayed with three decimal places. Typing

```
/RANGE
        FORMAT
                FIXED
                DECIMAL PLACES:  3
                RANGE:  G22..G22  [enter]
```

will accomplish this.

Enter the label **For a value of X =** in cell D24. The desired value of **53600** can be entered in cell D25. After entering the label **The estimated Y =** in cell D25, the formula can be generated in cell F25. Developing formulas in LOTUS 1-2-3 requires that you specify a value cell by entering a number or one of the key characters in the first position of the cell. In G22, the @ specified a value cell. For cell F25, you can start out with a + sign to specify a value cell. The formula will appear as

$$+F19*F24+G13$$

after it is entered. The relative cell references are generated by using the cursor arrows. The results are shown in Figure 8.21c.

To save the spreadsheet, type

```
/FILE
        SAVE
                A:\123\EX8_8A
```

which would save the spreadsheet in a directory named 123 on the diskette in drive A: under the name EX8_8A. The proper file extender is automatically added by LOTUS 1-2-3. If you have modified an existing spreadsheet, the name it is presently saved under will appear on the screen in the file name space. If you want to replace the old version with the new version, press **[enter]** and confirm that you want to replace the existing file by typing **R.** If you want to change the file name to be assigned to the file, you can press **[esc]** to clear the file name space, and then type the new file name. This will give you both versions saved under different names. To load the file from the diskette into the workspace, type

```
/FILE
        RETRIEVE
                A:\123\[enter]
```

This will produce a menu of all the LOTUS spreadsheet files in directory 123 of the diskette in drive A:. You can move the cursor to the desired file and load it by pressing **[enter].** Alternatively, you could type the whole file name with the command.

(EXAMPLE 8.8 □ Regression Analysis for Stress-Strain Data □ Continued)

After you have saved the file, you can print out the spreadsheet by using the commands

```
/PRINT
        PRINTER
                RANGE:  A1..G32
                ALIGN
                GO
                PAGE
                QUIT
```

The RANGE is the portion of the spreadsheet that you want printed out. 1-2-3 does not default to the whole spreadsheet. ALIGN tells the printer that you have positioned the paper at the top of a new page. It is necessary to select ALIGN each time you print a worksheet. GO starts the printing process. Since LOTUS 1-2-3 does not automatically go to the top of the next page after finishing printing, the subcommand PAGE should be executed before QUIT to move the printer to the top of the next page and to eject the last page of the printout. The QUIT command returns the spreadsheet to the ready mode.

The spreadsheet formulas can be printed out by the series of commands

```
/PRINT
        PRINTER
                OPTIONS
                        OTHER
                                CELL-FORMULAS
```

and will appear as shown in Figure 8.21d.

PROGRAM TESTING

The output of the LOTUS 1-2-3 model agrees with the output of the previous examples using these data. We will now continue the analysis by plotting the stress–strain data and the regression line. If necessary we will remove unrepresentative data and recalculate the spreadsheet.

Input values for X (stress) and Y (strain) are already in the spreadsheet. A third column, which will contain the values for plotting the regression line, will be added. The modulus of elasticity computation will also be added for the further analysis.

Once the values to be plotted are generated in the spreadsheet, you have to access the proper graphics commands to set up the graph. Then you have to save the graphics file, exit LOTUS 1-2-3, and run the PRINTGRAPH program that comes with the 1-2-3 package. If you are using Release 3.0 of LOTUS 1-2-3, you will have the capability of outputting graphs without leaving the spreadsheet module. Since many people are still using the other versions of 1-2-3, the PRINTGRAPH program will be described here. The student version also requires use of the PRINTGRAPH program.

To accommodate the estimated values of Y, you must add a column to the spreadsheet. Move the cursor to the column between Y and the regression output. Use the following commands:

```
/WORKSHEET
        INSERT
                COLUMN [enter]
```

to produce the blank column that you need. You will need to close up the underlines in rows two and eight using the /COPY command. You will also want to copy the headings from column B and edit them to produce Yest and ----- for column C. The numeric range (C12..C31) should be formatted for four decimal places using the RANGE FORMAT command.

The new column will contain the Y estimate for each given X value in the first column. You will need to use the slope and intercept of the regression equation; however, the reference to these parameters must be the same for each element of the column, while the reference to the X value must be relative to the row used. Therefore, the formula in column C must have both relative and absolute references. In cell C12, you can enter **+A12*$G$19+$H$13**, indicating that the reference to A12 is relative, while the references to G19 and H13 are absolute. This formula can now be copied into the range C13..C31 to produce all of the estimated Y values needed. We will place the label for the modulus of elasticity computation in cell E27. The formula

$$1/I19$$

in cell H27 will compute the modulus of elasticity. The cell is formatted for scientific notation with two decimal places. Figure 8.22a shows the modified spreadsheet.

Now you are ready to develop the parameters of the graph. Since the increments of the independent variable are not even, you need to develop an X-Y plot. The following series of commands will define the parameters of the graph. Explanatory notes are included in parentheses.

```
/GRAPH
    TYPE
        XY
    X
        X AXIS RANGE:  A12..A31
    A
        FIRST DATA RANGE:  B12..B31
    B
        SECOND DATA RANGE:  C12..C31
    OPTIONS
        LEGEND (legend for symbols)
            A
                ENTER LEGEND:  Measured Strain
        FORMAT
            A
                SYMBOLS (the A range [Y data] will be displayed as
                symbols)
            B
                LINES (The B range [Yest values] will be displayed as
                lines
            QUIT (exit to OPTIONS submenu)
        TITLES
            FIRST
                STRESS/STRAIN DATA
            X-AXIS
                Stress (psi)
```

(EXAMPLE 8.8 □ Regression Analysis for Stress-Strain Data □ Continued)

```
Y-AXIS
     Strain (in./in.)
GRID
     BOTH (X and Y grid lines will be added to graph)
SCALE
     Y-SCALE
          MANUAL (set scale manually)
          LOWER (lower limit for Y scale)
               0.0
          UPPER (upper limit for Y scale)
               0.0022
          QUIT (exit to next higher level of menu)
SCALE
     X-SCALE
          MANUAL
```

FIGURE 8.22   Scatterplot for stress-strain analysis.

(*a*) Spreadsheet with regression line (Yest) column added.

FIGURE 8.22 (Continued)

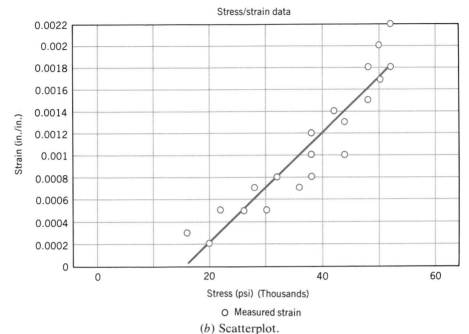

(*b*) Scatterplot.

```
                LOWER
                    0.0
                UPPER
                    52000
                    QUIT
            QUIT (exit OPTIONS submenu)
        VIEW (shows graph on screen) [esc]
        SAVE (saves the graph file for printing or plotting)
            A:\123\EX8_8B (file name for graph file)
        QUIT (exit to main 1-2-3 menu)
/QUIT (exit LOTUS 1-2-3)
    YES (confirms that you want to quit LOTUS 1-2-3 and return to DOS
C> PGRAPH (loads PRINTGRAPH program)
```

The first time you load the PRINTGRAPH program, you have to specify what type of output device you are using and what directory contains the graph file. If that has already been done, you can go directly to the IMAGE SELECT commands. A menu of the graphics files contained in the designated directory appears on the screen. Move the cursor to the desired field and press **[enter]** to load the graph file. The ALIGN and GO commands will start the printing or plotting process. When the graph is output, it will appear as shown in Figure 8.22*b*.

Examination of the scatterplot reveals that there are problems with the data. There is an obvious curvature in the data. Most of the data points on either end of the data are above the regression line, while most of the data points in the middle are below the regression line. The regression line also does not come close to going through zero. Further investigation shows that the modulus of elasticity for steel is approximately 30×10^6, while the value for our output data is only about 20×10^6. If you check any engineering strength of materials text, you will find that the yield strength of steel is approximately 36,000 psi. This

(EXAMPLE 8.8 □ Regression Analysis for Stress-Strain Data □ Continued)

means that with stresses above 36,000 psi, steel is no longer a purely elastic material. Some permanent deformation of the steel will occur with higher stresses, even though it will not actually break until much higher stresses are applied. In the higher range, the stress–strain relationship is no longer linear. The modulus of elasticity depicts the stress-strain relationship for the linear portion of the curve only.

It is obvious that the data that were supplied for this problem used applied stresses beyond the yield stress. Therefore, when we used the data to predict the modulus of elasticity, our value was not realistic. To make the value more realistic, we can eliminate all data values for stress above 38,000 psi, using the /MOVE or /COPY commands to close up the three 38,000 psi data points, and the /RANGE ERASE command to remove the unwanted data. The ranges for the independent and dependent variables in the /DATA REGRESSION command also have to be adjusted. After doing this, we will still have 11 data points, and the spreadsheet will appear as shown in Figure 8.23a. The modulus of elasticity is now just about what we would expect. However, after developing and printing the new scatterplot, Figure 8.23b, we see that the regression line still does not approach zero for zero stress. The most probable cause for this is a calibration error in the measuring instruments that caused a zero offset in all readings. Our only prudent course of action would be not to use the suspect data and to go back and figure out why the equipment was not calibrated properly.

FIGURE 8.23   Scatterplot after eliminating data outside elastic range.

```
        REGRESSION ANALYSIS AND RELATED STATISTICS
-----------------------------------------------------------------------
Enter any labels for this problem in rows 4-7
X = Stress (psi)
Y = Strain (in/in)
-----------------------------------------------------------------------

        X        Y       Yest
      -----    -----     -----
      20000   0.0002   0.0003              Regression Output:
      16000   0.0003   0.0002       Constant                   -0.00031
      22000   0.0005   0.0004       Std Err of Y Est           0.000149
      26000   0.0005   0.0005       R Squared                  0.767346
      30000   0.0005   0.0007       No. of Observations              11
      28000   0.0007   0.0006       Degrees of Freedom                9
      36000   0.0007   0.0009
      32000   0.0008   0.0007       X Coefficient(s)   3.31E-08
      38000   0.0008   0.0009       Std Err of Coef.   6.07E-09
      38000    0.001   0.0009
      38000   0.0012   0.0009       Correlation Coefficient       0.876
                                    For a value of X =     53600
                                    The estimated Y = 0.001452
                                    Modulus of Elasticity =    3.02E+07
```

(a) Spreadsheet.

FIGURE 8.23 (Continued)

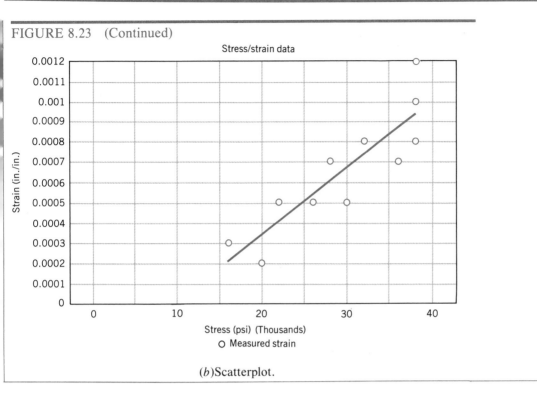

(b)Scatterplot.

We will now illustrate the matrix commands of LOTUS 1-2-3. Example 8.9 solves the planar truss problem of Example 7.4 using matrix inversion and multiplication.

EXAMPLE 8.9 □ Matrix Analysis of Truss (Matrix Commands)

STATEMENT OF PROBLEM

Develop a LOTUS 1-2-3 spreadsheet to solve the seven-member planar truss problem of Example 7.4.

MATHEMATICAL DESCRIPTION

Section 2.3.4 of Appendix F and Section 3.1 of Appendix G give the mathematical background for this problem. Figures 6.2 and G.1 give the schematic and free body diagrams for the truss under analysis.

ALGORITHM DEVELOPMENT

Input/Output Design

The coefficient matrix represents the directional sines and cosines of the truss members. Each row of the matrix corresponds to one of the equilibrium equations, and each column corresponds to one of the members. The matrix will be placed in a block of cells with appropriate labels identifying which equation and

(EXAMPLE 8.9 □   Matrix Analysis of Truss (Matrix Commands) □   Continued)

member each cell represents. The right-hand-side vector represents both the applied loads and the calculated reactions. It is a column vector and will be displayed as such in the spreadsheet.

Numerical Methods

LOTUS 1-2-3 has procedures for accomplishing most matrix operations, including inversion and multiplication. No information is given in the documentation as to what method is used for inversion.

Computer Implementation

The matrix inversion commands will be used to produce the inverse of the coefficient matrix, and matrix multiplication will be used to multiply the right-hand-side vector by the inverse to produce the solution vector of forces in each member.

PROGRAM DEVELOPMENT

The truss of Example 7.4 has seven members, so seven columns will be needed to store the matrix. If two more columns are used for the labels identifying the equilibrium equation that each row corresponds to, a total of nine columns will be required. This is too many columns to output on one sheet of output, unless you resort to using compressed print. However, you can easily get around this by adjusting the column width of the label columns and the matrix cells. Each row represents a summation of forces in either the X or Y direction at a node. The nodes of Figure 6.2 are labeled with letters. Therefore a label Xa could be used to represent the summation of forces in the X direction at node A, and so forth. If both columns A and B are set to a width of 3, you can enter the following labels in range A3..B12

```
       A    B
3      E
4      Q
5      U    Xa
6      A    Ya
7      T    Xb
8      I    Yb
9      0    Xc
10     N    Yc
11          Xd
12     #
```

To reset the column width of an individual column, you can place the cursor in the column to be set, and press

```
          /WORKSHEET
             COLUMN
                SET—WIDTH
                   3
```

Repeat this for both columns A and B. Since the coefficients of the truss matrix are all between $+1.0$ and -1.0, we could also make the remainder of the columns narrower than the default width of 9. You can reset the global column width to 8 by the following series of commands:

```
/WORKSHEET
        GLOBAL
                COLUMN-WIDTH
                        8
```

Any column widths that have been changed by previously outlined procedures will have precedence over the global setting, so columns A and B will remain at a width of 3.

The coefficient matrix will occupy cells C5..I11. To make all entries appear uniform in the spreadsheet, this range is formatted for four decimal places by the commands

```
/RANGE
        FORMAT
                FIXED
                        DECIMAL PLACES: 4
                RANGE:  C5..I11
```

Otherwise, the integer entries zero and one would appear as integers and the decimal values as decimals. The coefficient values can now be entered in the cells in the same order presented in Example 7.4. The sines and cosines have already been calculated, so you can enter them directly. However, spreadsheets have the capability of acting as calculators when you enter values. For example, the cosine of the angle that member AB makes with the horizontal can be calculated by superimposing a right triangle on node A, node B, and a point halfway between A and C. The cosine of the angle at A is equal to the length of the adjacent side divided by the length of the hypotenuse, member AB. The length of member AB in Figure 6.2 is 22.36 ft. If you enter **10/22.36** in cell D5 and press **[enter]**, the value 0.4472 would appear in the cell. However, you also have to accommodate the sign of the element of the equilibrium equation, which in this case is negative. This could be accomplished by entering **−10/22.36** in the cell.

You also should label each column of the matrix with the member force it represents. These labels will be placed in the fourth row as Fac, Fab, and so on, in the same order that they were developed for the equilibrium equations. In LOTUS 1-2-3 you have to enter the labels and then reformat them for justification. We want to center the labels, so we use the commands

```
/RANGE
        LABEL
                CENTER
                RANGE:  C4..I4
```

The top portion of Figure 8.24a shows the resulting coefficient matrix as entered in the spreadsheet.

The right-hand-side vector will be placed below the coefficient matrix to keep all output within one page width. Since each of the right-hand-side values also corresponds to a particular equilibrium equation, the same row labels can be used that were used with the coefficient matrix. Therefore, you can use the /COPY command to copy the range A3..B12 to A15 (the start of the right-hand-side block). The cells of this column are formatted for two decimal places, and the values will be entered in thousands of pounds (kips). The bottom half of Figure 8.24a shows the addition of the right-hand-side vector.

(EXAMPLE 8.9 □ Matrix Analysis of Truss (Matrix Commands) □ Continued)

It is now a fairly simple task to invert the coefficient matrix. The matrix commands of LOTUS 1-2-3 are contained within the /DATA commands. The following series of commands

```
/DATA
     MATRIX
          INVERT
               RANGE TO INVERT:  C5..I11
               OUTPUT RANGE:  C27 [enter]
```

will accomplish the inversion and place the inverted matrix in the spreadsheet with the upper left corner in cell C27. The following series of commands will multiply the inverse and the right-hand-side vector together:

```
/DATA
     MATRIX
          MULTIPLY
               FIRST RANGE:   C27..I33
               SECOND RANGE:  C17..C23
               OUTPUT RANGE:  C37
```

FIGURE 8.24   1-2-3 spreadsheet for matrix analysis of truss.

```
A15: [W3] 'E                                                          READY

       A B    C        D        E        F        G        H        I        J
1     MATRIX ANALYSIS OF A TRUSS
2
3      E       UNKNOWN MEMBER FORCES
4      Q       Fac      Fab      Fbc      Fbd      Fcd      Fce      Fde
5      U  Xa  1.0000  -0.4472   0.0000   0.0000   0.0000   0.0000   0.0000
6      A  Ya  0.0000   0.8944   0.0000   0.0000   0.0000   0.0000   0.0000
7      T  Xb  0.0000   0.4472   0.4472  -1.0000   0.0000   0.0000   0.0000
8      I  Yb  0.0000   0.8944  -0.8944   0.0000   0.0000   0.0000   0.0000
9      O  Xc -1.0000   0.0000  -0.4472   0.0000   0.4472   1.0000   0.0000
10     N  Yc  0.0000   0.0000   0.8944   0.0000   0.8944   0.0000   0.0000
11        Xd  0.0000   0.0000   0.0000   1.0000  -0.4472   0.0000  -0.4472
12     #
13
14
15     E       RIGHT-HAND SIDE VECTOR
16     Q       Forces and Reactions (1000 lbs)
17     U  Xa    0.00
18     A  Ya    5.00
19     T  Xb    0.00
20     I  Yb    0.00
21     O  Xc    0.00
22     N  Yc   10.00
23        Xd    0.00
24     #
25
28-Mar-90  10:31 AM          UNDO
```

(*a*) Input matrices.

FIGURE 8.24   (Continued)

```
MATRIX ANALYSIS OF A TRUSS
E        UNKNOWN MEMBER FORCES
Q           Fac      Fab      Fbc      Fbd      Fcd      Fce      Fde
U   Xa    1.000   -0.447    0.000    0.000    0.000    0.000    0.000
A   Ya    0.000    0.894    0.000    0.000    0.000    0.000    0.000
T   Xb    0.000    0.447    0.447   -1.000    0.000    0.000    0.000
I   Yb    0.000    0.894   -0.894    0.000    0.000    0.000    0.000
O   Xc   -1.000    0.000   -0.447    0.000    0.447    1.000    0.000
N   Yc    0.000    0.000    0.894    0.000    0.894    0.000    0.000
    Xd    0.000    0.000    0.000    1.000   -0.447    0.000   -0.447
#

E        RIGHT-HAND SIDE VECTOR
Q        Forces and Reactions (1000 lbs)
U   Xa     0.00
A   Ya     5.00
T   Xb     0.00
I   Yb     0.00
O   Xc     0.00
N   Yc    10.00
    Xd     0.00
#
                    INVERSE OF COEFFICIENT MATRIX
            1.000    0.500    0.000    0.000    0.000    0.000    0.000
            0.000    1.118    0.000    0.000    0.000    0.000    0.000
            0.000    1.118    0.000   -1.118    0.000    0.000    0.000
            0.000    1.000   -1.000   -0.500    0.000    0.000    0.000
            0.000   -1.118    0.000    1.118    0.000    1.118    0.000
            1.000    1.500    0.000   -1.000    1.000   -0.500    0.000
            0.000    3.354   -2.236   -2.236    0.000   -1.118   -2.236

                    MEMBER FORCES (1000 lbs)
M   ac     2.50
E   ab     5.59
M   bc     5.59
B   bd     5.00
E   cd     5.59
R   ce     2.50
    de     5.59
```

(*b*) Input and output matrices.

The range C27..I33 contains the inverse of the coefficient matrix, the range C17..C23 contains the right-hand-side vector, and the solution (member force) vector will be generated in a column beginning at C37. Figure 8.24*b* shows the resulting spreadsheet with identifying labels added. Also, you should note that the number of decimal places has been fixed at three for the matrices rather than the four decimal places included in the coefficient matrix of Figure 8.24*a*. This provides more definition for the columns without losing any essential information. The solution vector range is formatted for two decimal places.

PROGRAM TESTING

The solution is the same as generated previously. If you wanted to test various loading conditions, all you would have to do is make the appropriate modifications in the right-hand-side vector.

Next we will investigate the database management capabilities of LOTUS 1-2-3.

8.3.4.11 Database Management Commands for LOTUS 1-2-3

LOTUS 1-2-3 has a separate set of database commands under the general /DATA command. Also included under the /DATA command are the MATRIX and RE-GRESSION commands used in previous examples. The SORT and QUERY commands are specific to portions of a spreadsheet defined as a database. The method of specifying a portion of a spreadsheet as a database will be described in Example 8.10. The other subcommands of /DATA can be used with either databases or ranges of a spreadsheet.

Example 8.10 will now apply the database management commands of LOTUS 1-2-3 to the industrial production database of Example 7.1.

EXAMPLE 8.10 □ **Industrial Production Data (Sorting and Searching)**

STATEMENT OF PROBLEM

Develop a LOTUS 1-2-3 spreadsheet that contains the database of Example 7.1 in employee ID number order. Use the database commands of LOTUS 1-2-3 to sort the database into alphabetical order. Extract the data for a particular employee, and compute the total production, the total defects, and the production/defects ratio for that employee. Modify the database to include summary statistics for all employees. Use a conditional query to extract employees' names and ID numbers based on the quality of work that employees are doing.

MATHEMATICAL DESCRIPTION

To produce production and defect statistics, you will have to sum the monthly production and monthly defect data for each employee. To produce individual statistics, you will need to sum across the record for each individual.

ALGORITHM DEVELOPMENT

Input/Output Design

All data will appear on the spreadsheet, so you can design the database to appear as it would if you were entering it into tabular data sheets. This is both an advantage and a disadvantage. All the data are readily available. However, for large databases the spreadsheet can become cumbersome. The data for this example show a full year of entries. It would work equally well for a partial year's data. As the year progressed, you could enter monthly data. When you are entering data, you will want to use the WINDOW command of LOTUS 1-2-3 to freeze the name fields and scroll the production and defect columns. This will allow you to line up the column for the particular month next to the names.

Output of the full database requires multiple pages because of the number of columns. You should design your summary statistics and queries so that you can output just those portions of the spreadsheet in a single sheet.

Computer Implementation

The database will be developed in four stages. First the basic database will be entered and saved. Figure 8.25a shows the basic database. The database will then be sorted, as shown in Figure 8.25b. The third stage will be adding the criterion and output ranges and using the query command to extract data for a desired employee. The database commands will be used to develop summary production and defect data for the selected employee. Figure 8.25c shows the results of the query. This spreadsheet should be saved before undertaking the fourth stage of the example, which will involve modifications of the database.

FIGURE 8.25   1-2-3 spreadsheet for industrial production database.

	A	B	C	D	E	F	G	H	I	J
1	INDUSTRIAL PRODUCTION DATABASE									
2										
3	ID #	First Name	MI	Last Name	Pjan	Pfeb	Pmar	Papr	Pmay	Pjun
4	1001	James	S.	Johnson	155	170	190	188	175	166
5	1002	Brenda	A.	Thomas	130	132	110	129	108	142
6	1003	Pamela	B.	Dodd	145	153	115	135	100	134
7	1004	John	T.	Jones	143	133	127	140	99	131
8	1005	Keith	B.	Reynolds	125	140	122	167	190	122
9	1006	Elizabeth	R.	Thomas	160	155	153	182	145	78
10	1007	Rebecca	S.	Smith	130	125	135	123	157	133
11	1008	Joan		Carmen	132	115	132	135	135	167
12	1009	Thomas	K.	Flint	100	110	108	115	109	111

	K	L	M	N	O	P	Q	R	S	T	U	V
1												
2												
3	Pjul	Paug	Psep	Poct	Pnov	Pdec	Djan	Dfeb	Dmar	Dapr	Dmay	Djun
4	88	175	165	158	173	180	50	45	62	25	35	44
5	111	112	115	175	124	126	20	25	35	36	38	21
6	120	133	130	109	154	153	24	24	36	31	25	32
7	132	128	110	127	136	143	25	22	34	26	26	27
8	137	165	127	143	132	33	26	33	33	27	29	35
9	160	170	165	180	190	200	10	15	13	16	14	9
10	128	143	154	154	142	88	33	24	28	24	37	22
11	131	121	158	129	141	70	31	23	26	25	28	25
12	112	115	118	108	109	106	5	4	6	3	5	7

	W	X	Y	Z	AA	AB	AC	AD	AE	AF	AG	AH
1												
2												
3	Djul	Daug	Dsep	Doct	Dnov	Ddec						
4	40	21	36	43	36	28						
5	18	22	41	77	22	31						
6	33	24	39	21	18	27						
7	34	31	35	19	16	24						
8	37	28	29	18	18	1						
9	18	19	20	18	16	32						
10	32	26	22	22	12	33						
11	21	22	28	7	21	38						
12	4	3	6	4	5	7						

(a) Initial database entry.

(EXAMPLE 8.10   □   Industrial Production Data (Sorting and Searching)   □   Continued)

FIGURE 8.25   (Continued)

	A	B	C	D	E	F	G	H	I	J
1	INDUSTRIAL PRODUCTION DATABASE									
2										
3	ID #	First Name	MI	Last Name	Pjan	Pfeb	Pmar	Papr	Pmay	Pjun
4	1008	Joan		Carmen	132	115	132	135	135	167
5	1003	Pamela	B.	Dodd	145	153	115	135	100	134
6	1009	Thomas	K.	Flint	100	110	108	115	109	111
7	1001	James	S.	Johnson	155	170	190	188	175	166
8	1004	John	T.	Jones	143	133	127	140	99	131
9	1005	Keith	B.	Reynolds	125	140	122	167	190	122
10	1007	Rebecca	S.	Smith	130	125	135	123	157	133
11	1002	Brenda	A.	Thomas	130	132	110	129	108	142
12	1006	Elizabeth	R.	Thomas	160	155	153	182	145	78

	K	L	M	N	O	P	Q	R	S	T	U	V
1												
2												
3	Pjul	Paug	Psep	Poct	Pnov	Pdec	Djan	Dfeb	Dmar	Dapr	Dmay	Djun
4	131	121	158	129	141	70	31	23	26	25	28	25
5	120	133	130	109	154	153	24	24	36	31	25	32
6	112	115	118	108	109	106	5	4	6	3	5	7
7	88	175	165	158	173	180	50	45	62	25	35	44
8	132	128	110	127	136	143	25	22	34	26	26	27
9	137	165	127	143	132	33	26	33	33	27	29	35
10	128	143	154	154	142	88	33	24	28	24	37	22
11	111	112	115	175	124	126	20	25	35	36	38	21
12	160	170	165	180	190	200	10	15	13	16	14	9

	W	X	Y	Z	AA	AB	AC	AD	AE	AF	AG	AH
1												
2												
3	Djul	Daug	Dsep	Doct	Dnov	Ddec						
4	21	22	28	7	21	38						
5	33	24	39	21	18	27						
6	4	3	6	4	5	7						
7	40	21	36	43	36	28						
8	34	31	35	19	16	24						
9	37	28	29	18	18	1						
10	32	26	22	22	12	33						
11	18	22	41	77	22	31						
12	18	19	20	18	16	32						

(b) Database after sorting in alphabetical order.

The fourth stage will require the addition of columns to the database to sum production and defect data for each employee and to compute the production/defect ratio. In this manner, you are able to combine the spreadsheet functions (formulas and functions in cells) with the database functions. This point is not brought out in the 1-2-3 documentation, but it greatly adds to the versatility of LOTUS 1-2-3 as a database management tool. With the addition of individual

FIGURE 8.25   (Continued)

```
A20: [W6] 1009                                                    READY

      A       B        C      D      E     F     G     H     I     J
1  INDUSTRIAL PRODUCTION DATABASE
2
3  ID #   First Name MI Last Name  Pjan  Pfeb  Pmar  Papr  Pmay  Pjun
4   1008 Joan           Carmen      132   115   132   135   135   167
5   1003 Pamela      B. Dodd        145   153   115   135   100   134
6   1009 Thomas      K. Flint       100   110   108   115   109   111
7   1001 James       S. Johnson     155   170   190   188   175   166
8   1004 John        T. Jones       143   133   127   140    99   131
9   1005 Keith       B. Reynolds    125   140   122   167   190   122
10  1007 Rebecca     S. Smith       130   125   135   123   157   133
11  1002 Brenda      A. Thomas      130   132   110   129   108   142
12  1006 Elizabeth   R. Thomas      160   155   153   182   145    78
13
14 Criterion Range
15 ID #   First Name MI Last Name  Pjan  Pfeb  Pmar  Papr  Pmay  Pjun
16                      Flint
17
18 Output Range
19 ID #   First Name MI Last Name  Pjan  Pfeb  Pmar  Papr  Pmay  Pjun
20  1009 Thomas      K. Flint       100   110   108   115   109   111
21
22 Total Production          1321
23 Total Defects               59
24 Production/Defects       22.39
25
28-Mar-90  10:02 AM        UNDO
```

(*c*) Database with query and summary sections added.

statistics, conditional queries can be used to identify excellent or substandard employees. Figure 8.26 shows the modified database with a conditional query.

PROGRAM DEVELOPMENT

A database is entered into LOTUS 1-2-3 in much the same way as any other spreadsheet data. However, you have to include field definitions in the form of labels at the top of each column of the spreadsheet. Each row of the spreadsheet will represent a record of the database, and each column a field. The first field will contain the employee number. The second, third, and fourth fields will contain the first name, middle initial, and last name of the employee. The next 12 fields will contain the production data for January through December, followed by the defect data. Column A will be set at six spaces to accommodate the employee identification number. Column B will use 11 spaces for the first name, column C three spaces for the middle initial and period, and column D 11 spaces for the last name. The remainder of the spreadsheet will be set to a column width of six for the data. Figure 8.25*a* shows the database with the data entered for nine employees. The second and third groups of data actually appear in columns to the right of the first group. They have been made into a composite to fit on the figure.

(EXAMPLE 8.10 □ Industrial Production Data (Sorting and Searching) □ Continued)

FIGURE 8.26   Industrial production database with individual summaries and conditional query.

```
A19: [W6] 'ID #                                                        READY

         A       B       C     D        E      F      G     H     I     J
 1   INDUSTRIAL PRODUCTION DATABASE
 2
 3   ID #  First Name MI Last Name   TotP   TotD   P/D  Pjan  Pfeb  Pmar
 4   1008 Joan           Carmen     1566    295   5.31   132   115   132
 5   1003 Pamela      B. Dodd       1581    334   4.73   145   153   115
 6   1009 Thomas      K. Flint      1321     59  22.39   100   110   108
 7   1001 James       S. Johnson    1983    465   4.26   155   170   190
 8   1004 John        T. Jones      1549    319   4.86   143   133   127
 9   1005 Keith       B. Reynolds   1603    314   5.11   125   140   122
10   1007 Rebecca     S. Smith      1612    315   5.12   130   125   135
11   1002 Brenda      A. Thomas     1514    386   3.92   130   132   110
12   1006 Elizabeth   R. Thomas     1938    200   9.69   160   155   153
13
14   Criterion Range
15   ID #  First Name MI Last Name   TotP   TotD   P/D
16                                  +E4>1         +G4<5
17
18   Output Range
19   ID #  First Name MI Last Name   TotP   TotD   P/D
20   1003 Pamela      B. Dodd       1581    334   4.73
21   1001 James       S. Johnson    1983    465   4.26
22   1004 John        T. Jones      1549    319   4.86
23   1002 Brenda      A. Thomas     1514    386   3.92
24
28-Mar-90  10:17 AM        UNDO
```

Your first step will be to sort the database into alphabetical order. The first sort should be done on the last name. However, if you have two people with the same last name, there must be a way for the sort routine to differentiate between the employees. There are two employees with the last name of Thomas in this database. LOTUS allows you to specify a secondary sort key, so we will specify the first name for the secondary sort. This will put Brenda Thomas before Elizabeth Thomas. This also shows why it was important to put the first name, middle initial, and last name in separate fields. If all three were in the same field, you would have to enter the name as last name, first name, middle initial in order to do a proper alphabetical sort.

When you are specifying the range to be sorted, you *do not* include the field names. Therefore, you may start the sequence with the cursor in cell A4. The following series of commands will sort the database into ascending alphabetical order:

```
\DATA
    SORT
        DATA-RANGE:  A4..AB12
        PRIMARY-KEY: D4
            SORT-ORDER:  A
```

```
SECONDARY-KEY:  B4
      SORT-ORDER:   A
GO
```

The cell entered for the PRIMARY-KEY or SECONDARY-KEY can be any cell within the data range in the column that you want to use to establish the primary or secondary sort order. The data range for this example is from row 4 to row 12, so we chose D4 to indicate that the primary sort will be on the data in column D, and B4 to indicate that column B will be used for the secondary sort. Entering D8 and B10 would have accomplished the same end. Entering **GO** will resort the records into the order shown in Figure 8.25*b*. You can examine Figures 8.25*a* and 8.25*b* to confirm that all fields of each record have been sorted according to the keys. It is important to note that the order of the spreadsheet has been changed. If you need to keep the original order you should save the database spreadsheet prior to the sort. You also could do another sort on the employee identification numbers.

Now you have to develop a criterion range that can be used to search the database according to the criteria that you will specify. This is accomplished by copying the field names that you may want to use to set criteria. At least one field has to be included in the criterion range. It can contain up to all the fields available. You will select the actual fields for the criteria from the available range. The criterion range can be placed anywhere on the spreadsheet. For this example we will use all the fields in the criterion range and will place the criterion range below the database, starting in cell A15. The /COPY command is used to copy the whole row of field names from A3..AB3 to the range starting with A15. Figure 8.25*c* shows the first portion of the criterion range. Only the portion of the database that will fit on a single page has been included in the figure. This is the portion of the criterion range that we will use to set our criteria. A label, "Criterion Range", has been placed above the criterion range, and the word Flint entered under the Last Name field. This row under the field names establishes the criteria that will be used for the search. Entering Flint means that we will be searching for the record or records with a last name of Flint.

You also have to establish an output range for the query. The fields that you select for the output range establish which data from the records that satisfy the criteria will be copied to the output range. Again, you can choose any subset of the database field names that you want to, up to the whole list of field names. We will use the whole list in this example. Nothing will actually appear in the output range until you execute the query command. The following series of commands will execute the search:

```
/DATA
      QUERY
            INPUT
                  RANGE:   A3..AB12 [enter]
            CRITERION
                  RANGE:   A15..AB16 [enter]
            OUTPUT
                  RANGE:   A19..AB20 [enter]
            EXTRACT
            QUIT
```

The input range includes all the records of the database, plus the field names. Inclusion of the field names is required. You could leave off records at the end of

(EXAMPLE 8.10 □ Industrial Production Data (Sorting and Searching) □ Continued)

the database. The criterion range includes the row with the criteria field definitions, plus the next row that contains the actual criteria. If multiple records satisfy the criteria set, all those records will be copied to the output range. Thus, you must be careful not to have formulas or other data too close to the output range definition. If multiple records are copied, they will replace any information that is presently in the target cells. For our example, the output range A19..AB20 limits the output to one line. The first line (row 19) is taken up with the field definitions. Using this output range, you will get an error message if there are multiple records that satisfy the criteria. If you had specified A19..AB19 as the output range, all records that met the criteria would be copied to the output range. The EXTRACT command is what actually accomplishes the search and extracts the desired record or records. Our criterion was a last name of Flint, so the entire record of Thomas K. Flint would be copied to the output range. If you had wanted to extract the record for Elizabeth Thomas, you would have had to specify two criteria: the last name of Thomas, and the first name of Elizabeth. These could be typed under the appropriate criterion range field names, and then EXTRACT could be used to extract her record.

LOTUS 1-2-3 provides several functions that are particular to database applications. Among them is the function @DSUM(*input,offset,criterion*). This function finds the sum of the values in the column offset by the number of columns specified by the parameter *offset* from the leftmost column of the database. The parameter *input* specifies the input range for the database, and the parameter *criterion* specifies the criterion range to be used to select records from the defined range. However, this command will add down columns only. To produce the summary statistics that we want (total production and total defects for a year for the individual record extracted from the database) we can use the normal @SUM function. The formula @SUM(E20..P20) placed in D22 will sum the production data, and the formula @SUM(Q20..AB20) placed in D23 will sum the defect data. The product/defect ratio can be computed by the formula +D22/D23 in cell D24, after formatting D24 for two decimal places. Labels can be added in cells A22 through A24, resulting in the spreadsheet shown in Figure 8.25*c*.

Modifications will now be made in the database to generate the production and defect summaries for each employee in the database. You will now be able to extract information from the database based on the values for these parameters. After saving the original database, insert three columns to the right of the Last Name field. Label these as TotP, TotD, and P/D. The @SUM function is used to sum the production and defect data for each employee. The formulas are entered in E4 and F4, while the formula for the ratio is inserted in cell G4. These three formulas are then copied into the range from row 5 to row 12.

Conditional criteria will be used to select all employees with a total production greater than 1500 units per year and a production-to-defect ratio of less than 5. These could be considered as employees who have an acceptable production rate but are producing too high a percentage of defective products. For a conditional criteria, you have to enter the cell reference for any of the cells within the data range for the field to which the criteria will be applied. For example, the formula +E4>1500 placed in E16 will select the records only for employees with production totals greater than 1,500. The formula +G4<5 in G16 will select only records with a production/defect ratio less than 5. Inclusion of both criteria means that they both must be satisfied for a record to be selected. Note also that the criterion and output ranges have been modified to exclude all the actual production and defect data storage cells. Executing the DATA QUERY command with an output range of A19..G19 will result in the four records shown in Figure 8.26 being selected. You could also set different criteria to select employees who have both a high production rate and a high production/defect ratio, or any other criteria you wish to set.

PROGRAM TESTING

The individual data for the selected employee can be added up to confirm that the summary statistics are correct. You can also change the criterion. If you change it to Jones, the total production will become 1,549, the total number of defects will become 319, and the production-to-defect ratio will be 4.86.

8.3.4.12 Macros

A macro executes a series of keystrokes when you enter the appropriate macro command. Macros are useful for series of keystrokes that you will be executing many times. They save a great deal of time, since you do not have to stop to think about the series of keystrokes necessary to accomplish a particular task. However, you do have to remember the command keystrokes. Macros execute the same way every time you invoke them, so they help eliminate typing errors. They are similar in concept to command files that execute numerous operating system commands. Properly designed macros can improve your spreadsheet productivity.

When you define a macro for LOTUS 1-2-3, you have to define the keystrokes and the macro name. It is best to place macros in a section of the spreadsheet that you would not normally print out. The keystroke commands are entered as one or more labels in the spreadsheet. Macros can be edited just like the contents of any cell. You execute the macro command by selecting the **RUN** function key and typing the macro name. After you develop a macro, you have to debug it as you would debug a subprogram. If you wait to debug several macros at once, it may be difficult to determine which one is causing problems.

Macros are similar to FORTRAN subprograms in that once one has been developed it can be used in many applications. You can save macros and read them into new spreadsheets. Macros have been developed for most popular LOTUS 1-2-3 operations. Many of these are available commercially. In fact, books have been written just on macros. For example *101 Macros for LOTUS 1-2-3*, by Lunsford (1986), contains macros for operations such as setting currency formats, creating borders around spreadsheets, and creating and printing address labels.

LOTUS 1-2-3 also contains advanced macro commands that act as a powerful programming language. Conditional statements, branching, and subroutines are all supported. While it is true that you can develop useful spreadsheets without resorting to using macros, it is also true that even basic macros can help you to automate frequently used 1-2-3 operations. However, before developing a spreadsheet for a problem that will require advanced macro commands, you should evaluate the alternative use of a high-level language or equation solver.

8.3.5 Spreadsheet Summary

Spreadsheets are extremely versatile tools that can be used effectively in a whole range of engineering analysis and design problems. Any engineering problem that can be arrayed in a tabular format is a prime candidate for a spreadsheet solution. Spreadsheets are particularly useful for exploring different "What if" scenarios. Spreadsheets can easily be modified to solve many similar problems. Macros enhance the power of spreadsheets and can incorporate many of the features of high-level languages into spreadsheet applications.

8.4 INTEGRATED PACKAGES

8.4.1 Introduction

Early microcomputer applications packages each had their own way of storing and working with data, and there was virtually no way of sharing information among packages. As microcomputers began to be more widely used in the business and technical fields, the need for sharing of information, or integration of applications, was recognized. LOTUS 1-2-3 was the forerunner of integrated applications, with its spreadsheet, graphics, and database modules. The great success of LOTUS 1-2-3 led to the development of other more fully integrated packages and operating environments.

Integrated packages are single software programs designed to do several different types of tasks and to share data among the tasks. The most popular programs accomplish the tasks described in the following sections. Integrated packages started appearing in early 1984 and have been making steady progress since. But they have not dominated the software market, as some of their developers thought they would. Some of the reasons for this will be discussed in later sections. From an educational standpoint, integrated packages make a lot of sense. Students can use and become proficient in several different applications, without having to learn several different operating environments. The integrated package will also be less expensive than several stand-alone packages. Integrated packages allow you to spend more time on learning the concepts of the applications and less time on learning applications package syntax.

8.4.2 Functions

The most popular integrated packages include spreadsheet, word processing, database, graphics, and communications applications. By this time you should be familiar with the first four applications. The communications package provides terminal emulation software and file management software to allow you to connect your microcomputer to another computer system. You can interact with that computer and send files back and forth. Except for equation solvers, these are probably all the major applications that an engineer would ever have to use. However, when you are deciding what type of package to acquire, you have to ask yourself if you need all the functionality of an integrated package or if you would be better off having one or two specialized packages. You should carefully consider the advantages and disadvantages of integrated packages.

8.4.3 Advantages and Disadvantages

Although integrated packages are expensive, they are much less expensive than buying five separate packages. The popular top-end integrated packages sell for around $700, while LOTUS 1-2-3 alone retails for about $500.

Most integrated packages provide a consistency of commands and operations across the applications. The consistency involves syntax and menu structure, as well as common macro designs. There can never be complete consistency because the commands required for a word processing application are much different from the commands required for database management.

The ease of data transferal among modules is an important advantage of integrated packages. Although many stand-alone packages now have the capability of importing data from different types of applications, the process requires several steps, and the transfer is not always fully compatible. With integrated packages the transfer is made without having to leave the applications package, so compatibility is not a problem.

The word processing module is able to integrate data from the other modules to provide coordinated documentation for the applications of an integrated package. The better integrated packages have word processing capabilities close to those of the best stand-alone packages. The ability to incorporate data from the other four applications greatly increases the value of the word processing module.

A common complaint voiced by more sophisticated users about integrated packages is that none of the modules of the integrated package is as comprehensive as a stand-alone package for accomplishing similar tasks. Although this may be an accurate assessment, it should not be taken by users as condemnation of integrated packages. Some of the modules of the better integrated packages come very close to duplicating the functionality of stand-alone packages. Although the database management system in an integrated package might not satisfy a database specialist, its capabilities might be entirely adequate for a small engineering firm that is trying to increase its productivity. The spreadsheet module of an integrated package might not have some of the additional features provided in LOTUS 1-2-3, such as matrix operations, but most integrated package spreadsheets have all the essential elements that make spreadsheets powerful engineering tools.

Integrated packages do increase the learning time over a single stand-alone package. On the other hand, you will most likely observe an economy of learning time to master an integrated package over the time required to master several separate packages. This is because of the consistency of the command structure and syntax among the modules of the integrated package.

Integrated packages require a large amount of computer memory. Some of them partially get around this by having only one module in memory at a time. However, to undertake practical-sized problems most packages need a computer with at least 640K memory. Older IBM PCs came standard with 256K of memory, but can be expanded to 640K. The current IBM PS/2 line starts with 640K memory, and the more sophisticated models will accept several megabytes of memory. Therefore, the problem of memory consumption by integrated packages is less important than it once was.

Many users do not need all the modules. There is no need to buy an expensive software package if you are not going to use its capabilities fully. However, you may find that the compatibility between word processing, spreadsheet, and graphics applications may be enough to justify using an integrated package. You would then have the communications module available if you ever needed to copy files from another computer or access any of the information services available. Also, you may find that as your applications become more involved, or use larger amounts of data, you will develop a need for the database module.

We will now undertake a discussion of the integrated package ENABLE, followed by several applications examples, to illustrate some of the capabilities of integrated packages.

8.4.4 ENABLE

ENABLE was developed by ENABLE Software Inc., Ballston Lake, New York. It was designed from the ground up as an integrated package, whereas many of the other integrated packages are secondary products of companies that also produce stand-alone packages. ENABLE has steadily gained in popularity since its introduction and has an extensive support network. It has undergone several revisions, with the latest version called Office Automation, or OA.

8.4.4.1 Introduction

ENABLE is a modular program. The modules share common commands and structures, but are loaded into the computer one at a time instead of simultaneously. This saves some memory. However, the most recent version of the program still requires at least 512K, with 640K recommended. The documentation consists of five manuals: a "System Overview" manual; individual manuals for the word processing, database management, and telecommunications modules; and a combination manual for the spreadsheet and graphics modules. A student edition, a standard edition, and a local area network (LAN) edition are available.

The student edition contains all the functionality of the commercial version, but is limited in the size of problem that it can handle. In the student edition, word processing documents are limited to 10 single-space pages, spreadsheets cannot be larger than 64 columns by 256 rows, and database management files can contain no more than 100 records. These limitations are not a problem in the educational environment. Student exercises that are significantly smaller than the limitations can be developed. All the applications in this chapter were developed using the student edition of ENABLE.

Two separate manuals that cover all the modules are provided with the student edition (Spezzano 1990). They contain most of the information that a student will need to learn ENABLE. However, a set of the full documentation should be available to answer detailed questions that may arise. A student applications workbook and instructor's manual are also available. The price of the student version is comparable to what a student would pay for a course textbook. John Wiley & Sons, the publisher of the student version of ENABLE, provides technical support to instructors who have adopted ENABLE, but it does not supply support to students. It is expected that students will ask their instructor for help.

User support for the commercial versions of ENABLE is provided in several forms. ENABLE Software Inc. employs a trained technical staff for assisting users with problems. Access to this group is provided through one of several levels of subscription support services, depending on the type of computing environment the user is in and the amount of support needed. The subscription support also includes an electronic mail service that can be accessed through the ENABLE communications module. User group support is also provided by ENABLE Software Inc.

Two types of training in the use of ENABLE are supported by ENABLE Software Inc. A network of independent, authorized training centers has been developed. Although these centers are independent of the parent company, they receive a considerable amount of support from ENABLE Software Inc. These centers provide a variety of services, including classroom instruction and seminars, either at the training center or at a client company; training publications on ENABLE; and instructional video courses. ENABLE Software Inc. has developed ENABLE/Learn, which consists of 15 lessons for presentation in 20 to 24 hours of classroom instruction. The package includes the student version of ENABLE, student workbooks and instructor's manual, and a set of overhead transparency masters.

8.4.4.2 Description of Modules

ENABLE consists of five line modules: word processing, spreadsheet, graphics, database management, and telecommunications. A Master Control Module controls the other modules. On-line help and interactive tutorials are available in the package. Commands can be entered through the ENABLE menu structure or through shorter series of keystrokes called expert commands.

Master Control Module   The Master Control Module is a supervisory, or control, program similar to those that run large mainframe computers. All system-wide functions are controlled by this module. These functions include macros, windows, file management, an interface with DOS, a menu generator, a high-level language interface, and system profiles. The macro facility allows you to automatically record keystrokes and commands as they are executed for use as macros. Macros can be edited through the word processing module and can be accessed through any of the other modules. The window facility allows you to open up to eight separate windows, each containing a different application. Data can be cut and pasted from one module to another through windows. You can use the built-in menu generator to develop menus for user control of specific applications. The high-level language interface is designed for the programmer who wants to use routines in a high-level programming language to perform database management functions in ENABLE.

Profiles allow you to customize your use of ENABLE. Profiles specify printer and plotter hardware and set text attributes for screen display. They establish which word processing specifications will be displayed. Through profiles you can set default page margin and justification characteristics. You can establish printing characteristics and set telecommunications parameters. Profiles specify default spreadsheet and database management operations. You can also use them to define what function certain keys will have. Since all profile settings, except those specifying hardware, have default values, it is not necessary to go beyond the hardware specifications when you are learning ENABLE. As you become more sophisticated in your use of ENABLE, you may want to modify your profile. Different profiles can be saved to use for certain types of applications. All applications in this chapter will use the default profile settings.

Word Processing   Word processing is not covered in detail in this text because of the numerous word processing packages that are available for microcomputers. A word processing package was probably one of the first microcomputer applications that you were introduced to in high school or college. Therefore, it is assumed that you will already be proficient in at least one word processing package before undertaking a computer applications course. Some of the characteristics of the ENABLE word processing module will be discussed here. The text examples will include simple word processing applications, mainly to illustrate integration of data from the other modules. You can compare the characteristics of the ENABLE word processing module to the package or packages that you are already familiar with. You will find that the ENABLE package has nearly all the functionality of a stand-alone package.

Some of the advanced features of the ENABLE word processing module include outlining functions and text block operations. Tables of contents and indexes can be created. A spelling checker is included with the package. It contains an 80,000-word standard dictionary. The user has the option of adding words to the dictionary and can specify that duplicate words be identified to help eliminate typographical errors. The most recent version includes a thesaurus. An add-on thesaurus is available for earlier versions. Extensive search and find as well as search and replace capabilities are included. These include verification of matches of text in the document with the designated search character string, plus marking of text for later deletion. You can use wild cards that will produce matches between the document text and search string based on portions of words. Wild cards are helpful in locating words or phrases whose spelling you are unsure of. You can also specify the number of unknown characters in a word. A calculator function is available for use when you are in the word processing mode. Files from several other popular word processing packages can be translated into the ENABLE format and included in ENABLE documents.

Spreadsheet   The commands and syntax of the ENABLE spreadsheet closely mimic those of LOTUS 1-2-3, as you will see when you study Example 8.12. ENABLE can read and write files in 1-2-3 and several other formats and can translate most 1-2-3 macros. ENABLE recalculation is generally slower than recalculation with LOTUS 1-2-3.

The allowable size of ENABLE spreadsheets has been somewhat smaller than in 1-2-3. In older versions you can have spreadsheets with up to 255 rows and 255 columns, or 65,025 cells. Six row/column configurations are available for sizing worksheets. The Office Automation version has increased the allowable spreadsheet size to 1,024 columns by 9,999 rows.

You can work with up to eight spreadsheets at once in separate windows. Data can be copied between active spreadsheets. In the OA version of ENABLE, you can link spreadsheets together by referencing cells in other worksheets. Spreadsheet auditing is facilitated by a command that highlights all cells that refer to a target cell. There is a step feature for manually or automatically recalculating a critical variable until a target value is reached.

More specific capabilities of the ENABLE spreadsheet will be illustrated in Example 8.12. Commands and functions are not reviewed here because they would be repetitious of the LOTUS 1-2-3 commands and functions described previously.

Database Management The prominent features of the database management module of ENABLE will be discussed here. More specific commands and applications will be discussed in Example 8.11.

In older versions of ENABLE, database files are limited to 65,000 records, with 254 fields per record. The OA version can handle over one billion records. The number is limited only by disk size. You can use a default data entry form or design your own form using the data entry form design function. Input data strips can be placed anywhere on the screen. You can use ENABLE's word processing commands to add labels, instructions for data entry, or other information. Automatic duplication of data from one input record to subsequent records can be specified for data that is the same for several records. You can also change any of the duplicated values, and the new value will be carried forward. Fields may contain formulas for calculation, or they can be used to access data from records in another database.

ENABLE will automatically open and update a file that is related to another file. It also supports data interchange with many file formats and with other ENABLE modules. Sorting of a database can be based on up to eight fields. Record delete and undelete options are available. Records are marked for deletion, but are not actually deleted from the database unless you specifically request the elimination of records. The records marked for deletion are ignored during all database operations, but can still be undeleted if you desire.

Database searches and record retrieval can be accomplished by specifying conditional criteria in a "where clause." The criteria can be based on defined or derived fields of the database. A derived field is a field whose value is calculated using values from one or more other fields and a formula, similar to the use of formulas in spreadsheets. Selected records within a database can be marked for further processing, graphing, or transfer to a spreadsheet or word processing document. A query by example facility is available to find particular records for editing or data addition. You enter on your data input form what information you want to match with the target record, and ENABLE forms conditional criteria and accomplishes the search.

Report generation can be accomplished through the ENABLE report generator. You can use the default columnar format, or you can design an output form by placing the output fields on the screen and adding text information using word processing commands. You can specify that a report field is derived and supply a formula to derive the value from data in the database. However, a field derived for a report cannot be used to select records for inclusion in the report. Reports can also be developed through the ENABLE procedural language.

A high-level language interface for BASIC, C, and PASCAL programs is available. This allows programmers to customize ENABLE applications.

Graphics The graphics capabilities of ENABLE are similar to those of LOTUS 1-2-3. The commercial version of ENABLE comes with an independent presentation graphics package called PERSPECTIVE that can be used to prepare two- and three-dimensional graphs using ENABLE data.

Communications The communications module of ENABLE allows you to use a modem and telephone lines to connect your microcomputer with another microcomputer, a minicomputer, or a mainframe computer. The software will make your microcomputer act like a terminal. It also gives you the option of sending or receiving files or data to and from the computer you are communicating with. Automatic sign-on can be accomplished using stored specifications. The software supports auto-dial modems that will automatically telephone the remote computer system and transmit the first two commands after the connection. If there are additional steps in the sign-on sequence, you take over manually. ENABLE also provides the option of full-screen data editing during data transmission.

8.4.4.3 Operation

The ENABLE software is started by designating the disk directory where the software is stored, which would be C:\EN200 if you use the default directory created by the ENABLE INSTALL program, and typing **ENABLE (,,,,A:)**. The (,,,,A:) designates that the disk in drive A: will be used for all data storage. Instructions are given in the ENABLE documentation for designating another directory on your hard disk as the data directory, if you so desire. It is best not to store data in the same directory as the software.

Figure 8.27a shows the initial ENABLE screen. If your computer has a calendar and time function that continually updates the date and time, and you are going to use the default profile of ENABLE, you can bypass this screen by pressing **[end]**. The [◄┘] is used by ENABLE to indicate that you should press **[enter]**.

After you leave the initial screen, the first menu of Figure 8.27b appears. Pressing **[esc]** will return you to the previous screen. The top-line menu presently in effect is enclosed within the inner block. If there are expert command options, they will appear along the bottom border of the screen. Any of the top-line menu options can be accessed by moving the cursor to that selection and pressing **[enter]** or by typing the first letter of the option. If there are multiple options that start with the same letter on the same menu, numbers are used to differentiate among them. The submenu topics for the option presently under the cursor are shown on the line below the active menu line. For example, the first portion of Figure 8.27b shows the submenu options for the USE SYSTEM menu option. The second portion of Figure 8.27b shows the menu screen after USE SYSTEM has been selected and the active menu line has moved down to its submenu. The sub-submenu options line shows the selections for word processing, since the cursor is presently over the WORD PROCESSING option. The sequence of commands for the different modules will be discussed as they are used in the examples.

8.4.4.4 Applications

In the examples in this text, we will use default settings as much as possible, except in the case of hardware definition. User profiles were discussed in general in the section on the Master Control Module. Example 8.11 will illustrate the use of the database management functions of ENABLE to work with the industrial production data of Examples 8.10 and 7.1. All the data will be included in a single database file for this example. You can compare the methods of working with data in a true database management system with the methods used through a FORTRAN program or the database functions of LOTUS 1-2-3. Example 8.12 illustrates a spreadsheet and graphics solution to the particle trajectory problem of hitting a window solved with a FORTRAN program in Example 4.14 and with TK SOLVER in Example 8.2. Use of the ENABLE windows feature to integrate data from the spreadsheet, graphics, and word processing modules will be illustrated.

FIGURE 8.27   Initial ENABLE screen and menus.

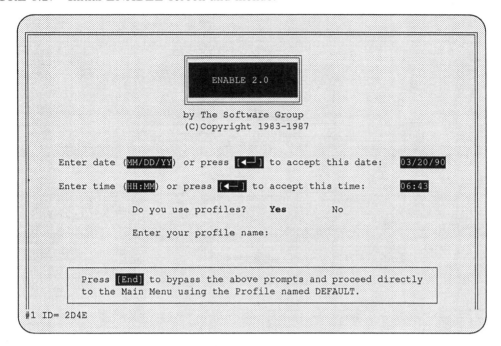

(*a*) Initial screen.

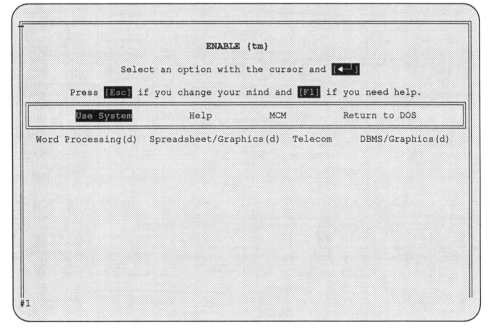

(*b*) Initial menu options.

FIGURE 8.27   (Continued)

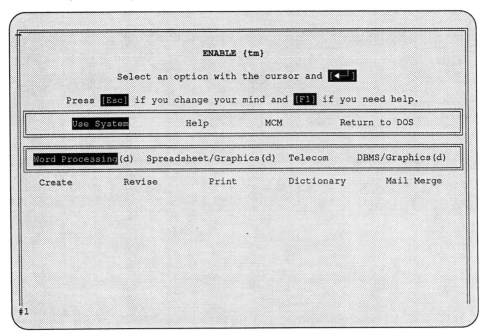

(*b*) Initial menu options (continued).

**EXAMPLE 8.11   □   Industrial Production Data
(Database Design, Entry, Sorting, Searching, and
Report Generation)**

STATEMENT OF PROBLEM

Develop an ENABLE application to create, enter, and work with the industrial production database presented in Examples 7.1 and 8.10. Use a single file for the employee database. The database should be entered in employee identification number order. Enter the necessary commands to sort the database into alphabetical order. Create a report form for extracting data for a particular employee and generating summary production and defect data for that employee. Also, develop a method to calculate summary data for all employees and to extract information based on the quality of work that employees are producing.

MATHEMATICAL DESCRIPTION

Production and defect data will be summed across each record to develop the summary statistics for each employee. The production-to-defect ratio will be used as a measure of the quality of employee work.

ALGORITHM DEVELOPMENT

Input/Output Design

The default ENABLE database input form will be used. This will place the defined data fields one under the other on the screen, with the field name to the left as a label. Data fields will be defined as they were in Example 8.10. The original database and the sorted database will be output using the default columnar format. Report forms will be developed for the individual data query and summary as well as for the production quality query. The format of the reports will closely approximate the format of the spreadsheet output of Example 8.10.

Computer Implementation

ENABLE's menu-driven structure will be used to develop all the applications of this example. The report forms will be saved as files so they can be used any number of times to produce reports based on user input criteria.

PROGRAM DEVELOPMENT

The first screen of Figure 8.28a shows the main menu with DBMS/GRAPHICS highlighted. After selecting DBMS/GRAPHICS, the second screen of Figure 8.28a would appear, with DESIGN highlighted. Since you first need to create a database definition, you press [enter] twice, and the database definition screen of Figure 8.28b will appear. The definition of the form of the database will be stored in a file separate from the actual database. A message on the screen indicates this. ENABLE distinguishes between the different kinds of files by the extenders placed on the file names. After entering the database name, **EX8_11**, the computer would respond with FILE NOT FOUND. You would respond by selecting the option NEW FILE, and the second screen of Figure 8.28b would appear. The description will be entered as Industrial Production Data. No entries will be made in the default input and output fields because the ENABLE default is being used. You can still define your own output forms later on.

After you complete the initial database definition screens, the database definition screen of Figure 8.29a will appear, asking for the first field name. **ID** is entered as the first field name, and the field description of Figure 8.29b will appear. The QUICK method of field definition will be used for this database. The DETAILED method will be illustrated for a later part of the example. After selecting QUICK, the data type and field length have to be entered. The screen of Figure 8.29a will again appear, but now with the field definition for ID shown, as in the first screen of Figure 8.29c. The source of data is listed as Keyboard, which means that you will enter the data into the database from your keyboard. The lengths for fields in this example will be the same as those used for Example 8.10. After repeating the definition process for all the fields, the second screen of Figure 8.29c will appear. The field definitions have scrolled up off the screen as more were entered. Note that the production and defect data for each month are defined one after the other, rather than defining all the production fields first, followed by all the defect fields. This order will facilitate addition of data for each month throughout the year. To save the database definition, you have to press the [F10] key to display the top-line menu and select SAVE.

Not all the characteristics of the fields are displayed in the database definition screen of Figure 8.29. You can select the PRINT option of the top-line menu to print out a complete description of each field. Before you do this the first time you have to use your operating system to copy the file DBMSDEF.$RF from the ENABLE tutorial disk to the disk or directory in which you are storing the database files. This is most easily accomplished before you start up ENABLE, but if you did not do it then, you can use the DOS Window through the Master

(EXAMPLE 8.11 □ **Industrial Production Data (Database Design, Entry, Sorting, Searching, and Report Generation)** □ **Continued)**

FIGURE 8.28 Initial database screens.

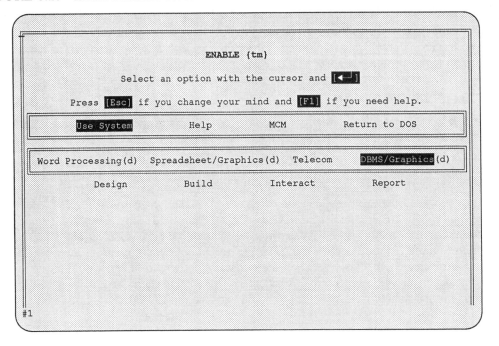

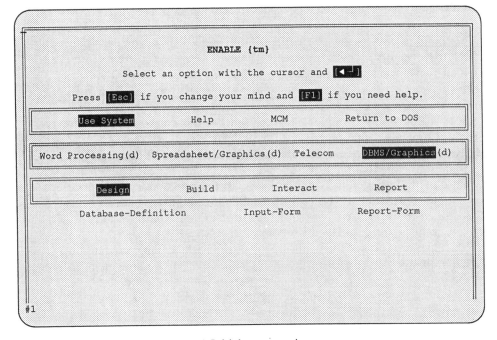

(*a*) Initial menu options.

FIGURE 8.28 (Continued)

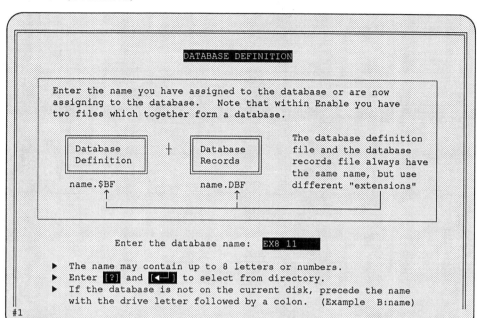

(b) Initial database definition screens.

(EXAMPLE 8.11 □ Industrial Production Data (Database Design, Entry, Sorting, Searching, and Report Generation) □ Continued)

FIGURE 8.29 Database definition.

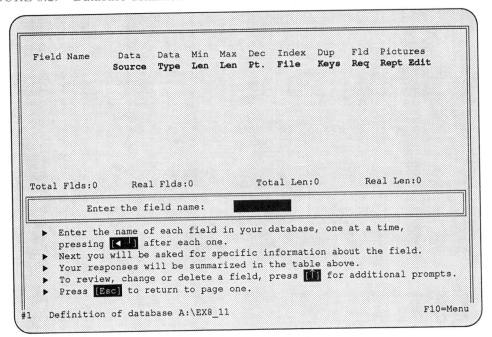

(*a*) Database definition screen.

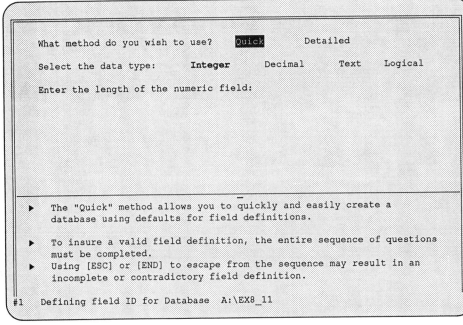

(*b*) Field definition screen.

FIGURE 8.29   (Continued)

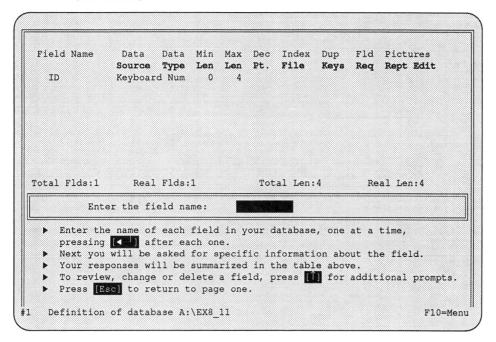

(c) Database definition screens with fields defined.

(EXAMPLE 8.11   □   **Industrial Production Data
(Database Design, Entry, Sorting, Searching, and
Report Generation)   □   Continued)**

FIGURE 8.29   (Continued)

```
DATABASE DESCRIPTION: INDUSTRIAL PRODUCTION DATA
DEFAULT INPUT FORM:
DEFAULT REPORT FORM:
-----------------------------------------------------------------
FIELD:  ID              METHOD:  Detailed    SOURCE OF DATA:  Keyboard
REQUIRED?: No  INDEXED?: Yes  ALLOW DUPLICATES?: No   INDEX FILE: EX10_1
DATA TYPE    NUMERIC TYPE    TEXT TYPE       LOGICAL TYPE    OTHER TYPE
Integer
MINIMUM LENGTH:  0  MAXIMUM LENGTH:  4  DECIMAL PLACES:
MINIMUM AND MAXIMUM VALUES?:  No
LIST OF ACCEPTABLE VALUES?:  No
EDIT PICTURE                            REPORT PICTURE


REPORT HEADING:
ERROR MESSAGE:
-----------------------------------------------------------------
FIELD:  FIRST           METHOD:  Quick      SOURCE OF DATA:  Keyboard
REQUIRED?: No  INDEXED?: No  ALLOW DUPLICATES?: No   INDEX FILE:
DATA TYPE    NUMERIC TYPE    TEXT TYPE       LOGICAL TYPE    OTHER TYPE
Text
MINIMUM LENGTH:  0  MAXIMUM LENGTH:  11  DECIMAL PLACES:
MINIMUM AND MAXIMUM VALUES?:  No
LIST OF ACCEPTABLE VALUES?:  No
EDIT PICTURE                            REPORT PICTURE


REPORT HEADING:
ERROR MESSAGE:
-----------------------------------------------------------------
FIELD:  MI              METHOD:  Quick      SOURCE OF DATA:  Keyboard
REQUIRED?: No  INDEXED?: No  ALLOW DUPLICATES?: No   INDEX FILE:
DATA TYPE    NUMERIC TYPE    TEXT TYPE       LOGICAL TYPE    OTHER TYPE
Text
MINIMUM LENGTH:  0  MAXIMUM LENGTH:  2  DECIMAL PLACES:
MINIMUM AND MAXIMUM VALUES?:  No
LIST OF ACCEPTABLE VALUES?:  No
EDIT PICTURE                            REPORT PICTURE


REPORT HEADING:
ERROR MESSAGE:
-----------------------------------------------------------------
FIELD:  LAST            METHOD:  Quick      SOURCE OF DATA:  Keyboard
REQUIRED?: No  INDEXED?: No  ALLOW DUPLICATES?: No   INDEX FILE:
DATA TYPE    NUMERIC TYPE    TEXT TYPE       LOGICAL TYPE    OTHER TYPE
Text
MINIMUM LENGTH:  0  MAXIMUM LENGTH:  11  DECIMAL PLACES:
MINIMUM AND MAXIMUM VALUES?:  No
LIST OF ACCEPTABLE VALUES?:  No
EDIT PICTURE                            REPORT PICTURE
```

(*d*) Printout of database definition.

Control Module to copy the file. The procedure for this can be found in the ENABLE documentation. The first page of the database definition printout is shown in Figure 8.29d. To exit from the database definition submodule, select QUIT from the top-line menu and confirm with YES, and you will be returned to the main menu.

From the main menu, you have to start all over with USE SYSTEM, DBMS/ GRAPHICS, and BUILD to specify that you want to build, or add data to, your database. The ADD screen of Figure 8.30a will appear. You would enter the database name as EX8_11. Press **[enter]** in response to the USING FORM prompt. This will indicate that you are using the default input form, and the STANDARD INPUT FORM screen shown in Figure 8.30b will appear. The nine records of industrial production data from Example 8.10 will be entered into the database. You can enter data in the fields of the record in any order. The cursor can be moved from one field to another using the up or down arrow keys. If you completely fill a field, the cursor automatically moves to the next field. If the data only partly fill the field, you can press either **[enter]** or one of the arrow keys to enter the data into the field and move to another field. Entered values can be changed by moving the cursor back to the field and retyping the entry. Once you have entered all that you want to into the record, press the **[F5]** function key to save the record and move on to the next record. You can go back later to edit or add data to saved records.

Once you have completed entering all nine records, use **[F10]** to bring up the top-line menu and select QUIT. This returns you to the ADD DATA screen. Using **[F10]** to bring up the top-line menu from this screen will provide the REPORT option. Selecting REPORT will bring up the REPORT screen shown in Figure 8.31a. For this initial report we will be using the default columnar report, so we do not have to enter anything in the USING FORM field. The INDEX field allows you to enter the field name for an indexed variable. This will cause the records to be output in the order of the index, rather than the order of entry in the database. For example, we could have indexed the last name field and output the records in alphabetical order of the last names. However, you can specify only one index variable on the report form, so you could not alphabetize on both the last and first names. The method for doing that will be illustrated shortly. The report can be sent to the screen, to a disk file, or to the printer, depending on the TO option that you choose. The WHERE field is for entering conditional criteria for selecting records for the report. Since we want to output all records, this is left blank. The FIELDS field designates what fields should be included in the columnar output. If you leave this blank, the default output form will try to print more fields than can appear on an 80 column output, and the data will appear on the page in a confusing array. Therefore, it is important that you enter only as many field names as can be printed on a single page. Figure 8.31a shows the fields ID, FIRST, MI, LAST, PJAN, DJAN, PFEB, DFEB, PMAR, and DMAR identified. These fields will fit across one page. The other fields will be output in separate reports. The field names are shown in the box near the bottom of the screen in case you cannot recall them. The operators that can be used in WHERE clauses are also shown. These will be used in a later report. Once you enter the TITLE, the screen shown in Figure 8.31b will appear. This lists print options. We will use the default options, so pressing **[alt/F2]** will cause the report to print.

Figure 8.32 shows a composite of the three reports required to output the

(EXAMPLE 8.11 □ Industrial Production Data (Database Design, Entry, Sorting, Searching, and Report Generation) □ Continued)

FIGURE 8.30   Adding data to the database.

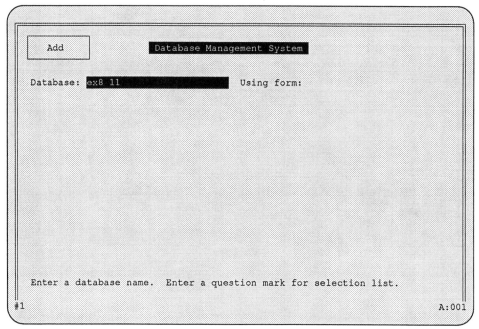

(*a*) ADD data form.

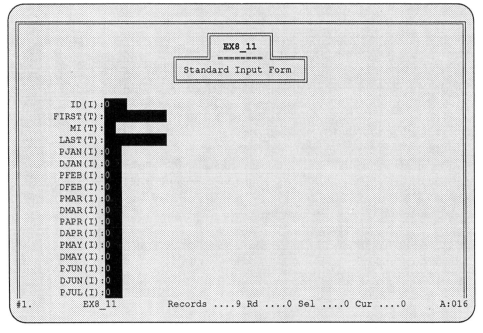

(*b*) Standard input form.

FIGURE 8.31 Default report generation.

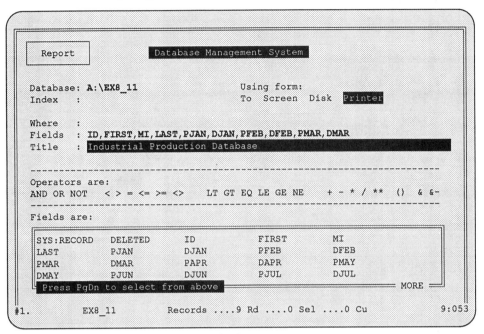

(*a*) REPORT screen.

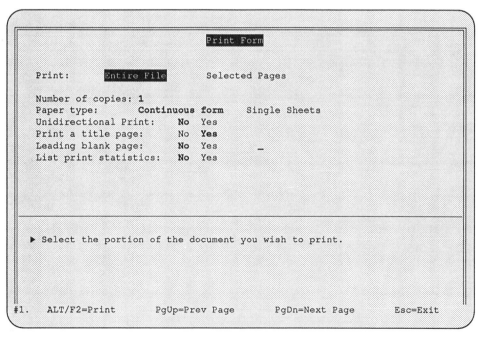

(*b*) Print form screen.

(EXAMPLE 8.11   □   **Industrial Production Data
(Database Design, Entry, Sorting, Searching, and
Report Generation)**   □   **Continued)**

FIGURE 8.32   Default report.

```
Header and Page Break
Industrial Production Database
ID.. FIRST...... MI LAST.......  PJAN  DJAN  PFEB  DFEB  PMAR  DMAR

1001 James        S. Johnson      155   50   170   45   190   62
1002 Brenda       A. Thomas       130   20   132   25   110   35
1003 Pamela       B. Dodd         145   24   153   24   115   36
1004 John         T. Jones        143   25   133   22   127   34
1005 Keith        B. Reynolds     125   26   140   33   122   33
1006 Elizabeth    R. Thomas       160   10   155   15   153   13
1007 Rebecca      S. Smith        130   33   125   24   135   28
1008 Joan            Carmen       132   31   115   23   132   26
1009 Thomas       K. Flint        100    5   110    4   108    6




** End of report **  Arrow keys to scroll sideways, RETURN to end.
#1.        EX8_11        Records ....9 Rd ....9 Sel ....9 Cur ....9      9:053
```

```
Header and Page Break
Industrial Production Database
ID.. PAPR DAPR PMAY DMAY PJUN DJUN PJUL DJUL PAUG DAUG PSEP DSEP

1001  188   25  175   35  166   44   88   40  175   21  165   36
1002  129   36  108   38  142   21  111   18  112   22  115   41
1003  135   31  100   25  134   32  120   33  133   24  130   39
1004  140   26   99   26  131   27  132   34  128   31  110   35
1005  167   27  190   29  122   35  137   37  165   28  127   29
1006  182   16  145   14   78    9  160   18  170   19  165   20
1007  123   24  157   37  133   22  128   32  143   26  154   22
1008  135   25  135   28  167   25  131   21  121   22  158   28
1009  115    3  109    5  111    7  112    4  115    3  118    6




** End of report **  Arrow keys to scroll sideways, RETURN to end.
#1.        EX8_11        Records ....9 Rd ....9 Sel ....9 Cur ....9      9:053
```

FIGURE 8.32   (Continued)

```
┌──────────────────────────────────────────────────────────────────────────┐
│ Header and Page Break                                                      │
│ Industrial Production Database                                             │
│ ID..  POCT  DOCT  PNOV  DNOV  PDEC  DDEC                                    │
│                                                                            │
│ 1001  158    43   173    36   180    28                                    │
│ 1002  175    77   124    22   126    31                                    │
│ 1003  109    21   154    18   153    27                                    │
│ 1004  127    19   136    16   143    24                                    │
│ 1005  143    18   132    18    33     1                                    │
│ 1006  180    18   190    16   200    32                                    │
│ 1007  154    22   142    12    88    33                                    │
│ 1008  129     7   141    21    70    38                                    │
│ 1009  108     4   109     5   106     7                                    │
│                                                                            │
│                                                                            │
│                                                                            │
│                                                                            │
│                                                                            │
│                                                                            │
│ ── End of report ──  Arrow keys to scroll sideways, RETURN to end.         │
│ #1.         EX8_11          Records ....9 Rd ....9 Sel ....9 Cur ....9  9:053 │
└──────────────────────────────────────────────────────────────────────────┘
```

entire database. The employee identification number is output with each report to identify the data records.

You should inspect the reports of Figure 8.32 to check that the data have been entered correctly. You can now develop the commands for sorting the database. Pressing **[F10]** will bring up the top-line menu. Selecting the DISPLAY OPTIONS option and the SORT suboption will bring up the SORT screen shown in Figure 8.33a. You can use this screen to sort the database on up to eight fields. Since we want to alphabetize on the last and first names, we would enter LAST,A in FIELD1, and FIRST,A in FIELD2. The A indicates ascending order. Moving the cursor to FIELD3 and pressing **[enter]** without typing anything in the field will execute the sort. A file named EX8_11.SS, which contains the sorted database, will be generated. The original database file is not changed by the sort procedure. The SS extender stands for select set. It is not itself a database, but just a file with pointers to various records so that the records will be output in the desired order. The select set is only a temporary file. If you do any other procedure that would select a subset of the database, a new select set will be formed and will replace the old one.

Pressing **[F10]**, selecting DISPLAY OPTIONS, and then selecting REPORT will bring up the REPORT screen. The default file indicated will be EX8_11.SS. Figure 8.33b shows the first page of the report for the sorted database. You can confirm that the sort accomplished the same end as the LOTUS 1-2-3 sort of Example 8.10. The big difference is that the order of the ENABLE database has not been changed, whereas the actual database was sorted with LOTUS.

Pressing **[F10]** and selecting QUIT will return you to the main ENABLE menu. Selecting USE SYSTEM, DBMS/GRAPHICS, and INTERACT brings up the screen shown in Figure 8.34a. From this menu you can access several of

(EXAMPLE 8.11 □ Industrial Production Data (Database Design, Entry, Sorting, Searching, and Report Generation) □ Continued)

FIGURE 8.33   Sorting database.

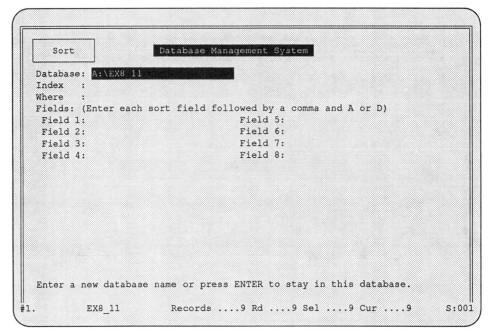

```
┌──────────────┬────────────────────────────────────┐
│    Sort      │     Database Management System      │
│              └────────────────────────────────────┘
│  Database: A:\EX8_11
│  Index   :
│  Where   :
│  Fields: (Enter each sort field followed by a comma and A or D)
│   Field 1:                      Field 5:
│   Field 2:                      Field 6:
│   Field 3:                      Field 7:
│   Field 4:                      Field 8:
│
│
│
│
│   Enter a new database name or press ENTER to stay in this database.
│ #1.      EX8_11        Records ....9 Rd ....9 Sel ....9 Cur ....9      S:001
└──────────────────────────────────────────────────────────────────────────┘
```

(*a*) SORT screen.

```
Header and Page Break
Industrial Production Database
ID..  FIRST......  MI  LAST.......  PJAN  DJAN  PFEB  DFEB  PMAR  DMAR

1008  Joan             Carmen        132    31   115    23   132    26
1003  Pamela       B.  Dodd          145    24   153    24   115    36
1009  Thomas       K.  Flint         100     5   110     4   108     6
1001  James        S.  Johnson       155    50   170    45   190    62
1004  John         T.  Jones         143    25   133    22   127    34
1005  Keith        B.  Reynolds      125    26   140    33   122    33
1007  Rebecca      S.  Smith         130    33   125    24   135    28
1002  Brenda       A.  Thomas        130    20   132    25   110    35
1006  Elizabeth    R.  Thomas        160    10   155    15   153    13



  ** End of report **  Arrow keys to scroll sideways, RETURN to end.
 #1.      EX8_11.SS      Records ....9 Rd ....9 Sel ....9 Cur ....9      9:053
```

(*b*) Report for sorted database.

FIGURE 8.34 Extracting employee data from the database.

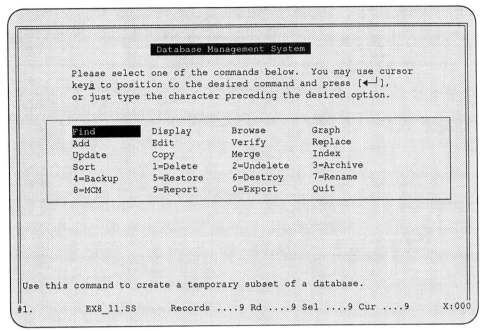

(a) Interactive menu.

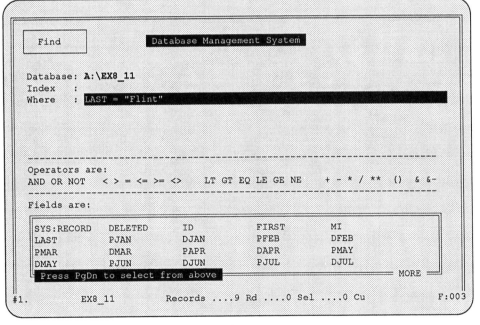

(b) FIND screen.

(EXAMPLE 8.11 □ Industrial Production Data (Database Design, Entry, Sorting, Searching, and Report Generation) □ Continued)

FIGURE 8.34 (Continued)

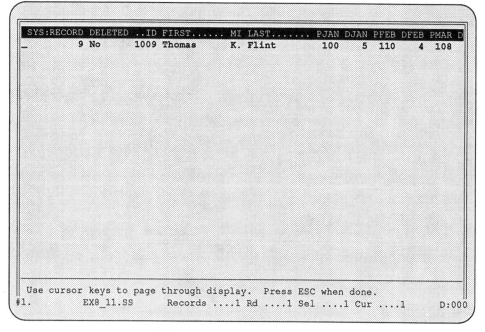

```
┌──────────────────────────────────────────────────────────────┐
│  ┌──────────┐        ┌─────────────────────────────┐         │
│  │ Display  │        │ Database Management System  │         │
│  └──────────┘        └─────────────────────────────┘         │
│                                                              │
│   Database: A:\EX8_11.SS                                     │
│   Index   :                                                  │
│   Where   :                                                  │
│   Fields  :                                                  │
│                                                              │
│                                                              │
│                                                              │
│                                                              │
│                                                              │
│   Enter a new database name or press ENTER to stay in this database. │
│  #1.        EX8_11.SS       Records ....9 Rd ....0 Sel ....0 Cur ....0     D:001 │
└──────────────────────────────────────────────────────────────┘
```

(*c*) DISPLAY screen.

```
┌──────────────────────────────────────────────────────────────┐
│ SYS:RECORD DELETED ..ID FIRST...... MI LAST....... PJAN DJAN PFEB DFEB PMAR D │
│  _         9 No     1009 Thomas      K. Flint      100   5  110    4  108    │
│                                                              │
│                                                              │
│                                                              │
│                                                              │
│                                                              │
│ Use cursor keys to page through display.  Press ESC when done. │
│ #1.        EX8_11.SS       Records ....1 Rd ....1 Sel ....1 Cur ....1     D:000 │
└──────────────────────────────────────────────────────────────┘
```

(*d*) Display of selected records.

the database management functions. We will use FIND and DISPLAY to extract information on a particular employee from the database. Placing the cursor over FIND and pressing **[enter]** brings up the screen shown in Figure 8.34*b*. Typing LAST = ''Flint'' in the WHERE field and pressing **[enter]** will direct ENABLE to search the database for any records with a LAST field containing Flint. The find operation will produce a new select set. Therefore, the database sort has now been removed.

Pressing **[esc]** will return you to the interaction menu, from which you can select DISPLAY. This will bring up the screen shown in 8.34*c*. Note that the default file is EX8_11.SS. Pressing **[enter]** three times will bypass all the options and display all the fields of the selected records. The screen display would now look like Figure 8.34*d*. The right arrow key can be used to scroll over to the fields that are beyond the right edge of the screen display. Pressing **[esc]** will bring you back to the interactive menu, and selecting QUIT will bring you back to the main ENABLE menu.

In order for you to develop the summary production and defect data for a selected employee, it is necessary to sum the production and defect data for each month for that employee. These summary data are shown in Figure 8.25*c* of Example 8.10. In the ENABLE database management system, the summing can be accomplished by defining a derived field. When you develop an output form, you can include derived fields in it. We will use the menu options

```
USE SYSTEM
     DBMS/GRAPHICS
          DESIGN
               REPORT-FORM
```

to enter the report form design module. You will be prompted to enter the file name for the form. You can call the form by the same name as the database, because it will have a different extender. After entering EX8_11, select NEW FILE, and enter EX8_11 for the name of the database file. After you enter the report description as ''Extracting Individual Employee Data,'' the screen of Figure 8.35 will appear. It provides instructions on how to design the output form. You can enter labels and other text information as if you were developing a word processing document. When you want to place an output field, you move the cursor to where you want the field to start and press **[shift/F9]**. Figure 8.36*a* shows the report form with a title and four labels entered. The output fields for the identification number, last name, and first name have also been defined. The cursor is now two spaces to the right of the colon after Middle Initial. When **[shift/F9]** is pressed, the screen of Figure 8.36*b* appears. Note that an inner window has been placed around a portion of the output form, and the rest of the screen is filled with instructions and information. You would enter the field name MI at the prompt. Since this is a field that has been defined in the database definition, the screen of Figure 8.36*c* appears. The default report picture for the MI field should be XX, designating two characters for the display field.

The field for total production will be called TOTPROD. When you enter this field name, ENABLE assumes that the field is to be a derived field because the field definition does not exist in the database definition. The screen of Figure 8.36*d* would appear to request additional information on the derived field. The formula in this case is the summation of each month's production data. The result data type is numeric, the number of decimal places is zero (the value is an

(EXAMPLE 8.11 □ **Industrial Production Data (Database Design, Entry, Sorting, Searching, and Report Generation)** □ **Continued)**

FIGURE 8.35 Report form instructions.

```
        You are now ready to revise or create your report form.

  1.  Lay out your report as you would have it appear on the screen
      or the printed page.  All of Enable's word processing features are
      available to do this.

  2.  Position the cursor where you want database information to
      be shown.  Press [Shift] and [F9] to tell Enable to "Put It Here".

  3.  Type a field name and answer the questions asked about the field.

  4.  Repeat the process for each field in your report form.  If you wish
      to delete a field after you have positioned it on the screen, or if
      you wish to revise a field's options, press [F10] for a menu.

  5.  When you're finished, press [F10] to save the form using the
      Save options or to return to the Main Menu.

  6.  Press any key when you're ready to begin and Enable will display a
      blank screen or your previously created report form.
          Note: If you wish to use headers and footers in your report,
          create a word processing document and use the Report Language.
#1   Definition of report form A:\EX8_11                          F10=Menu
```

integer), and the length of the derived field is four. NNNN is the report picture, which defines a numeric value with a displayed field length of four. The derived field for total defects can be defined in a similar manner. The third derived field, the production/defect ratio, is the ratio of the first two derived fields. Since all lines on the report form will be output each time the report is output, you need to remove any extra lines after the final output field. Otherwise, you will have unwanted blank lines in your report when multiple records are to be output. The blank lines can be removed by placing the cursor two lines below the last entry and repeatedly pressing **[alt/F3]** until the solid line at the bottom of the screen moves up to the cursor. After you have completed the output form, you can save it and call the REPORT screen up. The completed REPORT screen is shown in Figure 8.37*a*. We make sure that EX8_11 is displayed in the USING FORM: field. The WHERE field is again searching for records with the last name of Flint. After you complete the REPORT window, a whole series of PRINT FORMS are displayed. These are shown in Figure 8.37*b*. You can bypass all the PRINT FORMS and go directly to printing by pressing **[alt/F2].** The printed report is shown in Figure 8.37*c*.

Now it is time to use a conditional query to extract information from the database on the quality of employee work. In Example 8.10, we searched for a combination of employees who produced more than 1,500 units and also had a production/defect ratio of less than 5. The easiest way to accomplish this would

FIGURE 8.36   Report form definition.

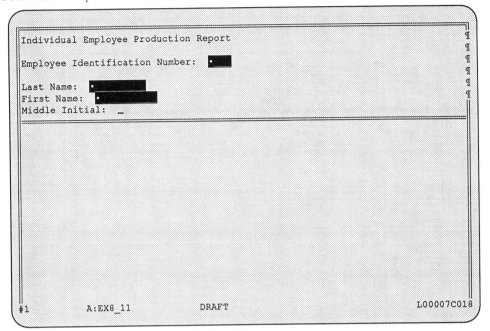

(*a*) Initial entries.

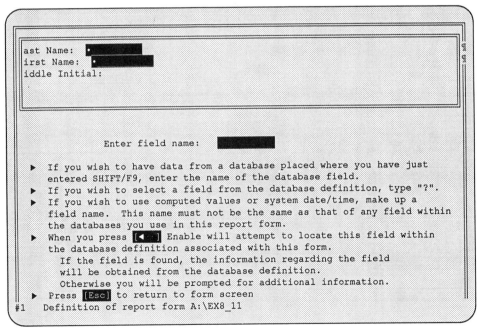

(*b*) Field name definition instructions.

(EXAMPLE 8.11   □   Industrial Production Data (Database Design, Entry, Sorting, Searching, and Report Generation)   □   Continued)

FIGURE 8.36   (Continued)

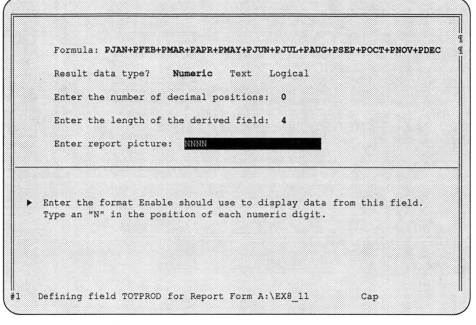

(c) Report picture definition.

(d) Field information for derived field.

FIGURE 8.37   Report using form EX8_11.

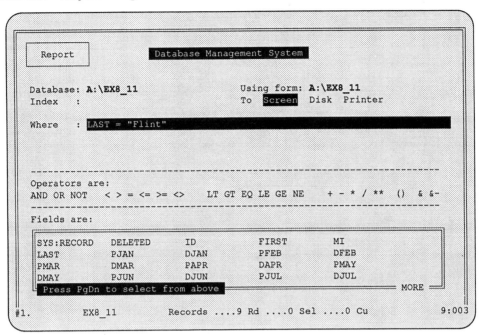

(*a*) REPORT screen.

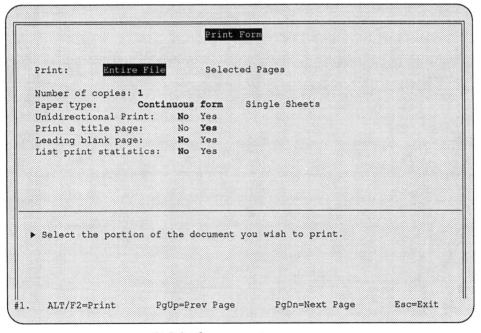

(*b*) Print form screens.

(EXAMPLE 8.11 □ Industrial Production Data (Database Design, Entry, Sorting, Searching, and Report Generation) □ Continued)

FIGURE 8.37   (Continued)

```
                              Print Form

    Date picture:      Standard   Military   Numerical   European Numerical

    Should Table of Contents entries be copied into text:    No   Yes

    Select style of Unit of Standard numbers:      1   2=A   3=a   4=I   5=i
    Select style of Chapter numbers:               1   2=A   3=a   4=I   5=i
    Select style of Section numbers:               1   2=A   3=a   4=I   5=i
    Select style of Minor Section numbers:         1   2=A   3=a   4=I   5=i




  ▶ This option determines the format a date appears in when you use
    the "%DATE" function in your text.  For example:
    Standard:  August 22, 1983          Military: 22 August 1983
    Numerical: 8/22/83          European Numerical: 22/8/83

#1.   ALT/F2=Print        PgUp=Prev Page        PgDn=Next Page        Esc=Exit
```

```
                              Print Form

    Enter printer code:  G

    Type of printer interface:  Parallel    Serial   File

    Which parallel port are you using?  1=LPT1:   2=LPT2:   3=LPT3:






    0 =Not listed           | Y =HP Laser Jet(Y Font)
    AB=Hermes               | AF=Diablo 630 (ECS)
    AC=IBM ProPrinter XL    |
    AC=IBM ProPrinter       |
    F =Epson RX80, RX100    |
    3 =Diablo 630           |
    G =HP Laser Jet         |
#1.   ALT/F2=Print        PgUp=Prev Page        PgDn=Next Page        Esc=Exit
```

(*b*) Print form screens (continued).

FIGURE 8.37 (Continued)

```
                        Page Form


        11.00 ... Length
        8.00  ... Width
        6     ... Lines per inch

        6     ... Top margin (number of lines)
        6     ... Bottom margin (number of lines)
        2     ... Blank lines between header and text
        2     ... Blank lines between text and footer

        0     ... Left margin offset
        0     ... Line spacing (0-9)


    ▶ Enter page length in inches.




#1.   ALT/F2=Print      PgUp=Prev Page      PgDn=Next Page      Esc=Exit
```

```
                        Page Form


    Should pages be numbered:   Yes   No

    Location of Footnotes:  Bottom of each page    End of document

    Font:  Pica   Elite   Compressed

    Should proportional spacing be used?  Yes   No

    Should letter quality be used?  Yes   No



    ▶ Pages can be automatically numbered in final form or at print time.




#1.   ALT/F2=Print      PgUp=Prev Page      PgDn=Next Page      Esc=Exit
```

(*b*) Print form screens (continued).

(EXAMPLE 8.11 □ **Industrial Production Data (Database Design, Entry, Sorting, Searching, and Report Generation)** □ **Continued)**

FIGURE 8.37   (Continued)

```
Individual Employee Production Report

Employee Identification Number:   1009

Last Name:  Flint
First Name:  Thomas
Middle Initial:  K.

Total Production: 1321
Total Defects:        59
PRODUCTION/DEFECT RATIO: 22.39
```

(c) Printed report.

be to set up a WHERE clause for TOTPROD > 1500 AND RATIO < 5 in your REPORT screen. However, you cannot do this with the existing database because TOTPROD and RATIO are derived variables for the output form. If you examine all the fields in the bottom window, you will find that only database-defined fields can be used in WHERE clauses. The SYS:RECORD and DELETED fields are fields included in every database by ENABLE. The first is a sequential index of fields in the database, while the second indicates records that have been marked for deletion. TOTPROD, TOTDEF, and RATIO do not appear in the list in the window.

You could try to put the original formulas for TOTPROD, TOTDEF, and RATIO in the WHERE clause, for example

```
PJAN+PFEB+PMAR+PAPR+PMAY+PJUN+PJUL+PAUG+PSEP+POCT+PNOV+PDEC>1500 AND
(PJAN+PFEB+PMAR+PAPR+PMAY+PJUN+PJUL+PAUG+PSEP+POCT+PNOV+PDEC)/
(DJAN+DFEB+DMAR+DAPR+DMAY+DJUN+DJUL+DAUG+DSEP+DOCT+DNOV+DDEC)<5
```

However, you would find that the available field for the WHERE clause is not large enough to hold the whole formula. Its maximum length is only 148 characters. You could produce two reports and manually compare the selected records for production greater than 1,500 with those for production/defect ratio less than 5, but this would be inefficient and cumbersome. There is a rather simple solution to this problem, which illustrates some of the powers of a database management system.

The solution is to create a new database definition that includes TOTPROD, TOTDEF, and RATIO as defined fields. You can do this without actually creating a new database. The new database definition can access the records from the old database. To link two database definitions, there must be an indexed field in the database from which data are to be extracted so that the new database definition can ascertain from which record to extract data. The employee identification number is the obvious choice since it is unique for each employee even if two employees have the same last name. When we originally defined the EX8_11 database, we could have defined ID as an indexed field. We can also do that now by using the following series of commands:

```
USE SYSTEM
    DBMS/GRAPHICS
        INTERACT
            INDEX
```

to bring up the INDEX screen shown in Figure 8.38. Identify the field name to be indexed as ID, and name the index file EX8_11. Any other index files would have to be given a unique name. Since the identification number is unique, specify that duplicates are not allowed. The ID index file is now created. Use QUIT to return to the main ENABLE menu, and use

```
USE SYSTEM
    DBMS/GRAPHICS
        DESIGN
            DATABASE-DEFINITION
```

FIGURE 8.38   INDEX screen.

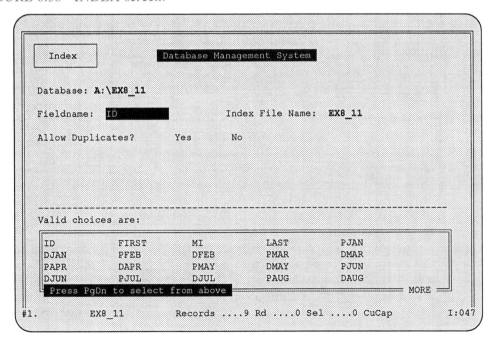

(EXAMPLE 8.11 □ Industrial Production Data (Database Design, Entry, Sorting, Searching, and Report Generation) □ Continued)

to bring up the database definition screen. This database definition will be called EX8_11A, the description will be Industrial Production Summaries, and the default output form will be EX8_11A. To link the field definitions with fields of another database, we need to use the detailed field definition method. Figure 8.39 illustrates the definition process. Figure 8.39*a* shows the field definition screen after entry of the first three fields. Note that the data source for ID is Keyboard. while the source for LAST and FIRST is External. ID will be the only field that is entered into this database. It is the field that is used to link the EX8_11 database with the EX8_11A database. Figures 8.39*b* and 8.39*c* show the detailed definition process. After you have entered MI for the next field name in the database definition screen, you are asked whether you want to use QUICK or DETAILED field definition. After you have selected DETAILED, the screen of Figure 8.39*b* will appear. The next line asks if you want to copy an existing field definition. You can copy a field definition from any database definition that has been saved to a disk. To illustrate this, we will copy the field definition for MI from EX8_11. If you select YES, you are prompted to enter the name of the database definition to copy from (EX8_11) and the name of the field to copy from (MI). It should be noted that the field characteristics, not the field name, are copied. You have to continue with the detailed field definition, because all that is copied from the old database definition is the type of data and the field length. This is because we used the QUICK method to define the fields in the original definition. The field will not be indexed since none of the fields in the new definition has to be indexed to accomplish our goal. However, the field is required. The source of data is another database, so at this prompt select ANOTHER DATABASE.

The screen of Figure 8.39*c* will now appear. Enter **EX8_11** as the database, **MI** as the lookup field in the other database, **ID** as the indexed link field in the other database, and **ID** as the link field in this database. ID can also be entered as the columnar report heading and the error message field can be left blank. This means that whenever MI is used in the EX8_11A database, the data will be looked up in the MI field of the corresponding record of EX8_11. It will be used in any reports generated with EX8_11A, but will not actually be copied into EX8_11A/.

The next three fields for TOTPROD, TOTDEF, and RATIO will be derived fields. Therefore, the definition process is slightly different. The first screen to appear will be the same as Figure 8.39*b*, but this time you will not copy a field definition from another database. You will select DERIVED for the source of data rather than ANOTHER DATABASE. This choice will bring up the screen of Figure 8.39*d*. The formula and other field definition information would be entered as shown.

Figure 8.39*e* shows the database definition after you have entered the derived fields and the first production data field (PJAN). Since all the rest of the fields will be very similar to the PJAN field, you can save time by copying your field definitions from PJAN in this definition. However, before you can do that, you have to save the definition for EX8_11A by invoking the top-line menu and selecting SAVE. Now you can type **EX8_11A** and **PJAN** in response to the copy

FIGURE 8.39   Detailed database field definition.

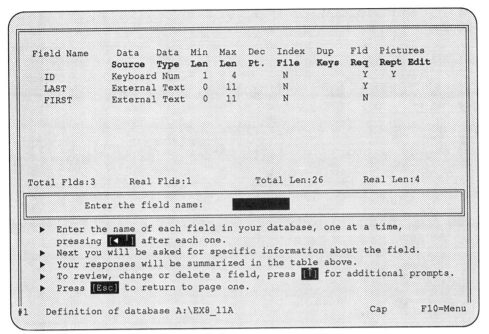

(*a*) Definition screen.

(*b*) First detailed definition screen.

(EXAMPLE 8.11   □   Industrial Production Data (Database Design, Entry, Sorting, Searching, and Report Generation)   □   Continued)

FIGURE 8.39   (Continued)

```
Enter the database name:  EX8_11██████████

Name of lookup field in other database:  MI

Name of indexed linking field in other database:  ID

Name of linking field in this database:  ID

Columnar report heading:  MI

Error msg:
────────────────────────────────────────────────────────────
▶  Enter the name of the database in which the field containing the
   data you want to include is located.
▶  The data will NOT be copied.  Enable will read data from
   this "other" database, using it as required in the database
   you are now defining.


#1   Revising field MI for Database  A:\EX8_11A            Cap
```

(*c*) Second detailed definition screen for data from another database.

```
Formula:  PJAN+PFEB+PMAR+PAPR+PMAY+PJUN+PJUL+PAUG+PSEP+POCT+PNOV+PDEC█

    Result data type?   Numeric   Text   Logical

    Enter the number of decimal positions:  0

    Enter the length of the derived field:  4

Enter report picture:  NNNN

Columnar report heading:  TOTPROD
────────────────────────────────────────────────────────────
▶  Enter the computation you want Enable to perform.   No error
   checking will be performed at this time. Error checking occurs when
   the database is opened.

▶  Up to 135 characters can be used to enter the formula (the line scrolls).
▶  You may use field names and any of the following operators:
            +   -   *   /    **    &    &-

#1   Defining field TOTPROD for Database  A:\EX8_11A        Cap
```

(*d*) Second detailed definition screen for derived fields.

FIGURE 8.39   (Continued)

Field Name	Data Source	Data Type	Min Len	Max Len	Dec Pt.	Index File	Dup Keys	Fld Req	Pictures Rept Edit
ID	Keyboard	Num	4	4		N		Y	Y
LAST	External	Text	0	11		N		Y	
FIRST	External	Text	0	11		N		Y	
MI	External	Text	0	2		N		Y	
TOTPROD	Derived	Num	0	4	0	N		Y	Y
TOTDEF	Derived	Num	0	4	0	N		Y	Y
RATIO	Derived	Num	0	5	2	N		Y	Y
PJAN	External	Num	0	3		N		Y	

Total Flds:15   Real Flds:1   Total Len:65   Real Len:4

Use the **[↑]** and **[↓]** to position the cursor on the desired field.

▶ To review or change the field definition, press **[◀┘]**
▶ To enter detail edit mode, use **[>]** or **[.]**
▶ To delete the field, press **[Del]**
▶ To give the field a new name, press **[R]**

▶ Press **[Esc]** to return

#1   Definition of database A:\EX8_11A

(*e*) Definition screen after addition of derived fields.

and existing field definition prompts. All the detailed definitions for PJAN will be copied. All you have to do is scroll down to the screen of Figure 8.39*c* and change the lookup field and columnar report heading field to the name of the field presently being defined, PFEB, PMAR, and so on. After you have defined all the remaining fields, the definition is saved again.

Report form EX8_11A now has to be defined. Most of the format of the report for EX8_11 can be used. The definitions for TOTPROD, TOTDEF, and RATIO are all that have to be changed. Enter the report form definition module, and enter **EX8_11** as the report form name. The old report form will be brought up. To remove the report form field definition for the TOTPROD, place the cursor in the first space of the output field on the form screen, and repeatedly press **[del]** until the output field disappears. Then press **[shift/F9]** to define a new field. This field will not be a derived field, since TOTPROD is a defined field in the EX8_11A database. After repeating the process for TOTDEF and RATIO, you can save the modified EX8_11 report definition as EX8_11A using the NEW NAME option. Thus, you will now have both the form definitions saved.

After you have entered the employee identification numbers into database EX8_11A, you are ready to produce your report. Figure 8.40*a* shows the REPORT screen with the WHERE clause entered, and Figure 8.40*b* shows the resulting report. Note that the report resembles the one that was produced in Example 8.10.

**(EXAMPLE 8.11   □   Industrial Production Data
(Database Design, Entry, Sorting, Searching, and
Report Generation)   □   Continued)**

FIGURE 8.40   Report for selected summary statistics.

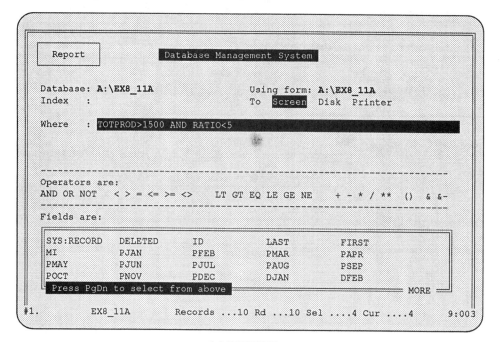

(*a*) REPORT screen.

FIGURE 8.40 (Continued)

```
Individual Employee Production Report
Employee Identification Number: 1001
Last Name:      Johnson
First Name:     James
Middle Initial: S.
Total Production: 1983
Total Defects:    465
PRODUCTION/DEFECT RATIO:  4.26
Individual Employee Production Report
Employee Identification Number: 1002
Last Name:      Thomas
First Name:     Brenda
Middle Initial: A.
Total Production: 1514
Total Defects:    386
PRODUCTION/DEFECT RATIO:  3.92
Individual Employee Production Report
Employee Identification Number: 1003
Last Name:      Dodd
First Name:     Pamela
Middle Initial: B.
Total Production: 1581
Total Defects:    334
PRODUCTION/DEFECT RATIO:  4.73
Individual Employee Production Report
Employee Identification Number: 1004
Last Name:      Jones
First Name:     John
Middle Initial: T.
Total Production: 1549
Total Defects:    319
PRODUCTION/DEFECT RATIO:  4.86
```

(*b*) Report.

PROGRAM TESTING

The database was edited and found to be correct. The summary statistics agree with those developed in Example 8.10.

Example 8.11 illustrated some of the database capabilities of ENABLE. Although virtually the same report was produced using the database commands of LOTUS 1-2-3, the treatment of the data was quite different. In ENABLE the data were stored in a database file. You can edit, add, or delete data from the file using special commands that allow you to search for records with particular attributes. You can mark records for deletion, and they will not be included in any subsequent operations or reports using the database. Records marked for deletion are not physically removed from the database unless you use a specific set of commands to do so. The deleted records can be undeleted as long as they have not been removed. When you use the database to produce reports, you do not do anything to the order or form of the data in the database. This is in contrast with LOTUS 1-2-3, in which the data are contained right in the spreadsheet, and any operations, such as a sort, involve the actual data. Although the ENABLE database management system is more difficult to learn than the database commands of LOTUS 1-2-3, it provides much more flexibility for manipulation of data and generation of reports. Also, the ability to extract data from two or more databases is a powerful characteristic of relational databases such as ENABLE.

Example 8.12 will now illustrate some of the integrating capabilities of ENABLE. Initially, the spreadsheet module of ENABLE will be used to develop a solution to the particle trajectory problem of hitting a target at a distance x and height H from the point of launch. This is the same problem that was used to illustrate a FORTRAN solution to a trial-and-error problem in Example 4.14, and the iterative, table, and graphics capabilities of TK SOLVER in Example 8.2. After we have developed the spreadsheet, the graphics module of ENABLE will be used to illustrate the solution process and the final solution. The word processing module will be used to develop a simple report that will import the data from the spreadsheet and one of the graphics images to produce an integrated document.

EXAMPLE 8.12 □ Particle Trajectory Problem (Integration of Spreadsheet, Word Processing, and Graphics)

STATEMENT OF PROBLEM

Develop an integrated ENABLE application that will determine what angle or angles of launch will cause a particle to hit a target. The particle is launched with an initial velocity of V_0. The target is at a distance x and a height H from the point of launch. The solution is to be integrated into a report containing spreadsheet and graphics data.

MATHEMATICAL DESCRIPTION

The portions of equations G.15 that are applicable follow:

$$y = \int V_y dt = \int (V_{0_y} - gt)dt = y_0 + V_{0_y}t - \frac{1}{2}gt^2$$

$$x = \int V_x dt = V_{0_x} \int dt = x_0 + V_{0_x}t \quad \text{or} \quad t = \frac{(x - x_0)}{V_{0_x}} \quad \text{(G.15)}$$

$$V_{0_y} = V_0 \sin \theta \qquad V_{0_x} = V_0 \cos \theta$$

in which θ is the initial angle of the trajectory, t is the time for the particle to reach a distance x from the launch point, g is the local acceleration due to gravity, and y is the height of the trajectory at distance x. If θ is in degrees, you have to divide it by 57.3 for use in ENABLE trigonometric functions. These equations can be combined into a single equation:

$$y = y_0 + V_0 \sin\left(\frac{\theta}{57.3}\right)\left(\frac{(x - x_0)}{V_0 \cos\left(\frac{\theta}{57.3}\right)}\right) - \frac{1}{2}g\left(\frac{(x - x_0)}{V_0 \cos\left(\frac{\theta}{57.3}\right)}\right)^2 \quad \text{(8.4)}$$

For this application, the origin is at the point of launch, so x_0 and y_0 are both zero. The FORTRAN expression for the right side of Equation 8.4 after eliminating x_0 and y_0 is

```
V0*sin(Theta/57.3)*x/(V0*cos(Theta/57.3))-0.5*g*(x/(V0*cos(Theta/57.3)))**2
```

This expression will be converted to an ENABLE spreadsheet formula by substituting cell references for the variables.

ALGORITHM DEVELOPMENT

Input/Output Design

Input data will include

V_0:	the initial velocity of the projectile
X:	the distance to the target
H:	the height of the target relative to the point of launch
g:	the local acceleration due to gravity

Also, the range of angles to be used to test the height of the trajectory at distance X vs. the target height of *H* must be input. Final output will include all the data of the spreadsheet, plus a graph of *y* vs. *x* for an angle that hits the target. These data will be incorporated into a word processing document. Interim spreadsheet and graphics output will help narrow down the range of angles used to compute spreadsheet data. This will increase the estimation accuracy of the angle or angles that cause the particle to hit the target.

Numerical Methods

The trial-and-error solution will be accomplished by developing a series of trajectory height values at X for a series of initial angles between the two limits specified. By changing the limits and recalculating, you can rapidly converge on the angle that will hit the target. At the same time, you will easily be able to identify trajectories that hit the target for two angles or that will not hit the target at all. No convergence criteria need be included in the spreadsheet. It is up to the judgment of the user to decide whether the trajectory comes close enough to the target to consider it a hit. As will be seen, this method can be used to develop much more accurate estimates than the program of Example 4.14 and can provide estimates as accurate as the TK SOLVER solution of Example 8.2.

Computer Implementation

To ensure that all the spreadsheet can be viewed on a single screen, 10 angle increments will be used between the limits input. Thus, the body of the spreadsheet table will be 11 lines long. Data will be developed for the height at distance X for the 11 angles, as well as *y* vs. *x* data for a selected angle. The latter can be used to provide additional information about angles that hit the window.

PROGRAM DEVELOPMENT

The initial step in developing the trajectory solution is to access the ENABLE spreadsheet module. This is accomplished through the following series of commands:

```
                  USE SYSTEM
                        SPREADSHEET/GRAPHICS
                              CREATE
```

**(EXAMPLE 8.12   □   Particle Trajectory Problem
(Integration of Spreadsheet, Word Processing,
and Graphics)   □   Continued)**

which results in the screen of Figure 8.41 appearing. EX8_12 is entered as the unique name. This file name will be given an extender of SSF by ENABLE to identify it as a spreadsheet file.

The initial screen of the ENABLE spreadsheet is virtually indistinguishable from the initial screen of LOTUS 1-2-3. Figure 8.42 shows the screen after the labels to be used with the spreadsheet have been typed in. The top-line menu is also shown, with the menu cursor positioned over WORKSHEET. The submenu commands for WORKSHEET are shown on the second line. This menu can be brought up by typing / or by pressing **[F10].** You will note some differences between the top-line menu of ENABLE and the top-line menu of LOTUS 1-2-3. ENABLE includes integrating commands, so some of the first-level commands available in LOTUS 1-2-3 have been moved to submenus in ENABLE. For example, the RANGE command is in the WORKSHEET submenu, as is shown in Figure 8.42.

The labels of the spreadsheet have purposely been kept to a minimum. Any additional text information can be added when the spreadsheet is brought into the word processing report. One difference between ENABLE and LOTUS 1-2-3 is that the default justification for labels is centered rather than left-justified. Another difference is that when you are typing in information, ENABLE is normally in the type-over mode. Therefore, if you move the cursor to a particular position and type in text, the new text will replace the old text. In order to insert text and move the remaining text to the right, the **[insert]** key has to be pressed. This will produce a flashing cursor, and you will be able to insert text. To return to the type-over mode, press the **[insert]** key a second time. This is important to remember when you are editing labels or are typing word processing documents.

At this point the spreadsheet was saved using the top-line menu **[F10]** and the SAVE command. When the session was restarted, the commands

```
USE SYSTEM
    SPREADSHEET/GRAPHICS
        REVISE
```

were selected, and **?** was entered for the request to enter the spreadsheet name. Entering the **?** brought up the directory listing shown in Figure 8.43. All spreadsheets on the data disk or directory will be listed. To choose a particular spreadsheet, move the cursor to the name and press enter. Since we only have one spreadsheet saved, we just press **[enter]** and the spreadsheet will be loaded into the active window. The prompts at the bottom of the screen indicate that there are several other file management options available when the directory listing is displayed.

The screen of Figure 8.42 will reappear once the EX8_12 file is loaded. The initial velocity, acceleration due to gravity, distance to target (X), and height of target (H) can be entered in cells B1 through B4. The initial and final angles can be entered in cells F1 and F2. Since we have decided to use 10 angle increments between the limits, the formula

$$+(F2-F1)/10$$

FIGURE 8.41 ENABLE spreadsheet module access from the main menu.

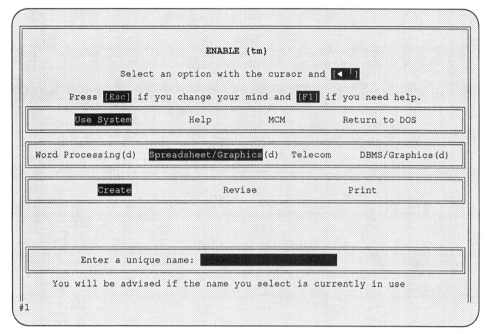

FIGURE 8.42 Spreadsheet screen with top-line menu and label entry.

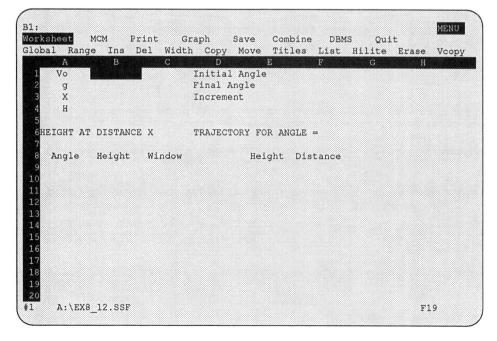

**(EXAMPLE 8.12 □ Particle Trajectory Problem
(Integration of Spreadsheet, Word Processing,
and Graphics) □ Continued)**

can be entered into cell F3. For the angle entries in cells A9 through A19, you
can start out with the formula +F1 in cell A9. That brings the initial angle into
A9. In cell A10 you can enter the formula

$$+A9+\$F\$3$$

which will add the increment to the initial angle and place that value in A10. The
formula from A10 can be copied into the range A11..A19. The copy sequence is

```
/WORKSHEET
       COPY
             FROM RANGE:   A10..A10
             TO RANGE:     A11..A19
```

which is similar to the LOTUS 1-2-3 sequence except that the COPY command
is accessed through the /WORKSHEET command rather than directly.

Since the reference to cell F3 was entered as an absolute cell reference, all the
formulas will reference the increment, while the relative cell reference A9 will
change with each formula so that it always refers to the value in the previous cell
in the A column. The ENABLE spreadsheet formula for the height of the trajectory at a distance X from the origin can be developed from Equation 8.4. It is
entered in cell B9 as

```
+$B$1*@sin(A9/57.3)*$B$3/($B$1*@cos(A9/57.3))
    -.5*$B$2*($B$3/($B$1*@cos(A9/57.3)))**2
```

You can easily see the similarities between this formula and the FORTRAN
expression for Equation 8.4. The formula can be copied into cells B10..B19
to estimate the height at distance X for the various initial angles. Note that
ENABLE uses ** to signify exponentiation, which is the same convention used
in FORTRAN. Most other spreadsheets use ^, which originated with the BASIC
programming language. Figure 8.44 shows the resulting data, which indicates
that there must be two angles that hit the target. In preparation for graphing, an
additional column has been added to the table that gives the height of the target.
This will facilitate plotting a reference line to compare against the height-at-X
data.

The trajectory-for-a-certain-angle table has also been filled out. For the data
shown, a + is typed in cell G6, and the arrow keys are used to move the cursor to
A14, which produces the formula +A14 in G6. For this table, the distance is the
independent variable. Zero can be entered in the first cell of the series, F9.
Succeeding values can be generated by adding X/10 to the value in the preceding
cell. To accomplish this, we will illustrate another capability of ENABLE, one
also available in LOTUS 1-2-3 and most other spreadsheets. A range of cells can
be given a name, and that name, rather than the cell references, can be used in
formulas. You can also still refer to the range in the normal manner even if it is
named. In this case, the range will be a single cell. We will name cell B1 V0, cell
B2 G, and cell B3 X. The sequence of commands to name a range of cells is

FIGURE 8.43   ENABLE file management screen.

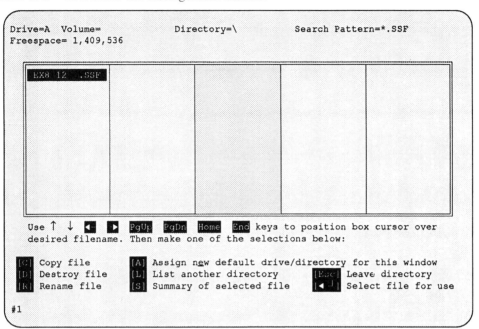

FIGURE 8.44 ENABLE spreadsheet for particle trajectory analysis.

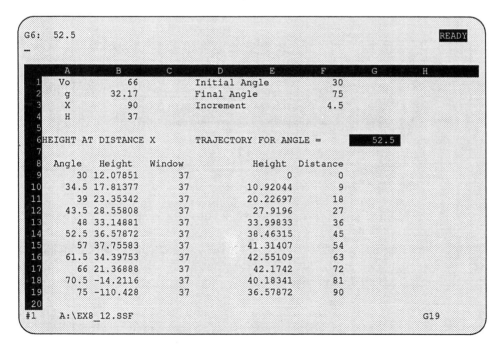

(EXAMPLE 8.12 □ Particle Trajectory Problem (Integration of Spreadsheet, Word Processing, and Graphics) □ Continued)

```
/WORKSHEET
     RANGE
          NAME
               CREATE
                    NAME:  X
                    RANGE:   B3
```

When a name is used in a formula, ENABLE treats it as a range rather than a particular name. Therefore, if a name is used in a formula, and the formula is copied into succeeding cells, the reference remains relative. The new cell references, not the name, will appear in the copied formula. For example, if we typed the formula

```
+F9+X/10
```

in cell F10, the correct distance of 9 ft would appear in that cell. The name X refers to the value contained in cell B3. However, if you copy this formula into cell F11, the formula would appear as

```
+F10+B4/10
```

and the value in F11 would be 9, not 18 as it should be. There is an easy solution to this problem. By placing a $ in front of the name, you can make it an absolute reference. Therefore, you should type the formula

```
+F9+$X/10
```

in cell F10 and copy this formula into F11..F19. This will produce the proper sequence of distance values, as shown in Figure 8.44. The formula

```
+$V0*@sin($G$6/57.3)*F9/($V0*@cos($G$6/57.3))
     -0.5*$G*(F9/($V0*@cos($G$6/57.3)))**2
```

can be typed in E9 and copied into the range E10..E19 to produce the correct height vs. distance data. $V0 and $G are used to make the range reference for the name absolute. Also, $G refers to the acceleration due to gravity, while $G$6 refers to the value contained in cell G6, which is the selected angle.

A printout of either ENABLE spreadsheet data or formulas can be accomplished in the same manner as in LOTUS 1-2-3. The formula output format is also the same, with formulas printed one under the other, starting with the first column, and continuing for the rest of the columns. Only cells that contain formulas are included.

The initial and final angles can be adjusted to refine the estimate for which angles might hit the target. Plotting the height vs. angle data will assist in identifying a refined interval. Therefore, we will generate and display such a plot. The commands are similar to those used in LOTUS 1-2-3. You do not have to leave the ENABLE spreadsheet module to print out your graph. The following series of commands will produce a plot of height at distance X vs. initial angle:

```
/GRAPH
    CREATE
        NAME:  EX8_12
        OPTIONS
            GLOBAL
                TYPE:  XY
                AXIS
                    X AXIS FORMAT
                        FIXED
                        1 (decimal place)
                    Y AXIS FORMAT
                        FIXED
                        2
                    X AXIS DATA
                        RANGE:  A9..A19
                HEADINGS
                    MAIN
                        Height at X vs. Angle
                    X-AXIS
                        Initial Angle (degrees)
                    Y-AXIS
                        Height @ Distance X
                XYL-FORMAT
                    BOTH (lines and symbols)
            DATA GROUP
                1
                    DATA:  B9..B19
                2
                    DATA:  C9..C19
        DISPLAY (displays graph on screen) [esc]
        PRINT (prints graph on printer)
```

They will produce the plot shown in Figure 8.45. From this plot it is plain that there are two angles that hit the target, and both of them are between 50° and 60°. Therefore, we will use 50 and 60 for our refined interval.

Figure 8.46a shows the spreadsheet after changing the interval to between 50° and 60°, and Figure 8.46b shows the corresponding graph. The graph is automatically updated with the new data. For this graph, the Y axis range has been specified to go from 34° to 38°. The default would be to go from zero to the maximum data value, which would make it more difficult to estimate the solution angles. After we examine the plot of Figure 8.46b, it can be determined that one of the solutions lies in the interval 53–54°. Figure 8.47 shows the spreadsheet with this 1° interval entered. It can be seen that the approximate solution is 53.3°, which is the same angle found in previous examples. Figure 8.46a shows that the other angle is approximately 59.0°.

Next, a second graph can be developed to show the arc of the trajectory for the angle that hits the target. The cell pointer is used to select 53.3° and the trajectory data appears as shown in columns E and F of Figure 8.47. For this graph we will select the options of displaying an X-Y grid, double-density printing, and not displaying symbols for the data points. Also, two different fonts will be demonstrated. Roman 1 will be used for the main title, and Script 1 for the X-Y axis titles. The plot will appear as shown in Figure 8.48.

The spreadsheet data and the second plot will now be incorporated into a

(EXAMPLE 8.12   □   Particle Trajectory Problem (Integration of Spreadsheet, Word Processing, and Graphics)   □   Continued)

FIGURE 8.45   Trajectory height at X vs. angle.

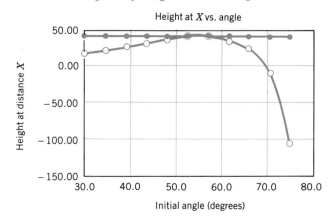

FIGURE 8.46   Trajectory spreadsheet for interval 50–60°.

```
F2:   60                                                              READY
 —

         A         B         C         D         E         F         G         H
 1    Vo        66              Initial Angle           50
 2    g         32.17           Final Angle             60
 3    X         90              Increment                1
 4    H         37
 5
 6 HEIGHT AT DISTANCE X        TRAJECTORY FOR ANGLE =             53
 7
 8  Angle   Height    Window              Height   Distance
 9     50  34.86414      37                    0          0
10     51  35.61588      37             11.11603          9
11     52  36.28198      37             20.58069         18
12     53  36.84875      37             28.39398         27
13     54  37.30016      37             34.55591         36
14     55  37.61732      37             39.06646         45
15     56  37.77799      37             41.92565         54
16     57  37.75583      37             43.13348         63
17     58  37.51956      37             42.68994         72
18     59  37.03191      37             40.59503         81
19     60  36.24827      37             36.84875         90
20
#1    A:\EX8_12.SSF                                                ' G19
```

(*a*) Spreadsheet.

FIGURE 8.46 (Continued)

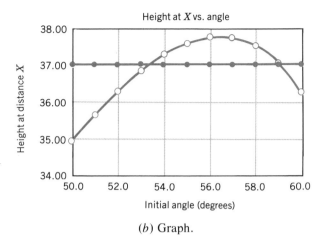

(*b*) Graph.

FIGURE 8.47 Trajectory spreadsheet for interval 53–54°.

(EXAMPLE 8.12 □ Particle Trajectory Problem (Integration of Spreadsheet, Word Processing, and Graphics) □ Continued)

FIGURE 8.48   ENABLE plot for trajectory at 53.3°.

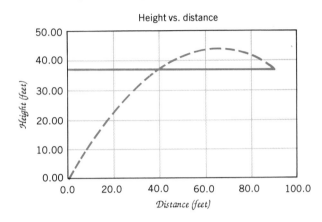

simple word processing document. To create a word processing document, use the series of commands

```
USE SYSTEM
       WORD PROCESSING
              CREATE
                     FILENAME:  EX8_12
```

The initial word processing screen is shown in Figure 8.49*a*. At the top is the default ruler, which controls left and right margins, tabs, and centering position. Information is also given as to how to change the ruler. After you press **[enter],** a window for entering a title for the document will appear. This will be printed out on a separate cover page. After you have entered the title, the cursor can be moved down to the document and you can enter the text. The start of the document is shown in Figure 8.49*b*. At this point we are ready to open windows and transfer the desired spreadsheet information and plot.

From the word processing module, windows can be opened by accessing the top-line menu, **[F10],** and selecting

```
MCM
       WINDOWS
```

which brings to the screen the pull-down menu shown in Figure 8.50. In this case we want to open a window, so the cursor is placed over that option and **[enter]** is pressed. This brings you back to the main ENABLE menu, but for window #2. The #2 will appear at the lower-left corner of the screen display, and the border of the screen will be a different color if you have a color monitor.

The ENABLE initial menu will come up on the screen, and the commands

```
USE SYSTEM
       SPREADSHEET
              REVISE
```

FIGURE 8.49   Word processing initiation.

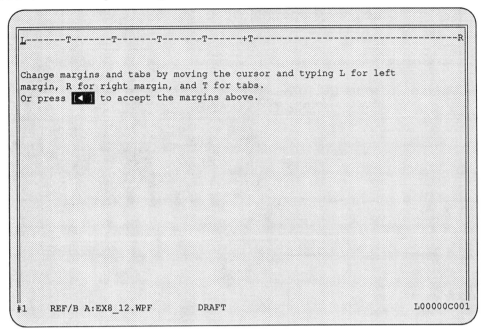

(*a*) Initial screen.

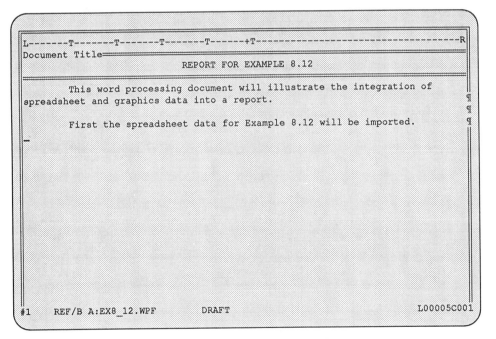

(*b*) Screen after entering title and text.

(EXAMPLE 8.12 ☐ Particle Trajectory Problem (Integration of Spreadsheet, Word Processing, and Graphics) ☐ Continued)

FIGURE 8.50 Pull-down menu for Windows commands.

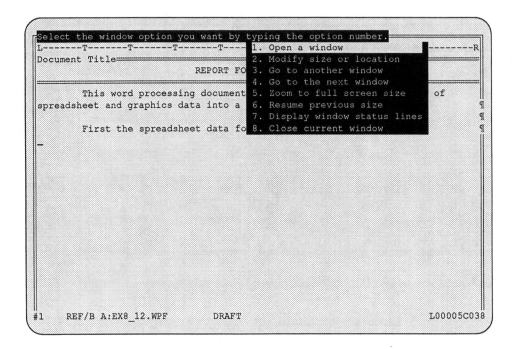

bring up the prompt to enter the spreadsheet name. Entering EX8_12 will load the spreadsheet into window #2. By creating a window for the plot from the spreadsheet, you will make the spreadsheet window unavailable for copying. Therefore, you should import the spreadsheet data into the word processing document before opening the third window for the plot. From the spreadsheet, the commands

```
/MCM
     OTHER WINDOWS
          GO TO 1
```

can be used to go back to the word processing document in window #1. Place the cursor where you want to start the imported data. Then, from the top-line menu **[F10]**, select

```
COPY
     COPY
          CHANGE OPTIONS
               BLOCK
                    OTHER WINDOW
```

This will bring up the screen shown in Figure 8.51*a*. All the windows that have been defined are shown. When you move the cursor down to window #2 and press **[enter]** once, the material to be copied will appear on the screen. When

FIGURE 8.51   Importing spreadsheet data.

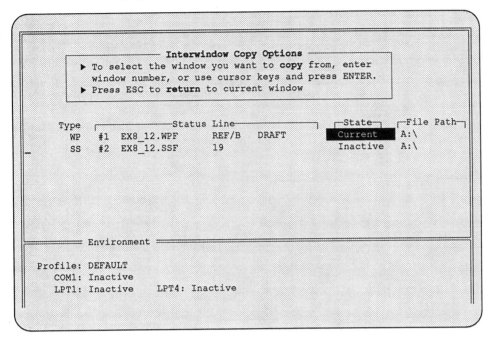

```
                  ┌────── Interwindow Copy Options ──────┐
                  │ ▶ To select the window you want to copy from, enter
                  │   window number, or use cursor keys and press ENTER.
                  │ ▶ Press ESC to return to current window
                  └

        Type   ┌──────────────Status Line──────────────┐  ┌─State─┐ ┌File Path┐
         WP   #1  EX8_12.WPF     REF/B   DRAFT           │ Current │ A:\
         SS   #2  EX8_12.SSF     19                        Inactive  A:\



        ══════ Environment ═══════════════════════════════════════════
   Profile: DEFAULT
     COM1: Inactive
     LPT1: Inactive      LPT4: Inactive
```

(*a*) Window selection menu.

```
L-------T-------T-------T-------T------+T---------------------------------------R
Document Title═════════════════════════════════════════════════════════════════
                            REPORT FOR EXAMPLE 8.12

         This word processing document will illustrate the integration of
spreadsheet and graphics data into a report.                               ¶
                                                                           ¶
         First the spreadsheet data for Example 8.12 will be imported.     ¶
                                                                           ¶
   Vo          66          Initial Angle        53                         ¶
    g        32.17         Final Angle          54                         ¶
    X          90          Increment            0.1                        ¶
    H          37                                                          ¶
                                                                           ¶
HEIGHT AT DISTANCE X        TRAJECTORY FOR ANGLE =         53.3            ¶
                                                                           ¶
  Angle  Height   Window             Height  Distance                     ¶
     53 36.84875     37                   0        0                       ¶
   53.1 36.89937     37              11.2354        9                      ¶
   53.2 36.94882     37              20.7962       18                      ¶
   53.3 36.99708     37              28.68241      27                      ¶
   53.4 37.04414     37              34.89401      36                      ¶
   53.5 37.08998     37              39.43102      45                      ¶
#1    REF/B A:EX8_12.WPF        DRAFT                         L00000C001
```

(*b*) Word processing document with spreadsheet data imported.

(EXAMPLE 8.12 □ Particle Trajectory Problem (Integration of Spreadsheet, Word Processing, and Graphics) □ Continued)

you press **[enter]** a second time, the data will be imported into the word processing document, as shown in Figure 8.51*b*. All the spreadsheet data have been imported and can be viewed by scrolling the cursor down.

The cursor is now moved to the bottom of the spreadsheet, and the text

```
''Now the graphics image for Height vs. Distance will be imported:''
```

is entered, the cursor is moved down two lines, and the top-line menu is brought up. The commands

```
                MCM
                        WINDOWS
                                GO TO ANOTHER WINDOW
```

will bring up the window selection screen. By moving the cursor to the window #2 line and pressing **[enter]** you will move to the spreadsheet window. You do not have to reload the spreadsheet, because it has already been brought into the window. If you wanted to, though, you could replace the spreadsheet in window #2 with any other ENABLE application by exiting the spreadsheet module, returning to the main ENABLE menu, and loading a new application into window #2. In this case, we want to create a new window for the graphics image of the second graph developed. The commands

```
                /GRAPH
                        SELECT
                                EX8_12A
                        DISPLAY
```

will automatically open up a third window to display the graph on the screen. The screen outline will again be a different color, and #3 will appear in the lower left corner. When you press **[F10]** while in the graphics window, a windows option menu will come up. This window includes the commands

OPEN—open a new window

MODIFY—change the size of the plot

GOTO—go to another window

NEXT—move to the next open window

ZOOM—return the graph to full-screen size

RESUME—return the window to its previous size

DISPLAY—display the system status screen

CLOSE—close the graphics window

Selecting GOTO will bring up the window selection screen. The screen will indicate that window #2 is not available. Moving the cursor to window #1 and pressing **[enter]** will bring you back to the word processing document at the place where you left off. When you enter the same series of commands for copying as before, except that you select window #3, you will bring the graph up on the screen when you press **[enter].** The bottom-line menu indicates that

FIGURE 8.52 Word processing document with data and graph imported.

This word processing document will illustrate the integration of spreadsheet and graphics data into a report.

First the spreadsheet data for Example 8.12 will be imported:

Vo	66	Initial Angle	53
g	32.17	Final Angle	54
X	90	Increment	0.1
H	37		

HEIGHT AT DISTANCE X   TRAJECTORY FOR ANGLE =   53.3

Angle	Height	Window	Height	Distance
53	36.84875	37	0	0
53.1	36.89937	37	11.2354	9
53.2	36.94882	37	20.7962	18
53.3	36.99708	37	28.68241	27
53.4	37.04414	37	34.89401	36
53.5	37.08998	37	39.43102	45
53.6	37.13458	37	42.29343	54
53.7	37.17791	37	43.48124	63
53.8	37.21996	37	42.99445	72
53.9	37.26072	37	40.83307	81
54	37.30016	37	36.99708	90

Now the graphics image for Height vs. **Distance** will be imported:

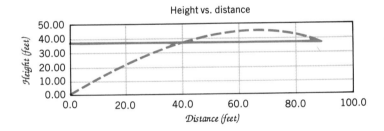

pressing **[alt/F5]** will copy the graph. After you have accomplished this, the graph will be copied but will appear as a shaded area in the word processing document. You can view the graph by switching to the graphics mode through the MCM module of ENABLE. The document can be saved and then printed using the commands in the top-line menu. Letter quality printing was selected as a print form option, and then **[alt/F2]** was used to start the printing process. The document will appear as shown in Figure 8.52. Note that the graph prints out correctly. After the document has been printed, you can close the two open windows and exit ENABLE.

PROGRAM TESTING

The answers found in this example agree with the answers found in the previous examples illustrating the particle trajectory application.

8.4.5 Summary of Integrated Packages

Integrated packages, such as ENABLE, offer a great deal of functionality and facilitate the sharing of information among their various modules. Integrated packages incorporate the different applications modules into a single software package with a consistent command syntax for data transfer and other general program operations. They are more difficult to master than a single package, but are cheaper and easier to master than several individual packages. If a person needs to use most of the applications included in the integrated package, an integrated package is probably a good choice.

8.5   INTEGRATION OF PACKAGES

Integration involves the carryover of data from one type of application to another, either through use of an integrated package such as ENABLE or through integration of separate packages. There was a time when integration among applications packages was almost nonexistent, but capabilities in this area have been increasing rapidly. Integration requires the sharing of data among two or more stand-alone packages through some integrating environment or through compatible file formats. People who do not want or need all the functionality and complexity of an integrated package may still need to integrate data from different applications packages.

8.5.1 File Formats

Integration among applications can be accomplished if data and/or program files from one application can be imported into another application program and used or modified by that program. File integration is still difficult across different operating systems. For example, file integration from Apple Macintosh to IBM PC and vice versa is difficult because the different operating systems store data on disks in different formats. There are some translation programs available, but their performance is generally less than satisfactory. Text files can be transmitted from one machine to another, independent of the operating system, by using a communications link and software. However, this is a slow process and is not efficient for routine integration.

Integration among different packages using the same operating system is dependent on compatible file formats. Since LOTUS 1-2-3 has become such a popular spreadsheet program, it has almost become a de facto standard. Competitors, in order to encourage potential users to try their product, have frequently either made their file formats compatible with LOTUS 1-2-3 or have included translation programs with their software. These developers also know that if a user has developed a large base of 1-2-3 applications, he or she would be hesitant to switch to any other package if that meant having to redevelop all the spreadsheet applications. Some of the compatibilities among spreadsheet programs have been described in the previous descriptions of the integrated packages. A similar trend either to compatibility or translation utilities is evident in database management systems and word processing programs.

Integration among different types of applications has also been increasing. For example, LOTUS now markets a whole range of products that can work together in a loose form of integration. Spreadsheet data from 1-2-3 can be transferred to their presentation quality graphics package, FREELANCE PLUS. Data and graphics from 1-2-3 and FREELANCE PLUS can be imported into MANU-SCRIPT, the LOTUS word processing package. MANUSCRIPT also has the capability of accessing a WORKSHEET at the time of printing and importing the latest updated information from the WORKSHEET. Since all the packages are produced by the same firm, the time to learn each package is reduced because there is some consistency in commands. This type of integration may be the answer for many users. However, there is some cost involved. The total cost for the three LOTUS products described above would be about twice the cost of ENABLE.

8.5.2 Desktop Organizers

Desktop organizers do not actually integrate applications, but they do provide useful accessories and utility programs that may improve a user's productivity. A good example of desktop organizers is Borland's SIDEKICK PLUS. It can either be called up from DOS or be resident in RAM along with other applications. Applications include a note pad, which is a word processor for fairly short documents up to about 11,000 words, and an outliner for organizing thoughts and reports. A phone book provides automatic communications capabilities with user-specified parties. Communications can be accomplished simultaneously with other applications. The time planner allows the entry of appointments in a calendar format, will remind you of upcoming appointments, and will automatically dial telephone numbers at a prearranged time. With SIDEKICK PLUS, you can search for blocks of free time in a busy schedule. A calculator module provides several different types of calculators for business or scientific/engineering applications. Many of the file management functions of the operating system can be accomplished from within SIDEKICK using the file manager module. Copy and paste services for transferring text blocks from one application to another are also provided. Any or all of these applications can be useful, but there will be a learning curve associated with program familiarization. SIDEKICK will take up about 1.5 megabytes of storage on your hard disk and will use between 72K and 150K of RAM when it is loaded into your computer.

8.5.3 Operating Environments

Operating environments allow loading of separate software packages into the computer simultaneously. Each can be accessed through a "window" on the screen.

Microsoft WINDOWS is one such operating environment. It allows you to work with several programs at once. You can switch back and forth between windows that are each using different programs. You can do this with a few keystrokes or by using a mouse pointing device. When you switch from one window to another, the application that you leave is not terminated, so you can return to it and continue from where you left off. Information can be copied from one part of a window to another or between windows.

The MS-DOS executive window is the launching point for many applications. It displays names of files and subdirectories in the current directory and can mark multiple files for collective copying or deletion. Microsoft WINDOWS applications or other applications programs and associated data files can be loaded simultaneously, provided sufficient memory (RAM) is available.

The different applications available directly through WINDOWS include:

Write: A scaled-down version of the Microsoft WORD word processing package.

Paint: A computer-aided drawing module.

Calculator: A basic calculator with memory. This is not a scientific calculator.

Clock: Shows current time as set by the computer system.

Terminal: Emulates a VT100 terminal.

PIF Editor: Program Information Files give the WINDOWS environment more information on an application, including directory location, memory requirements, display mode, etc. The editor creates and changes PIFs.

Note pad: A text editor that you can use to create, modify, and display text files. It is designed mainly for notes and short memos.

Card file: A filing application that you can use to keep track of names, addresses, phone numbers, directions, or any other information. It is like a set of index cards that sorts itself. Card file is a type of database.

Clipboard: Holds information to be copied or moved.

Other operating environments include IBM's TOPVIEW and Digital Research's GEM.

SUMMARY

You have been introduced to several different microcomputer applications packages in this chapter. Each has its own strengths and limitations. The repetition of earlier examples should give you some insight into which tool might be better for a particular application. Think about this as you restudy the examples in this chapter and as you study Chapter 9. We will discuss choosing the right tool in more depth at the end of that chapter.

REFERENCES

SIDEKICK PLUS User's Manual, Borland International, Scotts Valley, CA, 1989.

Borse, G. J., *FORTRAN 77 and Numerical Analysis for Engineers*, PWS Publishers, Boston, 1985.

ENABLE, Version 2.0, The Software Group, Ballston Lake, NY, 1986.

ENABLE Applications Development Guide, The Software Group, Ballston Lake, NY, 1988.

Growney, A. S., and Growney, W. J., "ENABLE in Action: A Student Applications Workbook," John Wiley & Sons, Inc., New York, 1986.

Konopasek, M., and Jayaraman, S., *The TK!SOLVER Book*, Osborne/McGraw-Hill, Berkeley, CA, 1984.

LOTUS 1-2-3, Release 2.01, Lotus Development Corporation, Cambridge, MA, 1986. Release 2.2 and Release 3.0, 1989.

Lunsford, E. M., *101 Macros for LOTUS 1-2-3*, Macropac International, Cupertine, CA, 1986.

Mathews, K. A., *Exploring Physics with TK Solver Plus,* Universal Technical Systems, Inc., Rockford, IL, 1989. Designed to accompany either *Physics* or *Physics: Classical and Modern*, by Gettys, Keller, and Skove, McGraw-Hill, New York.

MICROSOFT WINDOWS, User's Guide, Microsoft Corporation, Redmond, WA, 1986.

O'Leary, T. J., *The Student Edition of LOTUS 1-2-3*, 2nd ed., Addison-Wesley, Reading, MA, 1989.

O'Leary, T. J., *The Student Edition of LOTUS 1-2-3, Release 2.2*, Addison-Wesley, Reading, MA, 1990.

Red, W. E., and Mooring, B., *Engineering: Fundamentals of Problem Solving*, Brooks/Cole Engineering Division, Wadsworth, Inc., Monterey, CA, 1983.

Shanzer, H., *ENABLE Educational Version 2.0 Instructor's Manual*, John Wiley & Sons, Inc., New York, 1988.

Spezzano, C., *ENABLE 2.14: The Fundamentals and Advanced Topics*, John Wiley & Sons, Inc., New York, 1990.

TK SOLVER PLUS, College Edition, Universal Technical Systems, Inc., Rockford, IL, 1987.

TK SOLVER PLUS 1.1, Universal Technical Systems, Inc., Rockford, IL, 1989.

TK Solver System: Roark & Young Application, Universal Technical Systems, Inc., Rockford, IL, 1989.

Wiggins, E. G., "Building Linear Regression into TK SOLVER Models," *CoED Journal*, Vol. VI, No. 4, Computers in Education Division of ASEE, Wash., D.C., Oct.–Dec. 1986, pp. 37–42.

Wright, V. E.; *TK!SOLVER for Engineers*; Reston Publishing Company, Inc., Reston, VA, 1984.

Young, W. C., *Roark's Formulas for Stress and Strain*, 6th ed., McGraw-Hill, New York, 1989.

EXERCISES

Since it is not certain which of the applications discussed in this chapter you will cover in your course, the exercises for this chapter have been included with the instructor's supplement. Your instructor will make the appropriate exercises available to you.

CHAPTER NINE

DEVELOPMENT OF NUMERICAL METHODS AND OTHER APPLICATIONS

9.1 INTRODUCTION

This chapter presents several computer applications using various software tools. It discusses accuracy and numerical precision in more detail. We will develop computer applications for some useful introductory numerical methods. Different computational tools will be compared for particular applications. You will learn general guidelines for developing and documenting comprehensive applications packages.

The concepts of accuracy, error, and numerical precision were introduced in Chapter 1. In the ensuing chapters we have applied these concepts to the solution of problems using numerical techniques and the computer. For example, the principle behind regression analysis is to minimize the error between the regression function and the data values. As we pointed out in Chapter 1, the computer itself is a very precise computational tool. The same results will be achieved each time you run a particular program for a given set of conditions. However, the precision of the data used in analysis can affect the value of the results. Imprecise data will not produce as good a regression as precise data. On the other hand, inaccurate data that have a consistent error can produce a very good regression. You have also seen how the method of storing values in the computer can cause small errors. We will now see how repetitive computations can increase these numeric round-off errors. The errors can lead to inaccurate numerical methods solutions.

Numerical methods involve formulating mathematical problems so that they can be solved by arithmetic operations rather than through an exact analytical solution. Numerical methods require either repetitive calculations to develop increasingly accurate approximate results or a number of sequential steps to arrive at a solution. The number of calculations can become tedious and error-prone if you do them manually. Since computers can provide efficient, error-free arithmetic operations, they are a natural adjunct to numerical methods.

In Chapter 2 we discussed some of the considerations in choosing the best computational tool. Since then you have learned more about a high-level language and several types of generic computer applications software. You have studied examples of the same problem solved by two or more different computer software tools. You should now have a better feel for why it is important to carefully evaluate which resource is best for any particular problem. You should also understand the need to be familiar with more than one type of computer applications tool. Trying to fit one tool, such as FORTRAN, to every computational problem can be inefficient and frustrating. On the other hand, there may be problems for which writing a high-level language program is the only sensible option.

When you are using the computer as an analytical tool, your first consideration should be efficiency. You do not need to write a program for every application. If you are familiar with available applications packages, you can consider using or modifying one of these packages for your analysis. You should develop your own application package only if nothing appropriate is available. If you do develop your own package, be sure to consider the needs of whoever might be using your program. You should carefully design and structure your program, and develop clear and concise documentation for it. The latter part of this chapter will give you some guidelines to assist in developing good applications packages.

9.2 ACCURACY AND NUMERICAL PRECISION

Solutions by numerical methods are particularly sensitive to accuracy and precision. The accuracy of numerical results can be reduced by the cumulative error caused by the number of significant digits (precision) that are used for computations. Since these errors are truncation errors, they have a tendency to accumulate rather than to cancel each other out. Even a single computation involving small differences between larger numbers can adversely affect results. However, numerical round-off errors are usually only a problem in algorithms that employ a long series of sequential computational steps or a large number of iterations. A characteristic indicator of the onset of round-off error is an increase in inaccuracy as iterations progress. You may encounter increasing errors when you decrease the step size or increase the number of steps for iteration. You would think that either of these steps would increase accuracy. It is possible for numerical methods to fail to satisfy convergence criteria because of computer round-off error. It is also possible for numerical methods to converge to approximate answers that are not accurate, especially when the convergence criteria involve differences between successive approximations.

The TK SOLVER model shown in Figure 9.1 was developed to demonstrate errors caused by small differences between relatively large numbers. Fifteen values were set up, ranging from one to 15 decimal places. The corresponding values differed by one digit in the final place. The input values are shown in the top half of Figure 9.1. Two operations were undertaken with the data. First the difference between the two values was calculated and then the result was multiplied by 10^n, with n representing the decimal place in which the difference occurred. Each final result should have a value of 1.0. The actual results are shown in the bottom half of Figure 9.1. Only the first three differences, or up to three significant digits, gave exactly correct results. Starting with the fourth significant figure, the differences and products started showing a small error. This error increased as the number of significant figures increased. Up through eight significant figures, the product was accurate when rounded to the eighth place. After that the accuracy decreased. The final value, after taking the difference in the fifteenth significant digit and multiplying by 10^{15}, was accurate only when rounded to the second significant

FIGURE 9.1   TK SOLVER model for small differences.

```
==================== RULE SHEET ==========================================
S Rule-----------------------------------------------------------------------
* DIFF = A - B
* PROD = DIFF*10^n
==================== LIST: n ==============================================
Comment:
Numeric Format:
Display Unit:
Calculation Unit:
Element-- Value--------------
1         1
2         2
3         3
4         4
5         5
6         6
7         7
8         8
9         9
10        10
11        11
12        12
13        13
14        14
15        15
==================== LIST: A ==============================================
Comment:
Numeric Format:
Display Unit:
Calculation Unit:
Element-- Value--------------
1         .1
2         .11
3         .111
4         .1111
5         .11111
6         .111111
7         .1111111
8         .11111111
9         .111111111
10        .1111111111
11        .11111111111
12        .111111111111
13        .1111111111111
14        .11111111111111
15        .111111111111111
```

FIGURE 9.1 (Continued)

```
=================== LIST: B ====================================================
Comment:
Numeric Format:
Display Unit:
Calculation Unit:
Element-- Value--------------
1          .2
2          .12
3          .112
4          .1112
5          .11112
6          .111112
7          .1111112
8          .11111112
9          .111111112
10         .1111111112
11         .11111111112
12         .111111111112
13         .1111111111112
14         .11111111111112
15         .111111111111112

=================== LIST: DIFF =================================================
Comment:
Numeric Format:
Display Unit:
Calculation Unit:
Element-- Value--------------
1          -.1
2          -.01
3          -.001
4          -.000099999999999989
5          -9.99999999999612E-6
6          -.000001000000000001
7          -9.99999999889978E-8
8          -9.99999999473644E-9
9          -9.9999999947364E-10
10         -9.9999994396249E-11
11         -1.0000000827404E-11
12         -1.0000056338555E-12
13         -1.0000333894311E-13
14         -1.0005885009434E-14
15         -9.9920072216264E-16

=================== LIST: PROD =================================================
Comment:
Numeric Format:
Display Unit:
Calculation Unit:
Element-- Value--------------
1          -1
2          -1
3          -1
4          -.99999999999989
5          -.999999999999612
6          -1.000000000001
7          -.999999999889978
8          -.999999999473644
9          -.999999999473644
10         -.999999943962493
11         -1.00000008274037
12         -1.00000563385549
```

digit. A similar model was evaluated using a FORTRAN program on the VAX, with the following results:

Significant Digits	Difference	Product
1	−0.1000000	−1.000000
2	−9.9999979E−03	−0.9999998
3	−1.0000020E−03	−1.000002
4	−9.9994242E−05	−0.9999424
5	−9.9986792E−06	−0.9998679
6	−9.9837780E−07	−0.9983778
7	−1.0430813E−07	−1.043081
8	−7.4505806E−09	−0.7450581
9	0.0000000E+00	0.0000000E+00

The product for the differences in the sixth and seventh significant digit were both accurate to the second significant digit, but both had more error than the product for the fifteenth place with the TK SOLVER model. The error increased dramatically for the product of the difference in the eighth significant digit. For differences in the ninth and higher significant digits, the VAX returned a value of zero.

These results are not a comprehensive study of the effects of significant digits and differences on computer results. Nonetheless, they point up the need to watch for situations that may cause computer-generated numerical inaccuracies. If two operations can produce the errors shown, you can imagine what thousands or millions of repetitive calculations could produce.

One example we have used where either round-off or small differences between large values can cause problems is in the application of Equation 5.1

$$a = \frac{n\Sigma XY - \Sigma X \Sigma Y}{n\Sigma X^2 - (\Sigma X)^2} = \frac{\Sigma XY - n\bar{X} \cdot \bar{Y}}{\Sigma X^2 - n\bar{X}^2} \tag{5.1}$$

to find the slope of a regression line in regression analysis. Sometimes the X and Y data are of different orders of magnitude. Or the data may represent a relationship with either a very large or very small slope. In these cases, round-off or lack of sufficient significant digits when taking the differences can cause serious error in the computation of the slope. This in turn will adversely impact the calculation of the intercept.

It is also possible to have instabilities in numerical solution methods. In numerical methods algorithms that involve application of a correction factor, the correction will sometimes overshoot the actual answer, and the next iteration will oscillate back to the original side of the answer. If the oscillation is balanced, the algorithm may never converge to a solution, or it may converge very slowly. The instability may increase in magnitude until the algorithm fails because of computer numerical overflow or underflow. Another possibility is that the initial solution estimates used to start an algorithm may cause the solution to jump to another portion of the solution domain, and may converge to an entirely different solution than the one sought. These types of instabilities will be illustrated when we discuss the Newton-Raphson method.

In FORTRAN, problems with numerical round-off errors can sometimes be overcome by switching from single to double precision data storage. You may also be able to reformulate the model. A different order of computation may make the round-off problems less severe. Solution instabilities can sometimes be overcome by starting the solution process at a different point. You may be able to modify the iteration process. It may also be necessary to search for a different numerical solution technique.

You may be able to apply less stringent convergence criteria if they do not invalidate the solution. You have to be careful, though, when your convergence criteria involve differences between successive approximations. If the iterations are slowly converging to a solution, the differences between successive approximations will be small. If the convergence criteria are too large, you may get a false indication that the algorithm has converged to a solution.

From the previous discussion, you should realize that it is important to thoroughly understand any numerical technique that you employ. Otherwise you may be unsuccessful in your attempts to arrive at a useful solution. This concept will be reinforced as you study the examples in the rest of the chapter.

9.3 NUMERICAL METHODS

In pre-computer times, a great deal of effort was expended in finding analytical solutions to problems. It was often necessary to make simplifying assumptions that made the solution applicable only for special cases. Other problems were approximated graphically. These solutions were limited by the fact that we cannot effectively work with more than two-dimensional graphs. Also, a lot of time was spent implementing numerical methods with slide rules or calculators. These methods were tedious and prone to simple entry and transcription errors that could negate hours or days of work. The advent of computer-assisted numerical methods has allowed engineers to undertake much larger and more complex problems.

While you are students, you will find that computer-assisted numerical methods will allow you to undertake more realistic problems to reinforce the theory that you are learning. You will be able to investigate more alternatives to learn about trade-offs and optimization. However, since numerical methods are just solution techniques, you must still know the basic theory behind the application. Computer-assisted numerical methods can help reinforce your understanding of mathematics. They also provide an effective way of learning how to use computers.

Although there are many packages available for implementing numerical methods, they often require customizing for a particular application. Thus, you will need to be thoroughly conversant with computer programming if you are to avoid being dependent on others for your numerical solutions. At a minimum, you must understand the methodology for implementing computer solutions of numerical methods if you are to avoid the pitfalls of improper computer usage.

You have been exposed to various numerical methods through the examples in previous chapters. The regression analysis developed and used in Examples 5.4, 8.3, 8.5, and 8.8 is a numerical method used to fit curves when there is a significant amount of error associated with the data being analyzed. In regression analysis you derive a curve that represents the general trend of the data without necessarily matching any individual points. The resulting equation for the curve is used to predict values for a dependent variable based on values for the independent variable or variables. The solution of the particle trajectory problem of Examples 4.14, 8.2, 8.4, and 8.12 by trial and error is an example of a primitive numerical method for estimating a mathematical model solution. This method can be reasonably effective when it is used to solve problems with one independent variable, but the number of possible combinations of independent variable values becomes unwieldy for problems with multiple independent variables. The matrix inversion and multiplication procedures used to find the solution for a system of linear equations in Examples 7.3, 7.4, and 8.9 involve a series of computations. First you have to find the inverse of the coefficient matrix, and then you have to multiply the right-hand-side vector by this inverse matrix to find the solution. Although in theory the solution found is exact, the number of sequential computations can lead to a considerable amount of numerical round-off error. The modeling of unsteady flow through a reservoir developed in Example 8.6 is an example of a graphical technique that has been adapted for computer solution. Some type of interpolation is required to implement the method.

Several numerical techniques, including finding roots of polynomial equations, numerical integration, numerical differentiation, and searches for function minima and maxima will be developed in this chapter. Some of these will expand on the previously illustrated methods, while others will break new ground. Our objectives are to introduce you to the methodologies of numerical methods and to stress the importance of the computer as a tool for implementing numerical methods.

9.3.1 Finding Roots of Equations

You have seen several trial-and-error solutions for equations in both examples and chapter-end exercises. In this section we will develop two methods for estimating roots of an equation that are usually more efficient than using trial and error. The methods developed will be the Newton-Raphson technique and the secant method. Either of these methods is readily implemented by either a FOR-TRAN program or a spreadsheet. TK SOLVER uses a form of the Newton-Raphson technique in its solution scheme. Bisection is a third method that is sometimes used for finding roots for polynomial equations. Development of the bisection method can be found in Chapra and Canale (1986).

Any polynomial equation can be written in the form $f(X) = 0$ if all elements of the equation are moved to one side. The values of X that satisfy the equation are called roots of the equation. The number of roots that can be found depends on the expanded polynomial order of the equation. A fifth-order polynomial will have five roots and so on. The actual order is not always readily identifiable. However, if you have an approximate value for the desired root, and some knowledge of the behavior of the function in the region around the root, you should be able to find the desired root. You can use the approximate value for the starting point of your numerical technique. Graphing the function in the region of interest will assist in estimating the root. You may also have prior knowledge of the approximate behavior of the function in the region of interest. The importance of knowing a reasonably accurate first estimate of the root cannot be overemphasized. A close first estimate will reduce the number of iterations required and will generally result in convergence to the root desired. A haphazard first estimate may result in convergence to the wrong root, or nonconvergence of the solution.

You must also be aware that the solution to some functions will entail complex roots. Other problems may require the solution of systems of simultaneous, nonlinear equations. Methods are available for finding complex roots and for solving systems of nonlinear equations. These methods have evolved from the Newton-Raphson technique described in the next section, but their development is beyond the scope of this text. Methods for finding real roots of functions of a single independent variable will be covered. The references at the end of the chapter can be consulted for information on the more advanced methods.

Numerical techniques are invaluable in engineering work where it is difficult or impossible to solve for a variable explicitly. These types of equations arise frequently. One example of this type of equation would be a situation where you wanted to solve for an interest rate that gives a particular uniform series present worth factor. An examination of the equations in Appendix E will reveal that you cannot solve explicitly for this value, so some sort of approximation technique would be required. The Newton-Raphson method is one method for solving this type of problem. In the following section we develop this method and use three different computational resources to show its application.

9.3.1.1 Newton-Raphson

The Newton-Raphson method is an efficient method for finding real roots of nearly any nonlinear equation that would be of interest to engineers, provided that it is possible to find the derivative of the function. As is the case with most numerical approximation techniques, you must begin with an initial estimate of the root desired. The more accurate the estimate is, the more quickly the method will converge to the solution. If the initial estimate is too far off, there is a chance that the method could converge to a different root than the one desired, or not converge at all.

When the root has been determined correctly, the value of the function will be zero. Since this is an approximation technique, the value of the function will probably never be exactly zero. However, it is possible to estimate the root accurately to any number of decimal places desired.

If you take X as the independent variable and $f(X)$ as some polynomial function of X with all constants and powers of X on one side of the equation, then the function

$$f(X) = 0 \qquad (9.1)$$

can be used to find values of X that satisfy the equation. These values of X are the roots. If you take the derivative of the function and evaluate it for an initial estimate of the root, X_0, the result gives you the slope of the function at that point. If you divide the value of the function at X_0 by the value of its derivative at the same point and subtract the resulting value from the value of X_0, you get the X intercept of the slope line if it is extended down to the axis. This would be point X_1 shown in Figure 9.2a. Point X_1 is a closer estimate of the actual root than X_0. X_1 can now be used to generate another refined estimate, X_2. This process can be repeated until your estimate is as close to the actual root as you require it to be. If you happen to choose the initial estimate of the root on the opposite side of the actual root, as is depicted in Figure 9.2b, the first correction will bounce you over to the proper side, and the solution will converge toward the actual root from there. An estimation algorithm can be written

FIGURE 9.2 Graphical interpretation of the Newton-Raphson method.

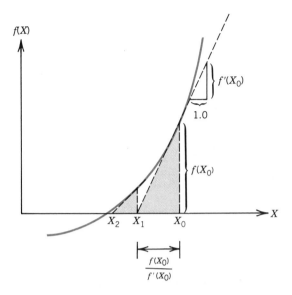

(a) Normal convergence.

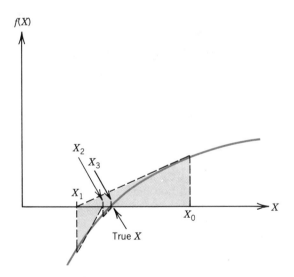

(b) Convergence from the opposite side of the root.

$$X_{n+1} = X_n - \frac{f(X_n)}{f'(X_n)} \qquad (9.2)$$

in which X_{n+1} = the next estimate of the root, X_n = the present estimate of the root, $f(X_n)$ = the function evaluated at the point X_n, and $f'(X_n)$ = the derivative of the function evaluated at the point X_n.

The initial estimate of the root (X_0) can be estimated by a rough graph or by knowledge of the approximate value expected. If a rough graph of the polynomial is available for the function, it will also tell you a great deal about the behavior of the function in the region of interest. You can use the graph to avoid situations that may cause the Newton-Raphson technique not to converge to the desired root.

It can be shown that the accuracy of the method is directly related to the absolute value of the difference between successive approximations of X. The formula

$$|X_{n+1} - X_n| < 10^{-a} \qquad (9.3)$$

reflects that decimal accuracy to a places is assured if the absolute value of the difference is less than 10^{-a}. This provides a mechanism for knowing when enough iterations have been accomplished. The desired decimal place accuracy is determined beforehand, and then iterations are performed until the above stated criterion is met. The estimate of X accurate to the desired number of decimal places is the value for X_{n+1}. The iterative solution technique is readily adaptable to computer solution and converges very quickly to the desired accuracy if the initial estimate is chosen wisely. It has to be remembered that the accuracy given by Equation 9.3 is numerical accuracy, not necessarily physical accuracy. You still have to determine how many significant figures can be reported.

As is the case with most numerical techniques, you have to be aware of times when the method might not provide adequate results. If you happen to choose an initial estimate of the root close to a minimum or maximum point for the function, the derivative will be very small and the correction factor $f(X_n)/f'(X_n)$ will be very large. This can cause the solution to jump to an entirely different root than the one that you are trying to find. If the initial guess happens to be right at the minimum or maximum point, the correction factor will be infinite, and the method will break down. Careful selection of your initial estimate is the key to avoiding this problem.

In the rare instance when a root of an equation is also a minimum or maximum point, the Newton-Raphson method will not successfully find the root. Both the function and its derivative will go to zero at that point. The secant method, described later in this chapter, will not encounter this problem.

Another instance when the method may not give adequate results is when there is a point of inflection in the function at or near the desired root. Figure 9.3 shows a case in which the successive approximations oscillate from one side of the root to the other because of a point of inflection. Convergence may be very slow in this case, or the solution might not converge at all. This problem does not arise often in practice, but you should be aware of it.

We will now explore three applications of the Newton-Raphson technique to estimate a root of a polynomial equation. The first will use a spreadsheet, the second a TK SOLVER model, and the third a FORTRAN program. The spreadsheet solution will be developed first because it closely parallels the hand application of the method. The results of that application can be used to check other applications. The polynomial function to be used has been created to give integer roots to facilitate accuracy checking. The function is

$$Y = X^4 - 14X^3 + 67X^2 - 126X + 72 \qquad (9.4)$$

It was developed by multiplying

$$(X - 1)(X - 3)(X - 4)(X - 6)$$

which has exact roots at $X = 1, 3, 4,$ and 6.

FIGURE 9.3   Oscillation of successive approximations about a point of inflection.

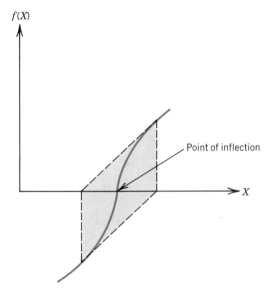

EXAMPLE 9.1 □ **Spreadsheet Approximation of Real Roots of a Polynomial Function**

STATEMENT OF PROBLEM

Develop an ENABLE spreadsheet that could be used to find the root of a polynomial function to a desired accuracy. Plot the function of Equation 9.4 in the interval $X = 0$ to $X = 7$. Use the spreadsheet to find the largest root in the interval.

MATHEMATICAL DESCRIPTION

Equation 9.2 will be used to develop the solution. The accuracy check, Equation 9.3, will be accomplished manually for this example.

ALGORITHM DEVELOPMENT

Input/Output Design

An initial estimate for X will have to be input. Also, the spreadsheet will be set up so that it could accept any expanded polynomial equation up to fifth order. Other polynomial forms could be input as formulas. Output will include values of all the intermediate calculations for checking and verifying the method.

(EXAMPLE 9.1 □ Spreadsheet Approximation of Real Roots of a Polynomial Function □ Continued)

Numerical Methods

Successive iterations will be accomplished until the error term is less than 10^{-a}.

Computer Implementation

At the top of the spreadsheet an area will be set aside to input the coefficients of the polynomial equation. These will be used to develop the function and function derivative values for a given value of X. They will also be used to generate values for plotting the function in a desired interval. The formulas for developing the polynomial function and derivative function values will be placed in the first row of the iteration table. The second cell of the second row will subtract $f(X_n)/f'(X_n)$ from the previous X value to find X_{n+1}. The remainder of the formulas can be copied from the first row. Successive rows can be copied from the second row until the correction term, $f(X_n)/f'(X_n)$, is less than 10^{-a}. You could also set up a number of iterations and manually select the row that meets the accuracy criterion.

PROGRAM DEVELOPMENT

Figure 9.4a shows the function plotted from 0.8 to 6.2. The function was not plotted for the full range zero to seven, because the value of the function becomes large near the end points. Inclusion of these large values would flatten out the definition of the plot in the vicinity of the roots. The plot shows that there is a root in the vicinity of $X = 6$, which we already knew. It also shows that there is a minimum point in the function somewhere between $X = 5$ and $X = 6$. To get a better idea of the behavior of the function in the vicinity of the root, the interval $X = 5.5$ to $X = 6.5$ was also plotted, and is shown in Figure 9.4b. It can be seen that the function is beyond the minimum point at $X = 5.5$, so we will choose that as the starting point. We chose a starting point somewhat away from the actual root of 6.0 to illustrate the convergence of the method.

Figure 9.5a shows the spreadsheet developed for determining a root of a polynomial equation. This particular spreadsheet is set up to estimate the roots for a polynomial in which each term contains the independent variable to a particular power. This is the standard form for a polynomial equation. Row 5 of the spreadsheet contains the coefficients of a polynomial of up to fifth order. If the polynomial to be evaluated is of a lower order, the unneeded coefficients can be set equal to zero. The solution table will contain all the elements of Equation 9.2. Row 8 is the initial evaluation of the function for X_0. Cell B8 contains the initial estimate of X, X_0, which will be used to evaluate the function and the derivative. The formula for evaluating the polynomial is defined by the formula in cell C8 of Figure 9.5b. This formula refers to the value of X_n using a relative reference, and refers to the coefficients of the polynomial with absolute references to facilitate copying of the formula into subsequent cells in the column. The value for $f'(X_n)$ is found in a similar manner. The formula is shown in cell D8 of Figure 9.5b. Note the mixed use of absolute and relative cell references in the formula to facilitate copying to other locations in the spreadsheet. The correction factor, $f(X_n)/f'(X_n)$, is computed by the formula +C8/D8 in cell E8. In

FIGURE 9.4 Polynomial function for finding roots.

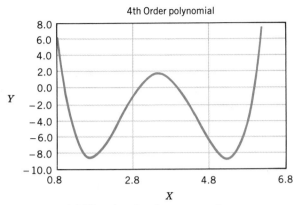

(a) Plot showing four approximate roots.

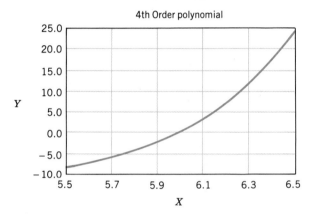

(b) Behavior of function in vicinity of root at $X = 6.0$.

the second row of the table, the formula $+A8+1$ is entered in A9 to compute the iteration number, and the formula $+B8-E8$ in cell B9 computes X_1 from X_0 and $f(X_0)/f'(X_0)$ according to Equation 9.2. Formulas for the last three cells in row 9 can be copied from the corresponding cell in row 8. Succeeding rows of the table can be copied from the row above.

PROGRAM TESTING

Figure 9.5a shows that the first step in the solution process flips to the other side of the root. You can see why this is so by examining Figure 9.4b. Because of the shape of the curve, the initial correction is fairly large and actually results in a second estimate that is farther away from the actual root than X_0. However, the solution converges nicely to the actual root in subsequent steps. Examination of Equations 9.2 and 9.3 reveals that the absolute value of the correction term, $|f(X_n)/f'(X_n)|$, is the error term for accuracy of the root estimate. Therefore, when the value of $f(X_n)/f'(X_n)$ is less than 10^{-a}, the estimate of X_{n+1} will be accurate to the number of decimal places defined by the value of variable a. In this case, we continued to add rows until the error term approached zero. This

(EXAMPLE 9.1 □ Spreadsheet Approximation of Real Roots of a Polynomial Function □ Continued)

FIGURE 9.5 ENABLE spreadsheet for finding roots using Newton-Raphson.

	A	B	C	D	E	F	G	H
1	FINDING REAL ROOTS OF A POLYNOMIAL EQUATION WITH NEWTON-RAPHSON							
2								
3	FUNCTION							
4	X^5	X^4	X^3	X^2	X	CONSTANT		
5	0	1	-14	67	-126	72		
6								
7	ITERATION	Xn	f(Xn)	f'(Xn)	f(Xn)/(f'(Xn)			
8	0	5.5	-8.4375	6	-1.40625			
9	1	6.90625	60.76491	113.8033	0.533947			
10	2	6.372303	16.00127	57.44753	0.278537			
11	3	6.093766	3.093865	36.08057	0.085749			
12	4	6.008017	0.242522	30.49901	0.007952			
13	5	6.000066	0.00197	30.00407	0.000066			
14	6	6	1.34E-07	30	4.46E-09			
15								

(a) Spreadsheet.

```
A9:    +A8+1
A10:   +A9+1
A11:   +A10+1
A12:   +A11+1
A13:   +A12+1
A14:   +A13+1
B9:    +B8-E8
B10:   +B9-E9
B11:   +B10-E10
B12:   +B11-E11
B13:   +B12-E12
B14:   +B13-E13
C8:    +$A$5*B8**5+$B$5*B8**4+$C$5*B8**3+$D$5*B8**2+$E$5*B8+$F$5
C9:    +$A$5*B9**5+$B$5*B9**4+$C$5*B9**3+$D$5*B9**2+$E$5*B9+$F$5
C10:   +$A$5*B10**5+$B$5*B10**4+$C$5*B10**3+$D$5*B10**2+$E$5*B10+$F$5
C11:   +$A$5*B11**5+$B$5*B11**4+$C$5*B11**3+$D$5*B11**2+$E$5*B11+$F$5
C12:   +$A$5*B12**5+$B$5*B12**4+$C$5*B12**3+$D$5*B12**2+$E$5*B12+$F$5
C13:   +$A$5*B13**5+$B$5*B13**4+$C$5*B13**3+$D$5*B13**2+$E$5*B13+$F$5
C14:   +$A$5*B14**5+$B$5*B14**4+$C$5*B14**3+$D$5*B14**2+$E$5*B14+$F$5
D8:    +$A$5*5*B8**4+$B$5*4*B8**3+$C$5*3*B8**2+$D$5*2*B8+$E$5
D9:    +$A$5*5*B9**4+$B$5*4*B9**3+$C$5*3*B9**2+$D$5*2*B9+$E$5
D10:   +$A$5*5*B10**4+$B$5*4*B10**3+$C$5*3*B10**2+$D$5*2*B10+$E$5
D11:   +$A$5*5*B11**4+$B$5*4*B11**3+$C$5*3*B11**2+$D$5*2*B11+$E$5
D12:   +$A$5*5*B12**4+$B$5*4*B12**3+$C$5*3*B12**2+$D$5*2*B12+$E$5
D13:   +$A$5*5*B13**4+$B$5*4*B13**3+$C$5*3*B13**2+$D$5*2*B13+$E$5
D14:   +$A$5*5*B14**4+$B$5*4*B14**3+$C$5*3*B14**2+$D$5*2*B14+$E$5
E8:    +C8/D8
E9:    +C9/D9
E10:   +C10/D10
E11:   +C11/D11
E12:   +C12/D12
E13:   +C13/D13
E14:   +C14/D14
```

(b) Formulas.

took six iterations for the function of Equation 9.4. It can also be seen that $f(X_n)$ is approaching zero as the iterations progress, which indicates that we are approaching the root. The final row of the table shows some round-off error, as the value of $f(X_n)$ should be zero for $X = 6$. The validity of the accuracy check can be verified by examining the values in columns B and E in Figure 9.5a. Note that the error term approaches zero quickly when the solution approaches the root. If you wanted to ensure two-decimal-place accuracy, you could stop at X_4, while the next estimate, X_5, gives four-decimal-place accuracy.

All that would have to be changed to search for another root would be the initial estimate for X_0 contained in cell B8. To analyze a different polynomial of the same form, all you would have to do is change the coefficients in row 5.

A TK SOLVER model to find the same root of Equation 9.4 will now be developed. TK SOLVER uses a modified form of Newton-Raphson iteration to solve either a single polynomial equation or a system of linear or nonlinear equations. As with the tabular Newton-Raphson method, it uses a convergence criterion to determine when the estimate is accurate enough.

EXAMPLE 9.2 □ TK SOLVER Solution for Real Roots of a Polynomial Equation

STATEMENT OF PROBLEM

Develop a TK SOLVER model that can be used to estimate real roots of a polynomial equation of up to order five.

ALGORITHM DEVELOPMENT

Input/Output Design

Polynomial coefficients and the initial guess of the root, X_0, will be input. Output will be the final estimate of the root accurate to the default convergence criterion of TK SOLVER. The default error tolerance is 0.000001. It can be reset in the GLOBAL sheet.

Numerical Methods

TK SOLVER employs a modified Newton-Raphson technique. This is a general method that is applicable to systems of linear or nonlinear equations.

Computer Implementation

Coefficients for the polynomial will be entered as variables to facilitate evaluation of any polynomial of the same form. Up to a fifth-order polynomial will be accommodated. TK SOLVER will not show the iterative steps in its output. It will deliver only the final estimate. However, you can set the maximum number of iterations to one, which will cause execution to pause after each iteration. The next iteration can be executed by pressing **[F9].** In this manner, you can view the convergence of the TK SOLVER solution.

(EXAMPLE 9.2 □ TK SOLVER Solution for Real Roots of a Polynomial Equation □ Continued)

PROGRAM DEVELOPMENT

The model was entered as shown in the RULE sheet of Figure 9.6a. Values for the coefficients of the polynomial were entered in the variable sheet, and X was specified as a guess, with an initial estimate of 5.5. When ! is pressed to solve the model, the output appears as shown in Figure 9.6b.

FIGURE 9.6   TK SOLVER model for finding roots of a polynomial equation.

```
=================== VARIABLE SHEET ========================================
St Input---- Name--- Output--- Unit----- Comment--------------------------
             Y
 G 5.5       X
    0        A5
    1        A4
  -14        A3
   67        A2
 -126        A1
   72        C
=================== RULE SHEET ============================================
S Rule--------------------------------------------------------------------
* Y = A5*X^5+A4*X^4+A3*X^3+A2*X^2+A1*X+C
* Y = 0.0
```
(a) Input model.

```
=================== VARIABLE SHEET ========================================
St Input---- Name--- Output--- Unit----- Comment--------------------------
             Y       1.603E-11
             X       6
    0        A5
    1        A4
  -14        A3
   67        A2
 -126        A1
   72        C
=================== RULE SHEET ============================================
S Rule--------------------------------------------------------------------
  Y = A5*X^5+A4*X^4+A3*X^3+A2*X^2+A1*X+C
  Y = 0.0
.PA
```
(b) Output model.

```
=================== VARIABLE SHEET ========================================
St Input---- Name--- Output--- Unit----- Comment--------------------------
    0        Y
             X       6
    0        A5
    1        A4
  -14        A3
   67        A2
 -126        A1
   72        C
=================== RULE SHEET ============================================
S Rule--------------------------------------------------------------------
  Y = A5*X^5+A4*X^4+A3*X^3+A2*X^2+A1*X+C
```
(c) Modified model.

PROGRAM TESTING

The error tolerance of TK SOLVER compares the left and right side of an equation; if the values agree within the prescribed tolerance, the iteration process is terminated. This is essentially the same error check employed with the tabular Newton-Raphson method. The default error tolerance results in the computer rounding the estimate to the exact root, as is shown in Figure 9.6*b*. However, the value of the function is not exactly zero; it is output as 1.6×10^{-11}. If you had formatted the output, this value would be rounded to zero for display. Also, if you eliminate the second rule, $Y = 0.0$, and specify that $Y = 0.0$ through an input value, the exact values would appear on the output, as is shown in Figure 9.6*c*.

If you wanted to solve for multiple roots of the equation, you could specify X as a list variable and store the various initial estimates of the roots in the X list. By specifying X as both the input list and a guess, and using the iterative list solver, you could generate the different root estimates as output in the X list.

We will now develop a FORTRAN program to implement the Newton-Raphson method to solve for the roots of Equation 9.4. The program will handle polynomial equations of up to fifth order. You would have to change the program for other polynomial forms. Example 9.5 will present a method that could be used to develop the program into a general subprogram.

EXAMPLE 9.3 □ FORTRAN Program for Finding Real Roots of a Fourth-Order Polynomial Using Newton-Raphson

STATEMENT OF PROBLEM

Develop a FORTRAN program that can be used to find a real root of a polynomial function given an initial estimate of the root and the required accuracy.

MATHEMATICAL DESCRIPTION

Equation 9.2 will be used to develop successive approximations of the root.

ALGORITHM DEVELOPMENT

Input/Output Design

Again, the program will be written in a format general enough to accommodate a polynomial of up to order five. Input will have to include the coefficients of the polynomial terms, the initial estimate of the root, and the decimal place accuracy desired. Output will include the iteration number, the final estimate of the root, and the accuracy criterion. The output could be modified to include each step of the iteration.

Numerical Methods

Iterations will continue until the convergence criterion of Equation 9.3 is satisfied to the specified number of decimal places.

(EXAMPLE 9.3 □ FORTRAN Program for Finding Real Roots of a Fourth-Order Polynomial Using Newton-Raphson □ Continued)

Computer Implementation

Figure 9.7 shows a flowchart for the Newton-Raphson solution of polynomial equations.

FIGURE 9.7   Flowchart for Newton-Raphson estimation of roots of a polynomial.

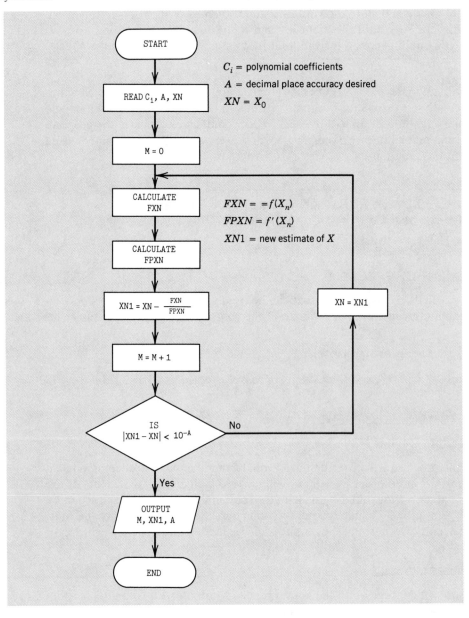

PROGRAM DEVELOPMENT

The program is developed directly from Figure 9.7.

```
********************************************************************
* NEWTON-RAPHSON SOLUTION FOR ROOTS OF A POLYNOMIAL EQUATION *
*   The program will solve for the roots of a polynomial      *
*     equation of up to fifth order                           *
*   Developed by:  T. K. Jewell            September 1975 *
********************************************************************
      REAL C(0:5), A, XN, FX, FPX, XN1, ACC
      INTEGER M
      PRINT *, 'Input coefficients for the polynomial equation'
      PRINT *, 'C(0) is the constant, C(1) the coefficient for'
      PRINT *, 'the first order term, etc.  Values for all six'
      PRINT *, 'coefficients have to be input.  Zero should be'
      PRINT *, 'entered for missing terms  '
      READ *, (C(I),I=0,5)
      PRINT *, 'Input decimal place accuracy desired  '
      READ *, A
      PRINT *, 'Input initial estimate of root  '
      READ *, XN
      M = 0
      CK = 1./10.**A
      ACC = 1000.
  100 CONTINUE
      IF (ACC.GE.CK) THEN
         FX = C(5)*XN**5+C(4)*XN**4+C(3)*XN**3+C(2)*XN**2+C(1)*XN+C(0)
         FPX=5.*C(5)*XN**4+4.*C(4)*XN**3+3.*C(3)*XN**2+2.*C(2)*XN+C(1)
         XN1 = XN - FX/FPX
         ACC = ABS(XN-XN1)
         M = M+1
         XN = XN1
         GO TO 100
      END IF
      PRINT 300,M,XN1,A
  300 FORMAT(//' AT ITERATION NUMBER',I3,' THE ROOT IS',F12.8,/,
     1          ' ACCURATE TO',F3.0,' DECIMAL PLACES')
      END
```

PROGRAM TESTING

Output for this program is the same as the data shown in Figures 9.5 and 9.6.

With any method of estimating roots of equations, it is wise to include a maximum number of iterations to guard against solutions that do not converge, or solutions that converge very slowly. It is also wise to substitute the final convergent estimate back into the function to make sure the function value is approximately zero. This is to guard against two slowly convergent estimates falling within the selected error bound.

The theory behind a second method of estimating roots of polynomial functions will now be investigated.

9.3.1.2 Secant Method

The secant method has many similarities to the Newton-Raphson method. The main difference is that you do not have to determine the derivative of the function to implement the secant method. You do, however, need two initial starting points.

The form of the algorithm is very similar to the form of the Newton-Raphson algorithm, as can be seen by Equation 9.5.

$$X_{n+2} = X_{n+1} - \frac{f(X_{n+1})}{Y_{n+1}} \tag{9.5}$$

The primary difference is that the term Y_{n+1} has replaced the derivative of the function. Y_{n+1} is actually an estimate of the derivative (slope) in the vicinity of the two points X_n and X_{n+1}. It is calculated by the formula of Equation 9.6.

$$Y_{n+1} = \frac{f(X_{n+1}) - f(X_n)}{X_{n+1} - X_n} \tag{9.6}$$

Figure 9.8 shows the application of Equations 9.5 and 9.6 for the points X_1 and X_0 to estimate the point X_2. The next step would be to estimate X_3 using X_2 and X_1. Iterations would continue until the change in the estimate of the root varies by less than the desired accuracy from one iteration to the next. Figure 9.9 shows the application of the method to find a root of Equation 9.5, with the initial estimates of $X = 5.5$ and $X = 5.6$. The solution converges to the root at $X = 6$. The method requires two more iterations than the Newton-Raphson technique to reach two decimal-place accuracy, but it is actually computationally more efficient. The reason for this is that at each iteration, the secant method evaluates only the function, not both the function and its derivative.

A variation of the secant method is the method of false position, which uses initial estimates on either side of the actual root. It is also referred to as the linear

---

FIGURE 9.8   Graphical interpretation of the secant method.

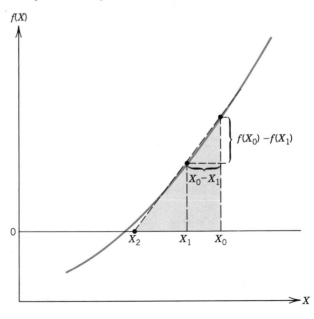

FIGURE 9.9   Application of the secant method.

```
A1:   FINDING REAL ROOTS OF A POLYNOMIAL EQUATION WITH SECANT METH        READY
 ─
          A         B         C         D         E         F        G       H
      1FINDING REAL ROOTS OF A POLYNOMIAL EQUATION WITH SECANT METHOD
      2
      3FUNCTION
      4    X^5       X^4       X^3       X^2       X     CONSTANT
      5    0         1        -14        67       -126      72
     ·6
      7ITERATION   Xn        f(Xn)       Yn      f(Xn)/Yn
      8    0       5.5       -8.4375
      9    1       5.6       -7.6544    7.831    -0.97745
     10    2     6.577449  29.69698  38.21314   0.777141
     11    3     5.800308  -4.83262  44.4316   -0.10877
     12    4     5.909073  -2.47895  21.63985  -0.11455
     13    5     6.023628   0.726288 27.97993   0.025957
     14    6     5.997671  -0.06971  30.66537  -0.00227
     15    7     5.999944  -0.00168  29.92611  -0.00006
     16    8     6          4.05E-06 29.99827   1.35E-07
     17    9     6         -2.3E-10  29.99997  -7.7E-12
     18
     19
     20
    #1      A:\ENABLE\FIG9_9.SSF                                          F94
```

interpolation method, or by its Latin name, regula falsi. The method of false position is described in Chapra and Canale (1988).

9.3.2 Numerical Integration

Integrating functions of one independent variable involves finding the area under the curve represented by the function between two given limits. Engineers study and learn the rules of calculus for integrating various functions. However, there are many functions for which it is extremely difficult or impossible to find analytical integrals. Determining centroids and moments of inertia for odd-shaped areas and finding the total flow from a hydrograph by summing incremental flow volumes are two examples for which it is difficult to apply calculus. These are situations where numerical integration is valuable. Any function that can be expressed in two dimensions can be approximately integrated by measuring the area under its curve. Numerical integration approximates the area as a summation of small differential area elements

$$A = \int_a^b f(x)dx \cong \sum_{i=1}^n \Delta A_i = \sum_{i=1}^n f(x_i)\Delta x \tag{9.7}$$

Different methods of numerical integration treat the function $f(x_i)$ differently and have varying amounts of error. Except for specialized situations, the differences between the methods are not significant. All these methods have the property of being more accurate the smaller the differential x approximation is. This approximation of dx is referred to as Δx. When you are using computers to accomplish the numerical integration, Δx can be made as small as is necessary to give the required accuracy. Texts on numerical analysis (James et al. 1985) contain error functions for evaluating the magnitudes of the errors for the different methods. Among the more popular methods of numerical integration are the trapezoidal rule and Simpson's rule. The trapezoidal rule will be illustrated here.

9.3.2.1 Integration by the Trapezoidal Rule

The trapezoidal rule treats the differential area (dA) approximation as a trapezoid, with the base equal to Δx and the sides equal to the value of the function $f(x_i)$ at each end of the differential, Δx. Thus

$$\Delta A_i = \left(\frac{f(x_i) + f(x_{i+1})}{2} \right) \Delta x \tag{9.8}$$

The approximation of the integral is the summation of all the differential areas. In Equation 9.8, $f(X_{i+1})$ will become the $f(X_i)$ for the next differential area. Because of this, the interior points will be used twice in the summation of areas. Using this characteristic, Equation 9.9 can be developed as

$$A = \sum_{i=1}^{n} \Delta A_i = (\Delta A_1 + \Delta A_2 + \cdots + \Delta A_n)$$

$$= \frac{\Delta x}{2} \{ [f(x_1) + f(x_2)] + [f(x_2) + f(x_3)] + \cdots + [f(x_n) + f(x_{n+1})] \}$$

$$= \frac{\Delta x}{2} \{ f(x_1) + 2f(x_2) + 2f(x_3) + \cdots + 2f(x_n) + f(x_{n+1}) \}$$

$$A = \frac{\Delta x}{2} \left\{ f(x_1) + 2 \sum_{i=2}^{n} f(x_i) + f(x_{n+1}) \right\} \tag{9.9}$$

FIGURE 9.10 $\operatorname{Sin}(\theta)$ for 0 to π radians.

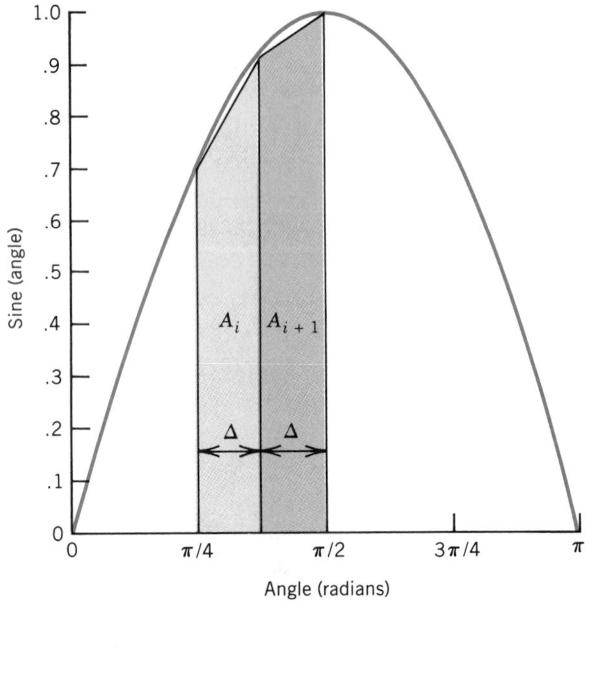

TABLE 9.1 EVALUATION OF
INTEGRAL OF $\sin(\theta)$ BY
TRAPEZOIDAL RULE

i	θ_i	$\sin(\theta_i)$	ΔA_i
1	0	0	
			0.0751
2	0.393	0.383	
			0.2140
3	0.785	0.707	
			0.3202
4	1.178	0.924	
			0.3778
5	1.571	1.000	
			0.3778
6	1.963	0.924	
			0.3202
7	2.356	0.707	
			0.2140
8	2.749	0.303	
			0.0751
9	3.142	0	
		Summation	1.974

which provides a computationally more efficient way of estimating the integral than adding up each individual incremental area.

To illustrate the trapezoidal rule method, the integral of $\sin(\theta)$ will be computed from 0 to π radians. By calculus,

$$\int_0^\pi \sin(\theta)\,d\theta = -\cos(\theta)\big|_0^\pi = 2.0 \qquad (9.10)$$

Figure 9.10 shows the function $\sin(\theta)$ from 0 to π radians. The figure also shows two representative ΔA strips. When Equation 9.10 is evaluated between these limits, it gives a numerical value of 2.0 for the area under the sine curve.

Table 9.1 shows the evaluation of this integral using the trapezoidal rule. The summation of ΔA_i is 1.974, which is fairly close to the actual value of 2.0. Even with the small number of intervals (fairly large θ increment), the numerical integral is a reasonably accurate estimate of the analytical integral.

Example 9.4 illustrates a spreadsheet application of the trapezoidal rule to find the centroid and moment of inertia for a triangle. This example is similar to Example 8.7, which illustrated the integration capabilities of TK SOLVER.

EXAMPLE 9.4 □ Centroid and Moment of Inertia for Triangle (Spreadsheet)

STATEMENT OF PROBLEM

Develop an ENABLE spreadsheet that can be used to approximate the y distance to the centroid and the moment of inertia of a triangle oriented as shown in Figure 9.11. Use this spreadsheet to approximate the centroid and moment of inertia for a triangle of base six and height nine.

MATHEMATICAL DESCRIPTION

The integral formulas for the y centroid and the moment of inertia about the X-X axis are shown in Equations 9.11 and 9.12. Equation 9.13 shows the expression for the differential area.

(EXAMPLE 9.4 □   Centroid and Moment of Inertia for Triangle (Spreadsheet) □   Continued)

FIGURE 9.11   Definition sketch for centroid and moment of inertia estimation.

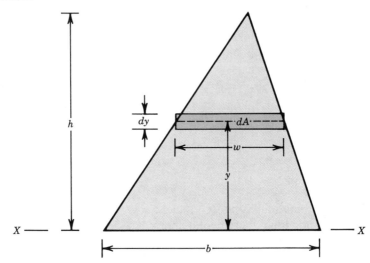

$$\bar{y} = \frac{\int_A y dA}{\int_A dA} \qquad (9.11)$$

$$I_x = \int_A y^2 dA \qquad (9.12)$$

$$dA = wdy = f(y)dy = \left[b - \left(\frac{b}{h} \right) y \right] dy \qquad (9.13)$$

ALGORITHM DEVELOPMENT

Input/Output Design

The base, height, and Δy values will be input variables. In addition to the computation table, output will include the centroid and moment of inertia.

Numerical Methods

The trapezoidal rule will be used to compute the differential areas of the triangle. Since the formulas also require taking differential moments, the y distance for the differential moment will be taken to the center of the differential area. This is the normal convention and is more accurate than taking the moment arm to either side of the differential area.

Computer Implementation

The spreadsheet will be set up as shown in Figure 9.12a. The computations for dA, ydA, and y^2dA will be offset on intermediate lines to indicate that they are

FIGURE 9.12 ENABLE spreadsheet for numerical integration.

```
NUMERICAL INTEGRATION TO FIND CENTROID AND MOMENT OF INERTIA OF TRIANGLE
 Base =      6
Height =     9
Delta Y =   1
    Point      Y        f(Y)        dA        YdA       Y**2dA
      1        0       6.0000
                                   5.6667    2.8333     1.4167
      2        1       5.3333
                                   5.0000    7.5000    11.2500
      3        2       4.6667
                                   4.3333   10.8333    27.0833
      4        3       4.0000
                                   3.6667   12.8333    44.9167
      5        4       3.3333
                                   3.0000   13.5000    60.7500
      6        5       2.6667
                                   2.3333   12.8333    70.5833
      7        6       2.0000
                                   1.6667   10.8333    70.4167
      8        7       1.3333
                                   1.0000    7.5000    56.2500
      9        8       0.6667
                                   0.3333    2.8333    24.0833
     10        9       0.0000
                                   -------   -------    -------
            Summations            27.0000   81.5000   366.7500
            Centroid    3.0185
Moment of Inertia 366.7500
```

(*a*) Spreadsheet.

```
B10:  +B8+$B$5
B12:  +B10+$B$5                    D19:  (C18+C20)/2*$B$5
B14:  +B12+$B$5                    D21:  (C20+C22)/2*$B$5
B16:  +B14+$B$5                    D23:  (C22+C24)/2*$B$5
B18:  +B16+$B$5                    D25:  (C24+C26)/2*$B$5
B20:  +B18+$B$5                    D28:  @sum(D9..D25)
B22:  +B20+$B$5                    E9:   (B8+B10)/2*D9
B24:  +B22+$B$5                    E11:  (B10+B12)/2*D11
B26:  +B24+$B$5                    E13:  (B12+B14)/2*D13
C8:   +$B$3-$B$3/$B$4*B8           E15:  (B14+B16)/2*D15
C10:  +$B$3-$B$3/$B$4*B10          E17:  (B16+B18)/2*D17
C12:  +$B$3-$B$3/$B$4*B12          E19:  (B18+B20)/2*D19
C14:  +$B$3-$B$3/$B$4*B14          E21:  (B20+B22)/2*D21
C16:  +$B$3-$B$3/$B$4*B16          E23:  (B22+B24)/2*D23
C18:  +$B$3-$B$3/$B$4*B18          E25:  (B24+B26)/2*D25
C20:  +$B$3-$B$3/$B$4*B20          E28:  @sum(E9..E25)
C22:  +$B$3-$B$3/$B$4*B22          F9:   ((B8+B10)/2)**2*D9
C24:  +$B$3-$B$3/$B$4*B24          F11:  ((B10+B12)/2**2*D11
C26:  +$B$3-$B$3/$B$4*B26          F13:  ((B12+B14)/2**2*D13
C30:  +E28/D28                     F15:  ((B14+B16)/2**2*D15
C31:  +F28                         F17:  ((B16+B18)/2**2*D17
D9:   (C8+C10)/2*$B$5              F19:  ((B18+B20)/2**2*D19
D11:  (C10+C12)/2*$B$5             F21:  ((B20+B22)/2**2*D21
D13:  (C12+C14)/2*$B$5             F23:  ((B22+B24)/2**2*D23
D15:  (C14+C16)/2*$B$5             F25:  ((B24+B26)/2**2*D25
D17:  (C16+C18)/2*$B$5             F28:  @SUM(F9..F25)
```

(*b*) Formulas.

(EXAMPLE 9.4 □ Centroid and Moment of Inertia for Triangle (Spreadsheet) □ Continued)

combinations of the widths, $f(y)$, at the top and bottom of the differential area.

PROGRAM DEVELOPMENT

Figure 9.12*b* gives the formulas for the spreadsheet. The function $f(Y)$ in the spreadsheet is the width of the triangle at the height Y. Note that absolute references are used to refer to the base and height so that when the formula is copied the reference will stay the same. Summations for the differential elements are at the bottom of the table. The rightmost column gives the approximate moment of inertia directly. The centroid is the ratio of the summation of the YdA and dA columns. The table could easily be expanded or shrunk for more or fewer differential elements by using the INSERT or DELETE ROWS, and COPY commands. The spreadsheet could also easily be adapted to do any numerical integration by modifying the formula for finding $f(Y)$ and using the appropriate summations.

PROGRAM TESTING

Integration shows that the actual centroid is three units above the X-X axis, whereas the actual moment of inertia about the X-X axis is 364.5. The TK SOLVER solution of Example 8.7 produced the actual moment of inertia. The present solution, using a Δy of 1.0, estimates the centroid to be 3.02 and the moment of inertia to be 366.75. If a Δy of 3.0 is used, the centroid estimate is 3.17, and the moment of inertia estimate is 384.75. From these results you can see that the increasing number of differential elements improves the estimates of the integrals.

When it is possible to perform an analytical solution for an integral, you are well advised to do so. However, numerical integration is an essential tool when there is no analytical solution for the integral. At other times numerical integration may be more efficient than trying to find the analytical solution. The next section will show that the computer is also a valuable tool in estimating derivatives.

9.3.3 Numerical Differentiation and Finding Minimum and Maximum Points of Functions

A first derivative gives the rate of change, or slope, of a function $Y = f(X)$ for a particular value of the independent variable X. The second derivative gives the rate of change of the slope. For trajectory or displacement problems, the first derivative is the velocity, and the second derivative is the acceleration. Your first course in calculus taught you how to find analytical derivatives and introduced you to the various properties of derivatives. You also learned that a value of X that makes the derivative equal to zero is a minimum or maximum point of the function. The second derivative will indicate whether that point is a minimum or a maximum. A positive second derivative indicates a minimum, whereas a negative second derivative indicates a maximum. For functions of two or more independent variables, the minimum or maximum points are found where the partial derivatives with respect to each independent variable are simultaneously equal to zero.

Some functions do not have analytical derivatives, and the derivatives for others may be difficult to find. Thus, there is a need for numerical differentiation, just as there is a need for numerical integration. You will also have to perform numerical minimization or maximization (optimization) of functions when you cannot evaluate the derivative. Numerical optimization may be a preferable method for multivariable problems in which numerical methods would have to be employed to solve the system of partial differential equations.

9.3.3.1 Numerical Differentiation

Numerical differentiation involves estimating the slope of a function at a point by using the linear slope between two points on the function. There are three different methods for doing this, all of which use finite divided differences. Figure 9.13a illustrates the forward finite divided difference method, in which the function is evaluated at X and $X + \Delta X$ and the slope is estimated by $\Delta Y/\Delta X$. The formula for doing this is

$$f'(X_i) \approx \frac{f(X_{i+1}) - f(X_i)}{X_{i+1} - X_i} = \frac{f(X + \Delta X) - f(X)}{(X + \Delta X) - X} \qquad (9.14)$$

For the function depicted in Figure 9.13a, the approximate derivative gives a slope somewhat steeper than the actual slope. The second method is the backward finite divided difference method, illustrated in Figure 9.13b. The application is similar, except that the second point is taken at $X - \Delta X$, and the formula is

$$f'(X_i) \approx \frac{f(X_i) - f(X_{i-1})}{X_i - X_{i-1}} = \frac{f(X) - f(X - \Delta X)}{X - (X - \Delta X)} \qquad (9.15)$$

This method gives an estimated slope somewhat less than the actual slope for the function depicted in Figure 9.13b. A third method is a combination of the forward and backward difference estimates, given by the formula

$$f'(X_i) \approx \frac{f(X_{i+1}) - f(X_{i-1})}{X_{i+1} - X_{i-1}} = \frac{f(X + \Delta X) - f(X - \Delta X)}{2\Delta X} \qquad (9.16)$$

provided that the ΔX increments are equal. The central difference method generally gives a better estimate of the slope than either of the other methods. Figure 9.13c shows this method for the function used to illustrate both the forward and backward estimates. You should also remember that generally the smaller ΔX is, the better the approximation will be.

Second derivatives can also be approximated numerically by taking second divided differences. Any of the three methods described for finding estimates of the first derivatives can be used, except that now the divided difference has to be taken for estimates of the slope. Only the central difference version of the second divided difference will be illustrated here. The forward or backward versions would use the same procedure, except that both ΔX increments would be taken on

FIGURE 9.13   Methods for numerically estimating derivatives.

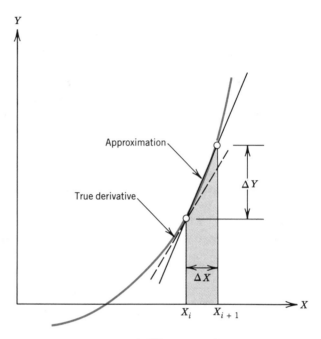

(*a*) Forward difference estimate.

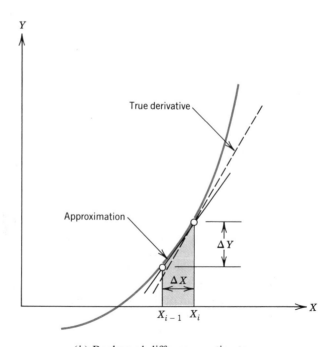

(*b*) Backward difference estimate.

FIGURE 9.13 (Continued)

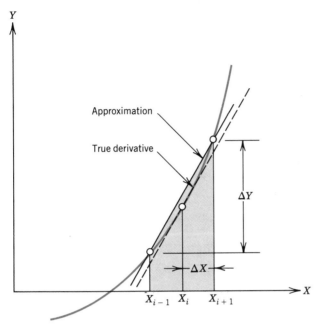

(c) Central difference estimate.

the same side of the selected point. As with the first divided difference, the central estimate generally gives more accurate results.

The central second divided difference is taken as the difference between the forward and backward first divided differences. In equation form

$$f''(X_i) = \frac{\dfrac{f(X_{i+1}) - f(X_i)}{\Delta X} - \dfrac{f(X_i) - f(X_{i-1})}{\Delta X}}{\Delta X}$$

$$= \frac{f(X_{i+1}) - 2f(X_i) + f(X_{i-1})}{\Delta X^2}$$

(9.17)

gives the numerical estimate of the rate of change of the slope of the function at X_i.

Example 9.5 illustrates a FORTRAN subprogram for finding the first and second derivatives using the central finite divided difference method.

EXAMPLE 9.5 □ Finding the Slope and Rate of Change of the Slope for a Polynomial Function (FORTRAN)

STATEMENT OF PROBLEM

Develop a general FORTRAN subprogram that will return estimates for the first and second derivatives of a function $Y = f(X)$ at any selected point X using the central finite divided difference method. Use this subprogram to estimate the slope and rate of change of the slope for Equation 9.4 when $X = 5.5$. Compare your estimates with the analytical solution values and compute the relative error for each.

MATHEMATICAL DESCRIPTION

Central divided differences will be calculated as per Equations 9.16 and 9.17.

ALGORITHM DEVELOPMENT

Input/Output Design

Input to the subprogram will include the value of X, the ΔX increment, and the reference to the function. Output will be the first and second derivative estimates.

Computer Implementation

An EXTERNAL specification will be used in the main program so that the function $Y = f(X)$ can be included as an argument in the subprogram. Section 5.10.2 of Chapter 5 described the syntax of the EXTERNAL statement. The function will be included in a FUNCTION subprogram after the main program. A loop will be included in the program so that multiple X points can be evaluated without exiting the program.

PROGRAM DEVELOPMENT

The general SUBROUTINE subprogram is as follows:

```
      SUBROUTINE DIFF(X,DELTX,FUNC,FIRST,SECOND)
**********************************************************************
*   SUBROUTINE for estimating first and second derivatives     *
*     using central finite divided differences                 *
*   Developed by:  T. K. Jewell              November 1988      *
**********************************************************************
C   X = Point at which first and second derivatives will be estimated
C   DELTX = X increment for divided difference
C   FUNC = f(x)
C   FIRST = Estimate of first derivative
C   SECOND = Estimate of second derivative
      REAL X,DELTX,FUNC,X1,X2,DIFF1,DIFF2,FIRST,SECOND
      X1 = X + DELTX
      X2 = X - DELTX
      DIFF1 = FUNC(X1)-FUNC(X)
      DIFF2 = FUNC(X)-FUNC(X2)
      FIRST = (DIFF1+DIFF2)/(2.*DELTX)
```

```
      SECOND = (DIFF1-DIFF2)/DELTX**2
      RETURN
      END
```

The main program and FUNCTION subprogram are as follows:

```
      EXTERNAL FUNC
      COMMON C(0:5)
      REAL FUNC,X,FIRST,SECOND,DELTX
C  FPX = Analytical first derivative for comparison
C  FDPX = Analytical second derivative for comparison
      PRINT *, 'Input coefficients for the polynomial equation'
      PRINT *, 'C(0) is the constant, C(1) the coefficient for'
      PRINT *, 'the first order term, etc.  Values for all six'
      PRINT *, 'coefficients have to be input.  Zero should be'
      PRINT *, 'entered for missing terms  '
      READ *, (C(I),I=0,5)
      PRINT *, (C(I),I=0,5)
      PRINT *, ' '
  100 CONTINUE
      PRINT *, 'Input X and DELTX'
      PRINT *, 'A value of zero for DELTX will terminate the loop  '
      READ *, X,DELTX
      IF(DELTX.EQ.0.0) STOP
      CALL DIFF(X,DELTX,FUNC,FIRST,SECOND)
C  Calculation of analytical derivative values
      FPX = 5.*C(5)*X**4+4.*C(4)*X**3+3.*C(3)*X**2+2.*C(2)*X+C(1)
      FDPX = 20.*C(5)*X**3+12.*C(4)*X**2+6.*C(3)*X+2.*C(2)
      ERR1 = (FIRST-FPX)/FPX
      ERR2 = (SECOND-FDPX)/FDPX
      PRINT 200, FIRST,FPX,ERR1
  200 FORMAT(' First Derivative',/,'   Estimate = ',F8.4,/,
     1       '    Actual = ',F8.4,/,'  Relative Error = ',F8.4)
      PRINT 300, SECOND,FDPX,ERR2
  300 FORMAT(' Second Derivative',/,'   Estimate = ',F8.4,/,
     1       '    Actual = ',F8.4,/,'  Relative Error = ',F8.4,/)
      GO TO 100
      END
      REAL FUNCTION FUNC(X)
      COMMON C(0:5)
      FUNC = C(5)*X**5+C(4)*X**4+C(3)*X**3+C(2)*X**2+C(1)*X+C(0)
      RETURN
      END
```

PROGRAM TESTING

The analytical derivatives for Equation 9.4 are

$$f'(X) = 4X^3 - 42X^2 + 134X - 126 \qquad (9.18)$$
$$f''(X) = 12X^2 - 84X + 134 \qquad (9.19)$$

When evaluated for $X = 5.5$, these give a first derivative of 6.0 and a second derivative of 35.0. These agree with the values found by the program.

It should be noted that this program could also be used to search for a minimum or maximum value by successively refining the estimate of X. Minimization and maximization of functions will be discussed in the next section.

9.3.3.2 Minimum and Maximum Points for Functions

You will frequently need to find a minimum or maximum point in a function of one or more independent variables. It could be a maximum point on a parabolic curve, a minimum point on a cumulative cost curve, or a peak on a three-dimensional surface. A special case would be a saddle point, which is a minimum point with respect to one axis and a maximum with respect to the other.

For easily differentiable functions, you can take the derivative and solve for values of the independent variable that make the derivative function equal to zero. Evaluation of the second derivative at a zero point of the derivative will show you if the original function is maximized or minimized for that value of the independent variable. A positive value for the second derivative implies a minimum. A negative value implies a maximum.

There is no infallible method for finding the minimum and maximum points of more complex functions. Even when you take analytical derivatives and solve for the roots, you have to ascertain whether you have found a global minimum or maximum, or just a local minimum or maximum. If you graph the function for two- and three-dimensional problems, you will stand a better chance of finding minimum or maximum points.

Functions that can be plotted in two dimensions have one dependent variable and one independent variable. A plot of $Y = f(X)$ will readily identify approximate minimum and maximum points for the function in the interval of interest. Successive refinement of the interval can be used to converge to the minimum or maximum. When you are searching for roots, you will be assured of at least one root in an interval if the function changes signs in the interval. When you are searching for a minimum or a maximum, you need to find a triplet of points such that the middle point is either less than or greater than both of the outer points. This confirms that there is a minimum or maximum point in the interval. Using a series of function values will assist you in finding a triplet with the required properties. The outer points of the triplet will help you refine the search interval.

Example 9.6 will illustrate a spreadsheet search for finding a minimum or maximum point for any function of X. A plot will identify approximate minimum or maximum points, then successive refinement of tabular values will be used to converge to the global minimum or maximum value.

EXAMPLE 9.6   □   Finding Minimum and Maximum Point for Polynomial Function (Spreadsheet)

STATEMENT OF PROBLEM

Use an ENABLE spreadsheet to determine how many minimum or maximum points occur in the interval between $X = 1$ and $X = 6$ for Equation 9.4. Next, use the spreadsheet to find the magnitude of the minimum or maximum point most distant from the X axis.

MATHEMATICAL DESCRIPTION

Equation 9.4 is the polynomial function

$$Y = X^4 - 14X^3 + 67X^2 - 126X + 72 \qquad (9.4)$$

FIGURE 9.14   Plot of Equation 9.4 from $X = 1$ to $X = 6$.

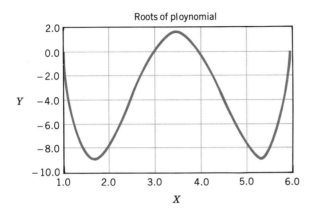

ALGORITHM DEVELOPMENT

Input/Output Design

Initially, the range $X = 1$ to $X = 6$ was plotted, and is shown in Figure 9.14. Input for the convergence iterations will include the beginning and ending X values for the test. Output will include 11 values of $f(X)$ and $f'(X)$ for 11 values of X evenly spaced in the test interval. The minimum and maximum values of $f(X)$ in the interval will also be output.

Numerical Methods

Successive refinement of the test interval will produce closer estimates of the minimum or maximum points.

Computer Implementation

This spreadsheet can be adapted from the spreadsheet of Example 9.1, which was used to find the roots of Equation 9.4. The input of the function and the computation of $f(X)$ and $f'(X)$ will be exactly the same. Eleven X values will be used. These values will be evenly spaced from the initial to final X values. Thus, the increment will be $(X_{final} - X_{initial})/10$. The minimum value for $f(X)$ in the interval will be found using the function @MIN(C10..C20). The maximum value will be found using the @MAX function for the same range. The derivative is included in the spreadsheet to help identify the approximate minimum point. The increment where the derivative changes sign is the interval in which the minimum or maximum value will occur.

PROGRAM DEVELOPMENT

Examination of Figure 9.14 shows that Equation 9.4 has two minimum and one maximum point in the interval $X = 1$ to $X = 6$. Both minima are farther from the X axis than the maximum, so their values will be estimated to see which one is the farthest from the X axis. It appears that the function is symmetrical in the interval, so the two minima may be equal. We will estimate both to confirm or disprove this hypothesis. Figure 9.15a shows the spreadsheet as adapted from

(EXAMPLE 9.6 □ Finding Minimum and Maximum Point for Polynomial Function (Spreadsheet) □ Continued)

FIGURE 9.15 ENABLE spreadsheet for determining minimum or maximum of a polynomial function.

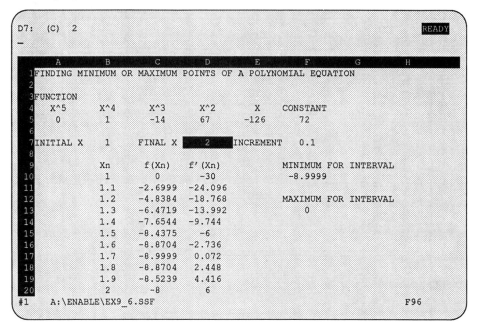

(*a*) Initial spreadsheet with interval $X = 1$ to $X = 2$.

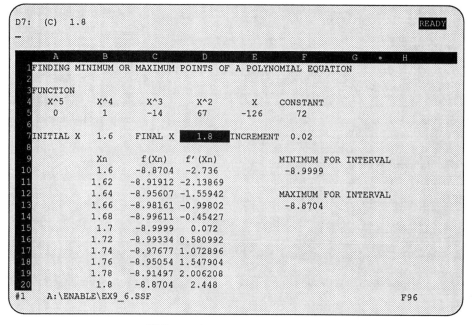

(*b*) Interval $X = 1.6$ to $X = 1.8$.

FIGURE 9.15   (Continued)

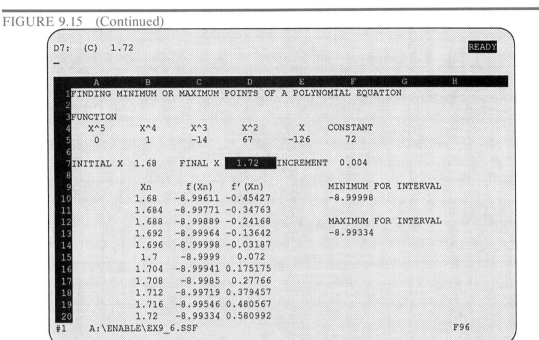

(c) Interval $X = 1.68$ to $X = 1.72$.

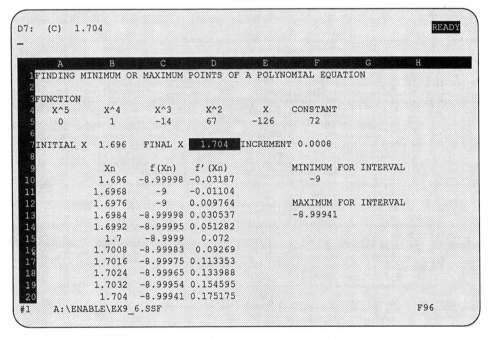

(d) Final interval for $X = 1.696$ to $X = 1.704$.

(EXAMPLE 9.6 □ Finding Minimum and Maximum Point for Polynomial Function (Spreadsheet) □ Continued)

FIGURE 9.15 (Continued)

| D7: (C) 5.310 | | | | | | | READY |

FINDING MINIMUM OR MAXIMUM POINTS OF A POLYNOMIAL EQUATION

FUNCTION

X^5	X^4	X^3	X^2	X	CONSTANT
0	1	-14	67	-126	72

INITIAL X 5.298   FINAL X 5.31   INCREMENT 0.0012

Xn	f(Xn)	f'(Xn)		MINIMUM FOR INTERVAL
5.298	-8.9997	-0.12367		-9
5.2992	-8.99983	-0.09269		
5.3004	-8.99993	-0.06164		MAXIMUM FOR INTERVAL
5.3016	-8.99998	-0.03054		-8.99932
5.3028	-9	0.000633		
5.304	-8.99998	0.031866		
5.3052	-8.99992	0.063161		
5.3064	-8.99983	0.094518		
5.3076	-8.9997	0.125937		
5.3088	-8.99953	0.157419		
5.31	-8.99932	0.188964		

#1   A:\ENABLE\EX9_6.SSF   F96

(e) Final interval for second minimum, $X = 5.298$ to $X = 5.310$.

the spreadsheet of Example 9.1, with an initial test interval of $X = 1$ to $X = 2$. This shows that the minimum point is in the vicinity of $X = 1.7$. Figures 9.15b, 9.15c, and 9.15d show successive estimates of the minimum point using narrower X intervals. From Figure 9.15d the approximate minimum point is equal to -9.0 and occurs for an X value of approximately 1.697. Figure 9.15e shows the last interval for estimating the second minimum point. This gives a minimum value of -9.0 for an X value of 5.303. The function is symmetrical.

PROGRAM TESTING

The minimum points found with this spreadsheet agree with the points where the analytical derivative is approximately equal to zero. This can be confirmed by solving for the roots of the derivative of Equation 9.4 using the Newton-Raphson or other root-finding algorithm.

9.4 CHOOSING THE BEST COMPUTATIONAL TOOL

You have now gained some experience in using various computer applications programs to solve engineering problems. You have also solved some problems with more than one type of applications program. We will now discuss some of the considerations for choosing a particular tool for an application. First we need to take one more look at the particle trajectory problem that we have solved several times to see if we have missed anything.

9.4.1 Particle Trajectory Revisited

We have looked at various forms of a particle trajectory problem in Examples 4.14, 8.2, 8.4, and 8.12. The objective in these examples was to find an angle or angles for a trajectory with a given initial velocity that would hit a target a distance X away from and a distance Y above (or below) the point of launch. The model

$$Y = V_{0_y} t - \frac{1}{2} g t^2 \qquad X = V_{0_x} t$$
$$V_{0_x} = V_0 \cos \theta \qquad V_{0_y} = V_0 \sin \theta \tag{9.20}$$

was developed, and substitutions were made to solve for Y in terms of the angle θ, which gave

$$Y = V_0 \sin \theta \left(\frac{X}{V_0 \cos \theta} \right) - \frac{1}{2} g \left(\frac{X}{V_0 \cos \theta} \right)^2 \tag{9.21}$$

At the time it was not apparent how this equation could be solved explicitly for θ, so a trial-and-error solution was undertaken. However, if we reach a little further into our repertoire of trigonometric identities, we may be able to find something that will simplify Equation 9.21. If all constants are grouped together, the equation becomes

$$Y = X \left(\frac{\sin \theta}{\cos \theta} \right) - \left[\frac{1}{2} g \left(\frac{X^2}{V_0^2} \right) \right] \frac{1}{\cos^2 \theta} \tag{9.22}$$

Since

$$\boxed{\frac{\sin \theta}{\cos \theta} = \tan \theta} \tag{9.23}$$

it would be helpful if we could find an expression for

$$\frac{1}{\cos^2 \theta}$$

in terms of $\tan \theta$. In fact we can, because

$$\boxed{\frac{1}{\cos^2 \theta} = \sec^2 \theta = 1 + \tan^2 \theta} \tag{9.24}$$

If Equations 9.23 and 9.24 are substituted into 9.22, it becomes

$$Y = X \tan \theta - \left[\frac{1}{2} g \left(\frac{X^2}{V_0^2} \right) \right] (1 + \tan^2 \theta) \tag{9.25}$$

or

$$\boxed{\left[\frac{1}{2} g \left(\frac{X^2}{V_0^2} \right) \right] \tan^2 \theta - X \tan \theta + \left[Y + \frac{1}{2} g \left(\frac{X^2}{V_0^2} \right) \right] = 0} \tag{9.26}$$

which is in the form of the quadratic equation

$$AZ^2 + BZ + C = 0 \tag{9.27}$$

so that the quadratic equation can be used to solve for the values of tangent θ that satisfy the equation. If

$$A = \frac{1}{2} g \left(\frac{X^2}{V_0^2}\right), \qquad B = -X, \qquad \text{and} \qquad C = Y + \frac{1}{2} g \left(\frac{X^2}{V_0^2}\right) \tag{9.28}$$

then the data for the original three cases of Example 4.14 give the following values:

Case	V0	X	Y	A	B	C
1	55	100	30	53.2	−100	83.2
2	66	90	37	29.9	−90	66.9
3	77	90	20	22.0	−90	42.0

Applying the quadratic formula to these data reveals that there are no real solutions for the first case because $B^2 - 4AC$ is negative. In other words, there are no angles that hit the target. The second case indicates there are roots at tangent θ equal to 1.667 and 1.342, which correspond to angles of 59.05° and 53.30°. The third case gives angles of 74.31° and 28.22°, the first of which is within the criteria established. These are the same angles that we found in the earlier examples.

This shows that searching our basic knowledge bank can often simplify our analysis tasks. In this case it would also save time. Although the search techniques we have used for the previously cited examples are all valid and give correct answers, we can get the same results with increased accuracy by making one simple substitution. Special thanks go to student William Earle for pointing this out to me.

9.4.2 Comparison of Text Examples

Examples 5.4, 8.3, 8.5, and 8.8 developed regression analysis to fit the best straight line through a set of X-Y data values. A FORTRAN program, TK SOLVER model, and LOTUS 1–2–3 spreadsheet were used. The number of data points used in the regression analysis may vary considerably for different problems. The flexible loop control structures of FORTRAN are useful for handling varying amounts of data. However, the ease of graphing the data is a definite advantage of TK SOLVER or LOTUS 1-2-3 applications.

Examples 4.14, 8.2, 8.4, and 8.12 involved solving by trial and error for the angle or angles that would cause a particle trajectory to hit a target some distance X away from and some height Y above or below the point of launch. The FORTRAN program for this problem was fairly involved and was only as accurate as the step size used for changing the angle. The application of TK SOLVER in Chapter 8 showed that the program would iterate quickly to the correct answer without having to write any type of procedural program. To efficiently apply an equation solver, you need to have some additional insight into the behavior of the problem to determine cases that will not hit the target for any angle, or that would hit the target only once. However, the ability of TK SOLVER to graph the solution will provide more than enough information on the behavior of the trajectory. With the ENABLE spreadsheet solution of Chapter 8, the path of the trajectory and the range of possible heights at the desired distance could readily be seen either through the tabular solution or by graphing the data. Thus it could easily be discerned how many times the trajectory would hit the target. The actual values for the angles could easily be found by successive convergence of the interval over which angles are taken. The section immediately preceding this one showed that if you dig a little deeper, you can modify the model so that you do not have to apply trial and error to solve it. This shows that you should not quickly decide to use a numerical method without spending a little extra time trying to simplify your model.

Examples 7.4 and 8.9 illustrated matrix inversion and multiplication. The FORTRAN application of Example 7.4 provided flexible data input. Separate FORTRAN subprograms could be used to develop the coefficients to be included in the matrix and the right-hand-side vector. The spreadsheet application of Example 8.9 was a quick and easy way of solving a system of equations when the coefficients were known.

Example 8.7 used one of the modules provided with TK SOLVER to perform numerical integration. The application found the moment of inertia for a triangle. The solution was straightforward, gave answers that agreed with the analytical solutions, and required only formulation of the differentials. Example 9.4 repeated the calculation of the moment of inertia using an ENABLE spreadsheet and the trapezoidal rule. This solution had the advantage of showing the magnitude of each step in the integration, but the number of integration intervals could not be changed as readily as would be possible with a FORTRAN program.

Examples 9.1, 9.2, and 9.3 showed ENABLE spreadsheet, TK SOLVER, and FORTRAN implementations of the Newton-Raphson method for finding roots of a polynomial equation. The spreadsheet was an excellent tool for showing the mechanics and convergence of the method. The TK SOLVER model was the quickest and easiest to develop, but it did not show you anything about the solution process. Development of the FORTRAN program was straightforward, but it took more time than the other methods. The FORTRAN program could be modified into a general subprogram for use with larger applications programs.

It can be seen from these brief discussions that different computational tools might be appropriate for the same application, depending on the type of information you are trying to develop. Generally, a FORTRAN program is better able to accommodate a variable number of data points or computation iterations. Spreadsheets provide a great deal of information on the dynamics of the solution process and respond quickly to changes in input data for sensitivity analysis. Equation solvers are extremely valuable if you require iterative solutions of functions for which you have a good understanding of the functional behavior in the region of interest. Being thoroughly familiar with these applications, as well as with database applications, will allow you to evaluate and choose the best tool for a particular problem application.

9.4.3 Comparison of Tools

The following is a brief description of the types of problems that are best suited to each of the computational tools that we have studied.

9.4.3.1 FORTRAN

FORTRAN 77 is a flexible language that is well-suited to problems that require many repetitive calculations, branching solution logic, or varying amounts of input and output. Since FORTRAN 77 is a structured language, it accepts user-developed modules that can be saved and used in other applications. You can also acquire modules that were written by others and incorporate them into your FORTRAN applications. Batch processing of FORTRAN programs is best suited for applications that will require dedication of a large portion of the computing resource available. The program may require most of the computer's resources because of its memory requirements, heavy input and output device usage, or large volumes of computation. Interactive FORTRAN programming is advantageous when you desire to input data as you run your application. Interactive programs are generally easier to use for people who have limited knowledge of computers and programming. However, interactive computer programs are more difficult to develop, because you have to concentrate on the user interface as well as the primary solution logic. If you are developing interactive applications, be sure to include error trapping and the option to save input data in a file.

You may have asked yourself, "Why have I learned high-level language programming when almost every application shown in this text could be done more efficiently with another tool?" You may have already answered the question for yourself.

It is indeed true that you have seen many cases in which equation solvers, spreadsheets, or database management systems could solve a particular problem more efficiently than a high-level language. However, you have seen other cases in which the flexibility and power of a high-level language are essential if a meaningful solution is to be found. This is especially true for larger problems and for problems that require many repetitive calculations. Knowing a high-level language helps you in developing the logic of solution algorithms, which in turn helps when you are developing applications for the other computer software tools. It will help you to understand the power and the limitations of the computer. Knowing a high-level language will give you the ability to modify or write programs when necessary. If you do not know a high-level language, how will you decide when that is the best tool to use for a particular application? High-level language competence will aid you in communicating with programmers, computer scientists, and other engineers. It is one component of what you need to establish your professional competence.

9.4.3.2 Spreadsheets

Spreadsheets are especially useful for manipulating data that can be arrayed in a tabular format. They are also well-suited to problems that require interactive iteration and recalculation, such as sensitivity analysis and repetitive design approximations. With spreadsheets, any data can be easily changed, and the effects of the changes on the output can be viewed almost instantaneously. Many spreadsheet applications can be developed and saved as templates for future use. Most spreadsheets also allow incorporation of elements from one spreadsheet into another, making it possible for you to save and recall procedures. The better spreadsheets also have preprogrammed procedures such as matrix operations, statistical analysis, and presentation graphics capabilities.

9.4.3.3 Equation Solvers

Equation solvers are excellent tools for the analysis of linear or nonlinear models that require iterative or trial-and-error solutions. If an engineer has a microcomputer conveniently available for use, an equation solver can function as an extremely powerful calculator. It will allow the engineer to solve many complex problems without having to do any programming. A set of commonly solved problems can be kept on disk, read in, and modified as necessary for new problems. Multiple problem files can be loaded together for more complex applications.

9.4.3.4 Database Management Software

As its name implies, database management software is best used for the storage, retrieval, and manipulation of large bodies of data. Related data in two or more databases can be accessed and used for analysis. Specialized commands are incorporated into the software to facilitate working with the data. Applications developers can produce user interfaces that allow people who know very little about the actual database management system to work with the data. Any number of customized reports can be developed for the database.

9.4.3.5 Integrated Packages

Integrated packages have several applications combined together. The most common applications are word processing, spreadsheet, graphics, database management, and communications. They are excellent for applications in which you need to pass data from one of these applications to another. Using an integrated package, you can easily develop reports containing spreadsheet data, graphics, or database material.

9.4.4 Commercial Applications Software

When searching for the best computational tool, we should give high priority to the identification of a previously developed program or application that is appropriate for our present problem. You should go to the trouble of developing your own application only if nothing else is available. This program or application may be one that you have developed, or it could have been developed by someone else in your firm. If there is nothing available within your firm, you should investigate commercial applications software.

Commercial applications software is any program or applications file that you acquire from a source external to your organization. Applications software can be free or can cost thousands of dollars. Since the quality of the software is not always a function of the price, you have to be extremely careful in evaluating the applicability of a particular package for a particular problem. Even so, it will often be advantageous for you to use commercial applications software.

9.4.4.1 Classification

Commercial applications software can be classified as proprietary and public domain. The rights to proprietary software have been retained by the developers. You must purchase a license to use the software for specific purposes. The listings of the source program are generally not made available for proprietary software. In public domain software, the developers have relinquished their rights to the software, and it is available for anyone to use. There may be a charge for user support. Most software developed by government agencies falls into the public domain classification.

9.4.4.2 Evaluation

Careful evaluation of commercial software is essential if you plan to use it for engineering applications. First of all, you should acquire and evaluate the program's user documentation. Poorly written documentation often indicates a poorly written source code. In your evaluation, try to anticipate problems that should have been illustrated but were not. These may reveal weaknesses in the program.

In evaluating a large program, you have to make sure that the documentation is complete and correct, and that the program is appropriate for its intended use. The program's reliability, uniqueness of solutions, and ease of use must also be considered.

9.4.4.3 Applicability to Problem

After defining your problem, developing your mathematical model, and considering solution algorithms, you have to determine whether the algorithms used in the program provide an acceptable engineering solution to the physical problem. If there are weaknesses in the solution, you have to look for a more applicable package or determine whether it is possible to write a program that would accomplish a more realistic solution.

9.4.4.4 Validation

Both types of software, proprietary and public domain, have to be validated through some means. Validation is establishing that the computer program does what it is supposed to do. As a user, you have to be careful not to assume false validation merely because the software has been used repeatedly. The fact that a program has been used in several projects does not validate its correctness. You have to be presented with concrete evidence of validation before you can accept the program without conducting your own validation program. Also, be careful not to accept developer-supplied output to check your output against unless you have evidence that the output has been validated.

Validation involves establishing the credibility of the program. Is the mathematical model used appropriate? Will the algorithms adequately solve the mathematical model? Does the computer programming code correctly translate the algorithms? These are questions that you should find answers for before you conclude that the program is the proper one for you to use.

9.4.4.5 User Support by the Developer

Carefully evaluate vendor support, including the vendor's quality assurance methods and training. The vendor's procedures for evaluating user or internally reported errors, their response time for fixing errors, and their method for disseminating error corrections should be assessed. The methods a vendor uses to disseminate revisions and updates to the program and manuals are important considerations if the applications package is to be used on a continuing basis. Technical support may or may not be available in the form of training, a technical service hot-line, or a newsletter. Although user support is important, it does not make up for a weak program.

9.4.4.6 Choice

Your final criterion for choice has to be economic. You are looking for a package that will meet your needs at minimum cost.

9.4.4.7 Calibration and Verification

Calibration of a computer program involves the comparison of computer results with field observations or the comparison of results with another previously verified model. Adjustments to the model are then made so that its output agrees with the field observations within some bound. Verification is the further testing of the calibrated model with an independent data set not used in the calibration procedure. This is a test of the generality of the calibration. Failure to be able to calibrate and verify the model is an indication that the model, solution algorithm, or computer program may not be valid. Under these circumstances, additional testing and evaluation would be required.

9.4.4.8 Use

When using commercial applications software, you should develop your own internal procedures for program support. You should always test those program features that you plan to use and then apply the program to your particular problem. If you encounter any errors, make sure that you report them and that you do not base any further analysis on the program until the errors are fixed. You also have a duty to make sure that the errors found do not invalidate any earlier designs for which the software may have been used.

9.5 DEVELOPMENT OF APPLICATIONS PACKAGES

When considering applications development, your first priority should be to find available software that is appropriate for the application being undertaken. Unfortunately, people tend to use an existing program to solve a new problem without sufficient regard for the appropriateness of the application. You have to guard against this tendency. If no appropriate software is available, you should try to modify an existing applications program. You may do this yourself, or you may contract with a software developer to do it, especially if you did not write the software in the first place. Because of the time and cost involved, your last option should be to develop a new applications program. This section will discuss some of the things you should consider if you do have to develop your own applications.

After defining your problem, mathematical model, and solution algorithm, you are ready to design and construct your applications package. An applications package consists of the software program and any necessary documentation for its use. The discussions, recommendations, and procedures that we present here apply equally to developing applications with any high-level language or any of the generic applications packages, such as spreadsheets and equation solvers.

An application is any computer program designed to accomplish some specific task. An applications package is generally a larger program designed to accomplish several related applications, in a serial and/or parallel structure. The word *program* is used here in the generic sense. It could equally be a spreadsheet or equation solver model.

Most programs are initially written for a specific purpose, but are later expanded to do other tasks. The problem with these expanded programs is that program errors are often introduced when a single program is developed to do too many tasks. As a program increases in size and complexity, the programmers seldom make enough executions to test all the branches in the program and to determine its full range of capabilities and limitations. In discussing civil engineering computer applications, Bowles (1981) stated that:

> Program errors can be introduced by attempting to write a single program to do too many tasks. A single program to solve all civil engineering problems would certainly be a laudable achievement, but very few persons, including the program authors, could ever use the program efficiently. The program would have tremendous memory and mass storage requirements, would have a large number of branches, and would take longer to execute than a smaller single purpose program. No single person could write such a program. Once the task is subdivided into areas of responsibility, the terminology, input format and general logic may become inconsistent. It would be extremely difficult to document the program adequately, Hence there would always be uncertainties in its use, except for certain clearly delineated problems.

9.5.1 Economic and Development Considerations

In deciding whether to use an existing application, modify an existing applications program, or create a new applications program, the criteria for making your final choice are largely economic. Given your problem definition, how can you perform the necessary analysis at minimum cost? Producing computer programs is expensive. Bowles (1981) has estimated that high-level language software development costs between $25 and $40 per line of code. Development of large programs containing full design, analysis, graphics, and database maintenance capabilities can require 25 man-years of labor at a cost of more than $1 million.

You should think of software development as a project that will produce a product. Think of development of the application as design; coding and producing it as construction; debugging it as troubleshooting; testing it as performance evaluation; and documenting it as final packaging with instructions. If the software is for commercial release, then the users will be evaluating your company on the overall usefulness of the product they have bought from you. Even if the software is only for internal use, the extra care taken in production and documentation will be worthwhile. You never know when you might want to revise the program. Also, someone who is unfamiliar with your program may want to use it.

As you would do with any programming project, you have to define your problem thoroughly and correctly. What are your objectives? Who will be the users of your package? What are the future development plans for the software? You will have to develop or gather appropriate mathematical models and algorithms. Consideration of the models and algorithms will help you to evaluate the proper computational tool to use for the application.

Practice streamlining your programs. Ask yourself if you would be able to read, understand, and use your program if you were someone else looking at it. A large portion of a program's usability is directly related to the quality of the program documentation. We have all been exposed to poorly documented programs or to a product's poorly written user's or assembly manual.

Many of the programmers developing engineering applications packages are engineers. This is because nonengineers often lack the appropriate technical skills to write engineering software. Always keep in mind, however, that the best-written applications packages are often a cooperative effort. Computer programmers and computer scientists can be invaluable in helping you design a well-structured and efficient program.

9.5.2 Program Design Considerations

9.5.2.1 Structured or Top-down Programming

The method of decomposition should be used to divide large problems into independent smaller parts. Step-wise refinement should then be used to develop the details of the computer algorithm. Once modules that accomplish one clearly defined procedure are identified, you can review available subprograms or modules to see if any might be applicable. Other subprograms or modules may be developed for this application and saved for future use. When developing high-level language applications, you should develop the main program as the coordinating procedure for the various subprograms. When using generic applications software, you have to make sure that all modules are properly linked together. You have to consider where, when, and how data will be input and output. In designing your application, be careful not to include options that will not be needed by the users of the program. This causes unnecessary expense in developing the program or model and makes it less attractive to users.

9.5.2.2 Portability

If you are developing FORTRAN or other high-level language applications, you should use only standard structures. This will minimize any problems when the program is transferred to a different computer system. Even such popular FORTRAN extensions as the DO WHILE should not be used, since they are not supported by all systems. When using generic applications software, you need to be aware of the different microcomputers supported by the software.

9.5.2.3 User Interface

The user interface refers to how the user will interact with the computer to load and execute an applications package. It includes the input of data and output of results. The user interface has a lot to do with the acceptance of a program by users. Excellent applications with poor user interfaces may be used by a small group who are willing to take the time to delve into the program. However, they will not gain widespread popularity. Data may be input through the keyboard or from files, or a combination of the two can be used. User input instructions should be clear and concise. Required units should be clearly indicated for all input. It may be advantageous to input large bodies of data from files, but then have the option within the program of changing selected data to test options. With high-level language applications, the user should be given the option of selecting a data file from within the program. That gives the user the capability of inputting different data if the program is executed multiple times. It also prevents you from having to use system file declaration commands. Output may be to the screen, to a printer, to a file, or to one of the other peripheral devices. Again, a combination may be necessary for more comprehensive applications packages.

Input data should be output for checking. This echoing of input data can be made optional for final output if it is not necessary for the presentation of the data.

Any default values that are used in the application should be explicitly stated. The user should have the option of changing default values. Whenever possible, range checks should be incorporated into the application to make sure that the input data values are within acceptable limits. Whenever significant data are to be entered through the keyboard, error trapping should be invoked for all input data variables. Error trapping is automatic when you are using generic software packages.

When developing programs that will use iterative approximations for solutions, you should include checks for divergence and excessive oscillation of approximations. In order to guard against very slow convergence, you should set a maximum number of iterations. When the approximation is for the root of an equation, you should always output the final value for the function to make sure that successive approximations with little change have not caused a false reading of convergence.

Every element of your output, including the units used, should be clearly identified. All tabular output should be well-designed, making it clear what is being presented. Additional output that will not be of interest to all users should be made available as an option.

9.5.3 Coding Specifications

The following comments refer mainly to high-level language applications. They also apply if you are using the programming options of generic applications software.

Comment statements should be used to provide introductory information for the program, to physically separate modules within the program, and to explain loops and conditionals or any other pertinent logic of the program. Indentation should be used to set off each level of nested logic.

All program variables should be declared in declaration statements and defined in comment statements at the beginning of the program. Variables for DO loop parameters and other counting variables can be typed using the default convention. Arrays should be dimensioned in declaration statements rather than DIMENSION statements.

If common blocks are used in subprograms, they should be labeled common blocks to facilitate definition of the variables to be passed to and from the subprogram by the calling program. You may find it easier and more general to pass variables and arrays through dummy and real arguments.

File references and processing should be accomplished using the standard FORTRAN OPEN, CLOSE, and REWIND commands. The use of these commands will make your application more transportable.

9.5.4 Validation and Testing

Validation implies that the program has been successfully tested and applied to a given set of sample problems similar to the intended applications. Validation means that the logic and coding of the program correctly implement the mathematical model for the problem. If validation is done properly, it also means that the computer solution agrees with solutions that would be considered correct by design engineers. The fact that it has been validated does not imply a guarantee that a given set of results from any particular application is correct. A validated program can still give incorrect answers if your input data are faulty. The use of validated programs for engineering computation in no way changes your responsibility for the accuracy of the computer-generated calculations when they are used in design.

A large portion of the development cost for applications packages may be spent on program validation and testing. Chien and Gilmore (1975) estimate that up to 40 percent of the development effort is spent confirming that the program does what it is supposed to do. Public safety is the reason for validation, since the program will in some way contribute to the design of some product that the public will use. Economics will govern the methods employed in validation. You need to validate the program, but you want to do it at minimum cost.

Validation establishes that the program correctly solves the mathematical model used in its development. It should also verify that the mathematical model is a valid representation of the physical system being modeled. Testing is the method by which validation is undertaken. It includes debugging of program errors, verification of output accuracy, and sensitivity analysis.

9.5.4.1 Testing

Thorough testing of a program would require that each statement of the program must be exercised at least once. Even this is not sufficient, however, since there may be multiple options available that use different sequences of instructions. Therefore, all options should be tested using known input and output. Each unique flow path through a program should be exercised at least once. Boundary or extreme values for input variables should be tested. Extreme values may cause a poorly designed computer algorithm to abort execution or to produce unreasonable answers. Correct data transfer to and from pretested modules must be verified. All this testing becomes a large and formidable task for complex programs with many options.

9.5.4.2 Debugging

We have developed numerous techniques for debugging in this text. Some of the more important concepts will be reviewed briefly here.

1. You should echo all input data. During the testing phase this will allow you to verify that data are getting into the computer correctly. After testing has been completed, echoing of input data will assist in identifying input data errors.
2. Use error messages provided by the generic software package or high-level language effectively. However, a lack of error messages does not mean that there are no further errors. You still have to verify application logic. User-defined error messages can be used to check for error conditions. Some of these may be specific to the debugging process. Others may be left in the application program to identify error conditions that can occur when the application is used incorrectly.
3. When debugging high-level language programs, use extra output statements liberally to check the specification and respecification of variable values during program execution. Any of these output statements that do not produce useful information for normal program applications can be removed after the program is completely checked out. Output statements can be used to check any data, including subscript indexes and DO loop

counters. They are also useful for showing you that all initializations and reinitializations have been made for loops and summations.

4. You will want to develop your application in stages. Make sure bugs are out of the present stage before progressing to the next logical element. For example, a program might contain the following stages:

INPUT
COMPUTATION STAGE 1
COMPUTATION STAGE 2
COMPUTATION STAGE 3
OUTPUT

5. Subprograms or modules can be developed and debugged separately and then interfaced as if they were previously developed modules. Avoid the temptation to write all the code before you begin debugging. This method works satisfactorily for small programs, but it only confuses the debugging process for large programs.

9.5.4.3 Validation

If you had to accomplish a detailed validation for each program that you use in design, much of the value gained through computer use would be lost. If the program has been tested and validated in detail in a controlled environment, you only have to check a selected set of critical output values to establish a high degree of confidence in the overall results. In the case of library subprograms, as a minimum you will have to check that your argument values are being transferred over to, and back from, the subprogram correctly. You can consider a program or subprogram to be validated if it meets one or more of the following criteria:

1. The computer program is a recognized program in the public domain and has had sufficient history of use to justify its applicability and validity without further documentation. However, you have to be careful of false validity assumed by repeated use of a program. You need to see evidence that the program has been adequately checked.

2. The computer program solution to a series of test problems agrees with the solution obtained with a similar and independently written and recognized program in the public domain. Again, this depends on documented validation of the other program.

3. The computer program solution to a series of test problems has been demonstrated to be substantially identical to those obtained from classical or analytical solutions.

Both methods one and two indicate that method three has been previously applied. Documentation for the program should substantiate this.

9.5.4.4 Sensitivity Analysis

Sensitivity analysis involves holding all parameters constant at their expected values, except for the one being tested. You would then vary the test parameter within reasonable expected limits and examine the change in the output. If relatively small changes in the parameter produce large changes in the output, the output is sensitive to that parameter. You will have to estimate values for sensitive parameters with more certainty. Information to this effect should be included in the user documentation. Insensitive parameters may not need to be in the model, or they may indicate problems in application development. If you find a variable that the output should be sensitive to, but is not, you may have a program error. You may also have uncovered a weakness in the mathematical model, or an error in the solution algorithm.

9.5.4.5 Validation Maintenance

Once testing is completed and the program is validated, you should take steps to ensure that changes are not made in the program by users who have not taken the changes through the testing sequence. In FORTRAN applications this safety feature can be accomplished by using compiled files for normal execution. Users who do not have access to the source code cannot make changes in the program. With a spreadsheet, you can lock cells that should not be changed. If you are a user who is modifying a program written by someone else, you will be responsible for the validation of your modifications. You also need to remember that your modifications may make it more difficult to implement updates that are provided by the original program developer.

9.5.5 Documentation of Applications Packages

The purpose of documentation is to provide enough information to users for them to use the application effectively. Users should not have to spend an inordinate amount of time studying the details of the program. As with any written communication, the tenor of the documentation has to take into account the intended audience. If the users will be engineers who understand the theory on which the program is based, you can give a brief overview of the theory and the solution algorithms used. You would have to include only enough information for the user to evaluate the applicability of the program. On the other hand, you would have to carefully spell out the input data instructions and user interface if the engineers are not familiar with computers. For students, more explanation of the theory and a discussion of the computer algorithm implementation might be in order. For any intended user group, the documentation should be clear, concise, and easily understood.

For short programs, comment statements may provide all the documentation that is required. Even if the comment statements are not the main documentation, they still provide valuable information that will make the program more understandable to a variety of users. At the beginning of the program, you should include an introductory block that explains the purpose of the program, the author, the date of development, and the date of the latest update. You could also include a brief description of the program if this is to be the only documentation. All variables that need definition should be defined in comment statements at the beginning of the program.

If the program is to be interactive, input/output instructions can be incorporated into prompts that will appear on the screen. If the program is to be run in the batch mode, but will not have separate documentation, input/output instructions can be included as comment statements.

Comment statements should be used to separate logic modules within the program and to explain the purposes of logic modules. The comments can also be used to explain the purpose of loops and conditional statements.

Longer programs and programs for which the source code will not be readily available to the user require preparation of a separate documentation manual.

This should be in the form of a user's manual and should provide all the information that a user will need. Preparation of separate documentation will be discussed in the next section.

9.5.5.1 Elements of Documentation

The following list shows the elements of thorough documentation. Some of the elements may be excluded from documentation for specific reasons. For example, proprietary software may not include a listing of the source code because the developers do not want users to modify the program.

1. Description of the problem that the program is designed to solve
2. Capabilities and limitations of the program
3. Theoretical basis for the program
4. Description of the program structure, including a flowchart
5. Special machine requirements
    Memory
    Tape and disk requirements
6. Definition of the variables
7. Data input requirements and instructions
8. Sample problem
    Statement
    Input and output
9. Program listing

9.5.5.2 Quality Assurance Program

A part of any program's documentation should be a description of the quality assurance program established by the developer for both the computer program and its documentation. The procedures followed in the validation of a computer program should be described. Documented proof of validation and samples of the test problems used should be presented. A record of any modifications that have been made, including the dates of these modifications, should be included.

9.5.6 Use (Production)

Engineering firms are professionally liable for the completeness and correctness of computer-aided design. The use of a computer applications package implies a knowledge and understanding of the package on the part of the firm's engineers. They are responsible for making sure that the package is appropriate for the problem being solved. They have to ensure data preparation accuracy, monitor the processing of the data by the computer, and interpret the results. In the interpretation of computer-aided results, there is no substitute for engineering knowledge, practical experience, and good common sense.

9.5.7 Continuation of Validation

Validation is an ongoing activity that continues as long as a program remains in productive use. A means of detecting and correcting errors found during program use must be considered to be part of the validation process. However, this is not an excuse for incomplete initial testing. If errors are found during the use of a program, you must determine whether those errors affect past calculations. If you believe that there may be any danger to the public welfare, you must take appropriate steps to inform the proper authorities of this danger.

9.5.8 Program Maintenance, Modifications, and Enhancements

One master copy of applications package code and documentation should be maintained for the purposes of updating and error correcting. Some efficient method must be implemented to disseminate modifications and enhancements on a timely basis. This might be on a yearly or a semi-annual cycle, depending on the volume of changes that the software undergoes and the importance of having an up-to-date version in the hands of users. Error corrections that might affect past results must be disseminated quickly, preferably as soon as an error has been verified and the correction implemented in the master file.

If a developer hopes to maintain control over a program, it is imperative that there be only one original from which the developer works. Each time the program is updated, that update becomes the new original. While there may be several versions available to the public, the developer will always maintain the most current version as the master copy. All updates should be fully compatible with applications files developed for earlier versions, so that users will have no trouble switching over to new versions.

When you modify an existing program for specific purposes, you have to be concerned about the updating of the base program by the original developer. When you modify programs, you should always keep a copy of the original. It will be up to you to implement changes in your modified version or to remodify the updated original version. With a program that is frequently updated by the developers, this can become a tedious and expensive process. It might be preferable to have the original developer work with you to implement the changes that you need.

9.6   INTEGRATION INTO PRACTICE

By now you should be convinced of the need for using computers as engineering tools. Properly using computers will increase your productivity and reduce the number of errors that you might make. Computers are state-of-the-art tools. Competent entry-level engineers are expected to be comfortable with the use of computers and all types of software as analysis and design tools. Computer hardware and software capabilities have advanced, and prices have decreased to levels that make computers practical for all engineering firms. Computers should be conveniently available in the workplace. If this is not the case in your firm, it may be up to you to demonstrate the need for up-to-date computer resources to your employers.

As a competent user of computers, you can help to avoid some of the pitfalls that may negate their effectiveness. Important among these is the problem of instant expertise, a phenomenon that may tempt engineers to use computer packages to accomplish analyses and designs for which they do not have the proper training. There is also the problem of becoming computer bound, causing engineers to turn to the computer for all analyses and designs, when other methods may be more efficient. You will need to emphasize the necessity for verifying computer results and will need to be able to give advice concerning whether or not the computer is the proper tool to use.

When using the computer, you must be acutely aware of the possible errors that can occur. You should always be suspicious of computer results until they have been validated by some other means. Pearson (1988) has put forth five tests for validity that are applicable to any computer analysis.

1. Are the results reasonable? You may have experience with similar problems and have a feel for the approximate range the answers should fall into. You can check values for the correct sign. Check to make sure that percentages add up to 100 percent and that probabilities add up to 1.0.
2. Estimate your results by some other method. We have been emphasizing throughout the text that it is essential to have a sample solution to check your computer results against. For complex problems, it may not be possible to develop a complete solution by hand. However, you may be able to develop an approximate solution that will give you an idea of the magnitudes of the correct answers.
3. Do small changes in your input data cause large changes in your results? This may indicate a problem with the mathematical model or solution technique you are using. If both the model and solution techniques are valid, then you have to make sure your input data for the sensitive parameters are sufficiently accurate.
4. Do large changes in the raw data produce little or no change in the results? If so, you probably have independent variables in your model that should not be there. Or there may be errors in your model or program that negate the influence of the one or more independent variables.
5. When you are using measured raw data to drive a computer program, examine the raw data graphically to see if any of the data values look suspicious. This can catch data acquisition errors. It can also catch errors in computer data entry. Incorrectly entered data are often orders of magnitude off.

You must keep in mind that the engineer still makes the decisions and has the responsibility for the correctness of solutions. The computer is only a tool. It is your responsibility to use it correctly. Make a personal commitment to continue to increase and update your computer competence throughout your career as an engineer.

9.7   IN THE FUTURE

In the future continue to review computer literature and to use the computer often. This is true both for the remainder of your time as a student and for when you enter and become established in the profession. You can join computer user groups to share ideas. Computer trade magazines will keep you informed about new developments. Make sure that you have a computer conveniently accessible in your workplace. You will also probably want to own a personal computer at home. This will greatly assist you in your efforts to continue developing your computer skills. Your employer should give you some time on the job to increase your professional skills, but the training will not all be computer-related. Since your employer is in business to make money, you will not be able to spend as much time as you would like to in experimenting with computer applications. Continue to learn new languages and applications packages, and keep studying documentation and package updates for packages you are using.

Remain a comfortable user of computers. Be open to new developments. The advances in microcomputer applications over the past ten years are a prime example of the need to be flexible. Similar advances in processing speed and computational power can be expected to occur over the next ten years. By remaining competent in computer applications, you will increase your value to your employer considerably. You will also make yourself more marketable when you seek new challenges with other employers.

SUMMARY

You have studied several specific numerical techniques in this chapter, including solving for roots of equations, numerical integration, numerical differentiation, and finding minimum or maximum points of functions. You have applied several different computational resources to help in implementing these numerical techniques. High-level language programs, spreadsheets, and equation solver models can all be used effectively under the proper conditions. As a user of numerical methods, you must first fully develop your problem definition and carefully define your mathematical model. The better your knowledge is of the behavior of a mathematical model in the region of interest, the more likely you will be to successfully find numerical solutions for the model. You must then evaluate which solution techniques are most appropriate for your mathematical model and which computational resources will be most efficient in implementing your solution technique. After using the computer to assist in finding solutions for your model, you must evaluate the appropriateness and accuracy of those solutions. The computer is actually a small part of the total process.

Do not stop trying to develop analytical solutions to problems just because you know numerical solutions are available. Analytical solutions that adequately model the phenomena being studied are still preferable to numerical solutions that only approximate the analytical solution. For problems that might be solved by either analytical or numerical methods, you need to ascertain which method will find an acceptable solution most efficiently.

If you do not have a previously developed application that is appropriate for a new problem, you may have to search for a commercial package or you may develop your own. Whatever resource you are using, you as the engineer are still solely responsible for the accuracy of your design. If you do develop your own applications packages, you should approach their development as you would any other design project. Proper problem definition, mathematical model development, and solution algorithm choice are more important than elegant computer programming. If you correctly translate your solution algorithm into a well-structured application and properly document it, you will make a valuable addition to the computational tools available to engineers. Before you release an applications package for others to use, you must feel confident that they will be able to read, understand, and use the program.

REFERENCES

Borse, G. J., *FORTRAN 77 and Numerical Analysis for Engineers*, PWS Publishers, Boston, 1985.

Bowles, J. E., "Is the Computer Giving Correct Answers?" *Civil Engineering*, American Society of Civil Engineers, October 1981, pp. 59–61.

Chapra, S. C., and Canale, R. P., *Numerical Methods for Engineers*, 2nd. ed., McGraw-Hill, New York, 1988.

Chapra, S. C., and Canale, R. P., *Introduction to Computing for Engineers*, McGraw-Hill, New York, 1986, Chapters 27, 28, 30–32.

Chien, E. Y. L., and Gilmore, M. I., "Providing the Profession with Design Programs," *Engineering Issues: Journal of Professional Activities*, Vol. 101, No. E12, Proc. Paper 11261, American Society of Civil Engineers, New York, April 1975, pp. 165–174.

James, M. L., Smith, G. M., and Wolford, J. C., *Applied Numerical Methods for Digital Computation*, 3rd. ed., Harper & Row, New York, 1985.

Pearson, R., "Lies, Damned Lies, and Spreadsheets," *BYTE*, McGraw-Hill, Petersboro, NH, December 1988, pp. 299–304.

Press, W. H., Flannery, B. P., Teukolsky, S. A., and Vetterling, W. T., *Numerical Recipes: The Art of Scientific Computing*, Cambridge University Press, Cambridge, 1986, Chapters 3, 4, 9, and 10.

Red, W. E., and Mooring, B., *Engineering: Fundamentals of Problem Solving*, Brooks/Cole Engineering Division, Wadsworth, Inc., Monterey, CA, 1983, Chapter 7.

EXERCISES

9.1 GENERAL NOTES

Throughout the text, chapter-end exercises have been organized by subject group. The exercises in this chapter will frequently refer to previous chapter-end exercises from the same subject group. You should use data from these previous exercises to test your applications for this chapter. Whenever they are appropriate, graphs should be used to illustrate your solutions.

General Topics for term projects are included with the exercises for this chapter, and are identified by [TERM PROJECT] at the start of the problem. These are comprehensive problems involving the development and documentation of an applications package. Either batch or interactive packages could be developed. You should concentrate on making your package as "user friendly" as possible. Many of the subprograms and procedures developed in previous chapters can be used in these applications packages. Your programs should incorporate such finishing touches as range checks and error trapping. Thorough user documentation should be developed. Only general topics are listed here. The actual form of your project will be up to your instructor.

9.2 MODIFICATION OF EXAMPLES

9.2.1 Example 9.1

Modify the spreadsheet of Example 9.1 to include a flag column, the value of which equals zero if the error check for the Newton-Raphson method is greater than 10^{-a} and equals one if the error condition is less than 10^{-a}. The logical structure should be designed so that the value would become one in the $(n+1)$st row, to indicate the correct estimate to 10^{-a} accuracy. A message should be included in the spreadsheet to indicate where the estimate of the root to the given accuracy can be found.

9.2.2 Example 9.2

Modify the TK SOLVER model of Example 9.2 so that it can find all the roots of the polynomial of Equation 9.4 in the range $X = 0$ to $X = 7$.

9.2.3 Example 9.3

1. Modify the program of Example 9.3 to include an upper limit on the number of iterations to guard against solutions that do not converge. If the number of iterations is greater than the limit, print an error message. Also, use the function value at the final estimate for convergent solutions to check the error estimate. Output a message either way.
2. Modify the program of Example 9.3 to include a general subprogram for computing real roots of a polynomial using Newton-Raphson and the procedures of Example 9.5 (EXTERNAL statement).

9.2.4 Example 9.4

1. Numerically confirm the parallel axis theorem (Section 1.3 of Appendix G) by applying numerical integration to the top and bottom portion of the triangle of Example 9.4 to find the moment of inertia about the centroidal axis. Compare this result with what you get by applying the parallel axis theorem to the estimate for the moment of inertia about the X-X axis of Example 9.4. The actual moment of inertia about the centroidal axis for this triangle is 121.5.
2. Modify the spreadsheet of Example 9.4 to use a Δy of 0.5, and compare the results with the previous results.

9.2.5 Example 4.14

Modify the program of Example 4.14 to incorporate the quadratic equation solution for the particle trajectory problem developed in Chapter 9. Use the quadratic equation subprogram developed in Example 5.7 to assist in your solution.

9.2.6 Menu-Driven Programs

Modify selected examples and/or exercises in this text to produce a comprehensive menu-driven applications package. Example 6.9 introduced you to a type of menu structure for FORTRAN programs.

9.3 GENERAL SYNTAX AND STATEMENT STRUCTURE

Develop either a FORTRAN program or a spreadsheet to implement the secant method for estimating the root of a polynomial equation. Use your program to estimate the highest-valued root of Equation 9.4.

Develop a FORTRAN program or a spreadsheet to verify the approximation for the integral of $\sin(\theta)$ found in Table 9.1. Compare the accuracy of this estimate with the accuracy of an estimate using 80 increments.

9.4 CHEMICAL ENGINEERING

9.4.1 [ALL] Ideal Gas Law

Estimate the slope and rate of change of slope of pressure vs. temperature from 0° to 100°C using either the van der Waals or the Redlich–Kwong equation.

[TERM PROJECT] Develop a menu-driven applications package to solve for temperature, pressure, volume, or number of moles using the ideal gas law, van der Waals equation, or the Redlich–Kwong equation.

9.4.2 [ALL] Chemical Kinetics

The following equation is used to predict the effect of mean ionic activities on the degree of dissociation of a weak monoprotic acid in aqueous solutions.

$$f(X) = AX^2 - B(1 - X)e^{-2.344\sqrt{AX}}$$

in which A is the molarity of the acid, B is the equilibrium constant of the acid, and X represents the degree of dissociation of the acid. Develop an application that can be used to find the degree of dissociation of an acid of given molarity and equilibrium constant. Test your application for a 1.0×10^{-2} molar solution of an acid with an equilibrium constant of 1.750×10^{-5}. Use your application to evaluate the sensitivity of the dissociation to changes in the molarity.

[TERM PROJECT] Develop an applications package that can be used to input chemical process data, determine the coefficients of either the first- or second-order kinetics models, select the best fit model, and use the model selected to predict concentration as a function of time.

9.4.3 [ALL] Depth of Fluidized-Bed Reactor

Develop an application that can be used to determine the rate of change of required bed height with change in particle density.

[TERM PROJECT] Develop a package that can be used to design a fluidized-bed reactor.

9.4.4 [ALL] Process Design, Gas Separation

Modify the previous gas separation applications to find the rate of change of the required number of plates and the removal efficiency for the secondary constituents when the desired removal for the primary constituent is changed.

[TERM PROJECT] Develop a package that can be used to design a gas separation column for up to five constituents.

9.4.5 [ALL] Dissolved Oxygen Concentration in Stream

Use numerical methods to estimate how far downstream a certain concentration would occur. You will want to plot the function first to see if the concentration reaches the value. The plot will also provide you with a first estimate of the distance.

[TERM PROJECT] Develop a package that can be used to evaluate the permissible concentration of a pollutant in a stream if the dissolved oxygen must remain above a critical lower limit at all points in the stream.

9.5 CIVIL ENGINEERING

9.5.1 [ALL] Equilibrium-Truss Analysis

Develop an application that will find the rate of change of reactions at A and C of the truss shown as the tension in the cable attached at D is increased. Continue until the reaction at C goes to zero.

[TERM PROJECT] Develop a comprehensive package for analysis of a planar truss of up to 25 members. Input should include the coordinates of the joints of the truss, starting and ending joints for each member, and loading data. All other information should be generated by the program. The truss will be supported on one end by a pin, and on the other end by a roller. The base of the truss will be in the horizontal plane.

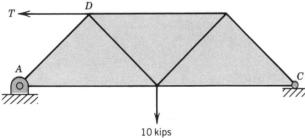

10 kips
Figure for Exercise 9.5.1

9.5.2 [ALL] Open Channel Flow Analysis

Use either the Newton-Raphson or secant method to estimate the depth of flow in a trapezoidal open channel for a given flow rate. Use the data of Exercise 8.4.2 (*Instructor's Manual*) to test your program. Compare this solution technique with using a TK SOLVER model to estimate the depth of flow.

[TERM PROJECT] Develop an applications package that can be used to design either a trapezoidal or circular open channel.

9.5.3 [ALL] Dam Stability Analysis

Develop an application that can be used to determine the rate of change of the required bottom width with change in the design height of water behind the dam.

[TERM PROJECT] Develop an applications package that can be used to analyze a concrete gravity dam for stability. The cross section can be a compound shape consisting of up to five standard shapes.

9.5.4 [ALL] Force on Submerged Gate

Develop a numerical integration application to verify that the formulas of Section 5.1 of Appendix G give the correct force and point of application for a 5-ft diameter circular window in the vertical plane if the top of the window is 10 ft below the surface of water in a tank.

[TERM PROJECT] Develop an applications package that can be used to find the magnitude and point of application of the resultant force on a submerged gate that is a combination of simple shapes.

9.5.5 [ALL] Volume of Excavation

Develop an application that can be used to find the incremental and total volume in a number of intervals between cross sections or the distance required for a given volume of excavation.

[TERM PROJECT] Develop an applications package that will determine the volume of cut and fill along a highway right-of-way. Include a function to calculate the cumulative net cut or fill as you progress from section to section.

9.6 DATA ANALYSIS AND STATISTICS

9.6.1 [ALL] Histogram

Develop an application to plot a histogram for any selected data, as well as histograms of the first and second derivatives of the data.

[TERM PROJECT] Develop an application that will produce a histogram for any data. Include appropriate labels and axis scaling.

9.6.2 [ALL] Sorting

Develop an application that will sort any set of data into ascending order; then take first and second differences between the sorted values to check for trends in the data. Test your application with the values 25, 16, 36, 9, 1, 4, 64, and 49.

[TERM PROJECT] Develop a general sorting routine for ascending or descending order. Include the option of saving the original sequence.

9.7 ECONOMIC ANALYSIS

9.7.1 [ALL] Economic Formulas

Develop an application to estimate the rate of change of the PWF, USPWF, and CRF with change in the interest rate.

[TERM PROJECT] Develop an applications package that will choose the proper economic formula to use based on the information supplied by the user. Use the package to determine the net present worth for a series of cash flows consisting of present worths, uniform series of payments, and future worths.

9.7.2 [ALL] Nonuniform Series of Payments

Develop an application that will estimate the rate of return for a nonuniform series of payments if the present value and the number of compounding periods are known.

[TERM PROJECT] Develop an application that will accept a nonuniform series of payments, compute the present worth of the series, and compute the equivalent uniform series of payments. Input should be interactive.

9.7.3 [ALL] Multiple Interest Rates or Number of Compounding Periods

Develop an application to estimate the rate of change of the PWF, USPWF, and CRF with the number of compounding periods.

[TERM PROJECT] Develop an interactive applications package to evaluate any combination of interest rates and compounding periods.

9.7.4 [ALL] Internal Rate of Return Analysis

Develop a FORTRAN function for determining the internal rate of return of a uniform series of payments given the present value P, the payment A, the number of compounding periods n, an initial estimate of the rate of return, and the decimal place accuracy desired. Test your program using $P = \$100,000$, $A = \$14,000$, and $n = 16$.

[TERM PROJECT] Develop an applications package that will find the internal rate of return for a combination of present worth, annual payments, and future worth.

9.7.5 [ALL] Mortgage Computations

Develop an application that will allow you to determine the interest rate necessary to repay a given principal in n payments of $\$A$ each.

[TERM PROJECT] Develop an application to find either i, A, or P for mortgage payments.

9.8 ELECTRICAL ENGINEERING

9.8.1 [ALL] Diode Problem

1. Section 5 of Appendix I presents a mathematical model for an electric circuit with two diodes and a voltage source. Develop an application that will plot the current vs. voltage curve and the slope of this curve for voltages from -0.05 to $+0.05$ V. Use $I_{s1} = 0.01$ amperes, $I_{s2} = 0.10$ amperes, $\lambda_1 = 38.10$ V^{-1}, and $\lambda_2 = 41.00$ V^{-1}.
2. Develop an application that can be used to find the required voltage to produce a desired current across the voltage source. Test your program for a current of 0.6 A. These exercises were adapted from an example by Borse (1985).

[TERM PROJECT] Develop an applications package that can be used to investigate the response of the two diode circuit of Figure I.5 of Appendix I to varying voltages.

9.8.2 [ALL] Capacitance and Resistance

Develop an application to estimate the first and second derivatives of the response of the circuit of Exercise 4.8.2.

[TERM PROJECT] Develop an applications package that will analyze a resistance circuit such as the one of Exercise 7.8.2 with up to 25 resistors and up to 5 output leads.

9.8.3 [ALL] Signal Processing Circuits

Develop an application that will find the first and second derivatives of the output of Exercise 4.8.3.

[TERM PROJECT] Develop an applications package that can be used to design a resonant circuit such as the one illustrated in Section 6 of Appendix I.

9.9 ENGINEERING MATHEMATICS

9.9.1 [ALL] Determinant and Cramer's Rule

Modify the application of Exercise 7.9.1 so that you can determine the change and rate of change of X_1, X_2, and X_3 as a function of the change in one of the right-hand-side values.

[TERM PROJECT] Develop an application that will find the determinant of up to a 5×5 matrix.

9.9.2 [ALL] Vector Cross Product, Unit Vector, Directional Cosines, Vector Components

Develop an application that will estimate the length of the position vector and the rate of change of that position vector for a particle that starts at a point in three-dimensional space and travels at a constant three-dimensional velocity. Test your application with a starting point of $(1,1,1)$ and a velocity of $2i + 1.5j - k$ units/s.

[TERM PROJECT] Develop a menu-driven program to find

a. A unit vector given either the coordinates of the vector end points or two points on the vector and the magnitude of the vector.
b. A position or force vector given the unit vector and magnitude or two points and the magnitude.
c. The vector cross product given a position vector and a force vector.
d. The summation of two vectors.

9.9.3 [ALL] Centroids and Moments of Inertia

Use numerical integration to estimate the centroid and moment of inertia about both the X and Y axes for a half-parabolic section whose base edges are coincident with the axes. Test your program using the formulas of Table G.1 of Appendix G.

[TERM PROJECT] Develop an applications package to find the centroid and moment of inertia of a compound area about either axis.

9.9.4 [ALL] Population Growth

Develop an application to take first and second differences for the data of Exercise 8.8.4 (*Instructor's Manual*) to assist in establishing the population trends. Use these results to estimate the population in the year 2000. Compare the results with your previous results.

[TERM PROJECT] Develop an applications package that can be used to analyze historic population data. Use data developed from this analysis to predict future population.

9.9.5 [ALL] Vertical and Horizontal Curves

Use numerical integration to find the volume and centroid of the solid parabolic arch shown. Use $h_{max} = 5$, $L = 8$, and $W = 4$ to test your application. Verify your answers through the methods of calculus.

[TERM PROJECT] Develop an applications package that can be used to analyze either vertical or horizontal curves.

$$h = -\frac{1}{2} ax^2 + bx$$

$$b = h_{max}\left(\frac{4}{L}\right) \qquad a = \frac{2b}{L}$$

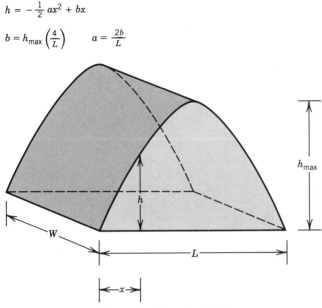

Figure for Exercise 9.9.5

9.9.6 Volume and Surface Area

Develop an application to use numerical integration to find the volume and surface area of a half cone on its side. Test your program for a cone whose base diameter is 8 ft and whose height along the x axis is 20 ft.

9.10 INDUSTRIAL ENGINEERING

9.10.1 [ALL] Project Management

Develop an application that will implement time/cost optimization for a project schedule. Use the example given in Section 1.6 of Appendix J to test your application.

[TERM PROJECT] Develop an applications package that incorporates all the project management techniques developed in previous project management exercises.

9.10.2 [ALL] Quality Control

Nondestructive testing can be accomplished for batches of parts coming off an assembly line. You can test any percentage of the batches that you want to. However, the cost of testing goes up at an increasing rate because of the disruption in the product flow caused by the testing. A low level of testing will result in extra costs because of recalls, and other disruptions. The testing costs can be represented by the function

$$C_T = C_c P^n$$

and the costs for defects can be represented by

$$C_D = C_0(1 - P)^b$$

in which P is the percentage of batches tested. Develop an application that will add these two costs together and search for the optimum level of testing, which is the minimum total cost. Plot all three curves of cost vs. percent checked. Test your application for $C_c = \$300,000$, $n = 1.2$, $C_0 = \$500,000$ and $b = 3$.

[TERM PROJECT] Develop an interactive package that can be used to accomplish the quality control analysis described in the quality control exercises of previous chapters. Your application should also compute the percentage of parts that have one to six defects.

9.10.3 [ALL] Managerial Decision Making

Due to economies of scale for an industrial storage complex, the storage costs per unit stored go down exponentially with increasing storage capacity. However, due to scarcity of available land, the site costs increase exponentially with increasing size. Develop a computer application that could be used to find the minimum point for the cumulative cost curve.

[TERM PROJECT] Develop an applications package for break-even analysis using all the features of the previous exercises.

9.10.4 [ALL] Queuing (Wait Line) Theory

Develop an application that can estimate the rate of change of the average number of down machines for variations in the total number of machines available.

[TERM PROJECT] Develop an applications package for finding the optimum number of machine repairmen for the queuing model of Section 3 of Appendix J. You should be able to change any of the parameters and rerun the simulation without exiting the program.

9.11 MECHANICAL AND AEROSPACE ENGINEERING

9.11.1 [ALL] Shear and Bending Moment

Develop an application that will find the shear and bending moment of Exercise 7.11.1 as a function of distance across the beam using the integration method. Numerically integrate the functions.

[TERM PROJECT] Develop an applications package that will analyze shear and bending moment for a beam cantilevered beyond one end, simply supported at both ends, or cantilevered from both ends. A uniform load and up to four point loads should be accommodated.

9.11.2 [ALL] Crank Assembly Analysis

Develop an application that can determine and plot the position, velocity, and acceleration of the piston as a function of the angular position of the crank for a given crank rotational speed. Use the data for Exercise 4.11.2 to test your program.

[TERM PROJECT] Develop an applications package that can be used to analyze user-selected portions of the crank problem of this exercise. Give the user the option of conducting sensitivity analysis.

9.11.3 [ALL] Friction

Develop an application that will estimate the force that will cause both slipping and tipping of the crate for a given point of application of the force on the crate.

[TERM PROJECT] Develop an interactive package to analyze a crate/dolly combination similar to the one used in Exercise 2.10.3.

9.11.4 [ALL] Variation of Atmospheric Pressure with Altitude

Develop an application that will estimate the decrease in temperature per foot of elevation, α, of Exercises 4.11.4 and 2.10.4. Use the secant method.

[TERM PROJECT] Develop an applications package that will solve for any of the variables of either of the atmospheric pressure variation models used in Exercises 2.10.4 and 5.11.4.

9.11.5 [ALL] Aerospace Physics

The trajectory of a particle is given by the equations

$$y = 200 + 50t - 16t^2$$
$$x = 50t - 3t^2$$

Use numerical differentiation to estimate the velocity and acceleration at $t = 10$ s. Use central difference and compare the values for $\Delta t = 0.01$, 0.1, and 1.0 s with the analytical solution for the velocity and acceleration. Develop a spreadsheet or FORTRAN program that will allow comparison at any time t.

[TERM PROJECT] Develop an applications package that will thoroughly analyze the trajectory for either a constant acceleration or constant thrust rocket.

9.11.6 [ALL] Energy Loss in Circular Pipeline

Develop an application that will estimate the rate of change of the energy loss with respect to pipe diameter for a given flow rate.

[TERM PROJECT] Develop a general design package for friction loss in a circular pipeline with either turbulent or laminar flow. The user should have the option of using the Chen, Swamee–Jain, or Colebrook equations to estimate f for turbulent flow. The program should be able to find friction loss, flow rate, or diameter given the other two.

9.11.7 [ALL] Torsion

Develop an application that will determine the rate of change of θ with increases in the applied torque. Use the results of Exercise 4.11.7 to test your application.

[TERM PROJECT] Develop a menu-driven package that will solve for any of the design variables of the shaft torsion problem.

9.12 NUMERICAL METHODS

9.12.1 [ALL] Interpolation

Develop an application that can be used to compare interpolation using linear interpolation, Lagrange quadratic interpolation, and cubic divided-difference interpolation for a given set of data.

[TERM PROJECT] Develop a general application for interpolation using linear interpolation, Lagrange quadratic interpolation, or cubic divided-difference interpolation.

9.12.2 [ALL] Solution of Polynomial Equation

Develop an application to solve the problem of Example 9.1 using the secant method.

[TERM PROJECT] Develop a general package for solving for any root of a polynomial equation. Use the EXTERNAL statement as described in Example 9.5.

9.12.3 [4,5,6,7,8] Saddle Point of a Hyperbolic Paraboloid

Add first and second divided differences to assist in the search for the saddle point of the hyperbolic paraboloid of Exercise 4.12.5.

9.12.4 [4,5,6,7,8] Polynomial Surface

Use first and second divided differences to assist in the search for the minimum or maximum point on a two-variable polynomial surface. Test the application using the surface of Exercise 4.12.6.

9.12.5 Matrix Inversion SUBROUTINE

[TERM PROJECT] Develop a FORTRAN SUBROUTINE for inverting a matrix of dimension up to 25×25. Appendix F contains the necessary background theory.

9.12.6 Interactive Package for Solution of Simultaneous Equations

[TERM PROJECT] Develop an interactive program for solving a set of simultaneous linear equations. Use either the matrix inversion subprogram developed in Exercise 9.12.5 or a resident subprogram on your computer system. Your program should be able to solve up to 25 equations with 25 unknowns.

9.13 PROBABILITY AND SIMULATION

9.13.1 [ALL] Simulation of Two Dice

Develop an application to plot the distribution of the frequency counts for the simulation of two dice for any number of simulation iterations.

[TERM PROJECT] Develop an applications package to analyze the simulation of throwing two dice n times.

9.13.2 [ALL] Normal Probability Distribution

Develop an application to estimate the maximum tolerance if a set percentage of the rods of Exercise 7.13.2 are to be acceptable.

[TERM PROJECT] Develop a computer application to find a probability for a given tolerance or the tolerance for a given probability.

APPENDICES

APPENDIX A

SOLUTIONS TO SELECTED PROBLEMS

CHAPTER 3

3.3.1 FORTRAN Constants

1. *Correct*   *Type*

   3.567   REAL
   1.56E6   REAL
   3.0   REAL
   'Mary'   CHARACTER
   '3-dimensional' CHARACTER

2. 'John Doe'
   3364185
   3.567856241894D0

3.3.2 Identification of Correct and Incorrect FORTRAN Variables

1. *Correct*   *Type*

   A   REAL
   I   INTEGER
   SMAX   REAL
   MAXT   INTEGER
   HUM   REAL
   LAMB   INTEGER
   X12   REAL

   Incorrect *Reason*

   1JUMP Starts with a number
   X-RAY   - not allowed in variable name
   C*30   * not allowed in variable name
   MAXIMUM More than six letters or numbers in the name
   WRITE   Reserved word in FORTRAN

3.3.3 Evaluation of Expressions

2. b. $s = 375$
   d. $yc = 0.499$
   f. $Z = 27.75$

4. b. Two operators, /-, immediately adjacent to each other.
   d. Cannot have an expression to the left of the = sign.

3.3.4 Evaluation of Program Sequences

1. $q = 68.$
   $r = 64.$
   $s = 21.33333$
   $t = 0.0$ (not defined)
   $u = 32.$
   $v = 10.$

3.3.5 Development of Expressions

2. `Q = 1.49/XN*A*Rh**(2./3.)*SQRT(So)`

4. `REAL i,n`
   `A = P*(i+i*((1.+i)**n-1))/((1.+i)**n -1.)`

6. `REAL k1,k2,Lin`
   `Di = k1/(k2-k1)*Lin*(EXP(-k1*t)-EXP(-k2*t)+Din*EXP(-k2*t)`

8. `REAL n`
   `p = R*T/(V/n-b)-a*(n/V)**2`

3.3.9 Subscripted Variables and Arrays

1. `REAL X(0:39)`

2. and 3.

Array	Capacity	Range of Index Values
Y(2,30)	60	1 to 2 and 1 to 30
A(2,3,6)	36	1 to 2, 1 to 3, and 1 to 6

5.

C(30)		30-Element array, with each element 8 characters long.
F(10,15)*6		150-Element array, with each element 6 characters long.
H*2		Character variable of length 2.

3.3.11 Finding Errors in Function Statements and References

b. $A(X,Y,2)$ Actual arguments in wrong order.

3.3.12 List-Directed Input and Output

2. 9.800000   4.890000   6.500000   14.69000   0.650000   9.000000   28.00000
   22.00000   13.00000   32.50000

CHAPTER 4

4.3.2 Developing DO Structures

2.
```
      BIG = -1.0E23
      DO 100 I=1,3
         DO 200 J=1,5
            IF(A(I,J).GT.BIG) BIG = A(I,J)
200      CONTINUE
100 CONTINUE
      DO 300 I=1,3
         DO 400 J=1,5
            A(I,J) = A(I,J)/BIG
400      CONTINUE
300 CONTINUE
```

4.
```
      REAL X(21),F1(21),F2(21),DIFF(21)
      X(1) = 0.0
      DO 100 I=1,21
         F1(I) = X(I)**3 -3.
         F2(I) = 3*X(I)**3 - X(I)**2
         DIFF(I) = F1(I) - F2(I)
         PRINT *, X(I),F1(I),F2(I),DIFF(I)
         IF (X.LT.21) X(I+1) = X(I) + 0.5
100 CONTINUE
```

4.3.3 Evaluating the Results of DO Structures

1. b. $Y = 90$

2. You could replace the IF with a BLOCK IF

```
      IF(X**2 - 2.0*X .EQ. Z) THEN
         K = I + J
         GO TO 800
      ENDIF
700 CONTINUE
800 CONTINUE
```

3. b.

I	J	K
1	1	1
2	4	4
3	9	9
4	16	0

4.3.4 Finding Errors in DO Structures and DO Statements

1. b. `DO 100 N = 1,20`   *N* and 100 backwards
   d. `DO 40 I = N,M,N`   Valid
   f. `DO 300 NA = 1,N4,1`   Mixed mode
   h. `DO 30 I = 20,1,-1`   Needs negative increment
   j. `DO 100 I = 1,20`   Ending statement reference left out

2. b. Improper nesting.
   d. First DO has improper negative increment.
   f. Improperly nested DO loops.
   h. I is used twice as an index in nested DO loops.

4.3.5 Evaluating Logical Expressions

1. `(x .lt. 3. .or. y .lt 4.)`   FALSE
   `(x .gt. y .and x .gt. z .or. y .lt. z)`   FALSE

2. b. T
   d. T
   f. F

CHAPTER 5

5.3.1 Calculation of Variable Values After Execution of SUBROUTINE or FUNCTION Subprograms

2. 4.000000   3.000000   5.000000

4. PCT = 80.0 and NWRONG = 10

6. b. First PRINT, I = 2 and J = 5
      Second PRINT, J = 2 and K = 6
      Within SUBROUTINE B: First CALL I = 1, and Second CALL I = 2

7. b. X = 0.529, Y = 4.0, and Z = 5.67
   d. X = 3.0, Y = 4.0, and Z = 7.67

5.3.2 Identify and Correct Errors in Calls to SUBROUTINE or FUNCTION Subprograms

1. `CALL ACT (Z,Y,N,X)`   *Y* and *J* do not agree in mode.
   `CALL ACT (X,I,M,W)`   *X* is declared as an array in the calling program, but not in the subroutine.

2. b. Function would return an integer value.
   d. Function names do not match.
   f. Correct.

5.3.3 Identifying and Correcting General SUBROUTINE and FUNCTION Subprogram Errors

1. b. A needs to be declared as an integer array in the subroutine.
   d. CALL statement should be CALL XLEST(A,B,X). You cannot use X(I) as a dummy argument in the subroutine. Change all X(I) to X.
   f. D is not one of the dummy arguments. A value for D would not be transmitted back to the calling program.

2. b. No dummy argument list.
   d. The P in the dummy argument list is not accomplishing anything.
   f. No function name.

5.3.4 Identify and Correct Errors in Passing Arrays To and From Subprograms

b. Array A cannot be larger than 40 in the subroutine.

d. A and B should be reversed in the dummy argument list. D should be used for the variable array size rather than I. The INTEGER statement in the subroutine should precede the REAL to declare D as an integer.

CHAPTER 6

6.3.1 Development of Input and Output Format Specifications

```
1.    100 FORMAT(3F8.0)
      110 FORMAT(3I4)
          READ(5,100) X,Y,Z
          READ(5,110) I,J,K
          READ(5,100) A,C

  123456789012345678901234567890
    65.739    .3578 3.00345
    30 6501000
  4.5E-05 656.789
```

6.3.2 Determination of Values Read In for Given FORMAT Specification

2. a. $A = 3.85, B = 6658.9, C = 66.781, D = 0.02233, E = 100.30, F = 40000.5$
   b. $A = 0.385, B = 10.030, C = 6.789, D = 123.456$
   c. $I = 3, J = 8500, K = 1$
   d. $A = 0.0385, B = 1.003, C = 6.789, D = 5E+06, E = 4.39E-02,$
      $F = 123456700., G = 987654300.$
   e. $A = 385006.1, B = 10030040.0, C = 6.7890000, D = 123456700.0$
   f. This would cause an error. The single I6 format would try to read an integer value from each line of the data file. The field on the third line of the data file contains a decimal point, which is not allowed in an I field.

4. $M = 5, N = 3$

   $A_{11} = 3.53, A_{12} = 6.98, A_{13} = 4.00, A_{14} = 4.98, A_{15} = 6.00$
   $A_{21} = 3.19, A_{22} = 6.50, A_{23} = 1.12, A_{24} = 2.15, A_{25} = 6.38$
   $A_{31} = 1.00, A_{32} = 0.98, A_{33} = 6.15, A_{34} = 50.1, A_{35} = 63.5$

6.3.3 Form of Output for Given FORMAT Specification

2. a. bbbbbb3.57bbbbbbb245.28bbb9987776.00
   b. bbbb3.57bb245.28********
      bbbbbbbbbbbbbbbbbbbbbbbbbb
      bb711bbb12bbb50
   c. b0.3568E+01
      b0.2453E+03
      b0.9988E+07
   d. bbbb3.57bb245.28
      bb711bbb12
      ********bbbb3.57

6.3.4 Determination of Number of Lines of Input or Output for Given FORMAT–Variable List Combinations

b. 4

d. 15

6.3.6 Data Statements

2. $B_{11} = 2.5$, $B_{21} = 3.0$, $B_{12} = 4.2$, $B_{22} = 5.0$, $B_{13} = 6.6$, $B_{23} = 7.5$, $B_{14} = 8.1$, $B_{24} = 9.0$

3. `DATA X,(A(I),I=1,3),Q,L/600*10.0,6.5,8.2,9.2,22.63,30/`

6.3.8 Identification of Errors in Formatted Input and Output Specifications

a. `(F6,2,F9 3,/)`   Comma between specifications

   `(3X,F6.2,A5,I6)`   Correct

   `(F6.2,F6.4)`   Remove literal from input format

   `(15F5.1)`   Trying to read 125 columns in one input record

b. `(4X,'Maximum Length = ',F8.3,' ft.')`   Need delimiters for literal

   `(1X,3I4,F6.2)`   Should have 1X at beginning

   `(1X,5F7.2,3F8.2,I5)`   Eliminate parentheses in 3F8.2

c. `FORMAT(20F4.0)`   No more than 80 columns per record

CHAPTER 7

7.3.1 Character Arrays

2.
```
      CHARACTER*1 A(11,11),AST,DASH
      DATA AST,DASH/'*','-'/
      DO 100 J=2,10
         A(1,J)=DASH
         A(6,J)=DASH
         A(11,J)=DASH
 100  CONTINUE
      JJ=1
      JJJ=11
      DO 200 I=1,11
         A(I,1)=AST
         A(I,11)=AST
         A(I,JJ)=AST
         A(I,JJJ)=AST
         JJ=JJ+1
         JJJ=JJJ-1
 200  CONTINUE
      DO 300 I=1,11
         WRITE(6,400) (A(I,J),J=1,11)
 300  CONTINUE
 400  FORMAT(1X,11A1)
      END
```

7.3.2 COMMON Blocks

b. Cannot repeat variable in COMMON and REAL statements. Mode of third variable in COMMON statement is integer in the main program and real in the subroutine.

d. Y(10) dimensioned twice.

APPENDIX B

ASCII
STANDARD
CODES

Seven-bit ASCII codes provide up to 2^7, or 128 characters. These are identified by the numbers 0 to 127. Codes 0–31 are used for special, nonprinting, 'control' characters, as is code 127. The remaining codes and the corresponding characters are given in Table B.1.

TABLE B.1   SEVEN-BIT ASCII STANDARD CODES

Numeric Code	Character	Numeric Code	Character	Numeric Code	Character
32	space	64	@	96	`
33	!	65	A	97	a
34	"	66	B	98	b
35	#	67	C	99	c
36	$	68	D	100	d
37	%	69	E	101	e
38	&	70	F	102	f
39	'	71	G	103	g
40	(	72	H	104	h
41	)	73	I	105	i
42	*	74	J	106	j
43	+	75	K	107	k
44	,	76	L	108	l
45	-	77	M	109	m
46	.	78	N	110	n
47	/	79	O	111	o
48	0	80	P	112	p
49	1	81	Q	113	q
50	2	82	R	114	r
51	3	83	S	115	s
52	4	84	T	116	t
53	5	85	U	117	u
54	6	86	V	118	v
55	7	87	W	119	w
56	8	88	X	120	x
57	9	89	Y	121	y
58	:	90	Z	122	z
59	;	91	[	123	{
60	<	92	\	124	|
61	=	93	]	125	}
62	>	94	^	126	~
63	?	95	_		

APPENDIX C

ADDITIONAL FORTRAN FUNCTIONS

Table 3.1 illustrates some of the most commonly used FORTRAN intrinsic functions. Table C.1 gives a complete list of the intrinsic functions. Generic functions are shown in bold type, and specific-type functions in regular type. The variable X represents any type argument, RX is a real argument, CX is a complex argument, DX is a double precision argument, CHAR is a character argument, LOG is a logical argument, and I is an integer argument.

TABLE C.1   FORTRAN INTRINSIC FUNCTIONS

Function Name	Function Produces	Function Type	Comments
ABS (X)	**Absolute value of X**	**Same as Argument**	
IABS (I)	Absolute value	Integer	Integer argument
DABS (DX)	Absolute value	Double precision	Double precision argument
CABS (CX)	Absolute value	Complex	Complex argument
AINT (X)	**Truncation of X**	**Same as Argument**	**Decimal portion of value truncated**
DINT (DX)	Truncation of DX	Double Precision	Double precision argument
ANINT (X)	**X rounded to nearest whole number**	**Same as Argument**	
DNINT (DX)	DX rounded	Double precision	Double precision argument
NINT (X)	X rounded to nearest integer value	Integer	
IDNINT (DX)	DX rounded	Integer	Double precision argument
ACOS (X)	**Arccosine of X**	**Same as Argument**	**Radians**
DACOS (DX)	Arccosine of X	Double precision	Double precision argument
ASIN (X)	**Arcsine of X**	**Same as Argument**	**Radians**
DASIN (DX)	Arcsine of DX	Double precision	Double precision argument
ATAN (X)	**Arctangent of X**	**Same as Argument**	**Radians**
DATAN (DX)	Arctangent of DX	Double precision	Double precision argument
ATAN2 (X,Y)	**Arctangent of X/Y**	**Same as Argument**	**Radians**
DATAN2 (DX,DY)	Arctangent of DX/DY	Double precision	Double precision arguments
COS (X)	**Cosine of X**	**Same as Argument**	**Radians**
DCOS (DX)	Cosine of DX	Double precision	Double precision argument

TABLE C.1   (Continued)

Function Name	Function Produces	Function Type	Comments
CCOS (CX)	Cosine of CX	Complex	Complex argument
COSH (X)	**Hyperbolic Cosine of X**	**Same as Argument**	**Radians**
DCOSH (DX)	Hyperbolic Cosine of DX	Double precision	Double precision argument
DIM (X,Y)	**X—(minimum of X and Y)**	**Same as Argument**	
IDIM (IX,IY)	IX—(min of IX and IY)	Integer	Integer arguments
DDIM (DX,DY)	DX—(min of DX and DY)	Double precision	Double precision arguments
DBLE (X)	**Double precision of X**	**Double precision**	
CMPLX (X)	**X + j0**	**Complex**	
CMPLX (X,Y)	**X + jY**	**Complex**	
CONJG (CX)	Conjugate of CX, a − jb	Complex	
EXP (X)	**e to the power X**	**Same as Argument**	
DEXP (DX)	e^{DX}	Double Precision	Double precision argument
CEXP (CX)	e^{CX}	Complex	Complex argument
INDEX (CHX,CHY)	Position of substring CHY in string CHX	Integer	Arguments are character strings
INT (X)	**Integer value of X**	**Integer**	
IFIX (RX)	Integer value of RX	Integer	Real argument
IDINT (DX)	Integer value of DX	Integer	Double precision argument
LEN (CHX)	Length of character string CHX	Integer	Argument is character string
LGT (CHX,CHY)	CHX is lexically greater than CHY	Logical	Arguments are character strings
LLE (CHX,CHY)	CHX is lexically less than or equal to CHY	Logical	Arguments are character strings
LLT (CHX,CHY)	CHX is lexically less than CHY	Logical	Arguments are character strings
LOG (X)	**Natural log of X**	**Same as Argument**	
ALOG (RX)	Natural log of RX	Real	Real argument
DLOG (DX)	Natural log of DX	Double precision	Double precision argument
CLOG (CX)	Natural log of CX	Complex	Complex argument
LOG10 (X)	**Log 10 of X**	**Same as Argument**	
ALOG10 (RX)	Log 10 of RX	Real	Real argument
DLOG10 (DX)	Log 10 of DX	Double precision	Double precision argument
MAX (X,Y,...)	**Maximum of (X,Y,...)**	**Same as Argument**	
MAX0 (IX,IY,..)	Maximum of (IX,IY,...)	Integer	Integer arguments
AMAX1(RX,RY,..)	Maximum of (RX,RY,...)	Real	Real arguments
DMAX1(DX,DY,..)	Maximum of (DX,DY,...)	Double precision	Double precision arguments
AMAX0(IX,IY,..)	Maximum of (IX,IY,...)	Real	Integer arguments
MAX1 (RX,RY,..)	Maximum of (RX,RY,...)	Integer	Real arguments

TABLE C.1   (Continued)

Function Name	Function Produces	Function Type	Comments
MIN (X,Y,...)	**Minimum of (X,Y,...)**	**Same as Argument**	
MIN0 (IX,IY,..)	Minimum of (IX,IY,...)	Integer	Integer arguments
AMIN1(RX,RY,..)	Minimum of (RX,RY,...)	Real	Real arguments
DMIN1(DX,DY,..)	Minimum of (DX,DY,...)	Double precision	Double precision arguments
AMIN0(IX,IY,..)	Minimum of (IX,IY,...)	Real	Integer arguments
MIN1 (RX,RY,..)	Minimum of (RX,RY,...)	Integer	Real arguments
MOD (X,Y)	**Remainder of X/Y**	**Same as Argument**	
AMOD (RX/RY)	Remainder of RX/RY	Real	Real arguments
DMOD (DX,DY)	Remainder of DX/DY	Double precision	Double precision arguments
REAL (X)	**Real value of X**	**Real**	**If X is complex, real part of CX is returned**
AIMAG (CX)	Imaginary part of CX	Real	Complex argument
FLOAT (I)	Real value of I	Real	Integer argument
SNGL (DX)	Convert DX to real	Single precision	Double precision argument
SIGN (X,Y)	**Transfer sign of Y to |X|**	**Same as Argument**	
ISIGN (IX,IY)	Transfer sign IX to |IX|	Integer	Integer arguments
DSIGN (DX,DY)	Transfer sign DX to |DY|	Double precision	Double precision arguments
SIN (X)	**Sine of X**	**Same as Argument**	**Radians**
DSIN (DX)	Sine of DX	Double precision	Double precision
CSIN (CX)	Sine of CX	Complex	Complex argument
SINH (X)	**Hyperbolic Sine of X**	**Same as Argument**	**Radians**
DSINH (DX)	Hyperbolic Sine of DX	Double precision	Double precision argument
SQRT (X)	**Square root of X**	**Same as Argument**	
DSQRT (DX)	Square root of DX	Double precision	Double precision argument
CSQRT (CX)	Square root of CX	Complex	Complex argument
TAN(X)	**Tangent of X**	**Same as Argument**	**Radians**
DTAN (DX)	Tangent of DX	Double precision	Double precision argument
TANH (X)	**Hyperbolic Tangent of X**	**Same as Argument**	**Radians**
DTANH (DX)	Hyperbolic Tangent of DX	Double precision	Double precision argument

APPENDIX D

DIMENSIONS, UNITS, CONSTANTS, AND CONVERSIONS

D.1 DIMENSIONS AND UNITS

Dimensions are unique quantities used to describe the physical attributes of some system or entity, whereas units are used to quantify the dimensions. Dimensions can be divided into fundamental and derived dimensions. Table D.1 gives the fundamental dimensions engineers use.

TABLE D.1
FUNDAMENTAL
DIMENSIONS

Fundamental Dimension	Symbol
Mass	M
Length	L
Time	t
Force	F
Temperature	T
Charge	Q
Electric current	I
Molecular substance	n

Derived dimensions are combinations of the fundamental dimensions that are used to describe attributes of a system or entity. For example, volume is L^3 and acceleration is L/t^2.

The fundamental dimensions of mass and force are related through Newton's second law

$$F = Ma$$

or force equals mass times acceleration.

There are several sets of units in common use. We will limit our discussion to two systems, the international system of units (SI) and the English system of units. The SI system is often referred to as the metric system. Base units for the two systems are given in Table D.2.

TABLE D.2   BASE UNITS

Fundamental Quantity	SI Unit	Symbol	English Unit	Symbol
Electric current	ampere	I	ampere	I
Force	newton	N	pound-force	lb
Length	meter	m	foot	ft
Mass	kilogram	kg	slug	slug
Molecular substance	mole	mol	mole	mol
Temperature	Kelvin	°K	Rankine	°R
Time	Seconds	s	Seconds	s

Table D.3 gives commonly used derived units for both SI and English systems.

TABLE D.3   DERIVED UNITS

Derived Quantity	SI Units	Symbol	English Units	Symbol
Acceleration	meters/second^2	m/s^2	feet/second^2	ft/s^2
Density	kilograms/meter^3	kg/m^3	slugs/foot^3	slugs/ft^3
Power	newton-meters/second	Watt	foot-pound/second	ft-lb/s
Pressure	newtons/meter^2	Pascal	pounds/foot^2	lb/ft^2
Specific weight	newtons/meter^3	N/m^3	pounds/foot^3	lb/ft^3
Velocity	meters/second	m/s	feet/second	ft/s
Work or energy	newton-meters	N-m	foot-pounds	ft-lb

Newton-meters are also called Joules.

Weight is a commonly used derived unit. It is actually a force. Weight is related to the mass of a body and the local acceleration due to gravity through Newton's second law. In SI units,

Weight (newtons) = Mass (kilograms) · acceleration due to gravity (g)

and in English units,

Weight (pounds) = Mass (slugs) · acceleration due to gravity (g)

The acceleration due to gravity near the earth's surface is 9.81 m/s^2, or 32.17 ft/s^2. From the above relationships it can be seen that the specific weight is equal to the density times the local acceleration due to gravity.

When you are using mathematical equations, you have to make sure that the units for each of the terms on the left of the equal sign agree with the units of each of the terms on the right. Otherwise, your equation will give incorrect numerical values even though the mathematical expression itself may be entirely correct.

D.2 CONSTANTS AND CONVERSIONS

$\pi = 3.14159265$

$e = 2.718281828$

1 radian $= 57.2958°$

Avogadro's number $= 6.023 \times 10^{23}$ molecules/mole

Degrees Fahrenheit: $°F = °R - 459.69$ R = Rankine

Degrees Celsius: $°C = °K - 273.15$ K = Kelvin

$°F = \left(\dfrac{9}{5}\right) °C + 32$

1 kilogram $= 1,000$ grams

1 kilogram $= 6.85 \times 10^{-2}$ slugs

1 Newton $= 0.2248$ pounds-force (lb)

60 seconds $= 1$ minute

60 minutes $= 1$ hour

24 hours $= 1$ day

365 days $= 1$ year

10 millimeters (mm) $= 1$ centimeter (cm)

100 centimeters $= 1$ meter

1,000 meters $= 1$ kilometer (km)

2.54 centimeters $= 1$ inch (in.)

12 inches $= 1$ foot (ft)

3 feet $= 1$ yard (yd)

5,280 feet $= 1$ mile

1 horsepower (hp) $= 550$ foot-pounds/second

1 kilowatt (kW) $= 1,000$ Watts

1 million gallons per day (mgd) $= 1.55$ feet^3/second (cfs)

1 U.S. gallon $= 0.1337$ foot^3

1 liter (ℓ) $= 10^{-3}$ meter^3

APPENDIX E

ECONOMIC ANALYSIS

E.1 INTRODUCTION

Economic formulas are used to estimate the value of money over time. To compare different alternatives you have to be able to compare the costs and returns in terms of common factors. Economic formulas allow you to do this. A total of five parameters have to be considered in economic evaluations. These are a present amount of money (P), an amount of money in the future (F), a series of equal payments over time (A), the interest rate (i), and the number of times the interest rate is applied (n). The parameter n is also called the number of compounding periods. All conditions can be evaluated by using a combination of these factors and the formulas developed from them. The elementary formulas will always use four of the five parameters, with values for three of the four parameters known. Formulas can also be developed for series of payments that are not constant over time.

Interest is money that must be paid for the use of borrowed money. If you are the borrower, then you must pay the interest as a rental fee. If you are the lender, you receive the interest as return on your investment.

E.2 CASH FLOW DIAGRAMS

Cash flow diagrams are used to graphically represent the economic parameters over time. For example, Figure E.1 shows a typical investment in a piece of equipment. Figure E.1a shows the cash flows for the initial cost P, the annual maintenance costs A, and the salvage value F, when the piece of equipment is retired. Note that the latter is treated as a return. Costs are considered to be negative cash flows, and returns are treated as positive cash flows. Figure E.1b shows the cash flows for the return you are expecting to receive by using the machine to accomplish work. The two diagrams can be combined to form Figure E.1c. The costs and returns for the individual periods can be subtracted to form a net cost or return for that period, but all the costs and returns cannot merely be added together because they occur at different times.

FIGURE E.1   Cash flows for a piece of equipment.

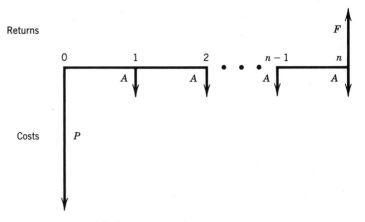

(*a*) Investment in equipment.

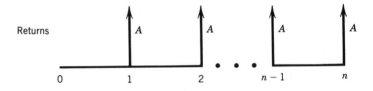

(*b*) Expected income from equipment.

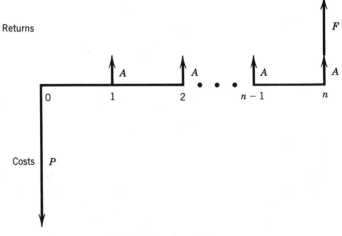

(*c*) Combined cash flows.

E.3 SINGLE-PAYMENT FORMULAS

Single-payment formulas involve only a single amount of money, rather than a uniform series of payments. A present amount of money P is compounded to give some estimated future amount of money F. Or a future amount of money is discounted back to its present worth. If a present amount of money is increased by the interest rate i over a compounding period, at the end of the first compounding period you would have

$$F_1 = P(1 + i)$$

If the new amount is subjected to interest for the second period, the interest is said to be compounding, and over n compounding periods the value of the initial P will grow to be

$$F = P(1 + i)^n \qquad \text{(E.1)}$$

The percentage increase, $(1 + i)^n$, is called the Compound Amount Factor, or CAF. It is customary to write the compound amount factor as (CAF, i, n) for the compound amount factor for i percent and n compounding periods.

   If you want to determine the present value of some amount of money in the future, all you have to do is take the inverse of the compound amount factor in Equation E.1. This is called the Present Worth Factor, or (PWF, i, n), and is shown in Equation E.2.

$$P = F \left[\frac{1}{(1 + i)^n} \right] = F(PWF, i, n) \qquad \text{(E.2)}$$

E.4   UNIFORM SERIES OF PAYMENT FORMULAS

Economic formulas involving uniform series of payments relate the series of payments either to a present worth P or a future worth F. In engineering analysis the usual practice is to assume that the payment A takes place at the end of the compounding period. The formulas presented in this section follow that practice. Similar formulas can be developed for payments at the beginning of a period.

E.4.1   Formulas Relating A to P

Many engineering projects are evaluated in terms of present worth. In order to do this, you have to have a formula for finding the equivalent present worth of a series of payments. The formula is derived by treating the present worth as a single payment that is subjected to n compounding periods. Each payment A is treated as a single payment that is subjected to a decreasing number of compounding periods as the payment comes later in the series. The first payment, since it occurs at the end of the first period, would be subject to $n - 1$ compounding periods. The last payment would not be compounded. If the sum of the payments and the compounded present worth are set equal to each other, you have

$$P(1 + i)^n = A_1(1 + i)^{n-1} + A_2(1 + i)^{n-2} + \cdots + A_{n-1}(1 + i) + A_n \quad \text{(E.3)}$$

This expression can be rearranged to give

$$P = A \left[\frac{(1 + i)^n - 1}{(1 + i)^n i} \right] = A(USPWF, i, n) \qquad \text{(E.4)}$$

The portion in brackets forms the Uniform Series of Payments Present Worth Factor $(USPWF, i, n)$. Conversely, the formula for finding the equivalent series of payments for a given present worth is the reciprocal of the $USPWF$, or

$$A = P \left[\frac{(1 + i)^n i}{(1 + i)^n - 1} \right] = P(CRF, i, n) \tag{E.5}$$

This factor is called the Capital Recovery Factor (CRF, i, n). In practical terms, the capital recovery factor can be used to determine what your payments would be on a loan of P dollars if you have n payment periods and the loan is at i percent interest.

E.4.2 Formulas Relating A to F

At times you will want to relate a series of payments to some future worth. For example, you might want to evaluate how much you would have to set aside each year to have a specified amount of money available in 18 years for your child's college education. Or if you are evaluating engineering projects in terms of annual payments, you will have to be able to convert a future worth into an equivalent series of payments. Formulas for doing this are developed in much the same way as the formulas relating A to P. Each of the payments A is an amount subjected to a decreasing number of compounding periods as the payment comes later in the sequence. The future amount F is the sum of the future worths of the payments. Therefore,

$$F = A_1(1 + i)^{n-1} + A_2(1 + i)^{n-2} + \cdots + A_{n-1}(1 + i) + A_n \tag{E.6}$$

which can be rearranged to give

$$F = A \left[\frac{(1 + i)^n - 1}{i} \right] = A(USCAF, i, n) \tag{E.7}$$

This formula for finding the future value F of a series of payments A is called the Uniform Series Compound Amount Factor $(USCAF, i, n)$. The formula for finding the equivalent series of payments for a given future worth is the reciprocal of the $USCAF$, and is called the Sinking Fund Factor (SFF, i, n). It is given by the formula

$$A = F \left[\frac{i}{(1 + i)^n - 1} \right] = F(SFF, i, n) \tag{E.8}$$

Table E.1 gives a summary of the relationships among the various economic parameters and provides some useful identities.

E.5 ECONOMIC FEASIBILITY OF PROJECTS

The economic feasibility of projects can be evaluated by comparing their costs and returns in terms of one of the economic parameters. This will normally require application of one or more of the economic factors discussed in the previous

TABLE E.1   SUMMARY OF ECONOMIC FORMULAS

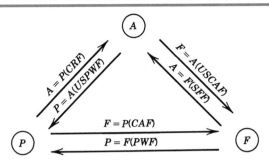

Definitions:

$\quad$ CAF = single payment compound amount factor = $(1 + i)^n$

$\qquad$ Known: i, n, P $\qquad$ Unknown: F

$\quad$ PWF = single payment present worth factor = $1/(1 + i)^n$

$\qquad$ Known: i, n, F $\qquad$ Unknown: P

$\quad$ USCAF = uniform series compound amount factor = $[(1 + i)^n - 1]/i$

$\qquad$ Known: A, i, n $\qquad$ Unknown: F

$\quad$ SFF = sinking fund factor = $i/[(1 + i)^n - 1]$

$\qquad$ Known: F, i, n $\qquad$ Unknown: A

$\quad$ CRF = capital recovery factor = $i(1 + i)^n/[(1 + i)^n - 1]$

$\qquad$ Known: P, i, n $\qquad$ Unknown: A

$\quad$ USPWF = uniform series present worth factor = $[(1 + i)^n - 1]/i(1 + i)^n$

$\qquad$ Known: A, i, n $\qquad$ Unknown: P

Identities:

$$SFF = CRF - i \qquad\qquad CRF = 1/PWF$$
$$SFF = (CRF)(PWF) \qquad USCAF = 1/SFF$$
$$USPWF = (USCAF)(PWF)$$
$$CRF = (SFF)(CAF) \qquad CRF = 1/USPWF$$

section to convert all amounts of money into the same form. The interest rate used should be the interest rate that could be realized by alternative investments.

E.5.1 Net Present Worth

Net present worth is the most widely used method of evaluating engineering investments. To evaluate the feasibility of an investment or project using net present worth, you first convert all series of payments into present worth using the uniform series present worth factor $(USPWF, i, n)$ and all future values into present worth using the present worth factor (PWF, i, n). Then you subtract the present worth of the costs from the present worth of the returns to get the net present worth. If the net present worth is positive, the project is more advantageous than making an alternative investment at an interest rate of i. If the net present worth is negative, the project is not economically attractive. You can compare competing alternative investments by comparing their net present worths. The alternative with the largest positive net present worth is the best investment.

E.5.2 Equivalent Uniform Net Annual Return

The process for evaluating investments and projects using equivalent uniform net annual return is the same as that for net present worth except that you convert all amounts into equivalent uniform series of payments. Present worths can be converted into equivalent series of payments using the capital recovery factor (CRF, i, n), and future amounts can be converted using the sinking fund factor (SFF, i, n).

E.5.3 Internal Rate of Return

The internal rate of return is an alternative way of evaluating investments. The internal rate of return is not used extensively by engineers, but it is favored by economists. Rate of return analysis is more involved than the other two methods already described. To find the internal rate of return you solve for the interest rate that makes the costs equal to the returns. However, you still have to do the comparison using all costs and returns in terms of P, A, or F. Therefore, you have to set up all the formulas that you would normally use for converting all the costs and returns into common terms. Then you equate the total cost to the total returns and solve for the interest rate that would make them equal. You have to use iteration to solve for i.

VECTOR
AND
LINEAR
ALGEBRA

F.1 VECTOR NOTATION AND OPERATIONS

F.1.1 Notation

Vectors are quantities in physics that possess both magnitude and direction, such as velocity and acceleration. Quantities that have magnitude but not direction are called scalars. A vector name is usually identified by a bold letter or letters. The vector itself can be represented in several different ways, as will be shown in the following sections. The fundamental way of expressing a vector is in terms of its components. In three-dimensional space the components are defined along the x, y, and z axes. They are usually labeled with the bold letters **i, j,** and **k.** Therefore, pertaining to the vector **A**

$$\mathbf{A} = a_x\mathbf{i} + a_y\mathbf{j} + a_z\mathbf{k} = 4\mathbf{i} + 3\mathbf{j} - 5\mathbf{k}$$

would represent a vector with components of 4 in the positive x direction, 3 in the positive y direction, and 5 in the negative z direction. You can define vectors of any dimension. However, at this point in your engineering career you will be working mainly with two- and three-dimensional vectors, so we will limit our discussion to those vectors. You can handle two-dimensional vectors in the same way as three-dimensional vectors, using the **i** and **j** components.

F.1.2 Addition and Subtraction

Vector addition and subtraction is best accomplished by components. If the vectors **A** and **B** are

$$\mathbf{A} = 4\mathbf{i} + 3\mathbf{j} - 5\mathbf{k}$$
$$\mathbf{B} = 3\mathbf{i} - 5\mathbf{j} + 8\mathbf{k}$$

then the vector **C** is

$$\mathbf{C} = \mathbf{A} + \mathbf{B} = 7\mathbf{i} - 2\mathbf{j} + 3\mathbf{k}$$

Like components are added together to give the sum. Vector subtraction is carried out in the same way except that the components of the **B** vector would be algebraically subtracted from the components of the **A** vector.

F.1.3 Norm (Length, Magnitude) of Vector

The norm of a vector is its scalar length and can be found by the formula

$$A = \sqrt{a_x^2 + a_y^2 + a_z^2} \tag{F.1}$$

The norm can also be used to find the distance between two points. Points are defined by a set of coordinates. In three-dimensional space the coordinates are given in the same order as the components of a vector. If you have two points, you can find the components of the vector joining those two points by subtracting the coordinates of the first from the coordinates of the second. If point one is (x_1, y_1, z_1) and point two is (x_2, y_2, z_2), the vector joining the two points is

$$\mathbf{r} = (x_2 - x_1)\mathbf{i} + (y_2 - y_1)\mathbf{j} + (z_2 - z_1)\mathbf{k} \tag{F.2}$$

and Equation F.1 can be used to find the length of **r**.

F.1.4 Characteristics of a Line in Two-Space

The length of a line segment joining two points in two-dimensional space, $P_1 = (x_1, y_1)$ and $P_2 = (x_2, y_2)$ can be found by the methods of the previous section. The midpoint of this line segment has the coordinates

$$\left(\frac{x_1 + x_2}{2}, \frac{y_1 + y_2}{2} \right) \tag{F.3}$$

The slope m of the line between P_1 and P_2 is given by

$$m = \frac{y_2 - y_1}{x_2 - x_1} \tag{F.4}$$

provided that x_1 is not equal to x_2. The slope-intercept equation for the line between P_1 and P_2 is

$$y = mx + b \tag{F.5}$$

The intercept b can be found from

$$b = y_1 - mx_1 \tag{F.6}$$

The perpendicular bisector of the line segment joining P_1 and P_2 is the line through the midpoint with a slope of $-1/m$, provided that m is not zero.

F.1.5 Unit Vectors and Directional Cosines

A unit vector, **u**, is a vector of magnitude 1 along a defined vector **r**. It is found by dividing the components of the vector by the magnitude of the vector. Therefore,

$$\mathbf{u} = \frac{\mathbf{r}}{r} = \frac{r_x}{r}\mathbf{i} + \frac{r_y}{r}\mathbf{j} + \frac{r_z}{r}\mathbf{k} = u_x\mathbf{i} + u_y\mathbf{j} + u_z\mathbf{k} \qquad (F.7)$$

The components of the unit vector, u_x, u_y, and u_z, are the directional cosines of the vector $\mathbf{r}$. They represent the cosines of the angles between the vector and the respective x, y, and z axes. It can be seen from Equation F.7 that the components of any vector can be found by multiplying its magnitude by its unit vector.

F.1.6 Dot (Scalar) Product

The dot or scalar product of two vectors can be used to find the angle θ between two vectors. It has the form

$$\mathbf{P} \cdot \mathbf{Q} = p_1q_1 + p_2q_2 + p_3q_3 = PQ \cos \theta \qquad (F.8)$$

F.1.7 Cross (Vector) Product

A moment is a force multiplied by the length of the moment arm between that force and the point about which moments are being taken. In three-dimensional space the moment arm is a vector that starts at the point about which the moment is being taken and terminates at some known point on the line of action of the force. The moment arm vector is called a position vector. If $\mathbf{r}$ is the position vector and $\mathbf{F}$ is the force vector, the vector moment, $\mathbf{M}$, is

$$\mathbf{M} = \mathbf{r} \times \mathbf{F} = \begin{vmatrix} \mathbf{i} & \mathbf{j} & \mathbf{k} \\ r_x & r_y & r_z \\ F_x & F_y & F_z \end{vmatrix} \qquad (F.9)$$

The determinant of the matrix shown gives the moment vector. Finding determinants will be discussed in the next section.

F.2 LINEAR ALGEBRA

F.2.1 Matrix Representation of Linear Equations

A system of linear equations

$$\begin{aligned} 2X_1 + 3X_2 + X_3 &= 4 \\ X_2 + 2X_3 &= 6 \\ 4X_1 + 2X_2 &= 12 \end{aligned} \qquad (F.10)$$

can be represented in matrix notation as

$$\begin{bmatrix} 2 & 3 & 1 \\ 0 & 1 & 2 \\ 4 & 2 & 0 \end{bmatrix} \begin{bmatrix} X_1 \\ X_2 \\ X_3 \end{bmatrix} = \begin{bmatrix} 4 \\ 6 \\ 12 \end{bmatrix} \qquad (F.11)$$

or in even more compact representation as $\mathbf{AX = B}$. $\mathbf{A}$ is called the coefficient matrix and $\mathbf{B}$ is the right-hand side. In general terms these matrices are

$$\begin{bmatrix} a_{11} & a_{12} & a_{13} \\ a_{21} & a_{22} & a_{23} \\ a_{31} & a_{32} & a_{33} \end{bmatrix} \begin{bmatrix} X_1 \\ X_2 \\ X_3 \end{bmatrix} = \begin{bmatrix} b_1 \\ b_2 \\ b_3 \end{bmatrix} \qquad (F.12)$$

The general term a_{ij} represents the matrix coefficient in the ith row and the jth column. $\mathbf{A}_{m \times n}$ represents a matrix of m rows and n columns.

F.2.2 Definitions

Square matrix: a matrix that has the same number of rows and columns.

Identity matrix: a matrix that has ones on the diagonal and zeros everywhere else. The identity matrix is labeled $\mathbf{I}$.

$$\mathbf{I} = \begin{bmatrix} 1 & 0 & 0 \\ 0 & 1 & 0 \\ 0 & 0 & 1 \end{bmatrix}$$

Augmented matrix: two or more matrices joined together. The augmented matrix helps in the solution of systems of linear equations or finding the inverse of a matrix. Finding the inverse will be discussed in a subsequent section.

$$(\mathbf{A}|\mathbf{B}) = \left[\begin{array}{ccc|c} 2 & 3 & 1 & 4 \\ 0 & 1 & 2 & 6 \\ 4 & 2 & 0 & 12 \end{array}\right]$$

is an augmented matrix.

Transpose: The transpose of a matrix interchanges the rows and columns.

$$\text{If } \mathbf{A} = \begin{bmatrix} 1 & 2 & 3 \\ 3 & 4 & 5 \\ 5 & 6 & 7 \end{bmatrix}, \text{ then } \mathbf{A}^{\mathbf{T}} = \begin{bmatrix} 1 & 3 & 5 \\ 2 & 4 & 6 \\ 3 & 5 & 7 \end{bmatrix}$$

F.2.3 Matrix Operations

F.2.3.1 Addition and Subtraction

Matrix addition requires the addition of corresponding elements of the two matrices. If $\mathbf{C} = \mathbf{A} + \mathbf{B}$, then $c_{ij} = a_{ij} + b_{ij}$. Matrices have to be the same size to be added together. Matrix subtraction uses the same operation, except corresponding elements are subtracted.

F.2.3.2 Multiplication by a Scalar

When a matrix is multiplied by a scalar, each element of the matrix is multiplied by the scalar value.

F.2.3.3 Multiplication of Two Matrices

Two matrices can be multiplied together if the number of columns of the first is equal to the number of rows of the second. For $\mathbf{C} = \mathbf{AB}$, the general formula is

$$c_{ik} = \sum_{j=1}^{n} a_{ij} b_{jk} \tag{F.13}$$

For example,

$$\begin{bmatrix} 3 & 4 & 5 \\ 1 & 2 & 3 \end{bmatrix} \begin{bmatrix} 1 & 1 \\ 0 & 1 \\ 2 & 0 \end{bmatrix} = \begin{bmatrix} c_{11} & c_{12} \\ c_{21} & c_{22} \end{bmatrix}$$

$$c_{11} = 3(1) + 4(0) + 5(2) = 13$$
$$c_{12} = 3(1) + 4(1) + 5(0) = 7$$
$$c_{21} = 1(1) + 2(0) + 3(2) = 7$$
$$c_{22} = 1(1) + 2(1) + 3(0) = 3$$

and

$$\mathbf{C} = \begin{bmatrix} 13 & 7 \\ 7 & 3 \end{bmatrix}$$

The product matrix will have the same number of rows as the first matrix and the same number of columns as the second matrix. If $\mathbf{A}_{m \times n}$ and $\mathbf{B}_{n \times p}$, then $\mathbf{C}_{m \times p}$. Generally, matrix multiplication is not commutative, that is, $\mathbf{AB} \neq \mathbf{BA}$, even if both multiplications are defined.

F.2.3.4 Division

Matrix division is not explicitly defined; however, under certain conditions there exists a matrix with the properties

$$\mathbf{A} \cdot \mathbf{A}^{-1} = \mathbf{A}^{-1} \cdot \mathbf{A} = \mathbf{I} \tag{F.14}$$

This is one of the few cases where matrix multiplication is commutative. $\mathbf{A}^{-1}$ is called the inverse of $\mathbf{A}$. Only square matrices have inverses. If a square matrix has an inverse, it is said to be nonsingular.

It is important to be able to find the inverse of a matrix because it can be used to find the solutions of systems of linear equations. If the system of equations

$$\mathbf{AX} = \mathbf{B}$$

is premultiplied by the inverse of $\mathbf{A}$, the result is

$$\mathbf{A}^{-1} \cdot \mathbf{A} \cdot \mathbf{X} = \mathbf{A}^{-1} \cdot \mathbf{B}$$
$$\mathbf{I} \cdot \mathbf{X} = \mathbf{A}^{-1} \cdot \mathbf{B}$$

$$\mathbf{X} = \mathbf{A}^{-1}\mathbf{B} \tag{F.15}$$

Thus, if the right-hand-side vector is premultiplied by the inverse of $\mathbf{A}$, the result will be the solution vector $\mathbf{X} = (X_1, X_2, X_3)$, which gives the values for the independent variables you wanted to solve for.

F.2.4 Determinant and Linear Independence

The determinant is an important property of a square matrix. We will first develop the methods for computing the determinant, and then discuss its application in operations such as testing the linear independence of systems of equations.

The determinant of a 2×2 matrix can be found by

$$|\mathbf{A}| = \begin{bmatrix} a_{11} & a_{12} \\ a_{21} & a_{22} \end{bmatrix} = a_{11}a_{22} - a_{21}a_{12} \tag{F.16}$$

You can compute the determinant for a 3×3 matrix by repeating the first two columns to the right of the third column, cross multiplying the augmented matrix, and adding up the results. For example,

$$|\mathbf{A}| = \begin{vmatrix} 2 & 3 & 1 & 2 & 3 \\ 0 & 1 & 2 & 0 & 1 \\ 4 & 2 & 0 & 4 & 2 \end{vmatrix}$$

$$= +(2)(1)(0) + (3)(2)(4) + (1)(0)(2) - (4)(1)(1) - (2)(2)(2) - (0)(0)(3)$$
$$= 0 + 24 + 0 - 4 - 8 - 0$$
$$= 12$$

A more general way of computing the determinant for any size matrix is the cofactor method.

$$|\mathbf{A}| = \sum_{i=1}^{n} a_{ij} A_{ij} = \sum_{j=1}^{n} a_{ij} A_{ij}$$

$$A_{ij} = cofactor = (-1)^{(i+j)} \begin{vmatrix} Determinant \\ of \\ minor \end{vmatrix}$$

(F.17)

A minor is the submatrix formed when row i and column j are eliminated from the matrix. For example, if

$$\mathbf{A} = \begin{bmatrix} 2 & 3 & 1 \\ 0 & 1 & 2 \\ 4 & 2 & 0 \end{bmatrix}$$

$$|\mathbf{A}| = (1)(2) \begin{vmatrix} 1 & 2 \\ 2 & 0 \end{vmatrix} + (-1)(3) \begin{vmatrix} 0 & 2 \\ 4 & 0 \end{vmatrix} + (1)(1) \begin{vmatrix} 0 & 1 \\ 4 & 2 \end{vmatrix}$$

$$= (1)(2)(-4) + (-1)(3)(-8) + (1)(1)(-4)$$
$$= 12$$

The determinant is used to compute the vector cross product, which was defined in Section F.1.7 of this appendix. It can also be used to solve for the unknowns in a system of equations using Cramer's rule. A subsequent section will develop this procedure. The determinant can test a system of equations for independence.

A system of equations is linearly independent if none of the equations can be written as a linear combination of some or all of the other equations. If the system of equations is not independent, you will not be able to find the inverse of the coefficient matrix. The coefficient matrix is said to be singular when this occurs. The determinant of the coefficient matrix for an independent set of equations will not be zero. Therefore, if the determinant is equal to zero, the coefficient matrix is singular, and the system of equations is not independent.

F.2.5 Solution of System of Linear Equations

F.2.5.1 Row Operations to Find Inverse

The inverse of a nonsingular square matrix can be obtained by applying elementary row operations to an augmented matrix containing the original matrix augmented with the identity matrix. These elementary row operations are

TABLE F.1   INVERSE AND LINEAR EQUATION
SOLUTION BY ROW OPERATIONS

INITIAL MATRIX

$$\left[\begin{array}{ccc|ccc|c} 2 & 3 & 1 & 1 & 0 & 0 & 4 \\ 0 & 1 & 2 & 0 & 1 & 0 & 6 \\ 4 & 2 & 0 & 0 & 0 & 1 & 12 \end{array}\right]$$

Step 1: Divide row 1 by 2.

$$\left[\begin{array}{ccc|ccc|c} 1 & 1.5 & 0.5 & 0.5 & 0 & 0 & 2 \\ 0 & 1 & 2 & 0 & 1 & 0 & 6 \\ 4 & 2 & 0 & 0 & 0 & 1 & 12 \end{array}\right]$$

Step 2: Add (-4) times row 1 to row 3. Note: If a_{21} had not been zero at this stage, some multiple of row 1 would be added to row 2 to eliminate a_{21}.

$$\left[\begin{array}{ccc|ccc|c} 1 & 1.5 & 0.5 & 0.5 & 0 & 0 & 2 \\ 0 & 1 & 2 & 0 & 1 & 0 & 6 \\ 0 & -4 & -2 & -2 & 0 & 1 & 4 \end{array}\right]$$

Step 3: Add (-1.5) times row 2 to row 1. Add (4) times row 2 to row 3. Note: If a_{22} had not been 1 at this stage, an additional step would be required to divide row 2 by a_{22}.

$$\left[\begin{array}{ccc|ccc|c} 1 & 0 & -2.5 & 0.5 & -1.5 & 0 & -7 \\ 0 & 1 & 2 & 0 & 1 & 0 & 6 \\ 0 & 0 & 6 & -2 & 4 & 1 & 28 \end{array}\right]$$

Step 4: Divide row 3 by 6.

$$\left[\begin{array}{ccc|ccc|c} 1 & 0 & -2.5 & 0.5 & -1.5 & 0 & -7 \\ 0 & 1 & 2 & 0 & 1 & 0 & 6 \\ 0 & 0 & 1 & -1/3 & 2/3 & 1/6 & 14/3 \end{array}\right]$$

Step 5: Add (2.5) times row 3 to row 1. Add (-2) times row 3 to row 2.

$$\left[\begin{array}{ccc|ccc|c} 1 & 0 & 0 & -1/3 & 1/6 & 5/12 & 14/3 \\ 0 & 1 & 0 & 2/3 & -1/3 & -1/3 & -10/3 \\ 0 & 0 & 1 & -1/3 & 2/3 & 1/6 & 14/3 \end{array}\right]$$

which is:

$$[\mathbf{I}|\mathbf{A}^{-1}|\mathbf{X}]$$

1. Interchanging two rows.
2. Multiplying a row by a scalar.
3. Adding a multiple of one row to another row.

If the original matrix is further augmented with the right-hand-side vector, the solution vector can be found directly by the same elementary row operations that are used to find the inverse. However, once you have found the inverse, you can solve for any number of right-hand-side conditions by matrix multiplication. Table F.1 shows the generation of the inverse of the coefficient matrix and the solution for Equations F.10 using row operations. This method is called Gauss-Jordan elimination. You can confirm that the inverse is correct by multiplying it by the original coefficient matrix to see if you get the identity matrix. You can also confirm the solution by multiplying the inverse by the original right-hand-side vector.

The element on the diagonal that you are working with at each stage of the transformation of Table F.1 is called the pivot element. An improvement over the method shown in Table F.1 is to rearrange the matrix before using a pivot element so that the element with the largest absolute value in that column is used as the pivot element. This reduces computer round-off error and prevents having zeros inadvertently appear as the pivot element. Both of these advantages are important for computer solutions. When you are working with the first row, you could rearrange any of the rows to put the largest element in position a_{11}. When you are working with the second row you want to have the largest element from the second column in a_{22}, but you do not want to disturb the first row because it has already been used to develop the first pivot element. Therefore, you would search for the maximum value for column 2 in row 2 and all subsequent rows. You would continue this procedure until the transformation is complete and the inverse is found.

F.2.5.2 Cramer's Rule

Solutions for systems of equations can also be found by using determinants and Cramer's rule. This method solves the system for one right-hand-side vector. A new right-hand-side vector would require a new solution. For a system of three equations and three unknowns

$$2X + 3Y - Z = 1$$
$$3X + 5Y + 2Z = 8$$
$$X - 2Y - 3Z = -1$$

the matrix of coefficients is

$$\mathbf{A} = \begin{vmatrix} 2 & 3 & -1 \\ 3 & 5 & 2 \\ 1 & -2 & -3 \end{vmatrix}$$

and the determinant of $\mathbf{A} = 22$. In applying Cramer's rule, you sequentially replace the column of the coefficient matrix corresponding to the variable being solved for with the right-hand-side vector, as shown below. You can confirm that this is the correct solution by finding the inverse and multiplying it by the right-hand side.

$$X = \frac{\begin{vmatrix} 1 & 3 & -1 \\ 8 & 5 & 2 \\ -1 & -2 & -3 \end{vmatrix}}{|\mathbf{A}|} = \frac{66}{22} = 3$$

$$Y = \frac{\begin{vmatrix} 2 & 1 & -1 \\ 3 & 8 & 2 \\ 1 & -1 & -3 \end{vmatrix}}{|\mathbf{A}|} = \frac{-22}{22} = -1$$

$$Z = \frac{\begin{vmatrix} 2 & 3 & 1 \\ 3 & 5 & 8 \\ 1 & -2 & -1 \end{vmatrix}}{|\mathbf{A}|} = \frac{44}{22} = 2$$

ENGINEERING MECHANICS

Most of you will have had college physics and an introductory mechanics course by the time you are using this text. This appendix will review the principles you will use to complete the exercises in the text. If you have not yet had a mechanics course, the material presented should be sufficient for you to undertake the computer applications.

G.1 PROPERTIES OF SHAPES

G.1.1 Area

Area can be found by integration by the general procedure

$$A = \int_{x_1}^{x_2} f(x)\,dx$$

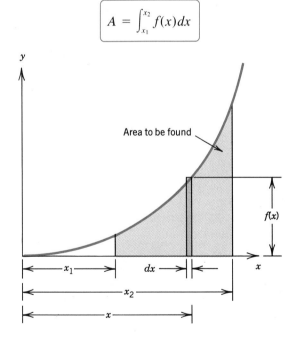

G.1.2 Centroid

The centroid is the center (middle) of an area. It is also the center of gravity for homogeneous bodies. The centroid of an area can be found using integration. Using the figure from the previous section,

$$x_{centroid} = \bar{x} = \frac{\overline{\int_{x_1}^{x_2} x f(x) dx}}{A}$$

If a shape is made up of several simple shapes, its centroid can be calculated by moments as shown in the following illustration:

$$\bar{y} = \frac{\bar{y}_1 A_1 + \bar{y}_2 A_2}{A_1 + A_2}$$

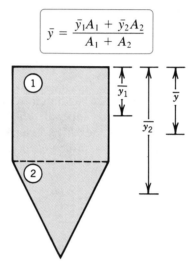

G.1.3 Moment of Inertia

The moment of inertia is the second moment of an area about a given axis. It can be found through integration. Using the definition sketch for area from Section G.1.1 of this appendix,

$$I_y = \int_{x_1}^{x_2} x^2 f(x) dx$$

The moment of inertia for compound shapes can be found by adding their individual moments of inertia about the same axis.

The parallel axis theorem provides a means of readily calculating the moment of inertia of a shape about an axis parallel to one of the centroidal axes of the shape. It can be found by the equation

$$I_y = I_c + x^2 A$$

In this expression, I_c is the moment of inertia about the y centroidal axis of the shape, and x is the distance between the axes.

G.1.4 Radius of Gyration

The radius of gyration of a hollow, circular column (k_x) is given by

$$k_x = \sqrt{\frac{J_0}{A}}$$

$$J_0 = I_x + I_y = \frac{1}{2}\pi(r_2^4 - r_1^4)$$

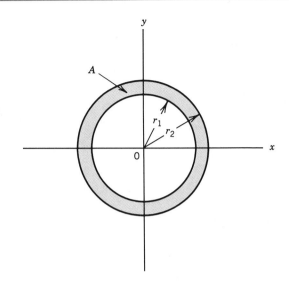

G.1.5 Table of Properties

Table G.1 shows the properties of common shapes that will be useful in chapter-end exercises.

TABLE G.1   PROPERTIES OF SHAPES

Rectangle

$$A = bh$$

$$\bar{x} = \frac{b}{2} \qquad \bar{y} = \frac{h}{2}$$

$$I_{x_c} = \frac{1}{12}bh^3$$

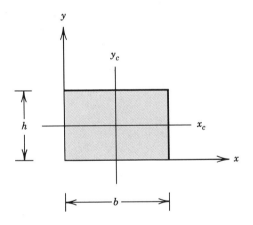

Triangle

$$A = \frac{1}{2}bh$$

$$\bar{y} = \frac{h}{3}$$

$$I_{x_c} = \frac{1}{36}bh^3$$

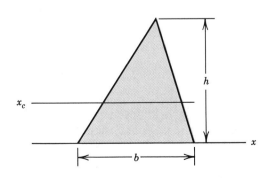

TABLE G.1   (Continued)

Circle

$$A = \pi r^2$$

$$I_x = \frac{1}{4} \pi r^4$$

$$S = r\theta \qquad (\theta \quad \text{in radians})$$

$$A_{sector} = \frac{1}{2} r^2 \theta$$

$$A_{ABS} = \frac{1}{2} r^2(\theta - \sin \theta) \qquad \text{(hatched area)}$$

Chord AB $Length = 2r \sin\left(\frac{\theta}{2}\right)$

$$\bar{r}_{sector} = \frac{r \sin(\theta/2)}{\theta/2}$$

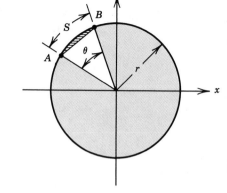

Parabola

$$\bar{x} = \frac{3a}{8}$$

$$\bar{y} = \frac{3h}{5}$$

$$A = \frac{2ah}{3}$$

$$I_x = \frac{2ah^3}{7}$$

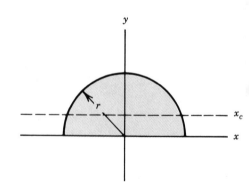

Semicircle

$$A = \frac{1}{2} \pi r^2$$

$$\bar{y} = \frac{4r}{3\pi}$$

$$I_x = \frac{\pi r^4}{8}$$

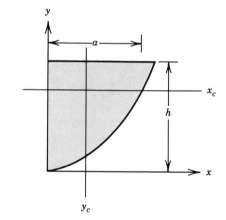

G.1.6 Parabolic Arch

Parabolic arches have many applications in engineering. The figure below shows a general parabolic arch with the key dimensions labeled.

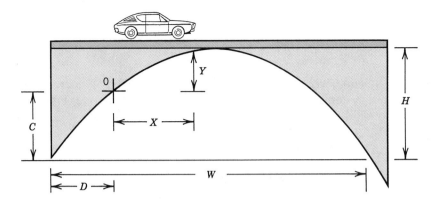

The equations necessary to analyze the arch are

$$Y = A(X + D)^2 + B(X + D) + C$$
$$A = -H/(W/2)^2$$
$$B = -AW$$
$$dY/dX = 2A(X + D) + B$$
$$@ \; X = W/2 - D; \; dY/dX = 0.0; \text{ and } Y = H - C$$

G.2 PROPERTIES OF VOLUMES

Table G.2 shows the properties of common solid shapes that will be useful in chapter-end exercises.

TABLE G.2 PROPERTIES OF SOLID SHAPES

Sphere

$$A = 4\pi r^2$$

$$V = \frac{4}{3}\pi r^3$$

Half Cone on Side

$$A = \frac{\pi r}{2}\sqrt{r^2 + h^2}$$

$$V = \frac{1}{6}\pi r^2 h$$

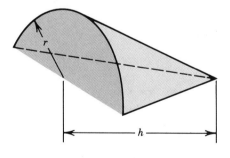

G.3 RIGID BODY MECHANICS (STATICS)

G.3.1 Equilibrium

Concurrent force systems are systems in which all the forces act through a common point. Since all forces act through a point, there are no moment equilibrium equations. Sums of forces along the axes are the equilibrium equations you can use to solve for unknown forces. In three dimensions,

$$\boxed{\Sigma F_x = 0 \qquad \Sigma F_y = 0 \qquad \Sigma F_z = 0} \qquad (G.1)$$

For two dimensions, only the sums of forces in the x and y directions would be used.

Rigid body equilibrium is used to analyze bodies that are subjected to nonconcurrent force systems. Both moment and force summations are applicable. For bodies in equilibrium in three-dimensional force systems, you have six equilibrium equations:

$$\boxed{\begin{array}{ccc} \Sigma F_x = 0 & \Sigma F_y = 0 & \Sigma F_z = 0 \\ \Sigma M_x = 0 & \Sigma M_y = 0 & \Sigma M_z = 0 \end{array}} \qquad (G.2)$$

In two-dimensional rigid body equilibrium problems you have three equilibrium equations available, as shown by Equations G.3. Points A and B are different points to take moments about. You can even take moments about three points, and not use any force summations, so long as the three points are not in a straight line with each other.

$$\boxed{\begin{array}{ccccc} \Sigma F_x = 0 & & \Sigma F_x = 0 & & \Sigma F_y = 0 \\ \Sigma F_y = 0 & \text{or} & \Sigma M_A = 0 & \text{or} & \Sigma M_A = 0 \\ \Sigma M_A = 0 & & \Sigma M_B = 0 & & \Sigma M_B = 0 \end{array}} \qquad (G.3)$$

Reactions are the external forces and moments supporting a body in equilibrium. Internal forces are the forces within the parts of the rigid body. For the chapter-end exercises and examples in the text we will be concerned with reactions for two-dimensional rigid bodies and the internal forces for a truss structure.

Reactions for two-dimensional rigid bodies can be found by taking a free body of the structure to be analyzed and applying the two-dimensional equilibrium equations to solve for up to three reactions, as shown below:

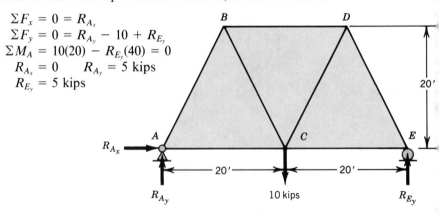

$$\Sigma F_x = 0 = R_{A_x}$$
$$\Sigma F_y = 0 = R_{A_y} - 10 + R_{E_y}$$
$$\Sigma M_A = 10(20) - R_{E_y}(40) = 0$$
$$R_{A_x} = 0 \qquad R_{A_y} = 5 \text{ kips}$$
$$R_{E_y} = 5 \text{ kips}$$

Trusses are structures that are made up of a set of long slender members that are connected at both ends by pins to other members. The members are arranged into geometric shapes to carry loads. The basic shape for a truss is the triangle. It is the combination of triangles that gives a truss its rigidity. Loads on trusses are assumed to act only at the joints between the members. Therefore, all members of the truss are subjected only to axial forces. Figure G.1 shows an expanded free body of the truss given above. Each member is separated from the pin connecting

FIGURE G.1   Free body for equilibrium analysis of truss.

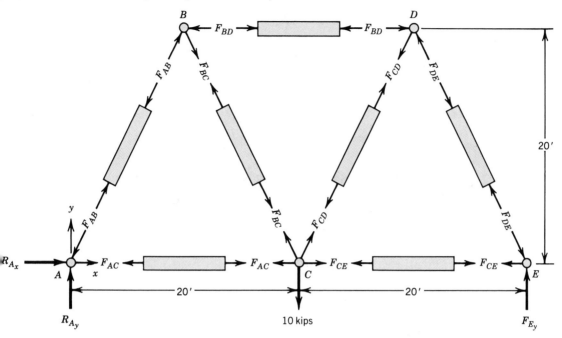

t to other members, and the forces acting on the member are added. Forces acting into the member free body show that you are assuming the member is in compression, while forces acting out of the member indicate the assumption of tension. Your final answers will tell you if you have assumed correctly. A negative value for a force shows that it acts in the opposite sense from your assumption.

The method of joints is a solution method that lends itself to computerized solutions for the forces in the members of a truss. You treat each joint as a concurrent force system and apply two equations of equilibrium to the joint. If you have n members, you will need n independent equations to solve for the forces. The truss of Figure G.1 has five joints, so you could write up to ten equilibrium equations. However, three of these will be redundant since you need only seven to solve for the forces in the seven members. You could pick any of the seven equations. Equations G.4 show the two equilibrium equations for joints A, B, and C, and one of the equilibrium equations for joint D.

At joint A:

$$\Sigma F_x = 0 = F_{AC} - \frac{1}{\sqrt{5}} F_{AB}$$

$$\Sigma F_y = 0 = -\frac{2}{\sqrt{5}} F_{AB} + 5$$

At joint B:

$$\Sigma F_x = 0 = \frac{1}{\sqrt{5}} F_{AB} + \frac{1}{\sqrt{5}} F_{BC} - F_{BD}$$

$$\Sigma F_y = 0 = \frac{2}{\sqrt{5}} F_{AB} - \frac{2}{\sqrt{5}} F_{BC} \qquad\qquad \text{(G.4)}$$

At joint C:

$$\Sigma F_x = 0 = -F_{AC} - \frac{1}{\sqrt{5}} F_{BC} + \frac{1}{\sqrt{5}} F_{CD} + F_{CE}$$

$$\Sigma F_y = 0 = -10 + \frac{2}{\sqrt{5}} F_{BC} + \frac{2}{\sqrt{5}} F_{CD}$$

At joint D:

$$\Sigma F_x = 0 = F_{BD} - \frac{1}{\sqrt{5}} F_{CD} - \frac{1}{\sqrt{5}} F_{DE}$$

These equilibrium equations can be rearranged into the form:

$$
\begin{aligned}
F_{AC} - \frac{1}{\sqrt{5}} F_{AB} &= 0 \\
\frac{2}{\sqrt{5}} F_{AB} &= 5 \\
\frac{1}{\sqrt{5}} F_{AB} + \frac{1}{\sqrt{5}} F_{BC} - F_{BD} &= 0 \\
\frac{2}{\sqrt{5}} F_{AB} - \frac{2}{\sqrt{5}} F_{BC} &= 0 \\
-F_{AC} \quad - \frac{1}{\sqrt{5}} F_{BC} \quad + \frac{1}{\sqrt{5}} F_{CD} + F_{CE} &= 0 \\
\frac{2}{\sqrt{5}} F_{BC} \quad + \frac{2}{\sqrt{5}} F_{CD} &= 10 \\
F_{BD} - \frac{1}{\sqrt{5}} F_{CD} \quad - \frac{1}{\sqrt{5}} F_{DE} &= 0
\end{aligned}
\tag{G.5}
$$

Equations G.5 can be written in matrix notation as

$$
\begin{bmatrix}
1 & -1/\sqrt{5} & 0 & 0 & 0 & 0 & 0 \\
0 & 2/\sqrt{5} & 0 & 0 & 0 & 0 & 0 \\
0 & 1/\sqrt{5} & 1/\sqrt{5} & -1 & 0 & 0 & 0 \\
0 & 2/\sqrt{5} & -2/\sqrt{5} & 0 & 0 & 0 & 0 \\
-1 & 0 & -1/\sqrt{5} & 0 & 1/\sqrt{5} & 1 & 0 \\
0 & 0 & 2/\sqrt{5} & 0 & 2/\sqrt{5} & 0 & 0 \\
0 & 0 & 0 & 1 & -1/\sqrt{5} & 0 & -1/\sqrt{5}
\end{bmatrix}
\begin{bmatrix}
F_{AC} \\ F_{AB} \\ F_{BC} \\ F_{BD} \\ F_{CD} \\ F_{CE} \\ F_{DE}
\end{bmatrix}
=
\begin{bmatrix}
0 \\ 5 \\ 0 \\ 0 \\ 0 \\ 10 \\ 0
\end{bmatrix}
\tag{G.6}
$$

or

$$\boxed{\mathbf{AF} = \mathbf{B}} \tag{G.7}$$

F is the vector of unknown member forces and **B** is the right-hand-side vector that contains the applied loads and the reactions.

Using the methods of Appendix F, the inverse of the coefficient matrix for this problem is

$$\mathbf{A}^{-1} = \begin{bmatrix} 1 & \dfrac{1}{2} & 0 & 0 & 0 & 0 & 0 \\[2ex] 0 & \dfrac{\sqrt{5}}{2} & 0 & 0 & 0 & 0 & 0 \\[2ex] 0 & \dfrac{\sqrt{5}}{2} & 0 & -\dfrac{\sqrt{5}}{2} & 0 & 0 & 0 \\[2ex] 0 & 1 & -1 & -\dfrac{1}{2} & 0 & 0 & 0 \\[2ex] 0 & -\dfrac{\sqrt{5}}{2} & 0 & \dfrac{\sqrt{5}}{2} & 0 & \dfrac{\sqrt{5}}{2} & 0 \\[2ex] 1 & \dfrac{3}{2} & 0 & -1 & 1 & -\dfrac{1}{2} & 0 \\[2ex] 0 & \dfrac{3\sqrt{5}}{2} & -\sqrt{5} & -\sqrt{5} & 0 & -\dfrac{\sqrt{5}}{2} & -\sqrt{5} \end{bmatrix} \qquad (G.8)$$

which when multiplied by the right-hand side gives the forces in the members

$$\mathbf{A}^{-1}\mathbf{B} = \mathbf{F} = \begin{bmatrix} F_{AC} \\ F_{AB} \\ F_{BC} \\ F_{BD} \\ F_{CD} \\ F_{CE} \\ F_{DE} \end{bmatrix} = \begin{bmatrix} 2.50 \\ 5.59 \\ 5.59 \\ 5.00 \\ 5.59 \\ 2.50 \\ 5.59 \end{bmatrix} \text{ kips} \qquad (G.9)$$

G.3.2 Friction

Figure G.2 shows the free body of a block subjected to dry friction forces. If the body is homogeneous, the weight W acts through the centroid. Otherwise you need to know where the center of gravity is. The force F_r is generated by the

FIGURE G.2   Free body for dry friction.

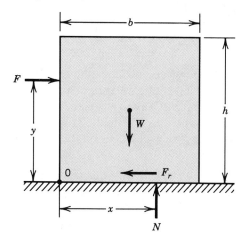

frictional resistance between the block and the surface it is in contact with. The maximum frictional force that can be generated is a function of the normal force and the coefficient of friction between the surfaces.

$$F_{r\max} = \mu N \qquad \mu = \text{coefficient of static friction} \qquad (G.10)$$

G.3.2.1 Analysis for Slipping

If the force F_r has to be greater than $F_{r\max}$ to maintain equilibrium, the block will slip along the surface.

G.3.2.2 Analysis for Tipping

The position of the normal force N is not necessarily under the weight of the block. Its position can be found by taking moments about point O. If the normal force location moves to the right of the edge of the block of Figure G.2, the block is no longer in static equilibrium. It will start to tip to the right.

G.3.2.3 Summary of Equilibrium Equations

Depending on the location of force F, the block may tip or slip first. The smallest force required will govern. The equilibrium equations are

$$
\begin{aligned}
\Sigma F_x &= 0 = F - F_r \\
\Sigma F_y &= 0 = W - N \\
\Sigma M_0 &= 0 = XN - \frac{b}{2}\, W - yF \\
F_{r\max} &= \mu N
\end{aligned}
\qquad (G.11)
$$

G.3.3 Shear and Bending Moment

Shear is an internal force transverse to the axis of a beam. If shear is large enough, it will cause the beam to fail in a scissor fashion. Bending moment will cause a beam to deflect, or bend, up or down. If an imaginary cut is made in the beam, the shear V and the bending moment M are positive according to the convention shown.

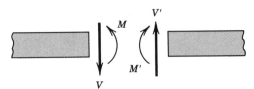

If you assume this convention when you solve the equilibrium equations, positive values for V or M will indicate positive shear or bending moment. Negative values will indicate the opposite. Another way to remember the convention is by remembering that external forces (reactions and applied forces) will cause positive shear and positive bending moment when they act on the beam as shown below.

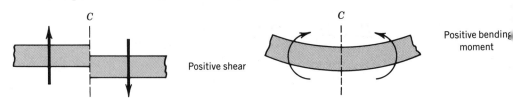

Positive shear

Positive bending moment

You can calculate shear and bending moment through either equilibrium or integration methods. Graphs of the resulting mathematical relationships are useful for determining maximum shear and bending moment, which are important design parameters. Figures G.3 and G.4 show the development of shear and bending moment diagrams for a point load and a uniform load using equilibrium methods.

You can also use integration methods to find the shear and bending moment for a section of a beam. The change in shear over an interval is the integral of the load, and the change in moment is the integral of the shear. For the beam of Figure G.4, the shear and bending moment would be

$$V = V_A - \int_0^x w\,dx = R_A - wx$$

$$M = M_A + \int_0^x V\,dx = \int_0^x (R_A - wx)\,dx = R_A x - \frac{1}{2} wx^2$$

(G.12)

FIGURE G.3   Shear and bending moment for point load.

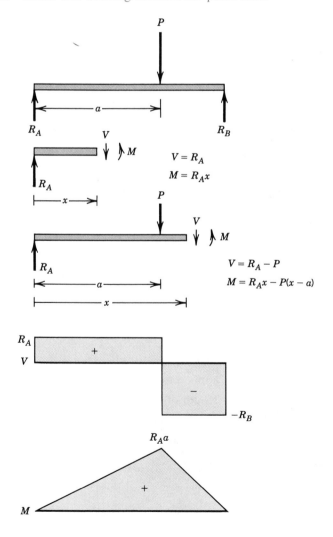

FIGURE G.4   Shear and bending moment for a uniform load.

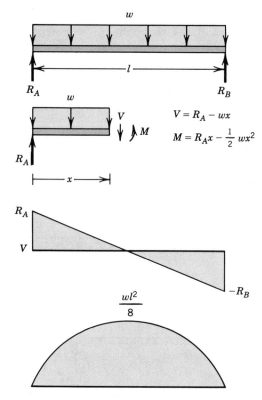

For multiple loadings, the total shear and bending moment are the sums of the shears and bending moments for the individual loads. This is the principle of superposition. It is useful for computer analysis of beams subjected to several loads.

G.3.4 Torsion

Torsion is the tendency of a torque to twist a shaft. Equations governing the torsion in a hollow circular shaft are

$$\theta = \frac{\Gamma \cdot L}{G \cdot J}$$

θ = angle of twist (radians)

Γ = applied torque (ft-lb)

L = length of shaft (ft)

G = torsional modulus of elasticity (lb/ft^2)

J = polar moment of inertia (ft^4)

$$T_{xy} = \frac{\Gamma \cdot OD}{2J}$$

(G.13)

T_{xy} = torsional shear stress (lb/ft^2)

$$OD = \text{outside diameter (ft)}$$

$$J = \pi \frac{(OD^4 - ID^4)}{32}$$

$$ID = \text{inside diameter (ft)}$$

G.4 PARTICLE MECHANICS (DYNAMICS)

G.4.1 Particle Motion with Acceleration in Both Axes

Particle motion in two dimensions with acceleration can be easily accommodated by remembering that position, velocity, and acceleration are all vector quantities, that velocity equals dx/dt or dy/dt, and that acceleration equals dV_x/dt or dV_y/dt. Integrations can be accomplished in each direction and the results combined by vector addition. Therefore, if **r** is the vector position of the particle at any time t,

$$\begin{aligned}
\mathbf{r} &= x\mathbf{i} + y\mathbf{j} \\
\mathbf{v} &= v_x\mathbf{i} + v_y\mathbf{j} \\
\mathbf{a} &= a_x\mathbf{i} + a_y\mathbf{j}
\end{aligned} \tag{G.14}$$

G.4.2 Particle Trajectory Under the Influence of Gravity

If air resistance is neglected, a particle trajectory under the influence of gravity is subject to the following governing equations:

$$\begin{aligned}
a_y &= -g \qquad a_x = 0.0 \\
V_y &= \int a_y dt = -g \int dt = V_{0_y} - gt \\
V_x &= \int a_x dt = V_{0_x} \\
y &= \int V_y dt = \int (V_{0_y} - gt) = y_0 + V_{0_y}t - \frac{1}{2}gt^2 \\
x &= \int V_x dt = V_{0_x} \int dt = x_0 + V_{0_x}t \\
V_{0_y} &= V_0 \sin\theta \qquad V_{0_x} = V_0 \cos\theta \\
&\quad y_{max} \text{ occurs when } V_y = 0.0 \\
t_{ymax} &= \frac{V_{0_y}}{g} = \frac{V_0 \sin\theta}{g} \\
x_{max} &= x_0 + V_{0_x}(2t_{ymax}) = x_0 + V_0 \cos\theta(2t_{ymax})
\end{aligned} \tag{G.15}$$

In the equations, g is the local acceleration due to gravity, and t is the time since launch. Figure G.5 gives a definition sketch for the other variables.

G.4.3 Rocket Trajectories

Constant thrust rocket motors produce constant thrust until the fuel is expended. Thus the rocket acceleration increases over time as the mass of the rocket de-

FIGURE G.5   Particle trajectory definition sketch.

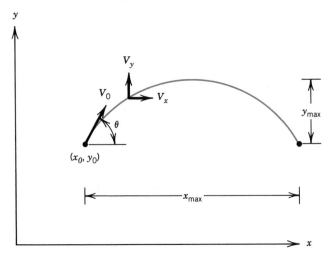

creases. In constant acceleration rockets, the thrust of the motor is varied over time to produce constant acceleration. If air resistance is ignored, the angle of inclination of the rocket thrust with the horizontal is assumed to be constant over the length of fuel burn, and the surface over which the rocket travels is assumed to be flat, the following equations describe the trajectories of constant thrust and constant acceleration rockets. Collectively, these equations are labeled as Equations G.16.

Constant Thrust

   Before Cutoff

$$v_x(t) = [v_e \cos(\theta)][-\ln(1 - bt)]$$
$$v_y(t) = [v_e \sin(\theta)][-\ln(1 - bt)] - gt$$
$$x(t) = [v_e \cos(\theta)]f(b,t)$$
$$y(t) = [v_e \sin(\theta)]f(b,t) - \frac{gt^2}{2}$$
$$f(b,t) = t + \frac{1}{b}(1 - bt)\ln(1 - bt)$$

The cutoff occurs at a time t_c defined by

$$bt_c = \Omega$$

   After Cutoff

$$x(t) = x_c + v_x(t_c)t$$
$$y(t) = y_c + v_y(t_c)t - \frac{gt^2}{2}$$

The optimum angle θ is determined from the root of

$$Q\sigma^3 - \left(\sigma^2 - \frac{1}{2}\right) = 0$$

where $\sigma = \sin(\theta)$, and   (G.16)

$$Q = [-\Omega + \ln(1 - \Omega)[\ln(1 - \Omega)]^2]\frac{g}{bv_e}$$

Constant Acceleration

Before Cutoff

$$v_x(t) = [v_e \cos(\theta)]bt$$
$$v_y(t) = [v_e \sin(\theta)]bt - gt$$
$$x(t) = [v_e \cos(\theta)]\frac{bt^2}{2}$$
$$y(t) = [v_e \sin(\theta)]\frac{bt^2}{2} - \frac{gt^2}{2}$$

The cutoff occurs at a time t_c defined by

$$bt_c = -\ln(1 - \Omega)$$

After Cutoff

$$x(t) = x_c + v_x(t_c)t$$
$$y(t) = y_c + v_y(t_c)t - \frac{gt^2}{2}$$

The optimum angle θ is determined from the root of

$$Q\sigma^3 - \left(\sigma^2 - \frac{1}{2}\right) = 0$$

where $\sigma = \sin(\theta)$, and

$$Q = \frac{1}{2}\frac{g}{bv_e}$$

For both constant thrust and constant acceleration, the rocket hits the ground at $t = t_{\text{hit}}$, the positive root of $y(t = t_{\text{hit}}) = 0$.

T = thrust of the engine (N)
θ = fixed angle of inclination of rocket thrust
m = current mass of the rocket (including fuel)
g = acceleration due to gravity (assumed to be constant)
v_e = velocity of the propellant relative to the rocket
M_0 = original mass of the rocket and fuel
Ω = ratio of the mass of the fuel to the total mass
β = burn rate of the rocket engines (kg/s)

b = ratio of the burn rate to the original mass

t_c = burn time for the engines

$x(t)$ and $y(t)$ = the x and y positions of the rocket at time t

x_c and y_c = the x and y positions at the time of engine cutoff

$v_x(t_c)$ and $v_y(t_c)$ = the x and y velocities at the time of engine cutoff

G.4.4 Motion of Crank Assembly

Figure G.6 shows a crank, connecting rod, and piston assembly, as well as the free bodies and vector diagram for the motion of the connecting rod. The crank AB

FIGURE G.6   Crank assembly.

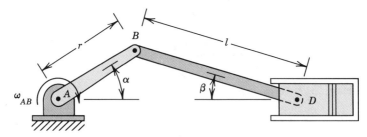

(*a*) Crank assembly.

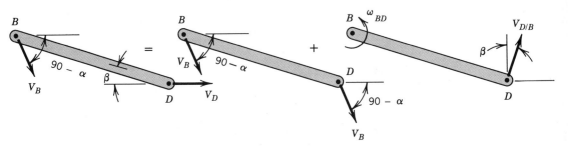

Plane motion = translation + rotation

(*b*) Free body of connecting link BD.

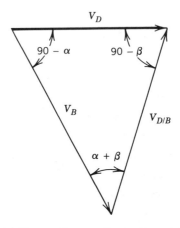

(*c*) Vector diagram for motion at D.

turns at a constant angular velocity. The connecting rod BD translates the circular motion of point B into the sliding motion of the piston at point D. The governing equations for the motion are

$$
\begin{aligned}
\frac{\sin \alpha}{l} &= \frac{\sin \beta}{r} \\[4pt]
V_B &= \omega_{AB} r \quad \text{(tangential velocity)} \\[4pt]
\omega_{AB} &= RPM \left(\frac{1 \text{ min}}{60 \text{ s}} \right) \left(\frac{2\pi \text{ rad}}{1 \text{ rev}} \right) \\[4pt]
\mathbf{V}_D &= \mathbf{V}_B + \mathbf{V}_{D/B} \\[4pt]
V_{D/B} &= \omega_{BD} l \\[4pt]
\frac{V_D}{\sin(\alpha + \beta)} &= \frac{V_{D/B}}{\sin(90 - \alpha)} = \frac{V_B}{\sin(90 - \beta)}
\end{aligned}
$$

$$(G.17)$$

G.5   FLUID MECHANICS

G.5.1   Forces on Submerged Surfaces

In evaluating forces on submerged surfaces, both the magnitude and the point of application are important. The figure for Exercise 2.4.4 shows a trapezoidal pressure prism acting on the side view of a vertical gate. The pressure prism represents the load placed on the plane surface by the liquid. The pressure prism is trapezoidal because the gate is submerged. At any point in a liquid the pressure is equal to

$$ p = \gamma h $$

$$(G.18)$$

in which γ = the specific weight of the liquid (lb/ft³ or kN/m³), and h = depth of the liquid above the point (ft or m).

At the top of the gate the pressure is γh_1. The pressure continues to increase linearly as the depth approaches the bottom of the gate. The total force on the gate is equal to the volume of the pressure prism. It acts through the centroid of the pressure prism. The force can be found by taking the integral of the pressure times the area acted on. The differential force is

$$ dF = p \cdot dA = p \cdot w(h)dh $$

$$(G.19)$$

in which $w(h)$ is the width of the differential area at depth h. If h_1 is the depth of the top of the surface and h_2 is the depth of the bottom of the surface, the force is

$$ F = \gamma \int_{h_1}^{h_2} h w(h)dh $$

$$(G.20)$$

The point of application h_p can be found by taking the first moment of the force and dividing it by the force, or

$$ h_p = \frac{\gamma \int_{h_1}^{h_2} h^2 w(h)dh}{F} $$

$$(G.21)$$

For common shapes, the foregoing equations simplify to

$$F = \gamma h_c A \qquad \text{(G.22)}$$

$$h_p = h_c + \frac{I_c}{h_c A} \qquad \text{(G.23)}$$

in which h_c = the depth of the centroid of the area acted on, A = the area acted on, and I_c = the moment of inertia about the horizontal centroidal axis of the area being acted on. Since γh_c is the pressure at the centroid of the area being acted on, the total force is the pressure at the centroid times the area being acted on. The force acts somewhat below the centroid because of the trapezoidal shape of the pressure prism. The analysis developed here is applicable to any plane shape in the vertical plane.

G.5.2 Energy Loss in a Circular Pipeline

Engineers have to be able to predict the energy loss in a pipeline flowing full if they are to design flow systems. The flow in a pipe is either laminar or turbulent, depending on certain characteristics of the flow and the pipe. Laminar flow travels along the pipe in concentric layers, or laminae. There is little mixing in the cross section of laminar flow. Turbulent flow is much less structured. There is considerable mixing in the cross section of turbulent flow. The type of flow can be predicted by a dimensionless parameter known as the Reynolds number:

$$Re = \frac{VD}{\nu} \qquad \text{(G.24)}$$

$$V = \frac{Q}{A} \qquad A = \frac{\pi D^2}{4}$$

in which D = diameter of the pipe (ft), A = cross-sectional area of the pipe (ft^2), V = average velocity in cross section (ft/s), Q = volumetric flow rate (ft^3/s), and ν = kinematic viscosity (ft^2/s).

In a pipe flowing under pressure, a Reynolds number up to about 2,000 indicates that laminar flow will occur in the pipe. A Reynolds number greater than 2,000 indicates that turbulent flow will occur.

A widely used equation for estimating energy loss in a circular pipe is the Darcy–Weisbach equation

$$h_L = f \left(\frac{L}{D} \right) \frac{V^2}{2g} \qquad \text{(G.25)}$$

in which h_L = energy loss per unit weight of fluid flowing (ft), g = local acceleration due to gravity (ft/s^2), L = length of pipe (ft), and f = the Darcy–Weisbach friction factor.

The formula used to estimate the friction factor is dependent on the type of flow in the pipe. For laminar flow ($Re \leq 2,000$).

$$f = 64/Re \qquad \text{(G.26)}$$

For turbulent flow ($Re > 2{,}000$), the friction factor is a function of both the Reynolds number and the relative roughness of the pipe, ε/D. The relative roughness is the average size of the roughness projections on the interior surface of the pipe ε divided by the pipe diameter D. The size of the roughness projections is a function of the pipe material. Larger projections make the pipe rougher, which increases the frictional energy loss for a given length of pipe.

There are several empirical equations for estimating the value of f for turbulent flow. One of these formulations is the Chen equation (Janna 1983), which estimates f as

$$f = \left[-2.01 \log \left(\frac{\varepsilon}{3.7065D} - \frac{5.0452}{Re} \log A \right) \right]^{-2}$$

$$A = \frac{1}{2.8257} \left(\frac{\varepsilon}{D} \right)^{1.1098} + \frac{5.8506}{Re^{0.8981}}$$

(G.27)

A second formula for estimating f is the Swamee and Jain equation (1976)

$$f = \frac{0.25}{\left[\log \left(\frac{0.27\varepsilon}{D} + \frac{5.74}{Re^{0.9}} \right) \right]^2}$$

(G.28)

Both the Chen and Swamee–Jain equations directly solve for f. A third formula is the Colebrook equation (Janna 1983)

$$\frac{1}{\sqrt{f}} = -2.0 \log \left[\frac{\varepsilon}{3.7065D} + \frac{2.5226}{Re\sqrt{f}} \right]$$

(G.29)

which contains $\sqrt{f}$ on both sides and cannot be solved explicitly for f. You must use an approximation technique to estimate f.

G.5.3 Open Channel Flow

The governing equation for flow of water in an open channel is Manning's equation

$$Q = \frac{u}{n} A R_h^{2/3} S_0^{1/2}$$

(G.30)

in which Q = volumetric flow rate (ft³/s or m³/s), $u = 1.0$ for metric units and 1.49 for English units, n = Manning's friction factor (dependent on channel material), A = cross-sectional area of flow (ft² or m²), R_h = hydraulic radius = A/P_w, P_w = wetted perimeter (length of cross-section perimeter that is wetted by the water), and S_0 = slope of the channel bottom in the direction of flow (ft/ft or m/m). The area and wetted perimeter are usually expressed as a function of the depth of flow, y, in the channel.

Figure G.7 provides a definition sketch for a trapezoidal channel with equal side slopes. For this geometry,

FIGURE G.7 Flow in trapezoidal open channel.

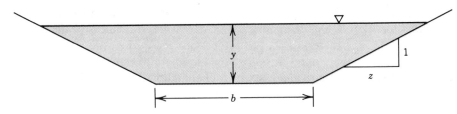

Side slope, $z = \Delta X / \Delta Y$

Cross-sectional area of flow, $A = yb + y^2 z$

Wetted perimeter, $P_w = b + 2y\sqrt{(1 + z^2)}$

FIGURE G.8 Flow in circular open channel.

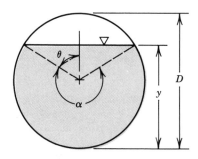

Figure G.8 provides a definition sketch for a circular channel flowing partly full. For this geometry,

Cross-sectional area of flow, $A = (\alpha - \sin \alpha)D^2/8$

Wetted perimeter, $P_w = \alpha(D/2)$

$$\alpha = 2(\pi - \theta)$$

for $y > (D/2)$, $\quad \theta = \sin^{-1}\left(\dfrac{\sqrt{(D/2)^2 - [y - (D/2)]^2}}{(D/2)}\right)$

for $y \leq (D/2)$, $\quad \theta = \pi - \sin^{-1}\left(\dfrac{\sqrt{(D/2)^2 - [(D/2) - y]^2}}{(D/2)}\right)$

G.5.4 References

Janna, W. S., *Introduction to Fluid Mechanics*, Wadsworth, Inc., Belmont, CA, 1983.

Swamee, K., and Jain, A. K., "Explicit Equations for Pipe-Flow Problems," *Journal of the Hydraulics Division*, American Society of Civil Engineers, New York, May 1976, pp. 657–664.

APPENDIX H

PROBABILITY, STATISTICS, AND SIMULATION

H.1 ERROR

Bias is the difference between the exact, or true, value and an estimate. It was referred to as accuracy in Chapter 3. Uncertainty is a measure of precision. We usually say that we are a certain percent sure that the true value lies within plus or minus a given amount from the estimated value. The total error is the sum of the bias and the uncertainty.

$$
\begin{aligned}
\text{Bias} &= \text{true value} - \text{estimate} \\
\text{True value} &= \text{estimate} \pm \text{uncertainty} \\
\text{Total error} &= \text{bias} \pm \text{uncertainty} \\
\text{True value} &= \text{estimate} + \text{bias} \pm \text{uncertainty}
\end{aligned}
$$

In terms of producing a product to certain specifications, we deal with a range of uncertainty, or tolerance.

$$
\text{True value} = \text{desired value} \pm \text{uncertainty (tolerance)}
$$

For example, if a diameter is to be 2.0 in. ± 0.01 in., it must be between 1.99 and 2.01 inches.

H.2 POPULATIONS AND SAMPLES

A population is the entire set of entities about which you want to develop some statistic. A sample is a subset of the population. Samples are used because it is often difficult, or impossible, to use the whole population to develop statistics. For example, you might want to develop data on which television programs people are watching. The population would be the total number of people in the United States watching television at a certain time. It would be impossible to poll them all, so you would choose a representative sample. Another example would be failure testing of a manufactured product. You could not test the whole population or you would have nothing left to sell, so you test a sample. From the sample you can make inferences as to the state of the population.

Equations for estimating the mean and standard deviation for a sample were given in Chapter 3, Equations 3.1 and 3.3. The relationships of these parameters with the corresponding parameters for a population are as follows:

Population

> Mean = μ
> Standard Deviation = σ

Sample

> Mean = $\bar{X}$
> Standard Deviation = S

Estimate of Population Parameter Using Sample

> Mean = $\bar{X}$
> Standard Deviation = $S(X) = S\sqrt{n/(n-1)}$ (also see Equation H.2)

H.3 HISTOGRAMS

A histogram is a graphical representation of the frequency of occurrence of particular data values when you take a sample. Each segment of the histogram may represent a range of data values, or it may represent a specific data value. For example, if you are analyzing the particle size of soil samples, one segment of the histogram might represent the number of times the particle size falls between 0.01 and 0.05 cm. If you are simulating the toss of two dice, one segment would be the number of times the sum came up six.

H.4 NORMAL PROBABILITY DISTRIBUTION

When repetitive measurements are made of some quantity that should theoretically be constant, the results are not generally identical. This is because the measurements are not completely precise, but vary over some range. The size of the range is dependent on the dispersion or scatter of the data. When the frequency of occurrence of each measured value is plotted as a histogram versus the range of values for the measured quantity, you will find a distribution similar to that shown in Figure H.1a. The most frequently occurring values are clustered around the mean. The general shape of the histogram is a bell-shaped curve, as is shown by the superimposed line. This is called the normal probability distribution, and it has properties that make it useful in predicting the probability of certain values being found. When a frequency distribution is normal, 68.27 percent of the values will lie within ± one standard deviation from the mean. Similarly, 95.45 percent will fall within ± two standard deviations and 99.73 percent of the values will fall within ± three standard deviations. If the total area under the probability curve is one, then the above percentages give the fraction of the area under the curve lying between the standard deviation ranges. Also, it can be seen that the area under the curve to the left of the mean is 0.5. This tells us that there is a 50

FIGURE H.1   Normal distribution.

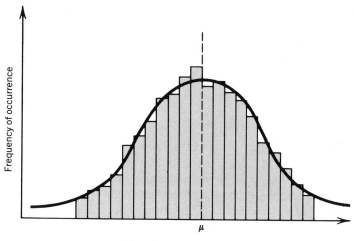

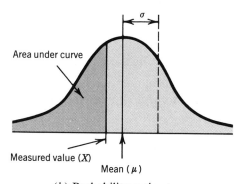

percent possibility that we will get a value that is less than or equal to the mean. The parameter

$$Z = \frac{X - \mu}{\sigma}$$ (H.1)

can be used to specify how many standard deviations away from the mean a particular value is. Z can be used to estimate the probability of occurrence of values less than or equal to that value by estimating the area under the curve to the left of Z. This requires interpolation in tables of Z.

H.5 STATISTICS

H.5.1 Measures of Central Tendency

Statistical measures of central tendency are related to what you think of as the "average" value. If the true value is known, then the measure of central tendency for your measured data will give you an indication of the accuracy of your measurements. The mean, illustrated in Equation 3.1, is the arithmetic average of a set of data. The median, illustrated in Equation 3.2, is the midpoint of a set of data values. In order to compute the median, you first have to order the values in either ascending or descending order. The median is less sensitive than the mean to extreme values. The mode of a set of data is the value that occurs with the most frequency. A data set may have more than one mode.

H.5.2 Measures of Dispersion

When you are measuring some quantity, there is always some amount of error in the measuring instrument. Therefore, you take several measurements and use the mean value as the estimate of the measured quantity. However, you should also know how precise the instrument is. Thus, you need some idea of the dispersion of the data around the mean value. The standard deviation, as presented in Equation 3.3, is one such measure of dispersion. The form given in Equation 3.3 is the standard deviation of a data sample. When you are estimating the population standard deviation using a sample, you would use the form

$$S(X) = \sqrt{\frac{\sum\limits_{i=1}^{n} (X_i - \bar{X})^2}{n - 1}} \tag{H.2}$$

The summation is the summation of the squared deviation of each data value from the mean value. $S(X)$ is said to be an unbiased estimator of the population standard deviation σ.

The coefficient of variation is a measure of the relative dispersion of data and is useful for comparing data sets with different mean values. It is given by

$$V = \frac{S(X)}{\bar{X}} \tag{H.3}$$

The range of a set of data is the difference between the two extreme values.

H.6 LINEAR REGRESSION

Regression is a method of fitting a predetermined mathematical model form to a measured set of dependent and independent variable values. The regression process is used to estimate the coefficients of the model. Regression is the process of estimating values for the coefficients to minimize the error between the observed data and the proposed functional relationship. The functional relationship is supposed to describe the phenomenon under study. The analyst must propose the form of the regression equation. Methods exist for estimating the parameters for linear regression equations with one or more independent variables, or for nonlinear regression equations. We will limit our discussion here to linear regression with one independent variable. This will allow you to fit a straight line function through the data.

FIGURE H.2 Linear regression.

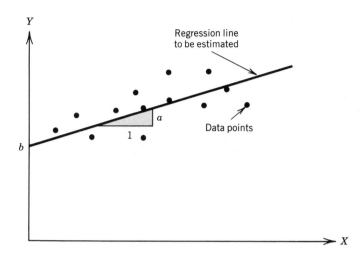

For linear regression with one independent variable, the model has the form

$$\hat{Y} = aX + b \qquad (H.4)$$

in which $\hat{Y}$ = the estimate of Y for a given X, a = the slope of the best fit line, and b = the Y intercept when $X = 0$.

The objective is to achieve the "best fit" by minimizing the cumulative error between data points and the regression line, as shown in Figure H.2. This is done by minimizing the sum of squared deviation between the measured data points Y_i and the values predicted by the model for the same X value, X_i.

$$\text{Minimize } SSD = \sum_{i=1}^{n} (Y_i - \hat{Y}_i)^2 = \sum_{i=1}^{n} (Y_i - b - aX_i)^2 \qquad (H.5)$$

It is important to note that SSD is a function of a and b, not X_i and Y_i. X_i and Y_i are now parameters. To minimize the cumulative squared error, the partial derivatives of SSD should be taken with respect to a and b. Both of these expressions can then be set equal to zero.

$$\frac{\partial SSD}{\partial a} = -2 \sum_{i=1}^{n} X_i(Y_i - b - aX_i) = 0$$

$$\frac{\partial SSD}{\partial b} = -2 \sum_{i=1}^{n} (Y_i - b - aX_i) = 0$$

These equations can be solved for a and b to give

$$a = \frac{\sum_{i=1}^{n} X_i Y_i - \bar{Y} \sum_{i=1}^{n} X_i}{\sum_{i=1}^{n} X_i^2 - \bar{X} \sum_{i=1}^{n} X_i} \quad \text{and} \quad b = \bar{Y} - a\bar{X} \qquad (H.6)$$

which with minor modifications give Equations 5.1 and 5.2 used in Example 5.4.

The correlation coefficient, calculated by Equation 5.3, is a measure of the goodness of fit of the regression line. Correlation coefficient values can vary between -1.0 and $+1.0$. The sign will be the same as the slope of the regression line, a. Values of plus or minus one would indicate perfect correlation. This means that the data points fall on a straight line. A correlation coefficient of zero means there is no relationship between the independent and dependent variables.

Another measure of goodness of fit is the coefficient of determination. It is the square of the correlation coefficient, or r^2. It is preferred by many statisticians because it gives the fraction of the total squared deviation that is explained by the regression. The coefficient of determination is always positive.

The standard error of estimate, $S_{Y,X}$ is a final measure of the goodness of fit of a regression line. It is given by

$$S_{Y,X} = \sqrt{\frac{\sum_{i=1}^{n} (Y_i - \hat{Y}_i)^2}{n - 2}}$$

(H.7

which measures the scatter of the data points with respect to the regression line

H.7 NONLINEAR REGRESSION

Some nonlinear functions can be linearized and then solved using the method of linear regression. For example,

$$Y = bX^a$$

can be linearized by taking logs of both sides,

$$\log Y = \log b + a \cdot \log X$$

You can use linear regression on the transformed data, and then convert the model back into its original form.

H.8 SIMULATION

Often it is desirable to simulate a process using the computer. It may be faste or easier to use the computer than to work with the actual process. One such process is the tossing of two dice. There are 36 possible outcomes for the tossing of the dice, and 11 separate sums of the two sides. If the outcomes of the dice throws are completely random, a large number of tosses would produce the fol lowing distribution:

Sum of Faces	2	3	4	5	6	7	8	9	10	11	12
Fraction	1/36	2/36	3/36	4/36	5/36	6/36	5/36	4/36	3/36	2/36	1/36

You can use a random number generator to randomly produce values betwee one and six to simulate one die, and use it again to simulate the second die. The sum of the two can be found and the process repeated any desired number o times. Your results should approximate the distribution given above.

Most computer systems have either an intrinsic function or a subprogram for random number generation. Just in case yours does not, an elementary random number generation subprogram is included here. It should be adequate for the applications in the text. In the subprogram, the argument SEED is a starting value for the generator. A different set of random numbers will be produced for a different seed value.

```
REAL FUNCTION RANDOM(SEED)
INTEGER A,M,TIME,SEED,X
REAL XM,XX
DATA A,M,TIME/1027,1048576,0/
IF (TIME.EQ.0) THEN
    X = SEED
    TIME = 1
    XM = M
END IF
X = MOD(A*X,M)
XX = X
RANDOM = XX/XM
RETURN
END
```

APPENDIX I

ELECTRIC CIRCUIT ANALYSIS

I.1 VOLTAGE, CURRENT, RESISTANCE, CAPACITANCE, INDUCTANCE

Voltage is the electrical potential difference that drives current through an electric circuit. Current is the rate of flow of electric charge through a circuit. Units for voltage are called volts, and units for current are called amperes, or amps.

Resistance is the constant of proportionality between voltage and current. Units of resistance are called ohms. The relationship between voltage E, current i, and resistance R, is referred to as Ohm's law:

$$E = iR \tag{I.1}$$

Capacitors are devices that store charge. They consist of two plates separated by an insulating material. Capacitance is measured in farads.

Inductors are coils of insulated wire through which current flows. They influence the rate of change of current through the circuit. Through inductors the voltage is approximately proportional to the rate of change of the current. The inductance is the constant of proportionality. The units of inductance are called henrys.

$$E \approx L \left(\frac{di}{dt} \right) \tag{I.2}$$

Figure I.1 shows the symbols used to represent the electrical circuit components used in the text.

I.2 PARALLEL AND SERIES RESISTANCES AND CAPACITANCES

Resistors and capacitors in series and parallel can be combined into a single equivalent resistor or capacitor by the formulas shown in Figure I.2.

FIGURE I.1   Symbols for electrical devices.

Voltage Source   12 V

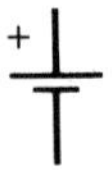

Resistor   3 Ω

Capacitor   2 μF

Inductor   3 H

FIGURE I.2   Parallel and series resistance and capacitance.

Series Resistance $R = R_1 + R_2$

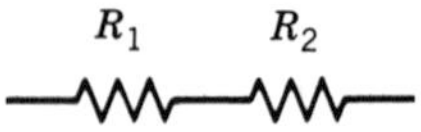

Parallel Resistance $\frac{1}{R} = \frac{1}{R_1} + \frac{1}{R_2}$

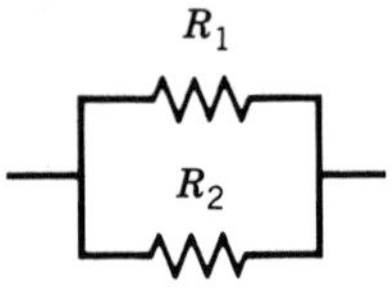

Series Capacitance $\frac{1}{C} = \frac{1}{C_1} + \frac{1}{C_2}$

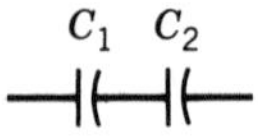

Parallel Capacitance $C = C_1 + C_2$

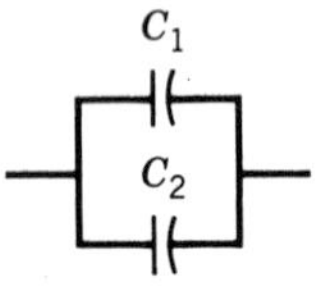

I.3 DC CIRCUIT ANALYSIS

I.3.1 Kirchoff's Laws

Direct current circuits are made up of branches, nodes and loops. A branch is
portion of a circuit between two terminals. Other branches may also connect to
terminal, or node. A loop is a closed path that is formed by connecting branches

Ohm's law, Equation I.1, is useful in analyzing DC circuits. Two other relationships applicable to DC circuit analysis are called Kirchhoff's laws. Kirchhoff's current law states that the algebraic sum of currents at a node must equal zero.

$$\sum_{i=1}^{m} I_i = I_1 + I_2 + \cdots + I_m = 0 \tag{I.3}$$

Currents flowing into a node are considered positive, whereas currents flowing away from a node are negative. Kirchhoff's voltage law states that the sum of the voltages around any loop must algebraically sum to zero.

$$\sum_{i=1}^{n} E_i = E_1 + E_2 + \cdots + E_n = 0 \tag{I.4}$$

Voltages include batteries and voltage drops across devices such as resistors.

I.3.2 Linear Equation Solution

A DC circuit is shown in Figure I.3. You desire to find the distribution of current in each branch of the circuit. Since there are ten unknown currents, you will need ten equations. There are seven nodes in the circuit. However, you can write node equations for only six of these. The seventh node equation would not be independent. It is up to you which six you choose. You can write loop equations for the four interior loops to make up the final four equations. The directions of the

---

FIGURE I.3   DC resistance circuit.

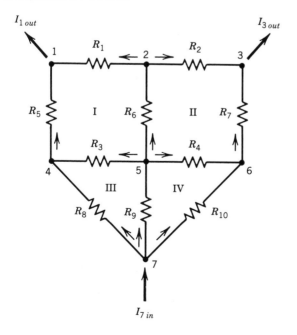

unknown currents are assumed. A negative value for current in a branch will indicate that the current was assumed in the wrong direction.

$$
\left.\begin{array}{ll}
\text{Node 1:} & I_1 + I_5 - I_{1OUT} = 0 \\
\text{Node 2:} & I_6 - I_1 - I_2 = 0 \\
\text{Node 3:} & I_2 + I_7 - I_{3OUT} = 0 \\
\text{Node 4:} & I_3 + I_8 - I_5 = 0 \\
\text{Node 5:} & I_9 - I_3 - I_4 - I_6 = 0 \\
\text{Node 6:} & I_4 + I_{10} - I_7 = 0 \\
\text{Loop I:} & E_3 + E_5 - E_1 - E_6 = 0 \\
\text{Loop II:} & E_2 + E_6 - E_7 - E_4 = 0 \\
\text{Loop III:} & E_8 - E_3 - E_9 = 0 \\
\text{Loop IV:} & E_9 + E_4 - E_{10} = 0
\end{array}\right\} \tag{I.5}
$$

You have to make the substitutions $E = IR$ in the last four equations. Then you can estimate the currents by moving the known currents, in this case I_{1OUT} and I_{3OUT}, to the right-hand side and solving the system of equations by matrix methods.

I.4 STEP RESPONSE OF RC CIRCUITS

A resistance-capacitance circuit is shown in Figure I.4. When the switch is closed, an initially uncharged capacitor will charge according to the relationships

$$
E = V_s \left(1 - e^{-\left[\frac{t}{RC}\right]}\right) \qquad I = C\frac{dE}{dt} = C\left(\frac{1}{RC}\right)V_s e^{-\left[\frac{t}{RC}\right]} = \frac{V_s}{R}e^{-\left[\frac{t}{RC}\right]} \tag{I.6}
$$

in which V_s is the voltage applied to the circuit, and t is the time in seconds. If at some time t_1 the switch is opened, the capacitor will discharge according to the relationships

$$
\begin{aligned}
E &= V_1 e^{-\left(\frac{t-t_1}{RC}\right)} \\
I &= -\frac{V_1}{R}e^{-\left(\frac{t-t_1}{RC}\right)}
\end{aligned} \tag{I.7}
$$

in which V_1 is the voltage across the capacitor at time t_1.

FIGURE I.4   Step response of resistance-capacitance circuit.

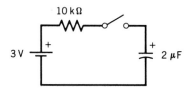

I.5 DIODE CIRCUIT

A diode circuit is shown in Figure I.5. Diodes are semiconductor devices that have the property of allowing current to flow readily in one direction but not the other. If I_0 is the desired current in the battery branch of the circuit, the governing equations are as follows:

$$
\begin{aligned}
I_0 &= I_1 + I_2 \\
I_1 &= I_{S1}(e^{\lambda_1 E} - 1) \\
I_2 &= I_{S2}(e^{-\lambda_2 E} - 1) \\
I_s &= \text{saturation current (amperes)} \\
\lambda &= \text{diode characteristic (volts}^{-1})
\end{aligned}
\tag{I.8}
$$

FIGURE I.5   Diode circuit.

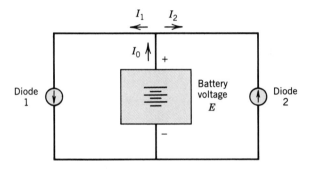

I.6 RESONANT CIRCUIT

A resonant resistance–capacitance–inductance circuit can be used as a tuner to select a specific frequency from a set of frequencies being received. Figure I.6 shows an RLC circuit which can be analyzed by the expressions

$$
\begin{aligned}
f_0 &= \frac{1}{2\pi}\left[\frac{1}{LC}\right]^{\frac{1}{2}} \\
\Delta f &= \frac{1}{4\pi}\left(\frac{L}{R}\right) \\
V_r(f) &= V_0\left[1 + \frac{(f^2 - f_0^2)^2}{f^2 \Delta f^2}\right]^{-\frac{1}{2}}
\end{aligned}
\tag{I.9}
$$

in which f_0 is the natural frequency of the circuit (the frequency to be tuned), f is the frequency of the driving voltage, Δf is selectivity of the circuit (a specified fraction of f_0), $V_r(f)$ is the voltage across the resistor, L is the inductance (henrys), and V_0 is the driving voltage.

FIGURE I.6   Resonant RLC circuit.

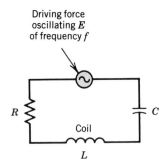

Driving force
oscillating E
of frequency f

R

C

Coil

L

I.7 COMPLEX ARITHMETIC AS APPLIED TO AC CIRCUITS

I.7.1 Imaginary and Complex Numbers

Imaginary numbers occur when you need to be able to take the square root of -1, but cannot do so by ordinary algebraic means. The letter j is used to represent the radical. Therefore

$$j3 = 3\sqrt{-1} \qquad -j2 = -2\sqrt{-1} \tag{I.10}$$

Complex numbers include both real and imaginary parts. For example, $3 + 4j$ is a complex number. Three is the real part and $4j$ is the imaginary part. Complex numbers frequently arise in the analysis of alternating current electric circuits.

I.7.2 Rectangular and Polar Forms

Complex numbers can be expressed in either rectangular or polar form.

$$\begin{aligned} a + jb & \qquad \text{Rectangular form} \\ Me^{j\theta} = M\angle\theta & \qquad \text{Polar form} \\ a + jb = M(\cos\theta + j\sin\theta) &= Me^{j\theta} \end{aligned} \tag{I.11}$$

The third part of Equation I.11 shows how to convert from one form to the other.

I.7.3 Rules of Complex Algebra

Two complex numbers are equal if and only if the real parts are equal and the imaginary parts are equal.

Two complex numbers in rectangular form are added or subtracted by adding or subtracting the real and imaginary parts separately.

$$(a + jb) + (c + jd) = (a + c) + j(b + d) \tag{I.12}$$

Multiplication of complex numbers is best accomplished when they are in the polar form. The magnitude of the product is the product of the individual magnitudes, and the angle of the product is the sum of the individual angles.

$$Ae^{j\alpha}Be^{j\gamma} = A \cdot Be^{j(\alpha+\gamma)}$$
$$A\angle\alpha \cdot B\angle\gamma = A \cdot B\angle(\alpha + \gamma) \qquad\text{(I.13)}$$

Division is also accomplished most readily when the complex numbers are in the polar form. The magnitude of the quotient is the quotient of the magnitudes, and the angle of the quotient is the difference of the angles.

$$\frac{Ae^{j\alpha}}{Be^{j\gamma}} = \frac{A\angle\alpha}{B\angle\gamma} = \left(\frac{A}{B}\right)\angle(\alpha - \gamma) \qquad\text{(I.14)}$$

I.7.4 Phasor Representation

In the analysis of alternating current circuits, currents and voltages will be continuously varying sinusoidal waves. To accomplish the additions and subtractions of currents and voltages required for the application of Kirchhoff's laws, the currents and voltages have to be represented by entities that portray the continuously varying magnitudes. Complex numbers supply the necessary characteristics. In polar form,

$$E = A\angle\theta \qquad\text{(I.15)}$$

would represent a voltage of A volts with a phase angle of θ. This representation is called a phasor. Resistances in phasor form are called impedances. Impedance, voltage, and current phasors can be multiplied, divided, added, and subtracted according to the rules described in the previous section. They can also be converted to and from rectangular form.

APPENDIX J

INDUSTRIAL ENGINEERING

J.1 PROJECT MANAGEMENT

J.1.1 General Concepts

Engineers need to be able to manage nonrepetitive projects made up of a large number of individual activities. A project might be the construction of a building. Some of the activities would include laying the foundation, completing the steel skeleton, finishing interior wall, and so forth. The activities are dependent on each other, in that some activities have to be completed before others can start. If you know these dependencies, you can display the project in a network form. Each of the activities will have a certain expected duration. The duration of a path through the network is the summation of the durations of the activities on that path. The path through the network that produces the longest total duration is called the critical path. The duration of the critical path is how long it will take to finish the project. There may be multiple paths that produce the same maximum duration. The critical path method, or CPM, is a commonly used method to find the longest path. There are two methods of displaying the activities of the project: either activity on arrow or activity on node. We will discuss both methods, since both are important in the development of computer algorithms to implement CPM.

J.1.2 Activity on Arrow Networks

Figure J.1a shows a project depicted with the activity on arrow notation. Each of the arrows of the network represents an activity. Activities leaving a node are dependent on the completion of the activities entering the node. For example, activity 2–5 cannot begin until activity 1–2 is completed. Activity 5–6 is dependent on the completion of both activities 2–5 and 4–5. Activity duration is written along the arrow. The dashed activities with zero duration are called dummy activities. They show dependency only. For example, activity 3–4 is dependent on the completion of both activities 1–3 and 1–2.

Since each link in this network represents an activity duration, and all activities have to be completed before the project is completed, the longest path through the network will set the minimum expected project completion time. This is the critical path. If any of the activities on this path are delayed, then the project completion time will also be delayed. A project will often have multiple critical paths.

Determination of project completion time and the critical path(s) is a three-step process. The first step is to go forward through the network and estimate the earliest expected completion time (T_E) for each node of the network. T_E is the

FIGURE J.1   Activity on arrow network.

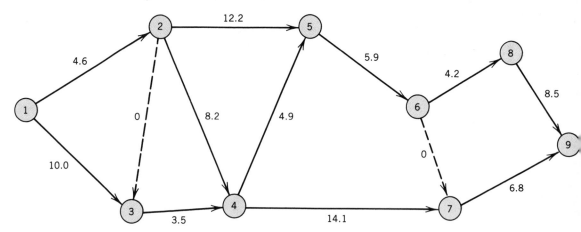

(*a*) Basic network (durations are in weeks).

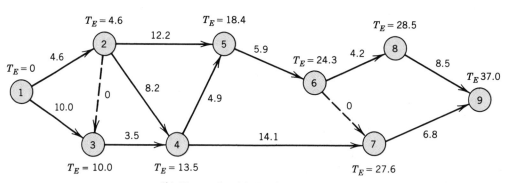

(*b*) Network with T_E times added.

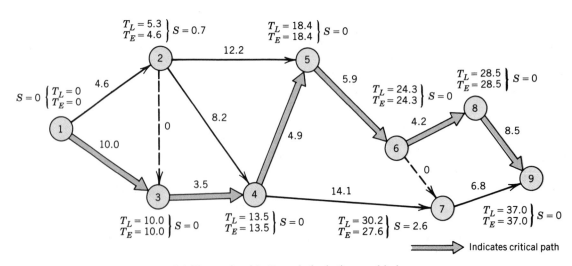

(*c*) Network with T_L and slack times added.

estimated time required to complete work up to a certain point in the network. Since the critical path is always the longest path, it is necessary to study the completion times of all the paths that enter each node. The T_E for that node would be the maximum estimated completion time among all activities entering the node. Completion time for a particular activity entering a node is determined by adding the duration of that activity to the T_E for the node from which the activity originates. For example, in Figure J.1b, at node 5 the choice is between the expected completion of activity 2–5 (12.2 + 4.6 = 16.8) and the expected completion time of activity 4–5 (4.9 + 13.5 = 18.4). Therefore, the T_E for node 5 would be 18.4 weeks. Other T_E values given in Figure J.1b can be found in a similar manner.

At each node on the forward pass, attention should focus on the entering activities. T_E for the ending event of the network is the estimated project completion time. Collectively, the T_E estimates for the network provide a schedule of when each event should be finished. They enable the manager to determine whether the project is on, behind, or ahead of schedule at any point during project execution.

After you complete the forward pass through the network, you make a pass in the other direction, from finish to start, to determine the latest allowable finish time (T_L) for each node. T_L is the greatest amount of time that can elapse between the beginning of the project and any point in the project without delaying the scheduled completion of the project. By definition, T_L for the ending event is equal to the estimated project completion time. The T_L value for each earlier node is the difference between the length of the longest path from that node to the end of the project and the estimated project completion time. However, as before, the computations can proceed from node to node. T_L is the minimum value of T_L for events directly successor to the event in question minus the intervening project duration. Figure J.1c shows the network of Figure J.1a with T_L times added. At node 8, T_L is equal to T_L for node 9 (37.0 weeks) minus the intervening activity duration for activity 8–9 (8.5 weeks). At node 6, T_L is the minimum between T_L of node 8 (28.5 weeks) minus activity 6–8 (4.2 weeks), and T_L of node 7 (30.2 weeks) minus activity 6–7 (0 weeks). The backward pass would continue back to the starting node.

The difference between T_E and T_L for any event is the slack time (S) for that event. Slack time is the amount of time a particular event could be delayed without delaying the project completion time. Critical paths have zero slack time.

J.1.3 Activity on Node Networks

The major advantage of the activity on node network representation is that it provides information on slack time for individual activities, not just the nodes. Figure J.2a shows the network of Figure J.1 redrawn in activity on node representation. Note that this representation eliminates the need for dummy activities.

Forward and backward passes are again made through the network. The method of analysis is essentially the same as the activity on arrow representation. However, slightly different definitions are given to the times found since they refer to activities, not nodes. On the forward pass,

EST = earliest start time of an activity = maximum EFT of operations immediately preceding the activity in question.

EFT = earliest finish time of an activity = EST + activity duration.

FIGURE J.2   Activity on node network.

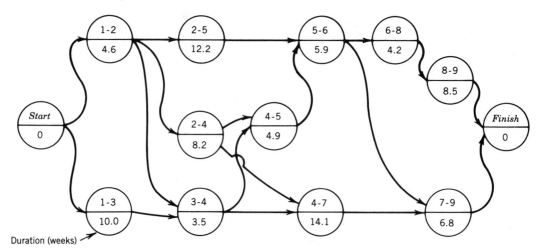

(*a*) Basic network.

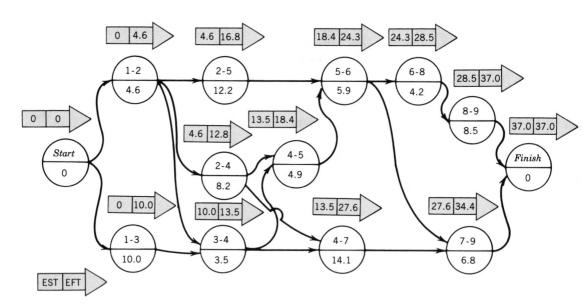

(*b*) Network with forward pass computations.

FIGURE J.2   (Continued)

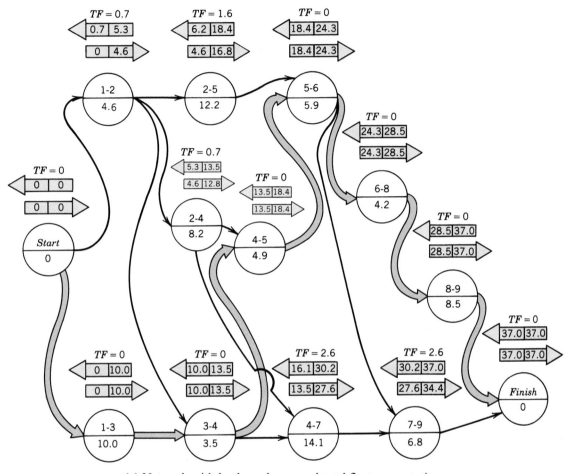

(c) Network with backward pass and total float computations.

The EFT is similar to T_E for the activity on arrow network. Forward pass times are usually identified by an arrow pointing to the right with the EST and EFT times placed in the indicated positions.

Figure J.2b shows the network of J.2a with the forward pass data recorded on it. The EFT for the ending event is again 37 weeks.

As is the case with the activity on arrow network, the backward pass for the activity on node network will establish latest times, but now they will be times for activities. The following definitions will be used:

LFT = latest finish time of an activity = minimum LST of activities immediately following the activity in question.

LST = latest start time of an activity = LFT − duration of the activity.

The LST is the latest time at which an activity may be started if the project is to be completed as scheduled. Similarly, the LFT is the latest time at which an activity may be finished without delaying project completion. The LFT is similar to T_L for the activity on arrow network. Figure J.2c shows the network of J.2a with backward pass calculations added. Backward pass times are usually identified by an arrow pointing to the left with the LST and LFT times placed in the indicated positions.

After completion of the backward pass, slack times for the activities can be calculated. The slack for the activity on node network is usually called the total float (TF). It is the amount of time a particular activity could be delayed without delaying project completion.

$$TF = LST - EST = LFT - EFT$$

The critical path passes through all activities with zero total float.

J.1.4 Bar Charts

A bar chart is a graphical representation of the activities of a project placed on a time scale. The length of the bar for each activity is its duration. In order to develop a bar chart, you have to know the EST and EFT for all activities. The bar chart is useful for seeing which activities overlap and for determining resource allocations. It does not display the dependencies among the activities. Figure J.3 shows a sample bar chart.

J.1.5 Resource Scheduling

If several activities of a project require the same resource, you can determine the requirements for the resource at any point in the project by adding the individual

FIGURE J.3   Bar chart.

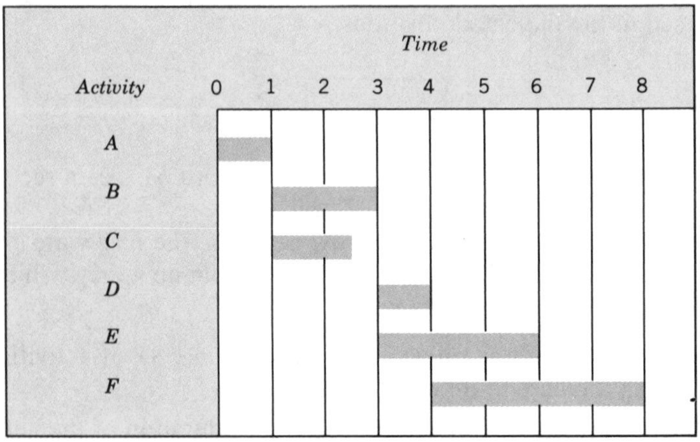

Activity	Time

requirements for activities that are being carried out simultaneously. A bar chart is useful for seeing which activities overlap.

J.1.6 Time–Cost Optimization

At some point during the progress of a project, it may be necessary to speed up the critical path activities in an effort to reduce the total project duration. One way of doing this is to apply more resources to selected activities on the critical path to speed them up. However, this will result in increased costs. Every activity has associated with it a normal time, which is the expected activity duration. The normal cost is the cost of completing the activity in the normal time. Most activities also have a crash time, which is the minimum possible activity duration. To reach this minimum time you may have to rent more equipment, add shifts, or work seven days a week. The crash time has a crash cost associated with it. The crash cost is larger than the normal cost. Between the two extremes, the marginal cost of reducing activity duration by one unit of time would increase as the activity is progressively compressed, because of decreasing returns on investment. This is shown as curve A in Figure J.4. However, to simplify computations, it is generally assumed that cost varies linearly with duration reduction between the normal and crash times (line B of Figure J.4).

The objective of time–cost optimization is to reduce the project completion time to a certain point with the minimum increase in project cost. The additional cost can then be compared with the cost of delaying project completion beyond the deadline. First it has to be determined if it will be possible to reduce project duration to the desired point. This can be accomplished by calculating the minimum possible project duration if all activities are crashed.

Table J.1 shows the normal and crash data for the network given in Figure J.5. Normal completion time for this project would be 23 weeks, and the normal cost would be $165,000. The marginal cost to crash is computed by the formula

$$\text{Marginal cost to crash} = \frac{\Delta C}{\Delta t} = \frac{\text{crash cost} - \text{normal cost}}{\text{normal time} - \text{crash time}} \qquad \text{(J.1)}$$

FIGURE J.4   Variation of cost with activity duration.

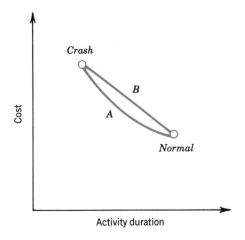

TABLE J.1   NORMAL AND CRASH DATA FOR PROJECT

Activity	Normal Time (weeks)	Crash Time (weeks)	Normal Cost	Crash Cost	Marginal Cost to Crash ($/week)
A	3	2	$20,000	$22,000	2,000
B	5	3	30,000	36,000	3,000
C	4	2	25,000	27,000	1,000
D	3	2	10,000	11,500	1,500
E	5	4	15,000	17,500	2,500
F	4	4	20,000	20,000	—
G	7	6	12,000	12,500	500
H	5	3	15,000	22,000	3,500
I	6	4	10,000	18,000	4,000
J	4	3	8,000	9,000	1,000

If all activities of the project were crashed, the network shown in Figure J.6a would result. The completion time has been reduced by 5 weeks to 18 weeks, but the project cost would now be $195,500. Completion time could not be reduced below 18 weeks. Crashing all activities is not an efficient way of reducing project duration. The following approach will identify the most economical way to reduce project duration to any completion time between the normal and all activities crashed completion times.

1. Find the critical path or paths.
2. Find the least expensive activity on the critical path. When multiple critical paths exist, it may be most economical to crash a combination of two activities on parallel critical paths.

FIGURE J.5   Network for time–cost optimization.

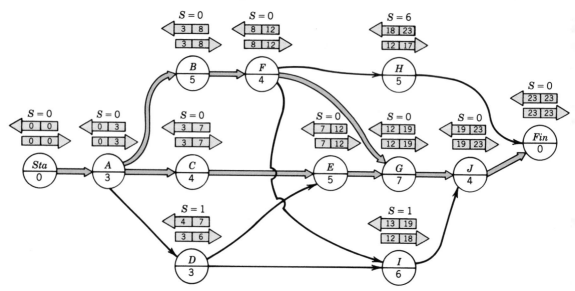

FIGURE J.6   Time–cost optimization.

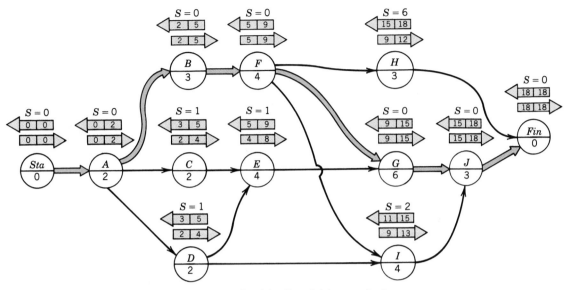

(*a*) Network with all activities crashed.

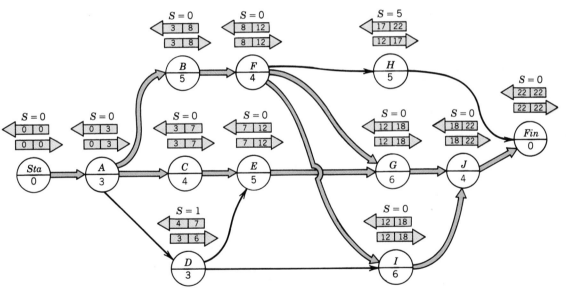

(*b*) Network with activity *G* crashed by one week.

FIGURE J.6   (Continued)

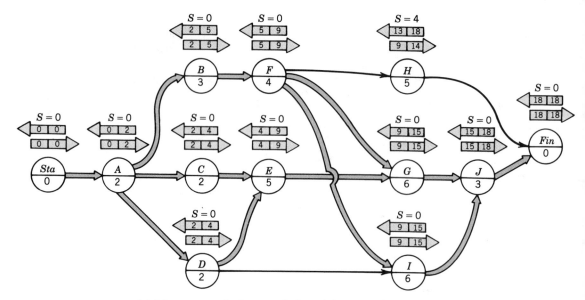

(c) Most economical network for minimum project duration.

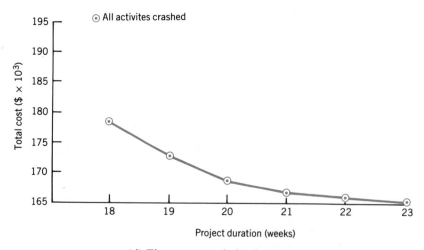

(d) Time–cost optimization curve.

3. Crash the most economical activity or activities by an amount equal to the minimum slack on the parallel paths, but not beyond its crash time.
4. Recompute network data and confirm critical path or paths.
5. Repeat steps 2, 3, and 4 until desired completion time is reached.

If the objective of network compression for the project depicted in Figure J.6a was to reduce project duration to 18 weeks in the most economical manner, the following steps would be required.

Step 1: Activities on the critical paths are A, B, C, E, F, G, and J. There are multiple critical paths at this step. Activity F cannot be crashed, so it will not be considered. Activity G has the lowest marginal cost to crash and is on the single portion of the critical path, so it is the most economical to crash. It can be crashed by 1 week. Minimum slack on parallel paths is also 1 week. Therefore, G should be reduced to 6 weeks and the critical path recomputed. This is shown in Figure J.6b. Note that activity I is now also on the critical path because of the 1 week reduction in the project completion time. Total project cost is now $165,500.

Step 2: Activity I has now joined the other activities on the critical path. G has been crashed to its limit, so it is no longer considered. F is also not considered. Of the remaining activities, C and J both have marginal costs to crash of $1,000/week. However, activity C is on a parallel critical path, and either activity B or F would have to be crashed with it to reduce project duration. Therefore, it is preferable to crash J. Five weeks of slack are available on the parallel path, but activity J can be crashed by only 1 week. This will reduce project duration by 1 week to 21 weeks, with a new total cost of $166,500. No new critical path was formed in this step.

Step 3: Next, activities A, B, C, E, and I should be considered. Activity C has the minimum marginal cost to crash. However, it is on a parallel critical path, so both B and C would have to be crashed to produce any project compression. This would give a total marginal cost to crash of $4,000/week. Activity A has a marginal cost to crash of $2,000/week, so it is the preferred activity to crash. Activity A has no parallel paths, so it can be crashed to its limit of 2 weeks' duration. Project duration will now be 20 weeks, and total cost will be $168,500. No new critical paths are created.

Step 4: Next, activities B, C, E, and I should be considered. Activities B and C have a combined marginal cost of $4,000, as does activity I. However, crashing I will not reduce project duration because of the parallel critical path. Activities B and C can be crashed by only 1 week at this step because of the slack on the parallel path through D. Project duration will now be 19 weeks, and the total cost will be $172,500. Activity D will now also be on the critical path; however, the path from D to I is not a critical path.

Step 5: Finally, activities B, C, D, E, and I should be considered. B and C can still be considered because they have not been crashed to their limits. Combinations of B, C, and D, or B and E could be crashed. The total marginal cost of each combination is $5,500. The choice would depend on project conditions and ease of crashing. In this case, B, C, and D will be crashed by 1 week, with a resulting project duration of 18 weeks and a total cost of $178,000. Additional crashing will not result in any further reduction in project duration. Figure J.6c shows the final network, while Figure J.6d shows a plot of project duration versus total cost. This plot could be used to determine the best way to reduce project duration to any time between the normal and crash limits. Note also that the total cost to reach the minimum possible project duration is $178,000 rather than the $195,500 required if all activities were crashed.

J.2 BREAK-EVEN ANALYSIS

Manufacturing processes involve fixed and variable costs. The fixed costs include items such as equipment and buildings that you would need to start production. The variable costs are the actual production costs. The total cost of producing a certain number of units of some product is the sum of the fixed and variable costs. The return is a function of the number of units produced. Figure J.7 shows a typical break-even analysis with linear cost and return functions. However, the costs and returns do not have to be linear. The break-even point tells us the minimum number of units that we should consider manufacturing. If the market will not support that many units, we should manufacture something else. Break-even analysis can also be used to determine if you should manufacture or buy a certain part. In that case the cost to buy replaces the return function. If the number of units you need is more than the break-even number, then you should manufacture them. Otherwise, you are better off buying.

J.3 QUEUING THEORY (WAIT LINE)

Wait line theory deals with the random arrival of people or objects to wait in a line (queue) to receive some service. It has many applications, including the hiring of repairmen to service machines. In this case the objects waiting in line are the broken machines. The objective is to determine how many repairmen should be hired to service a given number of machines.

FIGURE J.7 Break-even analysis.

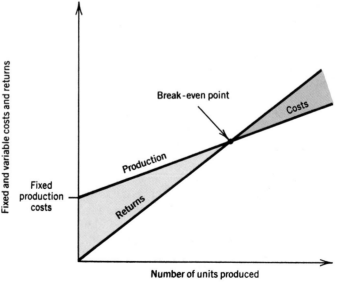

Assumptions made include:

1. All machines are identical.
2. Machines are subject to failure at random times.
3. The failure rate is known and is characterized by the average time between failures for an individual machine.
4. The average repair time is known.
5. A repairman can work on only one machine at a time.
6. The loss per machine while it is inoperative is $L/hr.

Good repairmen are expensive, but they also can fix machines at a faster rate. Too many repairmen will reduce the number of malfunctioning machines but may result in excessive personnel costs. You want to minimize the total costs of wages to the repairmen plus costs for downtime:

$$\text{Cost}(R) = (W)(R) + N_d L \tag{J.2}$$

in which W is the wage paid to the repairmen, R is the number of repairmen, and N_d is the average number of inoperative machines at any one time.

If P_0 is the probability that zero machines are down, and P_i = the probability that i machines are down, queuing theory can be used to develop the expressions

$$P_{i+1} = \rho \left(\frac{N - i}{i + 1} \right) P_i \quad \text{if} \quad i < R \tag{J.3}$$

$$\rho = \frac{\lambda}{\mu}$$

where

$$\lambda = \text{failures/machine/unit time}$$
$$\mu = \text{repairs/hour/repairman}$$
$$N = \text{total number of machines}$$

and

$$P_{i+1} = \rho \left(\frac{N - i}{R} \right) P_i \quad \text{if} \quad R \leq i < N \tag{J.4}$$

Each P_i can be related to P_0. After completing the series, P_0 can be found from the condition

$$\sum_{i=0}^{N} P_i = 1.0 \tag{J.5}$$

Since each P_i is proportional to P_0, you can find P_0 by first setting $P_0 = 1$, computing all the P_i on this basis, evaluating the sum in Equation J.5, and rescaling each of the P_i by replacing P_i by P_i/sum. The average number of down machines can then be estimated by the expression

$$N_d = \sum_{i=1}^{N} i \cdot P_i \tag{J.6}$$

APPENDIX K

CHEMICAL ENGINEERING PRINCIPLES

K.1 IDEAL GAS LAW

The ideal gas law, or equation of state, describes the relationship between the pressure, volume, temperature, and number of moles of an ideal gas. The relationship is

$$Pv = nRT \tag{K.1}$$

in which P is the pressure in atmospheres, v is the volume in liters, n is the number of moles, R is the universal gas constant, and T is the absolute temperature in degrees Kelvin. The ideal gas law succeeds fairly well in approximating the characteristics of real gases under relatively low pressures and temperatures. Researchers have developed empirical relationships to describe the behavior of gases when they vary significantly from the ideal. Two of those, the van der Waals equation and the Redlich–Kwong equation, will be discussed here.

Van der Waals equation models gases through the relationship

$$\left(P + \frac{a}{V^2}\right)(V - b) = RT \tag{K.2}$$

in which V = molar volume = v/n, and a and b are coefficients calculated through the equations

$$a = \left(\frac{27}{64}\right)\left(\frac{R^2 T_c^2}{P_c}\right) \qquad b = \frac{RT_c}{8P_c} \tag{K.3}$$

P_c and T_c are the critical pressure and temperature for the gas being considered.

The Redlich–Kwong equation has the form

$$P = \frac{RT}{V - b} - \frac{a}{T^{0.5}V(V + b)}$$

$$a = 0.427480 \frac{R^2 T_c^{2.5}}{P_c} \tag{K.4}$$

$$b = 0.086640 \frac{RT_c}{P_c}$$

A measure of the deviation from the ideal gas law is given by the compressibility factor,

$$Z = \frac{PV}{RT}$$

(K.5)

When Z approaches unity, the gas approaches the ideal state.

K.2 CHEMICAL KINETICS

The decomposition of a compound with time may follow first- or second-order kinetics. First-order kinetics are described by the model

$$C_t = C_0 e^{-k_1 t}$$

(K.6)

while second-order kinetics are described by the model

$$C_t = \frac{C_0}{C_0 k_2 t + 1}$$

(K.7)

in which C_0 is the initial concentration (moles/liter), C_t is the concentration at time t, and k_1 and k_2 are the rate constants.

K.3 DEPTH OF FLUIDIZED-BED REACTOR

Fluidized-bed reactors are used extensively in chemical engineering to provide more uniform contact for chemical reactions such as catalytic cracking in petroleum processing. The reactor contains a bed of granular material through which a fluid is flowing. The upward rate of fluid flow is such that the particles are suspended. When designing fluidized-bed reactors you have to make sure that the reactor is deep enough so that particles are not lost out the top of the reactor. The amount that the bed will expand is a function of both the particle and the fluid characteristics. The expanded height H_e of the bed is given by

$$H_e = H_0(1 + f) \sum_{i=1}^{n} \left[\frac{p_i}{(1 + e_i)} \right]$$

(K.8)

in which H_0 = static (unexpanded) bed height, f = void fraction of the unexpanded bed, p_i = fraction of bed particles with the diameter d_i, and e_i = porosity or void fraction of the expanded bed part that is made up of particles of size d_i. The individual void fractions e_i are functions of the flow rate and the particle size:

$$e_i = F^{1/3}(1 - e_i)^{1/3}$$

(K.9)

$$F = \left(\frac{180}{g}\right)\left(\frac{\mu}{(\rho_s - \rho)}\right)\left(\frac{q}{d_i^2}\right) \tag{K.10}$$

in which g = gravitational acceleration = 9.8 m/s², μ = fluid viscosity (N-s/m²), ρ_s = particle density (kg/m³), ρ = fluid density (kg/m³), and q = reactor bed flow rate (m³/s/m²) or (m/s). You can solve for F in Equation K.10, and then solve for e_i by successive substitution in Equation K.9. Start with an estimate of e_i, substitute that estimate into the right side of Equation K.9, and calculate the value of e_i on the left side. If the two values are within a small tolerance, your estimate is good enough. If they are not, substitute the new value into the right and repeat. You can use this procedure to find each of the individual e_i estimates, and then use Equation K.8 to estimate the expanded bed height.

K.4 PROCESS DESIGN, GAS SEPARATION

Chemical engineers use selective absorption of gases in liquids to separate a gas into its constituents. Differences in the absorption rates of the gases make this possible. The gas is passed through the liquid, and it is assumed that after each pass the gas will be in equilibrium with the liquid. Each pass is called a plate. A number of plates will be required to reach the desired separation efficiency for a particular constituent of the gas. This set of plates is referred to as a packed tower. The governing equation for a packed tower is

$$E = 1 - \frac{A - 1}{A^{n+1} - 1} \tag{K.11}$$

in which E = the desired absorption efficiency for a particular chemical constituent, A = the absorption factor for a particular constituent, and n = the number of plates. The liquid will absorb only the amount of each gas required to reach equilibrium. This amount is represented by the absorption factor

$$A = \frac{L}{kG} \tag{K.12}$$

in which L = mol/s of flowing liquid, G = mol/s of flowing gas, and k = experimentally determined equilibrium vaporization constant. Since the absorption factor is not really constant, an effective value is computed by the expression

$$A = A_e = \left[A_b(A_t + 1) + \frac{1}{4}\right]^{\frac{1}{2}} - \frac{1}{2} \tag{K.13}$$

in which A_b and A_t are the absorption factors appropriate for the bottom and top of the process. You can use a trial-and-error solution to solve for the number of plates, n, in Equation K.11 for the target constituent. The number of plates should be rounded to the next higher integer value. Then you can calculate the removal efficiency for each of the constituents based on that number of plates.

K.5 DISSOLVED OXYGEN CONCENTRATION IN STREAM

Dissolved oxygen concentration in a river downstream from a point source of pollution is given by the equation

$$DO = DO_s - D$$

in which DO = dissolved oxygen concentration at x (mg/ℓ), DO_s = saturation dissolved oxygen concentration for stream (mg/ℓ), and D = dissolved oxygen deficit for the stream. The time t (hours) to reach the point x ft downstream from the point of discharge is

$$t = x/V$$

in which V = average velocity of stream (ft/s). The dissolved oxygen deficit is given by the equation

$$D = \frac{k_d L_0}{k_r - k_d} (e^{-k_d t} - e^{-k_r t}) + D_0 e^{-k_r t} \tag{K.14}$$

in which k_d = oxygen use rate by natural pollution removal processes (time^{-1}), L_0 = initial concentration of pollutant at point of discharge (mg/ℓ), D_0 = initial dissolved oxygen deficit (mg/ℓ), and k_r = the reaeration coefficient, which can be estimated by the equation

$$k_r = \frac{(VS)^{0.5}}{H^{1.5}} \tag{K.15}$$

in which S = diffusivity of oxygen in water (ft^2/hr), and H = depth of flow (ft).

INDEX